19.28

Comparative Morphology
of Vascular Plants

A Series of Books in Biology

EDITORS
Donald Kennedy
Roderic B. Park

SECOND EDITION

Comparative Morphology of Vascular Plants

Adriance S. Foster
UNIVERSITY OF CALIFORNIA, BERKELEY

Ernest M. Gifford, Jr.
UNIVERSITY OF CALIFORNIA, DAVIS

W. H. FREEMAN AND COMPANY
San Francisco

COVER DRAWING:
Reconstruction of a specimen of *Medullosa noei*.
(Redrawn from Stewart and Delevoryas, *Bot. Rev.*
22:45, 1956.)

Library of Congress Cataloging in Publication Data

Foster, Adriance Sherwood, 1901–1973.
    Comparative morphology of vascular plants.

    Includes bibliographies.
    1. Botany–Morphology.   I. Gifford, Ernest M.,
joint author.   II. Title.
QK641.F6 1974   581.4   73–22459
ISBN 0–7167–0712–8

Printed in the United States of America

2 3 4 5 6 7 8 9

Contents

Preface

Although this edition represents a complete revision of the subject matter of the first edition, the essential purpose of our book remains unchanged, namely, to provide for upper-division and graduate-level college students a textbook combining the basic concepts of vascular-plant morphology with a detailed account of the organography, anatomy, and method of reproduction of all of the major groups of the Tracheophyta. To achieve this goal has proved an arduous and demanding undertaking because of the overwhelming number of research papers, reviews, and books on nearly every phase of morphology that have appeared during the past fifteen years. It would indeed be pretentious to imply that the new edition is in any sense a "complete" guide to the vast literature of plant morphology. But we entertain the hope that the new literature, incorporated in the references at the end of each chapter, will aid the teacher and the interested student in locating some of the most important publications concerned with recent advances in knowledge of vascular plants.

It will be instructive to enumerate some of the major advances in vascular-plant morphology that we have felt compelled to consider and to use in the preparation of this edition. Prominent among these advances are certain recent paleobotanical discoveries that have more-or-less drastically changed our previous scheme of classification of the vascular plants. As examples, we may mention (1) the revised treatment we have given in Chapter 7 to the Devonian "Psilophytales," which now seem to represent three parallel lines of evolutionary specialization among early vascular plants rather than a single taxon, and (2) our decision, as discussed in Chapter 14, to reject the "Pteropsida" as a valid taxon, in the light of recent studies and reconstructions of *Archaeopteris* and other Paleozoic progymnosperms. A second rapidly expanding field in vascular-plant morphology is *morphogenesis*, the experimental study of the factors that regulate ontogenetic processes and hence "condition" the expression of plant form. In this edition, increased attention is given to problems of morphogenesis in ferns (Chapter 13) and to the developmental processes in apical meristems concerned in the differentiation of strobili and flowers (Chapters 17 and 19). Since the publication of the first edition, there has been an enormous surge of interest in the ultrastructural aspects of cytology and morphology as revealed by the electron microscope. Illustrations of the value of the information obtained by the use of these instruments are given at several places in this edition, such as in the description of the varied types of fern sori, and in the new sections in Chapter 19 on pollen and floral organogenesis. Last, we have revised our previous descriptions of the morphology, systematics, and reproductive cycles in vascular plants in the light of new review articles and monographs dealing comparatively with such fundamental structures as sporangia, gametophytes, gametangia, and embryos and, for flowering plants, the endosperm tissue of the seed.

During the preparation of this edition, we have recognized the inevitable dangers that accompany the expansion of a textbook on morphology. To quote from the first edition, the student, in his initial approach to comparative morphology, may easily "miss the forest because of the trees." To avoid this dilemma as much as possible, we have retained, in revised form, the original "orientation chapters" (Chapters 1–6), which are intended to summarize (1) the salient morphological features of the tracheophytes as a whole, (2) the general organography and anatomy of the vegetative sporophyte, and

(3) the development and structure of sporangia, gametangia, and embryos. It is our belief that frequent cross-referencing between these introductory chapters and later portions of the book will help the student to relate morphological principles with the detailed discussions of structure and reproduction given in each of the chapters dealing with a specific group of vascular plants.

Although we have retained many of the illustrations from the first edition, our revision also includes many new photographs, line drawings, and diagrams essential to the expanded treatment of nearly all topics. The source of each figure is indicated in the accompanying legend. Many colleagues have generously donated illustrations for the present book, and they are either mentioned directly in the figure legends or in the acknowledgments that follow this preface.

We are happy to pay deserved tribute to our wives, Helen V. Foster and Jean D. Gifford, for their unfailing patience and encouragement during the writing of this book, and for their assistance in the preparation of the manuscript.

Adriance S. Foster
Ernest M. Gifford, Jr.

MARCH 1973
BERKELEY AND DAVIS

Acknowledgments

Although we assume full responsibility for the viewpoints expressed in this book, we wish to record our appreciation to many individuals associated with the University of California who, in a variety of ways, have assisted us during the preparation of this edition. We express our thanks to the following persons at the Davis campus: Professors John M. Tucker and Grady L. Webster, for their critical reading of several sections of the revised manuscript; Professor Richard H. Falk, for preparing a number of high quality scanning electron micrographs used as illustrations in our book; Professor D. I. Axelrod, for his assistance with the geologic time scale chart; Professor T. Elliot Weier, who donated photographs used in several chapters; Mr. Dennis Stevenson, for his critical reading of certain chapters and for editorial assistance; and Mr. Bijan Dehgan and Mr. Walter R. Russell, who collected some of the plant materials needed for photography. We express our gratitude to the following persons at the Berkeley campus: Professor H. G. Baker, who reviewed the section on pollination mechanisms in the angiosperms; Professor Lincoln Constance, for very helpful comments on the sections on fruits and inflorescences; Wayne L. Fry, for his assistance on many paleobotanical questions; Professor William A. Jensen, who was extremely helpful with the section on pollen tube growth and fertilization in angiosperms; Professors Donald R. Kaplan and Rudi Schmid, for valuable comments and discussions on various aspects of comparative and developmental morphology; Dr. Alan R. Smith, for his critical reading of the sections on fern systematics and angiosperm pollen; Mr. William Hirano, who assisted us by collecting many of the plant materials used in preparing certain illustrations for the present book. Our thanks are due to Mrs. Marie Dern, who assisted in the technical preparation of the manuscript.

We thank John Waller and Darwin Hennings and their assistants for the excellent preparation of the illustrations.

Finally, we would like to express our appreciation to Professor Ralph H. Wetmore and to Dr. Elizabeth G. Cutter for their critical reading of Chapter 13, especially the sections on morphogenesis.

Comparative Morphology
of Vascular Plants

1

The Science of
Plant Morphology

The extraordinary diversity in the form, stature, and habit of plants is a familiar fact of experience which is recognized by even the scientifically untrained observer. The "sea-weeds" of the ocean, the lowly "mosses" and graceful "ferns" of the woodlands, the towering cone-bearing trees of the northern forests, and the infinitely varied flowering plants of orchard and garden all are recognized as different kinds of plants by the layman, on the basis of more-or-less superficial criteria or earmarks.

Casual inspection of the *surface aspects* of plants, however, is a highly unreliable method either for separating plants into natural groups or for gaining a proper understanding of the nature and relationships of their parts. Thus, for example, the small green plants floating on the surface of ponds or garden pools are often commonly lumped together as "pond scum," "algae" or even

"moss," because of their small size and the absence of conspicuous flowers. However, rigorous scientific study of a population of such aquatic plants from the standpoint of morphology would show that it contains not only algae (in the scientific sense) but also aquatic ferns and minute flowering plants! Superficial observation of external similarities and differences among land plants often leads to equally incorrect conclusions. Frequently a wide variety of totally unrelated plants are called ferns by the layman, because they have divided or pinnatifid leaves. From a broad, comparative-morphological standpoint it is clear that the true ferns are remarkably diversified in leaf form and that their distinguishing characteristics are based on subtle but reliable similarities of structure and method of reproduction. And last there is confusion in the mind of the untrained observer in regard to flowering plants. With an understandable mental picture of a conspicuous and brightly colored garden or hothouse type of flower, the layman often fails to realize that the reproductive structures of grasses and of many trees and shrubs are flowers. This commonly leads to a wholly erroneous notion of the nature of reproduction in even the most common plants and to an astonishing underestimation of the diversity in form and habit of the flowering plants as a whole.

In marked contrast with such undisciplined regard of form and structure, the science of plant morphology attempts, by rigorous techniques and meticulous observations, to probe beneath these surface aspects of plants—in short, to explore and to compare those *hidden aspects* of form, structure, and reproduction which constitute the basis for the interpretation of similarities and differences among plants. One of the most fruitful results of early morphological studies was the recognition that a relatively few fundamental types of organs underly the construction of the plant body. Thus, the leaf, stem, and root were regarded as the principal types of vegetative organs, the size, form, proportions, and arrangement of which are subject to the most varied development or modification. As knowledge of the reproductive cycles of plants increased, sporangia and gametangia were added to this short list of major organ categories, and the importance of a broad comparative study of the resemblances, or homologies, of plant organs thus became established. Let us examine more closely the notion of homology as it is used in the interpretation of plant form and structure.

The Concept of Homology

The essence of the idea of homology was expressed in the writings of the great poet and philosopher Goethe, to whom we also

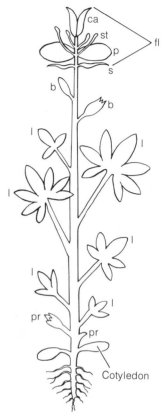

Figure 1-1
Diagrammatic representation of the organography of a flowering plant to illustrate Goethe's theory of metamorphosis. Note the "transitions" between primary leaves (pr) and the adult leaf type (l), and between the upper foliage leaves and the bracts (b). According to Goethe's theory, the parts of the flower (fl)—sepals (s), petals (p), stamens (st), and carpels (ca)—are also morphologically equivalent to "leaves." [Redrawn from *Organization and Evolution in Plants* by C. W. Wardlaw. Longmans, Green, London. 1965.]

owe the word "morphology" (literally, the science of form). Goethe sought for the nature of the morphological relationships among the various kinds of leafy appendages in higher plants. In his celebrated essay, *Metamorphosis in Plants,* published in 1790, he concluded that no real boundary exists between such organs as cotyledons, foliage leaves, bracts, and the organs of the flower—all are expressions of the same type of organ, i.e., the leaf (Fig. 1-1). Although Goethe's theory has been criticized as an example of idealistic morphology, it has proved an extremely astute viewpoint and indeed constitutes the theoretical basis for the current view that the flower is a determinate axis with foliar appendages (see Chapter 19).

The rapid expansion of botanical knowledge in the nineteenth century emphasized the importance of the concept of homology and the need for interpreting homologies in the broadest possible light. Goethe's ideas, and the earlier observations of C. F. Wolff (1774) on the origin of leaves at the growing point of the shoot, paved the way to a better understanding of *serial homology* in plants. With reference to a shoot, this term designates the equivalence in *method of origin* and *positional relationships* of the successive foliar appendages of a shoot. Thus, a bud scale, or a floral bract, is considered serially homologous with a foliage leaf because, like the latter, it arises as a lateral outgrowth from the shoot apex. Classical as well as modern ontogenetic studies have shown the very close resemblances in detail of origin and early histogenesis among the varied types of foliar organs of both vegetative and flowering shoots. Moreover, the different types of foliar appendages in the same plant are often interconnected by intermediate forms or transitional organs (Fig. 1-2). On the other hand, the concept of *general homology* in plants is much more difficult to demonstrate ontogenetically (see Mason, 1957, for a critical discussion). This is so because, unlike higher animals, plants are characterized by an open system of growth—a plant embryo is not a miniature of the adult, and hence homologies based on the resemblance in position, development, and form of two organs in different kinds of plants may be open to

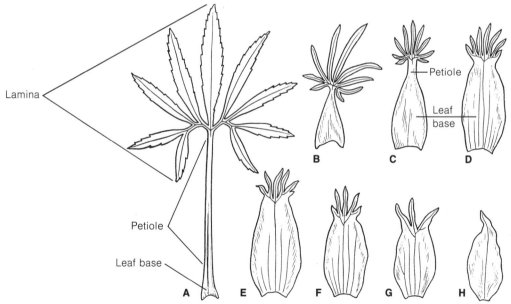

Figure 1-2
Serial homology between the foliage leaf (**A**) and the succession of floral bracts (**B–H**) in *Helleborus foetidus.* Note that the gradual suppression of a petiolar zone (bracts, **B–D**) and the progressive reduction of the lamina (bracts, **E–H**) is accompanied by a corresponding increase in the prominence of the leaf base. [Redrawn from *Vergleichende Morphologie der höheren Pflanzen* by W. Troll. Gebrüder Borntraeger, Berlin. 1935]

4

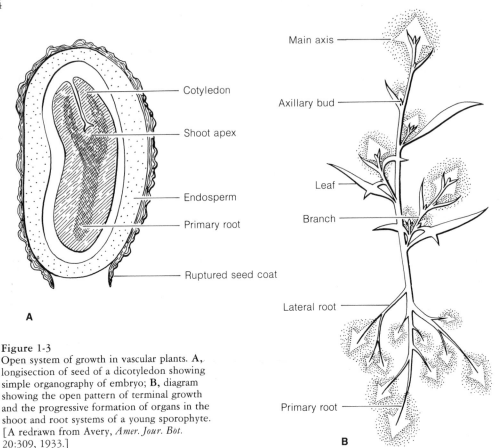

Main axis

Cotyledon

Axillary bud

Shoot apex

Leaf

Branch

Endosperm

Primary root

Lateral root

Ruptured seed coat

Primary root

A

B

Figure 1-3
Open system of growth in vascular plants. **A,**
longisection of seed of a dicotyledon showing
simple organography of embryo; **B,** diagram
showing the open pattern of terminal growth
and the progressive formation of organs in the
shoot and root systems of a young sporophyte.
[A redrawn from Avery, *Amer. Jour. Bot.*
20:309, 1933.]

serious question (Fig. 1-3). The cotyledons of
seed plants occur at the first node of the
embryo and in that respect may be held to be
homologous with one another. But whether,
for example, *all* foliage leaves in vascular
plants at large are homologous is a question
which is by no means easily resolved, either
from an ontogenetic or phylogenetic point
of view (see Chapter 3).

The question of homologies in plants was
placed in an entirely new position as the
result of the publication in 1859 of Charles
Darwin's classic, *The Origin of Species.* His
theory of the rôle of natural selection in producing the gradual adaptive changes in the
form and organography of both plants and
animals exerted a profound effect on all
questions of homologies. The goal of mor-

phology now became very clear: the interpretation of form and structure from a
historical (i.e., phylogenetic) point of view.
Resemblances or homologies between
organs were to be viewed as the result of
descent from a common ancestral "type."
Thus, the strong trend toward the phylogenetic interpretation of form and structure
which arose during the latter part of the
past century has continued to this day. In
addition to its effect on all concepts of homology, the phylogenetic approach to morphology has provided the basis for a more
realistic and natural classification of the
plant kingdom.

Although many of the widely recognized
similarities in basic structure between the
organs and tissues of *related plants* clearly

seem to be "homogenetic" i.e., due to the plants' origin from a common ancestor, it is also evident that there are remarkable structural resemblances between systematically *unrelated species* or groups of plants. In the latter case, the morphological correspondence is "homoplastic" and the result of what is termed convergent evolution. An excellent example of convergent evolution in vascular plants is provided by the presence of seeds in such widely divergent groups as the extinct seed ferns and the Cycadeoidales (fossil cycads of the Mesozoic Era), the modern conifers (e.g., pine, spruce, fir, and redwood) and the highly advanced and diversified taxa of modern flowering plants. The common possession of seeds by members of all these taxa—which otherwise are extremely dissimilar in morphological organization—is in all probability the result of evolutionary convergence or homoplasy.

Additional interesting examples of homoplastic developments in vascular plants are described and analyzed in great detail by Wardlaw (1952, 1955, 1965, 1968a, 1968b), who regards them as "homologies of organization" and worthy of intensive study from the combined viewpoints of comparative morphology, ontogeny, genetics, biochemistry, and biophysics (Fig. 1-4). It must be emphasized, however, that in actual practice it proves very difficult to separate homogenetic from homoplastic resemblances and, with our present knowledge, a rigid distinction is probably impossible. In this connection it must be constantly borne in mind that structural resemblances—whether they are interpreted as homogenetic or homoplastic—are the result of the interaction of genetical, physiological, and environmental factors which have been operating in *different ways* and to *different degrees* during the extremely long evolutionary history of vascular plants. Wardlaw (1965, p. 73) has admirably summarized the task as follows: "To obtain a balanced view,

it is therefore necessary to enquire to what extent prevalent homologies of organization can be accounted for in terms of genetical factors, on the one hand, and of what, for convenience, may be described as common, intrinsic, or not specifically genetical, factors, on the other."

Clearly, reliable interpretations require consideration of evidence that is derived from a wide variety of sources. Morphological theories increase in probability in relation to the extent to which collateral lines of evidence can be harmonized with one another. This chapter may therefore be most appropriately concluded by a brief, critical review of the sources of evidence which should be considered and evaluated in interpreting any problems of form and structure in plants.

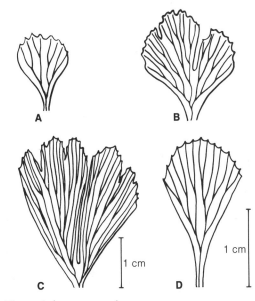

Figure 1-4

"Homology of organization" as illustrated by similar venation patterns in the leaves of unrelated plants. A, B, juvenile, dichotomously veined leaves of *Anemia adiantifolia,* a species of fern; C, dichotomous venation of a lamina segment of *Kingdonia uniflora,* a dicotyledon; D, open dichotomous venation in the lamina of *Circaeaster agrestis,* a dicotyledon. [Adapted from *Organization and Evolution in Plants* by C. W. Wardlaw. Longmans, London. 1965.]

Sources of Evidence in Morphological Interpretation

Adult Form and Structure

By far, the most voluminous data of comparative morphology have resulted from the study of the form of the adult plant.* Information derived from such study has contributed significantly to our knowledge of the wide variations in: (1) the form, venation, and phyllotaxy (arrangement on the axis) of foliar organs, (2) the patterns of branching of root and shoot systems, and (3) the morphological construction of such spore-producing structures as sporophylls, strobili, and flowers. During the second half of the nineteenth century, increasing emphasis was placed upon the study of the primary vascular system of the plant as the key to the interpretation of the morphological nature or homology of plant organs. The wide and continued use today of vascular patterns in morphology is based upon the fundamental assumption that the vascular system is more stable, or conservative, in a phylogenetic sense, than other tissue systems and hence is reliable as a criterion in morphological interpretation. (See Schmid, 1972, for a critique of the concept of vascular conservatism.) Considerable support for this assumption is provided not only from comparative study of living plants but also by the beautifully preserved patterns of vasculation in the vegetative and reproductive structures of extinct plants. Among the many examples that might be given of the use of vascular anatomy in the determination of homologies, the following are outstanding: the morphological interpretation of floral organs (see Chapter 19 for details); the interpretation of the phylogenetic development of the leaf traces in vascular plants (that is, the significance of the number of vascular strands which diverge into a leaf at a node); and the patterns of development of the primary xylem in the stem and root in primitive as compared with advanced plants. In addition to the emphasis on primary vascular systems, much attention has also been given to extensive surveys of the minute structure or histology of secondary xylem, or wood. (See Chapter 19.) The results of such surveys have been applied in the appraisal of the taxonomic aspects of genera and families in the seed plants, and particularly in the effort to determine the origin and trends of evolutionary specialization of tracheids and vessels (Metcalfe and Chalk, 1950; Bailey, 1954).

The Fossil Record

A salient problem common to all phylogenetic interpretations is the difficulty of determining the sequence in the evolutionary development of organs, tissues, and cells. A complete fossil record of the evolutionary history of the sporangium or the leaf would provide evidence of the origin and trends of specialization of these important structures in vascular plants; unfortunately, however, the known fossil record, as revealed by paleobotanical studies, is extremely fragmentary. Consequently, phylogenetic theories are still based largely on circumstantial or indirect evidence derived from the comparative study of living

---

*Strictly speaking, the term "adult" cannot have the same meaning for individual higher plants as it does for individual animals, e.g., vertebrates. In vertebrates the process of embryogeny yields a truly adult organism in which normally no additional organs are produced during the lifetime of the individual. But in vascular plants the continued activity of embryonic regions or meristems at the tips of shoots and roots results in an open system of growth that is characterized by the formation of new organs throughout the life span of the individual (Fig. 1-3). Moreover, in many vascular plants the vascular cambium makes more-or-less extensive periodic additions to the secondary vascular system of the older portions of stems and roots. For convenience in exposition therefore, adult will designate fully developed organs or plant tissues rather than the plant as a whole.

plants. The history of plant morphology is replete with examples of how the same series of morphological types has been interpreted by some investigators as a sequence of *advancing complexity,* and by others as a series in *progressive reduction.* In other words, the decision whether a given structure is primitive or advanced depends upon the interpretation of the apparently "simple" forms in the series; that is, these forms may be regarded as either the beginnings or as terminal specializations. Many simple forms which were regarded in the past as primitive now seem to be specialized because of profound evolutionary reduction. Therefore it is clear that inferences regarding the phylogeny of an organ must be based upon the wise evaluation of the evidence from extinct as well as living types of plants. New paleobotanical discoveries will continue to force morphologists to reconsider and revise many of the so-called classical viewpoints that were based solely on living plants.

To determine the age of rocks containing fossils it is necessary to establish geological chronology by converting the relative time scale of geological events (and the presence of fossils) into a quantitative scale having standard units of time. The procedures used have been termed the "hour-glass" methods. In the middle of the last century the hour-glass methods were based on the salinity of the oceans, sedimentation rates, and thickness of sediments and these measurements provided a framework for a rough time-scale, measured in years.

The historian and astronomer describe events in absolute units of time and it is now possible for geologists and paleobotanists to develop a satisfactory quantitative geological time scale (Table 1-1). This is an achievement of the present century and is based upon the radioactivity of isotopes of certain elements. The age of many minerals and rocks can now be determined within fairly narrow limits by radiometric methods. The dates of extremely ancient

rocks and strata can be established by analyzing the decay of uranium and thorium, for example, to lead and helium, or potassium (K^{40}) to argon (Ar^{40}) and rubidium (Rb^{87}) to strontium (Sr^{87}). Uranium-238 (U^{238}) disintegrates radioactively in a series of steps to a lead isotope (Pb^{206}) and half of a given amount of U^{238} will disintegrate in 4.5 billion years. If, in a sample of a uranium ore, all Pb^{206} has been formed by the process of radioactive disintegration since the original uranium-containing mineral was formed, the ratio Pb^{206}/U^{238} related to the half-life of U^{238} is a measure of this time. By use of radiometric methods with appropriate refinements, the age of the earth has been estimated to be 4.5 billion years.

Although these methods, using the isotopes mentioned, have proved to be essential in dating past Eras and Periods, they are of only limited value in the dating of recent strata because of the extremely long half-life of, for example, uranium-238.

The carbon-14 (C^{14}) method has proved to be extremely useful in dating biological specimens 30,000–50,000 years old. Cosmic rays slam into the earth's atmosphere at high velocities and produce nuclear particles, including some neutrons. Most of the neutrons are absorbed by nitrogen, changing it into the radioactive isotope carbon-14 (C^{14}). The C^{14} then becomes part of the world's reservoir of carbon. C^{14} decays with a half-life of about 5,570 years back to nitrogen-14 (N^{14}). Once an organism dies there is no longer an exchange with atmospheric radiocarbon and the C^{14} "clock" starts. The radiocarbon (C^{14}) begins decaying (reverting) to N^{14}. The ratio of the amount of carbon-14 in a fossil sample to the amount of C^{14} in modern wood or tissue indicates how long ago the organism died. There are several sources of error in this method but, when due precautions are taken, much useful dating information can be obtained for fossil remains and artifacts of prehistoric cultures.

Table 1-1 Geological time scale

ERA	PERIOD AND EPOCH (Years from beginning of time period to present)	CHARACTERISTIC PLANTS AND EVENTS	REPRESENTATIVE GEOLOGIC EVENTS AND ANIMALS
CENOZOIC (QUATERNARY)	RECENT		
CENOZOIC (QUATERNARY)	PLEISTOCENE About 1,500,000–2,000,000	Retreats and advances of vegetation coinciding with several advances and retreats of major continental ice sheets.	Medium elevation of continents and cordilleras. Rise of modern man.
CENOZOIC (TERTIARY)	PLIOCENE 7,000,000	Spread of grasslands. Local extinction or restriction of range of many species due to climatic change in temperate latitudes.	Elevation of Andes and continental uplift in many areas of the world. Mastodons, cats.
CENOZOIC (TERTIARY)	MIOCENE 26,000,000	"Modernization" of forest associations. Restriction of broad-leafed evergreens to lower latitudes. Gradual climatic cooling.	Alps commence to rise. Continental uplifts elsewhere. Climatic cooling, as shown by distribution of floras and faunas.
CENOZOIC (TERTIARY)	OLIGOCENE 37,000,000–38,000,000	Widespread occurrence of now relic taxa: *Metasequoia, Cercidiphyllum.* Dry climates spreading over southwestern North America.	Equable climates in high latitudes. Rodents, rhinoceroses.
CENOZOIC (TERTIARY)	EOCENE 53,000,000–54,000,000 PALEOCENE 65,000,000	Forests in northern and southern temperate latitudes very distinct. Many now extinct genera of woody angiosperms, though closely related to modern genera.	Continued inundation of Gulf and Atlantic States in U.S. Primitive horses.
MESOZOIC	CRETACEOUS 136,000,000	Angiosperms rise to dominance; numerous existing genera in Late Cretaceous: *Magnolia, Liriodendron, Quercus, Persea.* Pines well developed.	Beginning of uplift of Rocky Mts. and Sierra Nevada in U.S. Uniformity of climate resulting from widespread inundation of the continents. Spread of mammals. Last of dinosaurs.
MESOZOIC	JURASSIC 190,000,000–195,000,000	Origin of angiosperms (?) Ginkgoales and Coniferales world-wide. "Age of cycads." Greatest cosmopolitanism of floras in geological history.	Rise of higher insects and birds. Dinosaurs abundant.
MESOZOIC	TRIASSIC 225,000,000	Rise of cycadophytes and modern fern families. Spread of conifers from their Paleozoic ancestors.	Widespread savanna-type climates. Rise of dinosaurs. First mammals.

Era	Subdivision	Period / Age (years)	Plant life	Physical events and fauna
PALEOZOIC		PERMIAN 280,000,000	Rise of Coniferales–Voltziales. Gradual extinction of Carboniferous flora except for derivative forms: herbaceous lycopods, Equisetales.	Low uplift along Appalachian axis. Cooler and drier climate. Glaciation in Southern Hemisphere. Expansion of reptiles.
	CARBONIFEROUS	PENNSYLVANIAN 325,000,000	Widespread forest trees and coal swamps. Seed ferns, sphenophylls, calamites, lycopods, ferns, cordaites. Mosses, liverworts.	Glaciation in Southern Hemisphere. Widespread epicontinental seas. Mild, equable climate. Formation of great coal beds. Rise of reptiles. Insects abundant.
	CARBONIFEROUS	MISSISSIPPIAN 345,000,000	Seed ferns appear. Lycopods, cordaites, sphenophylls, calamites.	Widespread epicontinental seas and limestone deposition in interior U.S. Coal seams. Development of amphibia. Sharks.
		DEVONIAN 395,000,000	First seed plants(?) Rise of lycopods. *Rhynia, Horneophyton, Asteroxylon.* Primitive ferns. Progymnosperms.	Orogeny in New England. Abundance of fish in seas. Gastropods. Decline of trilobites.
		SILURIAN 430,000,000–440,000,000	Simple vascular land plants. *Cooksonia* (in Late Silurian). Fungi.	Arthropods. Echinoderms. Brachiopods.
		ORDOVICIAN 500,000,000	Algae common, bacteria.	First fish-like animals. Corals. Brachiopods.
		CAMBRIAN 570,000,000	Algae, bacteria. Evidence of fungi.	Climate warm and equable. Trilobites abundant. Other marine invertebrates representing most of the major groups.
		PRECAMBRIAN and earlier periods 1,500,000,000–3,200,000,000	Definite evidences of procaryotic algal and bacterial life.	Some questionable invertebrate microfossils.

Source: Time divisions from "Geological Society time-scale," *Quart. J. Geol. Soc.*, London 120(S): 260–262, 1964.

Fossils themselves have been used as indicators of the age of rocks in which they occur. This is based on the observation that certain strata around the world contain characteristic types of plants or animals (*index fossils*). Organisms that evolved rapidly, had wide geographical distribution, and lived for only a short period in geologic time before becoming extinct, constitute the best index fossils. The presence of index fossils in rocks of unknown age, together with certain geologic criteria, makes it possible to determine the age of the rocks by comparing them with rocks from regions where the age of the specific fossil-containing rocks are well known.

In concluding this very brief discussion of the use and limitations of the *known* fossil record in evaluating phylogenetic theories, it seems desirable to outline some of the principal modes of preservation of the organs and tissues of extinct plants. To the thoughtful student, our fragmentary knowledge of the structure of ancient plants should appear less important, at the moment, than the truly remarkable details of form and histology which have already been revealed by means of a variety of special techniques. During the past decade, new discoveries have been made which significantly affect our ideas on relationships among vascular plants, and, very probably, advances of equal or even greater importance in knowledge will be made in the future as paleobotanical research continues.

One of the most common types of preservation of fossil material is termed a *compression*. This results when such an organ as a leaf, for example, is flattened by the weight of the sediment deposited upon it. Although little or no internal cellular structure remains in a typical compression, the cuticle (i.e., the external layer of waxy material which covered the epidermis) is extremely resistant to decay and commonly displays, *in relief,* the pattern of arrangement and the form of the original epidermal cells and stomata. The organization of stomata varies, according to family, genus, or even species of plants, and hence proves of great value in the identification of the leaves and other organs of plants of the remote past (Florin, 1931, 1951). According to Harris (1956), cuticles have been isolated for study from fossils of Devonian age and thus reveal the organization of the epidermis of plants that existed on earth many millions of years ago.

Imprints of plant parts, devoid of organic material, which are formed when the sediment separates from the surface of a fossil, are called *impressions.* As Delevoryas (1962, p. 4) has indicated, an impression actually is the "negative" of a compression and at best may only show the contour and general venation pattern of such a structure as a leaf. Another type of fossil impression, only useful in reconstructing the form and gross organization of the parts (e.g., the stems) of extinct plants is termed a *cast.* In some cases, a cast is the result of the deposition of sediment in the cavity formed by the decay of some or all of the central tissues of a trunk or stem of a plant. Casts of this type may show markings which are the counterparts of the inner surface of the wood and are illustrated by the so-called "pith-casts" of the stems of *Cordaites* and *Calamites* (Arnold, 1947, p. 140, Fig. 61).

The most useful type of preservation, termed a *petrifaction,* reveals, often in *exquisite detail,* the cellular organization of such complex structures as pieces of secondary xylem (wood), stems, leaves, sporangia, and in a few recorded cases, even apical meristems and archegonia (Chapter 7). Petrifactions are formed by the infiltration and subsequent crystallization, in the lumina of the cells and in the intercellular spaces in plant tissues, of calcium, magnesium, or iron carbonate, and of silica or other mineral substances. Petrifactions can be prepared for examination by cutting out and mounting a small portion of the fossil on a slide and then grinding and polishing the surface until the specimen is thin enough to be studied under the microscope with the use of transmitted light. A less wasteful technique, which also makes it possible to obtain serial sections, has been used very successfully for petrifactions in which the carbonized plant material is embedded in a matrix of calcium or magnesium carbonate. The polished surface of the specimen is first treated with dilute hydrochloric acid, which removes the mineral matrix and leaves the cell walls of the tissue projecting above the rock surface; if the matrix is siliceous, hydrofluoric rather than hydrochloric acid must be employed. After the acid has been washed away, the surface of the specimen is flooded with acetone. Then a thin sheet of cellulose acetate is laid on the surface, allowed to dry and then peeled off. The "peel" contains a thin section of the organic remains of the fossil. The process just outlined may be repeated as many times as the specimen allows and in accordance with the number of serial sections which are desired. Each peel can be examined directly or mounted permanently on a slide for detailed microscopical study. (For more detailed information on types of plant fossils and the techniques for studying them, see Arnold, 1947; Andrews, 1961; and Delevoryas, 1962.)

Ontogeny

A highly important source of evidence for morphological interpretation is derived from the study of ontogeny—the actual development of a plant or of one of its component organs, tissues, or cells from the primordial stage to maturity. Histogenesis is a phase of ontogenetic study concerned with the origin of cells and tissues, and embryogenesis and organogenesis are concerned with the history of development of

embryos and organs, respectively. It should be emphasized, however, that the boundaries between these lines of ontogenetic study are drawn largely as a matter of convenience in dealing with varied aspects of development characteristic of the plant as a whole.

Ontogeny, like phylogeny, is thus concerned with origins and with stages in development. The attractiveness of ontogenetic study lies in the possibility of reconstructing, with considerable accuracy and completeness, the sequence of changes in form and structure which actually, occur between the primordial phase of an organ and its mature functional stage. Phylogeny—which is concerned with the sequence of changes in form and structure that have occurred in the past—is, as we have seen, limited in its reconstructions by the exceedingly fragmentary nature of the fossil record.

The importance of detailed ontogenetic information is clearly shown in the interpretation of the life cycles of vascular plants. Since the classical studies of Hofmeister (1862) on *Alternation of Generations,* it has been repeatedly verified that each of the two generations (sporophyte and gametophyte) begins ontogenetically as a single cell. The spore, which results from meiosis, is the primordial cell which develops into the gametophyte, while the zygote, resulting from gametic union or fertilization, is the starting point of the sporophyte. Thus Alternation of Generations, which is one of the most profound generalizations in plant morphology, rests upon the results of ontogenetic study. (see Fig. 2-1)

Ontogenetic studies have also proved essential to the solution of many special morphological problems. For example, the distinction between the two main types of sporangia in vascular plants is based primarily upon differences in their method of origin and early development (Chapter 4). Likewise, the numerous studies of the past century and of today on the ontogeny of foliar organs have shed important light on the morphological interpretation of such structures as stipules, bud scales, and particularly, floral organs (Foster, 1928, 1935). Within the past two decades there has been a notable renaissance of interest in the classical problems of structure and growth of the apical meristems of vascular plants (Gifford, 1954; Gifford and Corson, 1971; and Nougarède, 1967). These studies have not only greatly extended our knowledge of organogenesis and histogenesis but have also emphasized the need for an experimental approach to problems of development in plants (Cutter, 1965, 1966; Gifford and Corson, 1971; Wardlaw, 1952, 1968a).

Despite the demonstrated value of ontogenetic evidence, there are certain limitations to the ontogenetic interpretation of morphological problems which must be clearly understood. One of the most important of these limitations concerns the assumed occurrence of recapitulation in plants. According to the theory of recapitulation, the ontogeny of an organism tends — in abbreviated fashion — to repeat or recapitulate its evolutionary history. A frequently cited example of so-called recapitulation in plants is the development in seedlings of juvenile leaves which differ conspicuously in size, form, and venation from the foliage characteristic of the adult phase of the same plant. However, whether these juvenile leaves provide a reliable clue to the morphology of the ancestral leaf type in any given case is open to serious question. In general, ontogenetic sequences, whether in the succession of foliar types in the young plant or the stages in the ontogeny of an organ or tissue, do not fully or accurately depict the complex path of evolutionary history, and recapitulatory hypotheses should always be made with great caution (Sahni, 1925). This is true, because ontogenetic sequences vary so widely in their

extent and character. In some instances the ontogeny of a structure may be relatively protracted, thus permitting the phylogenetic evaluation of a well-defined series of stages. Vessel elements, for example, acquire their characteristic perforations late in ontogeny and their early development may closely resemble that of tracheids from which they undoubtedly have phylogenetically evolved. Very commonly, however, the ontogenetic history of a structure may be conspicuously abbreviated or telescoped and therefore may be of little or no value for phylogenetic questions.

In conclusion, it is well to realize the important interrelationship between the processes of ontogeny and phylogeny (Mason, 1957). Evolution or phylogeny involves historical change, but from our present point of view "change" is effected by factors which cause the gradual or abrupt modification of ontogenetic processes. A mature tracheid does not give rise to a vessel element, nor does a simple leaf give rise to a complex leaf. On the contrary, evolutionary changes in either tracheary cells or in foliar organs are the result of progressive modifications in the ontogeny of such structures. As Bailey (1944) has so clearly said: "Both comparative and developmental morphology will be more productive of valid generalizations when problems of mutual interest are attacked from a broadened viewpoint of the phylogeny of modified ontogenies in the vascular plants as a whole."

Physiology and Morphogenesis

If plant morphology is regarded as a phase of botanical science exclusively concerned with the description and phylogenetic interpretation of form and structure, it might seem to have little or nothing to do with those dynamic activities of plants that fall within the designation of plant physiology. But is not the wide separation which has developed between morphology and physiology not only artificial but highly undesirable? Goebel (1900) in his monumental *Organography of Plants* adopted the position that "the form and function of an organ stand in the most intimate relation to each other." A similar point of view was followed by Haberlandt (1914) in his attempt to classify and to characterize the tissue systems of plants on a functional basis. Abundant evidence exists that the interpretation of form and structure cannot logically be divorced from function. Sporangia and gametangia in vascular plants are complex multicellular organs, the structure, ontogeny and phylogeny of which constitute important morphological problems. But these organs are more than merely structural features of plants — they are also functionally essential, the sporangia producing spores which give rise to gametophytes and the gametangia forming sperms and eggs that are indispensable to the normal sexual reproduction of a plant. A further illustration of the interrelationships between structure and function is provided by the tracheids and vessel elements in the xylem of vascular plants. These two types of cells vary widely in their form and structure, and they furnish valuable criteria for the broad morphological interpretation of the xylem in vascular plants as a whole. But tracheids and vessels physiologically serve as the major water-conducting elements in plants, and their evolutionary specialization is clearly related to this important function (Bailey, 1953). As a final example of the correlation between structure and function, we may mention the foliage leaves of vascular plants. These organs — although highly diversified in shape, size, and details of anatomy — serve as the major photosynthetic structures of the majority of vascular plants.

It should be clear from these few illustrations that the morphological and functional aspects of plant organs and tissues are indeed interrelated. Although the adaptive or functional significance of many anatomical characters in plants has by no means been experimentally demonstrated, the rigid separation of form from function is, as Goebel remarked, a position which leads to "altogether unfruitful speculations."

We have attempted, in this chapter, to show that comparative morphology is fundamentally concerned with solving the complex problems of the *evolutionary relationships* — i.e., homologies — among plants revealed in their component organs and tissues. In contrast to this emphasis on phylogenetic questions, the rapidly developing field of *morphogenesis* is preoccupied with the *totality of factors*, genetical, biochemical, physiological, and environmental that *together* are responsible for the inception and development of form in plants. One of the many goals of morphogenesis is a better understanding of the "homologies of organization" — i.e., those structural resemblances, between unrelated organisms, which presumably have been produced by parallel and independent paths of evolution. Wardlaw (1968b, p. 99) considers the elucidation of homoplastic or parallel developments as "a major task, not only for the morphologist, but for botanists in general." In this connection, morphogenesis is more than simply the precise observation and description of the successive developmental stages of homogenetic or homoplastic structures: its goals are more comprehensive and are clearly indicated by the expression "causal morphology" — i.e., the *experimental study* of the genesis of form. Indeed, modern studies on morphogenesis use a very wide range of experimental techniques, such as microsurgical operations of various sorts on living apical meristems and the culture *in vitro* of isolated cells, callus tissue, embryos, and the terminal meristems of roots and shoots. (For details, see Cutter, 1965, 1966; Halperin, 1966; Soetiarto and Ball,

1969a, 1969b; Steward, 1968; Steward and Mohan Ram, 1961; Wardlaw 1968a, 1968b.)

In the future, the continued use of these and other techniques may begin to give answers to some of the following fundamental types of questions. (1) Why is the process of the initiation and differentiation of organs and tissues characterized by such an orderly and *integrated series* of developmental stages? Each step in the development of an embryo, for example, is *epigenetic*, i.e., dependent upon the preceding stage, and yet a growing embryo is an integrated "organism" at each and every phase of its development. (2) How and why do structurally similar primordia, which arise successively at the shoot apex, develop into such diverse appendages as bud scales, foliage leaves, and floral organs (e.g., sepals, petals, stamens, and carpels)? (3) And what factors control or "determine" the divergent paths of development of cells that arise from a common meristem? Answering such questions is a challenge for the future, and will require the coordinated efforts of experimental and evolutionary morphologists.

In concluding this chapter it seems appropriate to stress the great need, in the present age of excessive specialization in botany, for a broad multidisciplinary approach to the science of plant morphology. The great dangers of a limited or one-sided approach to morphological problems is very evident today and the dilemma is clearly indicated by the following quotation from Wardlaw's (1968a, p. 14) most recent treatise on plant morphogenesis. "All too often, the morphologist stops just at the point where the more searching study of the underlying physiological-genetical factors really begins; nevertheless, he performs a vital service: he indicates what is there to be investigated. But one can also think of excellently conceived physiological studies that have obviously been undertaken without an adequate knowledge of the *observable* morphogenetic developments. Somewhere between these extremes we ought to do better!"

References

Andrews, H. N., Jr.
  1961. *Studies in Paleobotany*. Wiley, New York.
Arnold, C. A.
  1947. *An Introduction to Paleobotany*. McGraw-Hill, New York.
Bailey, I. W.
  1944. The development of vessels in angiosperms and its significance in morphological research. *Amer. Jour. Bot.* 31:421–428.
  1953. Evolution of the tracheary tissue of land plants. *Amer. Jour. Bot.* 40:4–8.
  1954. *Contributions to Plant Anatomy*. Chronica Botanica Co., Waltham, Mass.

Cutter, E. G.
  1965. Recent experimental studies of the shoot apex and shoot morphogenesis. *Bot. Rev.* 31:7–113.
  1966. *Trends in Plant Morphogenesis*. Wiley, New York.
Darwin, C.
  1859. *The Origin of Species*. J. Murray, London.
Delevoryas, T.
  1962. *Morphology and Evolution of Fossil Plants*. Holt, Rinehart and Winston, New York.
Florin, R.
  1931. Untersuchungen zur Stammesgeschichte der Coniferales und Cordaitales. *Svenska Vetensk. Akad. Handl.* Ser. 5. 10:1–588.
  1951. Evolution in Cordaites and Conifers. *Acta Horti Bergiani*. Bd. 15. No. 11.

Foster, A. S.

1928. Salient features of the problem of bud-scale morphology. *Biol. Rev.* 3:123–164.

1935. A histogenetic study of foliar determination in *Carya Buckleyi var. arkansana. Amer. Jour. Bot.* 22:88–147.

Gifford, E. M., Jr.

1954. The shoot apex in angiosperms. *Bot. Rev.* 20:477–529.

Gifford, E. M., Jr., and G. E. Corson, Jr.

1971. The shoot apex in seed plants. *Bot. Rev.* 37:143–229.

Goebel, K.

1900. *Organography of Plants*, Eng. Ed., Pt. I. Clarendon Press, Oxford.

Goethe, J. W., von

1790. *Versuch die Metamorphose der Pflanzen zu erklären.* Gotha.

Haberlandt, G.

1914. *Physiological Plant Anatomy.* Macmillan, London.

Halperin, W.

1966. Alternative morphogenetic events in cell suspensions. *Amer. Jour. Bot.* 53(5): 443–453.

Harris, T. M.

1956. The fossil plant cuticle. *Endeavour* 15(60): 210–214.

Hofmeister, W.

1862. *On the Germination, Development and Fructification of the Higher Cryptogamia and on the Fructification of the Coniferae.* Published for the Ray Society by Robert Hardwicke, London.

Mason, H. L.

1957. The concept of the flower and the theory of homology. *Madroño* 14:81–95.

Metcalfe, C. R., and L. Chalk

1950. *Anatomy of the Dicotyledons.* 2 v. Clarendon Press, Oxford.

Nougarède, A.

1967. Experimental cytology of the shoot apical cells during vegetative growth and flowering. *Int. Rev. Cytol.* 21:203–351.

Sahni, B.

1925. The ontogeny of vascular plants and the theory of recapitulation. *Jour. Ind. Bot. Soc.* 4:202–216.

Schmid, R.

1972. Floral bundle fusion and vascular conservatism. *Taxon* 21:429–446.

Soetiarto, S. R. and E. Ball

1969a. Ontogenetical and experimental studies of the floral apex of *Portulaca grandiflora.* I. Histology of transformation of the shoot apex into the floral apex. *Can. Jour. Bot.* 47:133–140.

1969b. Ontogenetical and experimental studies of the floral apex of *Portulaca grandiflora.* II. Bisection of the meristem in successive stages. *Can. Jour. Bot.* 47:1067–1076.

Steward, F. C.

1968. *Growth and Organization in Plants.* Addison-Wesley, Reading, Mass.

Steward, F. C., and H. Y. Mohan Ram

1961. Determining factors in cell growth: some implications for morphogenesis in plants. Pp. 189–265 *in* Abercrombie, M., and J. Brachet (eds.), *Advances in Morphogenesis,* Vol. I. Academic Press, New York.

Wardlaw, C. W.

1952. *Phylogeny and Morphogenesis.* Macmillan, London.

1955. *Embryogenesis in Plants.* Methuen, London.

1965. *Organization and Evolution in Plants.* Longmans, London.

1968a. *Morphogenesis in Plants. A Contemporary Study.* Methuen, London.

1968b. *Essays on Form in Plants.* University Press, Manchester.

Wolff, C. F.

1774. *Theoria Generationis,* Editio nova. Halle.

2

The Salient Features of Vascular Plants

One of the most significant events in the long evolutionary development of the plant kingdom was the origin of the first land plants. These organisms, which in all probability arose from some ancient group of green algae, are believed by many students of evolution to have been extremely simple in organography. Perhaps they closely resembled such apparently primitive plants as the Devonian *Rhynia* and *Horneophyton* (Chapter 7). The conquest of the land must have been a long, hazardous, and costly process of slow adjustment to a new and inhospitable environment, and many of the steps in that early trial-and-error period may always remain unknown to science. But when the organography, anatomy, and spore develop-

ment of such ancient forms as *Rhynia* are compared with the morphology of a modern angiosperm, the magnitude of the changes which accompanied evolution on the land becomes evident. *Rhynia* was a rootless, leafless plant that was low in stature, and it had a simple and primitive vascular system. The reproductive structures of the sporophyte were crude sporangia located at the tips of the aerial branches, and they produced spores which were alike in form and size. In contrast, a present-day angiosperm, e.g., a modern dicotyledonous tree, is an extremely complex organism. In addition to its chief vegetative organs (roots, stems, and leaves), such a plant develops flowers, fruits, and seeds and exhibits a highly evolved vascular system which is periodically renewed and augmented by the activity of a vascular cambium. Obviously *Rhynia* and the angiosperm share certain general morphological features (photosynthetic, spore-producing, and vasculated sporophytes) but differ profoundly in the degree and type of development of other morphological characters. In this chapter an effort will be made to select and to describe briefly the salient morphological features that are common to most living vascular plants. Such preliminary analysis and orientation seem indispensable as an approach to the more detailed treatment of comparative morphology as presented in subsequent chapters. For convenience in exposition, the definitive features of sporophyte and gametophyte generations will be described separately.

Sporophyte Generation

The normal origin of this generation is from the zygote, a diploid cell which results from the fertilization of the egg by a sperm (Fig. 2-1). There is considerable variation in vascular plants with respect to the length of the period of attachment and physiological dependency of the young multicellular sporophyte upon the gametophyte generation. In certain lower vascular plants, for example *Lycopodium,* sporophytes may remain connected to the gametophyte for several years. By contrast, the period of dependency of the sporophyte in heterosporous plants, for example *Selaginella,* and seed plants is very brief and is usually limited to the phase of embryogeny. Ultimately, in *all types* of vascular plants the sporophyte gains physiological independence and develops into the dominant, typically photosynthetic, phase of the life cycle. *This physiological independence and dominance of the sporophyte constitutes one of the most definitive characters of all vascular plants.*

From an organographic viewpoint, the sporophyte typically consists of a shoot system (usually aerial) made up of stems, various types of foliar organs, and roots. The latter vary widely in their origin during embryogeny and later phases of growth—they are absent only from the Psilotaceae and certain highly specialized aquatic and parasitic plants. Shoots and roots are theoretically capable of unlimited apical growth, branching, and organ formation because of the maintenance at their tips of apical meristems. This capacity for continued indefinite apical growth is not found in the small dependent sporophytes of mosses and liverworts and hence represents a fundamentally important distinction between these organisms and all vascular plants (see Fig. 1-3).

From an anatomical viewpoint, one of the most definitive features of the sporophyte is the presence of a vascular system. Indeed, the presence in the xylem, or water-conducting part of this system, of cells known as tracheids is responsible for the designation of all vascular plants as "Tracheophyta." (See Bailey, 1953, for a discussion of the evolution of tracheary tissue.) As will become evident in subsequent chapters, the form and arrangement of the vascular system vary not only among different groups of tracheophytes, but among different organs of the same plant as well. Because of its remarkably well-preserved condition in fossils, the vascular system often provides important clues to relationships and to trends of phylogenetic development. Except for the vein endings, which in leaves or floral organs consist largely of tracheary cells, the vascular system consists typically of two distinctive tissues: *phloem,* the definitive conducting elements of which are the sieve elements; and *xylem,* the important conducting or tracheary elements of which are the tracheids and the vessel elements. The basic difference between the two types of xylem cells is that the tracheid is an imperforate cell, whereas the

18

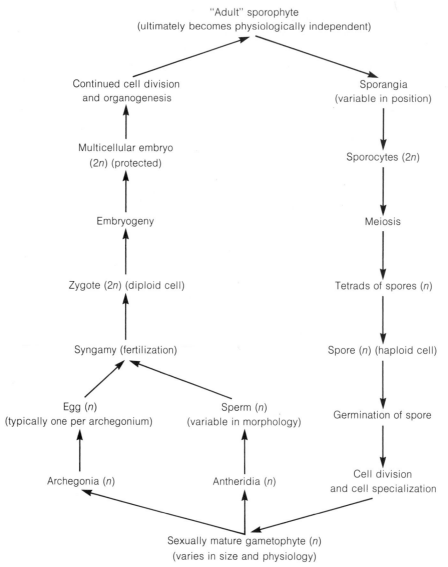

Figure 2-1
Alternation of sporophyte and gametophyte generations in the complete life cycle of lower vascular plants. The processes and structures represented are discussed in detail in the text.

ends of vessel elements ontogenetically develop one or more openings or perforations. A series of superposed vessel elements constitutes a vessel. Vascular tissues are essential to the conduction of water, mineral solutes, and soluble organic compounds — upon their coordinated functioning depends the normal physiology and growth of the sporophyte. Evidently, in the early occupation of the land by plants, the rapid evolution of an effective vascular system must have been of primary importance.

Although typical vascular tissue, in the sense already defined earlier in this chapter, is usually restricted to the sporophyte generation, the student should be aware that tracheary tissue has been observed in the gametophytes of certain plants. The best known example is the subterranean gametophyte of *Psilotum,* a vascular plant of uncertain phylogenetic origin and systematic relationships (see Chapter 8). The original discovery of vascularized gametophytes in *Psilotum* was made by Holloway (1939) in material collected from two widely separated localities on Rangitoto, a volcanic islet off the coast of Auckland, New Zealand. In about 30 gametophytes, which were one millimeter or more in diameter, a central conducting strand, demarcated by an endodermis, was present. A well-developed strand consists (in transection) of 1–3 tracheids with annular or scalariform thickenings, surrounded by elongated nucleated cells which occupy the position of "phloem" tissue but which lack sieve areas on their walls. The vascular strand arises at the apical meristem of the gametophyte but commonly is discontinuous, "fading out and reappearing usually several times in the length of a few millimeters" (Fig. 2-2).

Why a weak but usually definable central vascular strand tends to occur in robust gametophytes of *Psilotum* but is absent in gametophytes of comparable size in the closely related genus *Tmesipteris* is an interesting but unsolved problem (Lawson, 1917, Holloway, 1939, Bierhorst, 1968). On the other hand, several "explanations" for the occurrence of a vascular strand in *Psilotum* gametophytes have already been made by Holloway (1939) and deserve brief examination especially in the light of additional studies made since the publication of his paper.

First of all, it might be argued that the vascularized gametophytes of *Psilotum* are "abnormal" in chromosome number and that this cytological condition in some way "conditions" the development of a conducting

strand. Manton (1950, p. 236) made the necessary chromosome studies on gametophytes from Rangitoto supplied to her by Holloway and discovered that *all sizes* of gametophytes were diploid ($2n$) and that the sporophytes, to which they were related, were tetraploid ($4n$). Although chromosome number per se does not appear to "explain" the occurrence of vascularized gametophytes it is noteworthy that Bierhorst (1968, p. 241), in his intensive studies of *Psilotum,* only found vascular tissue in the larger diploid gametophytes of *P. nudum.* Haploid gametophytes of the same species, collected by Bierhorst from both Fiji and New Caledonia, were entirely devoid of a vascular system.

A second possible interpretation of vascularized gametophytes in *Psilotum* is that some type of physiological change—associated with the initiation of a procambial strand—occurs after a given gametophyte has reached a certain size in its growth. Holloway (1939, p. 332) states that as far as his studies are concerned, "the presence of the conducting strand in *Psilotum* is correlated with special robustness of growth." Bierhorst (1953, p.

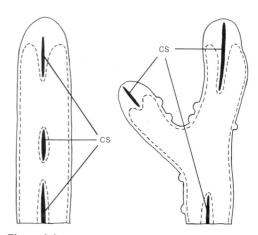

Figure 2-2
Psilotum triquetrum (= *nudum*). Diagrammatic longisections of two large gametophytes showing the interrupted character of the conducting strands (cs). [Redrawn from Holloway, *Ann. Bot.* n. s. 3:313, 1939.]

656) reached a similar conclusion and stated that "all the gametophytes are potential producers of vascular strands, and that a given gametophyte apex can, if the bulk of its meristem reaches a given threshhold, produce a strand."

A third possible interpretation of the vascular strand in the gametophyte — particularly in the light of recent paleobotanical discoveries — is that it represents the persistence of a type of vasculature that was characteristic of the prothallia of at least *some* of the ancient and primitive vascular plants of the Devonian Period, such as *Rhynia* (for a full description of the morphology of *Rhynia*, see Chapter 7). Pant (1962) made the suggestion that the axes of *Rhynia gwynne-vaughani*, which had previously been interpreted by Kidston and Lang (1921) as portions of the sporophyte, might actually represent vascularized gametophytes, comparable with those of the living *Psilotum*. Although Pant did not find convincing evidence of sex organs on the presumptive gametophytic axes of *R. gwynne-vaughani*, he maintains that the "hemispherical projections" which are present might be interpreted as being young *attached* sporophytes. Very recently, Lemoigne (1968, 1969, 1970) has discovered what appear to be well-preserved archegonia on certain of the vascularized axes of *R. gwynne-vaughani*. Curiously enough, none of the presumptive gametophytes appear to have borne antheridia, although the subterranean gametophytes of both *Psilotum* and *Tmesipteris* are bisexual, i.e., the same gametophyte forms both antheridia as well as archegonia (Chapter 8, p. 130).

It is evident from Lemoigne's interesting studies that much remains to be learned about the occurrence, systematic affinities, and phylogenetic significance of vascularized gametophytes in *Rhynia*. Perhaps, as Pant (1962, p. 298) cautiously suggested, the presumptive gametophytes of *Rhynia gwynne-*

vaughani may eventually prove to represent the sexual generation of *Rhynia major*. Lemoigne (1968, p. 3) seemed to admit this as a possible interpretation but later (1969) emphasized the need for more information about the branching pattern and other aspects of the general morphology of the gametophyte and sporophyte of *R. gwynne-vaughani* before a "reconstruction" of this "species" is possible.

The characteristic reproductive organ of the sporophyte is the sporangium. As will be shown in detail in Chapter 4, sporangia differ significantly in their position, methods of origin and development, and mature structure including sporangium wall and spore type. These differences provide reliable criteria, in many instances, for classifying vascular plants. From the standpoint of reproduction (i.e., spore formation) such diverse structures as the sporophylls of lower heterosporous plants and the stamens and carpels of the angiosperm flower are fundamentally similar. However, whether these structures are homologous in a morphological sense is a question which must be reserved for later discussion.

Gametophyte Generation

The normal origin of the gametophyte generation is from a spore, which, in turn, is a product of the meiotic (reductional) divisions of spore mother cells in the sporangium (see Fig. 2-1). In certain lower vascular plants, such as some species of *Lycopodium, Equisetum,* and most of the ferns, the gametophyte is a free-living photosynthetic plant; in others, such as *Psilotum, Tmesipteris,* and many species of *Lycopodium,* the gametophyte is subterranean, devoid of chlorophyll, and apparently dependent upon the presence of an endophytic fungus for its existence. In marked contrast, the gametophytes of gymnosperms and angiosperms are much smaller and are physiologically dependent upon the sporophyte generation. Aside from these morphological and physiological variations, the chief importance of the gametophyte is the production of male gametes (or sperms) and female gametes (or eggs). These gametes are developed in distinctive multicellular gametangia in all lower vascular plants, the sperms arising in antheridia and the solitary egg developing within the archegonium. In angiosperms, the male and female gametes are produced directly by the greatly reduced and modified gametophytes, and morphologically definable sex organs are not developed. The position, ontogeny, and structure of gametangia and gametes furnish important characters used in the comparison and classification of the groups of lower vascular plants. A full discussion of the comparative morphology of gametangia is presented in Chapter 5.

Alternation of Generations

Alternation is a consistent feature of all groups of tracheophytes and hence represents the basic pattern of reproduction in these dominant plants of the modern world. Figure 2-1 represents schematically the structures and processes common to the complete life cycle of lower vascular plants. The student must understand this generalized life cycle clearly before he becomes involved in the infinite variations of detail which occur in the cycle of reproduction of specific genera or groups of plants.

Because the sporophyte generation is the obvious and dominant phase of the life cycle, we may properly begin our analysis of Fig. 2-1 with the zygote. This diploid cell results from the union of a male gamete with the egg, a process occurring in the archegonia of lower vascular plants, including the majority of gymnosperms, and in the female gametophyte of angiosperms. The next event

is the process of embryogeny, which involves the production from the zygote of a multicellular embryo, the early development, form, and organography of which are often specific in a given group. By means of further growth and differentiation from the shoot and root apices of the embryo, the adult and independent sporophyte is developed. The general organography and vasculation of the sporophyte have already been outlined in this chapter. Ultimately as shown in Fig. 2-1, the sporophyte plant forms sporangia in which spore-mother cells, or sporocytes, are produced. These cells, like all normal cells of the vegetative sporophyte, are diploid. However, each sporocyte can by meiotic division give rise to a tetrad or group of four haploid spores. When circumstances are favorable a spore germinates and by cell division and cell specialization produces the gametophyte generation. As we have already seen, this phase in the life cycle varies considerably in size and physiology—its salient function is the production of sperms and eggs. The union—called fertilization or syngamy—of a male gamete and an egg restores the diploid chromosome number and produces a zygote from which a new sporophyte plant may develop.

It should be clear from this description that the alternation of sporophyte and gametophyte generations in the life cycle is normally coordinated with a periodic doubling followed by a halving of the chromosome number. The diploid zygote $(2n)$, from which the sporophyte arises contains twice the number of chromosomes typical of the spore (n) that produces the gametophyte. Is this difference in chromosome number of spore and zygote a clue to the remarkable morphological and functional differences between the generations which arise from these cells? Or, to ask the question in another way: Are syngamy and meiosis *always essential processes* in the production, respectively, of sporophyte and gametophyte?

These questions apparently deserve an answer in the negative because of certain deviations from a "normal" reproductive cycle, namely, *parthenogenesis, apospory,* and *apogamy.*

In certain angiosperms and a few ferns, the embryo arises from an unfertilized egg, a phenomenon designated as parthenogenesis (i.e., "virgin birth"). An example of parthenogenesis, discussed by Steil (1939, p. 447) is found in *Marsilea.* In this fern, certain of the female gametophytes, because of their origin from diploid sporocytes rather than from normal haploid spores, produce archegonia containing diploid eggs which in turn directly give rise to diploid sporophytes.

The phenomenon termed *apospory* is the development of gametophytes, without a haploid spore stage, from vegetative cells of the sporophyte. Aposporous gametophytes may develop spontaneously in nature but they can also be artificially induced by culturing detached leaves or pinnae under conditions favorable for regeneration. Very commonly, aposporous gametophytes bear functional antheridia and archegonia, and chromosome studies have shown the remarkable fact that their gametes are diploid and that a sporophyte resulting from their union is tetraploid (see Steil 1939, p. 446, and 1951 for further information on apospory).

A third deviation from the usual reproductive cycle of a vascular plant is the phenomenon of *apogamy,* which is the formation of a sporophyte, without the act of fertilization, from cells of the gametophyte (exclusive of the egg itself). Depending upon the chromosome number of the gametophyte (i.e., whether normal n or aposporous $2n$), the apogamous sporophyte may be either haploid or diploid in chromosome number.

In view of the extended discussion already included on the remarkable vascularized gametophyte of *Psilotum* (pp. 18–20), it is very interesting to note that Steil (1939,

pp. 439–440) states that "one of the first and best evidences of apogamy in ferns is the appearance of tracheids just back of the apical notch of the prothallium. Probably in all of the apogamous species, tracheids are produced at a comparatively early stage in the development of the embryo, usually before the gametophyte has grown to its maximum size." It should be emphasized, however, that several recent studies fail to support the idea that an *obligatory* relationship exists between the formation of tracheids in fern gametophytes and the occurrence of apogamy: DeMaggio, Wetmore, and Morel (1963), for example, by regulating the concentration of sugar in the culture medium, were able, in the absence of apogamy, experimentally to induce the formation of procambium and its subsequent differentiation into scalariform tracheids in the gametophyte of *Todea barbara,* a fern belonging to the family Osmundaceae (see Chapter 13). They concluded "that the formation of vascular tissue in the prothallus and the phenomenon of apogamy can be entirely independent of one another." A further example of the formation of vascular tissue in fern gametophytes, independent of any correlated apogamy, is provided by Klekowski's (1968, pp. 147–155) investigation on the fern *Osmunda regalis.* In one of his experimental cultures, which had already been used for genetical breeding tests, a vascular strand, sometimes unequally bifurcated, was found in most of the gametophytes. The vascular tissue was "comprised mainly of spirally thickened tracheids."

It seems evident, from this brief résumé, that, as with the diploid vascularized gametophytes of *Psilotum,* the significance – physiological, cytological, or phylogenetic – of the capacity of fern gametophytes to form vascular strands must for the present be considered an unsolved problem and one that invites much more study for its solution than it has received in the past. The genetic complexity, for example, that *may* be associated with vascularized fern gametophytes is strikingly illustrated by Morzenti's (1966) discovery of tracheids in many gametophytes of *Asplenium heteroresiliens*. This "species" is interpreted as a pentaploid ($5n$) hybrid between a tetraploid sexual plant of *Asplenium heterochroum* and an apogamous triploid form of *Asplenium resiliens*! According to Morzenti, tracheids were developed "whether or not the gametophyte had produced sporophytes."

Quite apart from the interesting questions raised by the phenomena of parthenogenesis, apospory, and apogamy, there remains the broader and still unsolved problem of the phylogenetic origin of the type of alternation of generations which is "normal" and prevalent throughout all living vascular plants. If, as is commonly believed, the remote ancestors of terrestrial plants were aquatic types of green algae, the central question at issue is the evolutionary origin of the diploid, vascularized, independent sporophyte typical of the vast majority of the Tracheophyta.

Nearly a century ago, Celakovsky (1874) recognized two principal forms of alternation which he distinguished by the terms "antithetic" and "homologous." In his opinion, antithetic alternation, which is characteristic of archegoniate plants (i.e., bryophytes and vascular plants with archegonia), arose, during evolution, by the interpolation of a new phase, i.e., the sporophyte, between successive gametophytes. In contrast, he limited his concept of homologous alternation to the succession of morphologically similar phases which occurs in the reproductive cycles of fungi and algae.

Today the terms "antithetic" and "homologous" are usually applied to two divergent theories which both seek to explain phylogenetically the characteristic type of life cycle in vascular plants. Bower (1935) has urged that in place of "antithetic" and "homologous" theories it would be more

appropriate to speak, respectively, of the "interpolation theory" and the "transformation theory." In his view, these substitute terms are explicit in that they convey the alternative methods of phylogenetic origin of the sporophyte generation. This point may be clarified by the following brief contrast between the two theories. For a more detailed analysis, the student is referred to Bower (1935), Fritsch (1945), and Wardlaw (1952).

According to the *interpolation theory*, which was strongly championed by Bower, the origin of the sporophyte generation was fundamentally the result of the postponement of meiosis and the development from the zygote of a new diploid vegetative phase. During subsequent evolution, this vegetative phase or sporophyte became increasingly complex, both organographically and anatomically but "meiosis and spore formation, though delayed, would still be the final result." Bower also maintained that the gradual specialization of an independent spore-producing diploid phase was closely connected, in a biological sense, with the transition from aquatic to terrestrial life by the algal-like "progenitors" of the first land plants.

The *transformation theory* of alternation, in contrast, postulates that the spore-producing and gamete-producing *individuals* of the *original* parental green algae were morphologically similar and hence that the life cycle was *isomorphic* in type (Fig. 2-3, A, B). This idea has been developed in great detail by Fritsch (1945) and some of his interesting ideas deserve brief attention at this point. According to Fritsch, the first step in the long evolutionary development of archegoniate plants from aquatic green algae was the attainment of a morphological distinction between *erect* branches and a creeping prostrate system. This resulted in what is termed a *heterotrichous habit* of growth. Further elaboration of the upright system by means

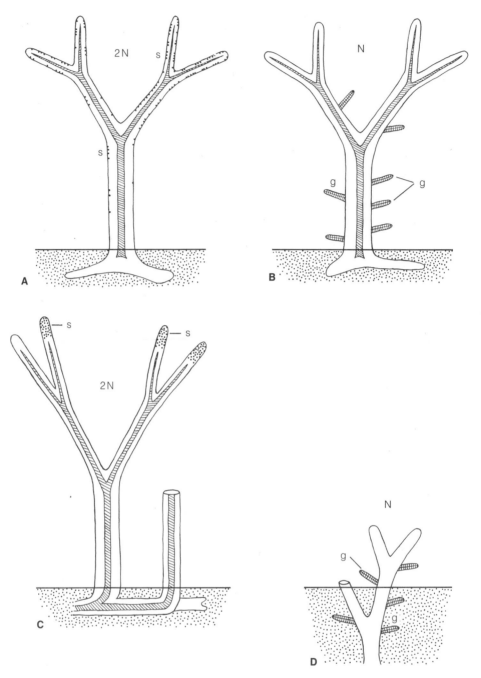

Figure 2-3
Diagrams illustrating the transformation theory of alternation of generations. **A, B,** isomorphic generations: **A,** sporophytic, **B,** gametophytic. **C, D,** heteromorphic generations: **C,** sporophytic, **D,** gametophytic. See text for further explanations. Abbreviations: g, gametangia; s, sporangia. [Redrawn from Fritsch, *Ann. Bot.* n. s. 9:1, 1945.]

of longitudinal divisions in some of the cells then produced a more-or-less "parenchymatous" type of organization. Fritsch regarded this second step a "most significant evolutionary advance, since in it lay the germ for the development of a plant body of almost unlimited size." Perhaps the remarkable terrestrial green alga, named *Fritschiella* by Iyengar (1932), might be regarded as typifying an early stage in development of a very primitive nonvascular land plant. Further evolutionary changes, according to Fritsch's theory, included (1) the appearance of dichotomous branching, (2) the origin of a primitive type of vascular system, (3) the formation of a surface cuticle and associated stomata, and (4) the structural specialization of sporangia and gametangia (Fig. 2-3, A, B). These postulated changes led to further morphological divergence between sporophytic and gametophytic individuals and the adoption of a *heteromorphic* type of life cycle (Fig. 2-3, C, D). In Fritsch's view, the gametophyte, at this stage in evolution, corresponded in form and dichotomous branching to the primary underground system of the sporophyte and had "lost the capacity for emergence overground." This conclusion is particularly interesting in the light of Lemoigne's (1968, 1969) recent discovery of "rhizome-like," vascularized gametophytes of the Devonian genus *Rhynia*.

It is hoped that future paleobotanical research will reveal additional examples of the morphological nature of truly primitive gametophytes and sporophytes and thus may aid in resolving the present divergence in opinion about the mode of origin of alternation of generations in vascular plants.

Classification of Vascular Plants

The dominant land plants of the earth, despite their extraordinary diversity in habit, organography, and method of repro-duction, share one highly important character, namely, the presence of a vascular system consisting of phloem and xylem that serves to conduct water, mineral solutes, and various organic materials throughout the plant body. This anatomical character has been widely accepted as the basis for designating all plants with *tracheary tissue,* i.e., xylem, as "Tracheophytes" (Sinnott, 1935), or more technically, "Tracheophyta" (Eames, 1936).

Certain modern authors, such as Bold (1967), have rejected the concept of "Tracheophyta" as a *major* division of the plant kingdom in favor of a polyphyletic scheme in which vascular plants are classified into nine "parallel" divisions. It must be emphasized, however, that all modern efforts to designate and to classify the major taxa of vascular plants are tentative and are subject to continued change or revision as new morphological types of organization continue to turn up in the fossil record. Therefore, in our opinion, the designation "Tracheophyta" is no more artificial than other categories proposed for higher land plants such as "Embryophyta," "Xylophyta," and "Stelophyta." (See detailed review of the problems of nomenclature and classification by Just, 1945, and Newman, 1947.) Certainly, from a broad evolutionary viewpoint, the origin of a vascular system must have been highly significant in the development of the sporophyte generation and in its continued adjustment to life on the land. For this reason, the use of the concept of "Tracheophyta" in this book seems to us fully justified from a physiological as well as a morphological standpoint (see Stewart, 1960, 1961).

A satisfactory phylogenetic classification of vascular plants into a reasonable number of subdivisions, classes, orders, and families is of course a formidable task which has not been completed to the satisfaction of contemporary morphologists and paleobotanists. One of the difficulties is the absence of convincing phyletic interconnections between the various groups of vascular plants which have been proposed. This tends to result in the rapid multiplication of "new" taxa, a result which may yield only temporary relief to the student of evolution and relationships among plants. Moreover, there is considerable confusion at the level of formal schemes of classification because of the continuous shifting of lower taxa (e.g., classes) to higher categories (e.g., subdivisions or divisions) or vice versa, merely by changing the suffix of certain names.

In view of these difficulties and nomenclatural problems, no attempt will be made in this chapter to contrast the various schemes of classifying vascular plants. But we have departed from the relatively simple classification, adopted in the first edition of this book, in order to use what we believe represents a significant advance in the treatment of vascular plants as a whole. In place of recognizing four major subdivisions of the Tracheophyta, namely, Psilopsida, Lycopsida, Sphenopsida, and Pteropsida, we have accepted, *in part,* some of the broad outlines of the recent classification of vascular plants proposed by Banks (1968). We believe that his taxonomic scheme displays the impact of new paleobotanical discoveries on taxonomy, particularly in that it recognizes the "Progymnospermopsida," a new class of tracheophytes from which all gymnosperms may have evolved (see Beck, 1960 and Chapter 14).

The conspectus shown below lists the twelve classes of the Tracheophyta which are recognized in the present text. The introduction of the conspectus at this point is intended simply to orient the student to the names and representative examples of the main classes of vascular plants. More detailed taxonomic treatment of each of the classes will be found in subsequent chapters.

CLASSIFICATION OF VASCULAR PLANTS
(TRACHEOPHYTA)

1. Rhyniopsida (extinct plants, e.g., *Rhynia*)
2. Zosterophyllopsida (extinct plants, e.g., *Zosterophyllum*)
3. Trimerophytopsida (extinct plants, e.g., *Trimerophyton*)
4. Psilopsida (2 living genera, *Psilotum* and *Tmesipteris*)
5. Lycopsida (extinct forms such as *Baragwanathia* and *Lepidodendron* and the living genera, *Lycopodium, Selaginella, Phylloglossum, Isoetes,* and *Stylites*)
6. Sphenopsida (largely extinct plants, e.g., *Calamites* and *Sphenophyllum;* the only living genus is *Equisetum.*)
7. Filicopsida (living and extinct ferns)
8. Progymnospermopsida (extinct plants, e.g., *Archaeopteris*)
9. Cycadopsida (seed ferns, cycadeoids, both extinct; the living cycads)
10. Coniferopsida (extinct and living conifers; *Ginkgo biloba,* a "living fossil")
11. Gnetopsida (*Ephedra, Gnetum, Welwitschia*)
12. Angiospermopsida (angiosperms or flowering plants)

References

Bailey, I. W.
    1953. Evolution of the tracheary tissue of land plants. *Amer. Jour. Bot.* 40:4–8.
Banks, H. P.
    1968. The early history of land plants. Pp. 73–107 *in* Drake, E. T. (ed.), *Evolution and Environment.* Yale University Press, New Haven.
Beck, C. B.
    1960. The identity of *Archaeopteris* and *Callixylon. Brittonia* 12:351–368.
Bierhorst, D. W.
    1953. Structure and development of the gametophyte of *Psilotum nudum. Amer. Jour. Bot.* 40:649–658.

1968. On the Stromatopteridaceae (Fam. Nov.) and on the Psilotaceae. *Phytomorphology* 18:232–268.

Bold, H. C.
1967. *Morphology of Plants,* Ed. 2. Harper and Row, New York.

Bower, F. O.
1935. *Primitive Land Plants.* Macmillan, London.

Celakovsky, L.
1874. *Bedeutung des Generationswechsels der Pflanzen.* Prag.

DeMaggio, A. E., R. H. Wetmore, and G. Morel
1963. Induction de tissue vasculaire dans le prothalle de Fougère. *Compt. Rend. Acad. Sci.* (Paris) 256:5196–5199.

Eames, A. J.
1936. *Morphology of Vascular Plants. Lower Groups.* McGraw-Hill, New York.

Fritsch, F. E.
1945. Studies in the comparative morphology of the algae. IV. Algae and archegoniate plants. *Ann. Bot.* n. s. 9:1–29.

Holloway, J. E.
1939. The gametophyte, embryo, and young rhizome of *Psilotum triquetrum* Swartz. *Ann. Bot.* n. s. 3:313–336.

Iyengar, M. O. P.
1932. *Fritschiella,* a new terrestrial member of the Chaetophoraceae. *New Phytol.* 31:329–335.

Just, T.
1945. The proper designation of the vascular plants. *Bot. Rev.* 11:299–309.

Kidston, R., and W. H. Lang
1921. Old Red Sandstone plants showing structure, from the Rhynie Chert Bed, Aberdeenshire. IV. Restorations of the vascular cryptogams and discussion of their bearing on the general morphology of the Pteridophyta and the origin of the organization of land plants. *Trans. Roy. Soc. Edinb.* 52:831–854.

Klekowski, E. J.
1968. Reproductive biology and evolution in the Pteridophyta. Ph.D. Diss., University of Calfornia, Berkeley.

Lawson, A. A.
1917. The gametophyte generation of the Psilotaceae. *Trans. Roy. Soc. Edinb.* 52: 93–113.

Lemoigne, Y.
1968. Observation d'archégones portés par des axes du type *Rhynia gwynne-vaughanii* Kidston et Lang. Existence de gamétophytes vascularisés au Dévonien. *Compt. Rend. Acad. Sci.* (Paris) 266:1655–1657.
1969. Contribution à la connaissance du gamétophyte *Rhynia gwynne-vaughanii* Kidston et Lang. Problème des protuberances et processus de ramification. *Bull. Mens. Soc. Linn. Lyon* 38(4):94–102.
1970. Nouvelles diagnoses du genre *Rhynia* et de l'espèce *Rhynia gwynne-vaughanii.* *Bull. Soc. Bot. France* 117:307–320.

Manton, I.
1950. *Problems of Cytology and Evolution in the Pteridophyta.* Cambridge University Press, London.

Morzenti, V. M.
1966. Morphological and cytological data on Southeastern United States species of the *Asplenium heterochroum-resiliens* complex. *Amer. Fern Jour.* 56:167–177.

Newman, I. V.
1947. The place of ferns and seed plants in classification. *Trans. Roy. Soc. New Zealand* 77:154–160.

Pant, D. D.
1962. The gametophyte of the Psilophytales. *Proc. Summer School Bot. Darjeeling, India.* (Held June 2–15, 1960) Pp. 276–301.

Sinnott, E. W.
1935. *Botany. Principles and Problems,* Ed. 3. McGraw-Hill, New York.

Steil, W. N.
1939. Apogamy, apospory, and parthenogenesis in the pteridophytes. *Bot. Rev.* 5:433–453.
1951. Apogamy, apospory, and parthenogenesis, II. *Bot. Rev.* 17:90–104.

Stewart, W. N.
1960. More about the origin of vascular plants. *Plant Sci. Bull.* 6:1–5.
1961. The origin of vascular plants: monophyletic or polyphyletic? *Rec. Adv. Bot.* 2:960–963.

Wardlaw, C. W.
1952. *Phylogeny and Morphogenesis.* Macmillan, London.

3

The Vegetative
Sporophyte

This chapter briefly reviews those features of the vegetative sporophyte of vascular plants that enter most commonly into morphological comparisions and interpretations. For the convenience of the student, the material is presented in five categories: (1) the contrasts between shoot and root, (2) the methods of branching or ramification of shoots, (3) the concept of microphylls and megaphylls, (4) the comparative anatomy of the sporophyte, and (5) the Stelar Theory. Each of these topics is extremely comprehensive in factual content and also is complex because of the conflicting theories which seek to unify and to explain its varied aspects. Hence the necessarily brief treatments are intended to orient the beginner in plant morphology rather than to provide a definitive résumé of the various topics.

Shoots and Roots

In the great majority of living vascular plants the developing embryo gives rise to a leafy stem, or shoot, and a primary root. Further development of the young sporophyte results, through the activity of apical meristems, in the formation of additions to the original shoot and root components, for example, the production of lateral branches in the primary shoot, the development of new roots from the stem, and the ramification of the primary root (See Fig. 1-3).

From the standpoints of organography, function, and anatomy, shoots and roots are very different types of systems. Roots develop no superficial appendages other than the absorbing root hairs and are to be regarded as naked axes; their chief functions are the absorption of water and solutes, and anchorage. In contrast, shoots have a jointed or segmental organography because the axis or stem bears conspicuous lateral appendages, or leaves; these structures are extraordinarily variable in size, form, and anatomy, and very probably have originated phylogenetically in at least two distinct ways, as will be discussed later in the chapter. The chief functions of shoots are photosynthesis, storage, and reproduction; the last is associated with the development from the shoot system of such structures as sporangia, strobili, and flowers.

In regard to apical growth and branching, roots and shoots differ in several important respects. The root apex consists of a root cap that functions as a protective buffer to the delicate meristem, which lies beneath it. This subterminal meristem of roots is the point of origin of two different patterns of cell formation. One adds new cells outwardly to the root cap, the other contributes the cells which become a part of the root body (Fig. 3-1, A). As we have already mentioned, roots are devoid of foliar appendages. Except for hairs, the only lateral appendages which may occur are lateral roots, and these structures, unlike the usually superficially developed branches of the shoot, originate deep within the tissue of the parent root. In marked contrast to the root apex, the apex of the shoot consists of the terminal meristem itself, and no cap of tissue comparable to a root cap is de-

veloped. Aside from giving rise to the primary stem tissues, a very important function of the shoot apex is the formation of new leaves (Fig. 3-1, B). Leaves originate as primordia (singular, primordium) by means of localized cell division and cell extension at discrete loci or nodes on the flanks of the shoot apex. Leaf primordia typically arise in acropetal sequence; that is, the succession is toward the apex, which means the youngest stages in leaf development are found nearest the summit of the meristem. In addition to their acropetal order of development, the primordia of leaves are laid down usually in an orderly and often distinctive arrangement or phyllotaxis with reference to the stem. In some plants the leaf primordia are formed in pairs, successive pairs being at right angles to one another (decussate phyllotaxis) or in groups of three or more (whorled phyllotaxis); perhaps most commonly, a single leaf occurs at each node—an arrangement designated alternate-spiral. Although consideration of such varied phyllotactic patterns is important, the causal factors responsible for them are poorly understood (See Wardlaw, 1952, 1965, for résumés of experimental studies on phyllotaxis.)

The differentiation between distinct root and shoot systems in most living tracheophytes is of considerable interest from an evolutionary viewpoint. It is now rather generally agreed that this organographic and anatomical differentiation did not exist in such Devonian land plants as *Horneophyton* and *Rhynia,* which were devoid of roots and also lacked foliar organs (Chapter 7). In these archaic organisms portions of the underground system of stems apparently served physiologically as roots. The presumedly modern survivors of an ancient line—*Psilotum* and *Tmesipteris*—are likewise rootless and, moreover, exhibit very primitive leaf-like structures. Both of these genera show absolutely no trace in their

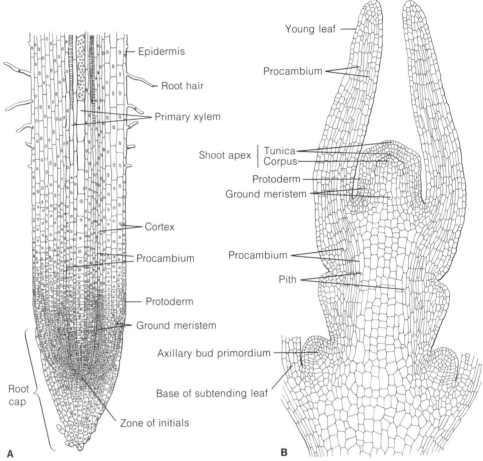

Figure 3-1
Longisections showing the apical meristems and the origin of the primary meristematic tissues in the root and shoot of angiosperms. **A**, root tip of *Hordeum sativum*; **B**, shoot tip of *Hypericum uralum*. [A redrawn from *A Textbook of General Botany*, Ed. 4, by R. M. Holman and W. W. Robbins. Wiley, New York. 1951; B redrawn from Zimmermann, *Jahrb. Wiss. Bot.* 68:289, 1928.]

embryogeny of a suppressed or abortive root, a fact that adds weight to the belief that they are primitively rootless plants (see Chapter 8). From such considerations it seems reasonable to postulate that roots originated phylogenetically from the leafless axes of early types of land plants. However, the steps in the further evolutionary divergence of roots from primitive shoots, which led to the acquisition of a root cap, a prevalent internal or endogenous origin of branches, and the retention of a primitive type of vascular system, are unfortunately completely obscure today. (See Chapter 19, for a more detailed account of root morphology and anatomy.)

Types of Branching in Shoots

The types of branching of the shoot system, in both extinct and living vascular plants, are highly varied and often definitive for the larger taxa, i.e., subdivisions or classes.

At this point in our discussion of the broad aspects of the vegetative sporophyte, brief consideration will be given to the major patterns of branching in the tracheophytes as a whole. Additional details on the types of ramification in the various groups of lower vascular plants and in gymnosperms and angiosperms will be found at appropriate points in later chapters of this book. The most comprehensive and critical review of the literature and theories dealing with branching patterns in the shoot was written by Troll (1937) whose monographic treatment of this intricate subject has been invaluable in preparing this section.

From the broadest possible morphological viewpoint, two principal types of ramification may be distinguished: dichotomous and lateral. The salient features of each of these may now be described.

Dichotomous Branching

Two of the widely accepted morphological characteristics of a dichotomously branched shoot system are (1) the absence of a dominant, or "major," axis and (2) the occurrence of a *sequence* of paired branches which are *not* associated with subtending leaves. These characteristics are the result of a particular type of apical ontogeny in which the terminal meristem itself divides into two more-or-less equal and divergent apices (Figs. 3-2, B; 3-3, B–D). This process of apical forking, or dichotomy, may be repeated on an indefinite scheme, leading in many cases to a very regular and distinctive kind of shoot system (Fig. 3-3, A).

If the successive bifurcations occur in one plane, a flattened dorsiventral system is formed. Troll (1937) terms this type of branching "flabellate dichotomy," citing as examples of it the extremely regular pattern in certain species of *Lycopodium* (see Chapter 9). In other types of plants, each dichotomy of the axis is approximately at right angles to the preceding bifurcation and the resulting "shrub-like" shoot system is termed "cruciate dichotomy" by Troll. An illustration of this kind of dichotomous branching is the aerial shoot system of *Psilotum nudum* (Chapter 8).

Although the interpretation of a given branching pattern as "dichotomous" has

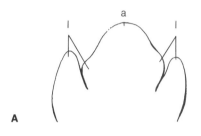

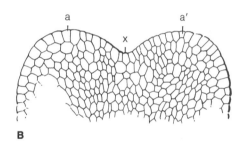

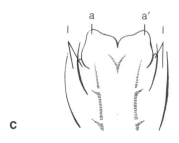

Figure 3-2
Dichotomous branching in *Lycopodium alpinum.*
A, vegetative shoot apex (a) with leaf primordia (1); B, dichotomous division of a reproductive shoot apex (x) into two new apices, (a and a'); C, later stage in dichotomy of vegetative shoot apex. [From *Vergleichende Morphologie der höheren Pflanzen* by W. Troll. Gebrüder Borntraeger, Berlin. 1937.]

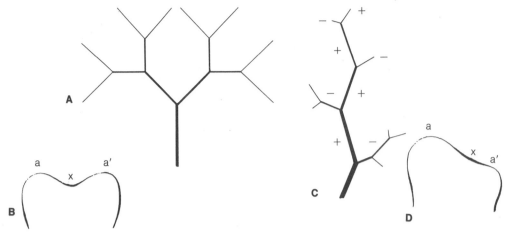

Figure 3-3
Isotomous and anisotomous types of dichotomous branching. **A,** isotomous branching, showing equal development following each dichotomy; **B,** origin of isotomous branch primordia (a, a′) by equal division of shoot apex (x); **C,** anisotomous branching, showing unequal development of stronger (+) and weaker (−) branches following each dichotomy; **D,** origin of anisotomous branch primordia (a, a′) by unequal division of shoot apex (x). [From *Vergleichende Morphologie der höheren Pflanzen* by W. Troll. Gebrüder Borntraeger, Berlin. 1937.]

very often in the past been based exclusively on the organization of the adult shoot system, it is also essential to study and to interpret the ontogenetic aspects of dichotomy. Many of the nineteenth century botanists limited their concept of dichotomy to those plants in which the initiation of branching could be referred to the longitudinal division of the apical cell in the shoot apex. Each of the two daughter cells produced by such a division was then observed to function as the new apical cell of a branch primordium (see Schoute, 1909; Siegert, 1965; and Bugnon, 1969, for critical discussions of the literature on dichotomous branching, with special reference to the rôle of apical cells).

However, if "dichotomy" is defined in the above restrictive sense, this type of branching would have to be regarded as extremely rare in the Tracheophyta because the shoot apex of the majority of vascular plants lacks a definable apical cell. On the other hand, in some lower vascular plants, such as *Lycopodium,* in which both flabellate and cruciate dichotomy occur, it now seems clear that *a group of initial cells,* rather than

a single apical cell, is concerned with the growth and bifurcation of the shoot apex. This has been demonstrated by the detailed histogenetic studies of Härtel (1938) on *Lycopodium complanatum.* In this species, the central cells in the "mother apex" form two new groups of initial cells that represent the points of origin of a new pair of branch primordia (Fig. 3-4).

In certain species of *Lycopodium,* the dichotomy of the shoot apex produces successive pairs of branches approximately equal in size and degree of development. Troll (1937) terms this kind of symmetrical branching *isotomous dichotomy* (Fig. 3-3, A, B). But in other species of *Lycopodium* the two members of a bifurcation develop unequally. One member — or "shank" — of each pair develops more strongly and soon overtops its weaker sister branch (Fig. 3-3, C, D). This derivative form of dichotomy, illustrated in both *Lycopodium* and *Selaginella,* is termed *dichopodial branching* by Bock (1962, pp. 11-14) who characterizes it, in its simplest expression, as the alternate promotion of the right and left shanks of successive dichotomies. The repetition of this *anisotomous dichotomy,* as Troll designates it, results in the formation of a zig-zag "axis" and shorter, more-or-less determinate "lateral branches." Bock regards the origin of dichopodial branching as a highly significant event in evolution that led to lateral branching, pinnate leaves, and the gradual formation of a midvein in the early phylogeny of venation patterns in leaves (for further details, consult Bock's 1962 and 1969 monographic treatments of dichotomy in vascular plants).

Lateral Branching

In contrast to dichotomy, lateral branching of a shoot system originates by the expansion of buds more-or-less distal to the shoot apex of the main or dominant axis (Fig. 3-1, B). This kind of ramification, which occurs in certain ferns, in *Equisetum,* and throughout the seed plants, is commonly termed "monopodial branching" to distinguish it from typical dichotomous or dichopodial branching.

Lateral buds, as shown by Troll's (1937) exhaustive survey, most frequently arise in some type of relationship to the leaves of a shoot. In some fern species that have dorsiventral rhizomes buds arise without reference to leaves ("acrogenous branching," according to Troll's terminology), but it is more common in ferns for buds to originate near or from the abaxial side of the leaf bases or from the petiole (Troop and Mickel,

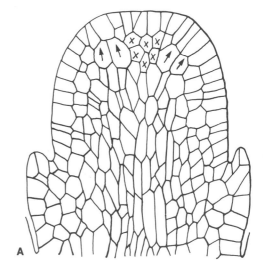

A

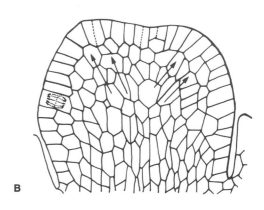

B

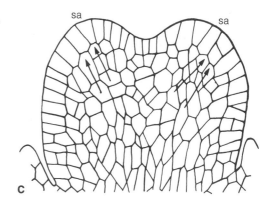

C

Figure 3-4
Early stages in dichotomous branching of the shoot apex of *Lycopodium complanatum.* **A,** preliminary stage of dichotomy: the arrows indicate the two new "centers" of cell extension in the "central tissue." Cells labeled with an "x" have ceased to divide; **B,** later stage: further active growth, in direction of arrows, has produced a slight depression in the contour of the original terminal part of the apex; **C,** well-defined stage of dichotomy showing the two new divergent shoot apices (sa). [Redrawn from Härtel, *Beitr. Biol. Pflanzen* 25:125, 1938.]

1968). *Equisetum* appears unique among vascular plants with monopodial branching in that the buds *alternate in position* with the fused leaves at each node (Chapter 10, p. 219). In seed plants, however, the dominant type of branching is *axillary* (Fig. 3-1, B), although it is very common for a portion of the lower leaves of an annual shoot to be devoid of axillary buds (e.g., such conifers as *Picea, Abies, Pseudotsuga, Taxus*) or to subtend rudimentary, dormant buds (e.g., many woody dicotyledons).

One of the most remarkable details of axillary bud formation in the angiosperms is the development, in a single leaf axil, of a series of *accessory buds* in addition to the main bud. Accessory buds, arranged in one or two vertical rows in the leaf axil are known as "serial buds" and are found in a considerable number of dicotyledons, e.g., *Juglans* and *Lonicera.* If the accessory buds are arranged in a crescentic pattern in the leaf axil, they are termed "collateral buds.' This type of arrangement is common in the monocotyledons. (Detailed treatments of accessory buds are given by Sandt, 1925, and Troll, 1937, pp. 531–540).

In many dicotyledons with normal axillary branching, it is common for the development of a terminal inflorescence or the abortion and subsequent abscission of the entire tip of a vegetative shoot to result in the continuation of the longitudinal growth of such shoots from one or more of the uppermost axillary buds (Garrison and Wetmore, 1961; Millington, 1963). If shoot-tip abortion occurs in a plant with decussate phyllotaxy, such as *Syringa,* the uppermost pair of axillary buds may subsequently expand into shoots and give the false impression that branching is dichotomous. Troll (1935, p. 104) designates this type of pseudodichotomy as a *dichasial sympodium* in contrast to the *monochasial sympodium,* which is produced when only a single axillary bud continues the development of the shoot. The latter structure is illustrated in such common dicotyledonous trees as *Ulmus, Tilia, Salix* and *Betula.*

Types of Branching in Monocotyledons

Although the usual method of branching in angiosperm shoots is obviously lateral and axillary, the ramification of certain arborescent monocotyledons has been interpreted as dichotomous. The most frequently cited example is *Hyphaene,* a genus of palms native to Africa and northeastern India. As is shown in Fig. 3-5, well-developed specimens of *Hyphaene* appear to branch in a cruciate-dichotomous manner. According to the classical morphological study of Schoute (1909), *only* the inflorescences in this palm arise in an axillary position, a fact which in his opinion, rules out the possibility of vegetative "pseudo-dichotomous" branching. Although he did not undertake an ontogenetic study, he concluded that the dichotomous pattern of branching in *Hyphaene* probably resulted from the division of the shoot apex into two separate terminal meristems. Certain modern authors (e.g., Eames, 1961), however, emphasize the need for ontogenetic and anatomical evidence before the remarkable branching pattern of *Hyphaene* can be accepted as a valid case of true dichotomy. On the other hand, Corner (1966) agrees with Schoute's interpretation and concludes that dichotomy in *Hyphaene* represents the retention of a primitive ramification pattern that "emphasizes the fundamental primitiveness of palms among the flowering plants."

Very recently Tomlinson has initiated a series of studies directed toward a developmental analysis of the extremely varied patterns of branching in the monocotyledons as a whole. His detailed investigation on

Figure 3-5
Dichotomous branching in the African palm *Hyphaene thebaica.* Calcutta Botanical Garden.
[Photo courtesy Dr. C. H. Lamoureux.]

36

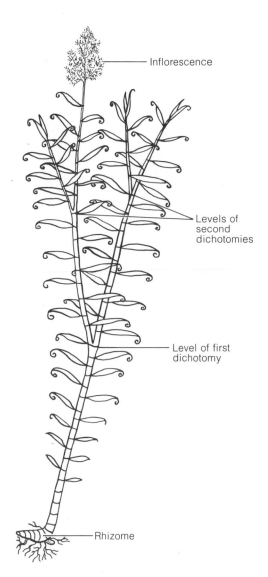

Inflorescence

Levels of
second
dichotomies

Level of first
dichotomy

Rhizome

Figure 3-6
Diagrammatic representation of an aerial, dichotomously branched shoot of *Flagellaria indica*. Note that
the distichous arrangement of leaves is continued,
without change, in the successive pairs of branches.
[Redrawn from Tomlinson, *Bot. Jour. Linn. Soc.* 63,
Supplement 1, 1970.]

Flagellaria indica, a tropical species, resulted
in what he regards "as the first unequivocal
demonstration of dichotomous branching
in a monocotyledon" (Tomlinson, 1970).
The aerial shoots of *Flagellaria,* which arise
from the axillary buds of the rhizome,
branch dichotómously in the plane of the
distichously arranged foliage leaves (Fig.
3-6). Bifurcation can occur at any level and
in some plants one of the sister branches of
a dichotomy may form a terminal inflorescence. Detailed histogenetic study revealed
that the dichotomous branching of the
aerial shoot results from the division of the
shoot apex, *above the level of the youngest
visible leaf primordia,* into two, more-or-
less equal apices; the latter then develop
into twin shoots identical in organization
and phyllotaxy to the parent shoot which
produced them.

A further example of dichotomous
branching in monocotyledons is provided
by Tomlinson's (1971) detailed ontogenetic
study of *Nypa fruticans,* a peculiar rhizomatous palm which grows in shallow
estuaries and sand flats in New Guinea.
Although *all* the inflorescences are axillary
in position, the rhizomes fork at irregular
and infrequent intervals, clearly suggesting
dichotomy (Fig. 3-7, B, C). Although Tomlinson did not succeed in finding shoot
apices in early stages of division, the evidence which he cites in support of dichotomous branching in *Nypa* may be summarized
as follows: (1) the two young daughter
shoots enclosed within but not subtended
by a leaf are in a plane perpendicular to the
dorsiventral axis of the enclosing leaf; this
arrangement is not that characteristic of
normal axillary branching (Fig. 3-7, A); (2)
the two daughter shoots at the tip of a rhizome are always at the *same developmental
stage* and possess equal numbers of leaves;
this suggests, according to Tomlinson, that
the paired shoots in all probability were
begun by the division of the shoot apex

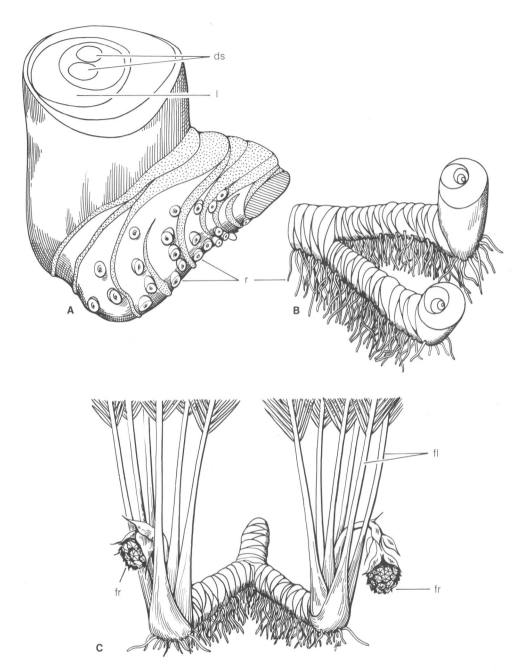

Figure 3-7
Nypa fruticans. **A,** drawing of a portion of a juvenile plant with terminal bud of rhizome cut transversely to show two daughter shoots (ds) enclosed by youngest leaf (l). Insertion of older leaves shown by stippling; note bases of numerous roots (r). **B, C,** diagrammatic reconstructions of a dichotomously branched rhizome: **B,** with older leaves removed and tip of each branch cut transversely to show leaf arrangement; **C,** the same showing crown of closely packed foliage leaves (fl) and axillary clusters of fruits (fr). [Redrawn from Tomlinson, *Ann. Bot.* 35:865, 1971.]

into two equal "growth centers;" and (3) there is no anatomical evidence to suggest the abortion of the main shoot apex and its replacement by two adventitious lateral shoots; all of the vascular bundles beneath the fork "are distributed more-or-less equally between the two daughter shoots."

Clearly, the same type of careful ontogenetic study which Tomlinson has applied to *Flagellaria* and *Nypa* is essential for the correct interpretation of other possible examples of dichotomous branching in monocotyledons. Greguss (1968) for example, cites *Hyphaene, Yucca* sp., *Dracaena, Draco,* and *Pandanus* sp. as examples of dichotomously branched monocotyledonous trees. However, he gives no ontogenetic analyses and Tomlinson et al. (1970) emphasize that with the possible exception of *Hyphaene,* these genera are all characterized by sympodial or "pseudodichotomous" branching. Striking examples of pseudodichotomy are provided by *Cordyline australis,* which, as shown in Fig. 3-8, A, superficially appears to branch dichotomously. A median longitudinal cut (Fig. 3-8, B through such a "dichotomy" reveals, however, the remains of the old terminal inflorescence in the crotch of the fork (for further details on sympodial branching in *Cordyline,* see Tomlinson and Fisher, 1971).

It remains for future investigations to determine the distribution and frequency of true dichotomous branching within the various taxa of monocotyledons. Tomlinson (1970, p. 12) summarizes the present status of the problem as follows: "The conclusion which should surprise us is not that dichotomous branching occurs in monocotyledons but that it has been overlooked for so long. This, however, merely reflects our relative ignorance of monocotyledonous morphology. My prediction is that when attempts are made to correct these deficiencies by careful examination of the growth habits of monocotyledons (un-

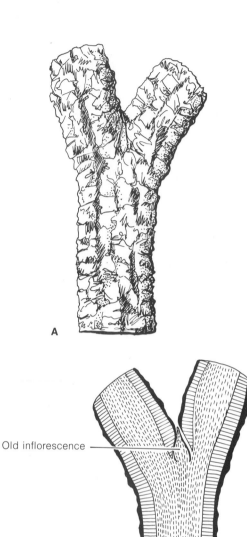

Figure 3-8
Pseudodichotomous branching of *Cordyline australis.*
A, external view of forked axis; B, median longisectional view of A, showing remains of the original terminal inflorescence in the crotch of the fork. Note continuity of primary tissues (stippled) between main and lateral axes. [Redrawn from Tomlinson, et al., *Phytomorphology* 20:36, 1970.]

fashionable though this kind of inquiry may be), dichotomous shoots may prove to be quite frequent in monocotyledons."

Microphylls and Megaphylls

Interpretation of the morphology and evolutionary history of the leaves of vascular plants constitutes a fundamental problem that has attracted much attention during the present century. Leaves, regardless of their size, form, or structure, arise as lateral protuberances from a shoot apex and at maturity represent the typical lateral appendages of the axis or stem (see Fig. 1-3). From an ontogenetic viewpoint, a leaf is a determinate organ. In contrast to the theoretically unlimited or open type of apical growth characteristic of the stem, the apical growth of the leaf primordium in most plants ceases early in ontogeny and is followed by a phase of tissue specialization and enlargement which culminates in the production of the final shape and structure of the adult foliar organ (Foster, 1936). The question is: How did the distinction between an indeterminate axis and its lateral foliar appendages arise during the evolutionary history of vascular plants? The answer to this question is particularly important in the light of paleobotanical studies on certain Devonian and Silurian plants, some of which represent the simplest vascular plants known to science. The sporophytes of *Rhynia* and *Horneophyton* are of exceptional interest because their aerial dichotomously branched axes were entirely devoid of foliar appendages (see Chapter 7).

Morphological Contrasts

MICROPHYLLS. It is unfortunate that the widely used term "microphyll"—literally "small leaf"—unduly emphasizes the small dimension of an appendage rather than its distinctive vascular anatomy and phylogenetic mode of origin. This point is well illustrated by comparison of leaf size in living members of the Lycopsida. In *Lycopodium* and *Selaginella,* for example, the leaves are indeed "small," measuring only a few millimeters in length. But in the genus *Isoetes,* the quill-like leaves attain considerable length, and in certain species (e.g., *Isoetes japonica*) may be 50 centimeters long. From a more precise morphological standpoint, one of the salient characters of a microphyll is its extremely simple and presumably primitive vascular system. As is shown in Fig. 3-9, D, the vascular supply of a typical microphyll consists of a single strand—the leaf trace—which diverges from the periphery of the *stele* (i.e., the vascular cylinder) of the stem and extends as an unbranched midvein through the leaf. It must be emphasized that the divergence of the single trace at the node of a microphyll is *not* associated with a corresponding break, or *leaf gap,* in the stele. In short, the development of microphylls is closely correlated with those taxa of living and extinct vascular plants in which the vascular cylinder of the stem is a *protostele,* i.e., a central column of vascular tissue devoid of a pith region (Fig. 3-22, A).

In certain genera of the Lycopsida a small scale-like outgrowth, termed the *ligule,* develops near the adaxial base of each of the vegetative microphylls and sporophylls (Figs. 9-21, 9-22). On the basis of the presence or absence of ligules, lycopsid plants can be arranged in two series: the "Ligulatae," which includes the living genera *Selaginella* and *Isoetes* as well as the members of the extinct order Lepidodendrales, all of which are characterized by ligulate leaves and ligulate sporophylls; and the "Eligulatae" in which ligules are absent, as in the living genera *Lycopodium* and *Phylloglossum* and possibly the extinct genus *Lycopodites.*

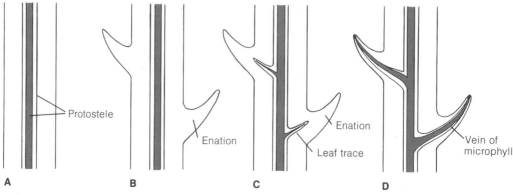

A   B   C   D

Figure 3-9
Longisectional diagrams showing the origin and evolutionary development of microphylls according to the enation theory. **A**, the leafless, protostelic axis of *Rhynia*; **B**, primitive enations, devoid of leaf traces, as illustrated by the shoot of *Psilotum*; **C**, portion of the "shoot" of the extinct lycopod *Asteroxylon*, showing the termination of leaf traces at the bases of the veinless enations; **D**, typical microphylls, in which a leaf trace extends as an unbranched midvein into each of the foliar appendages. The condition in **D** is the prevailing pattern in *Lycopodium*, *Selaginella*, and other members of the Lycopsida. [Redrawn from Lemoigne, *Bull. Mens. Soc. Linn. Lyon*, 37(9):367, 1968.]

The ligule appears very early in the ontogeny of a microphyll and is initiated from a single surface cell in *Isoetes* whereas in *Selaginella* it originates from several short rows of superficial cells (Harvey-Gibson, 1896; Hsü, 1937). At maturity the ligule in each genus is a surprisingly complex organ, considering its small size and short life. The base or foot region is to some extent sunken in the tissue of the microphyll and consists of the sheath, a layer of cells continuous with the epidermis of the leaf or sporophyll, and the glossopodium, which is particularly well-developed in *Selaginella* and consists of a wedge-shaped group of large, highly vacuolated prismatic cells (see Fig. 9-21). According to Dunlop (1949) the sheath cells of the ligule of *Isoetes macrospora* possess well-defined Casparian strips and thus resemble typical endodermal cells; a less conspicuous Casparian strip also occurs on the walls of the sheath cells of the ligule of *Selaginella rupestris*. In *Isoetes engelmanni*, investigated in great detail by Smith (1900), the foot region of the ligule becomes very massive and its lateral portions

grow upwardly and downwardly as pairs of horn-like processes. We have observed a comparable situation in other species, but a fuller investigation of the ontogeny of the ligule in *Isoetes* is highly desirable. In both *Selaginella* and *Isoetes* the free exserted portion of the ligule, prior to its death, is composed largely of relatively small, densely staining, polygonal-shaped cells. From a functional standpoint, it is believed that ligules are secretory organs which, by exuding water and possibly mucilage, serve to keep young leaves and sporangia moist. The frequent development of tracheid-like cells (transfusion tissue) between the ligule sheath and the adjacent vein of the microphyll indicates the possibility of direct conduction of water to the base of the ligule. Phylogenetically, the origin and homology of the ligule are very obscure problems. But the considerable antiquity of ligules is shown by their presence in certain of the extinct lycopods of the Paleozoic Era.

There are microphylls that, on the basis of vascular anatomy, may be taken as being typical in all members of the Lycopsida

(Banks and Davis, 1969) and possibly in members of the Psilopsida (see Chapter 8). The sterile (i.e., nonsporogenous) appendages of both *Psilotum* and *Tmesipteris* were interpreted by Bower (1935, p. 554) as microphylls, although the leaf of *Psilotum* is commonly evasculate or else provided with a very rudimentary trace which does not enter the appendage (Fig. 3-9, B). There was a comparable type of "incomplete" leaf trace in the extinct lycopod *Asteroxylon* (Fig. 3-9, C).

MEGAPHYLLS. The megaphyll (called "macrophyll" by some authors) is illustrated by the comparatively large pinnate leaves of ferns, although here too the salient feature is not size but rather morphological and anatomical organization. In contrast to a microphyll, the divergence of the leaf trace (or traces) in most of the living ferns is associated with the formation of parenchymatous areas, or *leaf gaps,* in the vascular cylinder of the stem (Figs. 3-10, 3-22, B). There are, however, notable exceptions to this anatomical correlation between leaf traces and leaf gaps in the leaves of sporelings (i.e., the young sporophytes) of many ferns, which have complex *adult* steles, and in the leaves of the adult phase of the sporophyte of *Lygodium, Cheiropleuria,* certain species of *Gleichenia,* and all members of the Hymenophyllaceae (Boodle, 1901; Bower, 1935). In all these exceptions, the vascular cylinder of the axis is a protostele, devoid of leaf gaps, as is shown, for example, by a series of transections cut through the node of the rhizome of *Lygodium japonicum* (Fig. 3-11). Near its point of separation from the stele, the xylem of the leaf trace is *continuous* with the peripheral xylem of the stele (Fig. 3-11, A). At a high level, the leaf trace separates from the vascular cylinder and appears as a central strand of xylem surrounded by phloem and

endodermis, thus closely resembling the histology of the protostele of the rhizome (Fig. 3-11, B).

Fern megaphylls differ from microphylls by their more complex patterns of venation. In contrast to the univeined microphyll, the lamina of a fern leaf, whether simple or pinnate in organization, develops a relatively complex system of branched vascular strands. It is now generally agreed that in ferns the most primitive venation pattern is the open-dichotomous type consisting of a series of repeatedly forked veins which terminate freely without anastomoses at the margins of the leaf blade (Bower, 1923, p. 93). This open type of venation is connected by transitional patterns (often beautifully displayed in the sequence of juvenile leaves in the young sporophyte) with the more highly evolved type of reticulate (i.e., anastomosed) venation found in the leaves of both eusporangiate and leptosporangiate ferns. (Chapter 13).

Phylogenetic Origin

As Bower (1935, p. 548) has clearly pointed out, the evolutionary significance of the morphological differences between microphylls and megaphylls "can only be solved by comparison between various types living and fossil, aided where possible by reference to individual development." From this broad outlook, Bower postulated that the microphylls and megaphylls of lower vascular plants are the results of separate paths of foliar evolution, i.e., the two leaf types are not homologous from a phyletic point of view.

Figure 3-9 shows diagrammatically the theoretical steps in the evolution of the microphyll. Beginning with the leafless type of axis found in such an ancient plant as *Rhynia,* the earliest stage in microphyll evolution may have been a simple emergence or "enation" devoid of a leaf trace

42

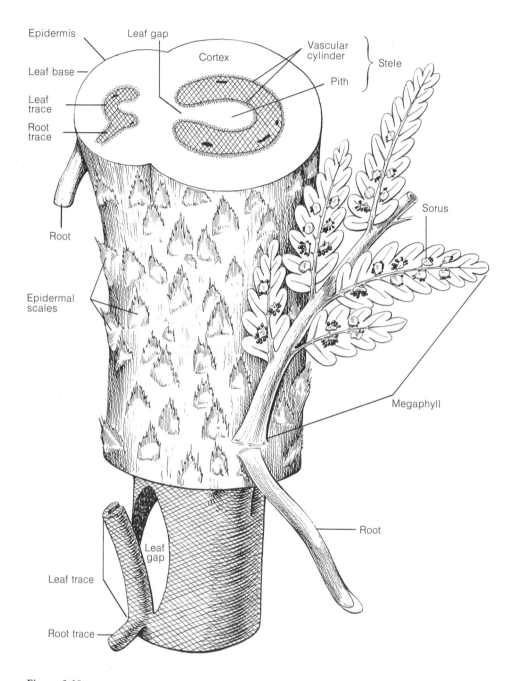

Figure 3-10
Organography and general vascular anatomy of a small portion of a fern shoot. A pinnatifid megaphyll with its abaxial sori is seen in surface view at right. Note that the divergence of a leaf trace into a megaphyll (shown at the top and bottom of the figure) is associated with a leaf gap in the stele of the stem. [From *The Anatomy of Woody Plants* by E. C. Jeffrey. University of Chicago Press, Chicago. 1917.]

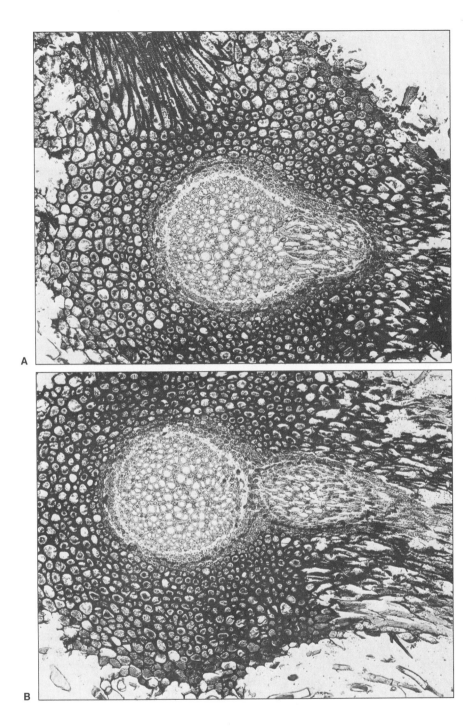

Figure 3-11
Transections of the rhizome of *Lygodium japonicum* showing the mode of divergence of the
leaf trace from the protostele. **A,** level just below the point of separation (at right) of trace
from stele. Note continuity between xylem of leaf trace and peripheral xylem of stele, and
the absence of a leaf gap; **B,** section at a higher level showing complete separation of leaf
trace and its oblique extension into the base of the megaphyll.

(Fig. 3-9, A). The further elaboration of primitive enations would be indicated by the initiation of "leaf traces" which diverged from the periphery of the protostele and terminated at the bases of the appendages. This theoretical stage in phylogeny is illustrated by the extinct lycopod *Asteroxylon* and possibly by the enation "leaves" of the living *Psilotum* (Fig. 3-9, B, C). In the final stage of microphyll evolution, a vascular strand, or leaf trace, was developed which continued as a single midvein into the microphyll (Fig. 3-9, D). This type of univeined microphyll is typical of all the living members of the Lycopsida. (For further details on the microphylls of the Lycopsida, see Chapter 9.)

In contrast to the progressive evolutionary specialization of microphylls from enations, megaphylls are considered by Bower to have evolved by the specialization of the distal regions of dichotomized branch systems. His theory, based on the earlier ideas of Lignier (1903), postulates that the fern megaphyll is a "cladode leaf" and that the first stage in its origin was a gradual change from equal dichotomous branching to a dichopodial type of growth. This consisted in the unequal development of the sister branches of dichotomizing axes, one continuing as the main axis or "stem," the other becoming laterally "overtopped" and representing the precursor of a megaphyll (Fig. 3-12, A, B). Subsequent flattening or "planation" occurred and finally the ultimate divisions of each of the overtopped branch systems became united, forming a simply dichotomously veined megaphyll (Fig. 3-12, C, D).

Although the evolutionary history of the megaphyll is now rather generally held to differ from that of the microphyll, it is an open question whether these terms — and their associated concepts — can also be applied to the leaves of seed plants. Jeffrey (1917) attempted to solve the problem by emphasizing the importance of the leaf gap,

a feature of stelar anatomy he believed was shared by all "megaphyllous" plants, i.e., ferns, gymnosperms, and angiosperms. He segregated these groups under the "Pteropsida" and placed the microphyllous "pteridophytes" in a separate category, the Lycopsida, which are distinguished by the absence of leaf gaps. However, as we have already noted, some of the living ferns clearly demonstrate that there is no *obligatory* relationship between leaf gap and leaf traces (see Fig. 3-11). In this connection, Bower (1935, p. 329) emphasized his view that "the existence of a leaf gap as a structural feature does not depend upon or indicate phyletic origin, but connotes a size relation of the leaf to the axis which bears it." It should also be noted that the validity of Jeffrey's phylogenetic concept of the "Pteropsida" — and the existence of leaf gaps in the vascular cylinder of seed plants — have been questioned in very recent years from both an anatomical and paleobotanical point of view (for detailed discussion, see Chapter 14).

Because of the difficulties and uncertainties inherent in reconstructing leaf phylogeny in seed plants, it seems best, for the present, to limit the terms "microphyll" and "megaphyll" to the contrasted "end products" of foliar evolution in lower vascular plants. Microphylls are believed to have evolved as enations from previously leafless axes whereas megaphylls are considered to represent the modification of flattened dichotomously branched systems of axes (compare Figs. 3-9 and 3-12). Although Wardlaw (1957) found no essential difference between the mode of initiation of microphylls and megaphylls at the shoot apex, ontogenetic information of this sort, because of the enormous time span which separates ancient and modern vascular plants, does not prove that microphylls and megaphylls *originated* by similar paths of evolution.

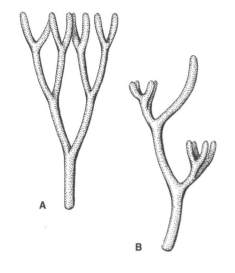

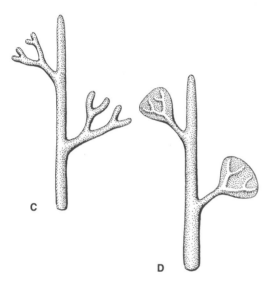

Figure 3-12
The phylogenetic origin and development of the
megaphyll according to the telome theory.
A, isotomous branch system, without distinction
between axis and megaphylls; **B,** unequal dichotomy
or "overtopping," the weaker branches representing
initial stages in megaphyll evolution; **C,** dichotomous
branching of primitive megaphylls in a single plane
("planation"); **D,** union between forked divisions of
megaphylls ("webbing") has produced a flat,
dichotomously veined lamina. [Adapted from
*Cryptogamic Botany, Vol. II. Bryophytes and Pterido-
phytes* by G. M. Smith. McGraw-Hill, New York.
1955.]

The Telome Theory

During the classical period in plant morphology, the study of homologies was very
largely "angiosperm centered;" i.e., it was
based on the belief that the organography
of the sporophyte in vascular plants *as a
whole* could be interpreted with reference
to such angiospermic organs as leaf, stem,
and root. Paleobotanical studies made during
the early part of the present century—
particularly the discovery by Kidston and
Lang (1917) of such morphologically simple
land plants as the Devonian psilophytes
(e.g., *Rhynia*)—suggested the need for a
"new outlook" in phyletic morphology,
especially a reappraisal of the evolutionary
history of the sporophyte. The most comprehensive synthesis of the major steps in
the evolution of the Tracheophyta was made
by Walter Zimmermann, who, in 1930,
originated the "Telome Theory," which he
later elaborated in great detail in an extensive series of publications (Zimmermann,
1953, 1959, 1965).

We will outline only the most salient
features of the Telome Theory. Students
interested in the historical aspects of this
theory and a more detailed analysis of it
should consult Zimmermann's (1965) most
recent monograph as well as the reviews in
English which are found in the papers by
Wilson (1953) and Stewart (1964).

Zimmermann selected *Rhynia* as a type
of very ancient "primordial" land plant
that provides a relatively simple example
of the application of his Telome Theory.
According to his interpretation, the dichotomously branched sporophyte of *Rhynia* was
composed of morphological "units," which
he designated as "telomes" and "mesomes."
Reference to Fig. 3-13 will serve to explain
his use of these terms. A telome, in the
broadest sense of the term, is one of the distal branches of a dichotomized axis. Each
telome ends at the point of forking of the

46

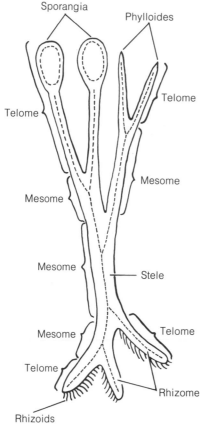

Figure 3-13
Simplified diagram showing the general organography of a primordial vascular land plant of the type of *Rhynia*. The morphological "units" of this elemental sporophyte (or "primordial syntelome"), according to Zimmermann's concept, are telomes (i.e., sporangia and vegetative phylloides) and mesomes. The telomes of the underground creeping rhizome bore rhizoids. See text for further explanation. [Adapted from *Die Telomtheorie* by W. Zimmermann. Gustav Fischer Verlag, Stuttgart. 1965.]

axis whereas a mesome represents the "internodal" region between two successive dichotomies of an axis. In ontogeny, a given telome becomes "converted" to a mesome if dichotomous branching continues (see Fig. 3-13, upper right). From a functional standpoint, telomes are classified as "fertile" when they terminate in sporangia, and as "vegetative" when they constitute "phylloides." A system of united telomes and

mesomes, as illustrated in Fig. 3-13, is termed a "primordial syntelome." Anatomically, the entire telome system is vascularized by a continuous protostele and is further morphologically differentiated into a creeping subterranean portion, composed of rhizoid-bearing telomes and an upright aerial portion terminating in sporangia and phylloides.

According to Zimmermann, the further phylogenetic development of the primitive *Rhynia*-type of telome system resulted from the operation of five "elementary processes" (Fig. 3-14). Independently—or more commonly in various combinations—these processes were responsible for the gradual evolution of the diverse types of leaves and sporophylls (sporangium-bearing structures) characteristic of the main taxa of vascular plants.

One of the most important of the elementary processes was "overtopping," i.e., the unequal development of certain parts of a dichotomously branched system, the subordinated overtopped lateral members representing the beginning of "leaves," and the overtopping portions, stem-like axes (Fig. 3-14, A). The additional processes of "planation" and "fusion" (syngenesis) are considered by Zimmermann to have been particularly significant with reference to leaf and sporophyll evolution. Planation resulted in the arrangement of groups of telomes and mesomes in a single plane (Fig. 3-14, B) while fusion, which at first only entailed parenchyma formation, connected these units into a flat dichotomously veined lamina. Subsequent fusion of the vascular strands in the leaf led to the formation of reticulate venation, and fusion in the stem resulted in the anastomosis of the originally separate steles, (Fig. 3-14, C, F).

According to the Telome Theory the elementary process of "reduction" accounts for the origin of the microphyllous type of leaf (Fig. 3-14 D). This interpretation rejects the "enation theory" of microphyll evolution (see pp. 41–44) and holds that *both* microphylls and megaphylls originated from subordinated and dichotomously branched portions of a primordial syntelome (see Zimmermann, 1965, p. 94, Fig. 58). Following planation and webbing, a group of telomes became progressively reduced to a single univeined "needle-leaf" or microphyll (Fig. 3-14, D). It is of interest to note that Zimmermann's phylogenetic concept of the origin of microphylls includes not only the foliar organs of the Lycopsida but also the univeined leaves of such members of the Sphenopsida as *Equisetum* and the needle-leaves of the Coniferales.

The elementary process of "recurvation" was responsible for the phylogenetic origin of the sporangium-bearing organs of the Sphenopsida. Recurvation, as conceived by Zimmermann, is a process resulting in the bending or "anatropous curvature" of the stalks of a group of sporangia, and is illustrated by the single curved fertile telome represented in Fig. 3-14, E).

Subsequent fusion of the bases of the curved sporangium-stalks led to the distinctive peltate sporangiophores of the strobilus of the modern genus *Equisetum* (see Zimmermann, 1965, p. 102, Fig. 63; and p. 237).

It was, of course, inevitable that the Telome Theory, because of its comprehensiveness, would produce decidedly mixed reactions from paleobotanists and morphologists. In his most recent monograph, Zimmermann (1965) vigorously defended his theory and rejected the numerous objections which have been raised against it. He concludes his book with the statement that the "universality" of his concept has dealt with all the "test-cases" and that, so far, no alternative theory has been proposed which so satisfactorily accounts for the enormous diversity of vascular plants as the Telome Theory. In this connection it is worthy of note that the distinguished paleobotanist, Rudolph

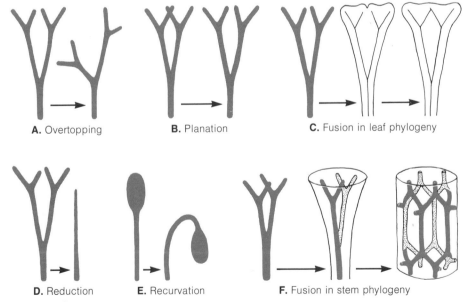

A. Overtopping   B. Planation   C. Fusion in leaf phylogeny

D. Reduction   E. Recurvation   F. Fusion in stem phylogeny

Figure 3-14
Diagrams illustrating the five elementary processes that, beginning with primordial system of telomes (see Fig. 3-13), were operative in the phylogenetic specialization of vascular plants. See text for further explanation. [Adapted from *Die Telomtheorie* by W. Zimmermann. Gustav Fischer Verlag, Stuttgart. 1965.]

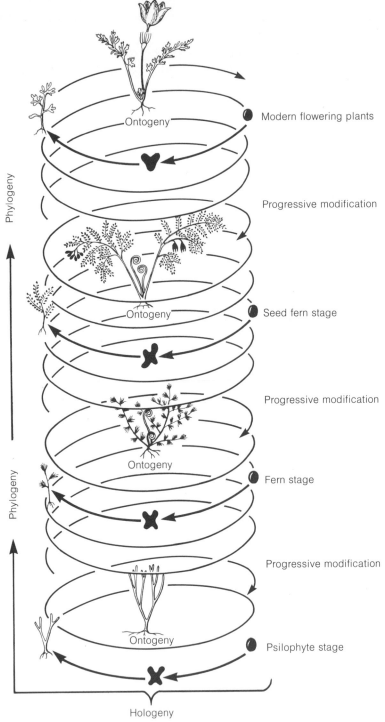

Figure 3-15
Diagram illustrating the progressive modification of ontogeny during the evolution of vascular plants, beginning with the "psilophyte" stage and progressing through the phylogeny of the ferns and seed ferns to modern flowering plants. The examples chosen are intended primarily to illustrate levels of specialization, not specific phylogenetic relationships between the taxa. [Adapted from *Die Telomtheorie* by W. Zimmermann. Gustav Fischer Verlag, Stuttgart. 1965.]

Florin (1951) adopted the Telome Theory in his interpretation of the evolution of the vegetative and reproductive structures of the Cordaitales and primitive conifers. Furthermore, although many morphologists such as Bailey (1949, 1956) have expressed skepticism about the value of the Telome Theory with reference to the angiosperms, Wilson (1937, 1942, 1950) utilized this concept in his interpretation of the origin and phylogeny of the stamen in flowering plants (see also Stewart, 1964, pp. 133–134).

New discoveries in paleobotany and the continued study of the morphology of the living angiosperms will doubtless test further the validity of the telome concept. But, for the present, it should be strongly emphasized that Zimmermann's concept of phylogeny is "hologenetic" — i.e., he regards evolution as the result of the *progressive modification* of ontogenetic processes which have been in operation over a time span of many millions of years (Fig. 3-15). In more explicit terms, an ontogenetically *mature* structure, such as a telome or a system of telomes, does not "give rise" to a more advanced type of mature organ, such as a microphyll or a megaphyll. On the contrary, from the standpoint of hologeny, new types of fertile and sterile organs are the result of *changes in the genotype* that in turn are expressed ontogenetically by the gradual modification in the organs and structures formed by the embryo and the apical meristems of the developing plant (for a more detailed discussion of hologeny, see Zimmermann, 1966).

Comparative Anatomy of the Sporophyte

It should be evident by now that the organographic evolution of the sporophyte has reached various levels in the different groups of living vascular plants. In the Psilotaceae and Lycopodiaceae, many ancient features such as dichotomous branching and microphylls are still in evidence. By contrast, the ferns and seed plants exhibit a much higher level of organ specialization, including the development of large complex leaves and the presence of monopodial branching. Paleobotanical evidence reveals that the evolution of cell types and tissues was similarly complex and variable. Indeed, one of the most important achievements in paleobotany has been the discovery and description of the well-preserved anatomy of several ancient groups of vascular plants. This has made possible at least the beginning of a true phylogenetic interpretation and classification of cell types and tissues (Hofmann, 1934; Foster, 1949, 1972). It seems reasonable to conclude this chapter with a brief review of the outstanding aspects of sporophyte anatomy. Of course, this résumé should in no sense be regarded as a satisfactory condensation of the vast subject matter of plant anatomy (Esau, 1965 a), but we hope that it will provide the indispensable orientation needed by the student in his introductory approach to the comparative anatomy of stem, leaf, and root.

Sachs' Classification of Tissue Systems

One of the most useful schemes for understanding the general topographical anatomy of the adult sporophyte was devised by the celebrated German botanist, Julius von Sachs (1875). The great merit of his classification is its simplicity and its wide applicability to the primary structure of the stem, leaf, and root. According to Sachs, the early phylogenetic development of vascular plants resulted in the differentiation of three principal systems of tissues: the external epidermal and cork layers collectively termed the *dermal system*; the strands of conducting phloem and xylem tissue which compose the *fascicular system*; and the remaining internal tissue or tissues designated the *fundamental* or *ground tissue system.* Sachs emphasized that each of these tissue systems may comprise the most varied cell types and that his scheme of classification was concerned with the broadest possible contrast between systems of tissues (see Foster, 1949, for a critical discussion of Sachs' scheme of classifying tissue systems).

Figure 3-16 shows diagrammatically the application of Sachs' classification and terminology to the gross anatomy of the stem, leaf, and root of flowering plants. This figure indicates plainly that the fascicular tissue system is the most variable of the three, from the standpoint of its pattern of development within plant organs. In stems the fascicular system appears either as a central cylinder of phloem and xylem or in the form of vascular bundles arranged in a cylinder or scattered throughout the ground tissue system (see Fig. 3-16, A; also Fig. 3-23, A). The form of the fascicular system of the leaf ranges from one to many bundles or a cylinder in the petiole, to a complex system of veins, usually arranged in a single plane, in the lamina (Fig. 3-16, B). In the root the pattern of the fascicular system is

very distinctive, consisting of a radial and alternate series of phloem and xylem strands; commonly the latter are joined at their inner edge to form a solid core of xylem as shown in Fig. 3-16, C.

In contrast with these diverse patterns exhibited by the fascicular system, the form and arrangement of Sachs' other tissue systems are comparatively simple. The dermal system, in all foliar organs and in young stems and roots, is represented by the epidermis (Figs. 3-16, 3-17). Typically this is a single layer of superficial cells which are tightly joined except for the stomatal openings; in some groups of dicotyledons the leaf may develop on one or both surfaces a multiple epidermis consisting of two or many layers of cells, all of which have originated from the subdivision of the original surface layer of cells (Fig. 3-18, C). In the stems and roots of many vascular plants the epidermis is eventually sloughed away by the development beneath it of *cork.* Like the cells of the epidermis, cork cells are compactly arranged and it is only at certain areas known as lenticels that well-developed intercellular air-space systems are found.

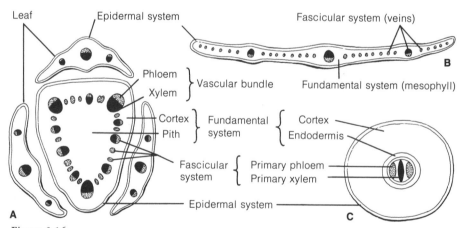

Figure 3-16

Diagrams (based on *Linum usitatissimum*) illustrating the positions and patterns of the epidermal, fascicular, and fundamental tissue systems in the vegetative organs of a dicotyledon. **A,** transection of stem and three leaf bases; **B,** transection of lamina of leaf; **C,** transection of root. [Redrawn from *Plant Anatomy* by K. Esau. Wiley, New York. 1953.]

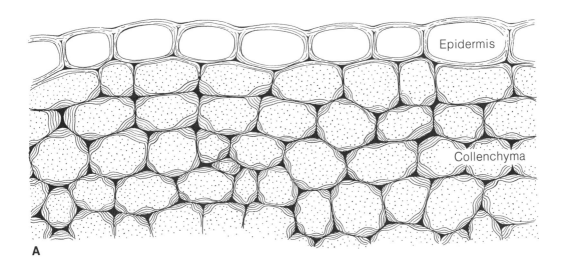

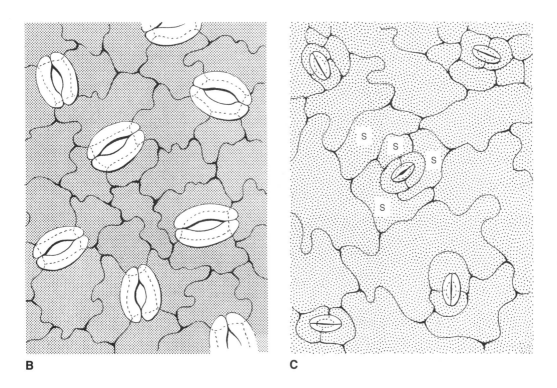

Figure 3-17
A, transection of epidermis and collenchyma tissue in stem of *Cucurbita*; B, surface view of stomata in lower epidermis of leaf of *Capsicum*; C, surface view of stomata with subsidiary cells (s) in lower epidermis of leaf of *Sedum*. [A-C redrawn from *Plant Anatomy* by K. Esau. Wiley, New York. 1953; B courtesy Artschwager.]

Situated below the dermal system and external to or surrounding the fascicular system is the fundamental tissue system (Fig. 3-16). In most stems this tissue system is represented by the cortex—a cylinder of tissues between epidermis and the phloem, and the pith—a central column of parenchyma. (see Fig. 3-16, A). The stems of the Lycopsida and most roots develop the cortical portion of the fundamental system but lack a pith (Fig. 3-16, C). In the lamina of leaves the photosynthetic mesophyll represents the fundamental tissue system (Fig. 3-16, B).

Structure and Development of Tissue Systems

In addition to its value in the topographical description and comparison of primary tissue systems, Sachs' scheme is very helpful in the morphological interpretation of *tissue development* at the apices of shoots and roots. Apical meristems, as shown by numerous researches, are extraordinarily variable in their histology (see Foster, 1949, pp. 42-48; Esau, 1965a; Gifford, 1954; Gifford and Corson, 1971; for literature on the subject). In some plants, for example, *Equisetum* and the leptosporangiate ferns, a single well-defined apical cell occupies the tip of the axis and represents the ultimate point of origin of all the meristematic tissue of the apex of root and shoot (see Fig. 19-33). However, the shoot apices of *Lycopodium*, of certain eusporangiate ferns, and of a great many gymnosperms possess several superficial apical initials. And finally, the shoot apices of the angiosperms have a typical stratified arrangement of cells, the outer layer or layers being the *tunica,* which surrounds a central mass of meristem designated the *corpus;* in this type of apex, the number and position of individualized apical initials is very uncertain in most instances (Figs.

3-1, B, 19-20). Despite the histological differences among the types of apices in vascular plants, an essentially similar pattern of early histogenesis or tissue formation, is common to all of them. This consists in the ultimate segregation, behind the apex of shoot and root, of three primary meristematic tissues: the *protoderm,* the *procambium,* and the *ground meristem* (Fig. 3-1). These tissues are the precursors of the epidermal, fascicular, and fundamental tissue systems, respectively, and their salient features may now be briefly examined.

THE EPIDERMIS. The protoderm, derived from the cells of the shoot or root apex, is a uniseriate layer of dividing cells that ultimately forms the epidermis of mature organs (Figs. 3-1, 3-17, 3-18). Certain protoderm cells enlarge, acquire cutinized outer walls, and differentiate as the typical epidermal cells of leaves and stems. In many plants large numbers of protoderm cells develop into the various types of epidermal appendages or trichomes; common examples of trichomes are root hairs and the various kinds of hairs so commonly seen on leaves and stems. The stomata are characteristic of the epidermal system of foliage leaves, and many types of stems, floral organs, and fruits (Fig. 3-17, B, C). Paleobotanical evidence shows clearly that stomata were present in such ancient and simple vascular plants as the Rhyniopsida. Stomata develop by the division and differentiation of certain protoderm cells into pairs of guard cells between which a stomatal opening or pore is formed. In many plants two or more of the cells bordering upon the stoma are distinctive in form and are termed subsidiary cells. (Fig. 3-17, C). The student should understand that the epidermis of fossil plants, particularly the arrangement and structure of its stomata, provides very important clues about the phylogeny and relationships of extinct tracheophytes. An

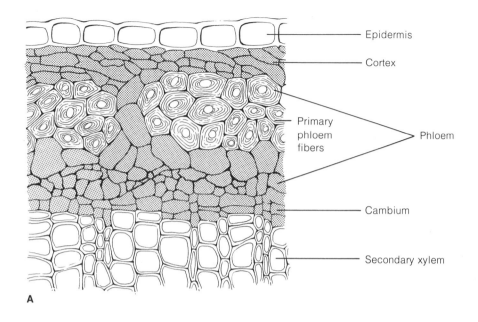

Epidermis

Cortex

Primary
phloem
fibers

Phloem

Cambium

Secondary xylem

A

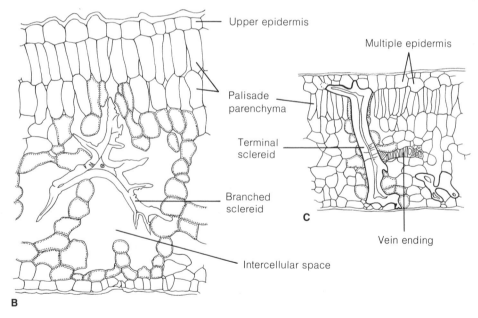

Upper epidermis

Multiple epidermis

Palisade
parenchyma

Terminal
sclereid

Branched
sclereid

Vein ending

C

Intercellular space

B

Figure 3-18
A, transection of stem of *Linum usitatissimum* illustrating position of strands of primary phloem
fibers; **B,** transection of portion of lamina of *Trochodendron aralioides* showing a branched sclereid;
C, transection of a portion of lamina of *Mouriria Huberi*, showing a columnar, ramified terminal
sclereid. [A redrawn from *Plant Anatomy* by K. Esau. Wiley, New York. 1953; B redrawn from
Foster, *Amer. Jour. Bot.* 32:456, 1945; C redrawn from Foster, *Amer. Jour. Bot.* 34:501, 1947.]

excellent illustration of the phylogenetic importance of epidermal structure is found in the detailed comparisons between the stomata of living and extinct gymnosperms made by Rudolf Florin (1931, 1951).

THE FUNDAMENTAL TISSUE SYSTEM.   This system of tissues originates from the ground meristem and, as mentioned before, is represented by the tissues found in the cortex of stems and roots, the pith of stems, and the mesophyll of foliar organs (Fig. 3-1). In contrast with the elongated and often spindle-shaped procambial cells, ground meristem cells, prior to differentiation, are polyhedral cells which closely approximate tetrakaidecahedra (14-sided bodies) in form. Cells of this type very commonly enlarge, become separated by intercellular spaces, and mature into the parenchyma tissue, which often is the principal component of the fundamental tissue system. But additional cell types and tissues may originate from unspecialized ground meristem and become part of the fundamental tissue system of plant organs. A common example of this is *collenchyma,* which is very commonly developed in the outer region of the cortex of stems and in the subepidermal region of petioles (Fig. 3-17, A). In the ontogeny of collenchyma the ground meristem cells divide and elongate, ultimately producing compact strands or a cylinder of living cells with unevenly thickened primary walls; very commonly the thickest portions of the wall are laid down at the angles or corners where several collenchyma cells meet. Regarded functionally, collenchyma tissue provides support and flexibility for growing organs and at maturity is characterized by considerable tensile strength.

Another extremely common type of tissue in the fundamental tissue system is *sclerenchyma,* which is composed of cells with thick, lignified secondary walls. Two fairly well-demarcated cell types are included under sclerenchyma: *fibers,* which typically are conspicuously elongated cells with pointed ends, and *sclereids,* which are polygonal (the so-called stone cells), columnar, or profusely branched in form. At maturity fibers very commonly are dead cells, devoid of protoplasts, and occur as strands or cylinders of tightly joined cells which evidently are of considerable importance in providing mechanical strength to plant organs (Fig. 3-18, A). Sclereids may occur in compact masses in various parts of the fundamental tissue system, but they also occur as isolated cells or *idioblasts.* In the leaves of many dicotyledons, branched idioblastic sclereids are frequently diffuse in their distribution in the mesophyll (Fig. 3-18, B), but in certain genera the sclereids are terminal, or restricted to the vein endings (Fig. 3-18, C). (See Foster, 1949, 1956, for further details and references to the literature on foliar sclereids.)

THE FASCICULAR TISSUE SYSTEM.   From what we have said regarding the variable patterns of the fascicular system, a corresponding variability is to be expected in the patterns of procambium formation in young organs. In the young, terminal regions of many stems (particularly the stems of gymnosperms and angiosperms) and in differentiating leaves the procambium consists of discrete cellular strands composed of elongated cells (Fig. 3-1, B); each strand matures as a vascular bundle composed of primary phloem and primary xylem tissue. In the stems of certain lower vascular plants (*Lycopodium,* and certain ferns) and in many roots, the procambium is a central core or column of tissue from which the vascular cylinder, devoid of pith, originates (Fig. 3-1, A). A discussion of the complex problem of the origin and development of primary vascular tissues from the procambium is entirely beyond the scope of this book (see Esau, 1965b, for a review of the subject and literature references), but certain aspects of vasculation in plants are essential to our presentation of comparative morphology and merit brief discussion here.

The first of these general problems concerns the general structure of primary xylem and the meaning of the terms *protoxylem* and *metaxylem*. According to the best of modern usage, protoxylem designates the pole of earliest developed primary xylem and includes all tracheary tissue which differentiates (i.e., completes its growth and secondary wall development) during the period of organ elongation. Tracheary elements of the protoxylem, as shown in Fig. 3-19, often develop their secondary walls as a series of rings (annular elements) or as one or more spiral bands (spiral elements). The metaxylem is the remaining portion of the primary xylem which completes its differen-

tiation, after the organ in which it occurs has ceased to elongate. Metaxylem cells, also shown in Fig. 3-19, usually have more extensively developed secondary walls which commonly appear as a series of connected bars (scalariform elements) or a network (reticulate elements), or else the wall is pitted.

It must be emphatically stated that the primary xylem may consist wholly of tracheids or of both tracheids and vessel elements. The type of secondary wall pattern is therefore not necessarily correlated with the presence or absence of vessel elements. The primary xylem of certain species of *Selaginella* and of such ferns as *Pteridium* and *Marsilea* contains well-defined vessels; these conducting structures also occur in

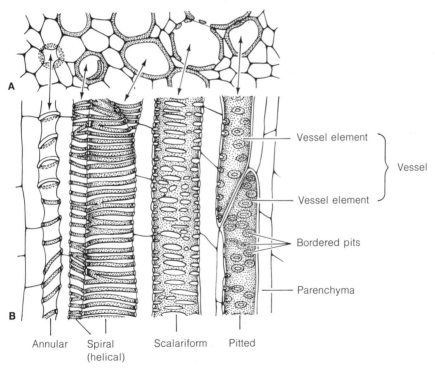

Annular   Spiral   Scalariform   Pitted
           (helical)

Vessel element
Vessel
Vessel element
Bordered pits
Parenchyma

Figure 3-19
A portion of the primary xylem of *Aristolochia* in transverse (**A**) and longisectional (**B**) view. Note diversity in types of secondary wall patterns of tracheary elements in progressing from protoxylem at left to metaxylem at right. [From *Plant Anatomy* by K. Esau. Wiley, New York. 1953.]

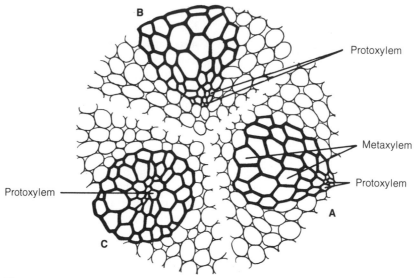

Figure 3-20
Directions of radial maturation of tracheary elements in the primary xylem of vascular plants. **A,** exarch primary xylem; **B,** endarch primary xylem; **C,** mesarch primary xylem. [Redrawn from *The Anatomy of Woody Plants* by E. C. Jeffrey. University of Chicago Press, Chicago. 1917.]

the primary xylem of many monocotyledons and dicotyledons. (See Bailey, 1944, 1957; and Cheadle, 1953, for a discussion of the evolution of vessels.)

A second problem involves the direction of radial maturation of successive procambial cells during the differentiation of the primary xylem. In the roots of all vascular plants, and in the stems of *Psilotum* and the Lycopsida, the first protoxylem cells to acquire secondary walls occur at the outermost edge of the procambial cylinder; these cells establish the future pattern of xylem differentiation which occurs *centripetally* or toward the center of the axis. Primary xylem of this type is termed *exarch* and is regarded as the most primitive condition in vascular plants (Fig. 3-20, A). In the stems of modern seed plants, however, the protoxylem begins its development from the innermost procambial cells—those situated next to the pith—and the remainder of the process of primary xylem differentiation occurs *centrifugally* or toward the

periphery of the stem (Fig. 3-20, B). This type of primary xylem is termed *endarch xylem* and is believed to be the most highly advanced type. Lastly, in the leaf and stem bundles of many ferns the primary xylem is *mesarch*. This means that the protoxylem begins development within the procambial strand and that further xylem formation occurs centripetally as well as centrifugally; consequently, at maturity the protoxylem cells are surrounded by the metaxylem, as can be seen in Fig. 3-20, C. The distinction between exarch, endarch, and mesarch xylem thus appears to be of considerable importance, not only among different groups of vascular plants but even between the root and stem of the same plant. Because the xylem of extinct plants is often well-preserved in fossils, the recognition of the position of the protoxylem in relation to the metaxylem is a matter of considerable significance in paleobotanical interpretation.

A third problem in regard to the structure of vascular tissues concerns the phloem. The

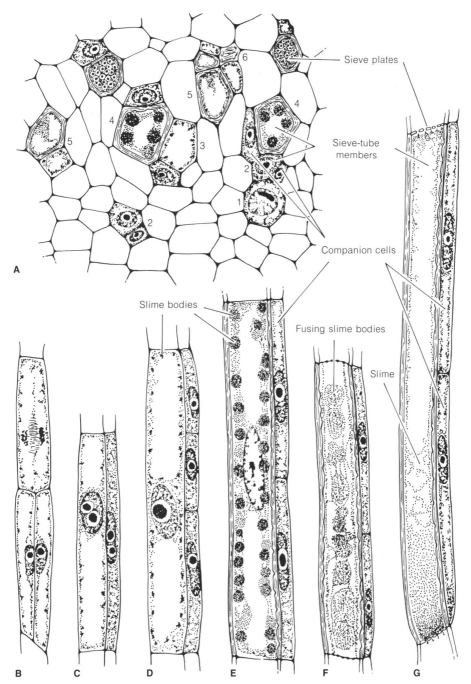

Figure 3-21

Differentiation of sieve-tube members and their associated companion cells in *Cucurbita.*
A, transection of phloem, the successive numbers (*1–6*) designating stages in development of
sieve-tube members and companion cells; **B–G,** corresponding stages in development as seen in
longisectional view. Note the disappearance of the nucleus (**E–F**) and the development of sieve
plates at the ends of the sieve-tube members. [Redrawn from *Plant Anatomy* by K. Esau. Wiley,
New York. 1953.]

primary phloem of vascular plants is characterized by the presence of specialized conducting cells known as sieve elements. The definitive feature of these elements, at least in angiosperms, is the absence at maturity of a nucleus and the presence of more-or-less specialized *sieve areas* or *sieve plates* in the walls (Fig. 3-21, A-G). A sieve area is a modified portion of the primary wall traversed by connecting strands of protoplasm, each of which at an early stage in development is surrounded by a cylinder of substance termed *callose.* Prior to the death and obliteration of sieve elements, the sieve areas (and the coarser sieve plates) are usually entirely covered with callose. In the angiosperms the sieve elements occur in superposed series collectively designated *sieve tubes.* Each sieve-tube member is associated with one or more nucleated *companion cells* (Fig. 3-21, A, C-G). In addition to the definitive sieve elements, primary as well as secondary phloem may contain parenchyma cells, sclereids, and fibers. For detailed treatments of the ontogeny and structure of both primary and secondary phloem, consult Esau, Cheadle, and Gifford, 1953; and Esau, 1960, 1965a, 1969.

Lastly, the concept of secondary vascular tissues and their demarcation from the primary vascular system needs to be considered. Secondary vascular tissues are produced from the vascular cambium, and since the cells of the secondary xylem are often arranged in radial rows this system of tissues often is clearly demarcated from the more irregular pattern of cells of the primary vascular system. (See Fig. 3-18, A). But the criterion of orderly versus irregular cell arrangement is not always valid; in some angiosperms and gymnosperms the tracheary cells of the primary as well as the secondary xylem are in regular radial alignment. As a consequence, the boundary between the primary and secondary vascular systems can be only approximately determined even when the entire ontogenetic development has been studied. Secondary growth by means of a vascular cambium has repeatedly arisen during the evolution of vascular plants. Many extinct groups of the tracheophytes (for example, *Lepidodendron* and *Calamites;* see Fig. 10-20, B) showed conspicuous secondary growth, and secondary growth is a prominent feature of all living gymnosperms and of a large number of the angiosperms. (See Chapters 17 and 19 for detailed descriptions of the secondary xylem of conifers and dicotyledons.) However, most of the lower vascular plants of today such as the Psilotaceae, the majority of the Lycopsida, *Equisetum,* and the majority of ferns, are devoid of cambial activity. In these organisms, as in most of the monocotyledons, the fascicular system is entirely primary and derived ontogenetically from the procambium.

The Stelar Theory

During the latter half of the nineteenth century the increasing emphasis placed on the importance of the vascular system in morphological interpretation led to the formulation of the Stelar Theory. This theory, which was developed by Van Tieghem and his students (Van Tieghem and Douliot, 1886) has had far-reaching effects on modern concepts of the morphology and evolution of the primary vascular system and hence deserves our attention. According to Van Tieghem, the primary structure of the stem and root are fundamentally similar in that each organ consists of a central *stele* enveloped by the cortex, the outer layer of which is the epidermis. The term "stele" was used in a collective sense by Van Tieghem to designate not only the primary vascular tissues but also the so-called conjunctive tissues associated with them: pericycle, vascular rays, and, when it occurs, the pith tissue (Fig. 3-22).

One of the critical—and in the light of modern studies, controversial—aspects of the Stelar Theory is the nature of the anatomical boundaries which separate the cortex from the stele. Van Tieghem considered that the inner boundary of the cortex is the endodermis, a cylinder of living cells which, from a strict histological viewpoint, are characterized by the presence of Casparian strips. These strips or bands are chemically modified portions of the radial and end walls of the endodermal cells and are thought to contain both lignin and suberin.

In roots and in the stems of many of the lower vascular plants an endodermis is present and represents a tangible boundary between cortex and stele (Fig. 8-4, A). But an endodermis, in the sense just defined, is not present in the stems of a large proportion of the seed plants, especially woody types, and in plants not having an endodermis the limits between cortex and stele are more difficult to establish. The pericycle likewise is not present in all vascular plants and the entire concept of pericycle and pericyclic fibers has been critically reexamined in recent years from an ontogenetic viewpoint (Blyth, 1958). Although the pericycle is a recognizable cylinder of cells at the outer edge of the stele of roots and the stems of lower vascular plants, ontogenetic studies have shown that the so-called pericycle in the stems of many angiosperms is actually the outermost portion of the primary phloem. In such instances the pericyclic fibers are morphologically a part of the protophloem and consequently there is no independent tissue zone separating the cortex from the stele (Fig. 3-18, A).

Despite the absence of consistent histological boundaries between the cortex and stele, the value of the Stelar Theory as a unified concept has been widely recognized and has led to efforts to classify and interpret

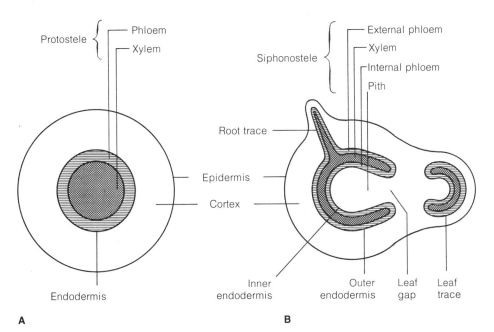

Figure 3-22
Types of steles in the stems of vascular plants. **A,** the protostele; **B,** the siphonostele.
[Redrawn from *The Anatomy of Woody Plants* by E. C. Jeffrey. University of Chicago Press, Chicago. 1917.]

phylogenetically the varied types of vascular cylinders that occur in stems and roots.

It is now rather generally agreed that from a phylogenetic as well as an ontogenetic standpoint the most primitive type of stele is the *protostele*, which is characterized by the absence of a central column of pith. In its simplest form the protostele is merely a central strand of primary xylem sheathed by a cylinder of phloem (see Fig. 3-22, A). This particular form of protostele is often termed a *haplostele* and was the type of conducting system in such primitive plants as *Horneophyton* and *Rhynia;* haplosteles are also common in the stems of young sporophytes of many ferns which develop later in ontogeny a medullated type of stele. In certain plants the contour of the core of xylem is lobed or starshaped in transectional view — this form is designated an *actinostele* (Chapters 8 and 9). *Psilotum* and various species of *Lycopodium* characterize this condition, and in other species of *Lycopodium* the xylem is a "sponge" and in transection seems to consist of separate plates of tissue between and around which occurs the phloem; this specialized type of protostele is called the *plectostele*.

A stele with a central column of pith is regarded phylogenetically as an advance in anatomical development and is termed a *siphonostele* (Fig. 3-22, B). Siphonosteles are characteristic of the stems of many ferns (see Chapter 13). Considerable difference of opinion has prevailed among morphologists about the method of phylogenetic origin and development of a pith. One theory, championed by the American anatomist E. C. Jeffrey (1917, pp. 283–291) holds that the pith is always *extrastelar* in origin — that it represents the inclusion of the fundamental tissue into the core of xylem. The opposed extreme view is that the pith is *stelar* in origin and represents degraded tracheary elements. Evidence for this view is furnished by the existence of protosteles in which the

central region consists of both tracheids and parenchyma cells; this type of vitalized protostele is presumed to typify a stage in the phylogenetic reduction of the centermost tracheary elements to parenchyma (Chapter 13). Today there is good reason to believe that medullated steles have originated in both of these ways. In other words, similar-appearing siphonosteles may have resulted either from the gradual inclusion of areas of cortical parenchyma or from the phylogenetic reduction of the original tracheids to parenchyma (see especially Bower, 1911).

The histological structure of the siphonostele varies widely. In certain fern genera (*Adiantum, Marsilea,* and *Dennstaedtia*) the cylinder of primary xylem is bordered internally as well as externally by phloem tissue. Furthermore, two endodermal cylinders are developed, one separating the cortex from the external phloem, the other situated between the internal phloem and the pith (see Chapter 13). This condition is termed *amphiphloic*, and a stele with this construction is specifically designated as a *solenostele* (Gwynne-Vaughan, 1901). The solenostele is often used to support the theory that the siphonostele has arisen by the invasion of the xylem by extrafascicular tissues; i.e., endodermis, pericycle, and phloem. In certain highly specialized angiosperm families such as Asclepiadaceae and Solanaceae, internal phloem is developed at the periphery of the pith. However, it is very probable that the amphiphloic condition in this case is a mark of extreme anatomical specialization rather than an example of the persistence of a primitive fern-like stelar structure. Most commonly, at least in seed plants, the stele of the stem consists of only an external cylinder of phloem enclosing the xylem; this condition is designated *ectophloic*.

Evolutionary modification of the solenostele in ferns is believed to have produced a type of stele with long overlapping leaf gaps (see Chapter 13). When visualized in three dimensions, a stele of this type is a *tubular network* of interconnected vascular strands enclosing the column of pith. Brebner (1902) designated this type of vascular cylinder as a *dictyostele*. In highly dissected dictyosteles, typical of many ferns, the parenchymatous areas between the vascular strands vary in number and vertical extent; doubtless some of these areas are not "leaf gaps" but are merely interfascicular strips of parenchyma not having a consistent relationship to the leaf trace system of the axis.

As viewed in a transverse section of the stem, a fern dictyostele appears as a "ring" of separate bundles, each of which is *amphicribral* in structure; i.e., it is composed of a central strand of xylem surrounded by a sheath of phloem tissue, followed by a layer of pericycle and an endodermis. A concentric bundle of this type is commonly called a "meristele" (Chapter 13).

In dicotyledons, the primary vascular cylinder also appears, in transections of the stem, as a "ring" of more-or-less discrete vascular bundles separated by areas of parenchyma (see Chapter 19, Fig. 19-22, A). This stelar type, termed the *eustele* by Brebner (1902), is distinguished from the fern dictyostele by the collateral, or bicollateral, organization of the vascular strands. Most commonly, each vascular bundle is collateral, consisting of a strand of xylem flanked externally by a strip of phloem tissue (Fig. 3-16, A). In certain taxa of the dicotyledons (e.g., *Cucurbita*) the vascular bundles are bicollateral with phloem on both sides of xylem strands. It should be emphasized at this point that the divergence of leaf traces from a eustele is associated with parenchymatous areas that are usually regarded as leaf gaps. However, it proves very difficult to demarcate the limits of each leaf gap and, as we will show in Chapter 14, the concept of leaf gap in the eusteles of both dicotyledons and conifers has been rejected by several recent investigators.

In our brief account of stelar types in the Tracheophyta, we have attempted to show the considerable value of the concepts of protostele, siphonostele, dictyostele, and eustele in the interpretation of the diverse patterns of the primary vascular system of the axis. A very difficult problem, however, is posed by the monocotyledons because the primary vascular system of the stems in this class of angiosperms usually does not conform to any of the stelar types found in other groups of vascular plants. In place of a concentric cylinder of primary xylem and phloem, or a tube of anastomosed strands, the vascular bundles in the stems of monocotyledons are dispersed or "scattered" in arrangement; no clear boundaries exist between cortex and pith, and the divergence of the numerous leaf traces at each node is obviously not associated with definable leaf gaps. Brebner (1902, p. 520) to whom we owe many terms used to designate stelar types, proposed the term *"atactostele"* (from the Greek root word "atactos" meaning "without order") for the stele in monocotyledons. He defined it as consisting "of a number of more or less irregularly arranged vascular bundles together with the ground tissue in which they are imbedded." Although the term "atactostele" has been widely adopted for descriptive purposes (Esau, 1960, 1965a; Fahn, 1967; Zimmermann, 1959), determining the significance — ontogenetic or phylogenetic — of the scattered arrangement of bundles in monocotyledonous stems has proved a most difficult and elusive problem. The magnitude of the problem is well illustrated by reference to bundle number in certain palms. According to Zimmermann and Tomlinson (1965, p. 165), about 1,000 bundles occur in the small stem (2–3 centimeters in diameter) of *Rhapis excelsa* and in stems of *Cocos*, a half a meter in diameter, there may be more than 20,000 vascular bundles "at any level"! In grasses, commonly used as examples of "typical" monocotyledons, the number of bundles is considerably less. Transections of the stem of *Secale,* for example, reveal a primary vascular system consisting of two "rings" of vascular bundles located in the outer peripheral region of the internode (Fig. 3-23, A). A contrasted pattern of bundle arrangement is illustrated by the stem of corn (*Zea*) in which numerous small bundles are closely

spaced at the periphery and larger, more widely spaced strands are distributed in the central region (Fig. 3-23, B).

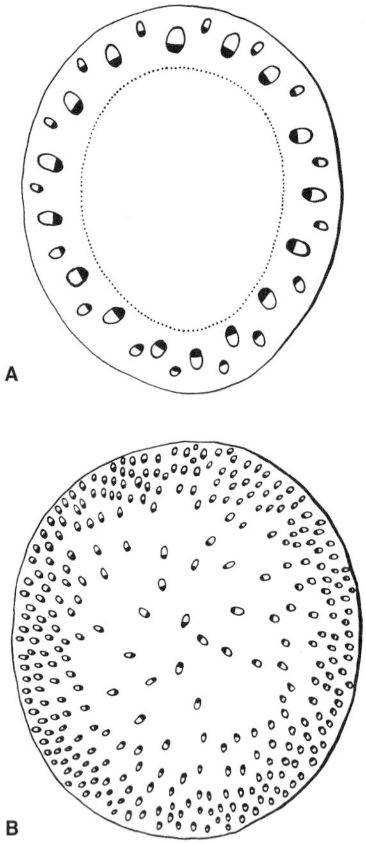

As Jeffrey (1917, pp. 192–196) pointed out many years ago, the large number of bundles seen in transections of monocotyledonous stems is correlated with the extremely large number of traces (often hundreds in palms) that vascularize the leaves. In his view, the number of leaf traces "is too large to be accommodated on the periphery as is the rule even in herbaceous dicotyledons." As a "consequence," the leaf traces are "displaced" into the pith region of the axis. This rather simplistic "explanation," however, fails to deal with the *vertical course* or the ontogeny of the "systems" of leaf traces in the stem. Classical studies on palm stems revealed that many of the centrally located bundles are leaf traces or leaf-trace complexes, which pursue an *undulate* course, diverging below a given node toward the stem center and, at a lower level, curving outwardly toward the periphery. This description merely outlines one aspect of the problem and it is only in recent years that Tomlinson and Zimmermann have initiated a series of studies designated to explain the course of the bundles in the stem and their complex behavior with reference to the vascularization of leaves and axillary inflorescences. These investigators devised an ingenious "optical shuttle method" for photographing, by means of a motion picture camera, successive transections of the stems of *Rhapis* (Palmaceae) and *Prionum* (Juncaceae). This method has resulted in considerable progress in reconstructing the extremely complex vascular system characteristic of these monocotyledons. A detailed review of the results of these new studies cannot be undertaken, however, in this text. Students wishing further information should consult the original papers by Zimmermann and Tomlinson (1965, 1966, 1967, 1968).

Figure 3-23
Transectional diagrams showing two patterns of vascular bundle arrangement in the stems of grasses. A, *Secale* (rye) — two "circles" of bundles at peripheral region of hollow stem; B, *Zea* (corn) — numerous bundles scattered throughout the transection. [Redrawn from *Allgemeine Botanik* by W. Troll. Ferdinand Enke Verlag, Stuttgart. 1959.]

References

Bailey, I. W.

1944. The development of vessels in angiosperms and its significance in morphological research. *Amer. Jour. Bot.* 31:421–428.

1949. Origin of the angiosperms: need for a broadened outlook. *Jour. Arnold Arboretum* 30:64–70.

1956. Nodal anatomy in retrospect. *Jour. Arnold Arboretum* 37:269–287.

1957. The potentialities and limitations of wood anatomy in the study of the phylogeny and classification of the angiosperms. *Jour. Arnold Arboretum* 38:243–254.

Banks, H. P., and M. R. Davis

1969. *Crenaticaulis*, a new genus of Devonian plants allied to *Zosterophyllum,* and its bearing on the classification of early land plants. *Amer. Jour. Bot.* 56:436–449.

Blyth, A.

1958. Origin of primary extraxylary stem fibers in dicotyledons. *Univ. Calif. Publ. Bot.* 30:145–232.

Bock, W.

1962. *Systematics of Dichotomy and Evolution.* (Geological Center Research Series. Vol. 2.) Geological Center, North Wales, Pa.

1969. *The American Triassic Flora and Global Distribution.* (Geological Center Research Series. Vols. 3 and 4.) Geological Center, North Wales, Pa.

Boodle, L. A.

1901. Comparative anatomy of the Hymenophyllaceae, Schizaeaceae and Gleicheniaceae. *Ann. Bot.* 15:703–747.

Bower, F. O.

1911. On medullation in the Pteridophyta. *Ann. Bot.* 25:555–574.

1923. *The Ferns (Filicales),* Vol. I. The University Press, Cambridge.

1935. *Primitive Land Plants.* Macmillan, London.

Brebner, G.

1902. On the anatomy of *Danaea* and other Marattiaceae. *Ann. Bot.* 16:517–552.

Bugnon, F.

1969. La notion de dichotomie: essai d'analyse sur l'example des Algues. *Mem. Soc. Bot. France* 115:149–155.

Cheadle, V. I.

1953. Independent origin of vessels in the monocotyledons and dicotyledons. *Phytomorphology* 3:23–44.

Corner, E. J. H.

1966. *The Natural History of Palms.* University of California Press, Berkeley and Los Angeles.

Dunlop, D. W.

1949. Casparian strips in *Isoetes macrospora. Bull. Torrey Bot. Club* 76:134–135.

Eames, A. J.

1961. *Morphology of Angiosperms.* McGraw-Hill, New York.

Esau, K.

1960. *Anatomy of Seed Plants.* Wiley, New York.

1965a. *Plant Anatomy.* Wiley, New York.

1965b. *Vascular Differentiation in Plants.* Holt, Rinehart, and Winston, New York.

1969. *The Phloem.* (Handbuch d. Pflanzenanatomie. Band V, Teil 5.) Gebrüder Borntraeger, Berlin-Stuttgart.

Esau, K., V. I. Cheadle, and E. M. Gifford, Jr.

1953. Comparative structure and possible trends of specialization of the phloem. *Amer. Jour. Bot.* 40:9–19.

Fahn, A.

1967. *Plant Anatomy.* Pergamon, Oxford.

Florin, R.

1931. Untersuchungen zur Stammesgeschichte der Coniferales und Cordaitales. *Svenska Vetensk. Akad. Handl.* Ser. 5. 10:1–588.

1951. Evolution in Cordaites and Conifers. *Acta Horti Bergiani* 15(11):285–388.

Foster, A. S.
  1936. Leaf differentiation in angiosperms. *Bot. Rev.* 2:349–372.
  1949. *Practical Plant Anatomy*, Ed. 2. Van Nostrand, New York.
  1956. Plant idioblasts: remarkable examples of cell specialization. *Protoplasma* 46:184–193.
  1972. Cell types: Spermatophytes. Table 12, Part I, pp. 132–135 in *Biology Data Book*, Ed. 2., Vol. I. Federation of Amer. Soc. for Exper. Biology, Bethesda, Maryland.
Garrison, R., and R. H. Wetmore
  1961. Studies in shoot-tip abortion: *Syringa vulgaris. Amer. Jour. Bot.* 48:789–795.
Gifford, E. M., Jr.
  1954. The shoot apex in angiosperms. *Bot Rev.* 20:477–529.
Gifford, E. M., Jr., and G. E. Corson, Jr.
  1971. The shoot apex in seed plants. *Bot. Rev.* 37:143–229.
Greguss, P.
  1968. Dichotomous branching of monocotyledonous trees. *Phytomorphology* 18(4): 515–520.
Gwynne-Vaughan, D. T.
  1901. Observations on the anatomy of solenostelic ferns. I. *Loxsoma. Ann. Bot.* 15:71–98.
Härtel, K.
  1938. Studien an Vegetationspunkten einheimischer Lycopodien. *Beitr. Biol. Pflanzen* 25:125–168.
Harvey-Gibson, R. J.
  1896. Contributions towards a knowledge of the anatomy of the genus *Selaginella* Spr. Part II. The ligule. *Ann. Bot.* 10:77–88.
Hofmann, E.
  1934. *Paläohistologie der Pflanze.* J. Springer, Wien.
Hsü, J.
  1937. Anatomy, development and life history of *Selaginella sinensis.* I. Anatomy and development of the shoot. *Bull. Chinese Bot. Soc.* 3:75–95.
Jeffrey, E. C.
  1917. *The Anatomy of Woody Plants.* University of Chicago Press, Chicago.

Kidston, R., and W. H. Lang
  1917–1921. On Old Red Sandstone plants showing structure, from the Rhynie Chert Bed, Aberdeenshire, Parts I–V. *Trans. Roy. Soc. Edinb.* 51–52.
Lignier, O.
  1903. Equisétales et Sphenophyllales. Leur origine filicinéenne commune. *Bull. Soc. Linn. Normandie* Ser. 5. 7:93–137.
Millington, W. F.
  1963. Shoot-tip abortion in *Ulmus americana. Amer. Jour. Bot.* 50:371–378.
Sachs, J., von
  1875. *Textbook of Botany.* Clarendon Press, Oxford.
Sandt, W.
  1925. *Zur Kenntnis der Beiknospen.* Bot. Abh., herausg. von K. Goebel. Heft 7. Gustav Fischer, Jena.
Shoute, J. C.
  1909. Über die Verästelung bei monokotylen Baümen. II. Die Verästelung von *Hyphaene. Rec. Trav. Bot. Neerl.* 6:211–232.
Siegert, A.
  1965. Morphologische, entwicklungsgeschictliche und systematische Studien an *Psilotum triquetrum* Sw. II. Die Verzweigung (mit einer allgemeinen Erörterung des Begriffes "Dichotomie"). *Beitr. Biol. Pflanzen* 41:209–230.
Smith, R. W.
  1900. The structure and development of the sporophylls and sporangia of *Isoetes. Bot. Gaz.* 29:225–258, 323–346.
Stewart, W. N.
  1964. An upward outlook in plant morphology. *Phytomorphology* 14:120–134.
Tomlinson, P. B.
  1970. Dichotomous branching in *Flagellaria indica* (Monocotyledons). (New Research in Plant Anatomy. Supp. 1.) *Jour. Linn. Soc. Bot.* 63:1–14.
  1971. The shoot apex and its dichotomous branching in the *Nypa* palm. *Ann. Bot.* 35:865–879.
Tomlinson, P. B., and J. B. Fisher
  1971. Morphological studies in *Cordyline* (Agavaceae). I. Introduction and general morphology. *Jour. Arnold Arboretum* 52:459–478.

Tomlinson, P. B., M. H. Zimmermann, and P. G. Simpson
  1970. Dichotomous and pseudodichotomous branching of monocotyledonous trees. *Phytomorphology* 20:36–39.

Troll, W.
  1935. *Vergleichende Morphologie der höheren Pflanzen.* Bd. 1, Lieferung 2. Pp. 101–107. Gebrüder Borntraeger, Berlin.
  1937. *Vergleichende Morphologie der höheren Pflanzen.* Bd. 1, Erster Teil. Pp. 465–660. Gebrüder Borntraeger, Berlin.

Troop, J. E., and J. T. Mickel
  1968. Petiolar shoots in the Dennstaedtioid and related ferns. *Amer. Fern Jour.* 58:64–70.

Van Tieghem, P., and H. Douliot
  1886. Sur la polystelie. *Ann. Sci. Nat. Bot.* Ser. 7. 3:275–322.

Wardlaw, C. W.
  1952. *Phylogeny and Morphogenesis.* St. Martin's Press, New York.
  1957. Experimental and analytical studies of Pteridophytes. XXXVII. A note on the inception of microphylls and macrophylls. *Ann. Bot.* n. s. 21:427–437.
  1965. *Organization and Evolution in Plants.* Longmans, Green, London.

Wilson, C. L.
  1937. The phylogeny of the stamen. *Amer. Jour. Bot.* 24:686–699.
  1942. The telome theory and the origin of the stamen. *Amer. Jour. Bot.* 29:759–764.
  1950. Vasculation of the stamen in the Melastomaceae, with some phyletic implications. *Amer. Jour. Bot.* 37:431–444.
  1953. The telome theory. *Bot. Rev.* 19:417–437.

Zimmermann, M. H., and P. B. Tomlinson
  1965. Anatomy of the palm *Rapis excelsa.* I. Mature vegetative axis. *Jour. Arnold Arboretum* 46:160–178.
  1966. Analysis of complex vascular systems in plants: optical shuttle method. *Science* 152:72–73.
  1967. Anatomy of the palm *Rapis excelsa.* IV. Vascular development in apex of vegetative aerial axis and rhizome. *Jour. Arnold Arboretum* 48:122–142.
  1968. Vascular construction and development in the aerial stem of *Prionium* (Juncaceae). *Amer. Jour. Bot.* 55:1100–1109.

Zimmermann, W.
  1953. Main results of the "telome theory." *The Paleobotanist* 1:456–470.
  1959. *Die Phylogenie der Pflanzen*, Ed. 2. Gustav Fischer Verlag, Stuttgart.
  1965. *Die Telomtheorie.* Gustav Fischer Verlag, Stuttgart.
  1966. Kritische Beiträge zu einigen biologischen Problemen. VIII. Die Hologenie. *Zeit. Pflanzenphysiol.* 54:125–144.

4

Sporangia

One of the salient and definitive features of the sporophyte generation in vascular plants is the production of more-or-less numerous spore sacs or sporangia. In marked contrast with a moss or a liverwort, wherein the entire sporophyte normally forms only a single, nonseptate sporangium, vascular plants are polysporangiate, and the number of sporangia and spores developed by a single individual may be enormous. For example, Bower (1908) has estimated that a single, well-developed male shield fern (*Nephrodium filix-mas*) may produce approximately 50,000,000 spores in a single season. Obviously, for a larger plant, such as a full-grown pine or fir tree, the total number of pollen grains would reach astronomical figures. The so-called sulfur showers which

are familiar phenomena each year in regions of coniferous forests are really countless yellow pollen grains that have been released from the sporangia of cones and are buoyed by wind currents. From a broad biological point of view, the apparently wasteful overproduction of spores, especially by lower vascular plants and wind-pollinated seed plants, actually tends to compensate for the high proportion of spores that do not survive after dispersal from the parent plant. Thus, as is true also of the prodigious development of gametes in animals and in many kinds of plants, it seems that overproduction of spores is a device for insuring the perpetuation of the race.

All normal sporangia of vascular plants share one important feature: they are *the* specific regions of the sporophyte where reductional division, or meiosis, occurs. This process can be localized further to the sporocytes, or spore mother cells, which are the essential components of any immature sporangium. Each sporocyte is normally a diploid cell which, as a result of meiotic divisions, yields a group of four spores known as a spore tetrad (Fig. 4-1). Aside from this functional identity, however, the sporangia in the various vascular plant groups differ widely in position, form, size, structure, and method of development. In a given group, or even in certain genera, the position, structure, and ontogeny of the sporangium provide consistent and useful criteria for morphological comparison and taxonomic utilization. Indeed, such major groups of the tracheophytes as the Psilopsida, Lycopsida, Sphenopsida, and Filicopsida are sharply distinguished from one another on the basis of sporangium morphology alone. Among the living members of the Filicopsida, the leptosporangiate ferns are a very vivid example of a large, biologically successful group that differs from all other vascular plants in the details of the development and structure of their sporangia (Fig. 4-1, A′–G′, and Fig. 13-13).

It should now be clear that sporangia are structures which, although conservative in their functional aspects, vary considerably in other ways. This points to a long, complex evolutionary history, the details of which may never become entirely clear. However, as paleobotanical research continues, fossilized sporangia in a fair state of preservation continually become available for study and comparison. Such sporangia throw considerable light on the morphology and evolution of the sporangia of modern vascular plants.

In this chapter the important topographical, ontogenetic, and structural features of the sporangia of the lower groups of vascular plants are emphasized. The purpose is to provide the student with a general con-spectus of the subject and an orientation, both of which will be indispensable in his survey of the comparative morphology of the lower groups of vascular plants. Later chapters will discuss sporangium morphology in the gymnosperms and angiosperms.

Position of Sporangia

The evidence provided by such extinct members of the Rhyniopsida as *Horneophyton* and *Rhynia* indicates that sporangia antedated leaves in evolution (Chapter 7). In other words, the most primitive and elemental type of sporangium appears to have been a *cauline* (i.e., belonging to the stem) structure and was merely a sporogenous tip of the axis. Among living vascular plants, cauline sporangia occur, according to recent interpretation, in *Psilotum* and *Tmesipteris*, the living representatives of the Psilopsida, and in *Equisetum*. In *Equisetum* the sporangia occur in circular groups attached to the under surface of peltate sporangiophores. These sporangium-bearing structures may represent condensed, highly modified fertile branches, but a completely satisfactory interpretation of their evolutionary history has not yet been offered (Chapter 10).

In other groups of vascular plants the sporangia are evidently related to leaves. A foliar structure which subtends or bears one or more sporangia is termed a *sporophyll*. Throughout the ferns, particularly the leptosporangiate types, the sporophylls are, to varying degrees, photosynthetic organs and are often large and pinnatifid like the sterile foliage leaves or fronds. In some genera, however, special areas of the frond, or even complete leaves, are devoted exclusively to spore production and are usually nonphotosynthetic. In the sporophylls of leptosporangiate ferns the sporangia typically are segregated into discrete groups, known as sori, at the margins or on the abaxial surface of the lamina (Chapter 13). There exists wide fluctuation in the number of sori produced by a fern sporophyll and the number of sporangia in each sorus. Sporangial number and spore output are of course tremendous in the huge sporophylls of tree ferns.

Certain groups of vascular plants are distinguished by the fact that the sporophylls, instead of being intermingled with ordinary foliage leaves, are aggregated into a compact cone-like structure termed a *strobilus*. The Lycopsida is a particularly instructive group in which various degrees of development and distinctness of the strobilus are strikingly displayed. In all lycopsid plants a solitary sporangium is associated with each sporophyll and is either situated in its axil, as in *Selaginella,* or attached adaxially to the basal region of the sporophyll, as in certain species of *Lycopodium* and all members of the genus *Isoetes.* Most species of *Lycopodium,* like members of the genus *Selaginella,* develop well-defined strobili which are borne at the tips of main or lateral shoots. But in *Lycopodium lucidulum* and *Lycopodium selago,* for example, the sporangia occur in poorly defined patches or zones which alternate with purely vegetative regions of the shoot system (see Chapter 9). The viewpoint is now generally held that the strobiloid forms of *Lycopodium* represent an evolutionary development from the morphological condition represented in such species as *L. selago.* In the genus *Isoetes* a definable strobilus distinct from the vegetative part of the plant likewise is absent. A high proportion of the quill-like leaves are fertile and bear large solitary microsporangia or megasporangia on their adaxial bases; however, many of the functionally sterile leaves actually develop rudimentary or abortive sporangia. Such an authority as Bower (1908, p. 165) states that "after the first sporangia appear, the whole plant may be regarded as a strobilus, imperfectly differentiated, as in the selago type, into fertile parts and parts sterile by abortion or by complete suppression."

The gymnosperms predominantly are a strobiloid group, and in the cycads and certain of the conifers the strobili (cones) are exceptionally large. The flowers of many angiosperms, particularly certain members of the Magnoliaceae and Ranunculaceae, are also remarkably similar to strobili in their general organography and function. However, a discussion of the strobili and sporangia of seed plants must be deferred until later chapters.

Structure of Sporangia

From the broadest possible structural standpoint, the mature sporangium in lower vascular plants consists fundamentally of one or of many spores enclosed by a protective wall (Fig. 4-1). In discussing the morphology and ontogeny of sporangia, "wall" is used to designate the layer or layers distinct in origin from the sporocytes, that constitute the sterile protective jacket of the sporangium. When a sporangium is sunken within the stem or the sporophyll the wall is merely an external multilayered cover which is not sharply demarcated from the adjacent sterile tissue. But in emergent sporangia the wall is much more clearly defined; furthermore, in this type the spore-producing portion of the sporangium is very often borne on a definite stalk (Fig. 4-1, G'). In addition to its protective rôle, the wall of most sporangia is definitely related to the method of dehiscence or splitting open of the spore case. Certain of the surface cells of the sporangia of many lower vascular plants are unevenly thickened and collectively form a distal plate, a ridge, or an incomplete ring; these varied cell patterns constitute dehiscence mechanisms, the biophysical and biochemical operation of which deserves much more investigation than has been accorded it in the past (Goebel, 1905). Undoubtedly, the most specialized and familiar dehiscence mechanisms are the annulus and stomium of the sporangium in advanced leptosporangiate ferns (see Fig. 13-13, I). In these plants the transverse rupture of the thin sporangium wall, and the slow recurving of the annulus and the upper half of the capsule, is followed by the return of the latter approximately to its former position with a sudden jerk that hurls the dry spores a distance of 1–2 centimeters from the parent plant (Fig. 13-16). (For detailed descriptions of the physical mechanism involved in the dehiscence of fern sporangia see Ursprung, 1915; Ingold, 1939; Meyer and Anderson, 1939, pp. 237–238.)

Besides the essential spores and the variously constructed wall, most sporangia develop a *tapetum* during the early or middle

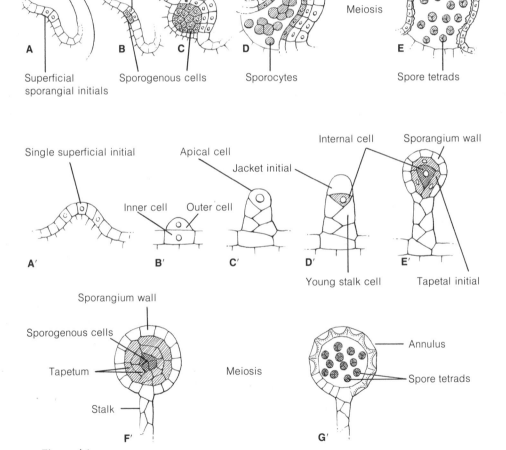

Figure 4-1
Ontogeny and structure of the two principal types of sporangia in vascular plants. A–E, the eusporangium; A′–G′, the leptosporangium. See text for detailed discussion of this diagram.

stages in their ontogeny. From a functional standpoint, the tapetum consists of cells which probably provide nourishment for the developing sporocytes and spores. According to Goebel (1905, pp. 596–597), there are two principal types of tapeta between which transition forms occur. In the sporangia of the Psilotales, in certain ferns (e.g., Ophioglossales), and in *Equisetum* the tapetum is a *plasmodial type* characterized by the breaking down of the cell walls and the intrusion of the protoplasts between the sporocytes and spores. By contrast, the sporangia of *Lycopodium* and *Selaginella* develop a *secretion type* of tapetum, the cells of which do not separate but remain in position and apparently excrete nutritive substances which are used by the developing spores.

The tapetum, whatever its physiological type, is structurally a continuous jacket of cells that completely surrounds the central mass of sporogenous tissue (Chapter 9). From an histogenetic point of view, however, the *outer part* of the tapetum (i.e., the portion toward the outer surface of the sporangium) and the *inner part* (i.e., the portion adjacent to the tissue of the sporophyll or the sporangial stalk) differ in their origin. In most eusporangia, the inner part of the tapetum develops from "sterile" tissue, i.e., from vegetative cells which have not arisen from the inner derivatives of the original sporangial initials. In contrast, a large portion of the outer tapetum most frequently is the innermost layer of the sporangium wall (e.g., *Lycopodium*) or is derived from the outer layer of potentially sporogenous cells (e.g., *Selaginella*). These two alternative methods of tapetum origin in eusporangia are represented diagrammatically in Fig. 4-2. Possibly, as has been suggested in the literature, the different methods of origin of the tapetum may somehow be correlated with the size of the sporangium. For example, in the small, delicate sporangia of leptosporan-

giate ferns, the wall is a single cell layer in thickness and the tapetal initials originate in a most precise manner from the divisions of a single large tetrahedral internal cell (see Figs. 4-3, 13-13). But in *Isoetes,* notable for the enormous size of its sporangia, the extensive tapetum originates in part from the innermost layer of the wall and in part from the sporogenous cells that lie in contact with the trabeculae, which incompletely partition the mature sporangial cavity (Figs. 9-34, D; 9-38).

Ontogeny and Classification of Sporangia

Our present terminologies and classification of sporangia in vascular plants are based upon the important researches of the great German morphologist Karl Goebel. In a paper published in 1881 he suggested that there are two principal types of sporangia: the eusporangium and the leptosporangium. The former type arises ontogenetically from several initial cells, and, at least before final maturation, develops a wall of several layers of cells. The eusporangium, as Goebel discovered, is the prevailing type throughout vascular plants. In contrast, the leptosporangium arises from a single parent cell or initial, and its wall is composed of but a single layer of cells. This more delicate type of sporangium, as Goebel correctly supposed, is restricted to the most highly advanced families of ferns which are now commonly called the leptosporangiate ferns.

Goebel's (1880, 1881) emphasis on ontogeny as a basis for the interpretation and classification of sporangia has had far-reaching effects, and the numerous researches since his original work have tended generally to support the distinction which he made between eusporangia and leptosporangia. In this regard the extensive and very detailed ontogenetic studies of F. O.

Bower, the distinguished British morphologist and student of the ferns, deserve particular mention. Although recognizing the distinctions which can usually be made between the eusporangiate and leptosporangiate methods of sporangial development, Bower (1889, 1891, 1935) has drawn attention repeatedly to the existence of intermediate patterns of sporangial ontogeny, as illustrated in the ferns by the family Osmundaceae. Such intermediate patterns, which will be described later in this chapter, justify, in Bower's opinion, the belief that the eusporangiate and leptosporangiate modes of development represent the end points of a continuous morphological series and that a rigid morphological distinction should not be made between them. This series, he believes, also represents the probable phylogenetic development of sporangia which, at least in the ferns, began with the eusporangiate type and terminated in the highly evolved leptosporangium. For convenience in presentation, the salient features of the ontogeny and structure of selected examples of eusporangia and leptosporangia will be treated separately.

The Eusporangium

In all lower vascular plants and in the microsporangia of certain conifers (Fagerlind, 1961) the parent cells or initials of the eusporangium are superficial in position (Chapter 17). Sometimes these initials are referred to as epidermal cells, but this seems an inappropriate term for the highly meristematic cells from which the sporangium arises. The first step in the development of the sporangium consists in the division of the initials by walls parallel to the surface (Fig. 4-1, B). Divisions in this plane are termed *periclinal,* and they result in the formation of an outer and an inner series of cells. In a very general sense, these two series of cells represent the starting points for the sporangium wall and the sporogenous tissue, respectively. Because of this, the outer series is commonly called jacket cells, or primary wall cells, or parietal cells, and the inner series primary sporogenous cells (Fig. 4-1, B). But the intensive studies of Bower and other investigators show that the first periclinal divisions in the sporangial initials do not always sharply define the future sporogenous tissue. On the contrary, further additions to the potential spore-forming cells may be made by periclinal divisions in the original outer cell series; moreover, the number of surface cells which function as parent cells for the sporogenous tissue may be somewhat variable even within a single genus. A few selected genera in the lower vascular plants will serve to illustrate the general eusporangiate pattern and some of its variants; see Fig. 4-2 for the general scheme of ontogeny. We may appropriately begin our discussion with the Lycopsida.

In the genus *Lycopodium,* which has been thoroughly investigated by Bower, the number of rows of surface cells which collectively function as sporangial initials varies from one in *Lycopodium selago* to as many as three in *Lycopodium alpinum.* Periclinal divisions in the initials result in a rather clear-cut distinction between (1) an outer cell layer which by further periclinal and anticlinal divisions (i.e., divisions resulting in new walls oriented perpendicular to the surface) builds up the several-layered wall, and (2) an inner fertile layer of potentially sporogenous cells from which by irregularly oriented planes of division the sporocytes ultimately arise (Fig. 4-1, B, C). Occasionally, according to Bower, periclinal divisions in the superficial cells resulting from the first periclinal divisions may contribute additional cells to the sporogenous tissue. The ontogeny of the sporangium in

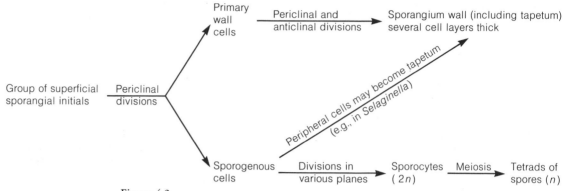

Figure 4-2
Summary of the ontogeny of the eusporangium in lower vascular plants.

Selaginella likewise conforms closely to the typical eusporangiate pattern and in all fundamental details resembles the early phases of sporangium development in *Lycopodium.* This is of particular interest since it shows that the mode of initiation and early development of a sporangium is not correlated with the condition of either homospory or heterospory. On the contrary, both young microsporangia and mega-sporangia in *Selaginella* are eusporangiate in type and become distinguishable only as the result of the breakdown of all but one or a very few megasporocytes in the future megasporangium (see Fig. 9-22). The differ-ence in method of origin of the tapetum in the sporangia of *Lycopodium* and *Selaginella* has already been mentioned, and it is il-lustrated in Fig. 4-2.

As might be expected on the basis of its enormous size at maturity, the sporangium of *Isoetes* arises from a relatively large num-ber of superficial initials; in certain species investigated by Bower four to five rows of surface cells of the sporophyll function as parent cells. Although all the internal cells produced by the earliest periclinal divisions are potentially sporogenous, extensive areas of sporogenous tissue eventually fail to reach their developmental possibili-ties and give rise to the remarkable trabec-

ulae, distinctive of both the microsporangia and megasporangia of *Isoetes* (Chapter 9). As in certain species of *Lycopodium,* addi-tions to the primary sporogenous cells may be made in *Isoetes* by subsequent periclinal divisions in the overlying surface cells.

In *Equisetum,* the only living member of the extremely ancient sphenopsid line, the sporangia develop by the eusporangiate method. However, according to Bower (1935, p. 183), a generous portion of the primary sporogenous tissue is referable in origin to the inner derivatives of a single surface cell. But the first periclinal division of the sporangial initial does not delimit all the future sporogenous cells; additions are made to the fertile area by further periclinal divisions in the original and adjacent surface cells (Fig. 10-11). Likewise, in the Ophio-glossaceae and Marattiaceae, which are designated as the eusporangiate ferns, a considerable portion or all of the primary sporogenous tissue of the sporangium origi-nates from a single inner derivative of a superficial initial (Bower, 1896; Campbell, 1911, 1918). But, as in *Equisetum,* surface cells adjacent to this initial contribute to the formation of the multiseriate wall of the sporangium.

Finally, mention should be made of the massive sporangia in the two living genera in the Psilopsida: *Psilotum* and *Tmesipteris*. In both genera, the sporangia develop by the eusporangiate method but the number of superficial initials concerned in the formation of the wall and the sporogenous tissue is difficult to determine (Bierhorst, 1968, p. 259, and Figs. 135, 136).

In concluding this discussion of the eusporangium, a few remarks about the mature wall and the ripe spores are in order. Although all typical eusporangia have a wall two or more layers thick at a middle stage in ontogeny, the *inner wall layers* (including of course the tapetum when this is of parietal origin) commonly become stretched, compressed, and ultimately destroyed. Thus, at maturity the walls of many eusporangia — for example, in *Equisetum*, *Lycopodium*, and *Selaginella* — may appear to consist of only a single layer of cells. But in *Psilotum*, in *Tmesipteris*, and very strikingly in *Botrychium* (a member of the eusporangiate ferns) the mature sporangium wall is multilayered.

The output of spores from eusporangia in both homosporous and heterosporous plants is variable but is frequently much greater than that in the leptosporangium. Estimates of spore number are made by counting (in serial sections) the number of spore-mother cells and multiplying by four. Such counts reveal that in certain homosporous types the numbers are in the hundreds, or in the thousands in various eusporangiate ferns. Apparently, the greatest spore output occurs in the microsporangia of *Isoetes*, where it is estimated that from 150,000 to 1,000,000 spores may develop in a single microsporangium.

The form of spores is quite variable and often is of considerable diagnostic value. Tetrahedral spores, with conspicuous triradiate ridges, are characteristic of most species in the Lycopsida, whereas bilateral spores are characteristic of *Psilotum* and *Tmesipteris*. In *Equisetum* the spores are unique because of four bandlike appendages,

or *elaters,* that are developed on the surface of each spore (see Chapter 10). These elaters are hygroscopic, coiling around the spore when the air is moist but uncoiling when it is dry. If the spores encounter alternating moist and dry air the elaters coil and uncoil repeatedly; this is believed to facilitate the dispersal of the spores from the open sporangium.

The Leptosporangium

In contrast to the multicellular origin of the eusporangium, the ontogeny of a typical leptosporangium begins with the transverse or oblique division of a single superficial initial (Fig. 4-1, A', B'). As is shown diagrammatically in Fig. 4-3, the inner of the two cells produced by the first division of the sporangial initial may (1) contribute cells which by further intercalary divisions produce a large part of the sporangial stalk or (2) remain inactive and play no role in the ontogeny of the sporangium. The latter pattern of development appears to be characteristic of leptosporangia with three-rowed stalks, an organization that Bower (1935, pp. 415–416) regarded as "the commonest of all in Leptosporangiate Ferns." A description will be given first of the main steps in development of a sporangium with a three-rowed stalk and the student should refer frequently to Figs. 4-1, A'–G', and 4-3 in order to follow the precise sequence of events. Following this description, a few comments will be added regarding the interesting divergence in the method of early development of the stalk in certain recently studied genera of leptosporangiate ferns.

As shown in Figs. 4-1, A'–G', and 4-3, the *outer cell* — formed by the division of the sporangial initial — may function as the parent cell from which both the stalk and the spore-containing capsule or "head" of the sporangium are derived. This outer cell, by means of three successive and

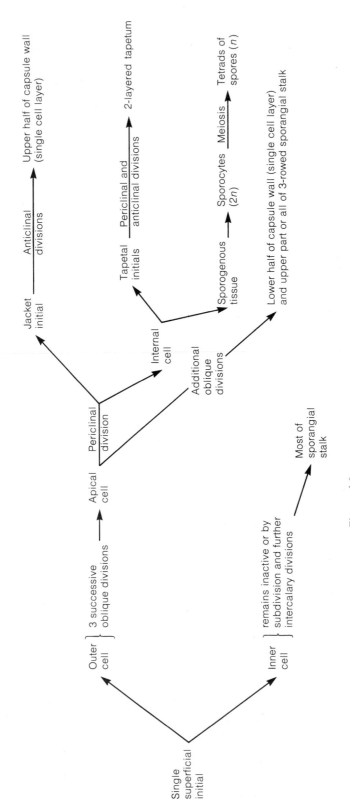

Figure 4-3
Summary of the ontogeny of the leptosporangium.

obliquely oriented divisions, forms a distal *apical cell,* which is tetrahedral in shape and which by means of further oblique divisions, parallel to its three lateral faces, produces additional segments. The lower segments give rise to stalk cells, and the three upper-most segments, by subsequent anticlinal divisions, form the lower portion of the capsule wall (Fig. 4-3).

A very critical stage in the development of the leptosporangium now occurs as the result of a distal periclinal division in the apical cell. This division yields an outer *jacket initial* and a pyramidal *internal cell* (Figs. 4-1, D'; 4-3). The jacket initial, by means of anticlinal divisions, completes the formation of the one-layered wall of the capsule (Fig. 4-3). Concomitantly the internal cell, by divisions parallel to its sides, produces four *tapetal* initials; from these cells, as a result of both periclinal and anticlinal divisions, there arises the two-layered *tapetum* characteristic of leptosporangia (Figs. 4-1, F'; 4-3). Following the formation of the four peripheral tapetal initials, the tetrahedral inner cell divides in various planes and forms a mass of sporogenous tissue from which in turn the *sporocytes* arise (Figs. 4-3; 13-13, F, G, H.).

Maturation of the leptosporangium includes the meiotic division of the sporocytes to form spore tetrads, the disintegration of the tapetum, the conspicuous elongation of the stalk cells, and the ultimate dehiscence of the capsule. In all the advanced families among the leptosporangiate ferns the *annulus* is vertical and the dehiscence of the capsule is transverse or brevicidal (Fig. 13-13, I). But in the phylogenetically less evolved groups, the annulus varies from apical to oblique, and the plane of dehiscence of the capsule is longitudinal or longicidal (Fig. 13-14).

As stated previously, the three-rowed type of sporangial stalk is very common in leptosporangiate ferns and its component cells are all apparently derived from lateral segments of the apical cell. In the genus *Diellia,* however, Wagner (1952) found that the first few divisions of the sporangial initial all take place in the transverse plane. After this 3–4 celled "filamentous stage" in development, the distal cell begins to function as an apical cell while the cells below it ultimately give rise, by intercalary divisions, to the characteristic one-rowed stalk; in *Diellia,* the stalk is composed of three cell rows only immediately below the capsule. A somewhat analogous divergence from the "usual" pattern of stalk initiation and development occurs in *Phlebodium aureum,* according to the study made by Wilson (1958). In this fern, the sporangial initial divides by the formation of an oblique wall into two cells: the inner cell divides transversely into a basal inactive cell and an upper cell that Wilson includes as part of the "sporangial primordium." According to his observations, the cells of the stalk are not produced by the activity of an apical cell but arise from intercalary divisions *within* "the sporangial segments" (see also Wilson, 1960).

It should be clear from the studies just reviewed that divergences in the *details* of early development of leptosporangia are to be expected, particularly with reference to the mode of origin of the *lower part* of the sporangial stalk. But on the other hand, the main steps in the ontogeny of the spore-producing capsule are very similar in the various genera of leptosporangiate ferns which have thus far been studied, and they provide a remarkable contrast with the developmental pattern characteristic of eusporangia (compare Figs. 4-2 and 4-3).

Although there is considerable variation in the average spore output of the lepto-sporangia in the various families, the number is, in general, less than for eusporangia. In none of the genera for which Bower made spore countings does the number of spores per sporangium equal the huge numbers produced by the sporangia in the Lycopsida

or in eusporangiate ferns. In typical leptosporangia the number of spores is a power
of two: 16, 32, 64, 128, 256, or 512. According to Bower (1935, p. 427) "in the vast
majority of the leptosporangiate ferns each
capsule contains 64 spores or less." Such
numbers as 24 and 48, though not powers
of two, do occur and are the result of
reductions in the number of cell divisions
which lead to the formation of sporocytes.
Mature fern spores often exhibit various
types of wall sculpturing which may be of
value in taxonomic differentiation. In certain highly evolved fern groups a special
deposit known as the perispore occurs on
the spore wall. This has been shown to have
considerable systematic value (Wagner,
1952).

As stated earlier in this chapter, development and structure of the stalk of the
sporangium in the Osmundaceae do not
easily fit the scheme of either a typical
leptosporangium or a typical eusporangium
(Williams, 1928). In the first place, as
shown in Fig. 4-4, A, B, the form of the
initial cell from which the sporogenous cell
and the tapetum originate is variable. Sometimes the cell is truncated at the base, thus
resembling the form of comparable cells in
a eusporangium; at other times, in the same
sorus, the cell is pointed at the base as in a
typical leptosporangiate fern. Secondly,
Fig. 4-4 reveals the fact that the entire
sporangium cannot be traced in origin to
a single cell, as can a typical leptosporangium. On the contrary, neighboring cells
contribute, as they do in a eusporangium, to
the development of the massive stalk which
contrasts strikingly with its more slender
counterpart in a typical leptosporangium
(compare Fig. 4-1, F' with Fig. 4-4, C, and
Fig. 13-13, I). Lastly, the spore output in
the transitional sporangia of the Osmundaceae is, in general, higher than in ordinary
leptosporangiate ferns; in *Osmunda regalis*,
for example, it ranges from 256 to 512.

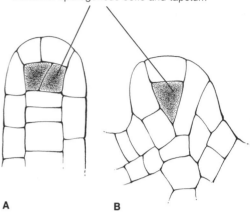

Initials of sporogenous cells and tapetum

A   B

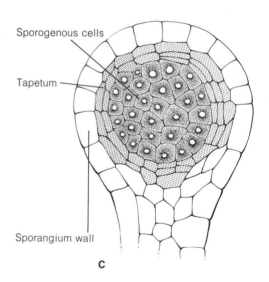

Sporogenous cells

Tapetum

Sporangium wall

C

Figure 4-4
Development of sporangia in the Osmundaceae.
A, B, *Todea barbara,* illustrating variations in early
stages of differentiation; C, *Osmunda regalis,*
longisection of sporangium showing wall, tapetum,
sporogenous tissue, and massive stalk. [From
Primitive Land Plants by F. O. Bower. Macmillan,
London. 1935.]

Summary and Conclusions

There are two principal types of vascular plant sporangia—eusporangia, and leptosporangia—and they are distinguished by their method of origin and development, their wall structure, and their spore output (Figs. 4-1, 4-2, 4-3).

Typical eusporangia originate in lower vascular plants from a series of superficial parent cells or initials which, by periclinal division, give rise to primary wall cells (external series) and primary sporogenous cells (internal series). To varying degrees, further additions from superficial cells may be made to the underlying sporogenous tissue. In most eusporangia the outer part of the nutritive layer or tapetum is developed either from the sporogenous tissue or from the innermost layer of the sporangium wall. The eusporangium, prior to complete maturity, develops a wall two or more layers thick, although at the time of dehiscence the inner wall layers may be crushed or obliterated. The spore output of eusporangia is high and variable.

By contrast, a typical leptosporangium originates from a single parent cell which first produces the stalk and then, by a curved periclinal division, is subdivided into a jacket or wall-producing cell and an internal primary sporogenous cell. The latter produces four tapetal initials, then gives rise to a relatively small number of sporocytes. The dehiscence of leptosporangia is either transverse or longitudinal and the spore output small and usually some power of two. The Osmundaceae is an example of ferns in which the ontogeny and structure of the sporangium is somewhat intermediate between that of typical leptosporangia and eusporangia (Fig. 4-4).

What general conclusions can be drawn from the existence of these two, well-defined morphological types of sporangia? Do they represent independent and parallel types, or can they be related in a phylogenetic sense? The widespread occurrence of the eusporangiate type throughout vascular plants, including two of the living orders of ferns, would suggest that the eusporangium is the more primitive form. That it indeed may represent the elemental and truly primitive type of spore case is suggested by its appearance in the living members of the Psilopsida and in Paleozoic ferns. As arranged by Bower (1935), a very persuasive series of ontogenetic patterns seems to exist which serves to bridge the gap between eusporangiate and leptosporangiate ferns and to demonstrate the derivation of the latter group.

One of the most important conclusions regarding the morphology of sporangia was expressed by Bower (1896, p. 8), who stated that the study of the sporangia of a given plant should be carried out in the light of a full knowledge of the segmentation of its apical meristems. This means that the mode of origin and pattern of development of sporangia should be viewed not as isolated phenomena but as a part of the general ontogeny of the plant—and an effort should be made to test the assumption that correlations between sporangial ontogeny and apical segmentation in the shoot do exist. It is now clear that Bower's hypothesis has considerable merit (see Fagerlind, 1961).

Throughout lower vascular plants the superficial cells of the shoot apex divide to varying degrees in the periclinal plane and thus do not collectively form a discrete surface layer. For this reason, internal cells of the apex and of the foliar primordia can be traced in origin to superficial cells or initials. As we have seen, this is precisely the pattern of origin of the sporangia, which differ ontogenetically only in the number of parent cells which produce them.

Much remains to be done in the way of testing Bower's idea with reference to the seed plants. In certain conifers the microsporangial initials, like those of eusporangia in lower groups, are superficial cells (Chapter 17). This is correlated with the relative frequency of periclinal divisions in the surface cells of the shoot apex of many different kinds of gymnosperms (Allen, 1946; Johnson, 1951; Clowes, 1961). In angiosperms the initials of the microsporangia are hypodermal cells in the anther, overlaid by a well-defined epidermis (Chapter 20). This condition, in turn, seems to be well correlated with the existence of a sharply defined external cell layer in the shoot apices of most angiosperms.

References

Allen, G. S.
  1946. The origin of the microsporangium of *Pseudotsuga. Bull. Torrey Bot. Club* 73: 547–556.
Bierhorst, D. W.
  1968. On the Stromatopteridaceae (Fam. Nov.) and on the Psilotaceae. *Phytomorphology* 18(2):232–268.
Bower, F. O.
  1889. The comparative examination of the meristems of ferns as a phylogenetic study. *Ann. Bot.* 3:305–392.
  1891. Is the Eusporangiate or the Leptosporangiate the more primitive type in the ferns? *Ann. Bot.* 5:109–134.
  1896. *Studies in the morphology of spore-producing members. II. Ophioglossaceae.* Privately printed, London.
  1908. *The Origin of a Land Flora.* Macmillan, London.
  1935. *Primitive Land Plants.* Macmillan, London.
Campbell, D. H.
  1911. *The Eusporangiatae. The Comparative Morphology of the Ophioglossaceae and Marattiaceae.* (Publication No. 140.) Carnegie Institution of Washington, D.C.
  1918. *The Structure and Development of Mosses and Ferns,* Ed. 3. Macmillan, New York.
Clowes, F. A. L.
  1961. *Apical Meristems.* (Botanical Monographs.) Blackwell, Oxford.
Fagerlind, F.
  1961. The initiation and early development of the sporangium in vascular plants. *Svensk Bot. Tidskr.* 55:299–312.
Goebel, K.
  1880. Beiträge zur vergleichenden Entwickelungsgeschichte der Sporangien. *Bot. Zeit.* 38:545–552.
  1881. Beiträge zur vergleichenden Entwickelungsgeschichte der Sporangien. *Bot. Zeit.* 39:681–694, 697–706, 713–719.
  1905. *Organography of Plants,* Part II, Eng. Ed. by I. B. Balfour. Clarendon Press, Oxford.
Ingold, C. T.
  1939. *Spore Discharge in Land Plants.* Clarendon Press, Oxford.

Johnson, M. A.
    1951. The shoot apex in gymnosperms. *Phyto-morphology* 1:188–204.
Meyer, B. S., and D. B. Anderson
    1939. *Plant Physiology.* Van Nostrand, New York.
Ursprung, A.
    1915. Uber die Kohäsion des Wassers in Farnannulus. *Ber. Deutsch. Bot. Ges.* 33:153–162.
Wagner, W. H., Jr.
    1952. The fern genus *Diellia. Univ. Calif. Publ. Bot.* 26:1–212.

Williams, S.
    1928. Sporangial variation in the Osmundaceae. *Trans. Roy. Soc. Edinb.* 55: 795–805.
Wilson, K. A.
    1958. Ontogeny of the sporangium of *Phlebodium (Polypodium) aureum. Amer. Jour. Bot.* 45:483–491.
    1960. The leptosporangium of the New Zealand fern *Anarthropteris dictyopteris. Contrib. Gray Herb. Harvard Univ.* 187:53–59.

5

Gametangia

Throughout the vascular plants sexual reproduction is achieved by the pairing of morphologically unlike gametes. In all lower vascular plants the male gamete, usually termed the sperm or spermatozoid, is a motile, flagellated cell that requires liquid water to reach the passive, nonmotile egg. From an evolutionary point of view such motile sperms, like their counterparts in liverworts and mosses, exemplify the persistence of the kind of gamete typical of many aquatic algae. In contrast with the somewhat casual dispersal of flagellated sperms in lower groups of the tracheophytes, modern seed plants, because of the development of a pollen tube by the male gametophyte, do not depend on water for

transportation of sperms—the pollen tube conveys the male gametes directly to the immediate vicinity of the egg. This is true of *Ginkgo* and the cycads, which have retained flagellated sperms, but it is particularly significant in all the other living groups of gymnosperms and angiosperms, for in these the male gametes are devoid of flagella.

In all the lower groups of vascular plants the sex cells are normally produced in separate organs or *gametangia*. Many instances have been recorded, in the literature, of bisexual gametangia that contain both eggs and sperms, and these remarkable structures will be discussed in more detail later in the chapter. But normal gamete development involves the production of distinctive, unisexual gametangia. The male gametangium, or *antheridium,* produces the sperms, which vary in number from four in *Isoetes* to several thousand in certain eusporangiate ferns. The female gametangium, or *archegonium,* is quite different in that typically it produces only a single, nonmotile egg which at the time of fertilization is situated at the base of the archegonial canal; the canal provides the channel through which the sperm must pass to reach the female gamete.

Antheridia and archegonia, as is true of the functionally equivalent sex organs of the bryophytes, are complex organs consisting of a jacket (usually one cell thick) of sterile cells which encloses and shelters the gametes. This is illustrated in Fig. 5-1. Such construction is quite unlike the simple unicellular sex organs that are typical of the algae, and it has led to speculation regarding the evolutionary history of gametangia and the possible homology between antheridia and archegonia. (These two problems will be discussed in some detail at the close of the chapter.)

As is true of sporangia, an exceptional amount of attention has been paid to the ontogeny and comparative structure of gametangia. One of the most interesting results of comparative research is the discovery that with respect to gamete production the archegonium is a remarkably uniform sex organ. Isolated cases have been reported of archegonia with several eggs, but generally only a single functional gamete is formed, as can be seen in Fig. 5-1, D'. On the basis of this standardized structure, plants with archegonia are very commonly designated collectively as Archegoniatae. Included in this group are the bryophytes, all lower vascular plants, including the ferns, and the majority of the gymnosperms. In the gymnosperms archegonia are modified in structure, but onto-

genetically and structurally they are strictly comparable with the archegonia of the lower groups. In angiosperms the egg is produced in a highly specialized embryo sac, and archegonia as definable organs are absent.

With reference to the evolutionary origin and modification of sex organs in vascular plants, it is interesting to note that antheridia, as definable multicellular organs, are restricted to the lower groups and, unlike archegonia, do not occur in living gymnosperms. This contrast between antheridium and archegonium seems to be correlated with the marked reduction in size of the male gametophyte generation in all heterosporous, lower vascular plants and in the seed plants. In the microgametophytes of *Selaginella* and *Isoetes,* for example, only a single antheridium is produced, whereas in gymnosperms the equivalent structure is further reduced to a few vegetative cells and usually two male gametes. This trend toward the elimination of sterile cells culminates in the male gametophyte of angiosperms, which normally consists of a single vegetative nucleus and two male gametes. Doubtless, the evolution of the pollen tube was a very significant factor in the drastic reduction of the antheridium to a few cells. On the other hand, the persistence of the archegonium as a distinct organ even among the relatively advanced gymnosperms is related to the larger size of the megagametophyte and to the prime biological importance of the archegonium as an organ which produces the egg and shelters the young embryo (Chapter 6).

This chapter is largely a general account of the position, structure, ontogeny, and homologies of the antheridium and the archegonium. As was the intent of the previous chapter on sporangia, this treatment is intended to provide orientation for the student by emphasizing those fundamental aspects (of gametangia) which are of biological and morphological importance. Further details of the morphology of gametophytes and sex organs will be given in later chapters that deal with specific groups of plants.

Position of Gametangia

It will be desirable first to contrast the gametophytes and sex organs of homosporous and heterosporous lower vascular plants. In the former the gametophyte is *exosporic*— i.e., free-living and not enclosed by the spore wall—whereas in all heterosporous groups the micro- and megagametophytes are *endosporic*—i.e., entirely or for the most part enclosed by the wall of the microspore or megaspore, respectively. These differences are important from a biological as well as a morphological standpoint, and there is little doubt that the relatively large exosporic type of gametophyte represents the original and primitive condition. We may therefore begin with this type.

Exosporic gametophytes are typically bisexual, which means they produce or have the capacity to produce both types of gametangia. In some instances the environmental conditions surrounding the developing gametophyte may be decisive with respect to the type or types of sex organs which are produced.

The gametangia of exosporic gametophytes are usually embedded in vegetative tissue. This is always the case with the venter or egg-containing portion of the archegonium, but the antheridium is less consistent. In the homosporous members of the Lycopsida, in *Equisetum,* and in the eusporangiate ferns the sterile jacket of the antheridium protrudes slightly above the surface, and the spermatogenous tissue is deeply sunken. But in the Psilotales and the leptosporangiate ferns the antheridia are conspicuously emergent organs. From the standpoint of ontogeny, however, the emergent antheridia of *Psilotum* and

Tmesipteris are quite like the embedded type, whereas in the higher groups of leptosporangiate ferns the antheridium is a more highly evolved and reduced organ, as will be shown later in this chapter.

The distribution pattern of antheridia and archegonia varies considerably and is correlated to some degree with the form and level of evolutionary development of the gametophyte. In the fleshy, radial, non-photosynthetic gametophytes of *Psilotum* and *Tmesipteris* the two kinds of gametangia are intermingled and tend to occur over the entire surface of the gametophyte (Fig. 8-10, A). An intermingling of sex organs also occurs in certain presumably primitive species of *Lycopodium*—for example, *Lycopodium lucidulum* (Spessard, 1922)—but in other members of this genus the antheridia and archegonia are formed in distinct patches or clusters on the upper surface of the gametophyte (Chapter 9). With reference to the thalloid, dorsiventral, photosynthetic type of gametophyte characteristic of the Marattiaceae and leptosporangiate ferns, several somewhat intergrading patterns of sex-organ distribution occur. In *Marattia* and *Angiopteris,* according to Campbell (1918), the archegonia are confined to the lower surface of the prominent midrib of the prothallus, whereas the antheridia, although more abundant below, develop on both surfaces. A more standardized plan of distribution of gametangia seems to prevail in the higher groups of leptosporangiate ferns. Here, both kinds of sex organs are commonly restricted to the lower surface with the archegonia limited to the cushion of tissue situated behind the notch of the heart-shaped prothallus; the more numerous antheridia occur near the basal end of the prothallus as well as on the wings (Chapter 13).

Endosporic gametophytes, as they are restricted to heterosporous plants, are usually strictly unisexual. As mentioned before, the male gametophyte is extremely reduced in structure, consisting in *Selaginella* and *Isoetes* of a single vegetative prothallial cell and one antheridium. The megagametophyte, by contrast, is more robust, consisting of a mass of food-storing tissue which usually fills the cavity of the megaspore and which, in *Selaginella* and *Isoetes,* is exposed by the cracking of the spore wall along the triradiate ridge (Fig. 9-28, A). The small archegonia are restricted in occurrence to the surface of the protuberant cushion of gametophyte tissue.

Structure of Gametangia

The initiation and ontogenetic development of gametangia will be outlined later in this chapter. But now, by way of orientation, we will describe the cellular structure of antheridia and archegonia as they approach the stage of functional maturity. Throughout this discussion we will refer frequently to Fig. 5-1.

Antheridium

In the embedded type of antheridium a well-defined sterile jacket or cover is always present. Because this layer is instrumental in the mechanism of sperm discharge, Goebel has proposed that it be designated the "opercular layer" (Fig. 5-1, D). This term is appropriate because, generally, one or more centrally located cells of this layer separate or become broken, thus creating an opening through which the sperms can escape. In the antheridia of many species of *Lycopodium* and in the eusporangiate ferns there is normally only a single cell in the antheridial jacket which is ruptured at maturity. This opercular cell (also termed the cap cell) is commonly triangular when seen in surface view, although it may be four-sided or irregular in contour. The situation in *Equisetum*

requires much further study. Some authors (for instance, Campbell, 1918) describe a single opercular cell, and others (Goebel, 1905) maintain that considerable variation occurs within the genus. In *E. limosum*, for example, a relatively large opening is produced by the pulling apart of a group of centrally placed cells. By contrast, in *E. pratense* there are, it is said, only two opercular cells "which then separate from one another in the middle somewhat after the manner of the guard-cells in a stoma" (see also Hauke, 1968). The mechanism of separation or rupturing of opercular cells is poorly understood—Goebel has suggested that the dehiscence may be attributable to the swelling of mucilaginous substances deposited in them. Most commonly, the jacket or opercular layer of embedded antheridia is only a single cell in thickness. But in *Botrychium* and *Helminthostachys*—members of the Ophioglossaceae—the jacket consists of two layers of cells except at the center where several cells, including the opercular cell, remain undivided. According to Campbell (1918), only the single opercular cell is destroyed when the sperms are released.

In the emergent type of antheridium the jacket is always a single layer of cells in thickness. Some variation occurs in the total number of cells which constitute this layer in the various families of the leptosporangiate ferns, although there is usually developed but a single cap or opercular cell. At one end of the series are the apparently highly specialized antheridia of the majority of the Filicales, the jacket here consisting of two ring-like cells and a terminal opercular cell. At the other extreme are the massive complex antheridia of the Gleicheniaceae, wherein the jacket may consist of as many as ten or twelve cells, one of which functions as an opercular cell (Stokey, 1950, 1951).

This discussion of antheridial morphology may be closed appropriately by a brief résumé of the structure and number of

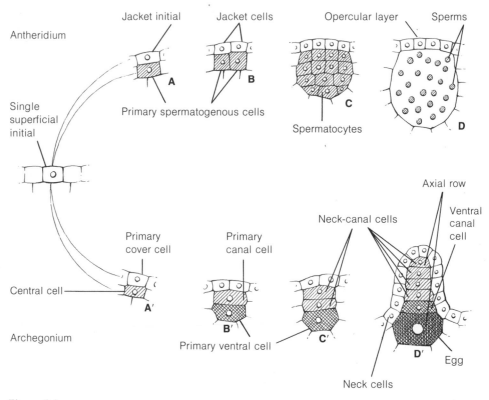

Figure 5-1
Ontogeny and structure of gametangia in lower vascular plants (exclusive of the leptosporangiate ferns):
A–D, the antheridium; **A′–D′**, the archegonium. Note the similarity in the mode of origin and the
first stage in development of the two types of sex organs.

sperms produced in the various groups of lower vascular plants. It is a well-established fact that there are two principal types of sperms, based upon the number of flagella developed. In *Lycopodium* and *Selaginella* the sperms are biflagellate and in this respect are quite unlike the prevailingly multi-flagellate sperms developed in all other groups including *Ginkgo* and the cycads. Since the spermatozoids of many algae are biflagellate, considerable interest, from the standpoint of evolution, is attached to the similar condition in *Lycopodium* and *Selaginella*. The taxonomic (and possibly phyletic) importance of the number of flagella developed by sperms is illustrated by the

genus *Isoetes*. On the basis of its solitary foliar eusporangia, its ligulate sporophylls, and its heterospory, *Isoetes* is commonly regarded as a member of the Lycopsida, allied in many respects to *Selaginella*. But the sperms of *Isoetes* are multiflagellate, and in this respect they are quite unlike the male gametes of either *Lycopodium* or *Selaginella*. As mentioned earlier in the chapter, the number of sperms formed by a single antheridium varies widely according to the genus or group of plants. *Isoetes* produces only four sperms from each microgametophyte. In contrast, the sperm output is in the hundreds in some leptosporangiate ferns (e.g., *Gleichenia*) and may reach several thousand, according to Campbell, in the antheridium of *Ophioglossum pendulum*.

In Bower's (1935) opinion, there is a close parallel between the output of sporocytes and the number of spermatocytes, as revealed by comparison of the sporangia and antheridia of eusporangiate and leptosporangiate ferns. In the first group there tends to be a relatively high number of spore mother cells and spermatocytes, but in the advanced leptosporangiates there is a marked reduction in number of both spores and male gametes. These differences, Bower believes, show that the progressive refinement in structure of spore-bearing structures has been extended in a comparable manner to the sex organs.

Archegonium

Just prior to its complete maturation a typical archegonium consists of the *neck*, which projects conspicuously above the surface of the gametophyte, and the so-called *axial row* of cells. The upper members of this row are enclosed by the neck, and the lowermost cells, including the egg, are sunk in the tissue of the gametophyte, as shown by Fig. 5-1 D'. The venter, or basal embedded portion of the archegonium, contains the *ventral canal cell* and the *egg*. In members of both the eusporangiate and leptosporangiate ferns the venter is demarcated from adjacent prothallial cells by a rather discrete cellular jacket, but in many vascular plants the cells bounding the lower portion of the axial row of the archegonium are not morphologically distinguishable from other cells of the gametophyte. The archegonial neck, which is the morphological and functional equivalent of the jacket of the antheridium, is structurally uniform, usually comprising four vertically arranged rows of cells. The number of cell rows is remarkably consistent, despite the wide variation in the total number of cells

which compose the neck. Without doubt, the longest necks are found in the archegonia of certain species of *Lycopodium*. These contrast markedly with the very short and scarcely projecting archegonial necks in eusporangiate ferns, in prothallia of *Lycopodium* grown *in vitro*, and in *Isoetes* and *Selaginella*.

From a morphological viewpoint, the most important and significantly variable part of the immature archegonium is the axial row (Fig. 5-1, D'). In certain species of *Lycopodium* the so-called axial row may actually consist of a partially or completely double series of cells. This seems to be the usual condition in *Lycopodium obscurum* var. *dendroideum*, according to Spessard (1922), and it has been recorded for other species by Lyon (1904). More commonly, however, the axial row is composed of a single series of cells, the function and morphological significance of which deserve full discussion at this point. As mentioned, the lowest cell in the axial row is the egg itself, and this normally is the only component of the row which survives, all the others disintegrating as the archegonium reaches full maturity. The cell located directly above the egg is known as the ventral canal cell; this cell ontogenetically is a sister cell of the egg and in exceptional cases apparently functions as a gamete. All the remaining cells of the axial row, situated above the ventral canal cell, are termed neck-canal cells and on theoretical grounds may be regarded as potential gametes also (Fig. 5-1, D').

From a comparative viewpoint it is possible to arrange the archegonia of lower vascular plants in a *reductional series*, beginning with those in which there are many neck-canal cells (e.g., *Lycopodium*) and culminating in plants that have only a single binucleate neck-canal cell (e.g., many leptosporangiate ferns). Whether this series actually corresponds to the general phylogenetic development of the archegonium is difficult to determine, since in admittedly primitive types like *Psilotum* and *Tmesipteris* there may be only two or possibly only a single binucleate neck-canal cell. The behavior of the ventral canal cell also is variable and hence of considerable comparative interest. In many lower vascular plants it is a well-defined cell, differing from the egg cell in its smaller size and more flattened form. But in the Ophioglossaceae, which also apparently represents an ancient stock of vascular plants, a distinct ventral canal cell (Campbell, 1911) proves difficult to demonstrate, even by careful ontogenetic study. This is true too in many gymnosperms where the ventral canal cell may be represented by only a short-lived nucleus. Furthermore, in these plants neck-canal cells have been entirely eliminated and the mature archegonium consists of a small neck of one or several cell tiers and a huge egg (Maheshwari and Sanwal, 1963).

Very little systematic investigation has been devoted to the mechanism of the opening of the archegonial neck. According to Campbell (1918), when the archegonium opens in leptosporangiate ferns "the terminal cells diverge widely and the upper ones are often thrown off." Goebel (1905) found that the four large terminal cells of the archegonial neck of *Equisetum* become strongly reflexed as they separate from one another, although still remaining attached to the adjacent tier of neck cells. As noted previously, all members of the axial row except the egg cell disintegrate, thus producing a canal through which the motile sperm can pass. In ferns it has been shown that malic acid exerts a chemotactic stimulus upon the spermatozoids; it appears that this organic acid or some other substance is present in the exudate from the open neck of the archegonium and this orients the sperm towards the neck and thence to the egg. Further experimental studies on the nature of the substances which may react in this way on male gametes

are highly desirable. (See Ward, 1954, for an interesting account of the opening of antheridia and archegonia in the leptosporangiate fern *Phlebodium*.)

Ontogeny of Gametangia

Despite the striking divergence between fully developed antheridia and archegonia, both kinds of sex organs are remarkably alike in their method of initiation and in the earliest stages of their ontogeny (Figs. 5-2, 5-3). In all investigated cases each type of gametangium originates from a single surface cell of the gametophyte, as can be seen in Fig. 5-1. Whether there are definable cytological differences between the parent cells of antheridia and those of archegonia is a question that has never been intensively studied. Spessard (1922) in his description of the gametophyte of *Lycopodium lucidulum*, where the sex organs are intermingled, states that "it is impossible to distinguish an antheridium from an archegonium in the very earliest stages of their development. There are some indications that the archegonial initial is slightly larger and longer than the antheridial initial, but this is so uncertain that it is useless as a criterion." A similar close resemblance between antheridial and archegonial initials is found also in other groups, such as *Psilotum* (Bierhorst, 1954) and the Marattiaceae; the latter are used by Goebel (1928) and Bower (1935) to illustrate the morphological equivalence of sex organs in vascular plants.

In addition to their identical mode of origin, antheridia and archegonia begin their development by the periclinal division of the parent cells (Figs. 5-1, A, A', 5-2, and 5-3). With reference to the embedded type of antheridium, this results in an outer *jacket cell*, the anticlinal subdivision of which produces the opercular layer, and an inner

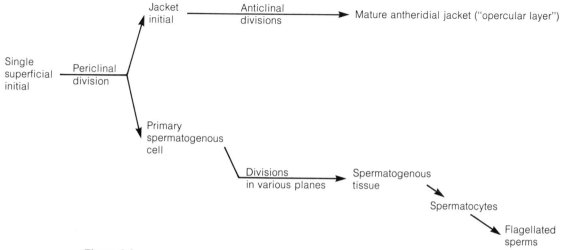

Figure 5-2
Ontogeny of antheridium in lower vascular plants (exclusive of leptosporangiate ferns).

primary spermatogenous cell (sometimes referred to as the primary androgonial cell) from which the usually numerous spermatocytes and ultimately the sperms originate (Figs. 5-1, A–D, and 5-2). There is a comparable distinction between outer sterile cells and inner potentially fertile cells, due to the first periclinal division of an archegonial initial (Figs. 5-1, A′, and 5-3). In this case, the outer cell or *primary cover cell*, by means of two successive anticlinal divisions at right angles to one another, produces four neck initials from which, by transverse divisions, the four vertical rows of the archegonial neck are developed. The inner of the two derivatives of the archegonial initial may first produce a *basal cell*, as in many ferns, and then serve as the parent cell of the axial row, or, as is more commonly the case, it may function directly without first forming a basal cell. In either instance, the parent cell of the axial row is usually designated as the *central cell* (Figs. 5-1, A′, and 5-3).

The presence or absence of a basal cell is not a consistent feature of archegonium development within a family or even a genus.

The Marattiaceae furnish a good example of the degree of fluctuation that may exist. In *Danaea*, for example, and in *Angiopteris evecta* a basal cell is consistently absent (Haupt, 1940). On the other hand, the presence or absence of a basal cell varies from species to species in the genus *Marattia* (Stokey, 1942). Apparently a basal cell is commonly developed in the young archegonia of leptosporangiate ferns; here, by means of a few anticlinal divisions, it contributes cells to the lower portion of the jacket of the archegonial venter (Fig. 13-31).

This description analyzes the early similarities and differences in the ontogenetic history of the gametangia. A brief discussion is in order of the method of origin of the component cells of the axial row in the archegonium and the comparative ontogeny of the emergent types of antheridia in the leptosporangiate ferns. Let us first examine the development of the cells in the axial row of the archegonia of various plants.

The central cell, in the archegonia of lower vascular plants, divides periclinally into an outer *primary canal cell* and an inner

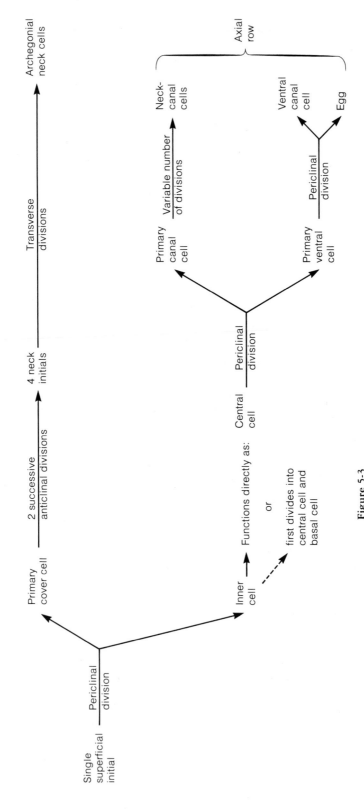

Figure 5-3
Ontogeny of archegonium in lower vascular plants.

primary ventral cell (Figs. 5-1, B', and 5-3). From the former, one or as many as sixteen neck-canal cells originate. As already explained, the most extensive development of neck-canal cells occurs in certain species of *Lycopodium* in which anticlinal divisions may result in a double series, or, if wall formation is inhibited, a single row of binucleate elements. When only a few neck-canal cells are formed, these usually are in a superposed row. But the archegonium of certain species of *Equisetum* quite regularly develops two laterally placed, boot-shaped, neck-canal cells (see Chapter 10). The primary ventral cell is believed to function in some cases as an egg. Perhaps more typically, however, it divides periclinally, forming the evanescent *ventral canal cell* and the basally situated *egg* (Figs. 5-1, D', and 5-3). In the gymnosperms, as mentioned, neck-canal cells are not developed. On the contrary, the central cell functions as the initial from which the large egg and the transitory ventral canal cell (or its nucleus) originate.

There remains for final consideration the ontogeny of the antheridium in leptosporangiate ferns, which differs in several interesting respects from that characteristic of embedded antheridia. Some of the differences parallel to a remarkable degree the ontogenetic differences between leptosporangia and eusporangia that have been described in Chapter 4. Perhaps one of the most definitive ontogenetic features of the emergent type of antheridium in leptosporangiate ferns is the fact that the first division wall of the initial does not set apart an outer sterile and an inner fertile cell. On the contrary, the formation of the primary spermatogenous cell is delayed until two or more sterile cells have been cut off by the antheridial initial. In such primitive fern groups as the Gleicheniaceae and Osmundaceae the antheridial initial, by means of alternating oblique divisions, first produces a series of basal cells which constitute the short stalk

and the lower portion of the antheridial jacket. Ultimately the initial becomes divided by a curved periclinal wall into an outer *jacket cell* and an inner *primary spermatogenous cell*. From the former, by anticlinal divisions, the upper part of the antheridial jacket including an opercular cell is formed, and the spermatocytes develop from the subdivision of the primary spermatogenous cell. The periclinal wall, which sets apart the primary spermatogenous cell from the sterile jacket initial, is a remarkable parallel with the similarly oriented wall delimiting the primary wall and primary sporogenous cells of a young leptosporangium.

The early ontogeny of the more delicate emergent antheridium of the highly specialized Polypodiaceae likewise results first in the production of sterile cells. These, however, are fewer and of unusual shape. According to the classical interpretation, the antheridial initial divides by the formation of a funnel-shaped wall into an outer "ring cell" and a central cell. The latter divides by the formation of a curved periclinal wall into an outer jacket cell and the internal primary spermatogenous cell. A curved anticlinal division in the jacket cell subdivides the latter into a second ring-shaped cell and a centrally located opercular cell. Certain details in this interpretation have been challenged by Davie (1951), who contends that the first division wall in the antheridial initial is straight or only slightly concave and not funnel-shaped (Fig. 5-4, B). By increasing turgor within the upper of the two cells this wall becomes bent downward until it nearly comes in contact with the original basal wall of the antheridial initial (Fig. 5-4, E, F). Meanwhile, the upper cell divides into the terminal cell and the primary spermatogenous cell (Fig. 5-4, D). Transverse division of the terminal cell produces a cap or opercular cell and the second ring cell (Fig. 5-4, E). As the primary spermatogenous cell enlarges, its upper convex wall

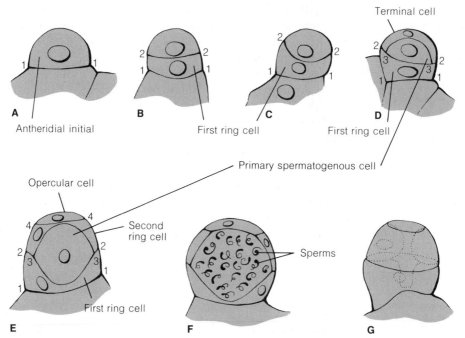

Figure 5-4

Ontogeny of the antheridium in *Pityrogramma calomelanos*. **A**, formation of the antheridial initial; **B**, two-celled stage; **C**, two-celled stage in which the first wall (2-2) is becoming depressed; **D**, three-celled stage, consisting of a terminal cell, a primary spermatogenous cell, and the first ring cell; **E**, transverse division of terminal cell by wall 4–4 has produced the opercular cell and the second ring cell; **F**, mature antheridium with sperms; **G**, three-dimensional view of antheridium (which has discharged its sperms) to show form of the ring cells. [Redrawn from Davie, *Amer. Jour. Bot.* 38:621, 1951.]

is forced into contact with the inner wall of the opercular cell (Fig. 5-4, E, F). At this stage the jacket of the antheridium consists of three cells: two rings cells and an opercular cell. But as a result of the contacts established between certain walls, the antheridium may appear in median longisectional view to have a five-celled jacket (see Fig. 5-4, F, G; also Fig. 13-31, I).

Although Davie's interpretation has been rather widely adopted, two recent studies have raised points of disagreement with his description of the orientation of each of the three successively formed walls of the developing antheridium (Fig. 5-4). Stone (1962), for example, in opposition to Davie's description (see Fig. 5-4, walls 2-2, 3-3, and 4-4), concluded from her survey that the

first wall (i.e., Davie's wall 2-2) is transverse, the second hemispherical and periclinal, and that the third wall, which cuts off the opercular cell, is invariably funnel-shaped. Verma and Khullar (1966), however, rejected Stone's conclusion regarding the plane of the third wall, which in their material was slightly concave or even flat, rather than funnel-shaped.

It should be clear from the studies that we have just briefly reviewed, that differences in the *details* of antheridial development are to be expected, not only between the various genera of ferns but, as was discovered by Verma and Khullar (1966), even between the antheridia of the same species. Continued research is obviously needed and should help eventually to resolve some of the present controversial aspects of antheridial ontogeny in leptosporangiate ferns as a whole (see, for example, Schraudolf, 1963).

Homology and Phylogenetic Origin of Gametangia

In the two preceding sections the structure and ontogeny of the sex organs in lower vascular plants have been described in considerable detail. Can any general conclusion be reached as to the evolutionary significance of the points of resemblance between antheridia and archegonia? Many morphologists believe that the close similarity in mode of initiation and early development of the two kinds of gametangia is evidence of their homology. Each arises from a single parent cell, and each, at a very early stage, exhibits a sharp distinction between external sterile cells and internal fertile ones. What then is the reason for the marked reduction in the number of functional gametes to a single basal cell in the archegonium?

Very possibly, as many writers have pointed out, this reduction indicates that the archegonium is the more specialized of the two gametangia and that it originated in evolution from a more generalized polygametic organ. The antheridium seems in many ways to correspond to such a hypothetical primitive gametangium. On this line of reasoning, the archegonium would be regarded as a highly modified and reduced organ in which a series of potential gametes —the entire axial row—are the morphological equivalents of the spermatocytes of the antheridium. But in modern archegonia all except one of these potential gametes are destroyed in the production of the archegonial canal. Some evidence that the archegonium formerly may have been more antheridial-like in structure is furnished by unusual or abnormal archegonia in the Bryophyta (Holferty, 1904; Meyer, 1911)

as well as in lower vascular plants (Lyon, 1904; Spessard, 1922). The archegonia of *Lycopodium complanatum*, for example, very often show from fourteen to sixteen neck canal cells, each of which is binucleate. This might be interpreted as an atavistic trend towards the multiplication of potentially fertile cells.

But the most remarkable structures are the bisexual gametangia that have been observed in various liverworts and mosses and also in certain species of *Lycopodium*. These organs resemble archegonia in their form but contain spermatogenous tissue as well as typical axial-row cells. In *Lycopodium lucidulum*, according to Spessard's investigations, the spermatogenous tissue may occupy either the neck or the venter of an abnormal archegonium. Great care must always be observed in the interpretation, especially from an evolutionary standpoint, of transitional forms between two kinds of organs. But taken in conjunction with the evidence from ontogeny, the aberrant archegonia lend further strength to the idea that both antheridia and archegonia have evolved from a rather general, undifferentiated, and probably polygametic sex organ. The details of the evolutionary process may always remain obscure, especially because it is difficult to derive the gametangia of living Archegoniates from the sexual "organs" of any of the green algae (Davis, 1903; Lyon, 1904). With reference to the possible relationship between the gametangia of bryophytes and lower vascular plants, Goebel (1905, p. 187) makes the following interesting conclusion: "The structure of the sexual organs is alike in its outlines in Bryophyta and Pteridophyta, but shows in the development and in the ultimate details so many differences that we have evidently here to deal with two phyletic series of which the higher has not been derived from the lower, but arising at an early period from simple similar primitive forms they have followed separate paths."

References

Bierhorst, D. W.
 1954. The gametangia and embryo of *Psilotum*. *Amer. Jour. Bot.* 41:274–281.

Bower, F. O.
 1935. *Primitive Land Plants*. Macmillan, London.

Campbell, D. H.
 1911. *The Eusporangiatae. The Comparative Morphology of the Ophioglossaceae and Marattiaceae.* (Publication No. 140.) Carnegie Institution of Washington, Washington, D.C.
 1918. *The Structure and Development of Mosses and Ferns*, Ed. 3. Macmillan, New York.

Davie, J. H.
 1951. The development of the antheridium in the Polypodiaceae. *Amer. Jour. Bot.* 38:621–628.

Davis, B. M.
 1903. The origin of the archegonium. *Ann. Bot.* 17:477–492.

Goebel, K.
 1905. *Organography of Plants*, Part II, Eng. Ed. by I. B. Balfour. Clarendon Press, Oxford.
 1928. *Organographie der Pflanzen*, Ed. 3. Erster Teil. G. Fischer, Jena.

Hauke, R. L.
 1968. Gametangia of *Equisetum bogotense. Bull. Torrey Bot. Club* 95:341–345.

Haupt, A. W.
 1940. Sex organs of *Angiopteris evecta. Bull. Torrey Bot. Club* 67:125–129.

Holferty, G. M.
 1904. The archegonium of *Mnium cuspidatum. Bot. Gaz.* 37:106–126.

Lyon, F.
 1904. The evolution of the sex organs of plants. *Bot. Gaz.* 37:280–293.

Maheshwari, P., and M. Sanwal
 1963. The archegonium in gymnosperms. A Review. *Mem. Indian Bot. Soc.* No. 4, 103–199.

Meyer, K.
 1911. Zur Frage von der Homologie der Geschlechtsorgane und der Phylogenie des Archegoniums. *Biol. Zeit.* 2:178–185.

Schraudolf, H.
 1963. Einige Beobachtungen zur Entwicklung der Antheridien von *Anemia phyllitidis. Flora* 153:282–290.

Spessard, E. A.
 1922. Prothallia of *Lycopodium* in America. II. *L. lucidulum* and *L. obscurum* var. *dendroideum. Bot. Gaz.* 74:392–413.

Stokey, A. G.
 1942. Gametophytes of *Marattia sambucina* and *Macroglossum Smithii. Bot. Gaz.* 103:559–569.
 1950. The gametophyte of the Gleicheniaceae. *Bull. Torrey Bot. Club* 77:323–339.
 1951. The contribution by the gametophyte to classification of the homosporous ferns. *Phytomorphology* 1:39–58.

Stone, I. G.
 1962. The ontogeny of the antheridium in some leptosporangiate ferns with particular reference to the funnel-shaped wall. *Aust. Jour. Bot.* 10:76–92.

Verma, S. C., and S. P. Khullar
 1966. Ontogeny of the polypodiaceous fern antheridium with particular reference to some Adiantaceae. *Phytomorphology* 16:302–314.

Ward, M.
 1954. Fertilization in *Phlebodium aureum* J. Sm. *Phytomorphology* 4:1–17.

6

Embryogeny

The term embryogeny designates the successive steps in the growth and differentiation of a fertilized egg or zygote into a young sporophyte. Although no sharp limits can be set, it will be convenient to distinguish between the early definitive stages in embryogeny and the later phases of growth during which the organs of the embryo arise and become functionally significant.

In all the Archegoniatae (including the majority of gymnosperms) fertilization and the first critical phases of embryogeny occur within the shelter provided by the venter of the archegonium and the adjacent gametophytic tissue. At this first period the polarity (i.e., the distinction between the apex and the base) of the embryo is determined as will be described in detail later in the chapter. The period of enlargement and organ development which follows results, in lower vascular plants, usually in the formation of a shoot apex together with one or more leaves, a root, a well-defined foot, and, in many genera, a suspensor. The foot serves as a haustorial organ which attaches the embryo to the nutritive tissue of the gametophyte. In the majority of archegoniate plants the apex of the young embryo faces inwardly toward the gametophytic tissue and away from the neck of the archegonium, as can be seen in

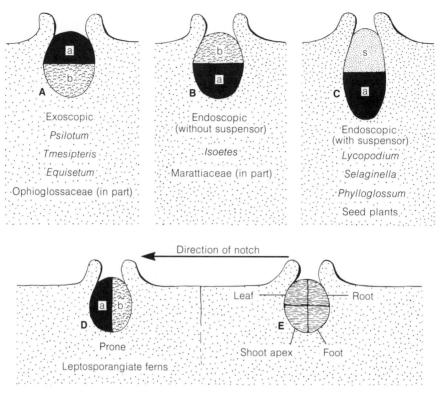

Figure 6-1
Main types of polarity in the development of the embryo in lower vascular plants. In all figures, the apical pole (cell a) is shown in black and the basal pole (cell b) is shaded. A, exoscopic polarity; B, endoscopic polarity without a suspensor; C, endoscopic polarity with a suspensor; D, orientation of two-celled embryo in leptosporangiate ferns; E, quadrant stage in embryogeny in leptosporangiate ferns illustrating points of future origin of leaf, shoot apex, root, and foot.

Fig. 6-1, B, C. An embryo of this type illustrates strikingly the intimate nutritive relationship between it and the gametophyte since, during enlargement and organogenesis, it must digest its way through a considerable volume of gametophytic tissue before emerging at the surface of the thallus. Ultimately, in contrast with the permanently dependent embryos of all the bryophytes, the embryo of lower vascular plants becomes free from the nursing gametophyte and grows into an independent sporophyte. In the majority of the gymnosperms and in angiosperms the embryo is shed from the parent plant in a somewhat dormant condition, is enveloped by a protective cover, and is often provided with a special nutritive tissue; here the embryo constitutes the essential part of that remarkable structure which we term a seed (see Chapters 14 and 21).

In this chapter the discussion will be restricted mainly to the principles of comparative embryogeny as illustrated by the lower vascular plants. As in the discussions of gametangia and sporangia, our intent is to provide orientation for the student and to illuminate those significant details of embryogeny which will be given later when we consider the morphology of special groups of tracheophytes. For an account of the embryogeny of seed plants, the student should consult Chapters 15-18 and Chapters 20-21, which include descriptions of the embryos of selected gymnosperms and angiosperms.

General Organography of Embryos

Before beginning a discussion of the origin and early development of embryos we must have a clear idea of the main categories of organs found in the embryos of vascular plants. From a broad morphological viewpoint, as has been repeatedly emphasized by Bower (1922, 1935), a very young embryo is somewhat filamentous or spindle shaped, and consists of two definable poles: the *apical pole*, which gives rise to the terminal meristem or shoot apex and one or more leaves, and the *basal pole*, which in many groups of the lower vascular plants is represented by the *suspensor*. Let us now examine these components of the embryo.

The apical pole is the portion from which the first or primary shoot takes its origin; it is a very consistent feature of the embryo of the large majority of vascular plants. In much of the literature dealing with the embryos of lower vascular plants the first leaf or leaves of the embryo are designated as cotyledons, a term which we, the authors, would prefer to reserve for the first foliar organs of the embryos of seed plants (Wagner, 1952, p. 45). The size of the first leaf or leaves of the embryo in lower vascular plants fluctuates widely; in some genera— for example, *Lycopodium*—they are very small and rudimentary, whereas in others— *Isoetes, Botrychium virginianum*, and many leptosporangiate ferns—the first leaf is large, precociously developed, and may even

be significantly photosynthetic (see Chapter 13). In seed plants, as will be described later, the number of cotyledons ranges from one to many.

As stated above, the base of the embryo in certain genera is represented by the suspensor; this structure, highly developed in most gymnosperms and found also in the embryos of angiosperms, is extremely variable in its occurrence in lower vascular plants (Fig. 6-2). The family Ophioglossaceae furnishes an excellent illustration, because in *Ophioglossum* and several species of *Botrychium* a suspensor is absent, whereas in other species of the latter and in *Helminthostachys* it is evident. Variation is found also in the Lycopsida, where two of the genera—*Lycopodium* and *Selaginella*—have embryos with suspensors, and *Isoetes* has none (Chapter 9). The morphological interpretation of the suspensor and its correlation with the type of polarity in the embryo will be discussed later, but here it should be noted that the suspensor is a temporary organ, the growth and enlargement of which serve to force the embryo into intimate contact with the nutritive tissue of the gametophyte.

In addition to the shoot apex, leaf primordia, and suspensor, the embryo of most tracheophytes develops a root. Conspicuous exceptions are found in *Psilotum* and *Tmesipteris*, two presumably primitive types of plants in which the sporophyte, at all stages in its growth, is devoid of roots (Chapter 8). Because there is no evidence in these genera of even a "vestigial" root during embryogenesis, the absence of roots at later phases in the ontogeny of their sporophytes is commonly regarded as a primitive rather than a derived condition.

As Bower (1935, pp. 540, 541) strongly emphasized, in lower vascular plants the root is consistently *lateral* with reference to the longitudinal axis of the embryo (Fig. 6-2).

Plants with this type of orientation of the first root of the embryo were termed *homorhizic* by Goebel (1930) in contrast to the *allorhizic* seed plants, in which shoot and root poles lie at opposite ends of the axis of the embryo (Chapter 19). A more detailed description of the general morphology and anatomy of the root of angiosperms will be found in Chapter 19, pp. 585–593.

There remains for brief consideration the so-called *foot* of the embryo, whose occurrence and degree of development vary more than that of the suspensor, even within species of the same genus (Fig. 6-2). Bower (1935) regards the foot as an "opportunist growth" which arises in a position convenient for performing its function as a suctorial or nursing organ. In some embryos,

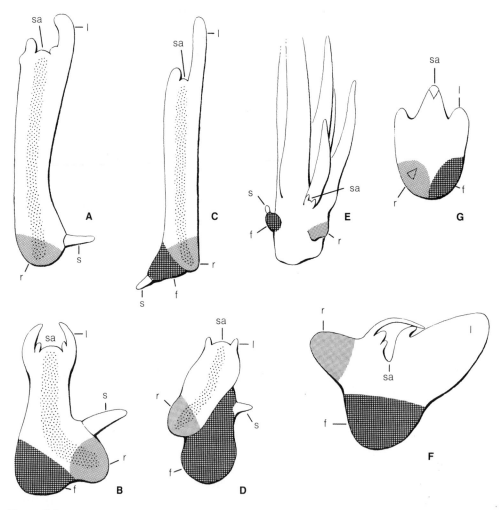

Figure 6-2
Variations in form and position of the organs of the embryos in lower vascular plants. **A,** *Selaginella spinulosa*; **B,** *Selaginella martensii*; **C,** *Lycopodium selago*; **D,** *Lycopodium clavatum*; **E,** *Lycopodium cernuum*; **F,** *Isoetes*; **G,** *Equisetum*. Abbreviations: f, foot; l, leaf; r, root; s, suspensor; sa, shoot apex. [Redrawn from *Primitive Land Plants* by F. O. Bower. Macmillan, London. 1935.]

such as those of *Selaginella, Isoetes, Equisetum, Ophioglossum,* and the leptosporangiate ferns, the foot is conspicuous, and its role in anchoring and conveying nutrients from the gametophyte is evident. Perhaps the most impressive example of a suctorial foot is found in the embryos of *Tmesipteris* and *Psilotum*. Here the entire lower half of the embryo develops into a haustorial structure which sends lobes and irregular processes deep into the tissue of the gametophyte (see Chapter 8).

Polarity and Early Embryogeny

During the middle and latter part of the nineteenth century much labor was devoted by morphologists to tracing in minute detail the origin and development of the organs of many types of embryos. One of the results of the close attention given the sequence and planes of cell division in growing embryos was the concept that there is a single or basic plan of segmentation throughout plant embryogeny. In particular, the embryos of *Equisetum* and leptosporangiate ferns were considered fundamental and typical of all embryos, since it was held that the major organs, leaf, stem, root, and foot could be traced back to specific quadrants (i.e., quarter sections) formed at the very beginning of embryonic development. In the light of our extended knowledge of embryos of all the major groups of vascular plants it is now evident that great variation exists in the definition of these so-called quadrants and their rôle in organogenesis, and that the study of embryogeny should be pursued in as flexible a manner as possible, with due regard to the effect of such external factors as gravity, light, and nutrition on the embryonic pattern (Ward, 1954; Wardlaw, 1955). In short, there is need for a broadened organographic attack on the problems of comparative embryology and a proper recog-

nition of the dangers of rigid histogenetic and organogenetic interpretations.

According to Bower (1908, 1922, 1935), who paved the way toward a more dynamic biological understanding of embryogeny, the first step in the development of the embryo is the definition of its axial polarity. Except for the leptosporangiate ferns, to be discussed later, the first division wall in the zygote is approximately transverse to the long axis of the archegonium and results in a definite distinction between the future apex and base of the embryo. However, two types of polarity arise from this first division of the zygote. In the more common polarity type, termed *endoscopic,* the apical pole — cell a in Fig. 6-1, B, C — is directed toward the base of the archegonium; in the opposed, or *exoscopic,* type the apex faces outward toward the neck of the archegonium — cell a in Fig. 6-1, A.

The occurrence of these two main types of embryo polarity in vascular plants at large poses some interesting morphological and phylogenetic questions. Embryos of such lower vascular plants as *Lycopodium, Selaginella,* and *Isoetes,* as well as all seed plants, are of the endoscopic type. It is particularly significant to note that all seed plants are of endoscopic polarity even though the mode of early development of the proembryo in gymnosperms is remarkably different from that in angiosperms. In the latter group, the first nuclear division of the zygote is followed by the formation of a wall; the only known exception to this generalization is provided by the free nuclear phase of proembryo development in the dicotyledonous genus *Paeonia* (see Chapter 20, Fig. 20-36, for further details). But in gymnosperms (with the exception of *Sequoia sempervirens* and *Gnetum* sp.) a more-or-less extensive free nuclear period of development is characteristic of the first phase of embryogenesis. Nevertheless, during the subsequent cellular phase of development of the proembryo, the future shoot apex region is directed away from the archegonial neck and the polarity is obviously endoscopic in type. (See Chapters 15–18 for detailed accounts of early embryogeny in gymnosperms.)

The less common exoscopic polarity characterizes the embryogeny of *Psilotum*, *Tmesipteris*, *Equisetum*, some members of the Ophioglossaceae, and, according to the recent studies of Bierhorst (1968), the leptosporangiate fern genus *Actinostachys*. Generally speaking, either endoscopic or exoscopic polarity characterizes the major groups within lower vascular plants, although both patterns may be encountered within the confines of a single genus, as in *Botrychium*. In a strict sense, the embryos of most of the investigated members of the leptosporangiate ferns do not fall into either the endoscopic or exoscopic categories because the first division wall in the zygote tends to be parallel to the long axis of the archegonium, as can be seen in Fig. 6-1, D. Consequently, the apical and basal poles of the embryo tend to be oriented laterally with respect to the archegonial axis, and the embryo is prone rather than vertical or curved with respect to the prothallus, as shown by Fig. 6-1, E. A further discussion of the distinctively oriented embryo of leptosporangiate ferns will be given after we describe the origin of the suspensor in various types of embryos.

The presence of a suspensor is invariably correlated with an endoscopic type of polarity. Figure 6-3 illustrates schematically the method of origin of the suspensor in the endoscopic type of embryogeny characteristic of *Lycopodium*, *Selaginella*, and certain of the eusporangiate ferns. The zygote divides by the formation of a transverse wall I-I into two cells as shown in Fig. 6-3, A. The upper cell (labeled s) is directed toward the neck of the archegonium and represents the parent cell of the future suspensor, which may enlarge without further division or give

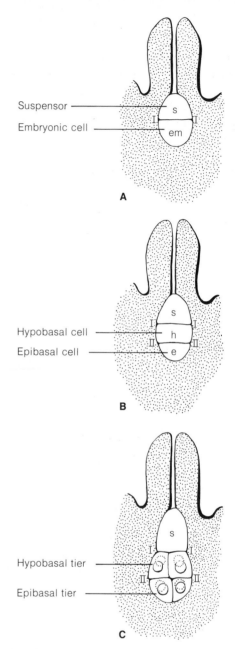

Figure 6-3

Successive stages in early development of an endoscopic type of embryo with a suspensor. **A,** the first division of the zygote is transverse (wall I-I) and yields a suspensor cell (s) directed toward the archegonial neck and an embryonic cell (em) directed inwardly; **B,** transverse division of embryonic cell (wall II-II) has produced an epibasal (e) and a hypobasal (h) cell; **C,** the octant stage, resulting from two successive vertical divisions of the epibasal and hypobasal cells.

rise to a short multicellular filamentous suspensor. The lower cell (labeled em), termed the *embryonic cell*, subsequently gives rise to the main body of the embryo. The plane of the first division of the embryonic cell may be transverse, yielding a *hypobasal cell* (in contact with the suspensor) and an *epibasal cell* (Fig. 6-3, B). Two successive vertical divisions in each of these cells results in an eight-celled embryo, consisting of an epibasal tier and a hypobasal tier (Fig. 6-3, C). Although this tiered arrangement of cells apparently is characteristic of the young embryos of many vascular plants, the sequence of wall formation in the embryonic cell appears to fluctuate. In various species of *Lycopodium*, for example, the plane of the first division wall in the embryonic cell is vertical rather than transverse; each of the two cells then divides in a similar plane yielding a group of four cells—a quadrant stage. Transverse divisions of each of these quadrant cells then results in the formation of an eight-celled embryo comprising an epibasal and a hypobasal tier. The variable rôle of the epibasal and hypobasal cell tiers in the subsequent development of the organs of the embryo will be reserved for later discussion.

In embryos devoid of a suspensor the zygote itself functions directly as the embryonic cell, and its first division results, as before, in an epibasal and a hypobasal cell (Wardlaw 1955, p. 22). But in contrast with the invariably endoscopic polarity of embryos with a suspensor, the suspensorless types vary widely in orientation of the epibasal cell or apical pole. *Isoetes* and certain genera in the Marattiaceae (*Christensenia, Marattia, Angiopteris*) lack suspensors, yet are characterized by endoscopic polarity. On the other hand, as we have mentioned, all exoscopic embryos (for example, those of *Tmesipteris, Psilotum, Equisetum,* and *Ophioglossum*) are devoid of a suspensor, and in them the epibasal region faces outward toward the neck

of the archegonium. To repeat, the orientation of the suspensorless embryo in the leptosporangiate ferns (with the apparent exception of *Actinostachys*) is neither strictly endoscopic nor exoscopic because the first or the basal wall is longitudinal rather than transverse, as can be seen in Fig. 6-1, D. In this case, the apical pole of the young embryo is directed toward the apex of the prothallus but is transverse with reference to the long axis of the archegonium. There appears to be no entirely satisfactory explanation for the anomalous orientation of the embryo in leptosporangiate ferns. Bower (1935) has pointed out that gravity may play a role in determining the plane of the first division of the zygote. In any event, an interesting result of the position of the embryo is that the first leaf and the first root emerge from beneath the lower surface of the gametophyte rather than penetrating the thallus as in endoscopic embryos (Chapter 13).

Origin and Development of Organs

Following the early definition of polarity and the establishment of an eight-celled embryo, the young sporophyte enters a phase of enlargement and organ formation. We may properly ask at this point whether the organs common to most embryos, such as root, first leaf, shoot apex, and foot, always arise in a uniform manner from specific sectors or regions of the embryo. If we restrict ourselves to the evidence furnished by classical accounts of the embryogeny of leptosporangiate ferns, this question would be answered in the affirmative. It is said that in these plants the shoot apex, first leaf, first root, and foot originate from specific quadrants of the embryo (Fig. 6-1, E). However, this interpretation, as will be discussed in more detail in Chapter 13, has been challenged in recent years on the grounds that it postulates a rigid correspondence between the position of organs of the embryo and the pattern of cellular differentiation arising from each of the original embryo quadrants. The foot, for example, as described by Ward (1954), originates in *Phlebodium* from cells derived from both of the anterior or epibasal sectors of the embryo. (See Wardlaw 1955, pp. 142–146 for further critical comments on the classical interpretation of embryogeny in leptosporangiate ferns.)

A precise type of organogenesis referable to definable quadrants is believed to exist also in the suspensorless embryo of *Equisetum*. According to the early investigations of Sadebeck (1878) on *Equisetum arvense* and *Equisetum palustre*, the outer epibasal tier of the embryo forms the shoot apex and the first whorl of leaves, and the foot and root originate from derivatives of the hypobasal tier. Although it is now considered doubtful that the root in *Equisetum* is always of hypobasal origin, there is little question about the discrete method of origin of shoot and foot, respectively, from the epibasal and hypobasal tiers (Chapter 10).

In contrast with *Equisetum* and the leptosporangiate ferns, organogenesis in the embryos of other lower vascular plants follows no clear-cut uniform or standardized pattern (Fig. 6-2). The shoot apex and the first foliar appendages show the greatest constancy of all the parts of the embryo because they invariably arise from the epibasal tier. According to the analysis of various embryonic patterns made by Bower, the shoot apex originates as nearly as possible at the geometrical center of the epibasal half of the embryo. In terms of cells, this center corresponds very closely to the point of intersection of the octant walls and is plainly defined in those plants in which a single definitive apical cell is produced by obliquely oriented divisions. Foliar organs, whatever their size, number, or arrangement, likewise

originate from the epibasal half of the embryo. In many plants, these primary leaves are arranged in a pair or a whorl around the embryo apex, thus resembling the relation of later leaf primordia of the sporophyte to the shoot apex. But in some embryos, particularly in *Isoetes* and in members of the Ophioglossaceae, the first leaf is unusually large and, because of its precocious growth, tends to displace the shoot apex to the side or even to cause a delay in its initiation.

With reference to its place and time of origin, the root is, without question, the most variable organ of the embryo. Sometimes it arises jointly with the first leaf and shoot apex from the epibasal half of the embryo. This is true of the embryos of *Isoetes* and many members of the Ophioglossaceae and Marattiaceae. In other embryos, such as those of certain species of *Selaginella*, the primary root develops from the hypobasal portion. In addition to its variable point of origin, the position of the first root with reference to other organs of the embryo is inconstant, as is shown very clearly in the genus *Selaginella*. In *Selaginella denticulata* the root lies on the same side of the embryo as the suspensor, and in *Selaginella poulteri* it lies on the side opposite the suspensor; finally, in *Selaginella galeotii* the root, although on the same side as the suspensor, lies between the suspensor and the shoot apex. These relationships are shown in Fig. 6-4.

With reference to the foot, it is evident that although this portion of the embryo arises from all or a part of the cells of the hypobasal tier, its form and degree of development fluctuate even within members of the same genus. This is illustrated in both *Lycopodium* and *Selaginella*, the individual species of which differ considerably in regard to the prominence of the foot (see Fig. 6-2).

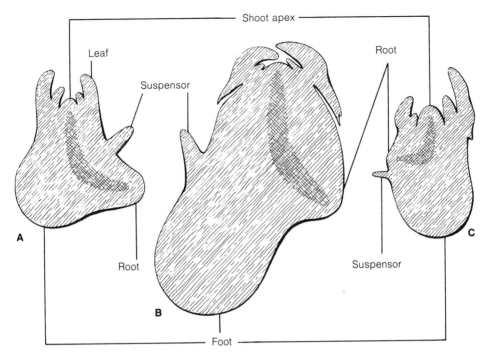

Figure 6-4
Embryos of three species of *Selaginella* showing variability in the position of the root with reference to other organs. **A**, *S. denticulata;* **B**, *S. poulteri;* **C**, *S. galeottii* [Redrawn from *Primitive Land Plants* by F. O. Bower, Macmillan, London. 1935.]

Embryogeny from a Phylogenetic Standpoint

As a conclusion to this chapter it is appropriate to examine briefly the phylogenetic implications of those various aspects of embryogeny which have been presented.

First, we may raise the question of the historical aspects and evolutionary significance of the two principal types of embryo polarity. Which is the more primitive form, the exoscopic or the endoscopic? It might be assumed, in agreement with the viewpoint of Bower, that an endoscopic embryo with a suspensor represents the "primitive spindle" form, from which the suspensorless types have taken origin. According to Bower, the abolition of the suspensor has taken place repeatedly and independently, as evidenced

by its variable occurrence in such families as the Marattiaceae and Ophioglossaceae and in the Lycopsida. Once the embryo became free of a suspensor its former obligatory endoscopic polarity might be retained, if this were selectively advantageous from the standpoint of nutrition, or a complete inversion to the exoscopic type of polarity might take place. Bower (1935, p. 533) states that such an inversion of polarity has occurred phylogenetically in *Ophioglossum* and *Botrychium lunaria*. Goebel (1928, p. 225) also holds that a rotation in polarity from endoscopic to exoscopic is admissable in some cases. On the other hand, the occurrence of exoscopic polarity in such primitive vascular plants as *Tmesipteris* and *Psilotum*, as well as throughout the Bryophyta, might equally well suggest that endoscopic polarity, which is typical of the majority of higher vascular plants, is the derived condition (MacMillan, 1898). At present there is absolutely no evidence that the embryos of *Tmesipteris* and *Psilotum* originally possessed suspensors and hence were endoscopic. Very likely, these phyletic questions are not answerable with our present knowledge of embryogeny and plant phylogeny.

There is great need for an intensive experimental attack on the problem of the polarity of embryos. In certain of the liverworts it seems clear that the orientation of the archegonium is not a determining factor because, regardless of whether it is upright, horizontal, or pendent, the embryo is strictly exoscopic. Probably, as Bower has suggested, the basal nutrition of the young embryo in these cases is an important factor in maintaining the exoscopic polarity. Almost no experimental work seems to have been done on the factors which might influence or control the polarity of the embryo in vascular plants (but see Chapter 13). Genera such as *Botrychium*, in which the polarity varies with the species, would be particularly interesting to study from a morphogenetic viewpoint.

With reference to the complex problems posed by organogenesis in embryos, several significant points deserve emphasis. During recent decades there has been a conspicuous awakening of interest in the cellular organization and patterns of growth and differentiation of apical meristems. The shoot apex in particular has been explored in a great variety of vascular plants, and the general conclusion reached is that there is no real basis for a rigid and highly deterministic interpretation of either histogenesis or organogenesis (see Foster, 1949; Esau, 1965; Gifford, 1954; Gifford and Corson, 1971). A similar flexible viewpoint seems entirely justified in the comparative and phyletic study of embryos. In other words, although correspondence may exist between the origin of organs and the pattern of cell divisions, this is not necessarily an obligatory relationship. Furthermore, it should be recalled that a plant embryo is very far from being a miniature of the adult organism. On the contrary, vascular plant sporophytes are characterized by an open type of growth and organogenesis because of the maintenance of embryonic areas or meristems at the tips of all shoots and roots. This open type of growth—or "continued embryogeny" as Bower terms it—has its beginning in the embryo itself, which thus foreshadows the morphology of the adult plant in only the most general fashion. For these reasons, phylogenetic interpretations of plant embryos should be made with due regard to the possible effects of gravity, light, and the source and type of nutrition on the polarity and pattern of development of the young encapsulated sporophyte (Wardlaw, 1955).

References

Bierhorst, D. W.
  1968. Observations on *Schizaea* and *Actinostachys* spp., including *A. oligostachys*, sp. nov. *Amer. Jour. Bot.* 55:87–108.

Bower, F. O.
  1908. *The Origin of a Land Flora.* Macmillan, London.
  1922. The primitive spindle as a fundamental feature in the embryology of plants. *Proc. Roy. Soc. Edinb.* 43:1–36.

1935. *Primitive Land Plants*. Macmillan, London.

Esau, K.
1965. *Plant Anatomy,* Ed. 2. Wiley, New York.

Foster, A. S.
1949. *Practical Plant Anatomy,* Ed. 2. Van Nostrand, New York.

Gifford, E. M. Jr.
1954. The shoot apex in angiosperms. *Bot. Rev.* 20:477–529.

Gifford, E. M., Jr., and G. E. Corson, Jr.
1971. The shoot apex in seed plants. *Bot. Rev.* 37:143–229.

Goebel, K.
1928. *Organographie der Pflanzen,* Ed. 3. Erster Teil. G. Fischer, Jena.
1930. *Organographie der Pflanzen.* Dritte Auf. Zweiter Teil. G. Fischer, Jena.

MacMillan, C.
1898. The orientation of the plant egg, and its ecological significance. *Bot. Gaz.* 25:301–323.

Sadebeck, R.
1878. Die Entwicklung des Keimes der Schachtelhalme. *Jahrb. Wiss. Bot.* 11: 575–602.

Wagner, W. H., Jr.
1952. The fern genus *Diellia. Univ. Calif. Publ. Bot.* 26:1–212.

Ward, M.
1954. The development of the embryo of *Phlebodium aureum* J. Sm. *Phytomorphology* 4:18–26.

Wardlaw, C. W.
1955. *Embryogenesis in Plants.* Wiley, New York.

7

The Early Vascular Plants: Rhyniopsida, Zosterophyllopsida, and Trimerophytopsida

In considering the history of early terrestrial plant forms, it is quite essential to distinguish between the first authentic vascular plants and the more primitive land plants that preceded them. An examination of Table 1-1 (pp. 8–9) reveals a gap of about 1–2 billion years between the first occurrence of algae in the fossil record and the first well-documented appearance of vascular plants. It may be concluded that the algae, bacteria, and fungi existed for an exceedingly long time before the origin of vascular plants (Schopf, 1970). Some of the alga-like plants may have been fairly large and, in being restricted to marshy areas, may have been in large part terrestrial, but they lacked characteristic vascular tissue as we know it today (see Andrews, 1959, 1961, for examples). The acquisition of a vascular system in ancestral forms probably led to rapid colonization of the land and then as forms diversified, there would have been opportunities for them to become established in the many unoccupied environmental niches. According to one view, the origin and diversification of vascular plants took place during a relatively short time— Late Silurian to Early Devonian (Banks, 1965, 1968a, 1968b, 1970). Other paleobotanists would suggest a longer developmental history for vascular plants (Axelrod, 1959; Leclerq, 1956). During much of the Paleozoic a climate prevailed that might be described as temperate and somewhat seasonal, like that of the coastal regions of California and Oregon. In the animal world, higher invertebrates were becoming specialized; fishes, amphibians, and reptiles made their appearance, and by the end of the Paleozoic amphibians were well established (Table 1-1).

Factual information concerning a member of a very ancient group of plants dates back to 1859 when Sir James W. Dawson described a plant (*Psilophyton princeps*) of Devonian age discovered on the Gaspé Peninsula, Quebec, Canada. Dawson described this fossil species as having dichotomous branches with interrupted ridges or closely appressed hairs or leaves; he described a vascular cylinder for the axes and stated that sporangia probably occurred in lateral groups on the stem.

In 1871, however, Dawson described the sporangia as being terminal. He called these plants *P. princeps* v. *ornatum*. Thus, the widespread concept that *P. princeps* had terminal sporangia is not based on Dawson's *original* description. Because his description was based upon rather fragmentary evidence (organic connection was not proved for any of the parts), some doubt existed about the morphological interpretation of the remains. It was not until almost sixty years later that the real significance of his discovery was appreciated. From 1917 to 1921 a series of articles was published by Kidston and Lang (1917) describing well-preserved plants from deposits of Devonian age in the Rhynie chert beds, Aberdeenshire, Scotland. These paleobotanists realized that there were some similarities between *Psilophyton* and their discoveries, and the order Psilophytales was established to include all of these presumably primitive extinct plants. Following these earlier studies additional genera were discovered in such widely separated places as Norway, China, and Australia. The Psilophytales became a catch-all group for extinct plants of Devonian

age that could not easily be placed with the lycopods or into any other lower vascular plant group. Examples can be found in standard texts on paleobotany (to mention but four: Arnold, 1947; Andrews, 1961; Gothan and Weyland, 1964; Mägdefrau, 1968).

With continuing research on fossil plants, however, it has become increasingly clear that there were at least two main lines within the Psilophytales as established by Kidston and Lang (1917). Hueber (1964) has pointed out that in one line the plants had terminal sporangia that were predominantly fusiform and dehisced longitudinally. The stems were naked (devoid of appendages). Examples are *Rhynia* and *Horneophyton* (Fig. 7-1, A, C). In the second group sporangia were lateral on axes, predominantly globose or reniform, and each sporangium dehisced along the distal edge. In most forms the axes had emergences or *enations,* but Banks (1965) has described some with lateral sporangia that lacked appendages. In line with the concept of two major groups we are adopting the system of Banks (1968b) in which he established the Rhyniophytina (Rhyniopsida as used in this text) for the forms with terminal sporangia. This group comprises the Psilophytales as established by Kidston and Lang (1917) and as generally recognized in current textbooks, including the first edition of this book. The second group is the Zosterophyllophytina (Zosterophyllopsida) (e.g., *Zosterophyllum,* Fig. 7-3, A). Banks (1968b) also has proposed a third subdivision, the Trimerophytina (Trimerophytopsida), to accommodate certain genera that do not seem to fit into the other two subdivisions. According to Banks (1968b), "this subdivision includes plants with a main axis that branches pseudomonopodially. Each lateral branches a number of times, either trichotomously or dichotomously and finally terminates in a mass of fusiform sporangia" (Fig. 7-4, B, C). These genera seem to be a natural outgrowth of the Rhyniopsida and foreshadowed the still more complex branching patterns found among the early ferns and progymnosperms.

Classification

The abbreviated outline of classification that follows is intended to serve as a guide for further elaboration of the morphology of selected members of each group. Many genera of Devonian vascular plants have been described, but we, the authors, will adopt the type method and describe only a representative number in detail. Remains of some of the genera have been found in a remarkably good state of preservation. Whether the plants in these groups represent the most primitive type of organization can be argued, and this is indeed questioned by some botanists. Nevertheless, they display by their morphology the type of organization typical of the three subdivisions — Rhyniopsida, Zosterophyllopsida, and Trimerophytopsida.

RHYNIOPSIDA: Extinct plants; dichotomous or modified dichotomous branching; sporangia terminal on main axes or laterals; stems naked; sporangia globose or fusiform, commonly dehisced longitudinally; xylem strand in stems, where known, terete in transverse section with central protoxylem; roots absent; sporangia eusporangiate and homosporous.

RHYNIALES.

RHYNIACEAE (Fig. 7-1).

Rhynia, Horneophyton, Taeniocrada, Hicklingia, Cooksonia.

ZOSTEROPHYLLOPSIDA: Extinct plants; dichotomous or modified dichotomous branching; sporangia borne laterally, globose or reniform in shape, and dehiscent along the distal edge; xylem strand, where known, elliptical in transverse section and exarch; roots absent; eusporangiate and homosporous.

ZOSTEROPHYLLALES.

ZOSTEROPHYLLACEAE: stems naked; sporangia aggregated into spikes.

Zosterophyllum (Fig. 7-3, A).

GOSSLINGIACEAE: Stems naked, spiny, or toothed; sporangia scattered along the stem.

Gosslingia (Fig. 7-3, B).

TRIMEROPHYTOPSIDA: Extinct plants; branching anisotomous; lateral branch systems trichotomously or dichotomously branched: sporangia terminal on ultimate dichotomies, fusiform and longitudinally dehiscent; xylem strand, where known, mesarch and elliptical to circular in transverse section; roots absent; eusporangiate and homosporous.

Psilophyton, Dawsonites (fertile branch system), *Trimerophyton* (Fig. 7-4).

Figure 7-1 (*facing page*)
Diagrammatic reconstructions of several members of the Rhyniopsida. A, *Rhynia major;* B, *Cooksonia* sp., C, *Horneophyton lignieri;* D, *Taeniocrada decheniana;* E, *Hicklingia.* [A, C redrawn from Kidston and Lang, *Trans. Roy. Soc. Edinb.,* Vol. 52, Part IV, 1921; B redrawn from Croft and Lang, *Phil. Trans. Roy. Soc. London* 231B:131–163, 1942; D redrawn from Kräusel and Weyland, *Abh. Preuss. Geol. Landesanstalt* 131:1–92, 1930; E redrawn from *Die Telomtheorie* by W. Zimmermann. Gustav Fischer Verlag, Stuttgart, 1965.]

A

B

C

D

E

Rhyniopsida—Rhyniales

Rhyniaceae: *Rhynia*

ORGANOGRAPHY. The two species of this genus, *Rhynia gwynne-vaughani* (but see page 112) and *Rhynia major,* are well known because numerous specimens of them have been found in the Scotland deposits. As mentioned previously, the preservation of these specimens is so excellent that they have yielded valuable information concerning the form and internal structure of the plants. *R. major* was a vascular plant that grew in marshy swamps to a maximum height of about 50 cm (see Fig. 7-1, A). Each stem was approximately 5 mm in diameter. The plant body consisted of an underground rhizome upon which tufts of absorbing rhizoids occurred. Roots were absent. The aerial portion (stem), which was merely a continuation of a part of the underground system, branched dichotomously and was devoid of appendages. Growth of the plant was apparently initiated by apical meristems located at the tips of the branches. Elongate cylindrical sporangia terminated many of the ultimate dichotomies.

ANATOMY. The stem of *Rhynia* possessed epidermal, fundamental, and vascular tissue systems (Chapter 3)—an organization that is characteristic of not only the members of the Psilopsida but also the more-advanced modern vascular plants.

The entire branch system was covered by an epidermis with a thick cuticle on its outer surface. Stomata were present on the aerial portion. Internal to the epidermis was a broad cortex comprising parenchyma cells which probably functioned as the photosynthetic tissue. A slender vascular cylinder occupied the center of the axes (Fig. 7-2) and was composed of a cylinder

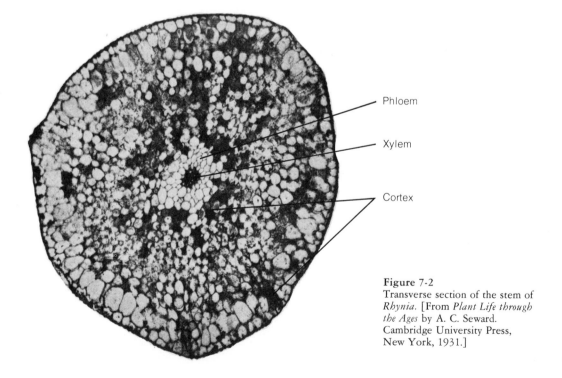

Phloem

Xylem

Cortex

Figure 7-2
Transverse section of the stem of *Rhynia.* [From *Plant Life through the Ages* by A. C. Seward. Cambridge University Press, New York, 1931.]

of primary xylem surrounded by a cylinder of primary phloem. This disposition of vascular tissue is designated specifically as a haplostele, as mentioned in Chapter 3. The xylem contained tracheids with annular thickenings, as interpreted by Lang, whereas the phloem consisted of thin-walled elongate cells.

The presumably indehiscent sporangia that terminated certain branches were, in some specimens, approximately 4 mm in diameter and 12 mm long. A rather massive sporangial wall, with a heavy cuticle on the outer surface, enclosed spores which were formed in tetrads. From all appearances, *Rhynia* was homosporous and the sporangium was of the eusporangiate type.

Rhyniaceae: *Horneophyton*

Considerable information has been obtained regarding *Horneophyton* (Fig. 7-1, C), another genus in the family, which is found with *Rhynia* and *Asteroxylon* — often in dense aggregations — in the Rhynie chert. The underground portion of *Horneophyton* consisted of a lobed rhizome from which arose the aerial dichotomously branched stems. The underground tuberous body lacked a continuous vascular system. Groups of long rhizoids located on the surface undoubtedly functioned in absorption. The aerial stem had a vascular system similar to that of *Rhynia*. Sporangial position and organization were very similar to those of *Rhynia*, differing only in the occurrence of a sterile columella in the sporangial locule.

Rhyniaceae: *Cooksonia*

At the moment *Cooksonia* appears to be the oldest known vascular plant (Obrhel, 1962). It has been dated as being of Late Silurian age. Species of *Cooksonia* are some of the smallest of vascular plants known

from the fossil record. They are slender, leafless, dichotomously branched plants with axes up to 6.5 cm in length and not exceeding 1.5 mm in width (Fig. 7-1, B). Vascular strands with annular tracheids have been identified. The sporangia were terminal and globose rather than being fusiform as in other members of the Rhyniopsida.

Zosterophyllopsida — Zosterophyllales

Zosterophyllaceae: *Zosterophyllum*

Generally *Zosterophyllum* is described as being a relatively small plant that branched dichotomously and was without leaves. An epidermal cuticle, stomata, and complete tracheids with annular to helical thickenings have been described (Lele and Walton, 1961). One species (*Zosterophyllum rhenanum*) from the Middle Devonian of the Rhineland may have depended upon water for support (Kräusel and Weyland, 1935). A branch system has been described for *Zosterophyllum myretonianum* (Walton, 1964) in which there was an upward directed system and a downward directed rhizome-type system (Fig. 7-3, A). Sporangia were aggregated into a terminal spike, were generally uniform in shape (2–4 mm wide), and opened by means of a slit along the distal edge.

Trimerophytopsida

In the same fossil beds with *Psilophyton*, Sir James W. Dawson discovered several plants with distinctive branching patterns and one of these is now known as *Trimerophyton robustius* (Hopping, 1956). The plant consisted of a rather robust main axis that branched anisotomously. Each lateral

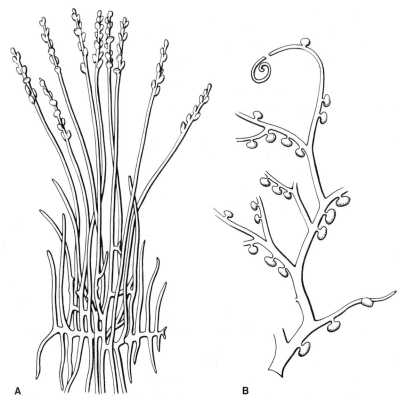

Figure 7-3
Diagrammatic representation of members of the Zosterophyllopsida. **A**, *Zosterophyllum myretonianum*; **B**, *Gosslingia breconensis*. [A redrawn from Walton, *Phytomorphology* 14:155–160, 1964; B redrawn from *Studies in Paleobotany* by H. N. Andrews, Wiley, New York, 1961.]

branched a number of times and there were fusiform sporangia at the tips of ultimate dichotomies (Fig. 7-4, C). The anatomy is known only for a portion of *Dawsonites*, a fertile branch (7-4, B). The axis supporting the fertile branch had a mesarch protostele and was elliptical in transverse section (Hueber, 1964). Plants of this group that produced large sporangial trusses could possibly have been derived from the Rhyniopsida by a change in growth habit. They occur in the fossil record at a later time than the earliest Rhyniopsida and could be ancestral to the primitive ferns and even the progymnosperms (See Chapters 11 and 14 and page 404).

Summary and Conclusions

In this chapter we have presented a discussion of the comparative morphology of certain extinct vascular plants. The classes Rhyniopsida, Zosterophyllopsida, and Trimerophytopsida are of particular interest and significance from an evolutionary standpoint because: (1) they include some of the most ancient, least specialized, and probably most primitive of known vascular plants; (2) some of the forms may well represent the type of early land plants from which the more highly developed sporophytes of ferns and other lower vascular plant groups have originated; and (3) the

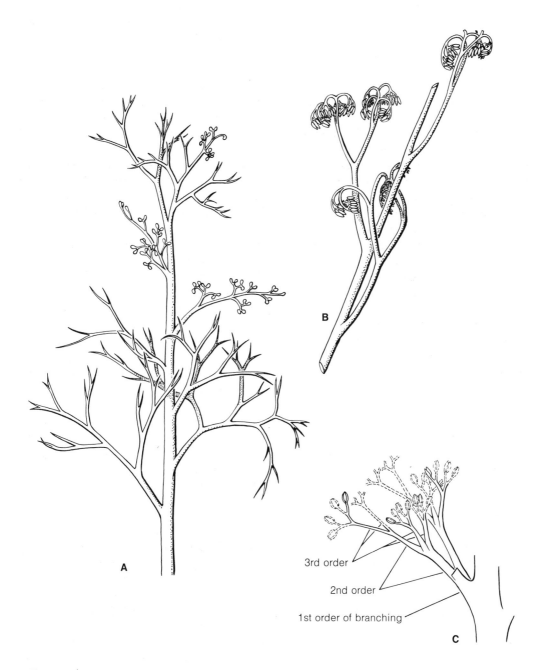

3rd order

2nd order

1st order of branching

C

Figure 7-4
Reconstructions of members of the Trimerophytopsida. **A**, *Psilophyton forbesii*; **B**, *Dawsonites arcuatus*;
C, *Trimerophyton robustius*. [A redrawn from Andrews, Kaspar, and Mencher, *Bull. Torrey Bot. Club*
95:1–11, 1968; B redrawn from Hueber, *Mem. Torrey Bot. Club* 21:5–9, 1964; C redrawn from Hopping,
Proc. Roy. Soc. Edinb. 66 (B):10–28, 1956.]

extremely simple vascular system of these ancient plants gives a clue to the arrangement and histology of primitive xylem in land plants. Well-preserved fossil remains of the Rhyniopsida were discovered in rocks of Devonian and Silurian age, and the reconstruction and description of these "Dawn Land Plants" is one of the most significant paleobotanical achievements in the twentieth century. The earliest known undisputed vascular plant is *Cooksonia* of the Late Silurian Period. For many years *Baragwanathia* (a lycopod) was considered to be the oldest known vascular plant but it has now been shown to be Devonian rather than Silurian in age (Jaeger, 1962). Another example of changes resulting from increased paleobotanical research is the shifting of the genus *Asteroxylon* from the "psilophytes" to the Lycopsida (See Lyon, 1964; Banks, 1968b; Chapter 9).

Returning to the main theme of the chapter, emphasis has been placed on three genera—*Rhynia, Zosterophyllum,* and *Trimerophyton*—each one representative of a certain type of growth pattern. All three genera were plants of relatively low stature and are believed to have been devoid of roots. In all three genera the aerial or upright portions of the sporophyte consisted of leafless, dichotomously or anisotomously branched axes or stems, some of which terminated in solitary, thick-walled, homosporous sporangia (*Rhynia*) or clusters of terminal sporangia (*Trimerophyton*). In *Zosterophyllum* the sporangia occurred in clusters near the tip of a branch but were attached laterally on the stem. The vascular system in these plants was a simple protostele consisting of a central strand of tracheids surrounded by a cylinder of phloem-like tissue.

An obvious omission from this discussion is an account of the gametophyte generation. It has usually been concluded that none of the known remains of early vascular plants are gametophytes. On the basis of reexamination of the original material of *Rhynia gwynne-vaughani,* Pant (1962) has suggested, however, that the axes of this species are more likely to be gametophytic than sporophytic. Furthermore, Kidston and Lang (1917) were never able to prove an actual connection between indisputable axes of *R. gwynne-vaughani* and sporangia; this would support the idea that these plants may not have been sporophytic at all. Pant (1962) admits that his evidence is indirect because he was not able to delineate well-preserved sex organs. In support of his belief Pant called attention to hemispherical projections and adventitious branches on the fossilized axes that possibly represent young sporophytes developing on gametophytes comparable to the gametophyte-sporophyte relationship observed in living *Psilotum* and *Tmesipteris* (Chapter 8).

More convincing evidence of Devonian gametophytes has been provided recently by Lemoigne (1968). Examination of sections of axes, presumed to be *R. gwynne-vaughani* (from the Rhynie chert), revealed rather convincing views of archegonia. The presumptive archegonia are embedded in the gametophyte but have protruding necks. Even the egg cell is apparent. It appears more and more likely that *R. gwynne-vaughani* was the gametophyte generation of *Rhynia major*. The axes of the former have vascular strands but this makes the discovery of even more interest because the gametophytes of the extant *Psilotum* may also possess vascular tissue. If the report of Lemoigne (1968) can be confirmed by other investigators, it will rank as one of the important discoveries in modern paleobotany. It will provide documentation for the Homologous Theory of the origin of alternation of generations and will stand as a monument to those botanists who many years ago predicted that sporophytes and gametophytes of primitive vascular plants were probably very similar in general morphology.

That paleobotany is not a "dead" study, but a lively subject, is reflected in the controversy regarding the time factor involved in the origin of land vascular plants. Some paleobotanists consider that vascular land plants first made their appearance in the Silurian and that this was followed by a rapid increase in numbers and diversification of the flora in the Devonian. Banks (1968a, b) and Chaloner (1970) contend that the stratigraphic occurrence of early vascular land plants should be reported very precisely and that it should be clearly shown that the fossils possess xylem and show evidence of other features such as cuticle, stomata, and spores with a triradiate mark. They argue that many errors are likely in a study of microfossils (e.g., spores, isolated cells)—for example, from the "contamination" of older rocks by spores from younger strata (also see Chaloner, 1964). Banks concludes that the first land plants were present in the Silurian and were simple, naked, dichotomously branched plants with terminal sporangia (e.g., *Cooksonia*). He asserts that sufficient time was available for the further elaboration of vascular plants by the end of the Devonian.

A different point of view has been expressed by other paleobotanists (e.g., Axelrod, 1959; Leclercq, 1956) to the effect that vascular plants originated in the Cambrian Period. This theory is based primarily upon the acceptance of reports of microfossils—spores and tracheids—being present in rocks of that horizon. There are also debatable reports of Cambrian macrofossils reputed to be vascular plants. According to the adherents of a Cambrian origin the presumed "rapid evolution" of the Devonian flora is due to a misinterpretation of the record. Rather than comprising successive evolutionary stages, the diversity of forms in the Devonian represents "only the replacement of lowland by more highly adapted upland floras which had evolved in areas of environmental diversity far removed from the lowlands where the record accumulated, and long before the Devonian" (Axelrod, 1959). Axelrod also emphasizes that not enough time had been available during the Silurian and Devonian for the diversification of the flora now known to have been present in the Devonian. The critics of a Cambrian origin of vascular plants attack the concept on the basis of the lack of positive evidence that the microfossils are actually from vascular plants of the Cambrian (Stewart, 1960; Banks, 1968a, b).

Whether a Silurian or Cambrian origin of vascular plants can be demonstrated conclusively remains a very open question. Indeed, plants and animals live today under conditions quite different from those which prevailed in the past. Appreciation of this fact combined with the possibility that minor genetic changes in simple primitive plants produced striking phenotypic changes should not be overlooked. Perhaps changes sufficient to produce a variety of basic forms no longer occur as freely as they did during the Early Devonian.

References

Andrews, H. N., Jr.
1959. Evolutionary trends in early vascular plants. *Cold Spring Harbor Symp. Quant. Biol.* 24:217–234.
1961. *Studies in Paleobotany.* Wiley, New York.

Arnold, C. A.
1947. *An Introduction to Paleobotany.* McGraw-Hill, New York.

Axelrod, D. I.
1959. Evolution of the psilophyte paleoflora. *Evolution* 13:264–275.

Banks, H. P.
1965. Some recent additions to the knowledge of the early land flora. *Phytomorphology* 15:235–245.
1968a. The stratigraphic occurrence of early land plants and its bearing on their origin. *In* Oswald, D. H. (ed.), *Proceedings of the International Symposium on the Devonian System,* Vol. I. Calgary, Canada.
1968b. The early history of land plants. Pp. 73–107 *in* Drake, E. T. (ed.), *Evolution and Environment.* (Symposium volume of papers at Centennial Celebration of Peabody Museum). Yale University Press, New Haven.
1970. Chipping away at early land plants: of people, places, and perturbations. *Plant Sci. Bull.* 16:1–6.

Chaloner, W. G.
1964. An outline of Pre-Cambrian and Pre-Devonian microfossil records: evidence of early land plants from microfossils. *In Abstracts, Tenth International Bot. Congress, Edinburgh.* pp. 16–17.
1970. The rise of the first land plants. *Biol. Rev.* 45:353–377.

Gothan, W., and H. Weyland
1964. *Lehrbuch der Paläobotanik.* Akademie-Verlag, Berlin.

Hopping, C. A.
1956. On a specimen of *"Psilophyton robustius"* Dawson, from the Lower Devonian of Canada. *Proc. Roy. Soc. Edinb.* 66(B): 10–28.

Hueber, F. M.
1964. The Psilophytes and their relationship to the origin of ferns. *Mem. Torrey Bot. Club* 21:5–9.

Jaeger, H.
1962. Das Alter der ältesten bekannten Landpflanzen (*Baragwanathia-flora*) in Australien auf Grund der begleitenden Graptolithen. *Paläeontol. Zeit.* 36:7.

Kidston, R., and W. H. Lang
1917–1921. On Old Red Sandstone plants showing structure, from the Rhynie Chert Bed, Aberdeenshire. Parts I–V. *Trans. Roy. Soc. Edinb.* 51–52.

Kräusel, R., and H. Weyland
1935. Neue Pflanzenfunde im rheinischen Unterdevon. *Palaeontographica* 80(B): 171–190.

Leclercq, S.
1956. Evidence of vascular plants in the Cambrian. *Evolution* 10:109–114.

Lele, K. M., and J. Walton
1961. Contributions to the knowledge of *Zosterophyllum myretonianum* Penhallow from the Lower Old Red Sandstone of Angus. *Trans. Roy. Soc. Edinb.* 64: 469–475.

Lemoigne, Y.
1968. Observation d'archégones portés par des axes du type *Rhynia gwynne-vaughanii* Kidston et Lang. Existence de gamétophytes vascularisés au Dévonien. *Compt. Rend. Acad. Sci.* (Paris), Ser. D, 266:1655–1657.

Lyon, A. G.
1964. Probable fertile region of *Asteroxylon mackiei* K. and L. *Nature* 203:1082–1083.

Mägdefrau, K.

1968. *Paläobiologie der Pflanzen.* G. Fischer Verlag, Jena.

Obrhel, J.

1962. Die Flora der Pridolí-Schichten (Budňany-Stufe) des mittelböhmischen Silurs. *Geologie* 11:83–97.

Pant, D. D.

1962. The gametophyte of the Psilophytales. (Pp. 276–301 *in* Maheshwari, P., B. M. Johri, and I. K. Vasil (eds.) *Proc. Summer School of Botany at Darjeeling* (1960). Ministry of Scientific Research and Cultural Affairs, India.

Schopf, J. W.

1970. Precambrian micro-organisms and evolutionary events prior to the origin of vascular plants. *Biol. Rev.* 45:319–352.

Stewart, W. N.

1960. More about the origin of vascular plants. *Plant Science Bull.* 6:1–5.

Walton, J.

1964. On the morphology of *Zosterophyllum* and some other early Devonian plants. *Phytomorphology* 14:155–160.

8

The Psilopsida

The class Psilopsida is made up of living plants comprising one order, one family, and two genera (*Psilotum,* Fig. 8-1; and *Tmesipteris,* Fig. 8-8). These plants are rather simple in organization and traditionally they have been grouped with the extinct "psilophytes" (Chapter 7). Indeed, there are some similarities between some of the extinct early vascular plants (Chapter 7) and the Psilopsida: (1) the sporophytes are dichotomously branched with an underground rhizome system and an upright system of branches; (2) there are no roots, that is, no water-absorbing structure with a root cap and root hairs; (3) the stems have a relatively simple vascular cylinder; (4) the sporangia are eusporangiate and homo-sporous; and (5) the sporangia may be interpreted as occurring at the ends of shortened axes. If this interpretation is correct, the Psilopsida must resemble in this respect the ancient vascular plants Rhyniopsida and Trimerophytopsida (Chapter 7).

Paleobotany is of very little value in attempting to prove the ancestry of *Psilotum* and *Tmesipteris* because of the lack of fossils of the two genera. If either genus is a relict form of some ancient or primitive vascular plant, we have no direct knowledge of it today. Approximately 400,000,000 years separate the Psilopsida and the earliest vascular plants and naturally, this has led to skepticism regarding the closeness of relationship. The two genera undoubtedly will be shifted from one major taxon to another in the years to come. For example, one botanist (Bierhorst, 1968a, b; 1969) would abolish a separate subdivision or class and remove the two genera, as a family, to the leptosporangiate ferns, primarily upon the basis of general gametophytic features and embryology. However, it should be pointed out that *Psilotum* and *Tmesipteris* are rootless and eusporangiate, whereas the fern (*Stromatopteris*) that is used for comparison has roots and is leptosporangiate.

Psilotin, a specific phenolic substance, has been found in *Psilotum* and *Tmesipteris* (Tse and Towers, 1967). This phenolic has not been found in the Lycopsida, which would support the conclusion that the two genera constitute a natural group.

Psilotales

Psilotaceae – Sporophyte Generation of *Psilotum*

DISTRIBUTION AND ORGANOGRAPHY. *Psilotum,* consisting of at least two species, *Psilotum nudum* (Figure 8-1) and *Psilotum complanatum* (Reed, 1966), is pantropical and subtropical in distribution, reaching north to Florida, Bermuda, and Hawaii. *P. nudum* has been reported in Nigeria (Savory, 1949), Basutoland (Morgan, 1962), Texas and Arizona (Mason, 1968), and for the first time in Europe—in Spain (Allen, 1966). Its presence in some of these new localities may be due to introduction by man, intentionally or inadvertently. Plants

Figure 8-1
Growth habit of *Psilotum nudum* under greenhouse culture.

occur as epiphytes on tree ferns, on coconut palm trunks, or at the base of trees, or they may be terrestrial, growing in soil or among exposed rocks. *P. nudum* (*P. triquetrum*) grows remarkably well under greenhouse conditions and is cultivated in most botanical gardens in the temperate regions. Depending upon its location and environment, the sporophyte of *P. nudum* may be pendent or erect and dwarfed (8 cm high) or as tall as 75–100 cm. The plant body consists of a basal branched rhizome system which is generally hidden beneath the soil or humus, and a slender, upright, green aerial portion which is freely and dichotomously branched

and which bears small appendages and sporangia (Fig. 8-2, A, B). The branched rhizome system, which bears numerous rhizoids, grows by means of apical meristems located at the tips of ultimate branches. According to the studies of Bierhorst (1954b), the degree of branching of the rhizome is directly related to the effects of obstacles which the apical meristem encounters in its growth through the soil. No structure is present which could be interpreted anatomically as a root, although the underground rhizome system anchors the plant and serves as an absorptive surface. A mycorrhizal intracellular fungus, gaining

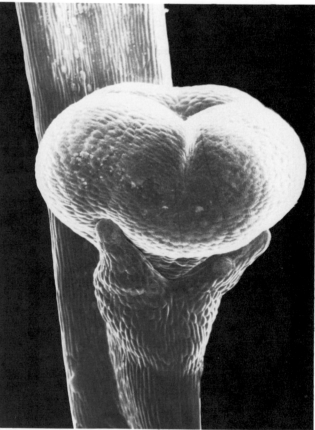

A

B

Figure 8-2

Organography of *Psilotum nudum*. **A**, portions of two branches showing three-lobed synangia and their associated forked appendages; **B**, scanning electron micrograph of a synangium and the forked foliar appendage.

entrance through rhizoids, is present in cells of the outer cortex (Verdoorn, 1938; Bierhorst, 1954b). This fungus may be related intimately to the physiology of the plant. Any one of the rhizome tips may turn upward and by apical growth produce an aerial branch system. The basal part of the shoot may be cylindrical with longitudinal ribs, whereas the more distal aerial stems are provided with three longitudinal ridges.

STEM ANATOMY. The apical meristem of rhizomes and aerial branches is reported to have a single, large apical cell (Bierhorst, 1954b; Marsden and Wetmore, 1954; Roth,

1963; Siegert, 1964) which divides repeatedly, giving rise to additional meristematic cells that differentiate eventually into tissues constituting the three primary tissue systems.

The lower portions of the aerial system as well as the more distal regions are covered by an epidermis, as shown in Fig. 8-3, in which the outer tangential cell walls are heavily cutinized and covered by a definite cuticle. Stomata are present mainly in areas between the longitudinal ribs (Zimmermann, 1927) and are without special subsidiary cells much like the haplocheilic type in certain gymnosperms (Pant and Mehra,

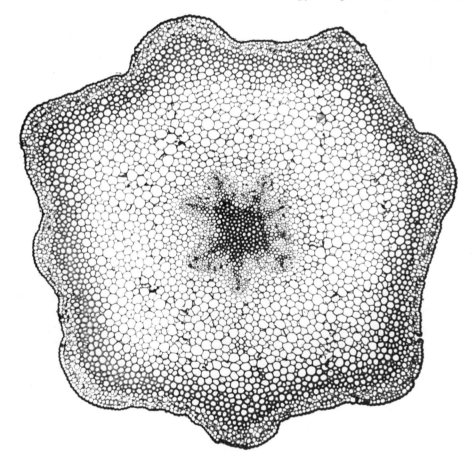

Figure 8-3
Transection (near base of aerial shoot) of stem of *Psilotum nudum.* Note differences in thickness of cell walls in outer and inner cortex, and xylem lobes and sclerenchymatous pith.

1963). Internal to the epidermis there is a rather broad cortex which can be resolved into three regions (Fig. 8-3). The outer portion, directly beneath the epidermis, consists of elongated, lobed parenchyma cells with intercellular spaces between the vertical rows. Starch grains are present in great numbers. Internal to this zone there is a band of vertically elongated and thick-walled cells, with small intercellular air spaces and few or no starch grains. In the lower portions of the aerial stems the walls of these cells apparently become lignified. In progressing from this zone to the vascular cylinder, the cell walls become thinner and thinner and less lignified with an increase in the number of starch grains per cell.

The boundary between the fundamental tissue (cortex) system and the vascular cylinder is marked by the endodermis (see Fig. 8-3; also Fig. 8-4, A), a very distinctive layer in that a conspicuous Casparian strip is present in the radial and end walls of the vertically elongated cells. Occupying the center of the rhizome in *P. nudum* is a slender cylinder of primary xylem which may be greatly reduced or even interrupted in small axes (Bierhorst, 1954b) and is a ridged or fluted cylinder in the aerial branches. Near the transition region from rhizome to aerial stem this cylinder may have as many as 10 lobes (Bower, 1935; Pitot, 1950), whereas a smaller number are present in the more distal parts of the aerial branch system (Figs. 8-3; 8-4, A). At levels where several xylem lobes are present, the center of the stem in *P. nudum* is generally occupied by elongate sclerenchymatous cells. Partially disorganized protoxylem tracheids, with helical or annular thickenings, occupy the extreme tips of the xylem lobes in aerial branches while the remainder is composed of metaxylem tracheids with predominantly scalariform or circular bordered pits. In summary, the rhizome is protostelic, becoming an exarch siphonostele throughout

a considerable portion of the aerial branch system; the upper branches may, however, be strictly actinostelic.

Internal to the endodermis is a cylinder of parenchyma-like cells, generally one layer thick, which is designated as the pericycle. The phloem is internal to the pericycle and occupies the regions between the lobes or flanges of the xylem. The smaller, somewhat angular cells are the sieve cells (Fig. 8-4, A). These cells are elongate, often lignified at the corners, and possess many spherical bodies as part of their cell contents (Ford, 1904; Stiles, 1910; Esau et al., 1953). Sieve areas occur on the oblique to nearly transverse end walls and on the lateral walls (Lamoureux, 1961). However, most of the tissue in the bays between the xylem arms is composed of elongate parenchyma cells.

Chloroplasts of the stem have been studied with the electron microscope and shown to be about 13 microns long and 3.6 microns wide. There are lamellated grana, similar to those in ferns, and the plastids contain relatively large osmiophilic bodies (Sun, 1961).

"LEAF" ANATOMY. As mentioned earlier, the "foliar" appendages in *Psilotum* are small scale-like structures which are helically arranged on the aerial stem (Fig. 8-2, A). Internally the appendage consists of photosynthetic parenchyma cells which are continuous, lower down, with similar tissue of the stem. There is no vascular bundle in the appendage of *P. nudum,* although in *P. complanatum* a "leaf" trace ends at the base of the foliar structure. Grouped generally on the upper portions of the stems are bilobed appendages, each of which is associated with a three-lobed synangium. (By definition, a synangium is the fusion product of two or more sporangia). A morphological interpretation of the foliar appendages is presented in Chapter 3.

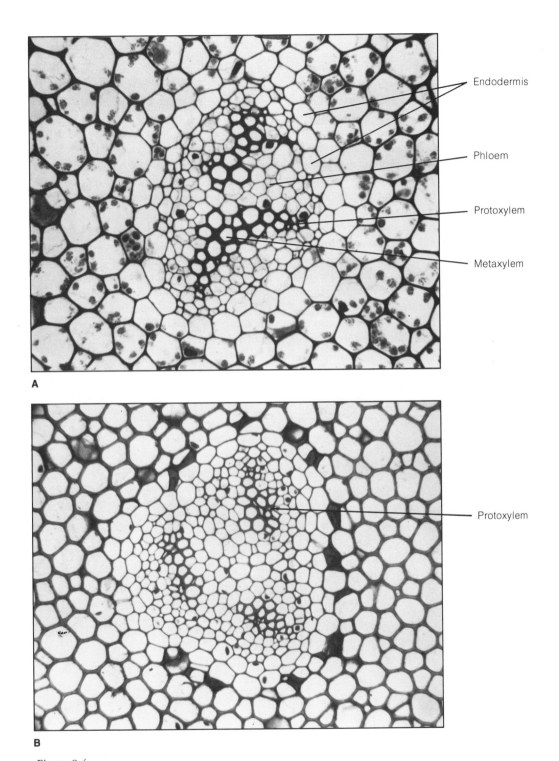

Endodermis

Phloem

Protoxylem

Metaxylem

A

Protoxylem

B

Figure 8-4
Stem anatomy of the Psilotaceae. **A**, transection of the vascular cylinder and adjacent cortex of a small aerial stem of *Psilotum nudum*; **B**, transection near the base of an aerial branch of *Tmesipteris* sp.

STRUCTURE AND DEVELOPMENT OF THE
SYNANGIUM. The morphological interpretations of the spore-producing structure
in the Psilotales are varied and controversial; see, for example, Solms-Laubach, 1884;
Bower, 1894, 1935; Eames, 1936; Smith,
1955; Campbell, 1940; Zimmermann, 1959;
Bierhorst, 1956; Roth, 1963. To comment
at length on all of the various theories is
beyond the scope of this book. Thus, only a
descriptive account, based upon the most
reliable sources, will be presented together
with selected interpretive theories.

The spore-producing structure of *Psilotum*
has been described as a trilocular sporangium and as a trisporangiate structure, i.e., a
synangium. The results of recent investigations support the latter interpretation,
which will, accordingly, be adopted in this
textbook.

The mature synangium of *Psilotum* is generally a three-lobed structure (Fig. 8-2, A, B;
and Fig. 8-5, B), 1–2 mm wide, located at
the tip of a very short axis, and closely associated with a forked, foliar appendage. Each
lobe of the synangium, corresponding to a
sporangium, exhibits loculicidal dehiscence
at maturity.

In early stages of development it is difficult to determine whether a lateral primordium will become simply a single vegetative
appendage or whether it will become eventually a "fertile" branch (Bower, 1894, 1935).
For *Psilotum* it was reported by Bower that
the lateral appendage is differentiated earlier
than its associated sporangial structure and

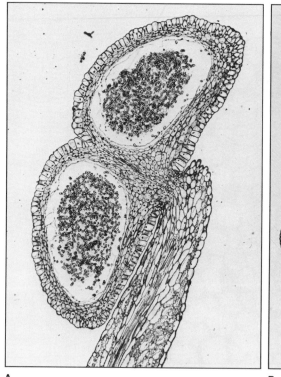

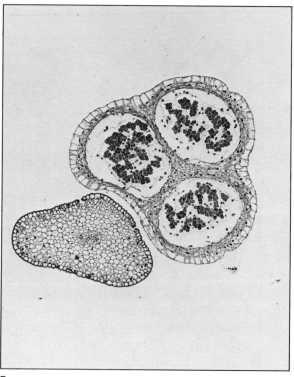

A   B

Figure 8-5
Synangia of the Psilotaceae. A, longisection, bilobed synangium of *Tmesipteris* sp.; B, transection, trilobed
synangium of *Psilotum nudum;* note that the sporocytes are surrounded by an irregular fluid-like tapetum.

that the latter arises on the adaxial side of the original appendage. Bierhorst (1956) concluded, however, that the original appendage is, in fact, the spore-producing apparatus and that the subjacent leaf-like structure is a lateral outgrowth on the "fertile axis."

Lastly, consideration should be given to the concept that perhaps each unit of the fertile axis (foliar and sporangial) represents, phylogenetically, a condensation of a more elaborate branch system (see Chapter 3; also Bierhorst, 1956). That the fertile axis may represent a condensed branch system has been strengthened recently by experimental work. *P. nudum* maintained on a long daily photoperiod (16 hours at 200–400 foot candles) produced aberrant fertile types in which the fertile axes proliferated into

definite branch-like structures. The investigator postulated that the fertile axis may have been derived, in the evolutionary sense, from a dichotomously branched ancestral form (Rouffa, 1967). Ancestral forms which might provide support for this conclusion are found in the extinct groups Rhyniopsida and Trimerophytopsida (Chapter 7).

Even more recently a variety of *Psilotum* from Japan, historically known as *Bunryu-zan* (personal communication by A. S. Rouffa, correcting a previously published name) or "green dragon's horns," has been described by Rouffa (1971). In this plant there are no sterile foliar appendages and the synangia are borne at the tips of branches (Fig. 8-6). Knowledge of this variety strengthens the concept that a synangium of

A   B

Figure 8-6
Two views of an appendageless form of *Psilotum* from Japan known as *Bunryu-zan:* note that the terminal synangia generally consist of more than three fused sporangia. [Courtesy Dr. A. S. Rouffa.]

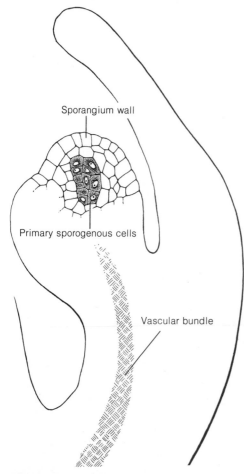

Sporangium wall

Primary sporogenous cells

Vascular bundle

Figure 8-7
A young developing synangium and its associated appendage in *Psilotum nudum*. The details of only one lobe (sporangium) are shown. The vegetative stem is to the left.

duced. Derivatives of the primary sporogenous cells divide in various planes to form the sporogenous tissue (Fig. 8-7). The entire sporogenous mass does not become converted into sporocytes; only irregular groups of sporocytes function, and the rest disintegrate and undoubtedly serve as a nourishing fluid in which the functional sporocytes and spores develop (Fig. 8-5, B). As is true of other plants with eusporangiate development, numerous spores are formed. Individual spores are bilaterally symmetrical with a reticulate wall pattern.

Unlike the foliar appendages, the fertile axis is vasculated (Fig. 8-7). A vascular bundle extends into the synangium and may become divided into three parts, corresponding to the three sporangia (Bierhorst, 1971).

Psilotaceae — Sporophyte Generation of *Tmesipteris*

DISTRIBUTION AND ORGANOGRAPHY. We have devoted considerable space to a detailed description of the sporophyte of *Psilotum* because we believe that there is more material of this genus generally available for teaching purposes in most institutions than there is of *Tmesipteris*. However, a descriptive treatment of the family would be incomplete without consideration of the interesting genus *Tmesipteris*.

In contrast with *Psilotum*, which is widespread in its distribution, *Tmesipteris* is confined to Australia, New Caledonia, New Zealand, and other islands of the South Pacific Ocean. *Tmesipteris* (Fig. 8-8) generally grows as a pendulous epiphyte, 5–20 cm long, on the trunks of tree ferns or other trees, but often it may be found on mounds of humus. For many years only two species of *Tmesipteris* were recognized. *Tmesipteris tannensis* was considered to be the more widespread species. More and more evidence has been accumulated which indicates

the more commonly known clones of *Psilotum* occurs at the terminus of a short branch, rather than being directly associated with a foliar appendage.

The development of each sporangium of the synangium in *P. nudum* is typically eusporangiate, i.e., separate groups of surface initials divide periclinally, setting apart primary-wall initials and primary sporogenous cells. By repeated periclinal and anticlinal divisions of the primary-wall initials, a sporangial wall of four or five layers is pro-

Figure 8-8
The pendant aerial branches of *Tmesipteris* sp. Synangia can be seen near the tip of the middle branch. In some of the appendages, the single unbranched midvein is evident.

that there are probably several forms which are sufficiently different to be considered as species. There are five recognizable populations in southeast Australia and at least another one in New Zealand (Barber, 1954, 1957). These differences are based on size and morphology of leaves, length and shape of sporangia, and spore length. The six species appear to be reproductively isolated from one another.

Organization of the *Tmesipteris* sporophyte is very similar to that of *Psilotum*, with a branching rhizome system and an aerial shoot portion. However, significant morphological differences do exist between the genera. Each aerial shoot of *Tmesipteris* may exhibit only one dichotomy. The "foliar" appendages are scale-like at the base, gradually increasing in size toward the tip. The majority of appendages are larger than those of *Psilotum*, and are flat and broadly lanceolate with a mucronate tip. The larger leaves are supplied with a single, unbranched vascular bundle. The bases of the leaves are strongly decurrent, and the distinction between stem and leaf is difficult to determine, particularly near the shoot tip, because a foliar appendage often terminates the axis. Roots are absent — a feature, it will be recalled, characteristic of ancient vascular plants (Chapter 7).

STEM ANATOMY. The rhizome of *Tmesipteris* is protostelic, gradually becoming siphonostelic in the aerial system (Fig. 8-4, B) with five or more protoxylem poles surrounded by metaxylem (Sykes, 1908). The primary xylem of aerial branches is mesarch in development, in contrast with *Psilotum* in which the primary xylem is exarch. More distally from the base of the stem of *Tmesipteris* there may be a smaller number of xylem strands which display an irregular arrangement because of the departure of leaf traces from the vascular cylinder. Protoxylem and metaxylem are composed of scalariform tracheids. The center of the stem consists of parenchyma-like cells which may have relatively thick walls. External to the strands of xylem is a cylinder of phloem in which the sieve elements are made evident by the presence of spherical inclusions and heavily lignified cell walls. These cells are reported as having tapering ends with numerous sieve areas on their lateral walls (Sykes, 1908). A characteristic endodermis is present in the rhizome, but in the aerial portions no such definable layer is evident. Between the phloem cylinder and an inner layer of cortex, the cells of which contain brown tanniferous or phenolic substances, is a zone two or three cells wide which physiologically may represent endodermis and pericycle (Fig. 8-4, B).

The cortex is composed of a compact tissue of parenchyma cells with evenly thickened, often lignified, cell walls. There may be small groups of photosynthetic parenchyma directly beneath the epidermis; the outer tangential walls of the latter are cutinized and covered by a definite cuticle.

LEAF ANATOMY. As mentioned earlier, the foliar appendages of *Tmesipteris* are larger than those of *Psilotum* and, moreover, exhibit a more diversified anatomy. The flattened appendage is covered by a uniseriate epidermis with cutinized outer tangential cell walls in which some of the thickening is laid down in the form of striations; stomata are conspicuously developed in both epidermal layers. The internal ground tissue is uniformly arranged, consisting of lobed parenchyma cells. The single concentric vascular bundle, located centrally, is composed of several protoxylem elements surrounded incompletely by metaxylem, which, in turn, is enclosed by phloem. As in the aerial stem, no definable endodermis is present, although a compact zone of parenchyma cells occupies the expected position of such a layer.

SYNANGIUM. As in *Psilotum*, the study of the spore-producing structure in *Tmesipteris* is beset with difficulties of interpreting the mature structure as well as the ontogenetic stages that are the bases for establishing phylogenetic sequences.

The mature two-lobed synangium of *Tmesipteris* is interpreted as occupying the terminus of a short lateral branch, although the axis tip is recurved and the synangium appears to be adaxial (Fig. 8-5, A). The two foliar appendages that are attached to the fertile axis just below the synangium extend some distance beyond the synangium. The single vascular bundle of the axis divides into three strands at the level of the foliar appendages. The lateral bundles traverse the appendages, whereas the median strand continues up the axis, and in some species ends in a trichotomy. The two lateral traces traverse the septum between the sporangia, and the central strand ends medially in the septum (Fig. 8-5, A; Sykes, 1908).

According to Bierhorst (1956), the development of the fertile axis is similar to that in *Psilotum:* appearance of a primordium near the vegetative shoot tip, apical growth of the primordium, appearance of separate groups of sporangial initials, and the formation of the two foliar appendages from a common outgrowth on the original fertile axis. Subsequent development also is similar to that in *Psilotum*, with the ultimate development of a synangium with two thick-walled sporangia devoid of a well-defined tapetum and containing a large number of spores. Dehiscence of each sporangium is effected through the formation of a longitudinal cleft.

Before this section on the sporophyte generation is closed, the studies on culturing tips of the aerial branches of *Psilotum* should be mentioned. If the tips of mature aerial branches are grown on agar substrate to which are added suitable nutrients, the axes become more like the rhizome both in ex-

Figure 8-9
Plant of *Psilotum nudum* grown from excised shoot tip that was placed on nutrient agar. [Courtesy of Dorothy Brandon.]

ternal morphology and anatomy (Marsden and Wetmore, 1954). This is interesting because it poses the question of causal relationships. It would seem that the absence of the endophytic fungus in the aerial stem is not sufficient evidence for the differential behavior of normal aerial and underground apices. The observed change may be due to the nutritional substrate on which the apex was grown, although, in some instances the aerial form may be retained in culture (Fig. 8-9).

Gametophyte Generation

The nature of the gametophyte generation long remained a serious gap in our knowledge of vascular plants; only in the twentieth century has it been discovered and described (*Tmesipteris*, Lawson, 1917a; *Psilotum*, Darnell-Smith, 1917, and Bierhorst, 1953).

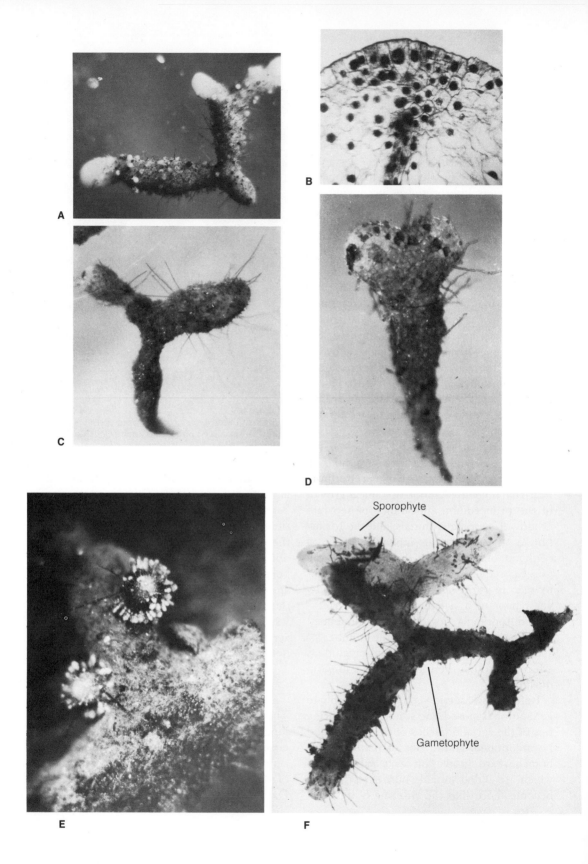

A

B

C

D

E

Sporophyte

Gametophyte

F

The gametophyte of both genera is a small structure, ranging from .5 mm to 2.00 mm in diameter, and is several millimeters in length, growing on the trunks of tree ferns or in the crevices of rocks; sometimes it is subterranean in its habitat. Original descriptions were based upon collections made at the natural habitat, and attempts to culture the prothallia were never successful. However, more recently (Moseley and Zimmerly, 1949; Zimmerly and Banks, 1950; Bierhorst, 1953) sexually mature gametophytes of *P. nudum* were discovered growing in pots at the base of ferns which had for many years been in a conservatory. Mature individual plants resemble pieces of the sporophyte rhizome in that they are brown, radially symmetrical, often dichotomously branched but frequently irregularly branched, and invested with rhizoids (Fig. 8-10, A). As is true of underground sporophytic axes, the branching of gametophytic plants is correlated to a considerable degree with apical injury, which results in the establishment of lateral axes (Bierhorst, 1953).

ANATOMY AND CYTOLOGY. Growth of a gametophyte is initiated by apical cells located at the tips of the ultimate dichotomies (Fig. 8-10, B). Branching is restricted, and configuration and growth of the gametophyte seemingly are determined in a large measure by surrounding objects. The gametophyte is devoid of chlorophyll, living a saprophytic existence, and is presumably aided by the presence of an endophytic fungus, which gains entrance through the rhizoids and invades nearly all cells of the plant (this can be seen in Fig. 8-12, A, B, D).

Cells of the gametophyte are parenchymatous; however, there are instances in *Psilotum* where annular and scalariform or scalariform-reticulate tracheids, surrounded by phloem and an endodermis, have been shown to occupy the center of the gametophyte (Holloway, 1938, 1939; Bierhorst, 1953). Such gametophytes are clearly the gamete-producing generation, because antheridia and archegonia are present on their surfaces. The presence of vascular tissue in the gametophytic plant is somewhat unusual among vascular plants (see Chapter 2). The external similarity of the gametophyte and the sporophytic rhizome, coupled with the presence of vascular tissue in the gametophyte, provides additional evidence for the Homologous Theory as it relates to the origin of the alternation of generations. Furthermore, confirmation of the report that the vascularized axes of the extinct *Rhynia gwynne-vaughani* actually may have been the gametophytes of *Rhynia major* (see Chapter 7) would constitute nearly conclusive evidence for the Homologous Theory. (Also see Chapter 2 for additional discussions on the origin of alternation of generations.)

Chromosome numbers are relatively high in *Psilotum* and *Tmesipteris*. A wild "diploid" sporophyte of *Psilotum* has been reported to occur in Ceylon (Manton, 1950), and to have a chromosome number $n = 52$ to 54. Bierhorst (1968b) has described similar races from Fiji and New Caledonia. Populations from India and elsewhere are tetraploids with $n = 104$ or 208. The form commonly grown in greenhouses for teaching purposes is tetraploid and it may have vascular tissue in the gametophyte. There is no

Figure 8-10 (*facing page*)
Gametophytes and rhizomes of *Psilotum nudum*. **A**, gametophyte showing meristematic apices (white) and prominent globular antheridia; **B**, longisection, apex of gametophyte (note large apical cell); **C**, young sporophyte of gemmaceous origin (gemmae are vegetative propagules formed both on rhizomes and gametophytes); **D**, Young gametophyte of gemmaceous origin (dark areas are presumably archegonia); **E**, clusters of gemmae on a sporophytic rhizome; **F**, gametophyte with attached sporophyte. [A–E courtesy Dr. D. W. Bierhorst.]

vascular tissue in gametophytes of diploid races. That tetraploidy has been operative in the genus is supported by the presence of quadrivalents during meiosis (Ninan, 1956). A similar type of polyploid series appears to exist for *Tmesipteris* (Barber, 1957) and is based upon the same basic theoretical number, probably $x = 13$.

GAMETANGIA. Sex organs (antheridia and archegonia) are scattered over the surface of the gametophyte and are intermingled (Fig. 8-10, A). Young sex organs begin development very close to the apices of the gametophyte; however, young gametangia may be found also among the more mature ones.

The first indication of antheridial development is the presence of a periclinal division in a single surface cell, which sets aside an outer jacket initial and an inner primary spermatogenous cell (Holloway, 1918; Lawson, 1917b; Holloway, 1939). By anticlinal divisions a single-layered jacket of several cells is produced enclosing a developing spermatogenous mass in which cell divisions occur in many planes. Ultimately the antheridium projects above the surface (see Fig. 8-12, A) much as in leptosporangiate ferns, although antheridial ontogeny in *Psilotum* is similar to that of *Lycopodium, Equisetum,* and eusporangiate ferns. Each spermatocyte or sperm-mother cell eventually becomes a

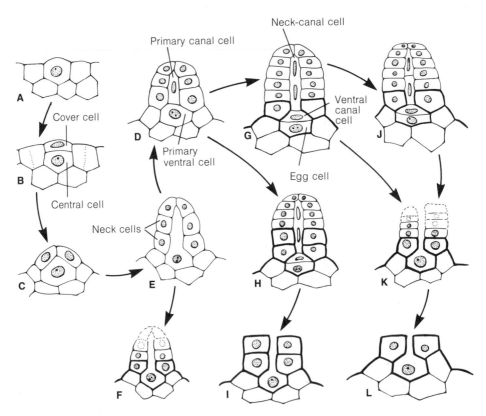

Figure 8-11
Archegonial development in *Psilotum nudum.* Although early development follows a uniform scheme (A–E), there is considerable variation in subsequent steps (F–L). A mature archegonium has only one or two tiers of neck cells, the others having been sloughed off during development. Consult text for details. [Redrawn from Bierhorst, *Amer. Jour. Bot.* 41:274, 1954.]

spirally coiled, multiflagellate sperm and escapes through an opercular cell on the side of the antheridium (Bierhorst, 1954a).

The archegonium likewise is initiated from a single superficial cell, as can be seen in Fig. 8-11, A. An initial periclinal division sets aside an outer cover cell and an inner central cell (Fig. 8-11, B). The cover cell undergoes two successive anticlinal divisions, and by further divisions of these cells and their derivatives, four rows of neck cells, consisting of from four to six tiers, are produced. The central cell divides periclinally to form the primary canal cell and the primary ventral cell (Fig. 8-11, D). The primary canal cell may divide to form two neck canal cells, although in most instances walls separating the two nuclei have not been observed (Fig. 8-11, G, J). There are also some indications that the central cell may function directly as the egg (Bierhorst, 1954a). The archegonial necks are straight, and the venter is embedded in the gametophyte. As the archegonia approach maturity the cell walls between certain tiers of neck cells become cutinized and the upper part of the neck is sloughed off (Fig. 8-11, F, K). In *Tmesipteris* the neck may break off flush with the surface of the gametophyte, whereas in *Psilotum* a smaller number of tiers may be lost. With the sloughing off of the upper portion of the neck and the disintegration of the axial row a passageway is created for the entrance of the motile sperms (Fig. 8-11, I, L). Bierhorst (1954a) reports that fertilization perhaps precedes the actual "decapitation" of the archegonium. Fertilization is accomplished by the union of sperm and egg.

The Embryo

In the method of early segmentation of the zygote, and in the structure and subsequent development of the embryo, there is close similarity between *Psilotum* and *Tmesipteris* (Holloway, 1921, 1939). The first division of the fertilized egg results in a wall being placed at right angles to the long axis of the archegonium. The cell directed toward the neck of the archegonium is the epibasal cell, the lower is the hypobasal cell. Since members of the Psilotales are illustrative of exoscopic polarity, the epibasal cell may be designated the apex, and the hypobasal cell the base (Chapter 6). The epibasal cell will ultimately give rise to the sporophyte shoot system (aerial and underground), while the hypobasal cell (base) will produce the foot — a structure that anchors the young sporophyte securely to the gametophyte.

By repeated cell divisions the shoot portion increases in size, and an apical cell is established at the distal end. Further vertical growth of the shoot is due in a large measure to the activity of this apical cell; frequently in *Tmesipteris* two apical cells are present on the flanks of the shoot, resulting in two precociously formed horizontal branches. Concomitant with embryonic development the gametophyte forms a calyptra-like outgrowth through which the young sporophyte eventually emerges (Fig. 8-12, D). While the shoot portion is assuming form, the foot enlarges by repeated cell divisions, sending haustorial outgrowths into the gametophytic tissue. The foot, by virtue of its position and organization, is well suited for the functions of anchorage and absorption until the shoot becomes physiologically independent.

To recapitulate, the young sporophyte consists of an axis (rhizome), with vascular tissue, which may be uniaxial or precociously branched, and an enlarged bulb-like foot embedded in the gametophyte. Ultimately the shoot becomes detached from the foot and the gametophyte through a separation layer in the vicinity of the original boundary between shoot and foot. Throughout all of the differentiation process this original boundary is clearly discernible. The rhizome continues to elongate and branch, and eventually ultimate dichotomies emerge above the soil or humus and develop into aerial branches.

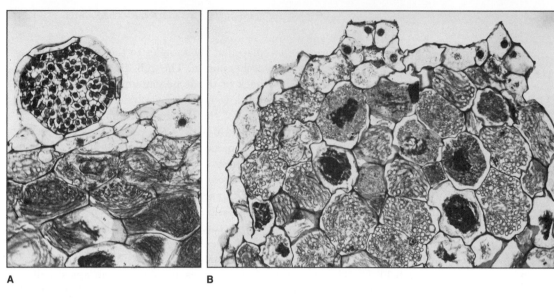

A   B

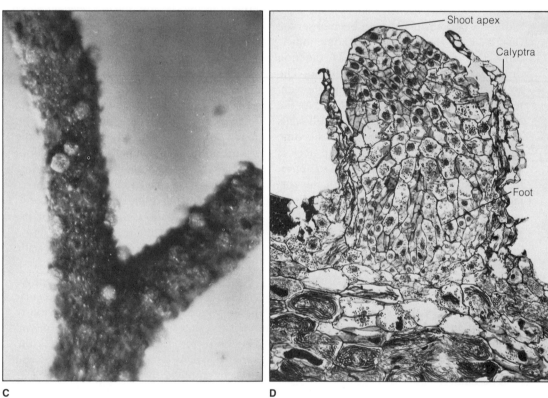

C   D

Figure 8-12
Gametangia and embryo of *Psilotum nudum*. **A,** section of nearly mature antheridium, showing jacket layer and spermatocytes; **B,** transection of gametophyte showing two mature archegonia (consult text and Fig. 8-11 for details of development; note that most cells of the gametophyte contain hyphae of an endophytic fungus); **C,** portion of a gametophyte with emergent antheridia; **D,** developing embryo attached to gametophyte by enlarged foot. [From slides prepared by Dr. D. W. Bierhorst.]

Summary and Conclusions

The greater part of the chapter has been devoted to the description and comparison of *Psilotum* and *Tmesipteris*, the only two genera in the Psilopsida. The sporophytes of these genera agree in certain respects—for example, in the absence of roots and in the dichotomous pattern of branching. But there are marked differences between them in other organographic and anatomical characteristics. The aerial vegetative stems of *Psilotum* bear small and inconspicuous scale-like appendages devoid of veins, whereas in *Tmesipteris* there are relatively large leaf-like appendages, with well-defined midveins, in the upper portions of the aerial shoots. The stele near the base of aerial stems of *Psilotum* consists of a lobed cylinder of exarch xylem (enclosing a central mass of sclerenchyma) and an external cylinder of phloem tissue. In contrast, the aerial stem of *Tmesipteris* develops a siphonostele consisting of a central pith surrounded by a dissected cylinder of mesarch xylem which, in turn, is enclosed by a cylinder of phloem.

Whether the spore-producing structure in either genus is (1) a tri- or bilocular sporangium or (2) is a synangium composed of fused sporangia is a debatable question, although the results of recent investigations favor the latter interpretation. According to one current idea, the synangium is terminal on a short "fertile" axis, a view supported by the presence of vascular tissue in the axis and within the septum between the fused sporangia. From an *evolutionary* standpoint the synangia in both genera probably have been derived from the fusion of once separate, terminal sporangia. Also, the morphological significance of the paired (*Tmesipteris*) or forked (*Psilotum*) appendages associated with a synangium has been the subject of various interpretations.

The mature gametophytes of *Psilotum* and *Tmesipteris* are very small, dichotomously branched, cylindrical structures which superficially resemble fragments of the rhizome of the sporophyte. Gametophytes are either wholly subterranean or they are buried in the crevices of rocks or in the mat of roots on the stems of tree ferns. In very recent years, intensive developmental and cytological studies made on the gametophytes of *Psilotum* have revealed and clarified many points of unusual interest. Chlorophyll is not found in the cells of the gametophyte, and this structure lives as a saprophyte apparently assisted by the presence of an endophytic fungus which gains entrance through the rhizoids and which is found within most of the interior cells. Unlike the nonvasculated gametophytes of most other vascular plants, certain gametophytes in *Psilotum* develop a central strand of xylem and sometimes phloem, enclosed by an endodermis. The significance of vascular tissue in the gametophyte generation of *Psilotum* with reference to the Homologous Theory of Alternation of Generations has been briefly discussed.

Sexual reproduction in the Psilotales is achieved by the union between the egg and a multiflagellate sperm. The gametophyte is bisexual, and the antheridia and archegonia are scattered and intermingled over its surface. After fertilization the zygote divides by a transverse wall and gives rise to an embryo with exoscopic polarity. The cells directed toward the neck of the archegonium initiate the shoot system, and the basal cells of the embryo produce a conspicuous foot which sends out haustoria into the adjacent tissues of the gametophyte. Ultimately the rootless young sporophyte becomes detached from the gametophyte and develops independently into a new plant.

References

Allen, B. M.
  1966. *Psilotum nudum* in Europe. *Taxon* 15: 82–83.
Barber, H. N.
  1954. New species of *Tmesipteris*. *Victorian Natur.* 71:97–99.
  1957. Polyploidy in the Psilotales. *Proc. Linn. Soc. New South Wales* 82:201–208.
Bierhorst, D. W.
  1953. Structure and development of the gametophyte of *Psilotum nudum*. *Amer. Jour. Bot.* 40:649–658.
  1954a. The gametangia and embryo of *Psilotum nudum*. *Amer. Jour. Bot.* 41:274–281.
  1954b. The subterranean sporophytic axes of *Psilotum nudum*. *Amer. Jour. Bot.* 41: 732–739.
  1956. Observations on the aerial appendages in the Psilotaceae. *Phytomorphology* 6: 176–184.
  1968a. Observations on *Schizaea* and *Actinostachys* spp., including *A. oligostachys,* sp. nov. *Amer. Jour. Bot.* 55:87–108.
  1968b. On the Stromatopteridaceae (fam. nov.) and the Psilotaceae. *Phytomorphology* 18:232–268.
  1969. On *Stromatopteris* and its ill-defined organs. *Amer. Jour. Bot.* 56:160–174.
  1971. *Morphology of Vascular Plants.* Macmillan, New York.
Bower, F. O.
  1894. Studies in the morphology of spore-producing members: Equisetinae and Lycopodineae. *Phil. Trans. Roy. Soc.* London 185B:473–572.
  1935. *Primitive Land Plants.* Macmillan, London.
Campbell, D. H.
  1940. *The Evolution of the Land Plants (Embryophyta).* Stanford University Press, Stanford, California.
Darnell-Smith, G. P.
  1917. The gametophyte of *Psilotum. Trans. Roy. Soc. Edinb.* 52:79–91.

Eames, A. J.
  1936. *Morphology of Vascular Plants. Lower Groups.* McGraw-Hill, New York.
Esau, K., V. I. Cheadle, and E. M. Gifford, Jr.
  1953. Comparative structure and possible trends of specialization of the phloem. *Amer. Jour. Bot.* 40:9–19.
Ford, S. O.
  1904. The anatomy of *Psilotum triquetrum. Ann. Bot.* 18:589–605.
Holloway, J. E.
  1918. The prothallus and young plant of *Tmesipteris. Trans. Proc. N. Z. Inst.* 50:1–44.
  1921. Further notes on the prothallus, embryo, and young sporophyte of *Tmesipteris. Trans. Proc. N. Z. Inst.* 53:386–422.
  1938. The embryo and gametophyte of *Psilotum triquetrum*. A preliminary note. *Ann. Bot.* n.s. 2:807–809.
  1939. The gametophyte, embryo, and young rhizome of *Psilotum triquetrum* Swartz. *Ann. Bot.* n.s. 3:313–336.
Lamoureux, C. H.
  1961. Comparative studies on phloem of vascular cryptogams. Ph.D. dissertation, University of California, Davis, Calif.
Lawson, A. A.
  1917a. The prothallus of *Tmesipteris tannensis. Trans. Roy. Soc. Edinb.* 51:785–794.
  1917b. The gametophyte generation of the Psilotaceae. *Trans. Roy. Soc. Edinb.* 52: 93–113.
Manton, I.
  1950. *Problems of Cytology and Evolution in the Pteridophyta.* Cambridge University Press, London.
Marsden, M. P. F., and R. H. Wetmore
  1954. In vitro culture of the shoot tips of *Psilotum nudum. Amer. Jour. Bot.* 41: 640–645.
Mason, C. T., Jr.
  1968. A new family of vascular plants (Psilotaceae) for Arizona. *Madroño.* 19:224.
Morgan, D.
  1962. *Psilotum triquetrum,* Swartz in Basutoland. *Nature* 195:1121.

Moseley, M. F., Jr., and B. C. Zimmerly
1949. *Psilotum* gametophytes matured under greenhouse conditions from self-sown spores. *Science* 110:482.

Ninan, C. A.
1956. Cytology of *Psilotum nudum* (L.) Beauv. (*P. triquetrum* Sw.). *Cellule* 57:307–318.

Pant, D. D., and B. Mehra
1963. Development of stomata in *Psilotum nudum* (L.) Beauv. *Curr. Sci.* 32:420–422.

Pitot, A.
1950. Sur l'anatomie de *Psilotum triquetrum* Sw. *Inst. Franc. d'Afrique Noire* (Paris) 12:315–334.

Reed, C. F.
1966. Index Psilotales. *Sociedade Broteriana, Boletin* 40:71–96.

Roth, I.
1963. Histogenese der Luftsprosse und Bildung der "dichotomen" Verzweigungen von *Psilotum nudum. Advan. Frontiers Plant Sci.* 7:157–180.

Rouffa, A. S.
1967. Induced *Psilotum* fertile-appendage aberrations. Morphogenetic and evolutionary implications. *Can. Jour. Bot.* 45:855–861.
1971. An appendageless *Psilotum.* Introduction to aerial shoot morphology. *Amer. Fern Jour.* 61:75–86.

Savory, H. J.
1949. A botanical discovery, *Psilotum. Nigeria* 30:317.

Siegert, A.
1964. Morphologische, entwicklungsgeschichtliche und systematische Studien an *Psilotum triquetrum* Sw. I. Allgemeiner Teil. Erstarkung und primäres Dickenwachstum der Sprosse. *Beitr. Biol. Pflanzen* 40:121–157.

Smith, G. M.
1955. *Cryptogamic Botany. Vol. II. Bryophytes and Pteridophytes*, Ed. 2. McGraw-Hill, New York.

Solms-Laubach, H. Grafen zu.
1884. Der Aufbau des Stockes von *Psilotum triquetrum* und dessen Entwicklung aus der Brutknospe. *Ann. Jard. Bot. Buitenzorg.* 4:139–194.

Stiles, W.
1910. The structure of the aerial shoots of *Psilotum flaccidum* Wall. *Ann. Bot.* 24:373–387.

Sun. C. N.
1961. Submicroscopic structure and development of the chloroplasts of *Psilotum triquetrum. Amer. Jour. Bot.* 48:311–315.

Sykes, M. G.
1908. The anatomy and morphology of *Tmesipteris. Ann. Bot.* 22:63–89.

Tse, A., and G. H. N. Towers
1967. The occurrence of psilotin in *Tmesipteris. Phytochemistry* 6:149.

Verdoorn, F.
1938. *Manual of Pteridology.* Martinus Nijhoff, The Hague.

Zimmerly, B. C., and H. P. Banks
1950. On gametophytes of *Psilotum.* (Abstract) *Amer. Jour. Bot.* 37:668.

Zimmermann, W.
1927. Die Spaltöffnungen der Psilophyta und Psilotales. *Zeit. Bot.* 19:129–170.
1959. *Die Phylogenie der Pflanzen,* Ed. 2. G. Fischer, Stuttgart.

9

The Lycopsida

The Lycopsida is a well-defined group of vascular plants consisting of fossil and living representatives. The known history of this group extends from the Paleozoic Era to the present. There are five living genera with more than 900 species which occur in various parts of the world under varied climatic conditions. The living genera consist of the "ground pine" or club moss *Lycopodium* (Figs. 9-1, 9-2); the club moss *Selaginella* (shown in Fig. 9-16); the small, tuberous plant *Phylloglossum* (Fig. 9-4), which is greatly restricted in its distribution; the quillwort *Isoetes* (shown in Fig. 9-33) and *Stylites* (Amstutz, 1957) found growing high in the mountains of Peru (Fig. 9-41, A, B).

All of these genera can be classified as small plants, some being erect, others living as epiphytes, or growing as creepers on the ground, or producing underground rhizomes. In contrast with these plants of mod-

est stature, many of the ancient lycopods (*Lepidodendron*) were good-sized trees, and their vegetative structures and spores constituted an important part of coal (see Fig. 9-29). The importance of this assemblage of vascular plants certainly cannot be measured in terms of the present economic value of living members, but rather by the striking morphological unity of the entire group and its value in the interpretation of phylogenetic trends in vascular plants.

The vegetative sporophyte is differentiated into a shoot, consisting of stems and leaves, and a root system. Reminiscent of the group Psilopsida, the shoot system of many forms is isotomously branched or modified by anisotomous branching (Chapter 3). Occasionally the axis may be unbranched (as in *Phylloglossum*). The arrangement of leaves is fundamentally helical with modifications (decussate, whorled) characteristic

Figure 9-1
Lycopodium sp. Note the sporangia (white structures) in the axils of certain leaves along the upper half of each branch and the clusters of gemmae on the uppermost portion of each shoot.

Figure 9-2

Lycopodium. **A**, *Lycopodium inundatum,* showing prostrate rhizomes and upright fertile shoots terminated by strobili. **B**, *Lycopodium obscurum,* portion of upright branched shoot and terminal strobilus. **C**, *Lycopodium clavatum,* strongly rhizomatous species with determinate, fertile side branches; a root is evident along lower edge of the main axis.

Figure 9-3
Lycopodium lucidulum. **A**, fertile region of branch showing sporangia in axils of leaves (sporophylls). **B**, enlargement, as seen with the scanning electron microscope; note line of dehiscence across top of each sporangium. [B courtesy of Dr. R. H. Falk.]

of certain species. Leaves of living genera are generally small, whereas those of certain extinct forms were considerably larger. Whatever the arrangement or form of the leaves, each one is generally traversed by a single unbranched vascular bundle. Such a leaf is designated a microphyll (Chapter 3). Each leaf of certain genera, such as *Selaginella, Lepidodendron, Isoetes,* and *Stylites* has a curious tongue-like appendage on its adaxial side termed the ligule (Figs. 9-22 and 9-34).

The vascular cylinder of the stem in most living species is protostelic. The primary xylem is generally exarch in development and consists primarily of tracheids with scalariform pitting. Whether the vascular cylinder is a protostele or a siphonostele,

there are no breaks in the vascular tissue at the point of departure of leaf traces. No leaf gaps exist (Chapter 3). In a stem with a siphonostele a branch gap is present in the vascular cylinder of the main axis only at the point of divergence of the branch trace. The Lycopsida are therefore only cladosiphonic (see p. 44). In most living genera the roots, arising from rhizomes (e.g., in *Lycopodium*) or other specialized structures (e.g., *Selaginella*), branch dichotomously. In *Isoetes* there is a definite, perennial, root-producing meristem. Although the formation of secondary tissues was very common in ancient lycopods, this feature is characteristic of only two living genera—*Isoetes* and *Stylites.* A feature that unifies the entire group is the position of the eusporangium. Each sporophyll is associated with a single sporangium which is either attached to the adaxial basal region of the sporophyll or is located in its axil. The conditions of homospory and heterospory are coexistent in the group; heterosporous forms always produce endosporic gametophytes, homosporous forms produce only exosporic gametophytes.

Classification

LYCOPSIDA: Sporophyte differentiated into leaf, stem, root and eusporangium; microphylls ligulate or eligulate; one sporangium attached to or associated with each sporophyll; no leaf gaps; exarch xylem predominates; protostelic or siphonostelic; some have secondary growth.

LYCOPODIALES: Living and extinct plants; sporophytes with primary growth only, no vascular cambium; leaves eligulate; majority have definite strobili; homosporous.

LYCOPODIACEAE: Living and extinct plants; herbaceous; microphylls not forked at tips; many with definite strobili; exosporic gametophytes; biflagellate sperms in the living genus *Lycopodium*.

Lycopodium, Phylloglossum, Lycopodites (extinct).

EXTINCT DEVONIAN LYCOPODS: of Devonian age; low, herbaceous plants; dichotomously branched upright shoots from rhizomes; vascular cylinder cylindrical to lobed; some leaves with one sporangium, located near axil or on adaxial side of leaf; no definite, compact strobili; homosporous.

Baragwanathia, Drepanophycus, Protolepidodendron, Asteroxylon.

SELAGINELLALES: Living and extinct plants; with primary growth only; no vascular cambium; microphyllous with ligule; definite strobili formed; heterosporous; gametophytes endosporic; sperms biflagellate in living members.

SELAGINELLACEAE: characteristics as in Selaginellales.

Selaginella, Selaginellites (extinct).

LEPIDODENDRALES: Extinct plants; tree-like and most, if not all, with secondary growth; microphyllous with ligule; large root stocks (rhizophores) formed; heterosporous; sporophylls grouped into strobili; some forming seed-like structures.

Selected genera: *Lepidodendron, Stigmaria* (form genus for rhizophores), *Sigillaria, Lepidostrobus* (form genus for strobili), *Lepidocarpon* (form genus for "seed-like" structures).

ISOETALES: Living and extinct plants; sporophytes with corm-like stems; secondary growth; perennial root-producing meristem; ligulate microphylls; heterosporous; endosporic gametophytes; sperms multiflagellate in living members.

ISOETACEAE: Characteristics as in Isotales.

Isoetes, Isoetites (extinct), *Stylites.*

PLEUROMEIALES: Extinct plant; upright unbranched stem with ligulate microphylls grouped at its upper end; upper end of axis terminates in a strobilus; subaerial rhizophore; heterosporous.

PLEUROMEIACEAE: Characteristics as in Pleuromeiales.

Pleuromeia.

Homosporous Forms in the Lycopsida

Lycopodiales — Lycopodiaceae

This family includes two living genera, *Lycopodium* and *Phylloglossum.* The former, commonly termed club moss, is worldwide in distribution, with species occurring in such varied climatic zones as arctic and tropic. Most of the species (about 200) of *Lycopodium* are tropical, but certain ones occur in the temperate regions of the world. *Phylloglossum,* a highly reduced and specialized monotypic plant, is restricted to Australasia (Fig. 9-4). *Lycopodites,* a fossil lycopod of the late Devonian and Carboniferous periods, resembled the modern club moss in many respects.

LYCOPODIUM. Although species of *Lycopodium* do not usually form a conspicuous part of the flora of temperate regions, this genus is very diversified in growth habit and abundantly represented in the American tropics (Haught, 1960).

Some species are erect shrubby plants (Figs. 9-1; 9-2, B), others have a trailing or creeping habit (Fig. 9-2, A, C) and still others grow as epiphytes. Some of the terrestrial prostrate types form "fairy rings" in open, undisturbed areas. Active rhizomatous growth takes place at the margins of such a circle, while that part of the colony produced in previous years decays. The ring may be in the shape of a circle, and increase in diameter, as a function of time, results in an exponential curve of surprising exactness. One ring, measuring 11.25 meters in diameter in 1964, was estimated to have originated in 1839 (Van Soest, 1964).

It has been traditional in earlier textbooks to include all club mosses under the single genus *Lycopodium.* Certain textbook writers, however, have recognized two subgenera — one in which well-defined strobili are not formed and branching is essentially isotomous (*Urostachya*). The second subgenus includes species that have definite cones and whose branching is anisotomous (*Rhopalostachya*). In another system two families are recognized — the Urostachyaceae with one genus and the Lycopodiaceae with three genera. All four genera are said to have different basic chromosome numbers (Löve and Löve, 1958) and the chemistry of their phenolic acids and lignins differ (Towers and Maass, 1965).

As pointed out by Hauke (1969) in the prospectus of *Flora North America,* and by Morton (1964), perhaps the collective genus *Lycopodium* should be retained until more information becomes available on gametophytes, life histories, chromosome numbers, etc. We will adopt this conservative viewpoint and recognize one genus with the realization that, in the future, *Lycopodium* may be split into at least three or four genera and two or more families.

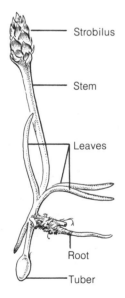

Strobilus

Stem

Leaves

Root

Tuber

Figure 9-4
Habit sketch, *Phylloglossum drummondii.* The "tuber" is a vegetative reproductive body and is capable of developing into a typical plant under favorable environmental conditions.

ORGANOGRAPHY. Whether a given species is erect or prostrate, branching is fundamentally dichotomous. The branches of a dichotomy may be equal, continuing to grow and to dichotomize, or the branches of a dichotomy may be unequal, with one branch overtopping the other. The weaker branch system may grow for a few years, generally becoming determinate, and these determinate branches may develop strobili (Fig. 9-2, C). This mode of branching is termed anisotomous, and it reaches its highest development in forms with a prostrate, rhizome-like main axis. The leaves of *Lycopodium* are eligulate microphylls which may be small (2–10 mm long) or in other species may become 2 or 3 cm long. The phyllotaxy is usually helical but may be decussate or whorled. In many forms the leaf bases are decurrent. In many of the creeping species dorsiventrality accompanied by anisophylly is characteristic of the lateral determinate branches.

Some species form vegetative reproductive structures — termed gemmae, or bulbils — which become detached from the plant and grow into new sporophytes (Fig. 9-1). These structures arise in the positions of leaves and consist of an organized bud and preformed root primordia (Takeuchi, 1962). The factors that favor bulbil formation are not well understood (Cutter, 1966).

Roots arise endogenously along the lower side of the stem in prostrate forms (Chapter 3; Fig. 9-2, C). In the upright forms roots may be initiated near the shoot tip and subsequently grow downward through the cortex, emerging at the base of the plant (Fig. 9-5, A). After the root emerges from the stem it may branch freely in a dichotomous fashion. No lateral endogenous roots are formed.

Sporangia always occur singly on the adaxial surface of the sporophylls or in their axils. The sporophylls may be aggregated into definite strobili and may be quite different from vegetative leaves (Fig. 9-2, B, C). In many species, however, "fertile" areas alternate with "sterile" regions along the stems, the sporophylls resembling ordinary foliage leaves (Figs. 9-1; 9-3, A).

STEM ANATOMY. The outermost layer of the stem is a uniseriate epidermis. The cortex is highly variable in thickness and structure (Fig. 9-5, A, C). In some species it remains parenchymatous, and in others the cells of specific regions undergo sclerification. Large air space systems may be present. Between the cortex and vascular cylinder there is an endodermis, the cells of which in the mature state are not easily recognized. During early development, however, these cells are reported to form Casparian strips and can be identified. Since the cell walls of the endodermis generally become thickened, the transition to the parenchyma cells of the pericycle is abrupt. The pericyclic cylinder may be two or three cells wide or constitute a rather broad tissue zone.

With the possible exception of the ferns, in no lower vascular plants is there such variation in the pattern of primary xylem and phloem in stems as is found in *Lycopodium*. The same species and even the same individual may show great variation during ontogeny (Hill, 1914, 1919; Wardlaw, 1924). In the mature plant body the vascular cylinder may be *actinostelic* with the primary phloem occupying the regions between the radiating arms of primary xylem (Fig. 9-5, B). In other species the primary xylem and phloem form strands of tissue which in transverse section appear as alternating bands of xylem and phloem; this type of vascular cylinder is designated a *plectostele* (Fig. 9-5, C, D). In still other species the central mass of xylem may be so modified as to form numerous strands of xylem and phloem (Ogura, 1938). It should be remembered that the seemingly isolated strands of xylem

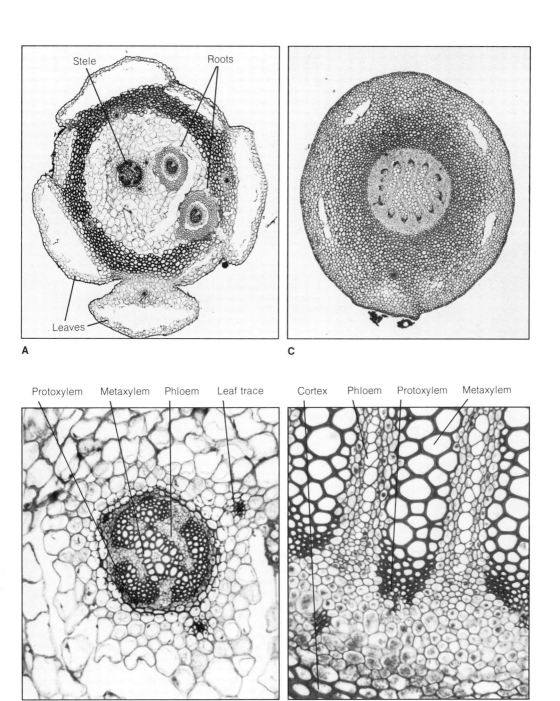

Figure 9-5
Stem anatomy in *Lycopodium*. **A**, transection, stem of *Lycopodium selago;* **B**, details of actinostele in *Lycopodium selago;* **C**, transection, stem of *Lycopodium* sp. showing plectostele; **D**, details, portion of stele in **C**.

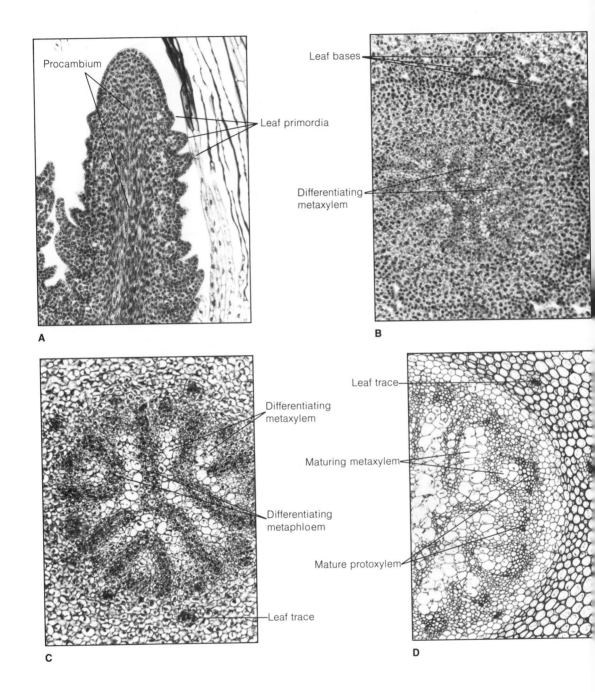

Figure 9-6
Development of the vascular cylinder (stele) in *Lycopodium*. **A**, longisection of shoot tip; **B–D**, transections of stem near shoot tip (**B**) and at increasing distances from the shoot apex (**C,D**). See text for details.

or phloem actually are interconnected. This can be demonstrated if their course is followed throughout the stem.

Ontogenetic studies have shown that the young sporophyte in most species is actinostelic. As growth of the sporophyte continues and the stem increases in size there is generally a change in the pattern of xylem and phloem. The actinostelic condition may persist with the formation of more protoxylem poles; or any of the configurations described above may result. In the smaller branches there may be a return to an actinostelic arrangement with only a few protoxylem points (Holloway, 1909; Jones, 1905; Chamberlain, 1917).

Xylem maturation generally has been accepted to be strictly exarch in *Lycopodium*, but the results of a recent reinvestigation of three species would suggest that xylem development is at least sometimes and possibly always mesarch in indeterminate branches (Wilder, 1970). Mesarchy is, however, inconspicuous in that there are only a few tracheids formed in the centrifugal direction from a protoxylem pole. The bulk of the xylem cylinder, of course, consists of centripetally formed tracheids of the metaxylem, the larger of which have scalariform or circular bordered pits.

The phloem consists of sieve elements and parenchyma. The sieve elements are elongate with sieve areas distributed over the lateral walls as well as on the long, oblique end walls.

Before leaving this discussion of stem anatomy it is important to describe the early formation and differentiation of primary vascular tissues in order to gain a fuller understanding of shoot development. The apical meristem of the shoot tip is reported to consist of a group of terminal initials (Turner, 1924; Härtel, 1938; Freeberg and Wetmore, 1967) which by periclinal and anticlinal divisions gives rise to the three primary meristematic tissues: protoderm,

ground meristem, and procambium, the derivatives of which differentiate into epidermis, cortex, and vascular tissue, respectively. The centrally located procambium extends very close to the shoot apex, a feature characteristic of many lower vascular plants (Wetmore, 1943; Freeberg and Wetmore, 1967). The cells of the procambium are elongate, and divide longitudinally, and frequently in the transverse plane.

To understand subsequent development, particularly vascular differentiation, an examination of transverse stem sections taken at successive levels from the apex is essential. A transverse section of a shoot very near the tip reveals that the future vascular cylinder is represented by a compact core of procambial cells. Very early, however, within the procambial cylinder of a plectostele, for example, there is the centrifugal blocking-out of the future stele. The first procambial cells to differentiate cytologically are the future cells of the metaxylem and metaphloem, followed by protoxylem and protophloem (Fig. 9-6, B, C). The future pattern may be established as close as 0.5 mm from the shoot apex. This initial centrifugal blocking-out is followed, at a lower level, by a stage of centripetal cellular maturation, the first elements to mature being the tracheids of the protoxylem (which form the secondary wall — Fig. 9-6, D), followed by sieve cells of the protophloem. Maturation then proceeds centripetally until all metaxylem and metaphloem elements are mature. Complete maturation of the vascular cylinder may be complete only at a distance of 4–6 cm below the shoot apex (Freeberg and Wetmore, 1967).

Radial differentiation and maturation have been stressed in the description above, but the student should realize that, during the differentiation stage, procambial cells are elongating as well as increasing in diameter, and that tissue maturation also occurs both longitudinally and radially in the stem.

LEAF ANATOMY. Initiation of the leaf is reported to take place in a single superficial cell on the flank of the apical meristem (Turner, 1924), although Bhambie (1965) reports that initiation occurs in several superficial cells for certain species. Growth in length, lateral extension of the midrib region, and maturation of tissues produce the mature leaf. The growth of a leaf is associated with the extension of a procambial strand into its base from the existing central core of differentiating vascular tissue of the stem (Härtel, 1938). Differentiation of cells within this original procambial tract produces a mesarch bundle, termed a leaf trace in its course through the cortex of the stem, and an unbranched midvein within the leaf itself. Leaf traces are generally attached to lateral flanges on the protostele of the stem axis. The mature leaf is usually small, ovate to lanceolate in outline, and without a definable petiole. A transverse section of the mature foliage leaf reveals a well-developed epidermal layer, with stomata occurring generally on both surfaces, although certain anisophyllous species have stomata restricted to one surface. The mesophyll in many species is composed of more-or-less isodiametric cells with a conspicuous intercellular air space system (Fig. 9-7, A).

Stomata are of the perigenous type (Bhambie, 1965; Pant and Mehra, 1964) in that only the guard cells are formed directly from a single protodermal initial, and are hence comparable to the haplocheilic type described by Florin for adult gymnospermous leaves (see Chapter 17).

ROOT. Except for the ephemeral primary root of the young sporophyte, the roots of actively growing plants arise from the stem very near the growing tip. These roots, which arise endogenously from the pericycle (Roberts and Herty, 1934), do not break through the cortex and epidermis immediately but often traverse the cortex for some distance before emerging. Since roots arise acropetally along the stem, many roots may be found in the stem cortex of the aerial portion of erect and epiphytic species (Fig. 9-5, A). In certain species (e.g., *Lycopodium pithyoides*) as many as 52 roots may be counted at one level (Stokey, 1907). Only near the base of the stem do these roots emerge. In prostrate forms the roots take a more direct course from the stem axis to the exterior. After emerging from the stem, a root branches dichotomously, often with great regularity; no lateral endogenous roots are formed.

In propagating species of *Lycopodium* it is important to (1) obtain a portion of the plant with intact roots, or (2) use the upper portion of a shoot (since roots are initiated near the tip), or (3) secure a portion of the stem with arrested roots which emerge from the stem cortex on contact with a moist surface. Arrested roots may be identified as mounds on the under side of the stem of a prostrate form (Roberts and Herty, 1934). Stokey (1907) has reported that four distinct groups of initials are present in the root apical meristem: a calyptrogen giving rise to the root cap; a tier of initials contributing to the developing protoderm; a group of initials giving rise to the cortex; and a set of initials for the vascular cylinder. Procambial differentiation and maturation result in a xylem strand that is crescent shaped (as seen in transverse section) and that partially surrounds a strand of phloem. Near the point of attachment to the rhizome a root may be polyarch (having several protoxylem poles) and maturation is exarch (Pixley, 1968). At this level the vascular cylinder (Fig. 9-7, B) of the root may resemble that of the stem and, except for size, it is sometimes difficult to distinguish between the two organs on the basis of stelar anatomy.

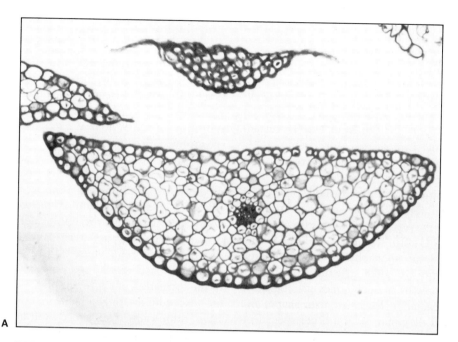

A

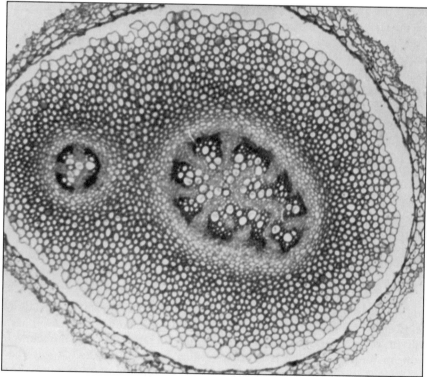

B

Figure 9-7
A, transection, leaf of *Lycopodium* sp; note stoma on adaxial (upper) side of leaf, and mid-vein; **B,** transection of large root of *Lycopodium* sp.: note general similarity in organization of stele to that of a stem in *Lycopodium*.

SPORANGIUM.   One of the definitive characteristics of the Lycopsida is the association of one sporangium with each sporophyll; each sporangium is located on the adaxial side of a sporophyll or in its axil. In certain species of *Lycopodium* (for example, *L. lucidulum and L. selago*) the sporophylls are similar to vegetative leaves (Figs. 9-1; 9-3, A). No definite strobili are formed, but rather there are "fertile" areas on the stem alternating with vegetative or "sterile" regions (Case, 1943). In species considered to be advanced the sporophylls are aggregated into definite cone-like structures or strobili; the sporophylls of such cones may be quite unlike vegetative leaves in size, shape, and color, and they may exhibit other specializations which are related to sporangial protection and spore dispersal. These strobili may occur on leafy stems or they may be elevated on lateral branches which have very small, scale-like leaves unlike those of the vegetative shoot (Figs. 9-2, B, C; 9-9, A).

Developmentally the sporangium is of the eusporangiate type; that is, its origin is from a group of superficial cells which divide periclinally (Fig. 9-8). The outer cells of such divisions form the multilayered wall, and the inner derivatives the sporogenous cells. The innermost layer of the sporangial wall functions as the tapetal layer. For a complete account of eusporangial development refer to Chapter 4.

Mature sporangia of most species are kidney shaped, are yellow to orange, and have a short stalk. As mentioned earlier, there are some very interesting relationships among position of mature sporangia, line of dehiscence, and specialization of the sporophyll. In certain species (e.g., *L. lucidulum*) the mature sporangium is axillary in association with a relatively unmodified sporophyll. Dehiscence is longitudinal and in the same plane as the expanded sporophyll (Fig. 9-3, B). In other species—those with definite strobili—the mature sporangia

are foliar in position, and the sporophylls are imbricated and have abaxial extensions (Fig. 9-9, B). Dehiscence in these species is modified, the opening being between the sporophyll and the abaxial extension of the sporophyll directly above. In still other species the sporangia are axillary, protected in the way just described by sporophyll modifications, and open in a similar manner. Whether the sporangium is protected or not, it is evident that the line of dehiscence occurs in such a position as to insure efficient dispersal of spores (Sykes, 1908).

Meiosis occurs in the sporocytes, resulting in the formation of spore tetrads. The spores are exceedingly small, light in color, and have thin walls. The spore wall in many species is sculptured, exhibiting various patterns. A triradiate ridge, present on the inner face of each spore, is indicative of the mutual contact between members of a spore tetrad (Fig. 9-9, C, D). Spore morphology is useful in delimiting subgroups within the genus. (See Wilce, 1972, for several interesting plates of photographs illustrating differences in spore wall ornamentation.)

The spores of certain species of *Lycopodium* are collected and sold as "lycopodium powder." This powder has been used in the manufacture of fireworks, but its use as a dusting powder on surgical gloves and pills has been discouraged; apparently the spores of *L. clavatum* cause inflammations in operative and other wounds (Whitebread, 1941).

CHROMOSOME NUMBERS.   Within the last several years some information has accumulated on chromosome numbers in lower vascular plants. In *Lycopodium* the haploid chromosome number (n) ranges from 14 to 24, 34, 78, 132 or even 170. In one species —*L. selago* (from Europe)—the diploid number was reported to be 260, and at meiosis there were many unpaired chromosomes,

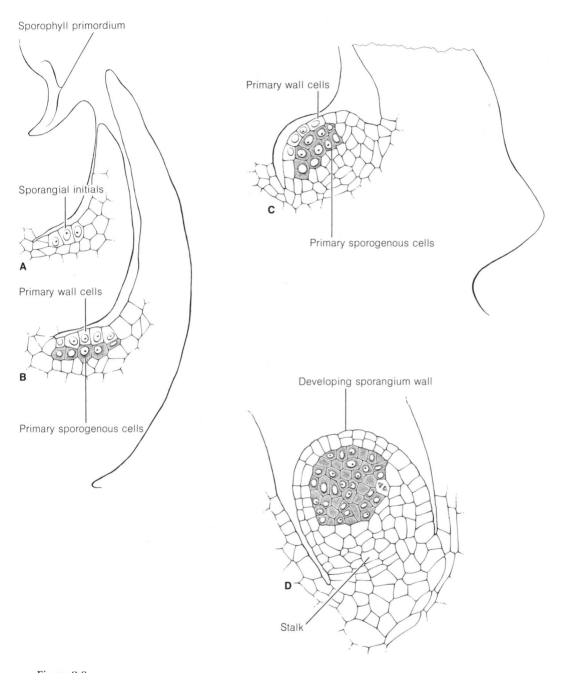

Figure 9-8
Ontogeny of the sporangium in *Lycopodium clavatum*. Note that initiation of the sporangium takes place in superficial cells by periclinal divisions, setting aside primary wall and primary sporogenous cells (**A, B**). The tapetum ultimately arises from inner cells of the sporangium wall.

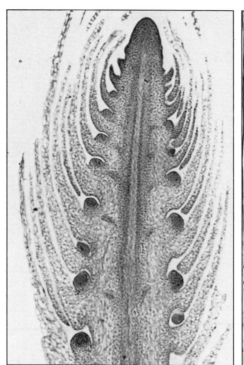

A

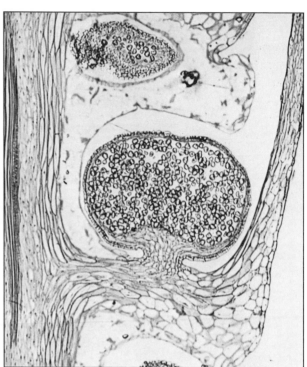

B

C

which indicated a history of hybridity to the investigators (Dunlop, 1949; Manton, 1950). Four cytological types of *L. cernuum* were reported from South India with numbers $n = 104$, 110, 136, and $2n = \pm340$. Here, again, cytology indicates that polyploidy and hybridization have been operative in the evolution of this species (Kuriachen, 1965; also see Mehra and Verma, 1957). It has been suggested that chromosome differences — numbers, size, and morphology — are so great in *Lycopodium* that the genus should be divided into several genera (Löve and Löve, 1958; also see page 141). *Phylloglossum drummondii*, the monotypic genus

referred to previously, has a $2n$ chromosome number of 502–510 with many univalents being present at metaphase I (Blackwood, 1953).

THE GAMETOPHYTE. Depending on the species of *Lycopodium*, the spores may germinate immediately or after a delay of several years. A gametophyte plant of the first type (*L. alopecuroides, L. cernuum, L. inundatum*), generally found on the surface of the substrate, is ovoid to cylindrical, with green aerial branches; the entire plant may not be over 3 mm long (Fig. 9-10, B). Rhizoids occur on the colorless basal end. An

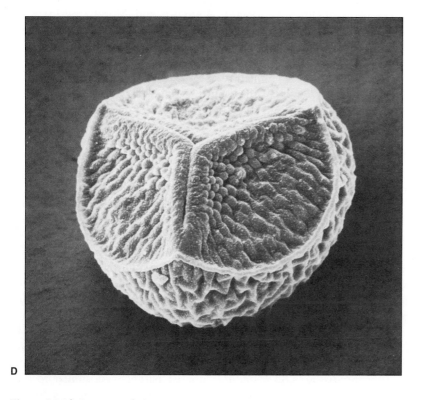

D

Figure 9-9 (*facing page and above*)
A, longisection of entire young strobilus of *Lycopodium clavatum*. Developing sporangia can be seen near the bases of sporophylls; **B**, a mature sporangium of *Lycopodium* sp. attached to sporophyll; note numerous spores and the overarching abaxial extension of the sporophyll above; the cone axis is to the left; **C, D**. Scanning electron micrographs of spores. C, spore tetrad of *Lycopodium reflexum* ($\times$ 2200); **D**, *Lycopodium inundatum*, note prominent triradiate ridge and contact faces with the other three spores ($\times$ 1500). [C, D courtesy of Dr. G. Breckon.]

endophytic fungus, entering the gameto-phyte plant early in development, is present in most species, occupying a definite region within the gametophyte. The sex organs generally occur near the bases of the aerial lobes. The time interval between spore germination and appearance of sex organs may vary from eight months to one year (Treub, 1884; Chamberlain, 1917; Koster, 1941; Eames, 1942).

After spore germination and when 6-8 cells have been formed, gametophytes of the second type enter into a rest period of a year or more. Apparently, further development is dependent on the entrance of a fungus. If this infection does not occur, all further growth ceases (Bruchmann, 1910). Physiologically, the fungus must supply certain substances vital for proper growth of the gametophyte plant. Subsequent development to a stage in which mature sex organs are present may require ten years or more (Eames, 1942). The development of these later stages takes place beneath the surface of the ground or within a layer of humus. The gametophyte (e.g., in *L. complanatum, L. clavatum*) is an oblong structure, ultimately becoming carrot-shaped or disc-shaped, with a convolute margin (Fig. 9-10, A, C). But in epiphytic species the gametophyte may be cylindrical, branched, and more attenuated. All of the subterranean gametophytes are colorless or yellowish to brown, developing chlorophyll only in those portions of the gametophyte that become exposed near the surface (Spessard, 1922).

The subterranean forms are long-lived, being continuously increased in size by a marginal ring of meristematic tissue. Old gametophytes may be up to 2 cm in length or width.

In species whose gametophytes are of the green, annual type, antheridia and archegonia are generally intermingled (Chamberlain, 1917), whereas in the subterranean forms the sex organs are segregated into definite groups (Fig. 9-13, A) except in certain species (Spessard, 1922). In the course of development, antheridia generally appear first near the middle of the crown of the gametophyte. Initiation of archegonia and more antheridia then occurs in the immediate derivatives of the meristematic ring (Fig. 9-13, A).

The dependence of *Lycopodium* species on the infection of the gametophyte by a fungus presents an interesting physiological problem. It has been possible to culture gametophytes, particularly the annual type, to maturity by sowing the spores on soil taken from the original habitat (see Koster, 1941). Wetmore and Morel (1951a) were able to culture to maturity, in the laboratory under sterile conditions, the gametophyte of *L. cernuum* (a green, annual type that in nature is associated with a fungus). After the spore coat had been sterilized with calcium hypochlorite, the spores were sown on a culture solution containing minerals and glucose. In some cultures the upright branches became club-shaped, while in others a filamentous "pin-cushion" type resulted (Fig. 9-11, A, B). After six months of continued growth, under regulated conditions, antheridia and archegonia were formed. Many sporophytes developed (Fig. 9-11, C, D). It may be assumed that the balanced nutrient solution contained substances that are supplied by the fungus under natural conditions. Through further experimentation it was found that under the action of the growth-regulating substance, naphthaleneacetic acid, a developing gametophyte is transformed into masses of undifferentiated parenchyma similar to that obtained in culturing the tissues of higher plants.

Not only has the small, annual green type of gametophyte (e.g., that of *L. cernuum,* just described) been cultured successfully *in vitro,* but so also have representative

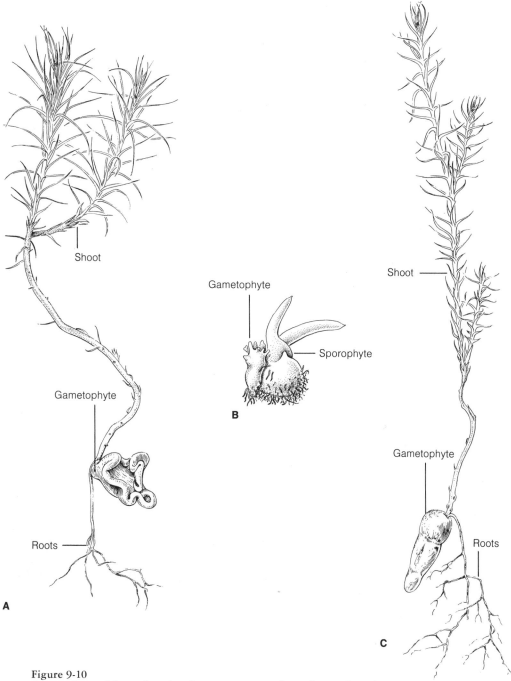

Figure 9-10
Gametophytes of *Lycopodium*. **A**, subterranean gametophyte of *Lycopodium clavatum*, with attached sporophyte; **B**, the subaerial or terrestrial type, *Lycopodium laterale*; **C**, subterranean type, *Lycopodium complanatum*. [A and C drawn from specimens supplied by Dr. A. J. Eames; B redrawn from Chamberlain, *Bot. Gaz.* 63:51, 1917.]

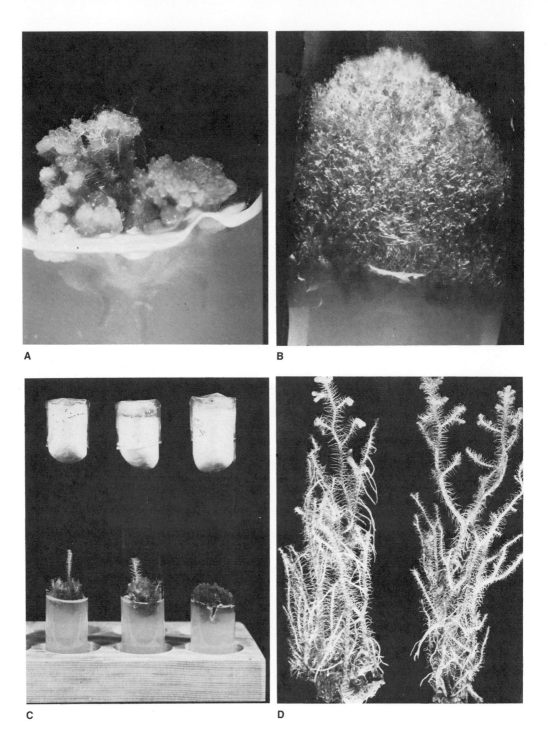

Figure 9-11

In vitro cultures of *Lycopodium cernuum*. **A,** gametophyte with club-shaped branches (note rhizoids on upright branches); **B,** an older pincushion-like gametophyte; **C,** gametophytes with young sporophytes attached, in two culture tubes at the left; **D,** sporophytes with young cones (tips of uppermost branches at left) and roots; a portion of the gametophyte can be seen at the base of each plant. [Courtesy Dr. Ralph H. Wetmore.]

species of the normally subterranean, long-lived types (e.g., *L. selago, L. complanatum*). For spores of the latter types to germinate it was necessary for them to be treated with H_2SO_4 or scarified; such treatment structurally modifies the spore wall, thus permitting the entrance of water and nutrients. The results of growth were somewhat unexpected. The gametophytic plants were green and, in general growth form, resembled the *L. cernuum* type (Fig. 9-12). After six to ten months, "sporophytic" shoots developed apogamously (without fertilization) or isolated sporophytic-like leaves developed directly from the gametophytes. When cultures were flooded with water, sperms were released, and fertilization and the formation of normal embryos followed. Microspectrophotometric measurements indicated no difference in DNA content between nuclei of the gametophyte and those of the apogamous sporophytes (Freeberg, 1957; Freeberg and Wetmore, 1957).

If portions of cultured gametophytes were transferred to a medium containing sucrose and autoclaved coconut milk, the new growth consisted of aggregations of cells and nodular outgrowths. Sporophytic buds appeared after two months in this medium. De Maggio (1964) concluded that the "genetic complement imposes no rigid restriction for growth and organization," and that the expression of a tissue as gametophyte or sporophyte may be evoked by manipulating the environmental conditions of growth. Similar conclusions have been reached for ferns (Chapter 13).

The ontogeny of gametangia in *Lycopodium* has been described in detail in Chapter 5. We will only re-emphasize here that a remarkable similarity in development exists in the early stages of ontogeny of the sex organs—namely initiation in a single superficial cell by a periclinal division, which sets aside the sterile jacket cell and the primary spermatogenous cell of the antheridium,

Figure 9-12
Gametophytes grown on nutrient agar. A, *Lycopodium selago;* B, *Lycopodium complanatum;* C, *Lycopodium cernuum.* Under natural conditions the gametophytes of *Lycopodium selago* and *Lycopodium complanatum* are subterranean, lack chlorophyll, and are of quite different form (cf. Fig. 9-10, A, C). [Courtesy of Dr. J. A. Freeberg.]

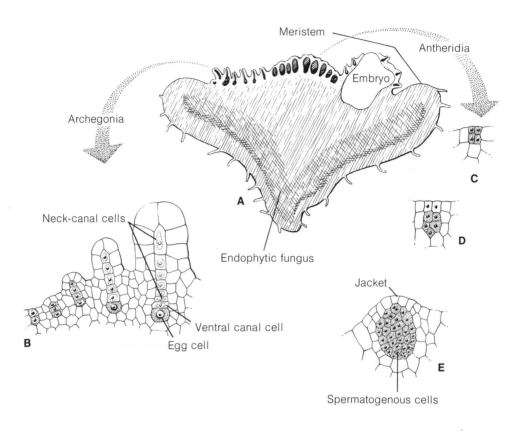

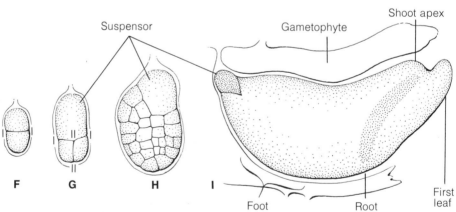

Figure 9-13
A, longisection of the gametophyte of *Lycopodium clavatum* showing the position of antheridia and archegonia, and one embryo; **B,** stages in development of an archegonium, *Lycopodium selago;* **C–E,** stages in ontogeny of an antheridium, *Lycopodium clavatum;* **F–I,** development of the embryo, *Lycopodium selago* (mouth of the archegonium is directed toward the top of the page). (Consult text for details.) [A redrawn from *Syllabus der Pflanzenfamilien* by Engler and Gilg, Berlin: Gebrüder Borntraeger, 1924; B, G, H, I redrawn from Bruchmann, *Flora* 101:220, 1910; C–E adapted from *Morphology of Vascular Plants. Lower Groups* by A. J. Eames. McGraw-Hill, New York, 1936.]

and a division which forms the primary cover cell and the central cell of the young archegonium. The latter cell is the progenitor of the axial row (Fig. 9-13, B, C-E). At maturity an antheridium consists of a sterile jacket or "opercular layer," one cell thick, enclosing many small sperm mother cells. Each spermatocyte matures into a biflagellate sperm which closely resembles the sperms of bryophytes and certain algae. The archegonia of surface-living, green, short-lived gametophytes have only three or four tiers of neck cells (Treub, 1884) and usually one neck-canal cell, whereas, according to Spessard (1922), archegonia of the subterranean forms have long necks with six or more neck-canal cells (Fig. 9-13, B). In either archegonium the venter is embedded in the gametophyte tissue. In certain forms a doubling of the axial row may occur (Spessard, 1922), or archegonia may be formed with exceedingly long necks (see Chapter 5 for further details).

With the degeneration of the neck-canal cells and the ventral canal cell a passageway is created for the entrance of the motile biflagellate sperms, which reach the archegonium by swimming through a film of water on the surface of the gametophyte. The results of one study indicate that free citric acid or salts of citric acid may play a role in the attraction of sperms to the archegonia (Bruchmann, 1909b).

THE EMBRYO. Embryogeny is correlated, to some degree, with the type of gametophyte, but closer examination reveals a common basic plan. To gain an understanding of embryogeny, we will begin with a species possessing an underground gametophyte (Fig. 9-13, F-I). The embryo in *Lycopodium* is endoscopic, that is, the shoot apex is directed away from the mouth of the archegonium. The first division of the zygote is transverse to the long axis of the arche-

gonium, setting aside an apex, cell *a,* and a base, cell *s* (see Figs. 6-1, C; 6-3). Cell *s* undergoes no further divisions, and becomes a suspensor. Cell *a* then divides at right angles to the original wall, and this is followed by two more divisions which result in two tiers of cells, each consisting of four cells. The tier of four cells next to the suspensor will give rise to the foot, and the distal tier will produce the first leaf and shoot apex. By further cell division and cell enlargement the foot becomes changed in orientation, since the first leaf and shoot apex grow laterally and upward, eventually breaking through the gametophyte near the meristematic zone (Fig. 9-13, I). The primary root is variable in its position, but commonly appears exogenously near the juncture of the foot and the primary leaf. The foot maintains close connection with the gametophyte and is important physiologically during growth of the young sporophyte. Sexually mature gametophytes may continue to live for some time, supporting one or more young sporophytes in different stages of development.

In the typically green, surface-living gametophytes (for example, in the tropical species *L. cernuum* and the temperate species *L. inundatum*) a foot is formed from the tier of cells next to the suspensor. The cells of the other tier, instead of developing immediately into the first leaf and shoot apex, undergo divisions in many planes forming a "neutral body," termed a protocorm, which pushes its way through the gametophyte. This green parenchymatous body develops leaf-like structures—protophylls—on the upper surface, and rhizoids on the lower surface. Eventually certain cells at the surface of the protocorm begin functioning as an apical meristem, giving rise to the shoot apex which will produce the "normal" type of shoot. The protocorm in these species has been interpreted as a stage of sporophyte development interpolated between the gametophyte and the "typical" adult sporophyte. (A detailed treatment can be found in Bower, 1935; Goebel, 1930.) For a more complete discussion of embryogeny, refer to Chapter 6.

Some Devonian Lycopods

The earliest plant with definite lycopod characteristics is *Baragwanathia* from the early Devonian in Australia (Jaeger, 1962). It branched dichotomously and the stem was covered with helically arranged leaves, 0.5–1 mm wide and as long as 4 cm. Sporangia have been found associated with some leaves, but it is unknown (because of poor preservation) whether they were on the adaxial or on the abaxial side of the leaves.

Another lycopsid plant from the Devonian is *Drepanophycus* (Fig. 9-14, C). As reconstructed, it exhibits a rhizome with upright dichotomous branches with falcate (curved) leaves. Stalked sporangia occurred singly on the adaxial side of some leaves midway between the enlarged leaf base and the leaf tip (Grierson and Hueber, 1968).

The genus *Protolepidodendron* is a commonly depicted lycopod from the Devonian (Kräusel and Weyland, 1932) and is characterized by leaves that forked near the tip. The plant was rhizomatous with upright dichotomous branches bearing helically arranged leaves, some of which had solitary sporangia located on the adaxial side of the leaf (Fig. 9-14, A).

The genus *Asteroxylon,* formerly classified with the "psilophytes" because it was assumed to have had terminal sporangia, is now considered to be a lycopod. It has been transferred to the Lycopsida because sporangia are now known to have been located in the axils of leaves (Lyon, 1964). *Asteroxylon mackiei* grew to a height of 0.5 meters. The plant had a naked, dichotomously branched subterranean rhizome. Small

branch systems were present that probably functioned in anchorage and absorption. At intervals the tips of some dichotomies became upright, bearing numerous, closely appressed, small leaves. The main upright axes exhibited irregular or anisotomous branching, whereas smaller side-branch systems displayed more regular dichotomous growth (Fig. 9-14, B).

The shoot of *Asteroxylon* was covered by an epidermis with thick outer walls, which was interrupted in places by stomata. The cortex was differentiated into three general regions: the outer, middle, and inner portions. The outer cortex was composed of homogeneous and compact parenchyma; the middle portion was highly lacunate; and the inner portion was compact and homogeneous (Fig. 9-15). Occupying the central region of the stem was the vascular cylinder, which may be designated as an actinostele (Chapter 3). Primary xylem in the form of a fluted cylinder occupied the center of the stem. Protoxylem occurred near the extremities of the lobes but was surrounded on all sides by metaxylem, making the xylem mesarch in development. "Leaf" traces, departing from the vicinity of the lobes, passed obliquely through the cortex and ended abruptly near the base of each leaf-like appendage. These traces were concentric; that is, each trace consisted of primary xylem surrounded on all sides by primary phloem. In one species, *Asteroxylon elberfeldense,* the vascular cylinder was siphonostelic in lower portions of the aerial stem.

Some of the Devonian lycopods just described were probably ancestral forms in the line leading to *Lycopodites* (a late Devonian and Carboniferous species) and then to the extant genus *Lycopodium.* The Devonian types were, in turn, probably derived from members of the Zosterophyllopsida—one of the groups of early vascular plants (Chapter 7).

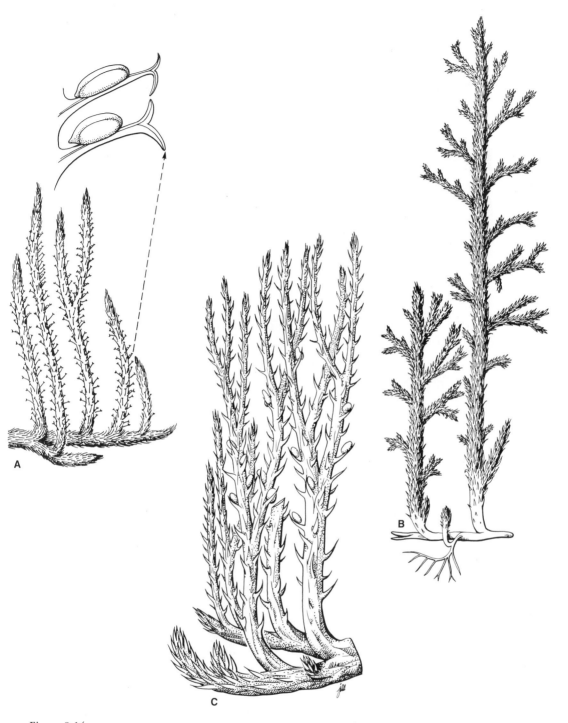

Figure 9-14
Diagrammatic reconstructions of extinct lycopods from the Devonian **A**, *Protolepidodendron scharyanum;* sporophylls and sporangia enlarged above; **B**, *Asteroxylon mackiei;* **C**, *Drepanophycus spinaeformis.* [A redrawn from Kräusel and Weyland, *Senckenbergiana* 14:391–403, 1935; B redrawn from Kidston and Lang, *Trans. Roy Soc. Edinb.* 52, Part IV, 1921; C redrawn from Kräusel and Weyland, *Palaeontographica* 80(B):171–190, 1935.]

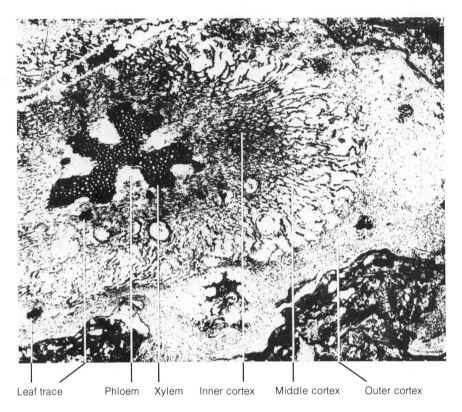

Leaf trace   Phloem Xylem Inner cortex Middle cortex Outer cortex

Figure 9-15
Transection, stem of *Asteroxylon mackiei.* [From Kidston and Lang, *Trans. Roy. Soc. Edinb.,* Vol. 52, 1920–21.]

Heterosporous Forms in the Lycopsida

Selaginellales — Selaginellaceae: *Selaginella*

GENERAL CHARACTERISTICS. The genus *Selaginella,* often referred to as the small club moss or spikemoss, is widely distributed over the earth. But even though the genus includes more than 700 species, it does not form a conspicuous part of the world's vegetation. Whereas many species of *Lycopodium* may be relatively large and coarse, most species of *Selaginella* are small and delicate. It is in the tropics that *Selaginella* is most abundantly represented, often being the dominant element of the forest floor in mesophytic tropical woodlands (Haught,

1960). Some species grow where climates are cold, and many others inhabit temperate zones, growing in damp areas or even occupying exposed rocky ledges. One species, *Selaginella lepidophylla,* cespitose in habit, has become adapted to existence on a Mexican desert and in the arid regions of southwestern United States. In arid places the entire plant forms a tight ball during periods of drought; in the presence of moisture the branches expand and lie flat on the ground. This species is commonly known as the "resurrection plant." *Selaginella* is a greenhouse favorite, and is often used as a border plant. Species of this genus growing together in a greenhouse present an array of color shades — dark to light green, bluish — and some are iridescent.

A

Figure 9-16
Two species of *Selaginella* displaying contrasting growth forms. **A**, *Selaginella kraussiana*, a creeping or scrambling type (note rhizophores along main shoot near points of branching); **B**, *Selaginella emelliana*, a form with strong rhizomes and large compound upright branch systems.

B

A

B

C

GROWTH FORM. In growth habit there is considerable variation, although most species can be referred to two or perhaps three growth types. Some species are erect and even shrubby; others form tufts or mounds and quite often possess leaves that are uniform in size and shape. The leaves are fundamentally spirally arranged, although they generally lie in four vertical rows. In other species the plant may be flat, creeping along the surface of the ground or scrambling over shrubs (Fig. 9-16, A). Still others have a strongly developed rhizomatous stem with large, frond-like side branches which stand erect (Figs. 9-16, B; 9-17, C). In the last two types anisophylly (the production of small and large leaves) is often a very prominent feature. In these species branching is fundamentally of the flabellate dichotomous type; the overall growth pattern is anisotomous.

Peculiar prop-like structures in certain species, originating from the stem at points of branching, are evident to even an untrained observer. These structures, termed "rhizophores," are of great morphological interest (see Figs. 9-16, A; 9-17, C also Fig. 9-18, B); more will be said of their morphological interpretation in a later section of this chapter.

STEM ANATOMY. The epidermis consists of cells in which the outer cell walls are cutinized. Stomata are said to be lacking. In many species there are several layers of thick-walled cells beneath the epidermis, which merge gradually with thin-walled chlorophyllous cells of the inner cortex. In most species showing a radial type of growth pattern, as well as in the dorsiventral species, the trailing stem or the prostrate rhizome is protostelic. Plants with a radial growth habit may have a simple, cylindrical protostele in the stem, whereas dorsiventral species may have two or more vascular strands that are either circular or ribbon-shaped as seen in transverse section (Fig. 9-18, A). The ribbon-shaped protostele in the rhizome of dorsiventral species (Fig. 9-18, D) may be replaced in the upright branches by a number of vascular bundles (meristeles). An even more complex stelar pattern has been described for the tropical species *Selaginella exalta.* In this species the stele of the very large erect stem is a three-lobed plectostele, termed an "actinoplectostele" (Mickel and Hellwig, 1969).

In still other species the rhizome may be solenostelic, and the upright branches may have as many as 10–15 separate meristeles. Experimentally it has been shown that if such an upright branch is placed in a horizontal position, the newly developed portion of the shoot will become solenostelic (Wardlaw, 1924). Whatever the stelar configuration, the composite vascular cylinder or each meristele, as the case may be, is supported in a large air-space system by radially elongated endodermal cells designated *trabeculae* (Fig. 9-18, A). These cells have the characteristic Casparian strips. If the air-space system is large, each support or trabecula may consist of several cortical cells as well as the endodermal cell.

The student should recognize the fact that, in descriptions of changing vascular patterns in the stem, stelar configurations at any given level are merely a reflection of differential growth patterns in relation to development of the entire shoot. The vascular system of a shoot is an interconnected network and should be visualized as such (Fig. 9-18, C).

Figure 9-17 (*facing page*)
A, B, branches of two species of *Selaginella* showing strobili at tips of determinate branches; C, *Selaginella mertensii* showing flabellate branches, prominent rhizophores, and roots. [C courtesy of Dr. T. E. Weier.]

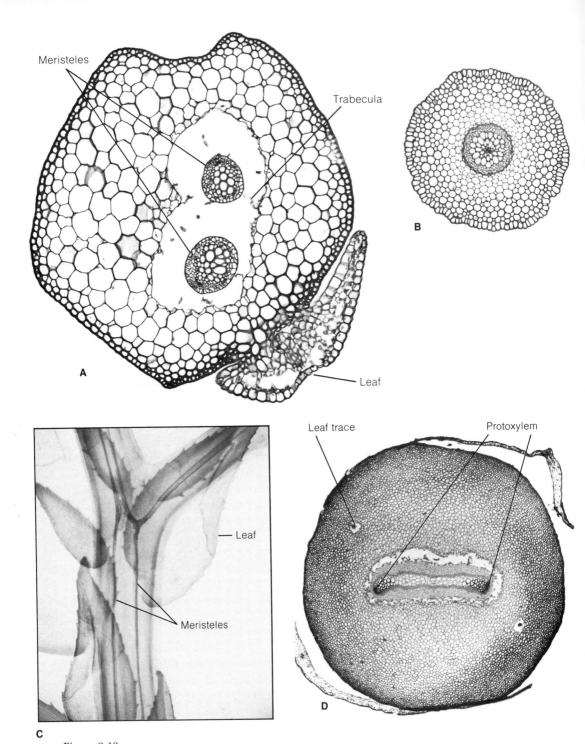

Figure 9-18
A, transection of stem of *Selaginella kraussiana;* only a few trabeculae appear in a transection of the stem; **B**, transection, rhizophore of *Selaginella* sp.; **C**, "cleared" and stained shoot of *Selaginella kraussiana* showing course of the vascular strands (meristeles) at a dichotomy of the stem; **D**, transection, stem of *Selaginella emelliana* near the base of a large branch.

Regardless of stelar organization, the primary xylem is fundamentally exarch in development, and the metaxylem consists primarily of tracheids with scalariform pitting. Ribbon-shaped steles may have more than one protoxylem pole (Zamora, 1958). Many years ago (Duerden, 1934) certain species were shown to possess vessels, a feature considered formerly to be one of the distinctive characteristics of angiosperms. Not only do certain species of *Selaginella* have vessels, but this type of tracheary structure is found in some ferns (Chapter 13) and in certain gymnospermous genera (for example, *Ephedra, Gnetum*). The phloem of *Selaginella* consists of sieve cells and parenchyma. The end walls of the sieve cells are usually transverse to slightly oblique and scattered single "sieve pores" occur on the lateral and end walls rather than typical sieve areas (Lamoureux, 1961).

LEAF ANATOMY.   Leaves of all species are small, attaining a length of a few millimeters. In form the leaves may be ovate, lanceolate, or orbicular. Although helical arrangement is a common feature in *Lycopodium,* most species of *Selaginella* have leaves that are arranged in four rows along the stem. As mentioned earlier, the shoots of most species are dorsiventral, and the leaves are anisophyllous (Fig. 9-16, A). There are two rows of small leaves on the dorsal side of the stem, and two rows of larger leaves on the ventral side. On close examination, a small tongue-like structure, the ligule, can be observed on the adaxial side of each leaf near the base; more will be said later concerning the development of this structure. Anatomically the mature leaf may vary considerably. The cells of the two epidermal layers may be similar, or in some species they may be somewhat different (Hsü, 1937). Some species have bristles or short hairs extending out from the epidermis. The mesophyll

may consist of a distinct palisade layer and spongy parenchyma, or the entire mesophyll may be a reticulum of lacunate parenchyma. Generally stomata are on the abaxial surface, although in certain species they are present on both surfaces (Harvey-Gibson, 1897). The small median vascular bundle of the leaf is collateral; the xylem consists primarily of tracheids with helical thickenings.

To gain an understanding of the growth of the entire shoot system (stem and leaves) it is necessary to examine the shoot apical meristem. In certain investigated species the shoot is terminated by an apical cell (Barclay, 1931; Hsü, 1937). This apical cell is tetrahedral and appears as an inverted pyramid in longitudinal sectional views of the shoot apex (Fig. 9-19). Through divisions of this apical cell, derivative cells are produced on the three "cutting" surfaces. Each of these cells (segments) undergoes a periclinal division, forming an outer and an inner cell. The outer cell, by further divisions, will produce the epidermis and cortex. Endodermis, pericycle, and vascular tissues are derived ultimately from derivatives of the inner cell (Fig. 9-19). Certain other species are reported to have two adjoining apical initial cells at the shoot tip (Williams, 1931) or a group of apical initials (Bhambi and Puri, 1963).

Leaves have their origin in superficial cells located along the flanks of the apical meristem. After leaf initiation a period of apical growth ensues, which is followed by generalized growth and cellular differentiation. The leaf is traversed by a procambial strand which is continuous with the vascular cylinder of the stem (Hsü, 1937). Procambial cells eventually differentiate into primary xylem and primary phloem of the leaf bundle.

The ligule (Fig. 9-22), located on the adaxial side of each leaf, makes its appearance through periclinal divisions in the embryonic cells of the protoderm. A detailed

166

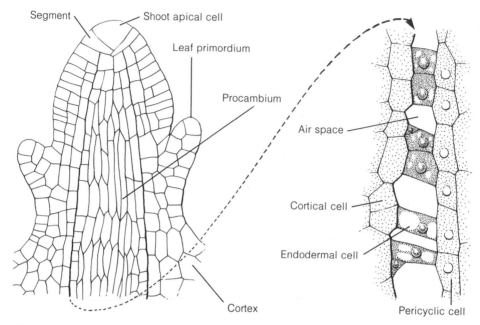

Figure 9-19
Stem development in *Selaginella sinensis*. Longisection of the shoot tip (left), and early development of trabeculae (endodermal cells) and of the air-space system surrounding the vascular cylinder (right); endodermal cells become separated from one another and undergo radial extension. [Redrawn from Hsü, *Bull. Chinese Bot. Soc.* 3:75, 1937.]

account of ligule development is presented in Chapter 3.

Of considerable interest is the large size attained by chloroplasts in this genus. Depending on the species, each mesophyll cell may have one large, cup-shaped chloroplast or from four to six smaller plastids (Ma, 1930).

Electron-microscopic examination of healthy plants of five species of *Selaginella* revealed that mature chloroplasts have distinct grana and fretwork that conform to the structure of chloroplasts of most gymnosperms and angiosperms (Jagels, 1970a, b). Jagels found that, although growth was halted when the plants were placed in complete darkness for several months, they remained green and chloroplast morphology

was similar (except for starch depletion) to that of normal, light-grown plants. This is rather unusual in that the plants would be expected to become bleached in the dark. Perhaps new chlorophyll was formed; some gymnospermous species are known to synthesize chlorophyll in the dark.

THE ROOT. Except for the primary root of the young sporophyte, the roots of most species are at the distal ends of rhizophores. The rhizophore, a structure that normally is without leaves and which arises at points of branching of the shoot, is an interesting organ (Figs. 9-16, A; 9-17, C). It originates from a meristem, termed an angle meristem (Cusick, 1954). The rhizophore possesses no root cap, and occasionally is transformed into a leafy shoot. Cusick considers the angle-meristem to be basically an embryonic shoot and believes that rhizophore formation involves a secondary change of growth pattern when the regulatory influence of the main axial apex is modified. Rhizophore formation can also be induced experimentally by applying indole-acetic acid to the cut end of a stem (Williams, 1937). The rhizophore is generally unbranched, but on contact with the substrate, roots are formed at the distal end. These roots branch dichotomously and have the recognized anatomical characteristics of roots.

The fact that rhizophores may be transformed into leafy shoots is not considered substantial enough evidence by some botanists for calling the rhizophore a stem rather than a root (Webster and Steeves, 1967; Mickel and Hellwig, 1969). These investigators argue that the angle meristem probably has the potential to develop into a leafy stem or rhizophore, but that it seems more reasonable to view the rhizophore as a root, based on its usual and normal behavior. It is argued that in a group of plants as primitive as the Lycopsida, root and shoot are probably not so distinct as in seed plants.

The use of the term rhizophore for *Selaginella* is deeply entrenched, however, in botanical literature and probably will not be displaced easily as a rather unique descriptive feature of *Selaginella*.

THE STROBILUS. Unlike *Lycopodium,* all species of *Selaginella* form strobili or cones. Strobili occur terminally on side branches, although in some forms the apical meristem of the cone may continue meristematic activity, producing vegetative leaves (Sykes and Stiles, 1910). All sporophylls of a strobilus are generally alike (although not differing from vegetative leaves as much as in certain species of *Lycopodium*) and are arranged in four distinct rows. The sporophylls may fit tightly together, or the vertical distance between sporophylls may be greater, the entire strobilus being a lax or open type of cone (Figs. 9-17, A, B; 9-20, A).

Because *Selaginella* is heterosporous, sporangia are of two types: microsporangia and megasporangia (Figs. 9-20, B; 9-21). The sporophylls associated with these two types of sporangia are termed, respectively, microsporophylls and megasporophylls. The one mature sporangium associated with each sporophyll is generally axillary in position, although its origin may be from cells of the axis or the base of the sporophyll (Lyon, 1901; Bower, 1935; Campbell, 1940). There is variation in distribution of sporangia within the strobili of different species. Strobili may consist entirely of microsporangia or of megasporangia (Mitchell, 1910). However, the mixed condition (either without order or arranged in definite patterns) is more common. The lower portion of a strobilus may consist of megasporangia and the upper portion of microsporangia, or the two types of sporangia may be mixed indiscriminately. A common arrangement is two vertical rows of each type (Fig. 9-20, B). In certain species (e.g., *Selaginella kraussiana*) only one megasporangium is present at the base of each strobilus.

The type and distribution of sporangia within the strobili of *Selaginella* have never been submitted to a detailed analysis by using mass collections. Recently, however, Horner and Arnott (1963) examined the pattern of distribution of mega- and microsporangia within strobili of species related taxonomically and geographically. Three major patterns of sporangia distribution in 30 North American species of *Selaginella* were recognized: Pattern I—strobili having a basal megasporangiate zone with an upper zone of microsporangia; Pattern II—strobili having two rows of microsporangia and two rows of megasporangia; Pattern III—strobili

A   B

Figure 9-20
Strobili of *Selaginella*. **A**, two enlarged, compact strobili comprising four rows of sporophylls; **B**, one-half of a strobilus that has been "cleared" and stained; microsporangia to the left and megasporangia to the right; note the four megaspores in each megasporangium, also the vascular bundle in each sporophyll that passes beneath the sporangium.

that are wholly megasporangiate. It was concluded that sporangial arrangement is a useful taxonomic tool in *Selaginella*. In general, a natural group or series of species is all of one type. Furthermore, Horner and Arnott (1963) concluded that Pattern I is more primitive since this arrangement was exhibited by Carboniferous tree lycopods and fossil species of *Selaginellites*. From this type other patterns may have evolved.

Mature microsporangia are generally obovoid or reniform and reddish to bright orange. Megasporangia are larger than microsporangia and frequently are lobed, conforming in outline to the large spores within them. The megasporangia are characterized by lighter colors: whitish-yellow or light orange.

The site of sporangial initiation, whether microsporangia or megasporangia are considered, is in superficial cells of the axis, directly above the sporophyll, or in cells near

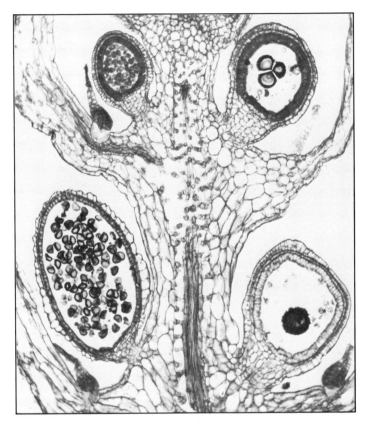

Figure 9-21
Portion of longisection of strobilus of *Selaginella* sp. showing late stages
in development of sporangia. At upper left, a microsporangium with
microsporocytes; note median sectional view of ligule. At lower left,
a mature microsporangium with numerous microspores. Three megaspores, surrounded by degenerating sporocytes, are evident in
megasporangium at upper right. The megasporangium at lower right is
nearly mature, and at this level of section a single megaspore is seen.
(Consult Fig. 9-22 for details of early ontogeny of sporangium in
Selaginella.)

the base of the sporophyll on the adaxial
side. Whether two, three, or more superficial initials are involved (Bower, 1935;
Campbell, 1940; Goebel, 1930), periclinal
divisions in these initials separate an outer
tier of cells—the primary wall cells—and an
inner tier—the primary sporogenous cells
(Fig. 9-22, A, B). By repeated anticlinal and
periclinal divisions of the primary wall
cells, a two-layered sporangial wall is formed.
The primary sporogenous cells divide periclinally, the outer cells eventually becoming
the tapetum; the inner cells, by dividing in
various planes, produce the sporogenous
tissue (Fig. 9-22, C). Undoubtedly cells
located at the base of the sporangium, not
identified with the original periclinal divisions, serve to complete the continuity of
the tapetum (Fig. 9-22, D).

In summary, a sporangium at this stage
consists of an immature sporangial wall of
two layers, a short stalk, and a conspicuous

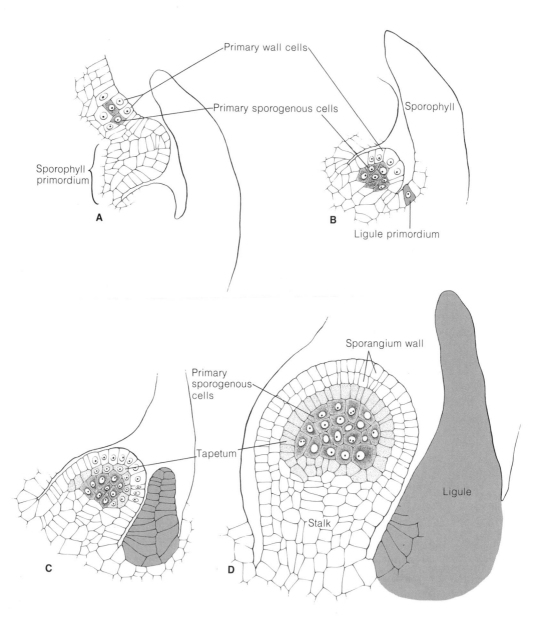

Figure 9-22
Early ontogeny of a sporangium in *Selaginella* sp. **A, B,** periclinal divisions in superficial cells separate primary wall and primary sporogenous cells. **C, D,** the tapetum is formed from outer sporogenous cells. Note precocious development of ligule. Whether the sporangium in **D** would have become a microsporangium or a mega-sporangium is not evident morphologically at this stage of development. However, physiological specialization may have occurred.

tapetal layer enclosing sporocytes which normally round off and separate from each other prior to the meiotic divisions (Fig. 9-22, D). Up to this developmental stage, microsporangia and megasporangia are indistinguishable, although the results of one study have shown that a pair of sporangia (micro- and megasporangium) at the same node in the cone exhibit different growth rates up to the premeiotic stage (French, 1970). As development continues, the two types become clearly defined. If a sporangium is to become a microsporangium, a large percentage of the sporocytes undergo meiosis to form tetrads of microspores (Fig. 9-21).

In a potential megasporangium the functional megasporocyte becomes distinct from the nonfunctional ones prior to meiosis. Nonfunctional megasporocytes develop large vacuoles and accumulate starch while the functional sporocyte retains a dense cytoplasm (is rich in RNA) and is without starch. The functional megasporocyte is generally in the central region of the sporocyte mass and undergoes meiosis, forming four megaspores; the nonfunctional sporocytes ultimately degenerate.

What determines why one sporangium will become microsporangiate and another one megasporangiate may be the late activation of sex determinants (Sussex, 1966). Cell "selection" in a megasporangium could, perhaps, be explained on a nutritional basis although this is only speculation.

Frequently one or more megaspores do not mature. In certain species more than one megasporocyte is functional; eight, twelve, and even more megaspores have been recorded (Duerden, 1929). An analysis of the stored food in megaspores of *Selaginella* indicates that 48 percent of the contents is fatty material, 0.43 percent is nitrogenous, and 1.26 percent is mineral matter (Sosa and Sosa-Bourdouil, 1940).

Spore morphology is useful in delimiting subgroups within the genus (Fig. 9-23). (See Tryon, 1949, for several interesting plates of photographs illustrating differences in spore wall ornamentation.)

CHROMOSOME NUMBERS. During the past 20 years the ferns have been the focus of cytological studies in lower vascular plants. *Selaginella* has remained neglected perhaps because of the assumptions of uniformity and smallness of chromosomes in the genus (1 micron long or less, in some species). Low chromosome numbers (e.g., $n = 9$) were recorded by Manton (1950). Perhaps *Selaginella* is not as simple, cytologically, as once thought. Some investigators recognize at least four basic chromosome numbers ($x = 8, 9, 10, 11$) with actual chromosome counts of $2n = 14, 16, 18, 20$, with a few being $2n = 50–60$ (Jermy, Jones, and Colden, 1967). Another investigator concluded that perhaps there are only two basic chromosome numbers, $x = 9$ and 10, although 10 may be basic and 9 an aneuploid derivative (Kuriachan, 1963). *Selaginella* remains as a genus which has not undergone spectacular increases in chromosome numbers as are shown by ferns. Polyploidy seemingly has played no major role in the evolution and distribution of the genus.

GAMETOPHYTES. In the genus *Lycopodium*, it will be recalled, spores do not germinate until their liberation from the sporangium. Also, most species of *Lycopodium* have a long and protracted growth period which is interpolated between germination and the formation of a mature exosporic gametophyte with sex organs. In *Selaginella,* however, germination of microspores and megaspores begins while they are still within their respective sporangia. In a microsporangium at a late stage of development the radial and inner tangential walls of the outer layer of the sporangium thicken; the

A

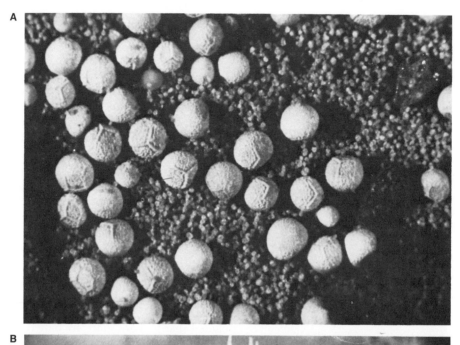

B

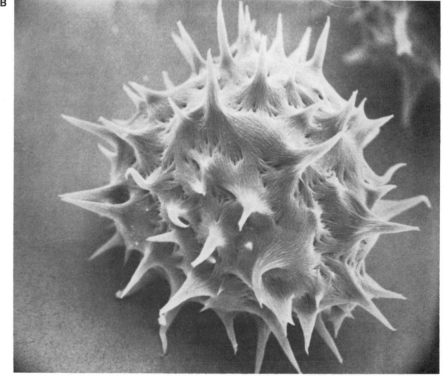

Figure 9-23
Spores of various species of *Selaginella*. **A**, spores
of *Selaginella* sp. illustrating dramatic difference in
size of large megaspores and numerous small
microspores ($\times$ 40); **B–D**, scanning electron
micrographs of spores; **B**, microspore of
Selaginella kraussiana ($\times$ 1200); **C**, microspores of
Selaginella flabellata ($\times$ 650); **D**, a megaspore of
Selaginella flabellata ($\times$ 200). [A, courtesy of
Dr. E. G. Cutter; **B–D**, courtesy of Dr. R. H. Falk.]

C

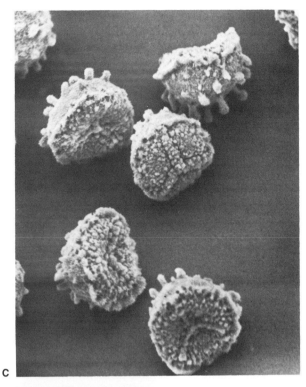

D

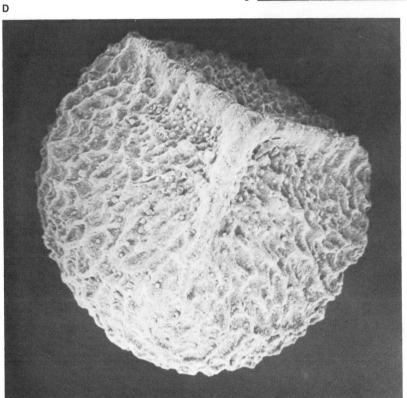

174

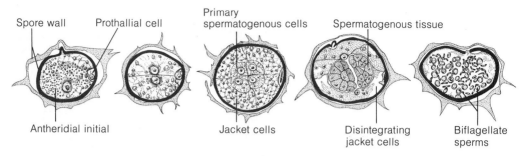

Spore wall   Prothallial cell   Primary
                                   spermatogenous cells   Spermatogenous tissue

Antheridial initial   Jacket cells   Disintegrating   Biflagellate
                                                     jacket cells   sperms

Figure 9-24
Development (from left to right) of the microgametophyte in *Selaginella kraussiana.* Early development may
occur within the microsporangium prior to sporangial dehiscence. [Redrawn from Slagg, *Amer. Jour. Bot.*
19:106, 1932.]

inner wall layer becomes stretched, and the tapetum may still be visible. Within the microsporangium the microspores (which may still be held together in tetrads) are tetrahedral and have developed a thick outer wall, which may be laid down in the form of varied spines or projections. The first division of the microspore results in the formation of a small prothallial cell and a large antheridial initial (Fig. 9-24). The antheridial initial becomes subdivided by several divisions in various planes. This results in the establishment of a sterile jacket enclosing four primary spermatogenous cells, all within the original microspore wall. Forceful ejection of the partially developed microgametophytes may occur at this time by splitting of the upper half of the sporangium along a line of dehiscence. This is followed by differential shrinkage of the lower portion. Whether microgametophytes are shed during this developmental stage or earlier, the final development occurs outside the sporangium wherever the spores may alight. According to Slagg (1932), the primary spermatogenous cells undergo several divisions, forming 128 or 256 spermatocytes which, on disintegration of the jacket cells and rupture of the spore wall along the triradiate ridge, are liberated as free-swimming biflagellate sperms (Fig. 9-24).

Just as early stages in formation of the microgametophyte begin while the microspores are still within the microsporangium, megagametophyte development begins while the megaspores are in the megasporangium. After meiosis in a megaspore mother cell the resulting megaspores soon develop a thick, multilayered cell wall. The raised portion on the spore surface is referred to as the perispore (usually forming a definite relief pattern as seen in face view, Fig. 9-23, D, or appearing as spines or ridges in sectional view, Fig. 9-25). Beneath this layer is a thick portion of the wall termed the exospore (or exine by some authors), which encloses the mesospore and endospore layers (which together are known as the intine). The exospore displays a rather unique organization in being in the form of packets of aligned granules, each packet forming a polyhedron (Martens, 1960a, b; Stainier, 1965). Very early a conspicuous vacuole develops within the enlarging protoplast of the megaspore. Associated with this enlargement is the division of the megaspore nucleus. Nuclear divisions continue without wall formation (a free nuclear phase), producing a thin layer of multinucleate cytoplasm surrounding a large central vacuole. In material processed for microscopic examination the intine may appear to be separate from the exine, but according to Pieniążek (1938) the apparent differential growth of the exine and intine is an artifact. He states that the spore coats remain in contact throughout development if the spores are not subjected to unusual physical or chemical changes. He was able to demonstrate that the spore protoplast can be plasmolysed and deplasmolysed.

Concomitant with an increase in the amount of cytoplasm and stored food is the process of cell-wall formation. The formation of walls around free nuclei takes place first at the apical end near the triradiate ridge (Fig. 9-25). In some species the cellular portion of the sexually mature megagametophyte may consist of only a few layers of cells separated from the large multinucleate vesicle. Only after the embryo begins development does the vesicle become cellular. In other species cell formation is a continuous process proceeding basipetally from the apical region. Irrespective of the timing of cell-wall formation, archegonia soon make their appearance in the apical region (Fig. 9-26, A). Each archegonium develops from a single superficial cell and at maturity consists of eight neck cells (arranged in two tiers of four cells each), a neck-canal cell, a ventral canal cell, and an egg cell (Fig. 9-26, B). The megaspore wall breaks along the triradiate ridge, apparently because of the growth of the cellular gametophyte (Figs. 9-25; 9-26, A; 9-28, A). In some species prominent lobes of tissue are produced, with certain cells forming tufts of rhizoids (Bruchmann, 1912).

It has been reported that a certain amount of chlorophyll is present in the upper part of the gametophyte, but this has never been confirmed. The amount of food produced through photosynthesis (if photosynthesis occurs at all) is probably small, and most of the nutritional requirements are met by the storage tissue in the lower part of the megagametophyte.

To achieve continuity in the description of gametophyte development, we have, to

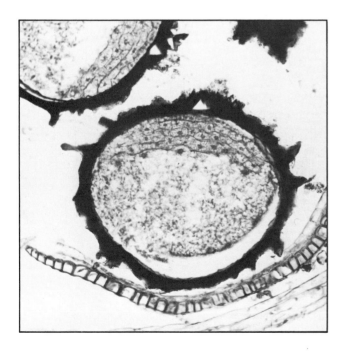

Figure 9-25
Section of developing cellular megagametophyte within thick megaspore wall of *Selaginella* sp. A large storage vesicle is present beneath the cellular tissue. The triradiate ridge is indicated by a triangular shaped space (white) at the upper end of the spore. The developing endosporic gametophyte is still enclosed within the megasporangium.

this point, made no mention of the time of natural shedding of the megagametophyte from the sporangium. The time of liberation is variable in the genus but may occur (1) at any time before the cellular stage, (2) after the archegonia appear, or (3), in two species, after development of the embryo. In species in which the time of liberation is (1) or (2), the final stages in megagametophyte development and fertilization take place while the megagametophyte rests on decaying portions of the sporophyte or on the soil or humus. In such species the rhizoids undoubtedly play an important role in the absorption of water and in anchorage. The partially developed microgametophytes complete their development while situated on the exposed megagametophyte tissue or in close proximity to it. After the sperms are liberated they swim to the archegonia in a thin film of dew or rain water.

EXPERIMENTAL STUDY OF THE MEGAGAMETOPHYTE. The successful experiments in culturing the gametophyte of *Lycopodium* have been described earlier in this chapter. The same workers (Wetmore and Morel, 1951b) have cultured the female gametophytes of two species of *Selaginella*. On the culture substrate the gametophytes remain alive for six months, and if vitamins are added large masses of undifferentiated tissue are produced, which are covered with rhizoids and archegonia.

THE EMBRYO.   After fertilization, the diploid or sporophyte generation is established. The first division of the zygote is transverse, separating a suspensor cell (that cell toward the archegonial neck) and the embryonic cell (labeled "apex" in Fig. 9-27, A) that will form the remainder of the embryo (see Chapter 6).

The embryo, as can readily be understood, is endoscopic. The suspensor cell may remain undivided or form several cells. In either instance the suspensor serves to thrust the developing embryo proper into the megagametophyte tissue. An apical cell is formed as a result of vertical and oblique

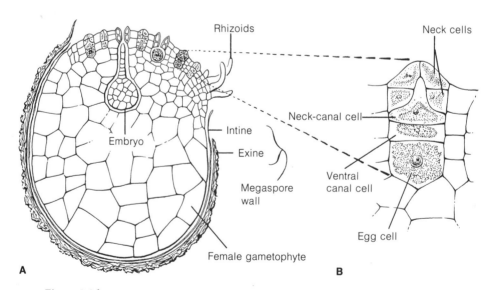

Figure 9-26
A, section of endosporic, cellular megagametophyte of *Selaginella denticulata;* B, details of archegonium. [Redrawn from Bruchmann, *Flora* 104:180, 1912.]

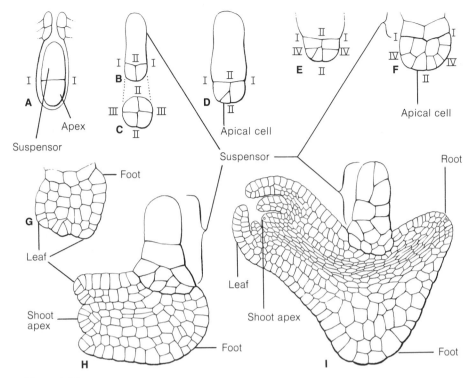

Figure 9-27
Embryogeny in *Selaginella martensii.* In all of the sketches the neck of the archegonium is directed toward the top of the page. The walls resulting from early cleavages are indicated as I–I, II–II, etc. An apical polar view of **B** is shown in **C.** (For details consult text.) [B–I redrawn from Bruchmann, *Flora* 99:12, 1909.]

divisions of the embryonic cell (Fig. 9-27, D). The apical cell of the future shoot apex is thus established early in embryogeny (Bruchmann, 1909a, 1912). Then, the first two foliar appendages, often referred to as "cotyledons," arise laterally and exogenously from the shoot apical meristem (Fig. 9-27, G). At approximately this stage of development the embryo shoot apex undergoes a 90-degree turn by differential growth (Fig. 9-27, H). Active cell division along the lower side of the developing embryo produces a definable foot (Fig. 9-27, I). Subsequently, periclinal divisions in superficial cells between the foot and the suspensor produce the root cap. Later, a definable root apical cell makes its appearance some distance from the surface (Bruchmann, 1909a). Con-

siderable variation exists in the origin and position of foot and root (Fig. 6-4).

By continued apical growth, cell enlargement, and differentiation of shoot and root, the young sporophyte emerges from the gametophyte tissue. That portion of the shoot below the first foliar appendages elongates rapidly (Fig. 9-28, B, C). In many species the first dichotomy of the shoot takes place immediately above the first pair of leaves. The primary root grows downward and enters the soil. The foot remains in close connection with the gametophyte, which, in turn, is surrounded by the megaspore wall. The complete separation of the sporophyte from the gametophyte may not occur until the sporophyte has undergone considerable growth.

Figure 9-28
Megagametophyte and young sporophyte of *Selaginella kraussiana*. **A**, gametophyte tissue protruding through cracked spore wall; **B**, young sporophyte attached to gametophyte, showing root (to the right), stem, and first pair of leaves (oriented in the plane of the page); **C**, older sporophyte (megaspore wall with enclosed gametophyte is still visible at the juncture of stem and root).

Carboniferous Relatives of *Selaginella*

Some herbaceous plants from the Carboniferous resembling *Selaginella* have been known for many years. The plants were small and were clearly heterosporous. Some of them were described as *Selaginellites,* others as *Selaginella.* One paleobotanical case history is of interest. Most species thought to be herbaceous and *Selaginella*-like, were described from compressions in a poor state of preservation. But in 1954 a ligulate lycopod was described from the Carboniferous, which had small stems and spirally arranged leaves. This plant was given the generic name *Paurodendron.* The stems were in an excellent state of preservation (Fry, 1954). In 1966 another plant of the same genus was found that had a basal root-producing portion, termed a rhizomorph (Phillips and Leisman, 1966). Complete knowledge of the plant was brought together in 1969 when bisporangiate strobili were found attached to stems. The entire plant was then described as a species of *Selaginella* because it resembled to a remarkable degree a certain modern-day species (*Selaginella selaginoides*) of the genus (Schlanker and Leisman, 1969). This is of great morphological interest because it shows rather convincingly that the genus *Selaginella,* as we recognize it today, probably existed in the Carboniferous, millions of years ago.

Lepidodendrales

In addition to the low-growing, herbaceous lycopods of the Carboniferous there were also large, tall trees with lycopsid features. These trees dominated the Carboniferous scene and are the best known of fossil lycopods. These tree-lycopods of the Carboniferous belong to the order Lepidodendrales. A lepidodendrid was arborescent, heterosporous, ligulate, and had distinctive markings on leaf cushions of the stem.

LEPIDODENDRON.   One genus about which considerable information has been accumulated is *Lepidodendron.* Well-preserved material, in particular the trunks, is especially common in coal beds in Great Britain and the central United States. Many of these dendroid plants attained a height of 100 feet or more and were 2–3 feet in diameter at the base. The trunk was unbranched for some distance and then terminated in a dichotomously branched, umbrella-like shoot system in which the branches were clothed with spirally arranged, linear or awl-shaped microphylls with decurrent leaf bases (Fig. 9-29, A). The leaves, upon abscission from

the larger branches, left a distinctive persistent scar (Fig. 9-30, A, B) that is used today as one feature in the separation of different genera in the Lepidodendrales. Elongate strobili terminated the ultimate branches (Fig. 9-29, A). The base of the trunk was divided into four large root-like structures (rhizophores), each of which was divided again to form two structures. Each of these prop-like structures branched repeatedly (Fig. 9-29, A).

Even though the trunk of one of these trees was very large, it had very little vascular tissue. Whether the primary vascular cylinder was protostelic or siphonostelic, the primary xylem was exarch in development. A vascular cambium produced a narrow cylinder of comparatively uniform secondary xylem consisting of scalariform-pitted tracheids and uniseriate rays. A relatively small amount of phloem was formed toward the outside. The entire vascular cylinder, even in quite large stems, was never more than several centimeters in diameter (Fig. 9-31).

The bulk of the stem consisted of a primary cortex and the "secondary cortex" or periderm produced centripetally by a cambium (phellogen). The small amount of xylem produced in relation to the large size of the trunk is remarkable since very little strengthening would be provided by primary cortex and periderm. This disparity between tree size and strengthening tissue may have literally led to the downfall of the tree-lycopods and their disappearance from the world's flora toward the end of the Carboniferous.

Preservation of lepidodendrids has been so good that something is known about the ontogeny of the plant. The results of certain studies indicate that the apical meristem and branches were reduced in size at each dichotomy until growth was no longer possible and apical growth became determinate (Andrews and Murdy, 1958; Eggert, 1961).

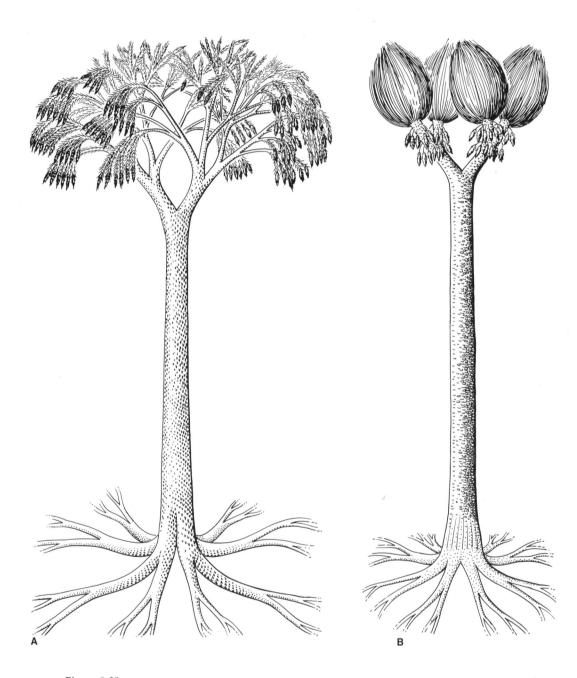

Figure 9-29
Diagrammatic reconstructions. **A**, *Lepidodendron* sp.; **B**, *Sigillaria elegans.* Note strobili and the large rhizophores at the base of the trunks. *Form* or *organ* genera exist for all basic parts of the plant bodies. (Consult text for pertinent information.) [Redrawn from *Handbuch der Paläobotanik* by M. Hirmer. R. Oldenbourg, Munich, 1927.]

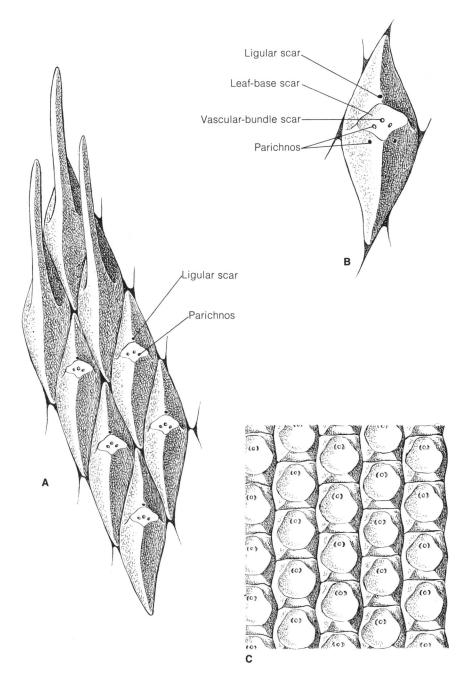

Figure 9-30
A, diagrammatic representation, portion of the surface of a branch of an arborescent lycopod (*Lepidodendron* sp.) showing three attached leaves and the scars left by the abscission of five others. B, one leaf scar of a lepidodendrid showing two sets of parichnos scars. C, surface of the stem of *Sigillaria* sp.

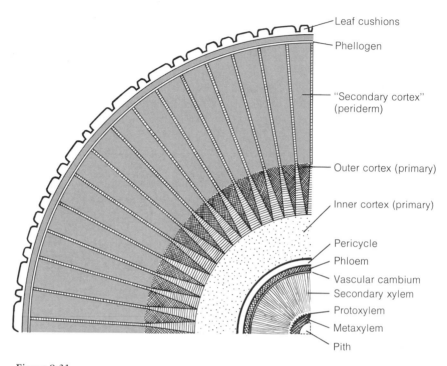

Figure 9-31
Schematic representation of the stem of *Lepidodendron vasculare* as seen in transverse section.
[Redrawn from *Handbuch der Paläobotanik* by M. Hirmer. R. Oldenbourg, Munich, 1927.]

FORM GENERA. From the account in the previous section the reader might conclude that the discovery of intact remains of this arborescent genus is indeed remarkable. However, the story is not as simple as we have recounted. To uncover a fossil plant with all parts intact is the dream of paleobotanists, but, unfortunately, at best only unconnected portions or separate organs of large plants are generally discovered. This is understandable if consideration is given to the amount of breakage possible during transport by water to sites of final preservation. Also the ontogeny of a plant has a bearing on the problem since some structures (strobili, spores, etc.) may be formed either at regular intervals or only near the end of the life cycle (Beck, 1969). Because plant fossils are usually only fragments,

generic names, as initially applied, rarely refer to whole plants. This dilemma was appreciated by the early paleobotanists and led to the establishment of *form genera* or *organ genera* for isolated plant fragments, which resulted in the accumulation of numerous, but necessary, generic names for leaves, stems, strobili, etc. For example, the genus *Lepidodendron* originally referred to portions of the trunk of lycopods with the distinctive leaf-scar pattern. The detached fossil leaves (sometimes found attached to stems) were placed in the organ genus *Lepidophylloides*. The massive, underground rootlike organ of lycopods was known as *Stigmaria,* another organ genus (Fig. 9-32, A, B). These dichotomizing axes bore "stigmarian" rootlets or appendages in a spiral pattern over their entire surface.

A

B

Figure 9-32
Tree stumps of lepidodendrids in "Fossil Grove," Victoria Park, Glasgow, Scotland.
The basal, dichotomously branched lobes, to which roots were attached, are designated as
Stigmaria (a form genus). These fossils are casts of the original tree. The stumps, which
measure 15–40 inches in their widest diameter, were exposed by carefully removing the
hard rock that encased them. [Photographs courtesy of Dr. E. G. Cutter.]

Usually the rootlets are detached and only the scars are visible. The internal structure of these rootlets is interesting because it resembles to a remarkable degree that of the roots of *Isoetes;* in both, the root has a vascular strand supported by a flange of the cortex in a large central air cavity (Fig. 9-37). By continued comparative studies and through fortunate discoveries of organs that were connected or in close proximity, paleobotanists were certain of the antici-pated conclusion that the organ genera just mentioned were merely parts of the same plant type, now designated as *Lepidodendron,* a natural genus (all parts known). Where all parts are known it has been customary to use the generic name originally applied to the form genus for stem and leaf cushions. Other natural genera have been described, e.g., the sparsely branched *Sigillaria* (Figs. 9-29, B; 9-30, C), which has terminal tufts of grasslike leaves, and *Lepidophloios.*

The strobili of *Lepidodendron,* which are found detached or occasionally in organic connection with a leafy stem, were desig-nated *Lepidostrobus.* These cones were long (up to 10–25 cm) with spirally arranged overlapping ligulate sporophylls. One spo-rangium was present on the adaxial side of each sporophyll. In some specimens there are microsporangia, enclosing several hun-dred microspores, in the upper portion of the strobilus, and megasporangia with fewer megaspores in the lower part of the cone (Brack, 1970). Megaspores were shed from the sporangia, and development of the mega-gametophyte was endosporic. Certain species of *Lepidodendron* progressed to an evolu-tionary level of producing seed-like struc-tures. *Lepidocarpon* is the organ genus represented by strobili in which typically only one functional megaspore, with its enclosed megagametophyte, was per-manently retained within the sporangium. Outgrowths from the sporophyll enclosed the entire structure except for a slit-like

opening along the top. At maturity the sporophyll and the megasporangium with the enclosed megagametophyte were shed as a unit. Archegonia occurred in the female gametophyte toward the opening at the tip; however, no embryos have ever been discovered. More detailed descriptions of this group of plants can be obtained from paleobotany books; see, for example, Darrah, 1939; Hofmann, 1934; Arnold, 1947; Andrews, 1961; Delevoryas, 1962).

Isoetales — Isoetaceae: *Isoetes*

The genus *Isoetes* is a most interesting, pro-vocative, and enigmatic vascular plant. The plant body of all species is relatively small with a greatly shortened axis, and has tufts of leaves and roots which in appearance resemble certain monocotyledonous plants (Fig. 9-33). This resemblance to seed plants caused Linneaus, as early as 1751, to characterize the plant as a seed plant; even the "fruit" and "seeds" were described! Even earlier, the herbalist Ray stated that "of this plant, one sees naught but leaves and roots, and knows not from what source it comes directly."

The first popular names recorded for *Isoetes* were "quillwort" and "Merllyn's Grass." The former name is applied to the genus in America; in Europe it is still known as Merlin's Grass. Economically the genus is relatively unimportant today, but there are records of the plants being eaten occasion-ally in Europe (Pfeiffer, 1922). Great quanti-ties of starch and oil are present in the plant body. Birds, pigs, muskrats, and ducks may eat the fleshy plant body, and cattle often graze on the leaves.

Isoetes includes more than sixty species, the majority of which are found, growing immersed in water (at least part of the year) or in swampy areas, in parts of the world having cooler climates. There are certain

Figure 9-33
Entire plant of *Isoetes* sp. showing crown of tightly packed leaves (sporophylls), short "corm" (dark) and roots.

species that have become adapted to existence in a dry environment at least during a part of the year.

ORGANOGRAPHY. The axial portion of *Isoetes* has been interpreted in various ways, and the complexity of descriptions of the species has been increased by botanists' use of various terminologies for it.

As mentioned earlier, the axis is a short, erect structure (commonly referred to as a corm) which, in the mature plant, may be divided into two, three, or, rarely, four lobes. Along the sides of the grooves or clefts are numerous roots. The upper part of the axis is covered with a dense cluster of leaves which have broad overlapping bases (Figs. 9-33; 9-34, A). The shoot apex is completely

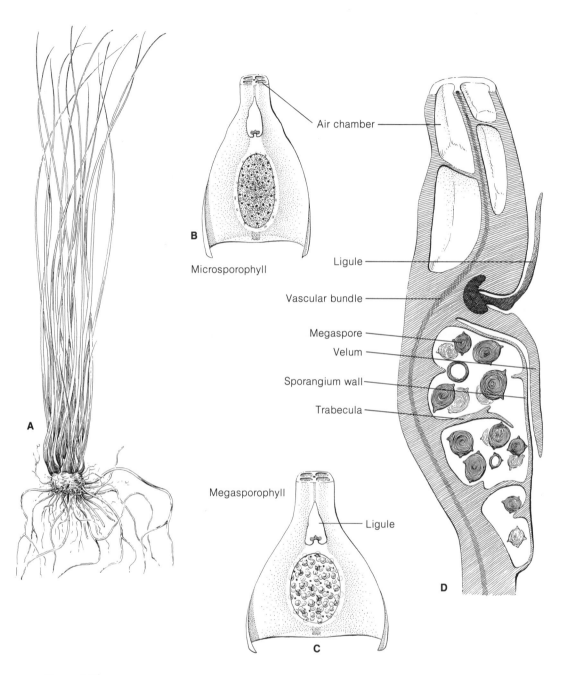

Air chamber

Microsporophyll

Ligule

Vascular bundle

Megaspore

Velum

Sporangium wall

Trabecula

Megasporophyll

Ligule

Figure 9-34
A–C, *Isoetes howellii*. **A**, habit sketch of sporophyte; **B**, adaxial view of base of microsporophyll showing
microsporangium as seen through the opening in the velum (spores are shown as black dots within the
sporangium, the ends of trabeculae as larger black asteroids); **C**, adaxial view of base of megasporophyll with
attached megasporangium; **D**, longisection of megasporophyll and megasporangium of *Isoetes* sp. (markings on
spore wall are entirely schematic).

hidden in a depression by the tightly imbricated leaves (Fig. 9-36, A). The sides of mature plants become very rough with layers of sloughing tissues. Each increment includes leaf bases and severed roots of previous growing seasons, and this sloughed tissue decays very rapidly.

ANATOMY OF THE CORM. Any morphological interpretation of the corm of this peculiar plant must necessarily be based partially on internal structure. The corm has been described variously as an erect rhizome; as a stock; as a stem; as a stem combined with a stigmarian type of rhizophore; and as an upper leaf-bearing part, the stem, and a lower root-bearing part, the rhizomorph (Scott and Hill, 1900; Stokey, 1909; Lang, 1915; West and Takeda, 1915; Osborn, 1922; Stewart, 1947; Bhambie, 1962; Paolillo, 1963). If a mature plant is cut longitudinally in the plane of the basal groove or furrow (considering that the plant is two-lobed), the shoot apex is seen at the bottom of a depression with surrounding leaf bases and leaf traces (Fig. 9-35, A). The xylem core is in the outline of an anchor or vegetable chopper—cylindrical in the upper part with the lower part extended horizontally with upturned arms. Roots and root primordia are evident below the xylem core.

Secondary growth is a characteristic feature of *Isoetes* and the specialized cambium comprises two parts: (1) A *lateral meristem* that gives rise to vascular tissue (prismatic layer) centripetally and to secondary cortical tissues, centrifugally; (2) a *basal meristem,* continuous with the lateral meristem, which adds to the xylem core and produces basally the surrounding ground tissue, in which root primordia become organized.

Both lobes of the corm are evident in a median longitudinal section cut at right angles to the furrow (Figs. 9-35, B; 9-36, B).

A transverse section through the leaf-bearing portion of a corm reveals the xylem core surrounded by the cylindrical prismatic layer and lateral meristem (Fig. 9-35, C). At a lower level the lateral meristem is evident and the basal meristem appears at two locations, reflecting the curved contour of the xylem core (cf. Figs. 9-35, C and D).

Of the many debatable features of *Isoetes,* interpretation of the secondary vascular tissue has probably caused the most discussion. The cells of this tissue, which are derived from a definite storied cambium (lateral meristem), are extremely short, being not much taller than they are wide. The form of these cells, combined with the difficulty of interpretation, led to the general acceptance of the noncommittal term "prismatic layer" for this part of the axis. Nevertheless, this tissue has been interpreted as secondary xylem (Stokey, 1909), as secondary phloem (West and Takeda, 1915), as a secondary tissue containing both tracheids and sieve elements (Scott and Hill, 1900), and as a tissue composed of (1) occasional tracheids, (2) considerable unmodified parenchyma, and (3) specialized parenchyma cells which are concerned with conduction (Weber, 1922).

Paolillo (1963) has provided a detailed study of the secondary vascular tissue (prismatic layer) of *Isoetes howellii.* In young plants of this species the derivatives of the lateral meristem differentiate as sieve elements. In older (larger) plants the inner derivatives of the lateral meristem differentiate as layers of sieve elements, alternating with layers of parenchyma (Fig. 9-36, C). Recognition of sieve elements is made possible by a combination of features: sieve areas, deposition of callose, and thicker walls than those of parenchyma cells (Esau et al., 1953). Nuclei do not always disappear entirely, as is generally true for sieve elements; the nuclei remain small and seem to be partially degenerate. In still older plants some

188

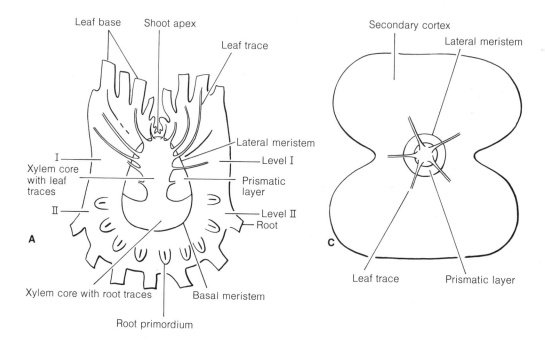

A

- Leaf base
- Shoot apex
- Leaf trace
- Lateral meristem
- Level I
- Prismatic layer
- Level II
- Root
- Xylem core with leaf traces
- Xylem core with root traces
- Basal meristem
- Root primordium

C

- Secondary cortex
- Lateral meristem
- Leaf trace
- Prismatic layer

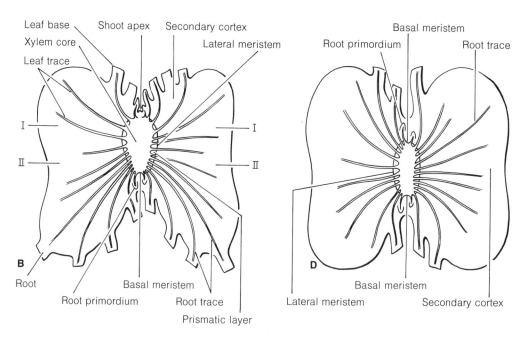

B

- Leaf base
- Shoot apex
- Secondary cortex
- Xylem core
- Lateral meristem
- Leaf trace
- Root
- Root primordium
- Basal meristem
- Root trace
- Prismatic layer

D

- Basal meristem
- Root primordium
- Root trace
- Basal meristem
- Lateral meristem
- Secondary cortex

Figure 9-35
Isoetes howellii. **A, B,** longitudinal sections of two-lobed plants in the plane of the furrow or groove (**A**) and perpendicular to the furrow (**B**); **C,** transverse section at high level (I–I in **A** and **B**); **D,** transverse section at low level (II–II, in **A** and **B**). [Redrawn from Paolillo, *Illinois Biol. Monogr.* 31, 1963.]

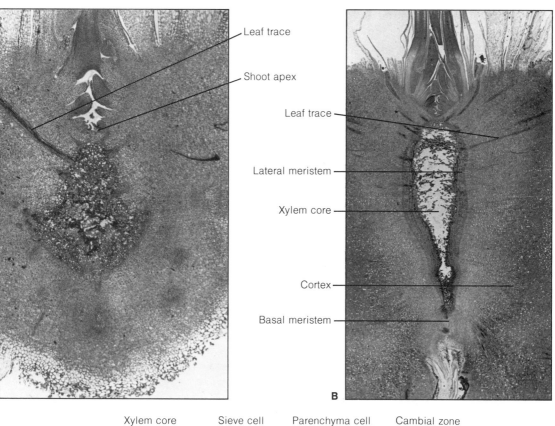

Leaf trace

Shoot apex

Leaf trace

Lateral meristem

Xylem core

Cortex

Basal meristem

B

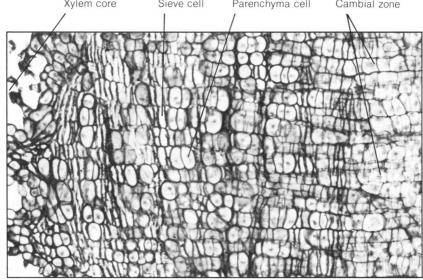

Xylem core   Sieve cell   Parenchyma cell   Cambial zone

C

Figure 9-36
Anatomy of the corm in *Isoetes howellii*. **A,** longisection, in the plane of the basal groove;
B, median longisection, at right angles to the basal groove (several root traces can be seen
at lower right); **C,** a portion of the secondary vascular tissues and adjacent tissues.

or many of the cells in the layers of parenchyma may differentiate as tracheids.

The lateral meristem also produces cells centrifugally which retain their meristematic activity, but ultimately mature into parenchyma cells — constituting a broad secondary cortex (Figs. 9-35, B–D); 9-36, B). The secondary cortex is added to regularly and the outer part sloughs off from the "shoulders" of the corm.

The basal meristem, as noted earlier, contributes to the xylem core and also forms ground tissue distally in which root primordia become organized. The roots eventually penetrate the ground tissue. Roots are produced on either side of the furrow — forming series (rows roughly parallel with the furrow) and orthostichies (rows running at right angles to the furrow). Series are more easily seen near the furrow (See Paolillo, 1963, for details).

Thus far, organography, basic organization, and functioning of lateral and basal meristems (collectively constituting a cambium) have been described. What constitutes primary growth (and tissues) of the corm? In a well-established plant the shoot apex is a low dome consisting of a group of apical initials. Leaves originate in the surface layer at the base of the small apical cone; the cells produced internally enlarge and divide, producing radiating files. It is within this tissue that axial procambium as well as leaf traces differentiate.

The primary plant body of the shoot, then, consists of a core of primary xylem, primary phloem, primary cortex, leaf traces, and leaves. The lateral meristem arises in the layer of procambium remaining outside of the primary phloem and gives rise to secondary vascular tissue (prismatic layer) from its inner face.

Paolillo (1963) regards the basal meristem as part of the cambium because the two meristems (lateral and basal) originate together and, during ontogeny, portions of the lateral meristem may be added to the basal meristem by conversion of initials from one function to another. Earlier, it will be remembered, the term xylem core was used to describe the central core of tissue. This term purposely was used because the origin of the central xylary tissue is from two quite different sources — primary xylem derived from procambium of the upper part of the plant axis and the internally produced cells of the basal meristem.

In summary, the question might be asked: "How is the mature plant form in *Isoetes* related to meristematic activity?" Longitudinal growth, accomplished through the functioning of the apical meristem of the shoot apex and of the basal meristem, is very slow, as evidenced by the extremely short axis and the crowding of appendages. The lateral meristem, which surrounds the erect portion of the vascular cylinder and encloses the sides of the rhizophore portion, gives rise yearly to small increments of secondary vascular tissue and a large amount of parenchyma tissue of the secondary cortex. This accounts for the fact that older plants are much broader than tall. Also, the formation of considerable secondary tissue on the sides of the rhizophore would tend to deepen the groove by the presence of overlying dead and dying secondary cortical tissue.

EXPERIMENTAL MORPHOLOGY. Expression of form in plants is determined largely by the occurrence and distribution of endogenous growth regulators. Perhaps the small, short corm-like form of *Isoetes* is due to a lack of certain endogenous regulating substances. In one experiment the investigator tested the effects of indoleacetic acid and gibberellic acid on *Isoetes*. Applied indoleacetic acid is known to stimulate or inhibit certain growth functions, and treatment with gibberellic acid is known to result in stem elongation in many plants. Except for some proliferation of cortical tissue in the upper part of the corm, stimulated by the indoleacetic acid, there were no other obvious effects on growth (Zinda, 1966). Additional experiments, using combinations of certain growth regulators, would be of great interest to the better understanding of control of form in *Isoetes*.

ROOTS.   Roots branch dichotomously after they emerge from the rhizophore ground tissue. For *Isoetes howellii* the root apical meristem is reported to consist of a layer of initials that gives rise to the cells of the outer cortex, epidermis, and root cap, and a group of initials common to the inner cortex and procambium (Paolillo, 1963).

A transverse section of a mature functioning root reveals a simple type of root of unusual interest. It consists of a cylindrical cortex which surrounds a large air cavity, and a vascular cylinder which is supported in a flange of the cortex in the cavity (Fig. 9-37, A). The primary xylem and phloem are collateral in arrangement, the phloem being oriented toward the cavity. An endodermis, with the usual Casparian strips, is present around the primary vascular tissues. The air cavity is formed by a breakdown of cortical cells throughout the length of that portion of the root which has emerged from

the rhizophore. Histologically the root of *Isoetes* resembles very closely an appendage on the rhizophore of *Stigmaria* (Fig. 9-37, B). The similarity is conclusive enough to support the belief, on the part of some botanists, that a phylogenetic relationship between *Isoetes, Stylites,* and some member of the Lepidodendrales is certain.

THE LEAF.   To this point foliar appendages have been treated simply as leaves, without reference to any special function. Actually each foliar appendage is a potential sporophyll, either a microsporophyll or a megasporophyll. (The terms leaf and sporophyll will be used interchangeably.) Each sporophyll has a thickened and expanded base, with a tapering upper portion which is awl-shaped and pointed. Leaves may be a few centimeters long or even as long as 50 cm (*Isoetes japonica*). The lower parts of the leaves may be buried in the soil and lack chlorophyll, and are commonly a glistening white. Some species have black leaf bases. In many species most of the leaves die and decay at the termination of the growing season. In aquatic species leaves may remain on the plant for some time. The upper portion of a leaf is traversed longitudinally by four large air chambers (Fig. 9-34, B, D) that may be partitioned into compartments

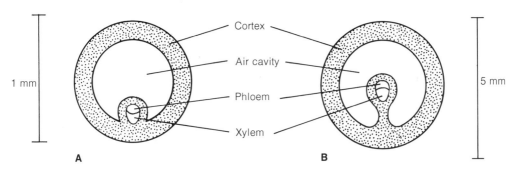

Figure 9-37
Schematic representations. A, transection of root of *Isoetes macrospora;* B, a stigmarian appendage of *Stigmaria*. [Redrawn from Stewart, *Amer. Jour. Bot.* 34:315–324, 1947.]

by transverse tissue diaphragms — the possession of large air cavities is a feature common to many water plants. Stomata occur on the leaves except those that are entirely submerged at all times. Running throughout the length of the leaf (a microphyll) is an unbranched vascular bundle.

Located on the adaxial side near the base of each leaf is a sporangium (Fig. 9-34, B, C). A triangular ligule is present just above the sporangium. Covering or partially covering the face of the sporangium is a protective flap of tissue, the velum (Fig. 9-34, D). All sporophylls have normal sporangia except several of the late-formed sporophylls of a growing season. At the end of a growing season a mature plant generally has an outer group of megasporophylls, which encloses a set of microsporophylls, which in turn encloses several poorly developed leaves with abortive sporangia.

SPOROPHYLL AND SPORANGIUM DEVELOPMENT.   Leaves have their origin in superficial cells of the apical meristem. At first they are crescent-shaped in outline, and growth in length is initiated by apical initials. When the primordium is only a few cells high, a conspicuous adaxial surface cell of the leaf near the base divides periclinally, giving rise to the ligule (Smith, 1900). By repeated divisions a filamentous structure is formed which soon overtops the leaf apex. The ligule eventually becomes tongue-shaped (Fig. 9-34, B, C) and shows a high degree of histologic specialization (a detailed description is given in Chapter 3). The sporangium and velum have their origin through periclinal divisions in surface cells below the ligule. The velum is first to take form, and is followed by growth of the sporangium. As a result of the first periclinal divisions, which localize the sporangial position, a central mass of potentially sporogenous cells is separated from outer layers

of peripheral cells. During development the cells of these outer layers may continue to add derivatives to the sporogenous mass (Smith, 1900). Ultimately the outer three or four peripheral layers constitute the sporangium wall. The cells of the sporogenous tissue divide rapidly in all planes.

Microsporangia and megasporangia are indistinguishable during early stages of development, and it is only after the potential microsporocytes or megasporocytes become apparent that the two types of sporangia can be distinguished. In a microsporangium irregular groups of deeply staining cells ultimately become the microsporocytes, and certain bands of lightly staining cells (originally potentially sporogenous) become the trabeculae, which traverse the sporogenous mass but do not divide it into compartments or locules (Fig. 9-38). Covering the trabeculae is a tapetum which may be biseriate or multilayered, the cells of which are derived also from potentially sporogenous cells. The tapetum of the trabeculae is continuous with a tapetal layer lining the sporangium wall. The microspore mother cells separate from each other prior to the meiotic divisions. Estimates of the number of bifacial microspores produced by a single sporangium range from 300,000 to 1,000,000. The spore number of each sporangium in *Isoetes* is probably greater than it is in any other vascular plant.

In a megasporangium, even before the trabeculae are distinguishable, certain cells are greatly enlarged over their neighbors and will become the megasporocytes (Smith, 1900; Goebel, 1930). Not all of the enlarged cells will become megaspore mother cells, but some will degenerate and be resorbed. It is only after megaspore mother cells are in evidence that trabeculae become apparent. As a result of meiosis, each sporangium contains approximately 100-300 tetrahedral megaspores which may range from 200 to 900 microns in diameter (Fig. 9-34, D).

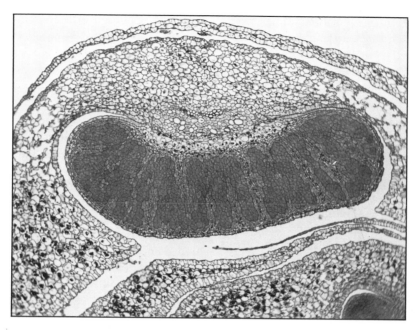

Figure 9-38
Transection of microsporophyll and developing microsporangium of *Isoetes howellii*. Trabeculae can be seen traversing the mass of potential microsporocytes. The bases of other sporophylls are evident.

Megaspores may be white, gray, or black, and they have distinctive protuberances on their thick cell walls. Microspores are very small, from 20 to 45 microns long, are ridged longitudinally, and have various characteristic wall patterns.

Sporangia are, in general, indehiscent, and liberation of spores is brought about by decay of sporophylls in the fall or winter seasons in cooler latitudes. Certain species of *Isoetes* growing in vernal pools, in which the corm is entirely buried, have a special means of exposing the sporangial contents. In these cases decaying sporophylls of the previous season are forced up by the expansion of mucilage cells at the base of the sporophylls, whereupon the spores are brought to the surface (Osborn, 1922). Distribution of spores is by wind, disturbance of mud in which certain species grow, or wave action in lakes; also, earth worms have been reported as carriers (Duthie, 1929).

CHROMOSOME NUMBERS. There is considerable variation in the reported chromosome number and size in *Isoetes*. In one species the chromosome number is $n = 10$, and the chromosomes are very small (1 micron, or less, in length), similar to the genus *Selaginella*. In other species the chromosomes may be 7–8 microns long, and the number may be from 54 to 56 (Dunlop, 1949; Manton, 1950). Additional cytological studies on still other species have led some investigators to postulate that polyploidy has been operative in the history of the genus. Chromosome counts of $n = 11, 22, 33, 54–56, 66$ would support this contention. A chromosome number of $n = 10$ (*Isoetes hystrix*) is considered to have been derived by aneuploidy (Abraham and Ninan, 1958; Ninan, 1958).

Characters such as spore color, extent of velum cover, and leaf length have generally been used to separate species. The results of certain population studies would indicate that these features are quite variable and are greatly modified by the environment. Therefore, some doubt has been cast on the choice of characters used to distinguish species of *Isoetes*. Chromosome numbers are the same $(2n = 22)$, except for one tetraploid, for populations of *Isoetes melanospora* and *Isoetes piedmontana* growing on certain granite outcrops of southeastern United States. Also, the populations have the same polyphenolic patterns (Matthews and Murdy, 1969). Additional studies of large populations of other species may result ultimately in a major revision of the genus.

GAMETOPHYTES. Spores may germinate immediately after being shed from the sporangium, but generally germination does not take place until winter in warm-climate species or spring in cold-climate species, after decay of the sporophylls. The gametophytes, as in *Selaginella*, are endosporic. The microgametophyte is retained entirely within the spore wall, though a portion of the megagametophyte may be exposed.

MICROGAMETOPHYTE DEVELOPMENT. The first division of the microspore forms a small prothallial cell, and a large cell that is interpreted as an antheridial initial. By several divisions the antheridial initial is subdivided into a layer of jacket cells and a total of four spermatogenous cells (the actual sperm mother cells, in this case). After a developmental period of about two weeks, the spore wall cracks along the flat surface. The jacket cells and prothallial cell degenerate, and four multiflagellate sperms, derived from the metamorphosed contents of the sperm mother cells, are liberated (Liebig, 1931). The general scheme of development is comparable to that in *Selaginella* (Fig. 9-24). The possession of sperms with many flagella contrasts strikingly with the bi-

flagellate condition of the sperms of other genera (*Lycopodium* and *Selaginella*) of the Lycopsida.

MEGAGAMETOPHYTE DEVELOPMENT. A mature megaspore contains a considerable amount of stored food surrounded by a spore wall of at least three layers. The primary nucleus is quite large and may be at the base or apex (toward the triradiate ridge) (Campbell, 1891; LaMotte, 1933). A period of free nuclear division ensues until about fifty free nuclei are distributed around the periphery of the protoplast. Wall formation then takes place rapidly around nuclei at the apex, proceeding basipetally and centripetally at a slower rate. Stored material is prominent in the basal end of the developing megagametophyte. No large central vacuole, as in *Selaginella*, is ever present. The megagametophyte may not become entirely cellular until the embryo is quite advanced in development (LaMotte, 1933). With an increase in volume of the megagametophyte, the megaspore wall breaks along the triradiate ridge (Fig. 9-39). The first archegonium appears at the apex (Liebig, 1931), and at maturity consists of four tiers of neck cells, a neck canal cell, a ventral canal cell, and an egg cell. If fertilization does not occur, many more archegonia may be formed among rhizoids that extend above the surface of the gametophyte (LaMotte, 1933).

THE EMBRYO. After fertilization, the first division of the zygote is transverse or oblique to the long axis of the archegonium (Fig. 9-40, A). The embryo becomes globose, and in some instances quadrants (Fig. 9-40, B) are recognizable (Campbell, 1891, 1940; LaMotte, 1937). The upper half (hypobasal) of the embryo (toward the neck of the archegonium) gives rise to the foot and root, and the lower half (epibasal)

Figure 9-39

Megagametophytes of *Isoetes howellii* protruding through cracked spore walls. In each case, the megaspore wall was ruptured along the triradiate ridge, and three portions of the wall are visible. The dark areas are archegonia, and one tier of four neck cells is apparent in an archegonium of the gametophyte to the left.

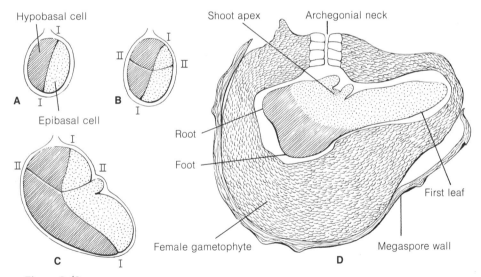

Figure 9-40

Early cleavages and subsequent growth of the embryo in *Isoetes lithophila*. **A**, the first cleavage (I–I) in this species is usually as shown, only rarely being at right angles to the neck of the archegonium; **B**, quadrant stage; **C–D**, later stages, each segment being multicellular, but not indicated by cells. More rapid growth occurs in the lower two quadrants, resulting in an apparent rotation of the embryo. [Redrawn from La Motte, *Ann. Bot.* n.s. 1:695, 1937; *Amer. Jour. Bot.* 20:217, 1933.]

produces the first leaf and shoot apex. The embryo is thus endoscopic by definition, but it will be noted that no suspensor is formed as in *Lycopodium* and *Selaginella.* Interpretation of subsequent embryonic development is indeed difficult, though the following description may represent a reasonably accurate account of development (LaMotte, 1933, 1937).

That portion of the embryo which will become the foot grows downward obliquely and into the storage tissue of the mega-gametophyte. At the same time, that portion of the embryo which will produce the first leaf and shoot apex grows laterally or perpendicular to the long axis of the arche-gonium. The primary root grows in the same plane but in the direction opposite from the leaf. Reorientation of the embryo is thereby achieved (Fig. 9-40, C, D). With further development the first leaf of the sporophyte breaks through the gametophytic sheath; the root emerges and turns downward. The young sporophyte may become firmly established on the substrate, but it remains attached for some time to the gametophyte and surrounding megaspore wall.

Isoetales – Isoetaceae: *Stylites*

The family Isoetaceae was considered to be monotypic until 1957 when another genus was established for plants which were found growing in dense cushions around the boggy margins of a small glacial lake high in the Andes of Peru (Amstutz, 1957). The plants resembled superficially certain species of *Isoetes* but the corms were much more elongate, up to several centimeters long, and branched dichotomously (Fig. 9-41, A, B). Secondary growth occurred, as in *Isoetes,* but roots generally were present only on one side of each branch of a dichotomy. The generic name *Stylites* was used because of the columnar, pointed

corm. The morphology, anatomy, and life history of the originally described species (*Stylites andicola*) and those of another species (*Stylites gemmifera*) are now well known (Rauh and Falk, 1959a, b). A reinvestigation of *Isoetes triquetra* from the Andes of Columbia revealed many similarities to *Stylites,* particularly with regard to position of sporangia and similarity of root anatomy (Kubitzki and Borchert, 1964). Kubitzki and Borchert believe that the systematic validity of the genus *Stylites* should be questioned. Whether the genus ultimately will be maintained is an open question. Perhaps the two described species of *Stylites* should be considered only as extremes of the morphological variation within the genus *Isoetes.*

Pleuromeiales

A representative member of the order Pleuromeiales is the extinct Triassic genus *Pleuromeia* (Fig. 9-41, C). The plant body consisted of an erect, unbranched stem about 1 meter or more in height and 10 cm in diameter. The stem axis terminated in a strobilus composed of closely overlapping sporophylls with apparently adaxial sporangia (Emberger, 1968; Delevoryas, 1962). Spirally arranged vegetative leaves were present beneath the bisporangiate strobilus. A very remarkable structural feature of the genus was the enlargement of the stem base. This basal part was divided commonly into four stigmarian-like lobes (a lobed rhizophore) upon which roots were produced in an orderly fashion.

Understandably, comparisons have been made between the large root-bearing structures of lepidodendrids (*Stigmaria*), *Pleuromeia, Stylites,* and *Isoetes.* It is quite likely that in *Pleuromeia* roots were produced in the same way as they are in *Isoetes.* Anatomically the roots of *Pleuromeia* were of the

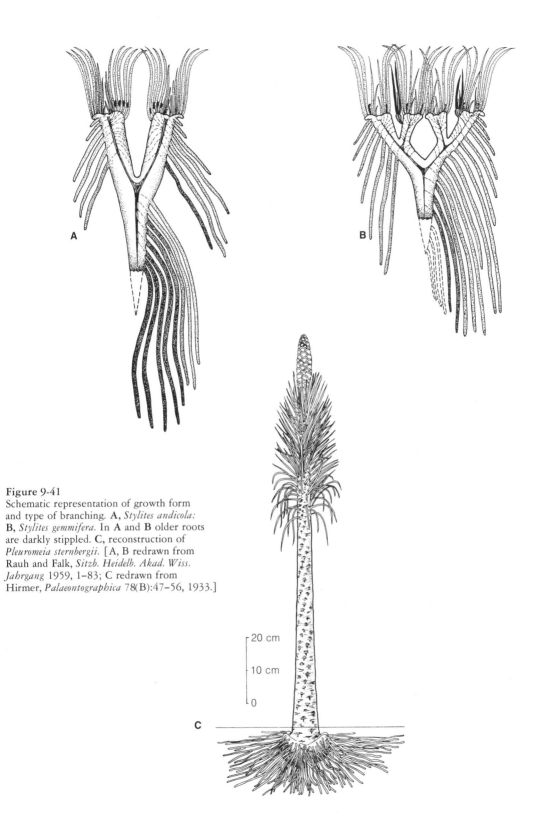

Figure 9-41
Schematic representation of growth form
and type of branching. **A,** *Stylites andicola:*
B, *Stylites gemmifera.* In **A** and **B** older roots
are darkly stippled. **C,** reconstruction of
Pleuromeia sternbergii. [A, B redrawn from
Rauh and Falk, *Sitzb. Heidelb. Akad. Wiss.
Jahrgang* 1959, 1–83; C redrawn from
Hirmer, *Palaeontographica* 78(B):47–56, 1933.]

stigmarian type (large central cavity with a flange of cortex supporting the vascular cylinder), a characteristic feature of *Isoetes* today. On the basis of root structure alone, some paleobotanists are convinced of the phylogenetic relationship between the lepidodendrids, *Pleuromeia,* and *Isoetes.* They further visualize the *Stylites* and *Isoetes* plant bodies as being of a *Pleuromeia* type in miniature, brought about by telescoping of the entire axis. An opposed view is that *Isoetes* has come from a smaller, more herbaceous plant (e.g., *Paurodendron,* p. 178). Plants with all the features of a modern *Isoetes,* and of Triassic age, have been found and given the generic name *Isoetites.*

Summary and Conclusions

The class Lycopsida includes a wide range of extinct and living forms, the study and comparison of which have shed important light on the morphology and evolutionary trends in primitive vascular plants. The morphological unity of the Lycopsida as a whole is remarkable, and the definitive characters of this subdivision are as follows: (1) the basic pattern of branching is dichotomous; (2) the leaves are typical microphylls, each of which is traversed by a single unbranched vein; (3) the primary xylem is predominantly exarch, the vascular system in the stem is commonly protostelic, and there are no leaf gaps; (4) in all lycopsid plants (extinct or living), a *solitary adaxial sporangium* (eusporangiate in type) is associated with each sporophyll, and is either situated in the axil of the sporophyll or attached adaxially.

Certain members of the Lycopsida that lived during the Carboniferous Period were very large tree-like plants, some of which (e.g., *Lepidodendron*) are believed to have attained a height of 100 feet or more. These giant lycopods exhibited secondary growth,

and their reconstruction has involved the study and synthesis of the fossil remains of trunks, leaves, root-like organs, and strobili. Many of these fossilized structures were at first described as separate "form genera," but subsequently were recognized as fragments of one main type of plant, the type, for example, now designated as *Lepidodendron.* Other Paleozoic lycopods were relatively small herbaceous plants which resembled rather closely certain living members of the Lycopsida (e.g., *Lycopodium* and *Selaginella*). The comparative morphology of extinct representatives of the Lycopsida has been briefly discussed at appropriate points in this chapter under the orders Lycopodiales, Selaginellales, Lepidodendrales, and Pleuromeiales.

The living members of the Lycopsida are worldwide in distribution, and are represented by five genera of comparatively small herbaceous plants: *Phylloglossum* (one species), *Lycopodium* (about 200 species), *Selaginella* (the largest genus with 600–700 described species), *Isoetes* (more than 60 species) and *Stylites.* Two of these genera, *Phylloglossum* and *Lycopodium,* are homosporous, with leaves and sporophylls devoid of ligules—they are classified under the family Lycopodiaceae. *Selaginella* and *Isoetes,* in contrast, are heterosporous and ligulate, but differ sufficiently in other characters to justify their segregation into the orders Selaginellales and Isoetales, respectively. The salient morphological features of living lycopsids may be conveniently summarized and compared under the following topics:

ORGANOGRAPHY.   In marked contrast with the Psilopsida, all living lycopsids form roots, which arise and develop in various ways. *Isoetes* is exceptional because a well-defined root system is produced from the basal end of the thick, fleshy corm. In *Selaginella* highly distinctive rhizophores (structures which arise at the points of branching of the shoot) are produced. The roots of *Lycopodium* arise endogenously within the stems and often traverse the cortex for some distance before emerging into the soil or substrate.

The microphyllous leaves of living lycopsids vary considerably in form and size. In *Phylloglossum* the fleshy tuber-like stem bears a cluster of narrow leaves, whereas the leaves of *Lycopodium* and *Selaginella* are small and often scale-like. *Isoetes* develops larger quill-like leaves consisting of a basal sheath and a cylindrical or angled lamina; each leaf in *Isoetes* is potentially fertile and bears a large solitary adaxial sporangium near its base.

The pattern of branching of the shoot is fundamentally dichotomous in *Selaginella* and *Lycopodium,* but in each genus asymmetrical development may result in an anisotomous system. In contrast, *Phylloglossum* and *Isoetes* are typically unbranched; occasional specimens of *Isoetes,* however, may be dichotomously forked. Repeated dichotomous branching is a regular feature of *Stylites.*

VASCULAR ANATOMY OF THE STEM. The vascular system in the short axis of *Phylloglossum* is a siphonostele with mesarch xylem; in the other genera, despite considerable variation in stelar type, the primary xylem is *exarch.* The structure of the stele in *Lycopodium* ranges from an actinostele (characteristic of the young stem of most species) to elaborate types of vascular cylinders in which the primary phloem and primary xylem occur as alternating interconnected strips (plectostele), or else the strands of xylem develop as a mesh-like mass with included groups of phloem. Stelar morphology within the species of *Selaginella* is likewise extremely variable; the vascular system may be protostelic or siphonostelic, or in some species a series of concentric meristeles may be formed. Unlike other genera in the Lycopsida, certain species of *Selaginella* develop vessels in the primary xylem. The primary vascular system of the corm of *Isoetes* is interpreted as a protostele, and its detailed form and structure have been discussed in this chapter. One of the many remarkable characters distinguishing *Isoetes* and *Stylites* from other living lycopsids is the development of a cambium. This meristem arises just outside the cylinder of primary phloem, and from it are derived externally the parenchyma constituting the so-called secondary cortex and internally the peculiar prismatic layer. The latter may consist of sieve cells, parenchyma, and tracheids in varying proportions and constitutes a tissue of debatable nature, the varied interpretations of which have been briefly considered.

SPORANGIA. As previously noted, the sporangium of the Lycopsida is eusporangiate in type and in an adaxial position with reference to the subtending leaf or sporophyll. In *Isoetes* each leaf of the sporophyte is potentially a sporophyll, and a distinct strobilus is not formed. On the contrary, in *Phylloglossum, Selaginella,* and most species of *Lycopodium* the sporangia are within strobili. *Phylloglossum* and *Lycopodium* are homosporous, and each functional sporangium produces a relatively large number of tetrahedral spores. In the two heterosporous genera, *Selaginella* and *Isoetes,* the early stages in development of the microsporangium (which forms numerous microspores) and the megasporangium (which forms fewer megaspores—commonly only four in each megasporangium of *Selaginella*) are quite similar. But as development continues in a microsporangium a large portion of the sporocytes undergo meiosis and produce spore-tetrads. In contrast, the further development of a megasporangium in *Selaginella* is generally characterized by the abortion of all but one sporocyte, which divides meiotically to form a single tetrad of large megaspores. The ontogeny of the megasporangium in *Isoetes* is comparable except for the survival of a larger number of functional megasporocytes. The sporangia of *Isoetes* and *Stylites* are extremely large, and are structurally distinguished from the sporangia of the other three genera by the development of sterile plates of tissue (trabeculae) which traverse but do not completely separate the mass of spores into compartments.

GAMETOPHYTES.   The exosporic bisexual gametophytes in *Lycopodium* vary, according to the species, in growth habit. Some may be wholly subterranean and devoid of chlorophyll, others are partially exposed with green aerial branches. In species with subterranean gametophytes an endophytic fungus enters the young gametophyte and becomes localized to certain regions of the mature thallus. Apparently the presence of such a fungus is essential to the normal development of the gametophyte in nature, although recently it has been possible to germinate and to grow gametophytes which, under sterile conditions in culture, formed sex organs. The subterranean types of gametophytes vary widely in form; some are conical, others are discoid with a convoluted or corrugated surface. Normal gametophytes produce abundant archegonia and antheridia and are apparently long-lived. The gametophytes of *Selaginella* and *Isoetes* are highly reduced and essentially endosporic structures. The male gametophyte in these two genera consists of only a few cells, namely, a prothallial cell, a sterile cellular layer, and the spermatogenous tissue. Each male gametophyte in *Selaginella* forms 128–256 sperms, whereas in *Isoetes* only four sperms are formed. The female gametophytes in *Selaginella* and *Isoetes* are more robust and begin development by a process of free nuclear divisions within the megaspores. This is followed by wall formation and the initiation of archegonia from certain of the superficial cells of the apical region of the megagametophyte. The megaspore wall ruptures along the triradiate ridge, exposing the female gametophyte and thus facilitating the fertilization of one or more of the archegonia by the sperms. In both *Lycopodium* and *Selaginella* the mature sperms are biflagellate, and in *Isoetes* the sperms are multiflagellate.

The polarity of the embryo in *Lycopodium, Selaginella,* and *Isoetes* is endoscopic, although the manner of subsequent growth of the embryo in *Isoetes* results in a reorientation of parts such that the shoot apex faces upward and the foot lies adjacent to the storage tissue of the megagametophyte. A suspensor is present in the embryos of *Lycopodium* and *Selaginella*, but not in the embryo of *Isoetes*.

References

Abraham, A., and C. A. Ninan
  1958. Cytology of *Isoetes. Curr. Sci.* 27:60–61.
Amstutz, E.
  1957. *Stylites*, a new genus of Isoetaceae. *Ann. Missouri Bot. Gard.* 44:121–123.

Andrews, H. N., Jr.
  1961. *Studies in Paleobotany.* Wiley, New York.
Andrews, H. N., Jr., and W. H. Murdy
  1958. *Lepidophloios* — and ontogeny in arborescent lycopods. *Amer. Jour. Bot.* 45:552–560.

Arnold, C. A.

1947. *An Introduction to Paleobotany.* McGraw-Hill, New York.

Barclay, B. D.

1931. Origin and development of tissues in stem of *Selaginella Wildenovii. Bot. Gaz.* 91:452–461.

Beck, C. B.

1969. Problems of generic delimitation in paleobotany. Pp. 173–193 in *Proceedings of the North American Paleontological Convention* Part C.

Bhambie, S.

1962. Studies in pteridophytes. II. A contribution to the anatomy of the axis of *Isoetes coromandelina* L. and some other species. *Proc. Indian Acad. Sci.* (Sect. B) 56:56–76.

1965. Studies in pteridophytes. V. The development, structure and arrangement of leaves in some species of *Lycopodium. Proc. Indian Acad. Sci.* (Sect. B) 61:242–252.

Bhambie, S., and V. Puri

1963. Shoot apex organization in Lycopodiales. *Mem. Indian Bot. Soc.* 4:55–63.

Blackwood, M.

1953. Chromosomes of *Phylloglossum drummondii. Nature* 172:591–592.

Bower, F. O.

1935. *Primitive Land Plants.* Macmillan, London.

Brack, S. D.

1970. On a new structurally preserved arborescent lycopsid fructification from the lower Pennsylvanian of North America. *Amer. Jour. Bot.* 57:317–330.

Bruchmann, H.

1909a. Von den Vegetationsorganen der *Selaginella Lyallii.* Spring. *Flora* 99: 436–464.

1909b. Von der Chemotaxis der *Lycopodium* — Spermatozoiden. *Flora* 99:193–202.

1910. Die Keimung der Sporen und die Entwicklung der Prothallien von *Lycopodium clavatum* L., *L. annotinum* L., und *L. Selago* L. *Flora* 101:220–267.

1912. Zur Embryologie der Selaginellaceen. *Flora* 104:180–224.

Campbell, D. H.

1891. Contributions to the life-history of *Isoetes. Ann. Bot.* 5:231–258.

1940. *The Evolution of the Land Plants (Embryophyta).* Stanford University Press, Stanford, California.

Case, I. M.

1943. Periodicity in the development of fertile and sterile zones in *Lycopodium Selago. New Phytol.* 42:93–97.

Chamberlain, C. J.

1917. Prothallia and sporelings of three New Zealand species of *Lycopodium. Bot. Gaz.* 63:51–65.

Cusick, F.

1954. Experimental and analytical studies of Pteridophytes XXV. Morphogenesis in *Selaginella Willdenovii.* II. Angle-meristems and angle-shoots. *Ann. Bot.* n. s. 18:171–181.

Cutter, E. G.

1966. Patterns of organogenesis in the shoot. Pp. 220–234 *in* Cutter, E. G. (ed.) *Trends in Plant Morphogenesis.* Longmans, Green and Co., London.

Darrah, W. C.

1939. *Textbook of Paleobotany.* Appleton-Century, New York.

Delevoryas, T.

1962. *Morphology and Evolution of Fossil Plants.* Holt, Rinehart, and Winston. New York.

De Maggio, A. E.

1964. Organization in a gametophyte callus of *Lycopodium* and its morphogenetic implications. *Proc. Nat. Acad. Sci. U.S.A.* 52:854–859.

Duerden, H.

1929. Variations in megaspore number in *Selaginella. Ann. Bot.* 43:451–457.

1934. On the occurrence of vessels in *Selaginella. Ann. Bot.* 48:459–465.

Dunlop, D. W.

1949. Notes on the cytology of some Lycopsids. *Bull. Torrey Bot. Club* 76:266–277.

Duthie, A. V.

1929. The method of spore dispersal of three South African species of *Isoetes. Ann. Bot.* 43:411–412.

Eames, A. J.

1942. Illustrations of some *Lycopodium* gametophytes. *Amer. Fern Jour.* 32:1–12.

Eggert, D. A.

1961. The ontogeny of Carboniferous arborescent Lycopsida. *Palaeontographica.* 108(B):43–92.

Emberger, L.
   1968. *Les Plantes Fossiles dans leurs Rapports avec les Végétaux Vivants,* Ed. 2. Masson et Cie., Paris.

Esau, K., V. I. Cheadle, and E. M. Gifford, Jr.
   1953. Comparative structure and possible trends of specialization of the phloem. *Amer. Jour. Bot.* 40:9–19.

Freeberg, J. A.
   1957. The apogamous development of sporelings of *Lycopodium cernuum* L., *L. complanatum* var. *flabelliforme* Fernald and *L. selago* L. *in vitro. Phytomorphology* 7:217–229.

Freeberg, J. A., and R. H. Wetmore
   1957. Gametophytes of *Lycopodium* as grown *in vitro. Phytomorphology* 7:204–217.
   1967. The Lycopsida—a study in development. *Phytomorphology* 17:78–91.

French, J. C.
   1970. Development of sporangia in *Selaginella*. M.S. thesis, University of California, Davis, Calif. 79 pp.

Fry, W. L.
   1954. A study of the carboniferous lycopod, *Paurodendron*, Gen. Nov. *Amer. Jour. Bot.* 41:415–428.

Goebel, K.
   1930. *Organographie der Pflanzen. Dritte Auflage. Zweiter Teil.* G. Fischer, Jena.

Grierson, J. D., and F. M. Hueber
   1968. Devonian lycopods from northern New Brunswick. Pp. 823–836 *in* Oswald, D. (ed.), *International Symposium on the Devonian System.*

Härtel, K.
   1938. Studien an Vegetationspunkten einheimischer Lycopodien. *Beitr. Biol. Pflanzen* 25:125–168.

Harvey-Gibson, R. J.
   1897. Contributions towards a knowledge of the anatomy of the genus *Selaginella*, Spr. III. The leaf. *Ann. Bot.* 11:123–155.

Haught, O.
   1960. Lycopods in the American Tropics. *Castanea* 25:127–129.

Hauke, R. L.
   1969. Problematic groups in the fern allies and the treatment of subspecific categories. *BioScience* 19:705–707.

Hill, J. B.
   1914. The anatomy of six epiphytic species of *Lycopodium. Bot. Gaz.* 58:61–85.
   1919. Anatomy of *Lycopodium reflexum. Bot. Gaz.* 68:226–231.

Hofmann, E.
   1934. *Paläohistologie der Pflanze.* J. Springer, Wien.

Holloway, J. E.
   1909. A comparative study of the anatomy of six New Zealand species of *Lycopodium. Trans. New Zealand Inst.* 42:356–370.

Horner, H. T., Jr., and H. J. Arnott
   1963. Sporangial arrangement in North American species of *Selaginella. Bot. Gaz.* 124:371–383.

Hsü, J.
   1937. Anatomy, development and life history of *Selaginella sinensis*. I. Anatomy and development of the shoot. *Bull. Chinese Bot. Soc.* 3:75–95.

Jaeger, H.
   1962. Das Alter der ältesten bekannten Landpflanzen (*Baragwanathia*-flora) in Australien auf Grund der begleitenden Graptolithen. *Paläontol. Zeit.* 36:7.

Jagels, R.
   1970a. Photosynthetic apparatus in *Selaginella*. I. Morphology and photosynthesis under different light and temperature regimes. *Can. Jour. Bot.* 48:1843–1852.
   1970b. Photosynthetic apparatus in *Selaginella*. II. Changes in plastid ultrastructure and pigment content under different light and temperature regimes. *Can. Jour. Bot.* 48:1853–1860.

Jermy, A. C., K. Jones, and C. Colden
   1967. Cytomorphological variation in *Selaginella. Jour. Linn. Soc. Bot.* 60:147–158.

Jones, C. E.
   1905. The morphology and anatomy of the stem of the genus *Lycopodium. Trans. Linn. Soc. London* (Ser. 2) 7:15–35.

Koster, H.
   1941. New *Lycopodium* gametophytes from New Jersey. *Amer. Fern Jour.* 31:53–58.

Kräusel, R., and H. Weyland
   1932. Pflanzenreste aus dem Devon. IV. *Protolepidodendron. Senckenbergiana* 14:391–403.

Kubitzki, K., and R. Borchert

1964. Morphologische Studien an *Isoëtes triquetra* A. Braun und Bermerkungen über das Verhältnis der Gattung *Stylites* E. Amstutz zur Gattung *Isoëtes* L. *Ber. Deutsch. Bot. Ges.* 77:227–233.

Kuriachan, P. I.

1963. Cytology of the genus *Selaginella*. *Cytologia* 28:376–380.

1965. Cytology of *Lycopodium cernuum* L. *Caryologia* 18:633–636.

LaMotte, C.

1933. Morphology of the megagametophyte and the embryo sporophyte of *Isoetes lithophila*. *Amer. Jour. Bot.* 20:217–233.

1937. Morphology and orientation of the embryo of *Isoetes Ann. Bot.* n.s. 1:695–716.

Lamoureux, C. H.

1961. Comparative studies on phloem of vascular cryptogams. Ph.D. dissertation. University of California, Davis, Calif.

Lang, W. H.

1915. III. Studies in the morphology of *Isoetes.* I. The general morphology of the stock of *Isoetes lacustris*. *Mem. Proc. Manchester Lit. Phil. Soc.* 59:1–28.

Liebig, J.

1931. Ergänzungen zur Entwicklungsgeschichte von *Isoëtes lacustre* L. *Flora* 125:321–358.

Löve, A., and D. Löve

1958. Cytotaxonomy and classification of lycopods. *Nucleus* (Calcutta) 1:1–10.

Lyon, A. G.

1964. Probable fertile region of *Asteroxylon mackiei* K. and L. *Nature* 203:1082–1083.

Lyon, F. M.

1901. A study of the sporangia and gametophytes of *Selaginella apus* and *Selaginella rupestris*. *Bot. Gaz.* 32:124–141; 170–196.

Ma, R. M.

1930. The chloroplasts of *Selaginella*. *Bull. Torrey Bot. Club* 57:277–284.

Manton, I.

1950. *Problems of Cytology and Evolution in the Pteridophyta.* Cambridge University Press, London.

Martens, P.

1960a. Sur une structure microscopique orientée, dans la paroi mégasporale d'une Sélaginelle. *Compt. Rend. Acad. Sci.* (Paris) 250:1599–1602.

1960b. Nouvelles observations sur la structure des parois mégasporales de *Selaginella myosurus* (Sow.) Alston. *Compt. Rend. Acad. Sci.* (Paris) 250:1774–1775.

Matthews, J. F., and W. H. Murdy

1969. A study of *Isoetes* common to the granite outcrops of the southeastern Piedmont, United States. *Bot. Gaz.* 130:53–61.

Mehra, P. N., and S. C. Verma

1957. Cytology of *Lycopodium. Curr. Sci.* 26:55–56.

Mickel, J. T., and R. L. Hellwig

1969. Actinoplectostely, a complex new stelar pattern in *Selaginella. Amer. Fern Jour.* 59:123–134.

Mitchell, G.

1910. Contributions toward a knowledge of the anatomy of the genus *Selaginella* Spr. V. The strobilus. *Ann. Bot.* 24:19–33.

Morton, C. V.

1964. New combinations in *Lycopodium. Amer. Fern Jour.* 54:71–73.

Ninan, C. A.

1958. Studies on the cytology and phylogeny of the pteridophytes. V. Observations on the Isoetaceae. *Jour. Indian Bot. Soc.* 37:93–103.

Ogura, Y.

1938. Anatomie der Vegetationsorgane der Pteridophyten. *Handbuch der Pflanzenanatomie.* Gebrüder Borntraeger, Berlin.

Osborn, T. G. B.

1922. Some observations on *Isoetes Drummondii*, A. Br. *Ann. Bot.* 36:41–54.

Pant, D. D., and B. Mehra

1964. Development of stomata in some fern allies. *Proc. Nat. Inst. Sci. India* 30(B): 92–98.

Paolillo, D. J., Jr.

1963. *The Developmental Anatomy of Isoetes.* (Illinois Biological Monographs. No. 31) University of Illinois Press, Urbana.

Pfeiffer, N. E.

1922. Monograph of the Isoetaceae. *Ann. Missouri Bot. Gard.* 9:79–232.

Phillips, T. L., and G. A. Leisman
1966. *Paurodendron*, a rhizomorphic lycopod. *Amer. Jour. Bot.* 53:1086–1100.

Pieniązek, S. Al.
1938. Über die Entwicklung und das Wachstum der Makrosporenmembranen bei *Selaginella*. *Sprawozdania Towarzystwa Nauk. Warszawskiego Wydziat 4 (Compt. Rend. Soc. Sci. Varsovie Cl. 4)* 31:211–230.

Pixley, E. Y.
1968. A study of the ontogeny of the primary xylem in the roots of *Lycopodium*. *Bot. Gaz.* 129:156–160.

Rauh, W., and H. Falk
1959a. *Stylites* E. Amstutz, eine neue Isoëtacee aus den Hochanden Perus. 1. Teil: Morphologie, Anatomie und Entwicklungsgeschichte der Vegetationsorgane. *Sitzb. Heidelb. Akad. Wiss. Jahrgang* 1959, 1–83.
1959b. *Stylites* E. Amstutz, eine neue Isoëtacee aus den Hochanden Perus. 2. Teil. Zur Anatomie des Stammes mit besonderer Berücksichtigung der Verdikungsprozesse. *Sitzb. Heidelb. Akad. Wiss. Jahrgang* 1959, 87–160.

Roberts, E. A., and S. D. Herty
1934. *Lycopodium complanatum* var. *flabelliforme* Fernald: its anatomy and a method of vegetative propagation. *Amer. Jour. Bot.* 21:688–697.

Schlanker, C. M., and G. A. Leisman
1969. The herbaceous Carboniferous lycopod *Selaginella fraiponti* comb. nov. *Bot. Gaz.* 130:35–41.

Scott, D. H., and T. G. Hill
1900. The structure of *Isoetes Hystrix*. *Ann. Bot.* 14:413–454.

Slagg, R. A.
1932. The gametophytes of *Selaginella Kraussiana*. I. The microgametophyte. *Amer. Jour. Bot.* 19:106–127.

Smith, R. W.
1900. The structure and development of the sporophylls and sporangia of *Isoetes*. *Bot. Gaz.* 29:225–258; 323–346.

Sosa, A., and C. Sosa-Bourdouil
1940. Sur la composition des macrospores et des microspores de *Selaginella*. *Compt. Rend. Acad. Sci.* (Paris) 210:59–61.

Spessard, E. A.
1922. Prothallia of *Lycopodium* in America. II. *L. lucidulum* and *L. obscurum* var. *dendroideum*. *Bot. Gaz.* 74:392–413.

Stainier, F.
1965. Structure et infrastructure des parois sporales chez deux Sélaginelles. *(Selaginella myosurus* et *S. kraussiana). Cellule* 65:222–244.

Stewart, W. N.
1947. A comparative study of stigmarian appendages and *Isoetes* roots. *Amer. Jour. Bot.* 34:315–324.

Stokey, A. G.
1907. The roots of *Lycopodium pithyoides*. *Bot. Gaz.* 44:57–63.
1909. The anatomy of *Isoetes*. *Bot. Gaz.* 47:311–335.

Sussex, I. M.
1966. The origin and development of heterospory in vascular plants. Pp. 140–152 *in* Cutter, E. G. (ed.), *Trends in plant morphogenesis*. Longmans, Green and Co., London.

Sykes, M. G.
1908. Notes on the morphology of the sporangium-bearing organs of the Lycopodiaceae. *New Phytol.* 7:41–60.

Sykes, M. G., and W. Stiles
1910. The cones of the genus *Selaginella*. *Ann. Bot.* 24:523–536.

Takeuchi, K.
1962. Study on the development of gemmae in *Lycopodium chinense* Christ and *Lycopodium serratum* Thunb. *Jap. Jour. Bot.* 18:73–85.

Towers, G. H. N., and W. S. G. Maass
1965. Phenolic acids and lignins in the Lycopodiales. *Phytochemistry* 4:57–66.

Treub, M.
1884. Études sur les Lycopodiacées. *Ann. Jard. Bot. Buitenzorg.* 4:107–138.

Tryon, A. F.
1949. Spores of the genus *Selaginella* in North America north of Mexico. *Ann. Missouri Bot. Gard.* 36:413–431.

Turner, J. J.
1924. Origin and development of vascular system of *Lycopodium lucidulum*. *Bot. Gaz.* 78:215–225.

Van Soest, J. L.

1964. Estimation of the age of a fairy circle (*Lycopodium complanatum* L. var. *chamaecyparissus* (A. Br.) Döll. *Acta Bot. Neer.* 13:623.

Wardlaw, C. W.

1924. Size in relation to internal morphology. No. 1. Distribution of the xylem in the vascular system of *Psilotum, Tmesipteris,* and *Lycopodium. Trans. Roy. Soc. Edinb.* 53:503–532.

Weber, U.

1922. Zur Anatomie und Systematik der Gattung Isoëtes. *Hedwigia* 63:219–262.

Webster, T. R., and T. A. Steeves

1967. Developmental morphology of the root of *Selaginella martensii* Spring. *Can. Jour. Bot.* 45:395–404.

West, C., and H. Takeda

1915. X. On *Isoëtes japonica. Trans. Linn. Soc. London* .8:333–376.

Wetmore, R. H.

1943. Leaf stem relationships in the vascular plants. *Torreya* 43:16–28.

Wetmore, R. H., and G. Morel

1951a. Sur la culture *in vitro* de prothalles de *Lycopodium cernuum. Compt. Rend. Acad. Sci.* (Paris) 233:323–324.

1951b. Sur la culture du gametophyte de Sélaginelle. *Compt. Rend. Acad. Sci.* (Paris) 233:430–431.

Whitebread, C.

1941. Beware of *"Lycopodium"! Amer. Fern Jour.* 31:100–102.

Wilce, J. H.

1972. Lycopod spores, I. General spore patterns and the generic segregates of *Lycopodium. Amer. Fern Jour.* 62:65–79.

Wilder, G. J.

1970. Structure of tracheids in three species of *Lycopodium. Amer. Jour. Bot.* 57:1093–1107.

Williams, S.

1931. An analysis of the vegetative organs of *Selaginella grandis* Moore, together with some observations on abnormalities and experimental results. *Trans. Roy. Soc. Edinb.* 57:1–24.

1937. Correlation phenomena and hormones in *Selaginella. Nature* 139:966.

Zamora, P. M.

1958. Anatomy of the protoxylem elements of several *Selaginella* species. *Philippine Jour. Sci.* 87:93–114.

Zinda, D. R.

1966. A preliminary report on the comparative morphology of the shoot apex of *Isoetes macrospora* Dur. and some effects of experimentally applied indole-3-acetic acid and gibberellic acid. *Jour. Minnesota Acad. Sci.* 33:107–116.

10

The Sphenopsida

The Sphenopsida constitutes a rather well-defined group of living and extinct plants. The oldest strata in which fossil remains of the Sphenopsida have been found are considered to be of Devonian age. During the Carboniferous the group attained almost worldwide distribution and significant diversity of growth form; a parallel situation, it will be remembered, is encountered for the Lycopsida. During the Carboniferous, arborescent and herbaceous sphenopsids were coexistent. By the end of the Triassic only a few representatives remained, and they were relatively small, herbaceous forms. Today only one genus, with about 25 herbaceous species, remains as a remnant of this once conspicuous and diversified group. How-

ever, this one genus, *Equisetum*, may itself have been present in the Carboniferous and may not have undergone any significant change in the course of time. If this is so, *Equisetum* may be one of the oldest living genera of vascular plants in the world today.

In characterizing the group, including living and extinct members, the most conspicuous external morphological feature is the subdivision of the shoot axis into definite nodes and internodes (Figs. 10-1, B; 10-2) — that is, the stem is jointed. At the nodes are whorls of relatively small leaves with buds between them. In addition to the stem joints there are definite, easily observed stem ribs. Reproductive structures of the sporophyte are terminal cones or strobili. The strobilus

A

B

Figure 10-1
Equisetum telmateia. **A,** colony growing along roadside embankment, Berkeley, California; **B,** series of plants arranged to show stages in the growth and expansion of vegetative shoots. [Courtesy Mr. Louis Arnold.]

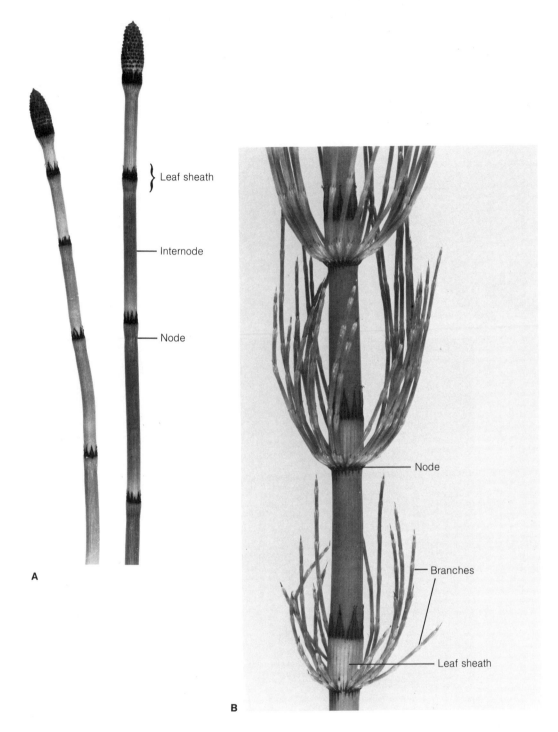

Figure 10-2
Equisetum shoots. **A,** *Equisetum hyemale,* portions of two unbranched shoots with terminal strobili.
B, *Equisetum telmateia,* portion of sterile (vegetative) shoot showing leaf sheaths and lateral branches.

consists of an axis with whorls of stalk-like structures (sporangiophores) to each of which a group of sporangia is attached (Figs. 10-9, A, B; 10-12, A).

Classification

The following outline of classification will serve as a guide to further descriptions. Only representative orders, families, and genera are included. In conformity with presentation in Chapter 9, the living member is placed first in the outline and is described in considerable detail later in the text. Descriptions of extinct members are necessarily brief, but their importance in the establishment of morphological unity within the group should not be minimized.

SPHENOPSIDA: Living and extinct plants; sporophyte differentiated into stem, leaf, root, and eusporangium; jointed, ribbed stems; buds not on same radii with leaves, but alternate with them; stem protostelic or siphonostelic; xylem exarch or endarch; secondary growth in some forms; sporangia borne on specialized stalks (sporangiophores) that are organized into strobili; largely homosporous with a few heterosporous forms (extinct).

EQUISETALES: Living and extinct plants; herbaceous sporophytes, no secondary growth; stem organized into nodes and internodes; siphonostelic and endarch; small microphyllous leaves basally united as a sheath at the node; roots developed from rhizome; sporangia-bearing structures (sporangiophores) organized into strobili. Gametophyte exosporic and green in living members; antheridium produces multiflagellate sperms; the embryo is exoscopic.

EQUISETACEAE: Characteristics as in Equisetales.

Equisetum, Equisetites (extinct).

HYENIALES: Extinct plants of Devonian age; rhizomatous plants with many upright branches having whorl-like arrangement of narrow, forked leaves; other axes with whorled, dichotomously branched sporangiophores having recurved terminal sporangia.

HYENIACEAE: Characteristics as in Hyeniales.

Hyenia.

SPHENOPHYLLALES: Extinct plants of Carboniferous and Permian age; small, either upright, climbing or trailing plants with some secondary growth; monopodial growth; small, lobed to wedge-shaped leaves, often notched along distal margin, arranged in whorls at nodes; strobili composed of whorls of bracts and sporangiophores, the latter often partially fused to whorls of united bracts.

SPHENOPHYLLACEAE: Characteristics as in Sphenophyllales.

Sphenophyllum; Sphenophyllostachys and *Litostrobus* (strobili).

CALAMITALES: Extinct plants of Carboniferous age; large plants with ribbed stems; endarch siphonostele in stem; considerable secondary growth; relatively small, linear or lanceolate leaves arranged in whorls at nodes; compact strobili, commonly consisting of (1) whorls of closely associated sporangiophores and bracts, or (2) alternate whorls of sterile bracts and sporangiophores.

CALAMITACEAE: Characteristics as in Calamitales.

Calamites (pith casts); *Calamodendron* and *Arthropitys* (stem petrifactions); *Annularia* (leaves); *Palaeostachya* and *Calamostachys* (strobili).

Equisetales–Equisetaceae: *Equisetum*

The sporophytes of *Equisetum*, with their characteristic jointed stems and rough texture, have earned several names, of which "horsetails" and "scouring rushes" are most popular. During the American colonial period, horsetails were used as scouring agents, as they probably are today in some regions of the world.

The horsetails are worldwide in distribution except for certain parts of Australasia. The majority of species occur in the Northern Hemisphere, with only a few appearing in the tropics and the Southern Hemisphere.

They generally grow in wet or damp habitats, being particularly common along the banks of streams or irrigation ditches (Fig. 10-1, A); some have become adapted to dry or mesophytic conditions, for at least a part of the year. In some localities they are a serious weed problem for farmers and a matter of concern for livestock owners because of the poisonous substances in the stems of some species. Horses are especially susceptible to their toxins. However, a certain species in Costa Rica is used medicinally, as a treatment for human kidney trouble (Hauke, 1969c).

Certain species have been used as indicators of the mineral content of the soil in which they grow (Vogt, 1942). These plants accumulate minerals in their bodies (including gold — up to 4½ oz per ton — Benedict, 1941), and the mineral content can be determined by either chemical or X-ray analysis of the plant tissues. This source of mineral information is therefore of some value in prospecting for new ore deposits, a most important activity in the present world struggle for resources.

The horsetails, despite the range of variation in, for example, branching and shoot dimorphism, are usually not mistaken for some other group and most botanists are content in recognizing the one genus *Equise-tum*. However, some specialists, dating even from the last century, have proposed that at least two genera be recognized. Consequently in some monographic treatments the genera *Equisetum* and *Hippochaete*, or even three genera, have been described. In the prospectus of Flora North America it was pointed out by Hauke (1969a) that there are some real and consistent differences between the two major groups of species, but that they are considered to be minor, and the number of similarities outweigh the differences. Therefore, he could see no gain in information or accuracy in establishing two genera instead of one. Two subgenera, however, are recognized to point up the differences. In this textbook we will use the generic designation *Equisetum* without attempting to specify the subgenus for any species. For those interested in the subject, earlier literature and taxonomic analyses can be found elsewhere (Rothmaler, 1944; Hauke, 1961, 1962a,b,c, 1963).

Organography

The shoot system of *Equisetum* consists of an aerial portion and an underground rhizome portion. The sporophyte is perennial, at least in the sense that the rhizome is perennial, even though the aerial system may die back each fall or winter. In the north temperate regions new aerial shoots are formed each year. Some species are small, particularly those of arctic and alpine regions. One such species grows within the Arctic Circle in Europe, in Siberia, and in North America. Several South American species are tall in growth habit (up to 10–12 meters); the stems are small in diameter and they are supported partly by the surrounding tall grass.

The stem of *Equisetum* has a very rough texture, and an irregular surface-pattern is especially evident in scanning electron micrographs of the stem surface (Fig. 10-3,

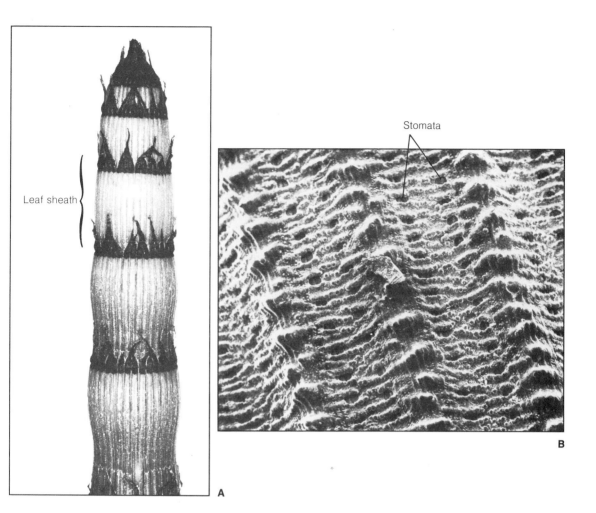

Figure 10-3
Equisetum hyemale. **A**, portion of as-yet-unelongated shoot showing successive leaf sheaths; stem not evident except for small portion at base of shoot; **B**, scanning electron micrograph of stem surface; note rows of stomata between "ridges." [A courtesy of Dr. T. E. Weier, from *Botany: An Introduction to Plant Biology* by T. E. Weier, C. R. Stocking, and M. G. Barbour. Wiley, New York, 1970; B courtesy of Dr. R. H. Falk.]

B). Stomata are arranged in longitudinal rows (1 or 2) in furrows between the ribs. The extremely rough texture is due to the deposition of silica (as opal) as discrete knobs and rosettes on the epidermal surface (e.g., in *Equisetum arvense*) or in a uniform pattern on and in the entire outer epidermal cell walls (e.g., in *Equisetum hyemale*, Kaufman, et al., 1971). Silica is known to be an essential element for normal growth of *Equisetum* (Chen and Lewin, 1969) and may play an important role in maintaining erect-ness of the plant, compensating for the very low lignin content of the cell walls. Other functions have been proposed: protection against pathogens and predators; prevention of excessive water loss (Kaufman, et al., 1971).

To reiterate, the shoot system of *Equisetum* consists of an underground rhizome and an aerial shoot, each system possessing the same fundamental organization. The stem is divided into definite nodes and internodes, and leaves are attached at nodes, united at least for a part of their length, forming a sheath around the stem (Figs. 10-2; 10-3, A). The number of leaves per node varies according to the species, and to the position on the stem in most species. Where adjacent nodes have the same number of leaves in the leaf sheath, there is a very precise type of symmetry. The number of leaves at a node corresponds to the number of ribs on the internode below. Confronting each rib is a vascular bundle, the upper part of which enters the leaf that is on a line with the rib (Figs. 10-5, A; 10-8). However, if a shoot is examined carefully, it will be noted that the number of leaves per node generally increases from the base for some distance and then decreases. With this pattern the type of vascularization described above is not obtained and adjustments in vascular connections occur in that some leaves of a whorl may have two vascular bundles instead of one (Bierhorst, 1959).

An internode may continue to grow for some time through the activity of a meristem located at its base. Lateral branches, when evident, are attached at the nodes and alternate with the leaves. During development, each branch breaks through the lower part of the leaf sheath in reaching the exterior. Roots arise at the nodes of the rhizome. Whether a species is highly branched or not, branch and root primordia are formed at each node. In the rhizome a root primordium develops into a root; a branch primordium may develop into an erect aerial shoot, remain as an arrested bud, develop into a new rhizome, or become abortive. In aerial shoots (if branching is a characteristic of the species) the branch primordia develop while the roots remain in an arrested condition unless the stems become procumbent and come to lie on a moist surface. Use can be made of this information in vegetative propagation. A plant may be propagated vegetatively also from arrested tuberous-like branches which occur at the nodes of rhi-

zomes. The propensity for this type of vegetative propagation is the major factor in *Equisetum* becoming a weed that is difficult to control in some localities. A small detached branch or tuber could be the start of the colonization of a very large area under favorable conditions for growth. For a farmer or cattleman to eradicate the last bit of rhizome is almost an impossible task, somewhat akin to catching the last two rats of a colony!

The rhizome, and the aerial stem, can be pulled apart into internodal lengths or "pipes." In one interesting study (Treitel, 1943), it was found that before a rhizome actually breaks under tension its elasticity compares favorably with that of muscle or rubber.

Abnormal growths or monstrosities (teratological forms) are always of interest, and large stands of *Equisetum* generally provide many examples. Certain shoots may have unusually short internodes, the internodal potentiality never having been realized. Other shoots may have flexuous (snake-like) stems or have continuous spiral leaf sheaths, exhibit dichotomous branching, or produce a vegetative shoot beyond the usual terminal cone (Schaffner, 1933). All of these abnormalities undoubtedly are the result of an unbalanced growth-regulating system.

Strobili may occur terminally on many of the main vegetative axes of a highly branched species, but more often a single strobilus terminates the main axis, whether the plant is branched or not (Fig. 10-2, A). Branching may be induced in a typically unbranched plant through injury of the main axis. In a few species there is a segregation of function: some shoots are green and purely vegetative (Figs. 10-1; 10-2, B) and others are unbranched (Fig. 10-4) and brownish in color and have terminal strobili (e.g., *E. arvense*). Still other species, in addition to having purely vegetative shoots, produce brownish, fertile shoots which may

A

B

Figure 10-4
Equisetum arvense. Photographs taken in early spring showing unbranched fertile shoots terminated by strobili; some branched sterile (vegetative) shoots are evident in **A** to the right, and in **B**. [Courtesy of Dr. Richard L. Hauke.]

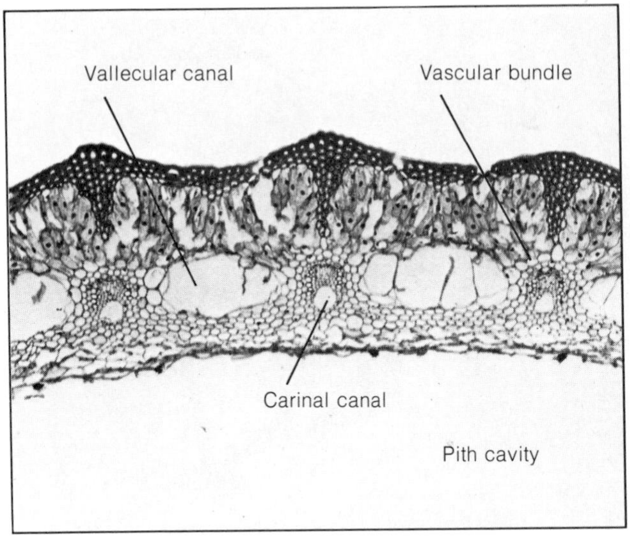

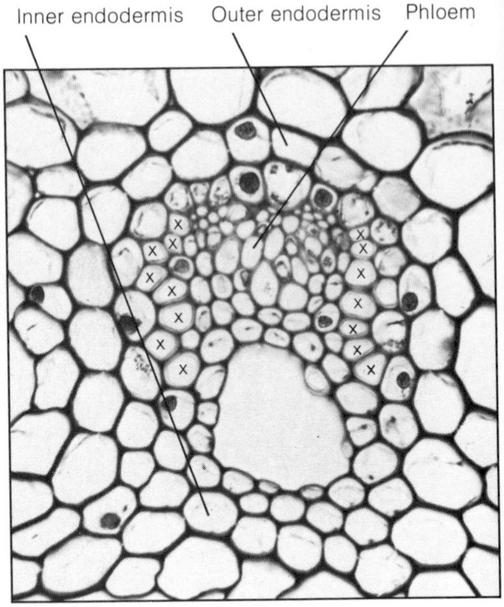

Figure 10-5

Stem anatomy in *Equisetum hyemale*. **A**, transection, portion of stem (note that carinal canals are opposite ridges on stem); **B**, structure of vascular bundle (one protoxylem tracheid is visible along edge of carinal canal at right; cross marks designate metaxylem tracheids.)

develop chlorophyll after the spores are shed; green branches then grow out from the nodes of the stem.

On the basis of external morphology alone, the evolutionary tendencies in the genus seem to be the following. The more primitive species have perennial, green shoots that are highly branched with green, pointed cones terminating several branches of the same shoot. The most advanced species have annual, branched aerial vegetative shoots that are green and have short-lived, unbranched fertile shoots which lack chlorophyll. Cones are rounded at the tip and are brownish. Naturally there are many species that show intermediate combinations of these characteristics (Schaffner, 1930). The more advanced species occur in the Northern Hemisphere and arctic regions, whereas the more primitive grow in South America, Cuba, and Mexico. In the words of Schaffner (1932), an authority on the genus *Equisetum*, "one immediately sees that the lowest [most primitive] species can be col-

lected as one stands on the equator, while the last and highest species [most advanced] can be gathered for a farewell bouquet as one steps from the farthest northern lands onto the ice of the Arctic Ocean."

Anatomy of the Mature Shoot

The stem of *Equisetum*, as seen in transverse section, is circular in outline with prominent ribs. In the majority of species the cell walls of the epidermis are thick and have a generous deposition of siliceous material as described earlier. Over the epidermis is a definite cuticle. Stomata often occur in two vertical rows in the furrows between the ribs. The stomatal apparatus, which may be in a depression in some species, consists of two guard cells and overarching subsidiary cells; the latter have ridges on their inner cell walls (Hauke, 1957). Beneath the epidermis is the cortex, which exhibits a rather complicated structure. The outer cortex is composed of sclerenchymatous tissue, which, in some species, is excessively developed opposite the ribs. Between adjacent ribs there occurs lacunate photosynthetic tissue (Fig. 10-5, A) with a single, large vertical air space (vallecular canal).

Opposite each ridge is a vascular bundle of unusual interest. A transverse section of the internode of a mature stem reveals a protoxylem lacuna (carinal canal) associated with a vascular bundle (Fig. 10-5, A). The carinal canal is formed during elongation of an internode due to the separation and disruption of protoxylem elements. In a mature stem one or more tracheids can be seen along the edge of the carinal canal (Fig. 10-5, B). The parenchyma cells lining the canal develop thick walls and the canal itself probably functions in water conduction (Bierhorst, 1958a). Experiments in which dyes were used support this conclusion. Opposite the carinal canal is the primary phloem flanked on each side by strands of xylem (Fig. 10-5, B), the ontogeny of which has been variously interpreted. In the past these strands have been considered to be the metaxylem of the bundle (Eames, 1909; Johnson, 1933; Golub and Wetmore, 1948a,b), but Bierhorst (1958a) considers this portion of the xylem to be separate and distinct (developmentally) from the carinal group; he refers to the flanking strands as "lateral xylem." Authors do agree that, despite some irregularities, maturation of the tracheary elements in the lateral xylem or metaxylem is centrifugal. Also, vessels have been reported to occur in the lateral xylem of the rhizomes of five species (Bierhorst, 1958b).

Depending on the species, one endodermal layer may be present external to the cylinder of vascular bundles, or there may be, in addition, an inner endodermis; in the latter instance one or both layers are often irregular in outline. Or, in certain species, an endodermis may completely surround each vascular bundle. A pericycle may or may not be evident. Sometimes the pericycle cells

are filled with starch. The vascular cylinder of *Equisetum* has been described as an ectophloic siphonostele of the *"Equisetum* type."

The stem nodes are unusual in that the metaxylem is extensively developed such that there is a conspicuous circular ring of xylem. There are no carinal or vallecular canals. In addition, a nodal plate of pith tissue separates one internode from another.

The system of vascular bundle connections will be discussed in a later section on shoot development.

The tips of underground rhizomes are covered by scale-like leaf sheaths, which may secrete mucilage from their abaxial epidermal cells. Trichomes may also be present on the adaxial side of the sheaths (Francini, 1942). There is sclerenchyma in the outer cortex. Vallecular canals and protoxylem lacunae are present. An endodermis around each vascular bundle is common (Fig. 10-6, C). There is often no consistency in the arrangement and number of endodermal layers (as seen in transverse section) in the aerial stem and rhizome of the same plant (Fig. 10-6, A, B).

Development of the Shoot

A large, four-sided inverted pyramidal cell with the rounded base uppermost occupies the apex of the apical meristem (Fig. 10-7). The so-called cutting surfaces are directed downward. The apical cell cuts off daughter cells in a continuous successive manner; each daughter cell or segment divides anticlinally to form two cells (segment cells). By tangential (periclinal) divisions in segment cells at their innermost ends, a pith meristem is formed. By further divisions (in various planes) of the segment cells, a definable circular bulge is formed below the shoot apex. This bulge consists of five or six layers of cells, the upper two of which will produce the future node and leaves; the lower layers will form cortical tissues of the internode.

216

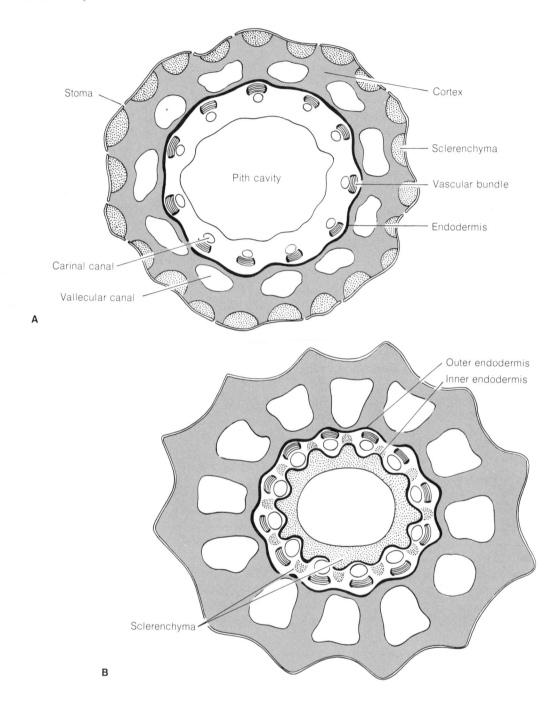

Figure 10-6
Schematic representations of variations in configuration and position of the endodermis in two species of *Equisetum*. **A**, aerial stem, *Equisetum sylvaticum;* **B**, rhizome, *Equisetum sylvaticum;* **C,** (*facing page*) rhizome, *Equisetum hyemale.*

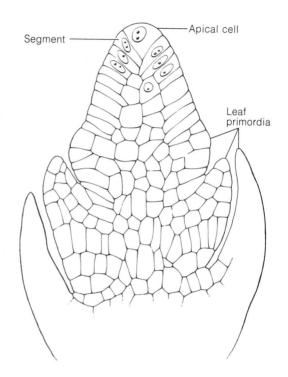

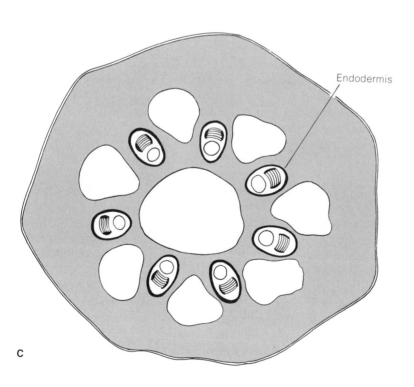

Figure 10-7
Longisection of shoot apex of *Equisetum* sp. showing
conspicuous apical cell and its derivative tissues.
The pith originates from inner cells derived from
periclinally dividing superficial cells. Leaf primordia
arise from superficial initials.

c

Very early in development, leaf-sheath initials can be distinguished as certain regularly placed cells of the upper two layers of the bulge. A leaf tooth then grows in length and widens at its base, eventually becoming fused to varying degrees with adjacent leaves, thus forming the characteristic leaf sheath (Figs. 10-2; 10-3, A). For a more complete discussion of apical growth the student is referred to the work of Golub and Wetmore (1948a).

At the time leaf initials are apparent, procambial cells become differentiated in the nodal and internodal regions near the pith. The results of one study (Golub and Wetmore, 1948b) indicate that the procambium, in one species at least, is acropetally continuous throughout the shoot. External to the procambium, and in internodal regions, an intercalary meristem gives rise, by oriented transverse divisions, to tissues that will differentiate into epidermis, cortex, endodermis, and pericycle.

Equisetum is a favorable plant for descriptive, histogenetic studies. In an investigation on the apical meristem of *E. hyemale*, Sinnott (1943) has emphasized that "what a given cell will do depends not upon some general principle of division, common to all cells, but upon the conditions which exist at that particular place and time," (related to the position of the cell in the organized developmental pattern of the plant). On the other hand, Taylor (1939) has described the embryo of *Equisetum* as exhibiting the general principle of the partitioning of cells. According to this concept, the laws of minimal surface and maximum stability are observed in meristematic cells. Sinnott points to the violation of this rule in the division of a segment (cell) in the apical meristem of *Equisetum*. Each segment divides anticlinally, but it should divide periclinally if the surface produced is to be minimal. Morphogenesis is, as yet, little understood, but *Equisetum*

and similarly favorable forms provide research tools for both descriptive and experimental studies.

Vascularization

There is considerable controversy regarding the vertical direction of differentiation of primary xylem and phloem (Gwynne-Vaughan, 1901; Queva, 1907; Meyer, 1920; Barratt, 1920; Browne, 1922; Johnson, 1933). From a study of *E. arvense*, Golub and Wetmore (1948b) have concluded that differentiation of protoxylem and protophloem within a procambial strand begins at about the fourth node. Differentiation then proceeds acropetally into the developing leaf and basipetally into the internode below. With elongation of the internode the protoxylem and protophloem elements become stretched and torn because the meristem (intercalary meristem) located at the base of each internode does not contribute new cells to the procambial strand. It is during this period of rapid elongation that protoxylem lacunae (carinal canals) are formed. After considerable elongation of an internode, metaxylem (lateral xylem) and metaphloem differentiation proceeds basipetally from a node through the internode below, and continuity is finally achieved with the same tissues in the node below.

During internodal elongation certain cortical cells do not accommodate themselves to this rapid extension, and they separate from each other, which results in the formation of vallecular canals. The pith cavity is reported to be formed by mechanical tearing of cells (Golub and Wetmore, 1948b).

In following a vascular bundle up through an internode it is found that at the level of leaf attachment a trichotomy of the protoxylem occurs. The median bundle enters the scale leaf, the two laterals diverge right and left, and each is joined laterally with an

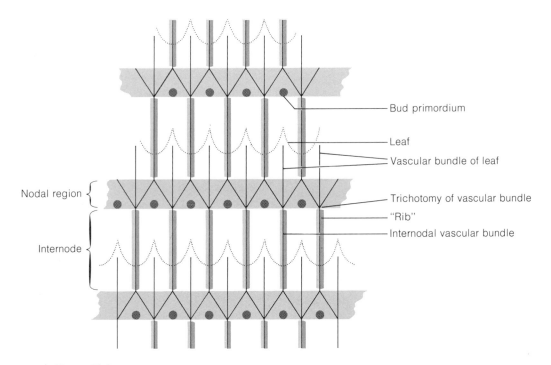

Figure 10-8
Schematic representation of a portion of the vascular system of *Equisetum,* shown in relationship to the "ribs," leaves, and buds.

adjacent strand to form one of the vascular bundles of the next higher internode (Fig. 10-8). As stated before, considerable metaxylem and metaphloem are formed within each bundle at the nodes.

At this point it is advisable to describe lateral branch development because branches occur at nodes on radii alternating with those of the leaves. At about the sixth node from the shoot tip, initiation of a lateral branch is in a single surface mother cell. An apical cell is soon established, and a branch bud with whorled leaves is formed. At about the time the first or second whorl of leaves is formed, a large deep-staining cell becomes apparent near the basal end of each branch. This endogenous cell is a root initial. The vascular tissue at the base of the branch shoot is in the form of a continuous cylinder (siphonostele) and is in continuity with the metaxylem and metaphloem of the nodal region. There is tissue continuity between the pith of the parent axis and the pith of the branch axis. These interruptions in the vascular tissue at the nodes then constitute branch gaps (Jeffrey, 1899). *Equisetum* is cladosiphonic (Chapter 3).

As mentioned earlier, the leaves at each node are united to varying degrees. The thick-walled cells of the abaxial epidermis have various types of ornamentation on their outer tangential walls. Stomata are arranged in longitudinal rows. Internally the mesophyll is lacunate and is traversed by one small median vascular bundle. Mesarch xylem has been reported by several workers.

One interesting feature of the *Equisetum* leaf is the presence of specialized "water stomata" (hydathodes) on the adaxial surface of the leaf along the midvein region (Johnson, 1937). *Equisetum* is a convenient plant for demonstrating the secretion of water as it is associated with the conditions of high moisture around the roots and a saturated atmosphere.

The Root

Similar to the type of ontogeny displayed by members of the Lycopsida, the primary root is ephemeral. All other functioning roots in *Equisetum* normally arise at the nodes of underground rhizomes. As seen in transverse section, the mature root has the following organization. The outer cortex is often sclerenchymatous, the inner parenchymatous. The xylem is triarch or tetrarch, or, in the smaller roots, may be diarch. Small roots may have one large metaxylem element in the center of the root (Walton, 1944). The cells of the endodermis and pericycle occur on the same radii, indicating a common origin. Lateral roots have their origin in the pericycle.

The root apical meristem contains an apical cell with four cutting surfaces. To the outside, a root cap is produced. The apical cell produces segments laterally, each of which by two divisions gives rise to three cells, which in turn produce the vascular tissue, cortex (including endodermis and pericycle), and epidermis (Johnson, 1933).

The Strobilus

Earlier it was noted that a mature strobilus terminates an axis, whether it be on a vegetative stem or a strictly fertile nonchlorophyllous axis. The strobilus is composed of an axis with whorls of stalked, peltate structures termed sporangiophores (10-9, A). Each sporangiophore is umbrella-like in shape, with pendant sporangia (five to ten in number) attached to the underside of the polygonal, disk-shaped shield (Fig. 10-12, A). The flattened tips of the sporangiophores

fit closely together, providing protection for the sporangia during development. At maturity the cone axis elongates, separating the sporangiophores, and the sporangia open by a longitudinal cleft that is formed down the inner side of each sporangium. Additional protection during early development is provided by a rudimentary leaf sheath, the so-called annulus, at the base of the cone in some species.

The vascular cylinder of the cone axis is a network of interconnected vascular bundles. No large canals are formed as described for the stem. Vascular bundles (sporangiophore traces) diverge from the vascular cylinder at regular intervals and enter the successive whorls of sporangiophores. At the distal end of the sporangiophore the bundle is branched; each strand is recurved and ends near the base of a sporangium (Fig. 10-9, B).

Early in development sporangiophore primordia arise in an acropetal manner on the flanks of the meristematic cone axis (Fig. 10-10). After enlargement of the sporangiophore primordium, sporangia are initiated in single superficial cells around the rim of the sporangiophore. The sporangium initial divides periclinally, setting aside an inner and an outer cell. The inner cell, by further divisions in various planes, produces sporogenous tissue (see Chapter 4, also Fig. 10-11). The outer cell, by anticlinal and periclinal divisions, gives rise to irregular tiers of cells, the inner of which may also become sporogenous; the outer tiers become the future sporangial wall cells (Bower, 1935). Superficial cells adjacent to the original initials may also contribute to the development of the sporangium (Chapter 4). Before the sporocytes separate and round-off prior to the meiotic divisions, two to three layers of cells adjacent to the sporogenous mass differentiate as the tapetum. In addition, not all of the sporogenous cells function as sporocytes;

Sporangiophores

Figure 10-9
Equisetum telmateia. **A**, fertile shoots
(the strobili are at different stages of
maturity, the youngest is to the left);
B, median longisection of one
sporangiophore showing the vascular
bundle in the stalk and its mode of
branching (crowded spores with elaters
are evident in the sporangia).

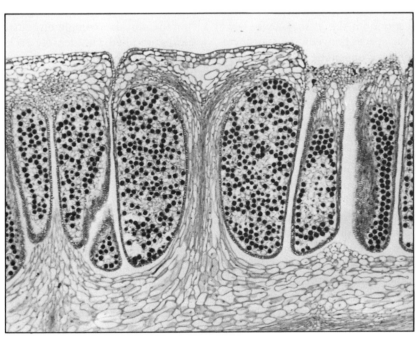

A

B

Figure 10-10
Scanning electron micrograph of a young strobilus of *Equisetum hyemale,* showing whorls of developing sporangiophores (× 60).

many degenerate and their protoplasm, together with that of the tapetum, forms a multinucleate nourishing substance which occupies the spaces between the sporocytes.

The results of one study indicate that the first sporangia to mature in a strobilus are sporangia situated in the widest part of the cone. Furthermore, within a single sporangium the sporocytes may be in various stages of meiosis. This may be related to the fact that the spore mother cells are separated into pockets surrounded by the multinucleate plasma (Manton, 1950).

After the meiotic divisions have taken place, the spore tetrads separate from one another, and each spore becomes spherical. The spore wall is said to be laminated (Beer, 1909). The outer layer is deposited on the spore in the form of four bands, derived presumably from the breakdown products of the nonfunctional sporocytes and tapetal cells. The four bands are attached to the spore wall at a common point and remain tightly coiled around the spore until the sporangium is completely mature. The spores are filled with densely packed chloroplasts, a feature quite uncommon to other lower vascular plants.

At maturity a sporangium consists of an outer wall composed of two layers of cells, the inner of which is generally compressed; the cells of the outer layer develop helical thickenings similar to tracheids. Internal to the wall is the mass of spores. At the time of dehiscence the free ends of the four bands, or elaters, separate from the spore wall. The elaters are hygroscopic, uncoiling as their water content decreases and recoiling with the addition of moisture (Fig. 10-12, B, C). It has been assumed that through this action the elaters assist in the dehiscence process and also bring about the dispersal of spores in large clumps from the sporangium.

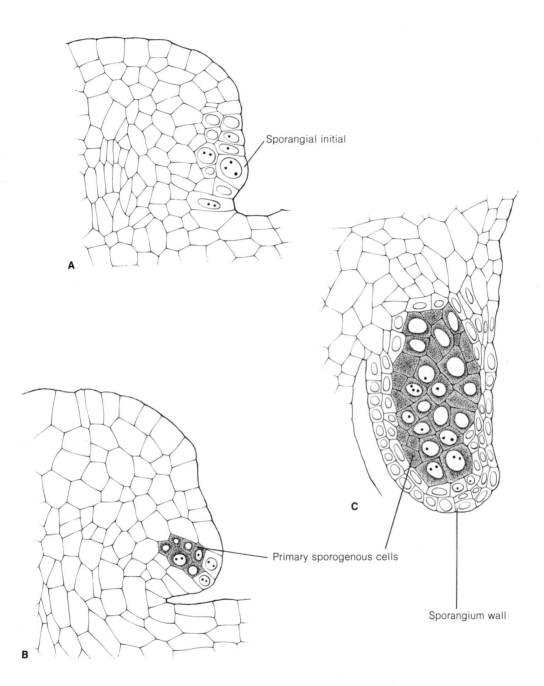

Figure 10-11
Early ontogeny of the sporangium in *Equisetum hyemale*. Initiation occurs in a single superficial cell (**A**), although lateral derivatives of the sporangial initial as well as adjacent superficial cells may contribute to the formation of primary sporogenous cells and primary wall cells.

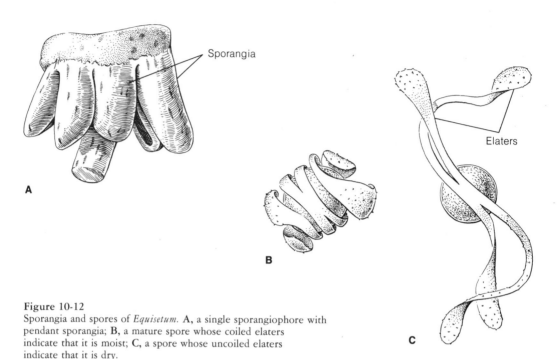

Figure 10-12
Sporangia and spores of *Equisetum*. **A**, a single sporangiophore with
pendant sporangia; **B**, a mature spore whose coiled elaters
indicate that it is moist; **C**, a spore whose uncoiled elaters
indicate that it is dry.

Chromosome Numbers

At this point it is appropriate to discuss the chromosomal constitution of *Equisetum*. In the several species that have been studied critically, the chromosome number has been determined to be $n = 108$ (Manton, 1950; Bir, 1960). Chromosomes vary in size and shape between the two groups (subgenera) in the genus, but the uniformity of the chromosome number may be indicative of the antiquity and conservativeness of the genus (Bir, 1960). However, a high degree of hybridity in the genus is suggested from the observed peculiarities at meiosis and from analyses of morphological variation of established sporophytes. In an intensive study of one subgenus, Hauke (1962c) described six hybrids between various pairs of seven species. Even if a hybrid produced nonviable spores, the sporophyte of the first generation could propagate itself very readily by the very efficient means of asexual reproduc-

tion mentioned earlier—fragmentation and growth of rhizomes (Hauke, 1969a). *Equisetum*, although an ancient genus dating from the Carboniferous, is not genetically static. Variability in chromosome number, however, seemingly has been lost in antiquity.

Gametophyte Generation

Under natural conditions the gametophytic plants may be found growing in damp areas, on mud, along creek banks, and even on the damp floors of abandoned mines and quarries (Matzke, 1941). Mature plants are dull green to brownish, and may range from 1 mm to 1 cm in diameter. In a tropical species (Kashyap, 1914) they may be even as large as 3 cm in diameter and from 2 to 3 mm in height. In some species the plant may be uniform in outline. Viewed from above, it resembles a miniature pin cushion (see Fig. 10-14, A).

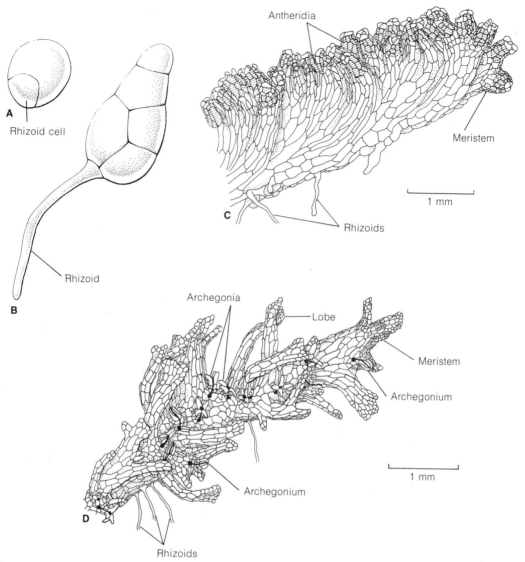

Figure 10-13
A,B, spore germination and early development of the gametophyte in *Equisetum hyemale;* **C,** apical portion, branch of male gametophyte of *Equisetum arvense;* **D,** branch of initially female plant. (See text for details of development). [C,D redrawn from Duckett, *Bot. Jour. Linn. Soc.* 63(4):327–352, 1970.]

The time interval between shedding of spores and germination is quite critical. Spores germinate very readily if they land in a suitable environment; however, the limiting factor seems to be the amount of available water. If the spores do not germinate at once their viability decreases rapidly, and the percentage still capable of germination drops greatly.

The first division of the spore results in two cells unequal in size. The smaller cell, which has fewer chloroplasts, elongates and forms a rhizoid (Fig. 10-13, A, B). The larger cell may divide transversely, or the division may be perpendicular to the original wall.

The results obtained from single spore cultures of five species (*E. arvense, E. telmateia, E. fluviatile, E. palustre,* and *E. sylvati-*

cum) have shown that within 15–20 days after germination a developing gametophyte becomes organized into two quite different regions: (1) a massive basal region with rhizoids on the lower surface and (2) an upright, bright green, leafy region. After 15–25 days in culture a localized meristem develops that gives rise to new basal cells of the "cushion" and produces derivatives which contribute to the development of upright lobes (Duckett, 1970a). Subsequent development and morphology is dependent on the type of sex organ produced, which leads us to a discussion of sexuality in *Equisetum.*

Although it has been known for more than a century that there is a gametophyte generation in *Equisetum,* sexual behavior in the genus has long been a standing controversy. Some investigators reported that for the species they studied, the gametophytes were either unisexual or bisexual, or were initially unisexual and then became bisexual. Heterospory was claimed to support sexual differentiation, but subsequent research failed to support this contention (Duckett, 1970b).

According to Duckett (1970a) the general picture to emerge from most investigations is that *Equisetum* gametophytes are initially either male or female and that the first sex organs to appear are antheridia on male plants. The older female plants eventually produce antheridia, generally in great numbers. There are, however, a few reports of archegonial formation on initially male plants. The following account of development is based on the research of Duckett (1970a).

MALE GAMETOPHYTES. The basal or cushion meristem, after producing a few upright lobes, suddenly begins forming antheridial tissue and the gametophyte often has a distinctive coloration (Fig. 10-13, C). Old males may have several hundred antheridia.

INITIALLY FEMALE GAMETOPHYTES. Archegonia are found on the cushion and at maturity lie at the bases of the upright lobes (Fig. 10-13, D). Female gametophytes sooner or later produce antheridia. When they do, archegonia and antheridia are both present, but in older gametophytes the number of antheridia produced increases. If a culture of a bisexual gametophyte (derived from one spore) is flooded with water, homozygous sporophytes are produced (also see Sporne, 1964).

Early sex determination, at least in some species, appears to be related to environmental conditions — being affected by temperature, light, and humidity as well as by the supply of nutrients (See discussions in Schratz, 1928; Wollersheim, 1957a,b; Hauke, 1967; Duckett, 1970a).

In contrast to the results of the experiments described earlier, Hauke (1969b) reported that gametophytes of *E. bogotense* were unisexual and never changed. For *E. giganteum* the gametophytes were normally bisexual, with archegonia and antheridia appearing at the same time. The pattern in *E. giganteum* was considered to be primitive by Hauke and could be correlated with the primitive features of the sporophyte of the species.

In the absence of clear-cut heterospory, correlated with the probability of a high degree of polyploidy, it is unlikely that any exact sex-determining mechanism is operative in *Equisetum*; rather, sexuality is controlled by a complex set of different interactions (Duckett, 1970a).

GAMETANGIA. For some species, as described earlier, spores germinate to give rise to male gametophytes, or to initially female plants which ultimately become bisexual. Antheridia first appear on the lobes of the male gametophyte or they may occur by the hundreds on the rounded, compact surface of old bisexual gametophytes at the

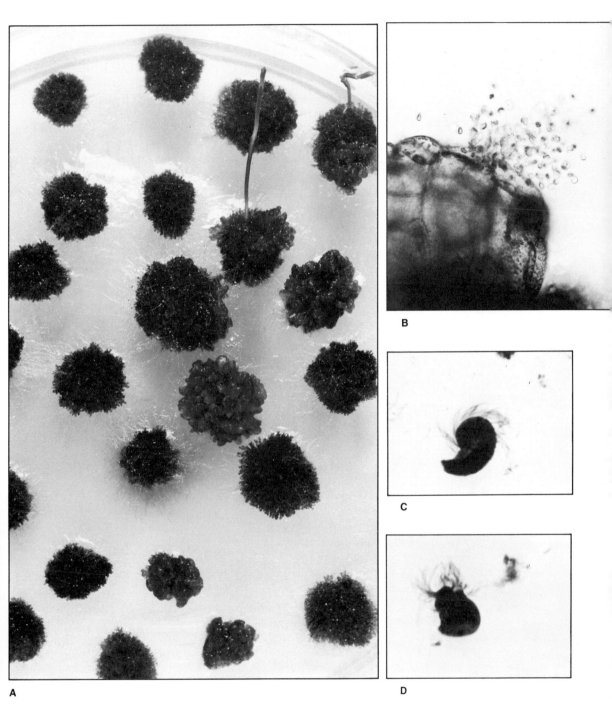

A

B

C

D

Figure 10-14
A, gametophytes of *Equisetum* sp. grown on simple inorganic nutrient medium; young sporophytes attached to gametophytes can be seen toward upper edge of figure; **B,** liberation of sperms from an antheridium; **C,D,** multiflagellate sperms, fixed with osmic acid fumes and stained with crystal violet.

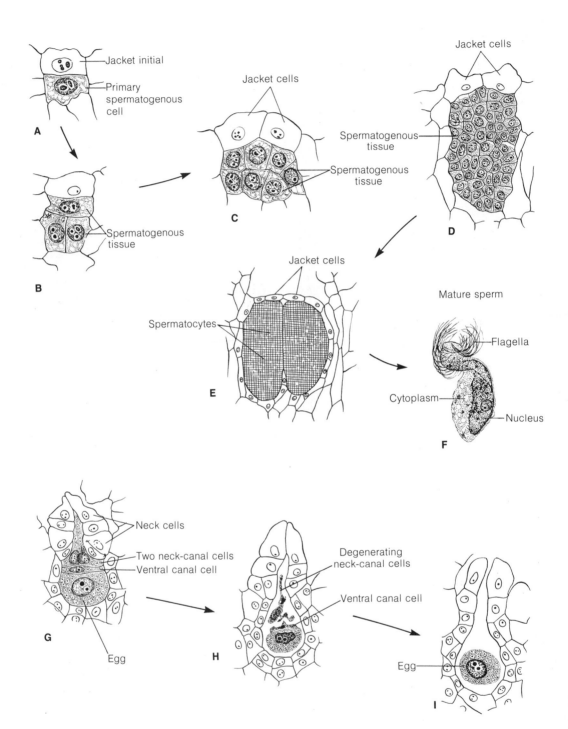

Figure 10-15
Equisetum ramosissimum Desf. subsp. *ramosissimum.* **A–F,** antheridial development; **G–I,** archegonial development. [Redrawn from Chatterjee and Ram, *Bot. Notis.* 121:471–490, 1968.]

"pincushion" stage. An antheridium is initiated by a periclinal division in a superficial cell setting aside a cover or jacket cell and a primary spermatogenous cell. The jacket cell divides anticlinally and at maturity the jacket layer may consist of two or more cells. The primary spermatogenous cell divides and the derivative cells then undergo a series of synchronized divisions, forming a packet of spermatocytes (Fig. 10-15, A–E). At maturity the large multiflagellate sperms escape through a pore (Fig. 10-14, B) created by the separation of the jacket cells (Chatterjee and Ram, 1968; Hauke, 1968). The sperms have numerous flagella attached to a linear flagellar band (Sharp, 1912; Chatterjee and Ram, 1968). A sac-like vesicle of cytoplasm may be attached to the posterior end of the sperm (Figs. 10-14, C, D; 10-15, F).

An archegonium follows the common type of ontogeny displayed also in other groups of lower vascular plants (Chapter 5; Fig. 10-15, G-I). At maturity an archegonium may have the following organization: a projecting neck which comprises two or more tiers of neck cells arranged in four rows; generally two adjacent neck-canal cells (boot-shaped cells), a ventral canal cell, and an egg at the base of the embedded venter (Jeffrey, 1899; Sethi, 1928, Chatterjee and Ram, 1968). At maturity all the cells of the axial row except the egg cell degenerate, and the upper tier of neck cells separate from one another (Fig. 10-15, I). In two species the neck cells have been reported to be greatly extended and spread apart at maturity (Slade, 1964; Hauke, 1968).

To achieve fertilization the gametophytes must be covered by at least a thin layer of water in which the motile sperms swim to the archegonia. Mature sperms are probably not released until the proper osmotic conditions are achieved. When gametophytes are flooded, water probably enters the cells of the mature antheridia because of an osmotic gradient. This would bring about swelling of the cells and is possibly the causal factor in the opening of the antheridium.

The Embryo

The origin of the diploid phase or sporophyte is the fertilized egg or zygote. Observations on fertilization have indicated that numerous sperms may enter the archegonium, and even penetrate the egg cytoplasm, but only one sperm actually fuses with the egg nucleus (Chatterjee and Ram, 1968). The first division of the zygote has been described as transverse (Sadebeck, 1878) for certain species; recently the first division in *E. arvense* has been described as being oblique (Laroche, 1968), setting aside an upper cell, the epibasal cell, and a lower cell, the hypobasal cell. The embryo is therefore, exoscopic in polarity (Chapter 6; Fig. 10-16).

There are conflicting accounts of subsequent embryogeny in *Equisetum,* which may be indicative of variation in the genus. Earlier descriptions were presented at a time when embryogeny was considered to be a precise series of unalterable events. Assignment of the first leaf, stem (future shoot apex), root, and foot to definite segments of the embryo in the quadrant stage was described by Sadebeck (1878). The first leaf and the shoot apex were said to develop from the epibasal portion, the first root and the foot from different quadrants of the hypobasal portion. Later workers reported variations in segmentation and origin of fundamental organs. Until more complete studies are made of more species, the following general outline of development may be taken as representative for *E. arvense* (Laroche, 1968) and some other species.

Following the first oblique division, the epibasal and hypobasal cells divide at right angles to the original wall. This establishes

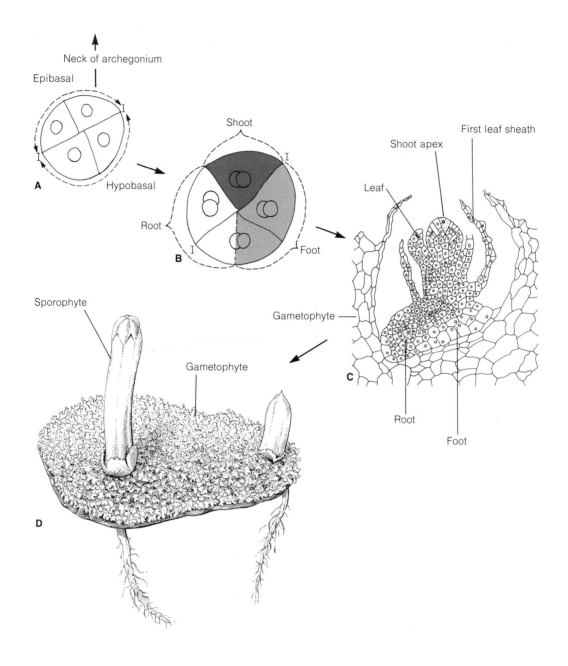

Figure 10-16

A–C, development of the embryo in *Equisetum* showing the first cleavage of the zygote (I–I in **A**) and derivation of shoot, root, and foot from the eight-celled embryo (**B**); **D**, "leafy" dorsiventral gametophyte of *Equisetum laevigatum* with two attached sporophytes. [A,B redrawn from Laroche, *Rev. Cytol. Biol. Veg.* 31:155–216; C redrawn from *Cryptogamic Botany*, Vol. II, Ed. 2 by G. M. Smith. McGraw-Hill, New York, 1955; D redrawn from Walker, *Bot. Gaz.* 71:378–391, 1921.]

the quadrant stage (Fig. 10-16, A). After subsequent divisions in the four cells, the future shoot apex is organized from derivatives of one quadrant of the epibasal hemisphere. The foot takes its origin from one quadrant of the hypobasal hemisphere and a portion of the other adjacent hypobasal quadrant. The first root is organized from one of the epibasal quadrants and a portion of the subjacent hypobasal quadrant (Fig. 10-16, B). The shoot then grows rapidly, forming whorls of three or four scale-like leaves (Fig. 10-16, C). Later the root penetrates the gametophytic tissue in reaching the soil or substratum (Fig. 10-16, D). Additional erect shoots are formed from buds on the primary axis of the sporophyte. Large, mature gametophytes may support several sporophytes (from multiple fertilizations) in varying stages of development.

Hyeniales–Hyeniaceae

As stated in the first part of this chapter, the Sphenopsida were widespread in their distribution and flourished particularly during the coal-forming age. Their representation today is but a vestige of that typical earlier. In the following section only brief descriptions will be given for extinct orders of the Sphenopsida.

The earliest known record of plants with sphenopsid features is from strata deposited during the early Devonian. The genus *Hyenia* is representative. It had a horizontal stem, with slender upright aerial branches, some sterile, some fertile. Both sterile and fertile appendages were somewhat whorled in arrangement, and the vegetative appendages branched dichotomously several times. The fertile appendages (sporangiophores) likewise dichotomized but had recurved tips with terminal sporangia (Fig. 10-17).

For the present, *Hyenia* is placed in the Sphenopsida, but it is possible that if more becomes known about its internal anatomy, it may become necessary to move it to another class (see Schweitzer, 1972). *Calamophyton,* a former companion of *Hyenia,* was removed from the sphenopsids when more information became available on its internal structure (Leclerq and Schweitzer, 1965).

Sphenophyllales – Sphenophyllaceae

This group was common in the Carboniferous, but is known to have existed as early as the late Devonian and as late as the Triassic. These plants are variously depicted as erect and self-supporting or as clambering over other vegetation. These plants are placed in the genus *Sphenophyllum,* although form genera exist for reproductive structures. Similar to other groups in the Sphenopsida, the stems were jointed with whorls of leaves at each node (Fig. 10-18, A). Unlike other groups, however, the leaves were commonly wedge-shaped, the distal margins were toothed or deeply notched, and the venation system was dichotomously branched (Abbott, 1958). The stem was an exarch protostele, which was usually triarch (Fig. 10-18, B). A cylinder of secondary xylem was formed which consisted of tracheids and xylem rays. Strobili of the group were structurally quite diversified and very complex in some instances (Fig. 10-18, C–E). Basically, there were alternating whorls of bracts and sporangiophores or, in some species the sporangiophores were positioned directly above a whorl of fused bracts and the recurved stalked sporangia were often fused with the bracts for some distance (Fig. 10-18, C, D). The majority of species were homosporous, but a few were heterosporous.

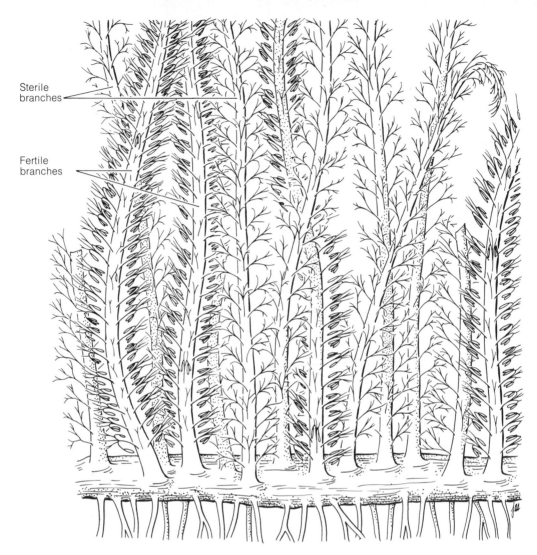

Sterile branches

Fertile branches

Figure 10-17
Schematic representation of portion of the rhizome and
upright branches of *Hyenia elegans*. [Redrawn from Leclercq,
Mem. Acad. Roy. Belg. Classe Sci. 12:1–65, 1940.]

Figure 10-18 *(facing page)*
Sphenophyllales. **A,** leaves and stem of *Sphenophyllum* sp.; **B,** xylem cylinder, stem of
Sphenophyllum sp. as seen in transverse section; **C,** longisection, strobilus of
Sphenophyllostachys dawsoni; **D,** same as **C,** viewed from above (stalked sporangia were
joined with the fused bracts for some distance); **E,** strobilus of *Litostrobus iowensis*
as seen in longisection. [B redrawn from *Les Plantes Fossiles dans leurs Rapports avec
les Végétaux Vivants* by L. Emberger. Masson et Cie, Paris, 1944; C redrawn from
An Introduction to the Study of Fossil Plants by J. Walton. Adam and Charles Black,
London, 1953; D redrawn from *Handbuch der Paläobotanik* by M. Hirmer.
R. Oldenbourg, Munich, 1927; E redrawn from Reed, *Phytomorphology* 6:261–272, 1956.]

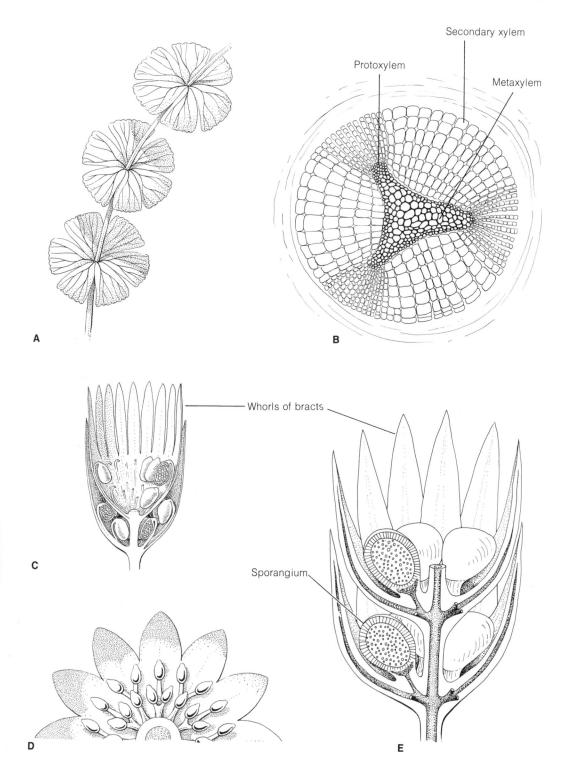

A

B

Protoxylem

Secondary xylem

Metaxylem

C

Whorls of bracts

D

E

Sporangium

Calamitales—Calamitaceae

Members of this group were of tree-like proportions (some plants being 1 foot in diameter and 60 feet or more in height) and constituted a significant part of the Carboniferous flora. It is in this group that establishment of *form genera* (explained in Chapter 9) is of great importance because only rarely are connected parts of a plant uncovered. If a complete plant could be assembled it would agree, in general, with the following description. The plant body consisted of an aerial branch system and an underground rhizome system similar to that of *Equisetum*. Roots, as well as aerial branches, originated at nodes along the rhizome. The aerial, articulated shoot exhibited limited or extensive branching. Whorls of linear, lanceolate or spatulate microphyllous leaves occurred at the nodes (Fig. 10-19). These leaves were considerably larger than those of *Equisetum* and were sometimes fused at their bases. In contrast to *Sphenophyllum*, leaves of *Annularia*, for example, developed only a single unbranched midvein. Strobili occupied the tips of side branches and were not placed terminally on major axes as is common in *Equisetum*.

As noted, form genera are very important in the Calamitales. There are form or organ genera for casts of pith cavities (*Calamites*), petrifactions of stems (*Arthropitys, Calamodendron*), leaf whorls (*Annularia, Asterophyllites*), and strobili (*Calamostachys, Palaeostachya*). Not all form genera have been listed here. It is conventional to use the generic name *Calamites* in speaking of entire plants (Fig. 10-19).

The remains of a calamite stem are of two types: pith casts and petrifactions. Pith casts were formed by infiltration of minerals into pith cavities and their subsequent hardening there (Fig. 10-20, A). When the organic material decayed, an image of the inner surface of the vascular cylinder re-mained. Many petrified stems of Calamitales reveal a remarkable similarity to those of *Equisetum,* except for the continuous cylinder of secondary xylem (Fig. 10-20, B).

Annularia is a genus for whorls of microphyllous leaves attached to small stems (Fig. 10-20, C). Strobili consisted of alternate whorls of bracts and sporangiophores (*Calamostachys*—Fig. 10-21, D, E) or whorls of peltate sporangiophores that are situated in the axils of sterile bracts (*Palaeostachya*—

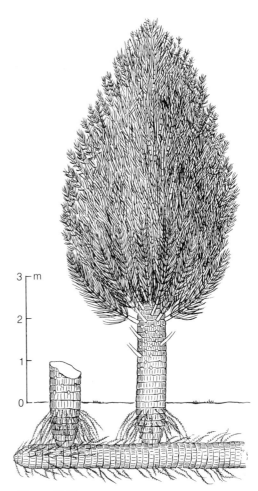

Figure 10-19
Schematic representation of *Calamites* showing rhizome and a tall, upright, branched portion. [Redrawn from *Les Plantes Fossiles* by L. Emberger. Masson et Cie, Paris, 1968.]

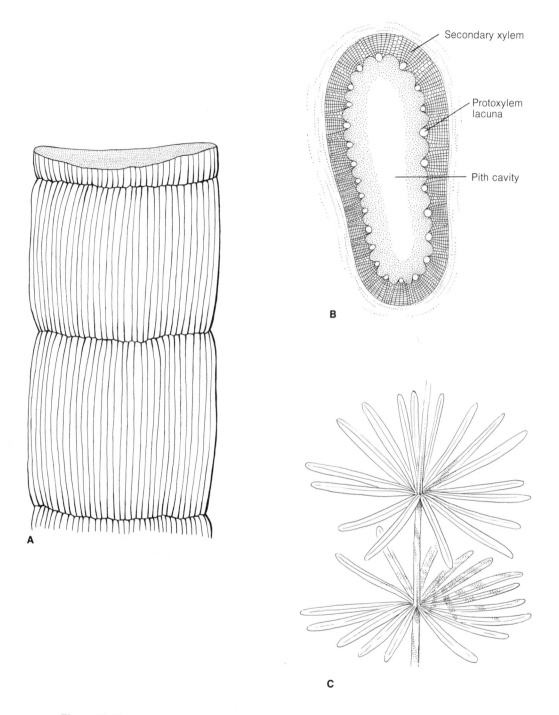

Figure 10-20
Schematic representations of Calamitales. **A**, a pith cast; **B**, a transverse section of stem; **C**, leaves and stem of *Annularia*.

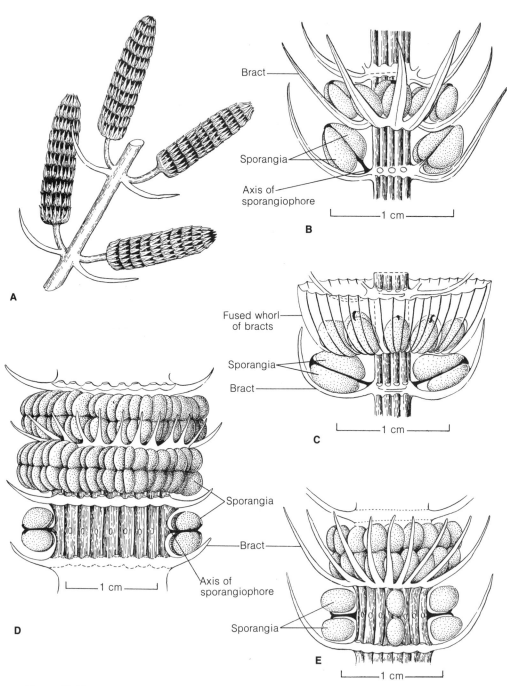

Bract

Sporangia

Axis of
sporangiophore

1 cm

A

B

Fused whorl
of bracts

Sporangia

Bract

1 cm

C

Sporangia

Bract

Axis of
sporangiophore

1 cm

D

Bract

Sporangia

1 cm

E

Figure 10-21
Diagrams of Calamitales strobili and portions of strobili showing position of sporangiophores and bracts.
A, *Palaeostachya decacnema;* **B,** *Palaeostachya ovalis;* **C,** *Palaeostachya aperta;* **D,** *Calamostachys interculata;*
E, *Calamostachys longibracteata.* [A redrawn from Delevoryas, *Amer. Jour. Bot.* 42:481–488, 1955;
B–E redrawn from Abbott, *Palaeontogr. Amer.* 6:1–49, 1968.]

Fig. 10-21, A–C). A few forms were hetero-sporous and in one reported instance attained the seed-like habit by retention of one megaspore in each mature megasporangium (Baxter, 1963). Some botanists consider this group to be a family coordinate with the Equisetaceae and place it in the order Equisetales.

Equisetales – Equisetaceae: *Equisetites*

Before leaving this discussion mention should be made of the genus *Equisetites*. As mentioned above, certain representatives of the Carboniferous resembled the living genus *Equisetum*. From the rocks of the Mesozoic and Cenozoic, plant remains have been discovered and given the name of *Equisetites*. It is extremely difficult to distinguish them from certain living members of the genus *Equisetum*.

An Appraisal

In this textbook we have recognized four orders in the class Sphenopsida. Other authors consider the class to comprise three orders: Hyeniales, Sphenophyllales, and Equisetales; the latter is composed of two families, the Calamitaceae and Equisetaceae. Additionally, the Sphenopsida is designated, for example, as the Articulatae or Arthrophyta by other authors.

The Sphenophyllales is quite distinct from other members of the Sphenopsida in leaf morphology, possession of a protostele, and the close association of sporangiophores with bracts. In some instances the sporangiophore-bract complex is exceedingly intricate.

The general morphology of *Equisetum* and that of an extinct calamite is similar except for the great size of a calamite and its formation of secondary vascular tissues.

Morphologists are frequently confronted with the question: Is the strobilar organization in *Equisetum* essentially primitive or is it specialized? In *Equisetum* the strobilus consists of sporangiophores without bracts. It is inviting to assume that this type of organization is derived from that of the Hyeniales. In the latter order the fertile appendages (sporangiophores?) were dichotomously branched and had recurved terminal sporangia. With some modification the sporangiophores of *Equisetum* could be derived from those of *Hyenia* (Fig. 10-22). If *Equisetum*, however, does truly represent

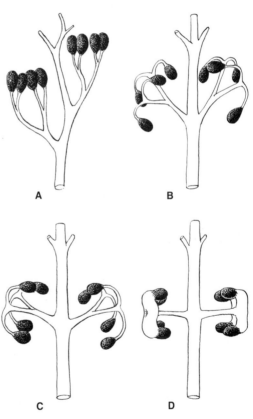

Figure 10-22
Possible evolutionary stages leading to the formation of sporangiophores typical of Hyeniales, Calamitales, and Equisetales. [Redrawn from Stewart, *Phytomorphology* 14:120–134, 1964; based on Zimmerman, 1949.]

a highly reduced calamite, then strobilar organization probably has involved the loss of bracts in the strobilus. In this case, the strobilus of *Equisetum* would be specialized and reduced — by loss of bracts in the course of evolutionary development. Another way of looking at the problem would be to consider sporangiophores and bracts as equivalent or homologous structures — perhaps both fertile in the past — the bracts having lost their reproductive structures (sporangia) in the course of time. There is no conclusive evidence for either theory (see Page, 1972, for additional information).

Heterospory in the Sphenophyllales has been reported and it has been definitely established for the Calamitales; heterospory in *Equisetum* has also been described, but has been disputed (see p. 226).

Despite differences in strobilar organization, the Calamitales still remains as the most likely, known, extinct group from which *Equisetum* (and *Equisetites*) was derived; derivation perhaps took place at an early stage in the evolution of the Calamitales.

Summary and Conclusions

This chapter has presented a discussion of the comparative morphology of the Sphenopsida, a subdivision of the lower vascular plants that flourished and attained its maximum diversity during the Paleozoic Era. The only surviving member of this ancient group is the genus *Equisetum*, which consists of about 25–30 species of widely distributed herbaceous perennials popularly known as horsetails or scouring rushes. Some of the ancient sphenopsids were large trees with pronounced secondary growth (e.g., *Calamites*), but the modern horsetails are plants which rarely exceed a meter or so in height and which are devoid of secondary vascular tissues. Both the extinct and living representatives of the Sphenopsida, especially the Sphenophyllales, Calamitales,

and Equisetales, differ from other groups of lower vascular plants in the following respects: (1) the shoots are conspicuously jointed with well-defined nodes and elongated internodes; (2) the leaves occur in whorls at each node; (3) branches arise at the nodes in marked contrast with the dichotomous plan of branching characteristic of the *Psilopsida* and of such genera as *Lycopodium* and *Selaginella* in the Lycopsida; (4) the internodal regions of the stem are commonly traversed by conspicuous longitudinal ribs; and (5) the sporangia are borne on whorled appendages termed sporangiophores. These organographic features of the sporophyte have been discussed briefly with reference to the morphology of the extinct orders of the Sphenopsida — that is, the Hyeniales, Sphenophyllales, and Calamitales — and in connection with the detailed description of the morphology and reproduction of *Equisetum,* the living representative of the order Equisetales.

The vegetative sporophyte of *Equisetum* consists of an underground system of rhizomes, which form buds and roots at their nodes; certain of the buds develop into the aerial shoots of the plant. The foliar organs of *Equisetum* are small univeined scales which occur in whorls at each node. The scale leaves are basally united into a cylindrical sheath, and the nodal branches, which alternate in position with the leaves, must break through the foliar sheath during their early expansion. In some species of *Equisetum* two kinds of shoots arise from the rhizome: unbranched, brownish, fertile shoots bearing large terminal strobili, and profusely branched, green, sterile shoots. Other species do not exhibit such dimorphism, and the strobili are borne terminally on the main axis or the lateral branches of green shoots. The presumable trends in evolutionary specialization of the varied types of shoots have been briefly discussed.

The anatomy of the stem in *Equisetum* is distinctive, and a detailed description has been given with particular emphasis on the structure and ontogeny of the vascular system. In contrast with the predominence of the protostele (with either exarch or mesarch xylem) in the stems of the Psilopsida and Lycopsida, the vascular system in the stem internodes of *Equisetum* consists of a cylinder of bundles with endarch xylem; each of these bundles develops a prominent carinal canal (i.e., a protoxylem lacuna). At the nodes the bundles are joined into a vascular girdle which is interrupted only by the gaps associated with the nodal branches. *Equisetum* is thus cladosiphonic and devoid of leaf gaps. The number of bundles in a given internode may agree with the number of external longitudinal ribs—each rib lies radially opposite a vascular bundle.

The mature strobilus of *Equisetum* consists of an axis to which are attached whorls of peltate sporangiophores. Each sporangiophore bears on the lower surface of its disk-shaped portion a ring of 5–10 sporangia. The sporangia are eusporangiate in method of development, and when ripe contain a large number of spores that are unique in wall structure. The outermost wall layer consists of four strips or bands known as elaters. These structures are hygroscopic, uncoiling as the spores dry and recoiling with the addition of water. Possibly the movements of the elaters may aid in the escape of the spores after the sporangium has dehisced.

Spores retain their viability for only a short time, but if moisture is available they germinate readily and give rise to distinctive cushion-shaped gametophytes, which produce thin erect photosynthetic lobes. Both antheridia (which produce multiflagellate sperms) and archegonia may develop on older gametophytes. For several species the gametophytes are initially unisexual but the female plants may later become bisexual.

The embryo of *Equisetum* is exoscopic in orientation and early in development three regions of the embryo become established— the future shoot, first root, and foot.

References

Abbott, M. L.
    1958. The American species of *Asterophyllites, Annularia,* and *Sphenophyllum. Bull. Amer. Paleontol.* 38 (No. 174):289–372.
Barratt, K.
    1920. A contribution to our knowledge of the vascular system of the genus *Equisetum. Ann. Bot.* 34:201–235.
Baxter, R. W.
    1963. *Calamocarpon insignis,* a new genus of heterosporous, petrified calamitean cones from the American Carboniferous. *Amer. Jour. Bot.* 50:469–476.
Beer, R.
    1909. The development of the spores of *Equisetum. New Phytol.* 8:261–266.

Benedict, R. C.
    1941. The gold rush: a fern ally. *Amer. Fern Jour.* 31:127–130.
Bierhorst, D. W.
    1958a. The tracheary elements of *Equisetum* with observations on the ontogeny of the internodal xylem. *Bull. Torrey Bot. Club* 85:416–433.
    1958b. Vessels in *Equisetum. Amer. Jour. Bot.* 45:534–537.
    1959. Symmetry in *Equisetum. Amer. Jour. Bot.* 46:170–179.
Bir, S. S.
    1960. Chromosome numbers of some *Equisetum* species from the Netherlands. *Acta Bot. Neer.* 9:224–234.
Bower, F. O.
    1935. *Primitive Land Plants.* Macmillan, London.

Browne, I. M. P.
1922. Anatomy of *Equisetum giganteum*. *Bot. Gaz.* 73:447–468.

Chatterjee, J., and H. Y. M. Ram
1968. Gametophytes of *Equisetum ramosissimum* Desf. subsp. *ramosissimum*. I. Structure and development. *Bot. Notis.* 121:471–490.

Chen, C., and J. Lewin
1969. Silicon as a nutrient element for *Equisetum arvense*. *Can. Jour. Bot.* 47:125–131.

Duckett, J. G.
1970a. Sexual behavior of the genus *Equisetum*, subgenus *Equisetum*. *Bot. Jour. Linn. Soc.* 63:327–352.
1970b. Spore size in the genus *Equisetum*. *New Phytol.* 69:333–346.

Eames, A. J.
1909. On the occurrence of centripetal xylem in *Equisetum*. *Ann. Bot.* 23:587–601.

Francini, E.
1942. La struttura dell'apice del rizoma in confronto alla struttura dell'apice del fusto aereo in "*Equisetum ramosissimum*" Desf. *Nuovo Gior. Bot. Ital.* 49:337–357.

Golub, S. J., and R. H. Wetmore
1948a. Studies of development in the vegetative shoot of *Equisetum arvense* L. I. The shoot apex. *Amer. Jour. Bot.* 35:755–767.
1948b. Studies of development in the vegetative shoot of *Equisetum arvense* L. II. The mature shoot. *Amer. Jour. Bot.* 35:767–781.

Gwynne-Vaughan, D. T.
1901. Remarks upon the nature of the stele of *Equisetum*. *Ann. Bot.* 15:774–776.

Hauke, R. L.
1957. The stomatal apparatus of *Equisetum*. *Bull. Torrey Bot. Club* 84:178–181.
1961. A resume of the taxonomic reorganization of *Equisetum*, subgenus *Hippochaete*. I. *Amer. Fern Jour.* 51:131–137.
1962a. A resume of the taxonomic reorganization of *Equisetum*, subgenus *Hippochaete*. II. *Amer. Fern Jour.* 52:29–35.
1962b. A resume of the taxonomic reorganization of *Equisetum*, subgenus *Hippochaete*. III. *Amer. Fern Jour.* 52:57–63.

1962c. A resume of the taxonomic reorganization of *Equisetum*, subgenus *Hippochaete*. IV. *Amer. Fern Jour.* 52:123–130.
1963. A taxonomic monograph of the genus *Equisetum*, subgenus *Hippochaete*. *Beihefte Nova Hedwigia* 8:1–123.
1967. Sexuality in a wild population of *Equisetum arvense* gametophytes. *Amer. Fern Jour.* 57:59–66.
1968. Gametangia of *Equisetum bogotense*. *Bull. Torrey Bot. Club* 95:341–345.
1969a. Problematic groups in the fern allies and the treatment of subspecific categories. *BioScience* 19:705–707.
1969b. Gametophyte development in Latin American horsetails. *Bull. Torrey Bot. Club* 96:568–577.
1969c. The natural history of *Equisetum* in Costa Rica. *Rev. Biol. Trop.* 15:269–281.

Jeffrey, E. C.
1899. The development, structure, and affinities of the genus *Equisetum*. *Mem. Boston Soc. Natur. Hist.* 5:155–190.

Johnson, M. A.
1933. Origin and development of tissues in *Equisetum scirpoides*. *Bot. Gaz.* 94:469–494.
1937. Hydathodes in the genus *Equisetum*. *Bot. Gaz.* 98:598–608.

Kashyap, S. R.
1914. The structure and development of the prothallus of *Equisetum debile*, Roxb. *Ann. Bot.* 28:163–181.

Kaufman, P. B., W. C. Bigelow, R. Schmid, and N. S. Ghosheh
1971. Electron microprobe analysis of silica in epidermal cells of *Equisetum*. *Amer. Jour. Bot.* 58:309–316.

Laroche, J.
1968. Contribution à l'étude de l'*Equisetum arvense* L. II. Etude embryologique. Caractères morphologiques, histologiques et anatomiques de la première pousse transitoire. *Rev. Cytol. Biol. Veg.* 31:155–216.

Leclercq, S., and H.-J. Schweitzer
1965. *Calamophyton* is not a sphenopsid. *Bull. Acad. Roy. Belg.* 11:1394–1402.

Manton, I.

1950. *Problems of Cytology and Evolution in the Pteridophyta.* Cambridge University Press, London.

Matzke, E. B.

1941. Gametophytes of *Equisetum arvense* L. *Torreya* 41:181–187.

Meyer, F. J.

1920. Das Leitungssystem von *Equisetum arvense. Jahrb. Wiss. Bot.* 59:263–286.

Page, C. N.

1972. An interpretation of the morphology and evolution of the cone and shoot of *Equisetum. Bot. Jour. Linn. Soc.* 65: 359–397.

Queva, C.

1907. Histogénèse et structure du stipe et de la fronde des *Equisetum. Bull. Soc. Hist. Natur. d'Autin* 20:115–152.

Rothmaler, W.

1944. Pteridophyten-Studien. I. *Repert. Spec. Nov. Regni Veg.* 54:55–82.

Sadebeck, R.

1878. Die Entwicklung der Keimes der Schachtelhalme. *Jahrb. Wiss. Bot.* 11: 575–602.

Schaffner, J. H.

1930. Geographic distribution of the species of *Equisetum* in relation to their phylogeny. *Amer. Fern Jour.* 20:89–106.

1932. Diagnostic key to the species of *Equisetum. Amer. Fern Jour.* 22:69–75; 122–128.

1933. Six interesting characters of sporadic occurrence in *Equisetum. Amer. Fern Jour.* 23:83–90.

Schratz, E.

1928. Untersuchungen über die Geschlechterverteilung bei *Equisetum arvense. Biol. Zentralbl.* 48:617–639.

Schweitzer, H.-J.

1972. Die Mitteldevon—Flora von Lindlar (Rheinland). 3. Filicinae—*Hyenia elegans* K. & W. *Palaeontographica* 137(B):154 –175.

Sethi, M. L.

1928. Contributions to the life-history of *Equisetum debile* Roxb. *Ann. Bot.* 42: 729–738.

Sharp, L. W.

1912. Spermatogenesis in *Equisetum. Bot. Gaz.* 54:89–119.

Sinnott, E. W.

1943. Cell division as a problem of pattern in plant development. *Torreya* 43:29–34.

Slade, B.

1964. Gametophytes of *Equisetum fluviatile* L. in agar culture. *Phytomorphology* 14: 315–319.

Sporne, K. R.

1964. Self-fertility in a prothallus of *Equisetum telmateia,* Ehr. *Nature* 201:1345–1346.

Taylor, T. M. C.

1939. Some features of the organization of the sporophyte of *Equisetum arvense* L. *New Phytol.* 38:159–166.

Treitel, O.

1943. The elasticity, breaking stress, and breaking strain of the horizontal rhizomes of species of *Equisetum. Trans. Kansas Acad. Sci.* 46:122–132.

Vogt, T.

1942. Geokjemisk og geobotanisk malmleting. III. Lift om planteveksten ved Rörosmalmene. [Geochemical and geobotanical ore prospecting.] III. Some notes on the vegetation at the ore deposits at Röros. K. *Nosske Vidensk. Selskab Forhandl.* 15:21–24.

Walker, E. R.

1921. The gametophytes of *Equisetum laevigatum. Bot. Gaz.* 71:378–391.

1931. The gametophytes of three species of *Equisetum. Bot. Gaz.* 92:1–22.

Walton, J.

1944. The roots of *Equisetum limosum* L. *New Phytol.* 43:81–86.

Wollersheim, M.

1957a. Untersuchungen über die Keimungsphysiologie der Sporen von *Equisetum arvense* und *Equisetum limosum. Zeit. Bot.* 45:145–159.

1957b. Entwicklungsphysiologische Untersuchungen der Prothallien von *Equisetum arvense* und *Equisetum limosum* mit besonderer Berücksichtigung der Frage nach der Geschlechtsbestimmung. *Zeit. Bot.* 45:245–261.

11

The Filicopsida: Ferns

For the groups of plants discussed thus far (Chapters 7–10) the *microphyll,* with some exceptions (e.g., in *Sphenophyllum*), is the prevailing type of foliar appendage. In microphyllous plants, a considerable portion of the primary vascular system of the shoot is cauline and in some lower vascular plants (e.g., *Lycopodium, Selaginella*) is not closely related in its development to the initiation and vascularization of the leaves (cf. Freeberg and Wetmore, 1967). Rather, each leaf trace appears to represent merely a small strand of vascular tissue that separates from the periphery of the protostele at the node. Ferns, on the contrary, have *megaphylls.* A megaphyll, whether large or small, usually has a branched venation system, and its leaf-trace system in ferns is generally associated with one or more *leaf gaps* in the vascular cylinder of the stem unless the stem is protostelic (Refer to Chapter 3 for a more complete discussion of microphylls and megaphylls.) In megaphyllous plants the character of the primary vascular system of the stem is apparently affected by the development of the larger leaves and their associated traces. It should be pointed out, however, that regardless of the stelar pattern of the shoot axis, the primary vascular system of the root is exarch and radial in organization as in all groups of vascular plants. This remarkable conserva-

Figure 11-1 (*facing page*)
Tree ferns growing in Golden Gate Park, San Francisco, California. Young leaves, which exhibit circinate vernation, are seen in various stages of growth. The mature fronds are large and compound pinnate.
[Courtesy Dr. T. E. Weier, from *Botany, An Introduction to Plant Science,* Ed. 2, by W. W. Robbins, T. E. Weier, and C. R. Stocking. Wiley, New York, 1957.]

Figure 11-2
Greatly enlarged crozier, or fiddle head, of a fern. Note that the pinnae and further subdivisions exhibit circinate vernation. [Courtesy Dr. T. E. Weier.]

tism of the root seems to be correlated with the absence of foliar appendages, and hence lends further support to the idea that the vascularization of stems in most ferns and seed plants is significantly correlated with the development of megaphylls and their traces.

Today approximately 10,000 species of ferns are widely distributed over the earth's surface. Some species are restricted to narrow environmental niches and are endemic to certain localities. The common bracken fern (*Pteridium aquilinum*), for example, is worldwide in distribution in the tropics and temperate zones and is a troublesome weed in some regions. Ferns are quite numerous and are most diverse in the tropical rain forests, many of them becoming trees 20–40 feet high or growing as epiphytes. However, even desert areas and mountains of the temperate regions may have a fern population.

What general characteristics do we associate with a common field, garden, or house fern? Naturally we think of a large fern leaf, commonly called a *frond* in everyday usage —a term also used by many fern specialists. The fern frond may be a simple expanded blade or lamina with a petiole or *stipe*—the latter term used by some students of fern morphology—or, which is more common, the frond may have incisions in the blade, resulting in a pinnatifid leaf (*pinnatus*, from the Latin meaning feather-like, or with parts arranged along the two sides of an axis). The pinnate plan of organization reaches its highest degree of development in pinnately compound leaves. In the latter type of organization the petiole (stipe) is devoid of any expanded blade, and its continuation as the main axis of the frond is called the *rachis*. Attached to the rachis by petiolules, and approximately opposite each other, are pairs of leaflets, each called a *pinna* (plural, *pinnae*) in fern terminology. Each pinna may likewise be subdivided into pairs of *pinnules,* and there may be further subdivisions. Thus, a frond may be once

pinnate, bipinnate, tripinnate, and so on (Fig. 13-1). These plans of organization do not describe all of the variations of pattern in fern fronds, but do describe those of a large number of ferns.

In most ferns the stem is an underground rhizome and is not apparent except in stocky, erect species. The large trunks of tropical ferns, however, compare in size with the trunks of moderately large palms. Roots usually arise from the lower part of an aerial stem or from the lower surface of a rhizome, often characteristically related to each leaf (Fig. 3-5).

The manner in which a young fern frond expands is a familiar matter to the more careful observer. Unrolling fronds of many ferns are often referred to as "fiddle heads," "monkey tails," or "croziers" (Figs. 11-1, 11-2).

Sporangia

Almost everyone has discovered the brownish to black splotches on the lower surface (abaxial side) of a fern frond and perhaps at first has interpreted these structures as being parasitic insects or their eggs. Actually each "spot" is technically a *sorus,* i.e., a collection of sporangia that is, in some species, protected by an outgrowth from the leaf surface called an indusium (Fig. 13-9).

The ferns, particularly members of the Filicales, provide the best and most striking examples of abaxial sporangia in vascular plants. In contrast with the adaxial, solitary sporangia of the Lycopsida, the sporangia of higher ferns are either marginal (probably the primitive position) or, more commonly, on the abaxial surface of the fertile pinnae (Fig. 13-7). Abaxial sporangia, frequently fused into synangia, are also typical of many members of the Marattiaceae. In the Ophioglossaceae, often regarded as the most primitive family in the ferns, the sporangia are either free and terminal on the ultimate divisions of the fertile "leaf segment" (*Botrychium* and *Helminthostachys*) or are embedded in the leaf tissue (*Ophioglossum*). In the light of these facts, it may be concluded that in the Filicopsida three more-or-less intergrading sporangial positions occur: terminal, marginal, and abaxial. The first position is likewise characteristic of certain microphyllous plants (e.g., Rhyniopsida, Psilopsida, and Sphenopsida) whereas the abaxial position is common in higher ferns and serves to help demarcate this group of plants.

From the standpoint of the structure and method of development of their sporangia, the Filicopsida are either *eusporangiate* or *leptosporangiate*. The eusporangiate method of sporangial development is characteristic not only of the more primitive orders of ferns (i.e., Ophioglossales and Marattiales) but also of the other classes of lower vascular plants as well as of seed plants. The more specialized ferns (Filicales, Marsileales, Salviniales) are remarkable among all vascular plants by virtue of the presence of leptosporangia, one of the most distinctive morphological features of these plants. As pointed out in Chapter 4, the leptosporangium is evidently an extreme modification of the more archaic eusporangium, and its presence in the fern orders mentioned above, is a striking example of the result of an independent trend of specialization in spore-producing structures.

With reference to the kinds of spores produced, the ferns are characterized by both homosporous and heterosporous types. *Homospory* is typical of most ferns; *heterospory* occurs in certain "water ferns" (Marsileales and Salviniales) as well as throughout the seed plants. It should be emphasized that just as for the eusporangium of certain other lower vascular plants (e.g., Lycopsida), there is no strict correlation between sporangial type and type of spores produced.

Gametophytes and Embryos

As is true of the other groups of vascular plants, the type and relative prominence of the gametophytes are closely correlated with the conditions of homospory and heterospory in the ferns. Thus, in the homosporous ferns the gametophyte is a freely developed, independent plant (*exosporic*) that is photosynthetic or (as in many of the eusporangiate groups) subterranean and associated with a fungus. In contrast, the male and female gametophytes of heterosporous ferns are *endosporic* and much smaller than those of homosporous ferns. All known living ferns produce multiflagellate sperms.

No single type of polarity characterizes the embryogeny of ferns. The pattern in which the first division of the zygote is longitudinal, and hence results in a lateral orientation of the apical and basal poles, characterizes the largest number of ferns (See Chapter 6). Exoscopic and endoscopic embryos also are found in ferns. It is remarkable that the endoscopic type of embryo with a suspensor, which Bower (1935) regarded as the primitive form in lower vascular plants, has become "standardized" in seed plants. The intermediate or "prone" type of embryo, devoid of a suspensor, is another feature that separates the highly specialized leptosporangiate ferns (except for a few species) from other vascular plants.

Ferns of Past Epochs

The ferns are well known from the fossil record that extends back to Middle Devonian times. Some of these fossils are not too different from those of certain members of the Trimerophytopsida. In fact, representatives of the familiar so-called "true ferns" are found in strata of the Permian. In addition, certain groups of ferns apparently existed for a period without leaving any direct descendants in today's fern flora. Even so, the Filicopsida, particularly the Filicales, is a highly successful group in the present epoch, having overcome the rigors of existence in a changing world much better than their frequent associates, the lycopods and horsetails.

In the opening of this chapter a general description of a modern fern was presented for orientation purposes. However, the student of fern morphology is interested not only in living ferns but also in ferns of the past. Deviating from the organization of previous chapters, the earliest fossil forms will be introduced first. This departure should not be difficult for the student because he is, by now, familiar with the ancient group, the Trimerophytopsida (Chapter 7), and has gained an insight into the possible evolutionary development of the megaphyll (Chapter 3). In a book of limited scope it is impossible to consider the vast amount of information available on extinct ferns, and we are compelled simply to present certain types which illustrate important morphological features and steps in the evolution of the group as a whole.

The search for earliest fern records immediately becomes a complicated study, but one not without some degree of hope. The bulk of Paleozoic fern foliage, originally thought to be exclusively that of spore-producing ferns and to which generic names were assigned, was shown to represent actually the leaves of a great many seed-producing ferns — the Pteridospermales (Fig. 15-6). Thus form genera (see the discussion in Chapter 9) exist for the foliage types of both the ferns and pteridosperms. Identification keys that use shape, method of attachment of pinnules, and type of venation have been established for Paleozoic fern leaves (Fig. 15-1). In certain instances a "natural plant" can be synthesized from the form genus for foliage and from the numerous form genera for other parts of the plant which were originally found as isolated fragments.

The problem, then, in tracing the history of the Filicopsida is to separate those fossil forms that may represent morphological steps in the evolution of the fern type of organization, but at the same time to recognize that seed ferns may have shared a common ancestry with the Filicopsida.

As a result of extensive studies on ferns, Bower (1935) proposed certain features that would, in his opinion, characterize a primitive fern. The fern archetype would be an upright, dichotomizing plant, if branched at all, in which the distinction between leaf and axis would be either absent or ill-defined. The leaf, where recognizable, would be long-stalked and dichotomously branched with the shanks of the dichotomies free from one another. Sporangia would be relatively large, solitary, and located at the distal ends of the subdivisions of leaves. The sporangial wall would be thick, opening by a simple dehiscence mechanism, and the sporangia would contain only one type of spore.

Cladoxylales

One assemblage of ancient fern-like plants is the order Cladoxylales (sometimes elevated to the rank of class). These Paleozoic plants are of interest because of their complex stelar configurations and forms of leaves. Most species were relatively small plants with irregularly branched main axes and dichotomously branched smaller ones. The smaller appendages were forked and the fertile leaves, occurring on terminal branches, were flat, fan-shaped, and bore many terminal sporangia on partially fused ultimate segments (Fig. 11-3). The vascular tissue was organized as a system of interconnected strands (each strand, on many specimens, being elongated radially as seen in transverse section of a stem). In some

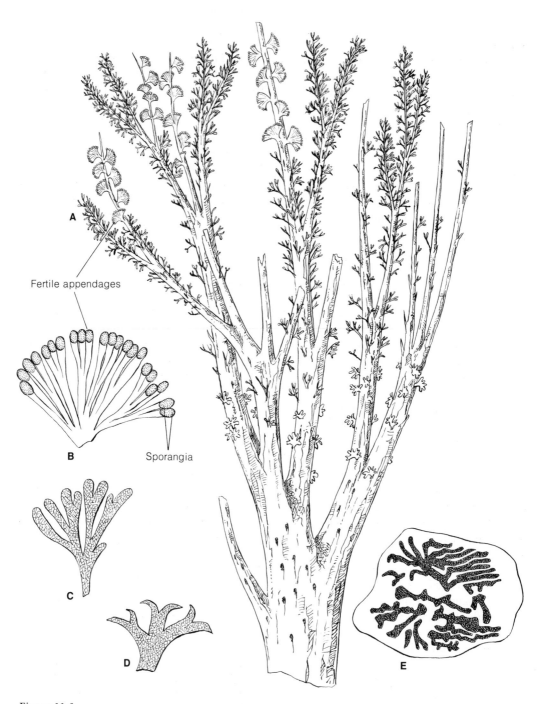

Fertile appendages

Sporangia

Figure 11-3
Cladoxylales—*Cladoxylon scoparium*. **A**, reconstruction, portion of a plant; **B**, fertile appendage; **C,D**, sterile appendages from upper (**C**) and lower (**D**) portions of the plant; **E**, xylem of stem as seen in transverse section. [Redrawn from Kräusel and Weyland, *Abh. Senckenberg. Naturforsch. Ges.* 40:115–155, 1926.]

species secondary growth took place around each primary vascular bundle.

The Cladoxylales is an interesting collection of plants, perhaps derived from the Trimerophytopsida, but one that probably ended without giving rise to any other groups.

Coenopteridales

Another rather enigmatic group of fern-like plants was the extinct Coenopteridales (Fig. 11-4). Some specimens have been identified as being of Devonian age, but the order was more common in the Carboniferous. These fern-like plants cannot be assigned with any certainty to modern groups of ferns, with perhaps one exception (see below). Admittedly, the order is somewhat artificial and is used as a matter of convenience.

In general, the Coenopteridales can be characterized by the general lack of distinction between the shoot axis and the leaf at the level of frond attachment. They exhibit their primitiveness by the fact that the stem is protostelic, by the three-dimensional branching of the frond in some, and by the terminal position of sporangia on ultimate segments of the frond in some. Specialization has led (1) to the formation of elaborate vascular cylinders in the frond axis and (2) to planate megaphylls with sporangia being located near the margins or even on the lower surface of pinnules. Tracheids were scalariformly pitted in some species; other species had circular bordered pits (see Eggert, 1964 for general discussion of the Coenopteridales).

We will now discuss representative genera of the Coenopteridales.

STAUROPTERIS. The two species assigned to this genus were apparently small, bushy plants with fronds that exhibited three-dimensional branching (Fig. 11-4, A). No laminar tissue was formed and the sporangia were terminal on ultimate branches of a frond. The petiole (stipe or "phyllophore") had four strands of vascular tissue, each one being pear-shaped. Traces from these bundles vascularized subdivisions of the frond (Fig. 11-5, C).

BOTRYOPTERIS. In the genus *Botryopteris* the lower part of the frond was three-dimensionally branched. In one example the distal portion was planate (branched in one plane) and the ultimate segments developed laminae (Fig. 11-4, C). The genus is distinct anatomically in that the xylem cylinder in the petiole was in the form of the letter W, as seen in transverse section (Fig. 11-5, B). Sporangia were in clusters or tufts on ultimate segments.

ANKYROPTERIS. The form of the vascular cylinder in the petiole is an important feature in separating coenopterid fossils. In *Ankyropteris* the xylem as seen in transverse section is in the form of the letter H (Fig. 11-5, A). The vegetative body of *Ankyropteris glabra* was described by Eggert (1963). In this species, unlike other coenopterids, all orders of branching of the frond were in the same plane, much like present-day ferns. Fertile fronds were discovered and given the name *Tedelea* (Eggert and Taylor, 1966). Sporangia occurred on the lower surface at the termination of lateral veins of pinnules (Fig. 11-4, D). Eggert concluded that this particular fern does not seem closely related to any completely known coenopterid and he therefore placed it in the order Filicales.

It is quite possible that as more becomes known about members of the Coenopteridales, some of them will be shifted to other groups of ferns. The reader is directed to the article by Eggert (1964) for additional information on other genera and for a general discussion of the taxonomy of the order.

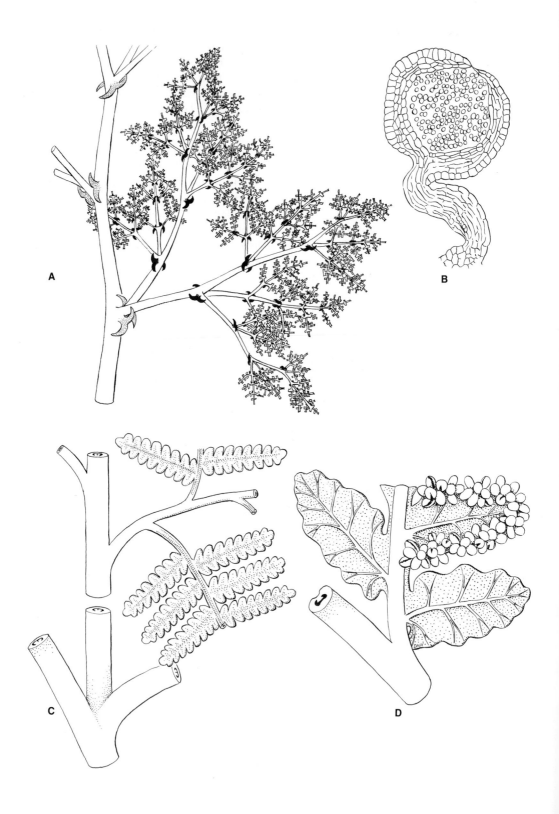

A

B

C

D

Figure 11-4 (*facing page*)
Reconstructions of portions of fronds of representative genera of the Coenopteridales. **A,B,** *Stauropteris oldhamia:* **C,** *Botryopteris* sp.: distal part of a frond; **D,** *Ankyropteris glabra:* portion of fertile frond (form genus *Tedelea* showing sporangia on abaxial side of one pinnule. [A,B redrawn from *Paläobiologie der Pflanzen* by K. Mägdefrau. Gustav Fischer Verlag, Jena, 1968; C redrawn from Delevoryas and Morgan, *Amer. Midland Natur.* 52:374–387, 1954; D redrawn from Eggert and Taylor, *Palaeontographica* 118(B):52–73, 1966.]

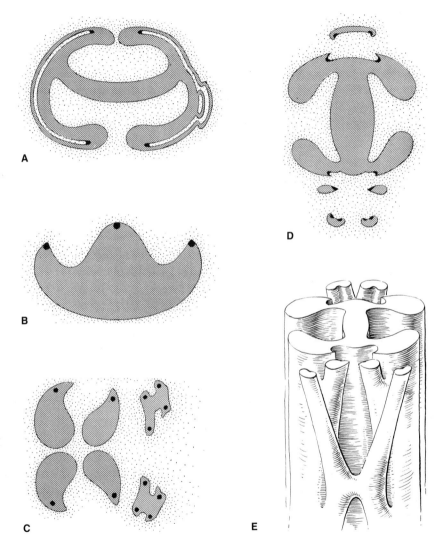

Figure 11-5
Petioles (stipes, phyllophores) of the Coenopteridales. **A–D,** xylem cylinders only, as seen in transection (solid black areas represent positions of protoxylem). **A,** *Ankyropteris;* **B,** *Botryopteris:* **C,** *Stauropteris;* **D,** *Etapteris;* **E,** stereogram of xylem cylinder in *Etapteris.* Smaller, peripheral strands in all diagrams are traces departing to subdivisions of the frond. [A, D, E redrawn from *Handbuch der Paläobotanik* by M. Hirmer. R. Oldenbourg, Munich, 1927; B, C redrawn from Anatomie der Vegetationsorgane der Pteridophyten by Y. Ogura. In *Handbuch der Pflanzen-anatomie.* Gebrüder Borntraeger, Berlin, 1938.]

In summary, the order, imperfectly defined as it is, probably did not give rise directly to other fern-groups, but may, however, have shared a "pre-fern" ancestry (from the Trimerophytopsida?) with other fern taxa. Most of the coenopterids probably represent end points in evolutionary experimentation.

Classification

The following outline of classification will serve as a guide for discussions of the various orders of ferns. The Coenopteridales has been included with the orders that have both living and extinct members.

FILICOPSIDA: Living and extinct ferns; plants showing a conspicuous alternation of generations (in modern representatives at least); sporophyte most conspicuous and often elaborately developed; megaphylls present; stems protostelic or siphonostelic, often with complex vascular cylinders; sporangia terminal on ultimate axes, terminal on veins, marginal, or on abaxial surface of fronds; eusporangiate, or more commonly leptosporangiate in living species; homosporous, a few heterosporous; gametophytes: (1) majority exosporic and green, (2) others exosporic, nonchlorophyllous, and subterranean, and (3) endosporic (restricted to heterosporous types); multiflagellate free-swimming sperms; embryo exoscopic, endoscopic, or intermediate.

COENOPTERIDALES: Extinct ferns; only sporophytes known; shoots erect, climbing, or creeping; fronds with pinnate plan or with modified dichotomous type of branching; stem and frond stipe or phyllophore anatomically similar at level of separation; stem protostelic; vascular cylinders of petioles (phyllophores) often complex; eusporangia grouped in clusters on pinnules or terminal on ultimate branches of a frond and often with a simple dehiscence mechanism; homosporous (one known to be heterosporous); form genera common; gametophyte and embryo unknown.

Representative genera: *Botryopteris, Stauropteris, Etapteris* (form genus for fronds), *Zygopteris, Ankyropteris* (*Tedelea,* fertile frond).

OPHIOGLOSSALES (Chapter 12): Living ferns, fossil record uncertain; sporophyte axis usually short and fleshy; stem siphonostelic; fronds simple or pinnately compound; vernation noncircinate; each fertile frond consists of a fertile segment or spike, bearing sporangia, and a sterile or vegetative segment; eusporangiate, producing great quantities of spores per sporangium; homosporous; gametophytes subterranean, bisexual, nonchlorophyllous, tuberous or worm-like with endophytic fungus; embryo exoscopic or endoscopic.

OPHIOGLOSSACEAE: Characteristics as in Ophioglossales.

Ophioglossum, Botrychium, Helminthostachys.

MARATTIALES (Chapter 12): Living and extinct ferns; sporophyte stem in most erect and short, or may be dorsiventral; stem with complex dictyostelic vascular cylinder; fronds commonly large, simple pinnate to tripinnate, and circinate in vernation; paired, clasping stipules at base of each leaf; eusporangia free and grouped into elongate sori or united into synangia on abaxial surface of fronds; many spores formed per sporangium; homosporous; gametophyte terrestrial, green, cordate to ribbon shaped, bisexual, and with endophytic fungus; endoscopic embryo.

MARATTIACEAE: Characteristics as in Marattiales.

Representative genera: *Angiopteris, Marattia, Psaronius* (form genus for stems), *Asterotheca, Scolecopteris* (form genera for sori).

FILICALES (Chapter 13): Living and extinct plants, of diverse growth habits and habitats; stems vary from protostelic to intricately dictyostelic; fronds simple to compound pinnate; sporangia scattered or grouped into sori; sori marginal or on abaxial side of fronds; sori with or without a protective structure, the indusium; leptosporangiate, most with a definite dehiscence mechanism, the annulus; spores numerous to few per sporangium, tetrahedral or bilateral; homosporous; gametophytes primarily green, exosporic, commonly thalloid, some filamentous; embryo "prone." Representative families are Osmundaceae, Schizaeaceae, Gleicheniaceae, Hymenophyllaceae, Dicksoniaceae, Matoniaceae, Cyatheaceae, and Polypodiaceae. Certain of these families are described in greater detail in Chapter 13.

MARSILEALES (Chapter 13): Living and extinct ferns; grow on damp soil or submerged in water; solenostelic; lamina consisting of four pinnae, bipinnate, or without pinnae; circinate vernation; sori enclosed in sporocarps; leptosporangiate; heterosporous; microsporangia and megasporangia in same sporocarp; endosporic gametophytes.

MARSILEACEAE: Characteristics as in Marsileales.

Marsilea, Regnellidium, Pilularia.

SALVINIALES (Chapter 13): Living and extinct ferns; small plants that float on surface of water; no true roots; leaves dimorphic, some serving as "roots"; sori enclosed in specialized indusia ("sporocarps"); leptosporangiate; heterosporous; microsporangia and mega-

sporangia in separate sporocarps; endosporic gametophytes.

SALVINIACEAE: Characteristics as in Salviniales.

Azolla, Salvinia.

Before proceeding with a detailed account of the living orders of ferns, which we have just outlined, it is important for the student to have a clear idea of the morphological features that are used for comparative purposes. It was the celebrated British morphologist, F. O. Bower, who realized the importance of exploring and exploiting the totality of morphological features before a reasonable phylogeny of ferns could be achieved. Bower (1923, 1935) concluded that there are at least twelve major morphological and anatomical criteria that should be utilized. These are listed here because discussions of these points are unavoidable in ferns, and because most of them are of great importance in later discussions.

1. External morphology and habit of plant
2. Apical meristem organization
3. Architecture and venation of the leaf
4. Vascular system of the shoot
5. Morphology of hairs and scales
6. Position and structure of the sorus
7. Protection of the sorus by an indusium
8. Development and mature structure of the sporangium including form of and markings on spores
9. Number of spores
10. Morphology of the gametophyte
11. Position and structure of sex organs
12. Embryology of the sporophyte

To this list should be added the more recent publications on important criteria to be considered in fern classification (Wagner, 1964) and the results from studies in the fields of anatomy (White, 1963), cytology and cytogenetics (Manton, 1950, 1959; Abraham et al., 1962; Fabbri, 1963), and experimental physiology and morphogenesis (Wardlaw, 1952, 1968a,b; Wetmore and Wardlaw, 1951).

References

Abraham, A., C. A. Ninan, and P. M. Mathew
1962. Studies on the cytology and phylogeny of the pteridophytes. VII. Observations on one hundred species of South Indian ferns. *Jour. Indian Bot. Soc.* 41:339–421.

Bower, F. O.
1923. *The Ferns,* Vol. I. Cambridge University Press, London.
1935. *Primitive Land Plants.* Macmillan, London.

Eggert, D. A.
1963. Studies of Paleozoic ferns. The frond of *Ankyropteris glabra. Amer. Jour. Bot.* 50:379–387.
1964. The question of the phylogenetic position of the Coenopteridales. *Mem. Torrey Bot. Club* 21:38–57.

Eggert, D. A., and T. N. Taylor
1966. Studies of Paleozoic ferns: on the genus *Tedelea,* gen. nov. *Palaeontographica* 118(B):52–73.

Fabbri, F.
1963. Primo supplemento alle tavole cromosomiche delle Pteridophyta di Alberto Chiarugi. *Caryologia* 16:237–335.

Freeberg, J. A., and R. H. Wetmore
1967. The Lycopsida—a study in development. *Phytomorphology* 17:78–91.

Manton, I.
1950. *Problems of Cytology and Evolution in the Pteridophyta.* Cambridge University Press, London
1959. Chromosomes and fern phylogeny with special reference to "Pteridaceae". *Jour. Linn. Soc. London Bot.* 56:73–91.

Wagner, W. H., Jr.
1964. The evolutionary patterns of living ferns. *Mem. Torrey Bot. Club* 21:86–95.

Wardlaw, C. W.
1952. *Phylogeny and Morphogenesis.* Macmillan, London.
1968a. *Essays on Form in Plants.* University Press, Manchester, England.
1968b. *Morphogenesis in Plants: A Contemporary Study.* Methuen, London.

Wetmore, R. H., and C. W Wardlaw
1951. Experimental morphogenesis in vascular plants. *Ann. Rev. Plant Physiol.* 2: 269–292.

White, R. A.
1963. Tracheary elements of the ferns. II. Morphology of tracheary elements; Conclusions. *Amer. Jour. Bot.* 50: 514–522.

12

The Eusporangiate Ferns

Ophioglossales — Ophioglossaceae

It is most unfortunate that the order Ophioglossales has no fossil history; nevertheless some morphologists consider the order to have been derived from the Coenopteridales; others believe that it originated from the Trimerophytopsida. Our earliest recorded knowledge of the Ophioglossales dates back to the year 1542 and to one of the herbalists, Leonhard Fuchs (Clausen, 1938).

The family comprises three or four recognized genera, of which two are more commonly seen and known by botanists. One genus, *Botrychium* (grape fern, moonwort), having about 30 or more species, is restricted mainly to the North Temperate Zone. *Ophioglossum* (adder's tongue), with 40 or more species, is widely spread throughout the habitable world but is more abundant in the tropics (Campbell, 1948). Although members of the Ophioglossaceae are typically terrestrial, two species of *Ophioglossum,* which grow in the American tropics, are epiphytic. Commonly the stem is short and erect and has a frond that is divided into a flattened, vegetative portion (sterile segment) and a sporangium-bearing portion (fertile segment). However, the genera *Ophioglossum* and *Botrychium* can be separated easily (with few exceptions) by examining the fronds. The lamina in *Ophioglossum* is characteristically simple in outline with reticulate venation; it is pinnate in plan in *Botrychium,* with open dichotomous venation. One other genus, *Helminthostachys,* which is native to the Indo-Malayan regions, can be recognized by its creeping rhizome and palmately compound leaves, which have an open dichotomous venation system. The genera *Botrychium* and *Ophioglossum* will now be described in detail.

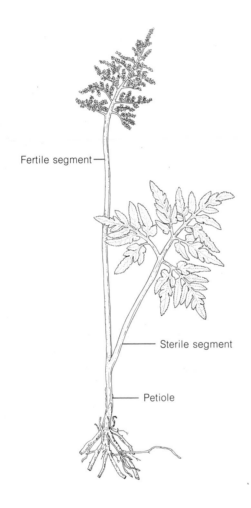

Figure 12-1
Habit sketch of sporophyte of *Botrychium dissectum* var. *obliquum,* showing pinnate fertile and sterile segments of one leaf.

Labels on figure:
Fertile segment
Sterile segment
Petiole

Botrychium (Grape Fern, Moonwort)

The plant axis is a short, stocky, subterranean rhizome from which roots arise at the bases of leaves (Fig. 12-1). Generally one frond matures each year in temperate-climate species; the decaying leaves of previous seasons surround the base of the plant. Each frond, all of which are fleshy, has a sheath at its base that encloses the next younger leaf. Although only one leaf matures each year, there are several immature leaves of future seasons in varying stages of development within the terminal bud. During development the leaf does not exhibit circinate vernation—a feature so common to many other ferns. The frond is fleshy and consists of two parts: the vegetative lamina, which usually exhibits a pinnate pattern of branching, and a fertile segment. The fertile segment likewise is constructed on the pinnate plan, bearing eusporangia in two rows on ultimate axes (Fig. 12-2, A, B). There is some degree of correlation between the amount of pinnate branching of the fertile and sterile portions of the frond. In certain species the vegetative lamina may be multipinnate and the fertile segment may be of the same degree of branching; in others, the blade may be simple and entire with the fertile segment showing the same degree of reduction. This reduction series is considered to be an evolutionary sequence. Morphological interpretation of the frond in *Botrychium* is a lively and controversial subject which will be discussed later in this chapter.

ANATOMY OF THE STEM. Growth of the shoot is reported to be initiated (Campbell, 1911) by an apical cell from which derivative cells are formed; these derivative cells provide initials for leaves, and some of them contribute to tissues of the axis. The vascular cylinder of well-established plants is an ectophloic siphonostele. The gaps do not overlap, consequently a continuous vascular

Figure 12-2
Botrychium californicum subsp. *silicifolium.*
A, growth habit; B, enlarged view, portion of
fertile segment showing sporangia; line of
dehiscence can be seen along top of some
sporangia.

cylinder or a cylinder with one gap may be
present in a stem as seen in a representative
transverse section (Fig. 12-3, A). The xylem
is endarch in development. The tracheids of
the metaxylem have circular or oval bordered pits in contrast with the scalariform
pitting that is so common in the vast majority of lower vascular plants. A definite
endodermis is present around the vascular

cylinder. In contrast with all other living
ferns, the stem of *Botrychium* is reported
to have a vascular cambium, which gives
rise to some secondary xylem and a limited
amount of secondary phloem (Fig. 12-3, A).
A critical developmental study of stem
anatomy has not been made to date. The
sieve cells of the primary phloem are conspicuously thickened except for sieve areas.

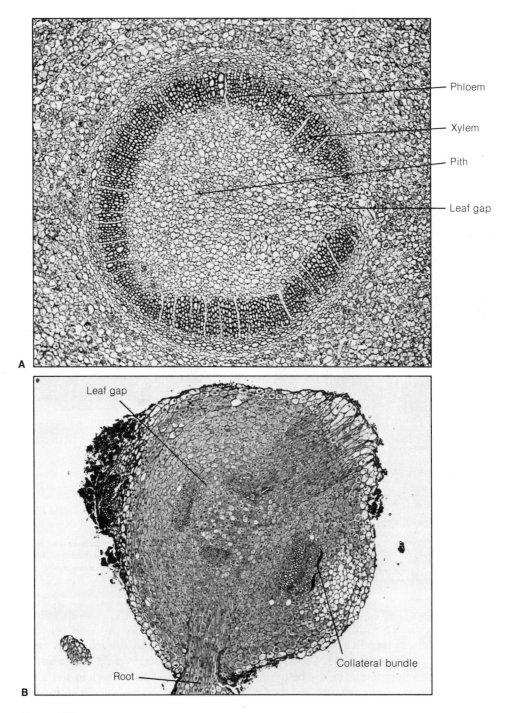

Figure 12-3
Stem anatomy in the Ophioglossales. **A,** transection, stem of *Botrychium* sp. with ectophloic siphonostele; **B,** transection, stem of *Ophioglossum* sp. showing ectophloic siphonostele.

However, the presence of callose has not been detected in the sieve areas (Esau, Cheadle, and Gifford, 1953). A cork layer is generally present at the periphery of the broad cortex.

THE ROOT. Attached near the bases of leaves are hairless, fleshy roots that generally exhibit limited monopodial branching. It is reported that a tetrahedral apical cell is present in the root tips, but cleavages which give rise to the root cap may not be as precise as they are in leptosporangiate ferns (Campbell, 1911; Bower, 1926). Mature roots are commonly tetrarch to diarch.

THE FROND (LEAF). As stated earlier in this chapter, the vegetative portion of the frond in *Botrychium* has an open dichotomous venation system with a rather homogeneous mesophyll organization. Transcending the mere knowledge of leaf organization is the question of the interpretation of the frond. A widely accepted view is that the fertile segment represents two pinnae which have become fused during the course of evolution and which now stand erect (Chrysler, 1925). This hypothesis is based on a study of the vascular system. In general, the leaf trace consists of one vascular bundle which may be dissected into two bundles where it enters the frond. Within the stipes of some species there are two crescent-shaped vascular bundles. This arrangement would appear to be a basic type in the genus *Botrychium*. At the level of the fertile segment the two bundles are branched; one small vascular strand from each of the original two bundles extends into the fertile segment. The two remaining larger vascular strands are continuous up the rachis of the vegetative segment, each one supplying lateral vegetative pinnae with a vascular bundle. Thus the presence of two vascular bundles in the rachis of the fertile segment is the evidence for the assumed evolutionary fusion of two basal fertile pinnae. The occasional presence in some species (for example, in *Botrychium lanuginosum*) of fertile pinnae occupying the position of otherwise normal vegetative pinnae tends to support this idea (Chrysler, 1925).

More recently the frond has been reinterpreted. The newer theory also is based on the branching of the vascular tissue in the frond. The repeated bifurcations of the vascular tissue are interpreted as being indicative of a once free dichotomous branch system. The first dichotomy is considered to be at a level below the union of the fertile spike and vegetative lamina (or even at the level of the departing leaf trace). Thus this interpretation adds even greater weight to the idea that the "frond" is a modified, reduced dichotomous branch system (Zimmermann, 1942; Chrysler, 1945).

As a natural extension of this concept some authors consider the petiole or stipe of the "leaf complex" of present-day species of *Botrychium* to be cauline and think that the sterile and fertile segments are phylogenetically equivalent to fronds, and are therefore homologous (Nozu, 1950, 1955; Nishida, 1952, 1957).

THE SPORANGIUM. The fronds of several future seasons are in varying stages of development within the bud. Therefore the stages of development of the fertile spikes are also variable. Eusporangia are initiated from one or from several superficial cells (Bower, 1935) in an acropetal manner on the pinnae. If only one initial superficial cell is evident early in development, sooner or later adjacent superficial cells may divide periclinally, which finally results in the separation of sporogenous cells and wall cells (Chapter 4). The tapetum becomes several layers thick, and its cells break down very early, their contents permeating the spaces between the spore mother cells.

At maturity each sporangium is large and

260

has a vascular strand that extends to the base of the capsule. Each sporangium, which dehisces by means of a terminal slit, produces 2,000, or more, tetrahedral spores.

CHROMOSOME NUMBERS. The results of several studies have shown that the haploid number of chromosomes in a large number of species of *Botrychium* is 45 (Manton, 1950; Wagner, 1963; Fabbri, 1963). One species, *Botrychium virginianum,* is presumably a tetraploid with $n = 90$.

THE GAMETOPHYTE. In *Botrychium* the gametophyte is subterranean, tuberous, usually somewhat longer than wide, and may be very short or as long as 2 cm. Although a gametophyte is fundamentally radial in organization, it may be somewhat flattened and dorsiventral and is covered with rhizoids (Fig. 12-4; 12-5, A). It is without chlorophyll, but it does have an endophytic fungus (Fig. 12-5, A). Chlorophyll, however, has been reported in some species under special conditions (Boullard, 1963). Antheridia are generally on a dorsal crest with archegonia along each side (Jeffrey, 1897; Nozu, 1954; Bierhorst, 1958). Antheridia are sunken and produce multiflagellate, free-swimming sperms. Archegonia have protruding necks of a few cells, and each one has a binucleate neck-canal cell, a ventral canal cell, and an egg cell.

THE EMBRYO. There is some variation in the early development of the embryo in *Botrychium,* but the first division of the zygote is transverse or slightly oblique. Depending on the species the embryo is either exoscopic or endoscopic. If endoscopic, a suspensor may or may not be formed. Quadrants are usually formed but the derivation of future parts of the embryo is not easily determined. Critical studies of embryogeny, using modern techniques, have not been undertaken, due in part to the

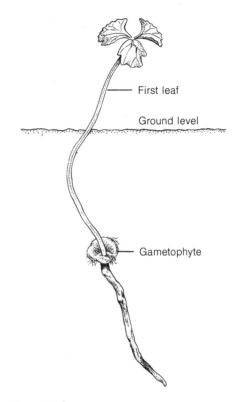

First leaf

Ground level

Gametophyte

Figure 12-4
Subterranean gametophyte of *Botrychium dissectum* with attached sporophyte. [Drawn from specimen, courtesy Dr. A. J. Eames.]

difficulty of obtaining suitable material. The relationship of a young sporophyte to the gametophyte is shown in Fig. 12-5, B. Additional information on embryogeny in *Botrychium* can be found in the following references: Jeffrey, 1897; Lyon, 1905; Nishida, 1955; Wardlaw, 1955; Rao, 1962; Bierhorst, 1971.

Ophioglossum (Adder's Tongue)

Most species are terrestrial; the exceptions are certain tropical epiphytic forms. The general organization of the plant body is similar to that of *Botrychium,* namely, there is a short rhizome, roots, and the formation

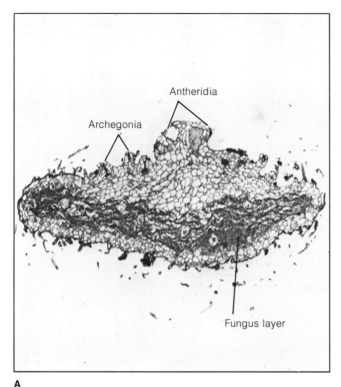

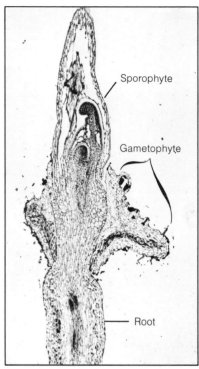

A   B

Figure 12-5
A, section, gametophyte of *Botrychium* sp. showing antheridial ridge and subjacent archegonia; B, section, young sporophyte attached to gametophyte of *Botrychium* sp. The sporophyte has several immature leaves in the bud; the first root is large and is toward the bottom of the page.

of a single mature frond each year in temperate species (Figs. 12-6, 12-7). Here also, a leaf sheath encloses the next younger leaf in the bud. The vegetative lamina is generally simple and entire in outline, or it may be palmately lobed or dichotomously branched in epiphytic species. In contrast with *Botrychium*, leaf venation in *Ophioglossum* is of the reticulate type. In general the fertile spike or segment consists of an axis with two lateral rows of embedded sporangia. In one tropical epiphytic species (*Ophioglossum palmatum*) there may be several "fertile spikes" near the base of the lamina. A reductional series (considered to be an evolutionary reductional series) also can be established in the genus beginning with the large epiphytic species and culminating with the minute and inconspicuous types a few centimeters in height (Eames, 1936).

ANATOMY OF THE SHOOT. It has been reported that there is a definite apical cell, which may be variable in form, at the shoot tip (Bower, 1926). For *Ophioglossum lusitanicum* L. subsp. *lusitanicum*, Gewirtz and Fahn (1960) reported that there is a group of initials at the shoot apex. The vascular cylinder of the shoot, enclosed by fleshy storage cortical tissue, is described as an ectophloic siphonostele (Fig. 12-3, B). The

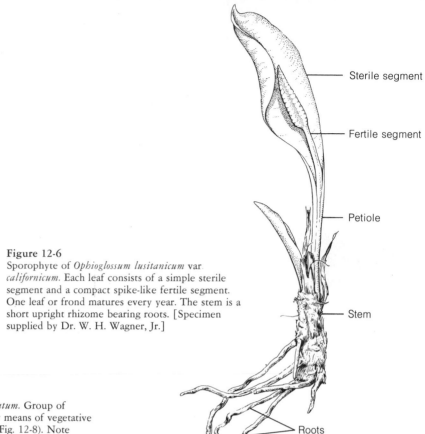

Sterile segment

Fertile segment

Petiole

Figure 12-6
Sporophyte of *Ophioglossum lusitanicum* var.
californicum. Each leaf consists of a simple sterile
segment and a compact spike-like fertile segment.
One leaf or frond matures every year. The stem is a
short upright rhizome bearing roots. [Specimen
supplied by Dr. W. H. Wagner, Jr.]

Stem

Roots

Figure 12-7
Ophioglossum petiolatum. Group of
plants produced by means of vegetative
reproduction (See Fig. 12-8). Note
sterile and spike-like fertile segments.

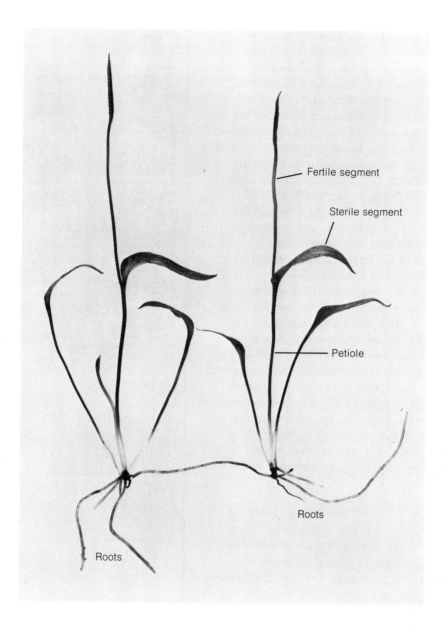

Figure 12-8
Entire plants of *Ophioglossum petiolatum*. Note that the short erect stems arise from roots; only one leaf is fertile on each shoot.

vascular bundles are collateral and are composed of primary phloem and of endarch primary xylem; the last formed metaxylem consists primarily of scalariformly pitted tracheids. The occurrence and position of the endodermis are seemingly variable in *Ophioglossum* (see Gewirtz and Fahn, 1960, for details). The entire vascular system of the stem axis, not only in *Ophioglossum* but also in *Botrychium*, has been interpreted as a meshwork of leaf traces and root traces, no part of it being considered as strictly cauline (Baas-Becking, 1921; Campbell, 1921; Maheshwari and Singh, 1934), although the basal part of a stem may be protostelic (Gewirtz and Fahn, 1960; Sen, 1968).

The base of the frond or leaf is traversed by several vascular bundles which are subdivisions of the single leaf trace (in many species). Vascular bundles on the adaxial side of the stipe are continuous into the fertile spike; others traverse the vegetative lamina and are branched, forming a reticulate venation system. The morphological interpretation of the fertile spike with its inherent problems is comparable to the situation in *Botrychium* (p. 259). The presence in *Ophioglossum palmatum* of several fertile spikes, arising on the margins of the lamina, is used as evidence for the branched nature of the frond.

THE ROOT. As in *Botrychium* there is generally one fleshy mycorrhizal root attached to the stem near each leaf. Roots arise endogenously, having their origin from cells near the phloem of vascular bundles in the shoot axis. Growth of the root is initiated by a tetrahedral cell; the root becomes monarch or diarch, or even more protoxylem poles are formed (Petry, 1914; Joshi, 1940; Chrysler, 1941; Sen, 1968).

Vegetative propagation in *Ophioglossum* is especially interesting in that new shoots take their origin directly from roots (Fig. 12-8). Almost all plants of a colony are formed in

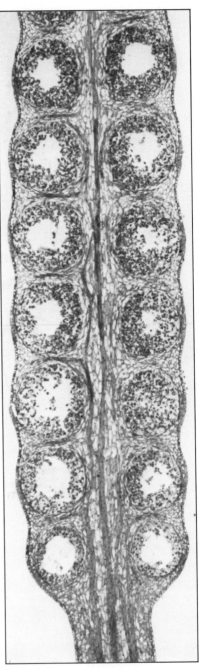

Figure 12-9
Longisection, fertile segment of *Ophioglossum lusitanicum* var. *californicum*, showing two rows of embedded sporangia. Branches of the main vascular system are evident between sporangia at left.

this fashion. The buds arise endogenously from cortical cells of the root and they also can be induced experimentally to arise from the pith of decapitated shoots (Wardlaw, 1953, 1954).

THE SPORANGIUM. Eusporangia, which originate from superficial cells along two sides of the young fertile spike, are in varying stages of development before the frond emerges for the season. The sporangia always remain embedded in tissues of the fertile spike; small vascular strands are present between the sporangia and often are turned toward them (Fig. 12-9). At maturity numerous tetrahedral spores are formed which are liberated through a slit in the sporangial wall, perpendicular to the wide surface of the fertile spike.

CHROMOSOME NUMBERS. Chromosome numbers are known for about 36 species in the Ophioglossales. The species having the *largest* number of chromosomes of all living plants is in the genus *Ophioglossum*: $2n = 1260$ in *Ophioglossum reticulatum* (Ninan, 1958). As pointed out by Stebbins (1971), it is nothing short of miraculous that so many chromosomes could sort themselves out at the time of meiosis, each one "seeking" a specific mate in the formation of hundreds of bivalents in each sporocyte. The lower end of the range of chromosome numbers in the genus is $2n = 240$ (Kurita and Nishida, 1965; Verma, 1965; Löve and Kapoor, 1967). It must be concluded that there has been an extremely high degree of polyploidy in these ancient ferns. The theoretical basic chromosome number is considered to be $x = 15$. If this is true, then some species are 48- and 84-ploid!

EXPERIMENTAL STUDIES ON THE DEVELOPMENT OF THE FERTILE SEGMENT. A study of the mechanism of elongation of the *fertile segment* in *Ophioglossum petiolatum* was un-

dertaken by Peterson and Cutter (1969a). In this species the fertile segment undergoes considerable elongation during development. The fertile segment originates as a primordium on the adaxial side of the sterile segment and exhibits apical growth initially. Apical growth ceases rather early in ontogeny at a time when the sporangial region becomes organized. Meristematic activity then becomes more pronounced in an intercalary meristem located at the base of the sporangial region. Marking experiments (using India ink) indicated that subsequent growth of the upper portion of the stalk of the fertile segment accounted for most of the final length of the stalk. These results were confirmed by the determination of mitotic rates, cell length, and by the use of H^3-thymidine to label DNA.

Additional experiments were performed, designed to determine the causal basis for elongation of the fertile segment. Surgical removal of the sterile segment of the leaf had no effect on the elongation of the fertile segment. Excision of the sporangial portion early in development resulted in cessation of elongation of the stalk of the fertile segment. Replacing the excised sporangial portion with auxins resulted in elongation of the stalk but gibberellic acid alone or kinetin alone had very little effect. These studies have shown the controlling influence that sporogenous tissue in general may have on adjacent tissues (Peterson and Cutter, 1969b).

THE GAMETOPHYTE. Spores germinate slowly, and growth may be extended over a period of years during which time the gametophytes become buried by several inches of humus or soil particles. Growth by an apical cell results in the development of a wormlike, sometimes branched, non-chlorophyllous gametophyte, from 1/8 inch to 2 inches in length (Fig. 12-10). An endophytic fungus, essential to the nutrition of

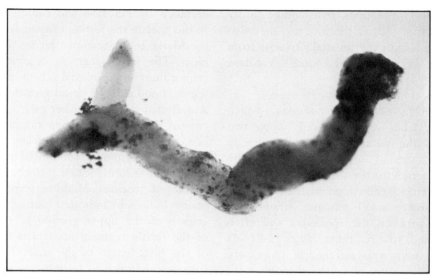

Figure 12-10
Gametophyte of *Ophioglossum vulgatum,* enlarged. Prominent root of young attached sporophyte is seen at left. [Specimen supplied by Mr. Dennis Stevenson.]

the gametophyte, gains entrance early in development and is present in cells of the older parts of the plant body. Sex organs take their origin from meristematic cells distal to the apical cell and mature within the region of the endophytic fungus. Antheridia and archegonia are scattered and intermingled over the surface of the gametophyte. The gametangia are very similar to those in *Botrychium.*

THE EMBRYO. As is true of most groups of lower vascular plants, the first division of the zygote (in *Ophioglossum*) is transverse, that is, perpendicular to the long axis of the archegonium. The embryo is exoscopic in polarity (Fig. 6-1). A cell division at the apical pole (epibasal cell) and at the basal pole (hypobasal cell) results in the formation of four cells—the quadrant stage of embryogeny. Subsequent cell divisions are irregular, but with little doubt the first leaf and future shoot apex are derived from the epibasal portion of the embryo and the foot is derived from the hypobasal portion. Ori-

gin of the root is uncertain, but seemingly it arises near the middle of the embryo and enlarges rapidly in a lateral direction before other parts of the embryo become conspicuous (Fig. 12-10).

Appraisal of the Ophioglossales

In this textbook we have included the Ophioglossales as an order in the Filicopsida along with other ferns; this is a commonly accepted procedure. Some botanists, however, consider the members of the Ophioglossales to be sufficiently different from other ferns to warrant making them a separate class. Some of the important features distinguishing them are: (1) the peculiar fertile segment, (2) collateral vascular bundles, (3) roots with endophytic fungus and without root hairs, (4) noncircinate vernation of leaves, (5) lack of sclerenchyma in the plant body, and (6) subterranean gametophytes with an associated fungus. Although some of these features are not confined to the Ophioglossales, no other group of lower vascular plants has all of these attributes. The Ophioglossales is undoubtedly an ancient group, but the lack of fossils continues to shroud its phylogeny. In the absence of a fossil record the living representatives undoubtedly will be shifted from one taxon to another in the future, depending upon the concepts of fern specialists.

Marattiales–Marattiaceae

This is a group of plants whose members resemble more closely, in general aspect, the Filicales; many of them possess large pinnate fronds with sporangia located on their lower surfaces. It is an ancient group of ferns and has a well-preserved fossil record that extends back to Middle Carboniferous. It is a tropical group and is generally known in temperate zones only through specimens in conservatories and dried speci-

mens in herbaria. According to one student of ferns, there are six genera and perhaps a hundred or more species (Copeland, 1947). The two better-known genera are *Angiopteris* and *Marattia*; the former is distributed throughout the South Sea Islands and north to Japan, the latter is pantropical in distribution.

In growth habit a marattiaceous fern characteristically has an upright, unbranched tuberous stem or short trunk to which large, pinnately compound leaves (circinate in vernation) and thick, fleshy roots are attached (Fig. 12-11). A pair of clasping fleshy stipules occurs at the base of each leaf, covering a part of the trunk; they persist, along with the leaf base, even after the frond abscises. On the lower surface (abaxial surface) of the fronds, which may be several feet in length, sporangia commonly occur along veins; the venation is of the open dichotomous type. In *Angiopteris* the

Figure 12-11
Angiopteris evecta, showing short, tuberous stem and large bipinnate leaves.

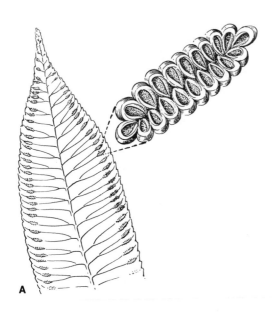

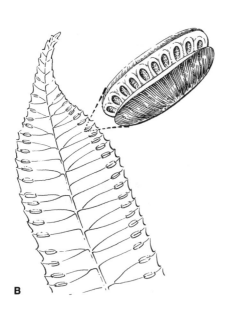

Figure 12-12
Abaxial views of fertile pinnae. **A**, *Angiopteris;*
B, *Marattia.* One sorus of each is enlarged.
The sporangia in *Marattia* form a definite
synangium. [Pinnae redrawn from *The Ferns,*
Vol. II, by F. O. Bower. Cambridge University
Press, London, 1926.]

sporangia are crowded together in two rows
along each side of a vein, and each sporangium dehisces by a longitudinal slit on the
side facing the other row of sporangia (Figs.
12-12, A; 12-13). A more specialized condition is observed in *Marattia* where the two
rows of sporangia are united into a compact
soral group surrounded by a common wall
(Fig. 12-12, B). This structure is termed a
synangium. At maturity the synangium
opens, much like a clam shell, exposing the
sporangia which dehisce by longitudinal slits.
There are exceptions to the above frond and
synangial types. In some genera the fronds
may be simple to once pinnate (*Danaea*), or
palmately compound with reticulate venation and scattered synangia (*Christensenia*).

ANATOMY OF THE SHOOT. The apical
meristem of a mature or adult shoot is reported to have a group of shoot apical initials
or an apical cell that is not regular or precise
in its divisions (Charles, 1911; Bower, 1923).
The apical meristem gives rise to derivative
meristematic cells from which leaves take
their origin; other derivative cells will differentiate into procambium and pith. A transverse section of a young stem may reveal a
cylinder of vascular bundles, but, at a higher
level in older plants, the vascular cylinder
becomes dissected into a complex polycyclic dictyostele consisting of two or more
concentric vascular cylinders. It should be
emphasized that the inner vascular cylinders
are continuous with the original outer cylinder at a lower level in the stem. It has been
shown that the majority of the vascular
bundles (meristeles) enter leaves at some
level of the shoot axis; however, some of the
bundles are reported to be simply interconnections between leaf traces (Farmer and
Hill, 1902; Charles, 1911; West, 1917). Additional detailed studies in this group by
interested botanists would be desirable.
Root primordia have their origin in tissue
near the vascular bundles of either the inner

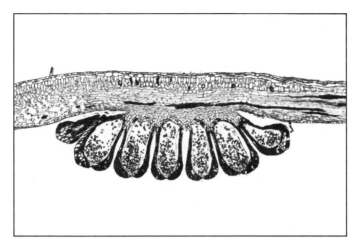

Figure 12-13
Section, lamina and sorus of *Angiopteris* sp. Note the presence
of numerous spores within each thick-walled sporangium.

vascular cylinders or the outermost cylinder; after their initiation the roots bore their way through the cortex to the surface (Farmer and Hill, 1902). Therefore root traces appear in transverse sections of a stem. In the outer cortex of the stem there are mucilage ducts as well as cells filled with tannin.

The leaves of marattiaceous ferns are circinate in vernation during development. The ultimate frond segments in most species have a single midvein with lateral dichotomous veins. The mesophyll in most forms is differentiated into an adaxial palisade tissue and an abaxial spongy mesophyll. Stomata occur on the abaxial surface. Mucilage cavities, hypodermal sclerenchyma, or collenchyma are often present in the petiole.

THE ROOT. In the primary root and the first-formed roots on the stem there is a definite apical cell; later-formed roots are reported to have a group of about four equivalent initials (West, 1917). Roots become large and fleshy and contain mucilage cavities. Typically the vascular cylinder is polyarch — a feature not generally found in other ferns.

THE SPORANGIUM. Sporangia are of the eusporangiate type and commonly originate from mounds of tissue paralleling the veins of developing fronds. At maturity each sporangium has a broad base and a sporangial wall that consists of several layers of cells. When mature the sporangia may be separate from each other (*Angiopteris*), or the sporangial walls may become confluent during development so that each sporangium is actually a pocket or loculus in a compact structure, the synangium (Figs. 12-12, 12-13). Dehiscence of individual sporangia in a synangium is brought about by the drying out of wall cells, which results in longitudinal splitting of each sporangium (after the halves of the synangium separate in *Marattia*), or by the formation of a pore at the tip of each sporangium as in *Danaea*. Spore output is large; spore numbers range from a minimum of 1,000 up to a maximum of 7,000 spores formed by each sporangium (Bower, 1935). In *Marattia sambucina* the spores are yellowish to tan, bilateral and elongate, with a longitudinal ridge (Stokey, 1942). In contrast, tetrahedral spores are formed by other members of the group.

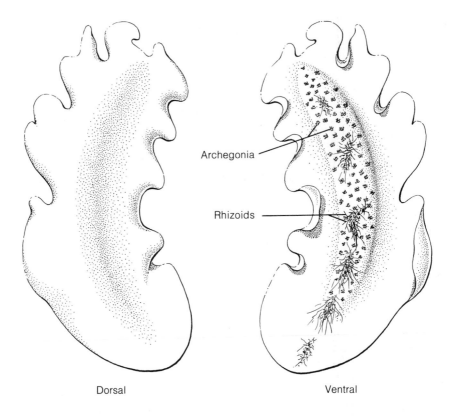

Archegonia

Rhizoids

Dorsal

Ventral

Figure 12-14
Ribbon-shaped gametophyte of *Marattia douglasii*. Most of the antheridia (not shown in drawing) are on the ventral surface. [Drawn from a specimen supplied by W. H. Wagner, Jr.]

CHROMOSOME NUMBERS. Some information is available on chromosome numbers in the order. In one study of species from South India the chromosome number in *Marattia fraxina* was shown to be $n = 78$, while that of *Angiopteris evecta* was $n = 80$. From one cytological study of ferns, including the Marattiales, it was concluded that the theoretical basic chromosome number of ancient ferns is probably $x = 13$, with polyploidy and aneuploidy having been operative in the groups (Ninan, 1956; Chiarugi, 1960).

THE GAMETOPHYTE. The gametophyte is a large, green, dorsiventral ribbon-shaped or heart-shaped structure with a prominent ventral midrib or cushion and thin, lateral wing-like extensions (Fig. 12-14). The thallus, which may be 2 cm, or more, in length is long-lived and has an endophytic fungus which, however, must play only a minor role in the nutrition of the gametophyte because of the presence of chlorophyll. There are absorbing rhizoids along the ventral midrib. Gametangia show the following pattern of distribution: Antheridia are on the ventral surface but may occur on the dorsal side of the thallus (Nozu, 1956); archegonia are restricted to the projecting ventral midrib (Haupt, 1940). Both antheridia and archegonia are sunken, and the main stages in their development are shown in Fig. 12-15. Mature sperms are coiled and multiflagellate.

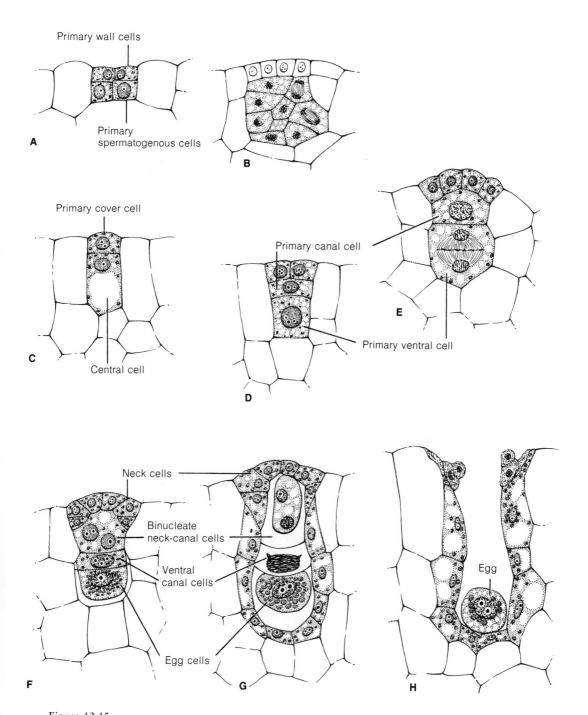

Figure 12-15
Stages in the development of an antheridium (**A, B**) and of an archegonium (**C–H**) of *Angiopteris evecta*.
[Redrawn from Haupt, *Bull. Torrey Bot. Club* 67:125, 1940.]

Figure 12-16
Reconstruction of the tree fern *Psaronius*, considered to have been approximately 20–25 feet tall. [Redrawn from Morgan, *Illinois Biol. Monogr.* 27:1–108, 1959.]

THE EMBRYO. Just as in the Ophioglossales, the first division of the zygote is transverse, resulting in a two-celled embryo. In contrast with most members of the Ophioglossales, the Marattiaceae exhibit endoscopic polarity (see Fig. 6-1). This means that the future shoot apex and first leaf have their origin from the cell (epibasal cell) directed away from the neck of the archegonium. The cell (hypobasal cell) toward the archegonial neck gives rise to a multicellular foot. The root meristem, appearing late in embryogeny, is endogenous in origin, and in one study it has been reported to be derived from the epibasal portion of the embryo (Campbell, 1911). With subsequent growth of the embryo the young shoot grows up through the gametophyte, emerging from the upper surface. The presence of a suspensor has been reported in some species of *Angiopteris* (Land, 1923) and other genera (Campbell, 1940). The first vascular bundle of the embryo is continuous between the root and first leaf. The vascular bundle of the next leaf is joined to the first vascular strand. The vascular strands of later leaves continue to form essentially an interconnected network of leaf traces and root traces (Campbell, 1911, 1921).

FOSSIL MEMBERS OF THE MARATTIALES. Considerable information is available about the extinct Carboniferous relatives of living members of the Marattiales (Morgan, 1959). They were tall trees, large in diameter at the base, tapering toward the top and ending in a collection of large pinnately compound fronds (Fig. 12-16). The presence of large masses of roots at the basal region of the trunk would account for a specimen's large size at that level. In many ways these extinct ferns resembled present-day tree ferns in general growth habit (cf. Fig. 11-1).

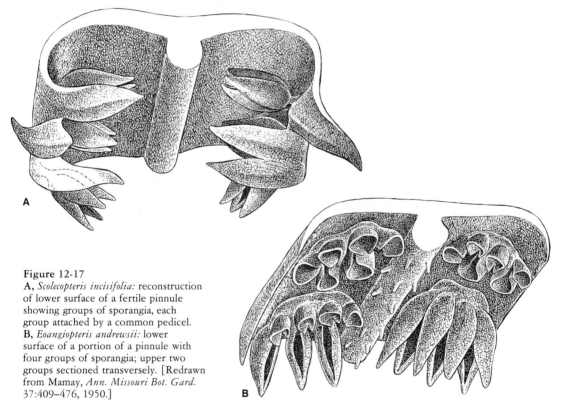

Figure 12-17
A, *Scolecopteris incisifolia:* reconstruction of lower surface of a fertile pinnule showing groups of sporangia, each group attached by a common pedicel. B, *Eoangiopteris andrewsii:* lower surface of a portion of a pinnule with four groups of sporangia; upper two groups sectioned transversely. [Redrawn from Mamay, *Ann. Missouri Bot. Gard.* 37:409–476, 1950.]

The vascular cylinder of the stem, probably rather simple at the base, became exceedingly complicated farther up the stem or trunk, being a polycyclic dictyostele of several concentric cylinders. Roots were initiated high up on the stem and then grew down through the outer part of the trunk, or on the surface of the trunk, some of them reaching the soil at the base. The mantle of roots was probably one of the main means of support of the stem. *Psaronius* is a form or organ genus for stems.

There is a wealth of information concerning sporangiate structures (Mamay, 1950). Pinnae have been found with sporangia organized as sori and in some specimens the sporangia were fused, forming synangia. In one form genus, *Scolecopteris*, four sporangia were attached to a common pedicel, but were free from each other beyond this point (Fig. 12-17, A). In other forms there were several sporangia in each sorus, but they were essentially separate (Fig. 12-17, B) as in the present day *Angiopteris* (cf. Fig. 12-12, A and Fig. 12-13).

There are records of the order in the Mesozoic and Tertiary, all of which points to a nearly complete lineage from the Carboniferous to the present time.

References

Baas-Becking, L. G. M.
    1921. The origin of the vascular structure in the genus *Botrychium*; with notes on the general anatomy. *Rec. Trav. Bot. Neerl.* 18:333–372.
Bierhorst, D. W.
    1958. Observations on the gametophytes of *Botrychium virginianum* and *B. dissectum*. *Amer. Jour. Bot.* 45:1–9.
    1971. *Morphology of Vascular Plants*. Macmillan, New York.
Boullard, B.
    1963. Le gamétophyte des Ophioglossacées. Considerations biologiques. *Bull. Soc. Linn. Normandie* 4:81–97.
Bower, F. O.
    1923. *The Ferns*, Vol. I. Cambridge University Press, London.

    1926. *The Ferns*, Vol. II. Cambridge University Press, London.
    1935. *Primitive Land Plants*. Macmillan, London.
Campbell, D. H.
    1911. *The Eusporangiatae*. Carnegie Institution, Washington, D.C.
    1921. The eusporangiate ferns and the stelar theory. *Amer. Jour. Bot.* 8:303–314.
    1940. *The Evolution of the Land Plants (Embryophyta)*. Stanford University Press, Stanford, California.
    1948. Notes on the geographical distribution of ferns. *Amer. Fern Jour.* 38:122–125.
Charles, G. M.
    1911. The anatomy of the sporeling of *Marattia alata. Bot. Gaz.* 51:81–101.
Chiarugi, A.
    1960. Tavole chromosomiche delle Pteridophyta. *Caryologia* 13:27–150.

Chrysler, M. A.

1925. *Botrychium lanuginosum* and its relation to the problem of the fertile spike. *Bull. Torrey Bot. Club* 52:127–132.

1941. The structure and development of *Ophioglossum palmatum. Bull. Torrey Bot. Club* 68:1–19.

1945. The shoot of *Botrychium* interpreted as a series of dichotomies. *Bull. Torrey Bot. Club* 72:491–505.

Clausen, R. T.

1938. A monograph of the Ophioglossaceae. *Mem. Torrey Bot. Club* 19:1–177.

Copeland, E. B.

1947. *Genera Filicum (The Genera of Ferns)*. Chronica Botanica, Waltham, Mass.

Eames, A. J.

1936. *Morphology of Vascular Plants. Lower Groups*. McGraw-Hill, New York.

Esau, K., V. I. Cheadle, and E. M. Gifford, Jr.

1953. Comparative structure and possible trends of specialization of the phloem. *Amer. Jour. Bot.* 40:9–19.

Fabbri, F.

1963. Primo supplemento alle tavole cromosomiche delle Pteridophyta di Alberto Chiarugi. *Caryologia* 16:237–335.

Farmer, J. B., and T. G. Hill

1902. On the arrangement and structure of the vascular strands in *Angiopteris evecta*, and some other Marattiaceae. *Ann. Bot.* 16:371–402.

Gewirtz, M., and A. Fahn

1960. The anatomy of the sporophyte and gametophyte of *Ophioglossum lusitanicum* L. ssp. *lusitanicum. Phytomorphology* 10:342–351.

Haupt, A. W.

1940. Sex organs of *Angiopteris evecta. Bull. Torrey Bot. Club* 67:125–129.

Jeffrey, E. C.

1897. The gametophyte of *Botrychium virginianum. Trans. Canadian Inst.* 5:265–294.

Joshi, A. C.

1940. A note on the anatomy of the roots of *Ophioglossum. Ann. Bot.* n.s. 4:663–664.

Kurita, S., and M. Nishida

1965. Cytotaxonomy of Ophioglossales. III. Chromosome numbers and systematics of *Ophioglossum. Bot. Mag.* (Tokyo) 78: 461–473.

Land, W. J. G.

1923. A suspensor in *Angiopteris. Bot. Gaz.* 75:421–425.

Löve, A., and B. M. Kapoor

1967. The highest plant chromosome number in Europe. *Svensk Bot. Tidskr.* 61:29–32.

Lyon, H. L.

1905. A new genus of Ophioglossaceae. *Bot. Gaz.* 40:455–458.

Maheshwari, P., and B. Singh

1934. The morphology of *Ophioglossum fibrosum. Schum. Jour. Indian Bot. Soc.* 13: 103–123.

Mamay, S. H.

1950. Some American Carboniferous fern fructifications. *Ann. Missouri Bot. Gard.* 37:409–476.

Manton, I.

1950. *Problems of Cytology and Evolution in the Pteridophyta*. Cambridge University Press, London.

Morgan, J.

1959. The morphology and anatomy of American species of the genus *Psaronius. Illinois Biol. Monogr.* 27:1–108.

Ninan, C. A.

1956. Studies on the cytology and phylogeny of the Pteridophytes. I. Observations on the Marattiaceae. *Jour. Indian Bot. Soc.* 35:233–239.

1958. Studies on the cytology and phylogeny of the Pteridophytes. VI. Observations on the Ophioglossaceae. *Cytologia* 23: 291–316.

Nishida, M.

1952. Dichotomy of vascular system in the stalk of Ophioglossaceae. *Jour. Jap. Bot.* 27:165–171.

1955. The morphology, gametophyte, young sporophyte and systematic position of *Botrychium japonicum* Und. *Phytomorphology* 5:449–456.

1957. Studies on the systematic position and constitution of Pteridophyta. 10. A further investigation on the vascular dichotomy in the phyllomophore of Ophioglossales, with special references to phylogeny. *Jour. Coll. Arts Sci. Chiba Univ.* 2:179–211.

Nozu, Y.

1950. On the so-called petiole of *Botrychium*. *Bot. Mag.* (Tokyo) 63:4–11.

1954. The gametophyte and young sporophyte of *Botrychium japonicum* Und. *Phytomorphology* 4:430–433.

1955. Anatomical and morphological studies of Japanese species of the Ophioglossaceae. I. Phyllomophore. *Jap. Jour. Bot.* 15:83–102.

1956. Notes on gametophyte and young sporophyte of *Angiopteris suboppositifolia* de Vries. *Bot. Mag.* (Tokyo) 69:474–480.

Peterson, R. L., and E. G. Cutter

1969a. The fertile spike of *Ophioglossum petiolatum*. I. Mechanism of elongation. *Amer. Jour. Bot.* 56:473–483.

1969b. The fertile spike of *Ophioglossum petiolatum*. II. Control of spike elongation and a study of aborted spikes. *Amer. Jour. Bot.* 56:484–491.

Petry, L. C.

1914. The anatomy of *Ophioglossum pendulum*. *Bot. Gaz.* 57:169–192.

Rao, L. N.

1962. Life-history of *Botrychium lanuginosum* Wall. ex. Hook et Grev. *Proc. Indian Acad. Sci.* (Sect. B) 55:48–64.

Sen, U.

1968. Morphology and anatomy of *Ophioglossum reticulatum*. *Can. Jour. Bot.* 46:957–968.

Stebbins, G. L.

1971. *Chromosomal Evolution in Higher Plants*. Addison-Wesley, Reading, Mass.

Stokey, A. G.

1942. Gametophytes of *Marattia sambucina and Macroglossum smithii*. *Bot. Gaz.* 103:559–569.

Verma, S. C.

1965. Chromosomes of *Ophioglossum coriaceum* A. Cunn. (Correction of previous paper.) *Cytologia* 30:173–174.

Wagner, W. H., Jr.

1963. A biosystematic survey of Unites States ferns—preliminary abstract. *Amer. Fern Jour.* 53:1–16.

Wardlaw, C. W.

1953. Endogenous buds in *Ophioglossum vulgatum* L. *Nature* 171:88–89.

1954. Experimental and analytical studies of pteridophytes. XXVI. *Ophioglossum vulgatum*: Comparative morphogenesis in embryos and induced buds. *Ann. Bot.* n.s. 18:397–406.

1955. *Embryogenesis in Plants*. Methuen, London.

West, C.

1917. A contribution to the study of the Marattiaceae. *Ann. Bot.* 31:361–414.

Zimmermann, W.

1942. Die Phylogenie des Ophioglossaceen-Blattes. *Ber. Deutsch. Bot. Ges.* 60:416–433.

13

The Leptosporangiate Ferns

Filicales

Members of this group are often called the "true ferns" and are the types most commonly grown in home gardens. Almost everyone can recognize a familiar name in the following brief list: cinnamon fern, Boston fern, Christmas fern, male fern, maidenhair fern, sensitive fern, and bracken fern.

Economic Importance

Aside from the purely aesthetic value of ferns, they do have moderate economic importance throughout the world; perhaps the most notable use is that of providing food for certain groups of peoples. Young leaf tips of some ferns are eaten throughout the Malayan and adjacent regions. In fact, in the same areas certain ferns are grown commercially in gardens with a value sometimes greater per plot than that of rice (Copeland, 1942). Not only are ferns enjoyed as food in such tropic regions, but "fiddlehead greens" have been imported from Maine to the best hotels in New York (Blake, 1942). One gourmet reputedly remarked, "They taste, simply and beautifully, like the soul of spring." During devastating famines in Europe in past times, precious stores of barley and rye were conserved by mixing small amounts of the grains with dry, ground-up male fern or bracken fern—the mixture being termed *"pain de fougère"* in France (Coquillat, 1950).

Commercially, ferns are used extensively by florists in bouquets and floral arrangements, the maidenhair fern being among the most common. A fern with better keeping qualities is the American-shield fern ("fancy fern"). In greenhouses and conservatories orchids often are grown on pieces of the trunk and roots of tree ferns or on *Osmunda* "fiber." In some regions of the world the fibrous material obtained from ferns is used as a stuffing for mattresses.

The male fern (*Dryopteris filix-mas*) has been used medicinally in the treatment of tapeworm infestation since its beneficial effect was discovered by Dioscorides in the days of Nero. Two more uses of the male fern might be cited: tissues of the fern have been substituted for hops in brewing ale, and juices of the fern are said to have been used by witches in concocting love philters!

Distribution and Growth Habit

There are approximately 8,000–10,000 species of "true ferns" and about 300 genera. The number of recognized families varies widely depending on the morphological concepts of fern specialists (Bower, 1935; Copeland, 1947; Holttum, 1947, 1949). The Filicales reach their greatest numbers and are most diversified in the tropics, where both epiphytic forms and tree forms are common. Some mountains may have hundreds of species. As an example, a certain mountain in Borneo has at least 437 species (Copeland, 1939). As was pointed out in Chapter 11, ferns are not restricted to the tropics; they may be found within wide climatic ranges. In temperate regions ferns are largely terrestrial, the shoot being erect or more commonly a prostrate rhizome with no aerial stem, the leaves being the only structures visible above the ground.

It is interesting to note that ferns constitute about one-fiftieth of the total species of vascular plants in California, whereas in Guam (Mariana Islands) one-eighth of the total species are ferns. How can we account for this unequal distribution of ferns? According to Copeland (1939), who was one of the leading scholars of fern distribution and systematics, the group undoubtedly had its evolutionary beginnings in the Australasian region and spread thence to the north and also across the Pacific Ocean.

Morphology and Anatomy of the Shoot

It is impossible to describe all of the morphological types and endless variations in the anatomy of the fern sporophyte. We can only look at certain common features, leaving complete surveys to monographic treatments. However, near the end of the chapter the student will find summaries of the important features of selected families and their interrelationships. In addition there are brief discussions of certain morphological and morphogenetic problems concerning ferns.

If the aerial stem or rhizome of a fern is short and erect, the plant appears to be a collection of leaves. If, on the other hand, the rhizome is prostrate, the leaves tend to be somewhat farther apart, and nodal and internodal regions can be more easily seen. Branching of a prostrate rhizome is usually irregular, but in some ferns it is dichotomous. The majority of true ferns have leaves that are basically organized on the pinnate plan, but many are of the simple type (Fig. 13-1). Members of all families of ferns possess epidermal appendages on the leaf and stem, and frequently they are present on the root. These appendages may be only simple hairs, or they may be developed into large chaffy scales called paleae (see Fig. 13-6, E, G–I). Roots usually arise at definite places along the rhizome; most commonly this position is near the base of leaves.

INITIATION AND DEVELOPMENT OF THE LEAF. Inasmuch as probably everyone has seen a fern frond at some time and noticed the associated reproductive structures, we will begin the organographic account of filicalian organization with the morphology of the leaf.

A description of the initiation and growth of a foliage leaf should include a description of the shoot apical meristem because of the

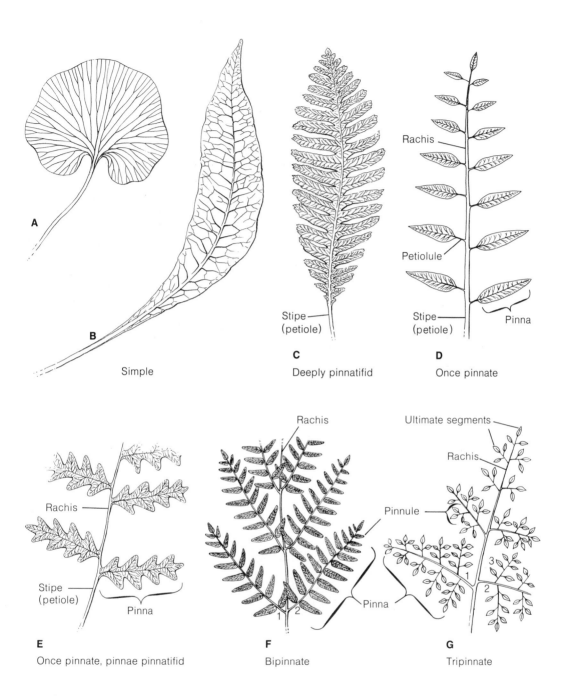

A   **B**

Simple

C

Deeply pinnatifid

Rachis

Petiolule

Stipe
(petiole)

Pinna

D

Once pinnate

Rachis

Stipe
(petiole)

Pinna

E

Once pinnate, pinnae pinnatifid

Rachis

Pinna

F

Bipinnate

Ultimate segments

Rachis

Pinnule

Pinna

G

Tripinnate

Figure 13-1
Form and organization of fern fronds. **A,** *Trichomanes,* simple leaf with open dichotomous venation;
B, *Paraleptochilus,* simple leaf with pinnate reticulate venation; **C,** *Trichomanes;* **D–G,** schematic only,
numbers 1, 2, 3, indicate degree of compoundness. [A,C redrawn from *The Ferns,* Vols. I, II, respectively,
by F. O. Bower. Cambridge University Press, London, 1923, 1926; B redrawn from E. B. Copeland,
Genera Filicum. A Chronica Botanica Publication. Copyright 1947, The Ronald Press Company.]

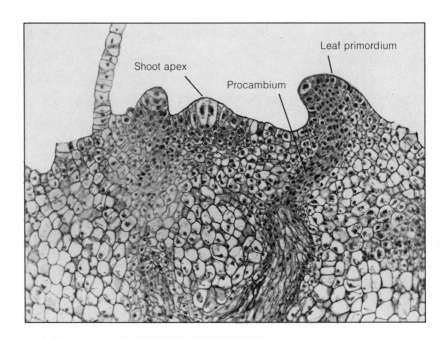

Figure 13-2
Longisection of shoot apex and leaf primordia of *Asplenium bulbiferum*.
Note procambial tissue extending into leaf primordium at right.
[Photomicrograph courtesy of K. Esau.]

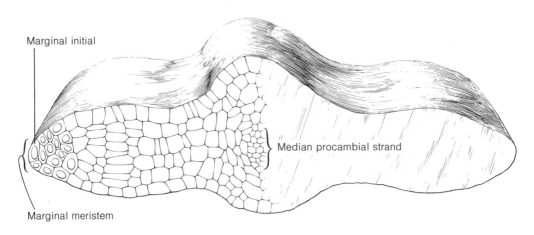

Figure 13-3
Transectional view of a young developing lamina of *Dicksonia*. The epidermal and mesophyll cells have
their origins from derivatives of the marginal initials. Only one cell of a row of marginal initials is shown.

close ontogenetic relationship of stem and leaf. At the tip of a fern stem is an obvious apical cell which may vary in shape according to the species. In some genera (e.g., *Pteridium, Polypodium*) the shoot apical cell has three faces, from which cells are cut off on two sides. In others (*Osmunda, Lygodium, Asplenium*) the apical cell is a tetrahedral, pyramidal cell; new cells are initiated from the three oblique basal sides (Fig. 13-2). The apical cell itself may divide infrequently, but the derivative cells divide at a rapid rate.

At a rather precise distance from the shoot apex, one cell among a group of dividing cells becomes conspicuous and functions as the initial of a new leaf. This initial soon becomes an apical cell and gives rise to derivative cells (Fig. 13-2). The leaf then undergoes a period of apical growth which may continue for varying lengths of time depending on the ultimate size of the frond. A developing vascular bundle may be evident in a leaf that is only a fraction of a millimeter in height. The lamina or pinnae (if the leaf is compound) are formed as lateral outgrowths on the original peg-like structure; the latter becomes the stipe (petiole) and rachis. The blade portions of the leaf are built up by the activity of marginal meristems (Fig. 13-3). The derivatives of the marginal initials continue to divide in a definite pattern, resulting in the formation of a characteristic lamina for each species. During the later phases of growth the typical fern leaf displays circinate vernation, not only of the main axis but also of its subdivisions. Circinate vernation is undoubtedly related in part to the presence of naturally occurring growth hormones (see Steeves and Wetmore, 1953, and Briggs and Steeves, 1959, for detailed discussions).

SIZE, FORM, AND TEXTURE OF THE LEAF. Some ferns have fronds that are dichotomously branched; the majority, however, have pinnatifid, pinnate, or simple leaves (Fig. 13-1). In size, fronds may vary from the enormous, almost branchlike leaves of tree ferns to the small leaves of certain water ferns. In living forms the large, compound, pinnate leaves are considered more primitive, whereas the small, simple leaves are considered to have been reduced in size during evolution. The latter type is therefore interpreted as a derived form. There is considerable variation in the texture of fern leaves: they may be thick and leathery, crisp, or very delicate (as in the so-called filmy ferns).

A survey of living ferns suggests that there have been two principal trends in the evolution of leaf form, namely, (1) the evolution of a simpler leaf form from a more complex one and (2) the evolution of all other principal leaf forms from the pinnate-determinate type (Tryon, 1964). An example of the pinnate-determinate form is shown in Fig. 13-4, A. Approximately 85 percent of the species and genera of ferns have this type. Specialization has led, for example, to the helicoid, radiate, and furcate types (Fig. 13-4, C–E). Tryon (1964) has concluded that there have been two principal adaptive evolutionary developments, one leading to the simplification of the lamina and reduction of leaf surface, such as in ferns of tropical xeric environments and the other to the climbing types with pinnate-indeterminate leaves (Fig. 13-4, F).

VENATION OF THE LEAF. The stipe, rachis, and axes of frond subdivisions have prominent veins. The smaller veins of the lamina usually show an open dichotomous type of venation. The dichotomies in some ferns are obscure, the venation pattern being of the reticulate type (Fig. 13-5). Any degree of vein fusions is considered an evolutionary advancement. The first steps toward vein fusions are expressed in various ways: sometimes by the appearance of marginal loops which connect the tips of vein dichotomies, or sometimes by the formation of a series of meshes adjacent to the midrib (Fig. 13-5, E) or throughout the pinna (Fig. 13-5, F).

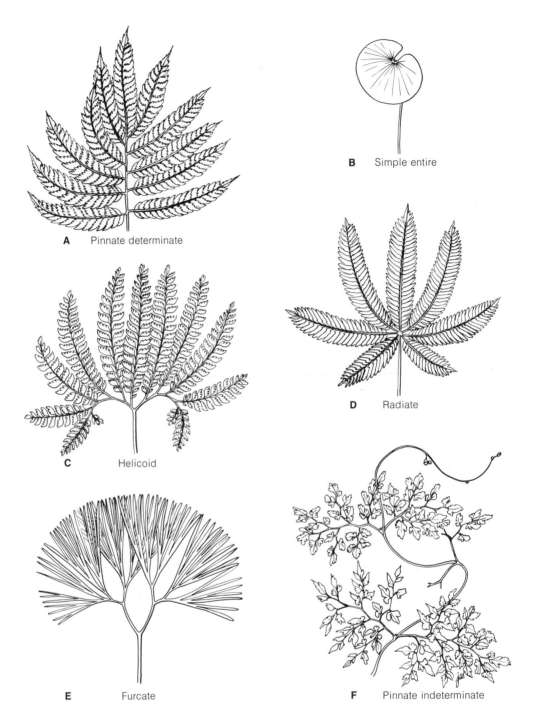

A   Pinnate determinate

B   Simple entire

C   Helicoid

D   Radiate

E   Furcate

F   Pinnate indeterminate

Figure 13-4
Diversity and evolution of leaf form in ferns. **A,** *Adiantum pulverulentum:* lamina has pinnately arranged parts and determinate growth; **B,** *Adiantum reniforme:* lamina is simple and not lobed or laciniate; **C,** *Adiantum pedatum:* parts of lamina are borne on only one side of rachis branches; **D,** *Cheilanthes radiata:* parts of lamina radiate from one point; **E,** *Schizaea dichotoma:* parts of lamina branch in a dichotomous manner; **F,** *Lygodium japonicum:* leaf has indefinite growth; the parts are arranged in a pinnate manner. [Based on Tryon, *Mem. Torrey Bot. Club* 21:73–85, 1964.]

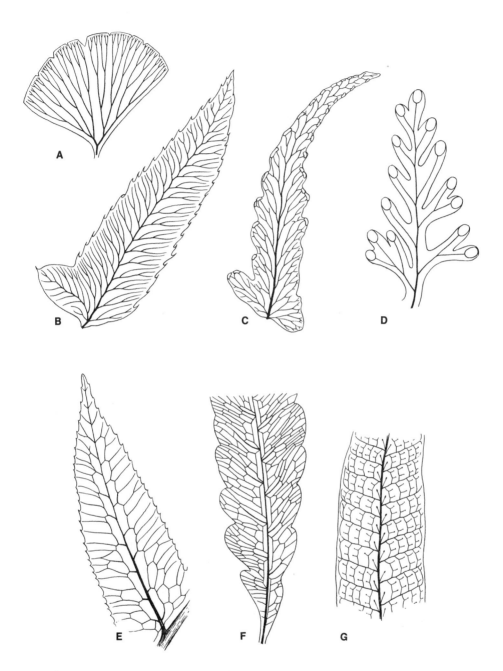

Figure 13-5
Leaf venation in ferns, illustrating the basic type of open dichotomous venation system in some
ferns as contrasted with the highly complex reticulate patterns in others. **A,** *Adiantum;*
B, *Polystichum;* **C,** *Diellia;* **D,** *Davallia;* **E,** *Woodwardia;* **F,** *Onoclea;* **G,** *Polypodium.* [A redrawn from
Organographie der Pflanzen. Dritte Auflage. Zweiter Teil, by K. Goebel. Gustav Fischer, Jena, 1930;
B redrawn from *Morphology of Vascular Plants. Lower Groups,* by A. J. Eames. McGraw-Hill, New
York, 1936; C, D redrawn from Wagner, *Univ. Calif. Publ. Bot.* 26:1, 1952; E–G redrawn from
The Ferns, Vols. I, II, III, by F. O. Bower. Cambridge University Press, London, 1923, 1926, 1928.]

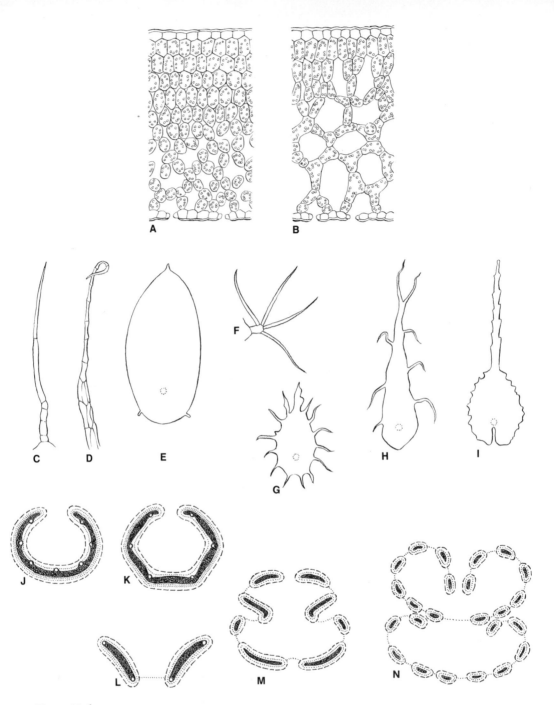

Figure 13-6

A, B, representation of internal leaf structure of ferns; C–I, representative types of dermal hairs and scales in ferns; the large peltate scales — E, G–I — are the more specialized forms; J–N, vascular cylinders of frond stipes as seen in transection; the primitive condition is seen in J, and a specialized condition in N; xylem is black, phloem stippled, protoxylem represented by white dots in J–L. A, B, E, G–I, *Polypodium;* C, *Matonia;* D, *Dipteris;* F, *Trichomanes;* J, *Osmunda;* K, *Gleichenia;* L, *Asplenium;* M, *Histiopteris;* N, *Cyathea.* [A, B, E, G–N redrawn from Anatomie der Vegetationsorgane der pteridophyten by Y. Ogura. In *Handbuch der Pflanzenanatomie.* Gebrüder Borntraeger, Berlin, 1938; C, D, F redrawn from *The Ferns,* Vol. I, by F. O. Bower. Cambridge University Press, London, 1923.]

ANATOMY OF THE LEAF. The typical fern leaf is dorsiventral. Epidermal cells have thickened, outer tangential walls and usually contain chloroplasts, a feature not shared by most other vascular plants. Stomata generally occur on the lower or abaxial surface of leaves. The mesophyll may be uniform in organization, consisting of homogeneous parenchyma with chloroplasts, or the cells may be organized into definite adaxial palisade and abaxial spongy parenchyma layers (Fig. 13-6, A, B). Considerable evolutionary importance is attached to the form and arrangement of the vascular tissue in the stipe and rachis. There may be a simple arc, as seen in transverse section, or the vascular bundle may be convoluted, or the vascular tissue may be broken up into a cylinder of separate bundles (Fig. 13-6, J–N). The last arrangement is considered to be a derived condition. As an added complication, the pattern of the vascular tissue may change in passing from the stipe through the rachis to the axes of the pinnae and pinnules.

THE SORUS. Keeping in mind the main aspects of shoot organography and anatomy makes it easier to understand the organization and position of the sorus. In the vast majority of ferns the foliage leaves serve in both photosynthesis and reproduction. In some species there may be a distinct separation of these functions: certain leaves function in photosynthesis, whereas others are strictly "sporophylls" and nonphotosynthetic. In other species there is an intermediate arrangement: For example, sporangia may be restricted to certain specific portions of a photosynthetic leaf. For the most part, sporangia are crowded into compact groups on leaves, each group being termed a sorus (Fig. 13-7). Sori may be circular or linear in outline. If sporangia are not grouped into definite sori they may form marginal tassels along narrow, reduced leaf segments or be scattered over the lower surface of expanded leaves, along and sometimes between veins.

In those species having definite sori the sorus is along or near the frond margin or away from the margin on the abaxial side of the frond.

ORGANIZATION OF THE SORUS. Both marginal sori and superficial sori (that is, ones on the abaxial side of the frond) are most commonly found over a vein or at the terminus of a vein (Fig. 13-7, C, F). That portion of the leaf surface to which sporangia are attached is termed the *receptacle* (Fig. 13-12). It may be a slight protuberance, it may be a definite bulge, or it may even be an elongated cone. It is from the superficial cells of the receptacle that sporangia originate while the leaf is still in a very young developmental stage. Undoubtedly the protection provided by circinate vernation is very important for the delicate sporangia during their ontogeny.

Sporangia may or may not be protected by a covering termed an *indusium*. If a sorus lacks an indusium it is often referred to as a naked sorus (Figs. 13-7, E; 13-9, D). If an indusium is present it may be formed by extensions from the adaxial and abaxial sides of the lamina; this results in the formation of a cup or pouch-like structure (Fig. 13-49, A, D, E, G). In some species a reflexed marginal portion of the lamina itself is associated with an indusium, forming a pouch-like structure (Figs. 13-8, B; 13-10, A). In forms that have sori on the abaxial surface some distance from the leaf margin, the indusium is an outgrowth from the epidermis of the lamina or of the receptacle. The form of the indusium is variable: it may be a delicate, linear flap attached along one side only (unilateral indusium); it may be horseshoe shaped or circular and elevated (peltate); it may be cup-shaped; it may be a collection of scale-like structures overarching the sporangia; or the leaf margin may be turned back upon itself ("false indusium"), with the sorus borne on it, to function in the protection of the sporangia (see Figs. 13-7, 13-8, 13-9, 13-10, 13-11 for types).

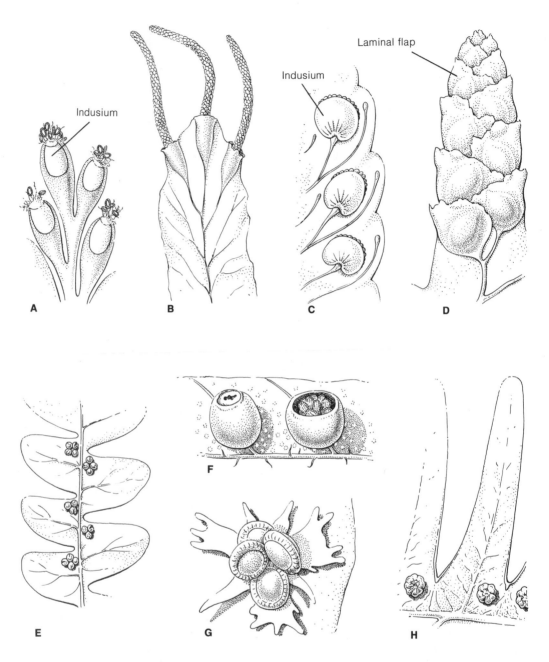

Figure 13-7
Variation in position and form of fern sori. **A,** *Davallia,* pouch-like indusium, joined with lamina, open at laminal margin; **B,** *Trichomanes,* marginal, receptacle elongate; **C,** *Nephrolepsis,* indusium attached at one side; **D,** *Lygodium,* each sporangium covered by a laminal flap; **E,** *Gleichenia,* superficial position, no indusium; **F,** *Cyathea,* cup-shaped indusium; **G,** *Woodsia,* basal membranous indusial segments; **H,** *Matonia,* peltate indusium. [C redrawn from *The Ferns,* Vol III, by F. O. Bower. Cambridge University Press, London, 1928; F adapted from *Morphology of Vascular Plants. Lower Groups,* by A. J. Eames. McGraw-Hill, New York, 1936.]

Figure 13-8
Portions of fertile fern leaves. **A,** *Pteridium*, young pinna left, older right (the reflexed margin of the lamina covering the coenosorus is evident at left); **B,** *Dicksonia*, each "indusium" consists of a reflexed laminal flap tightly joined with true abaxial indusium (also see Fig. 13-10, A); **C,** *Lygodium*, individual sporangia covered by a flap of leaf tissue; **D,** *Asplenium*, unilateral indusia.

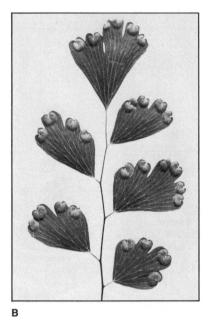

A

B

C

D

Figure 13-9
Morphology of fern sori. **A,** *Dryopteris*, reniform indusia; **B,** *Adiantum*, each sorus occurs on a reflexed portion of the lamina ("false indusium"); **C,** *Polystichum*, peltate indusia (sporangia can be seen extending beyond the margin of each indusium); **D,** *Polypodium*, naked sori ("exindusiate").

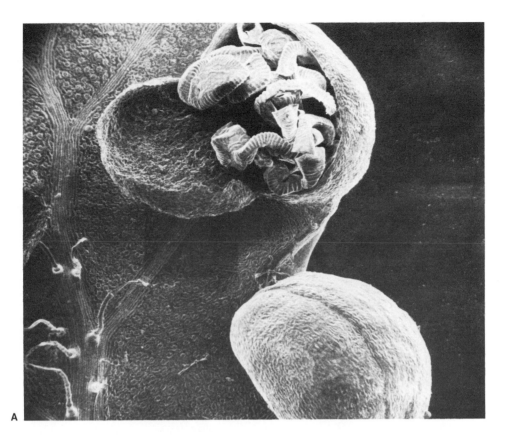

A

Figure 13-10
Scanning electron micrographs of fern sori. **A,** *Dicksonia antarctica*: the sporangia are protected by
the reflexed margin of the lamina, which is tightly appressed to the indusium (sorus toward the lower
edge of the figure); the indusium is depressed in the sorus toward the upper edge of the figure,
revealing the sporangia; note veins and numerous stomata on abaxial side of lamina (× 50);
B, *Dennstaedtia cicutaris*: indusia are cup-shaped and attached along the margin of the lamina (× 110).
[Courtesy of Dr. R. H. Falk.]

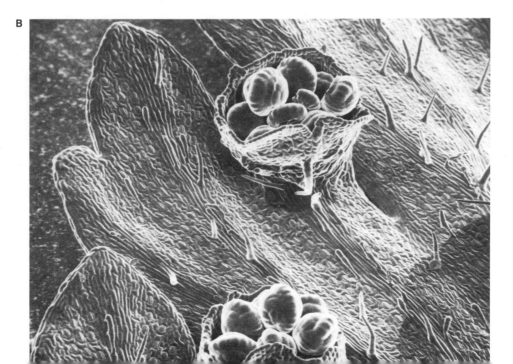

B

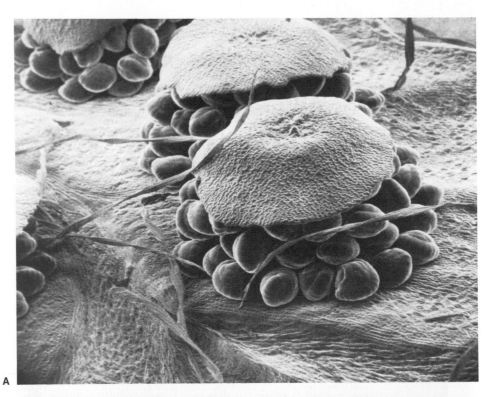

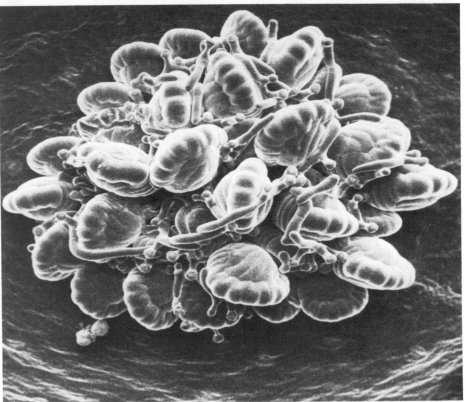

A

B

Figure 13-11 (*facing page*)
Scanning electron micrographs of fern sori. **A,** *Polystichum setosum:* note peltate indusia (× 35); **B,** *Polypodium heterophyllum:* note prominent annuli and branched paraphyses (× 80). [Courtesy of Dr. R. H. Falk.]

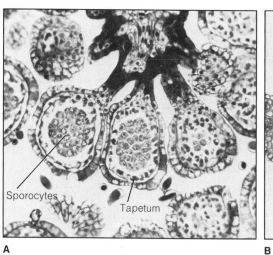

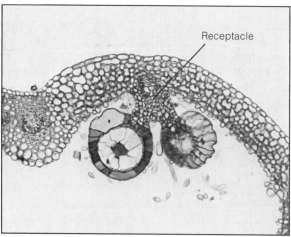

A
B

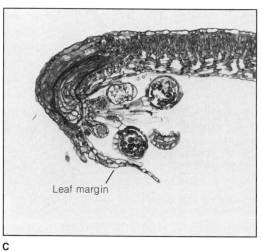

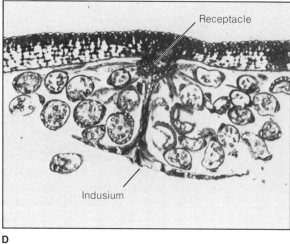

C
D

Figure 13-12
Sori and sporangia of ferns as seen in sectional view. **A,** *Osmunda,* section of leaf segment showing large sporangia, each one with many sporocytes and a massive tapetum; **B,** *Gleichenia,* showing large sporangia and lack of indusium; **C,** *Pteris,* intramarginal sorus with protective reflexed leaf margin (maturation of sporangia is of the mixed type); **D,** *Polystichum,* peltate indusium.

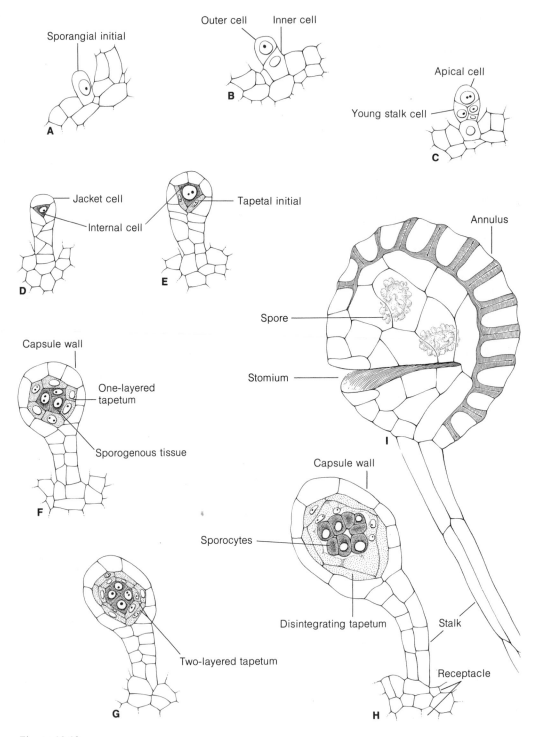

Figure 13-13
Stages in the development of a leptosporangium. Dehiscence has occurred in **I. A–H**, *Polypodium*; **I**, *Polystichum*.

DEVELOPMENT AND STRUCTURE OF THE SPORANGIUM. The student is referred to Chapter 4 (pp. 70–76), where a detailed description of the development of both the eusporangium and leptosporangium is presented. It will be recalled that the eusporangium has its origin from several superficial initials and that it has a massive wall and thick base or stalk at maturity. In the Filicales (leptosporangiate ferns) a single surface cell functions as the sporangial initial (Fig. 13-13, A, B). It will also be recalled that the sporangium grows for a time by the activity of an apical cell and by divisions in subjacent cells. The apical cell eventually forms a jacket cell and an internal cell. The latter cell gives rise to the tapetal initials and then to the sporocytes (Fig. 13-13). The wall remains one cell in thickness throughout development. Space does not permit a more detailed discussion of these features, and the remaining portion of this section will be devoted to an analysis of certain sporangial characteristics of evolutionary and phylogenetic significance.

The two features that are regularly considered in comparing sporangia are final length of the stalk, and the number of cells (or rows of cells) making up the stalk. Short, thick stalks are considered primitive, and long, delicate stalks (frequently consisting of three rows of cells or even one) are derived.

The sporangial wall is typically one cell in thickness at maturity. The main point of interest, however, is in the means of dehiscence and in the mechanisms involved. In most eusporangiate ferns dehiscence is brought about by a slit which opens down the side of the sporangium, which is fundamentally longitudinal. In the Filicales there are various methods of dehiscence, depending on the position of the annulus. It is interesting to note that in the Osmundaceae a group of thick walled cells (the annulus), located near the tip but to one side, is responsible for the formation of a cleft that runs over the top of the sporangium and

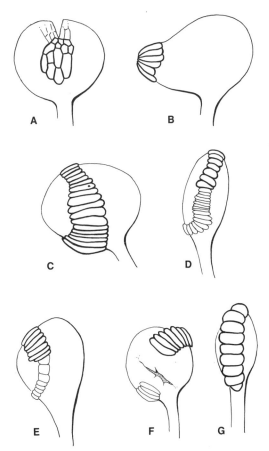

Figure 13-14
Variation in position of the annulus in leptosporangia. A, *Todea*, annulus subapical or lateral, which results in longitudinal dehiscence; B, *Lygodium*, annulus apical; C, *Gleichenia*, annulus oblique; D, *Plagiogyria*, annulus oblique; E, *Loxsoma*, annulus oblique, not all cells thickened; F, *Hymenophyllum*, annulus oblique, oblique dehiscence; G, *Leptochilus*, annulus vertical, which results in transverse dehiscence. [Redrawn from *Primitive Land Plants* by F. O. Bower. Macmillan, London, 1935.]

down the opposite side. The sporangium opens like a clam. In other ferns the annulus may form a cap at the distal end of the capsule (Fig. 13-14, B), be obliquely placed (Fig. 13-14, E), or run over the top of the capsule in line with the stalk (in a vertical or longitudinal position). These three positions result in longitudinal, oblique, and transverse dehiscence, respectively (Fig. 13-14). These three types of dehiscence represent a progressive evolutionary series as regards one morphological feature.

MATURATION OF SPORANGIA WITHIN A SORUS. The simplest way in which fern sporangia are borne is singly along the margins of leaf segments—each with a vascular bundle leading to its base (as in some extant ferns such as *Botrychium* and *Osmunda*). This arrangement is the primitive condition. In the majority of Filicales, sporangia are aggregated to form sori, which are disposed upon the sporophyll as described earlier in the chapter. A sorus in which all of the sporangia originate, grow, and mature at the same time is termed a "simple" sorus (Fig. 13-15). That the simple sorus is primitive is attested by the fact that members of those extant families, considered to be primitive in the Fili-

cales on other grounds have this type of development.

If sporangia are initiated over a period of time in a definite sequence the so-called "gradate" sorus is produced. The order of sporangial initiation and development is basipetal; the oldest sporangium is near the summit of a receptacle with successively younger sporangia toward the base (Fig. 13-15). When the fossil record is considered and compared with that of living ferns having the simple type of sorus, the gradate maturation is clearly a derived type. Certain fossils of the Mesozoic had gradate sori. It must be emphasized that all ferns with gradate sori are not necessarily closely related. In summary, gradate maturation of sporangia represents an evolutionary level of specialization that has been reached by certain species.

The most advanced evolutionary level of development is achieved in the sorus that has intermingled sporangia, all in different stages of growth (Fig. 13-15). This is the "mixed" sorus. The more highly specialized and evolved families have this mode of soral development. Bower (1923, p. 271; 1935) has emphasized the physiological importance of extending sporangial development over a longer period of time during the reproductive phase of the sporophyte. Neither should the adaptive value of prolonged spore production be underestimated.

SPORANGIAL DEHISCENCE. The dehiscence of the fern sporangium is ingenious and has attracted much attention from interested students. The annulus is the structural feature associated with dehiscence and the forceful ejection of spores. As a sporangium matures, water is lost from the cells of the annulus by evaporation. There is a powerful adhesion between the cell walls and water. The continued loss of water from each cell of the annulus results in the thin outer tangential wall of each cell being drawn inward, while the ends of the radial walls are pulled

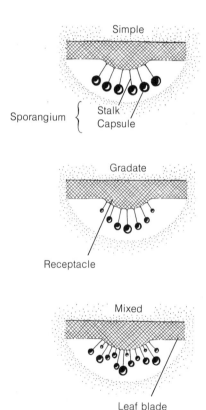

Figure 13-15
Three types of sporangial maturation in sori of leptosporangiate ferns.

toward each other. This results in the tearing open of the sporangium on the weak side and eventually in the complete inversion in the position of the annulus. The annulus is now under tremendous tension. Eventually the cohesive force of the water in cells of the annulus is exceeded, the water vaporizes into a gas phase, and the annulus returns suddenly to approximately its original position. In the process the spores are thrown out forcefully for a distance of a centimeter or so (Fig. 13-16). The tensions built up prior to dehiscence are equivalent to about 300 or more atmospheres of pressure. The advantage of such a dispersal mechanism is easily understood.

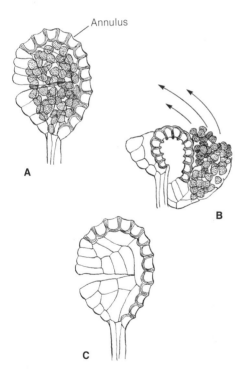

Figure 13-16
Behavior of a fern sporangium during drying and dispersal of spores. See text for discussion. [Redrawn from *Plant Physiology*, Ed. 1, by B. S. Meyer and D. B. Anderson, © 1939 by Litton Educational Publishing, Inc. Reprinted by permission of Van Nostrand Reinhold Co.]

A SURVEY OF SORAL TYPES AND THEIR DEVELOPMENT.   Variations in soral morphology are endless. At best only a few of the more common soral types can be described. Examples most often used for teaching purposes are those that illustrate a generalized type or represent steps in an evolutionary series. Whenever possible, as material is available, the student should not only gain an understanding of soral positions but should also keep in mind the several morphological features of the sorus as described earlier.

In subsequent discussions we will be describing the positions of sori on fern leaves, namely, the (1) marginal position, (2) intra-marginal (near the margin) position, and (3) the abaxial or "superficial" position in which sori are at some distance from the leaf margin on the lower (abaxial) side of a frond. It will be recalled that one stage in leaf development is the formation of a simple lamina, pinnae, pinnules, etc., by marginal growth (p. 281). The origin and final position of a sorus are correlated with the activity of marginal meristems and their derivatives (Wardlaw, 1958, 1962; Wardlaw and Sharma, 1961).

SORI OF MARGINAL POSITION.   For a sorus to qualify as truly marginal the receptacle and its sporangia must originate strictly from the margins of the developing pinnae or pinnules (Fig. 13-17, A–C). An indusium, if present, is formed by submarginal outgrowths around the receptacle. The indusium in its final form may be funnel-shaped or two-lipped. The outgrowth from the adaxial side is called the adaxial or outer indusial flap or lip; the abaxial outgrowth is the abaxial or inner flap or lip (Fig. 13-17, A-C). The two parts of the indusium may be structurally quite different. Often, the adaxial flap is thicker and histologically may resemble the lamina of the leaf. As can readily be understood, the receptacle is actually a

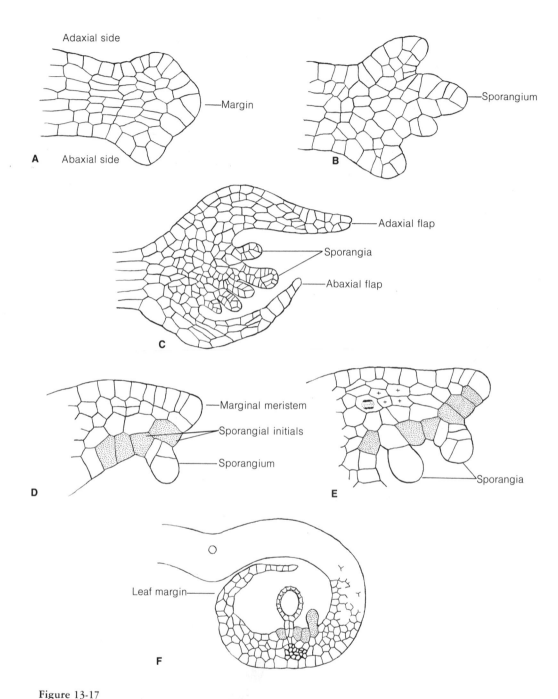

Figure 13-17
A–C, *Lindsaea linearis,* transverse sections through margins of fertile portions of leaves showing stages in development of marginal sorus; **D–F,** *Cryptogramma crispa,* sections through leaf margin showing development of intramarginal sorus. Margin of lamina becomes extended as a protective flap (**F**). [A–C redrawn from *The Ferns,* Vol. III, by F. O. Bower. Cambridge University Press, London, 1928; **D–F** redrawn from Wardlaw, *Ann. Bot.* 25:481, 1961.]

continuation of the marginal meristem and retains its activity in the production of sporangial initials. Wardlaw (1958, 1962) has recommended that the term "sporogenous meristem" should replace the term receptacle, since it is more descriptive of its function. Examples of the marginal type are: *Hymenophyllum, Trichomanes, Lindsaea* (some species).

SORI OF INTRAMARGINAL POSITION. The sori of many ferns are described as being marginal, because that is how they appear on the adult lamina, but actually they only originate near the margin during ontogeny. In these ferns, the cells at the margin suddenly or gradually lose their meristematic potentialities and become parenchymatous, often forming a thin extension of the margin (Fig. 13-17, D–F). Submarginal cells on the abaxial side of the developing lamina in close proximity to the margin continue their meristematic activity — forming the receptacle or sporogenous meristem. Associated with this change in activity, an inner or abaxial indusial flap also may be formed as a thin outgrowth from surface cells near the receptacle. The final width and length of the receptacle depends upon the length of time that the sporogenous meristem is active and upon how much of the laminal margin participates in the processes just described, which are, of course, under genetic control. In some instances individualized sori may be formed along veins near the leaf margin; in other instances the receptacle (and sporangia) may extend parallel with and along the entire length of the leaf segment. Examples of the intramarginal type are: *Pteris, Cryptogramma, Pellaea, Pteridium,* and *Dicksonia* (the last two also have a true abaxial indusium).

SORI OF ABAXIAL OR "SUPERFICIAL" POSITION. In ferns whose sori are in the abaxial position, a young receptacle has its inception from submarginal cells on the abaxial side rather early during marginal growth of the lamina. However, in contrast to the intramarginal type of development, the margin of the lamina remains actively meristematic and adds new tissue to the lamina. Therefore, as growth of a young leaf segment proceeds, the young sorus occupies positions progressively farther from the margin (Figs. 13-18, 13-19). An indusium, if characteristic of the species, is formed from superficial cells near the developing receptacle and eventually overarches it (Fig. 13-18). The indusium, in its final form, varies in shape from elongate and attached along one of the longer sides (unilateral indusium), to halfmoon shaped or reniform and attached at the sinus (*Dryopteris*, Fig. 13-9, A). In some species the indusium is an outgrowth from the top of the receptacle, resulting in the formation of a stalk and a radially symmetrical cap (peltate type; Figs. 13-9, C; 13-11, A; 13-12, D).

ANATOMY OF THE STEM. A discussion of fern stem anatomy should logically begin, as was done for the leaf, with a description of the apical cell. A variety of structures and tissues take their origin from the meristematic derivative cells of the apical initial: leaves, the protoderm of leaves and of the axis, epidermal hairs and dermal appendages, ground meristem, and procambium. From available studies it appears that procambium in ferns, as is true of the majority of investigated vascular plants, develops in relation to the appearance of leaves. (For a review see Esau, 1954). Very soon after a leaf is initiated a procambial strand becomes differentiated at its base and soon becomes apparent in the leaf itself (Fig. 13-2). The development of procambium is acropetal and apparently always in continuity with existing procambium at a lower level in the stem. Within a procambial strand, phloem differentiation also takes place in an acropetal direction

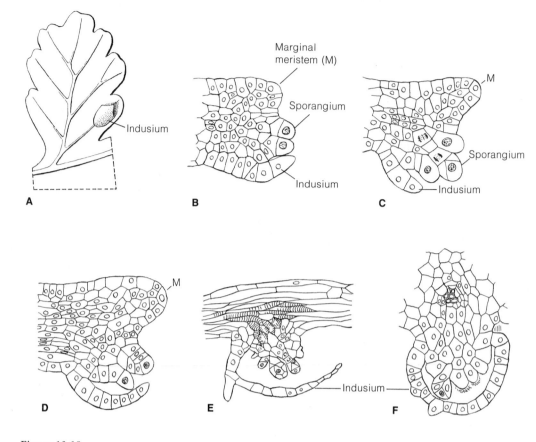

Figure 13-18

Cystopteris bulbifera. **A,** pinnule showing position of sorus; **B–E,** transverse sections through margins of pinnules, showing development of sorus; **F,** developing sorus as seen in a section that cuts across a vein. [Redrawn from Palser and Barrick, *Bot. Gaz.* 103:172, 1941, University of Chicago Press.]

(upward into the leaves). In some ferns there is evidence that the differentiation of tracheary elements in the xylem may be discontinuous before the leaf has its normal amount of mature xylem. In general, though, the wave of vascular differentiation is in an acropetal direction.

Throughout the Filicales, including the fossil forms, there is no indication of cambial activity resulting in the formation of secondary vascular tissues. There is, however, the development of considerable sclerenchyma in the axes of some ferns, in contrast with its absence in the Ophioglossales and its limited occurrence in the Marattiales. The radial maturation of xylem in a vascular bundle of the stem or leaf is char-

acteristically mesarch but exceptions do occur. The large tracheids of the metaxylem generally have tapered ends, and scalariform pitting is present on all sides of the cell. It has been demonstrated that certain ferns (e.g., *Pteridium aquilinum*) possess vessels; actual scalariform perforations instead of pits are present in the oblique end walls between two vessel members (Bliss, 1939; also see p. 361).

An extensive survey of the length of tracheary elements and their structure in ferns revealed that the length of tracheary elements varied with location in the plant, age of the plant, length of an internode, polyploidy and habitat (White, 1963a,b). In spite of these variations the following

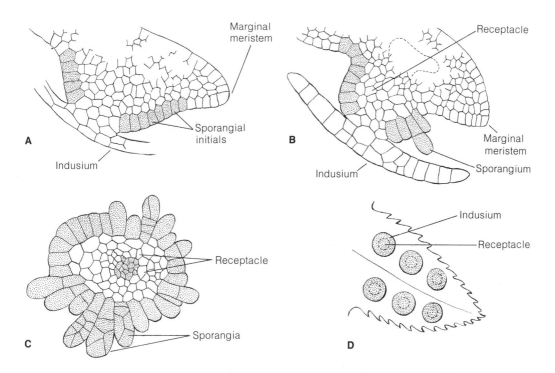

Figure 13-19
Polystichum aculeatum. **A,B,** transverse sections through margins of fertile leaf segments showing origin and development of the sorus; note displacement of sorus from marginal region by continued activity of the marginal meristem. **C,** transverse section of a young sorus at a level below the indusium; note origin of sporangia from surface cells (sporogenous meristem) of the receptacle. **D,** schematic representation showing position of mature sori on abaxial side of a fertile leaf segment. [Redrawn from Wardlaw, *Ann. Bot.* 25:482, 1961.]

changes appear to be correlated with specialization from the primitive condition for each feature: (1) shortening of tracheary elements, (2) increase in the occurrence of a modified type of scalariform pitting (opposite and alternate), (3) increase in the occurrence of slightly oblique to transverse end walls and (4) sporadic occurrence of vessels, e.g., in *Pteridium* and *Marsilea* (see p. 361).

TYPES OF VASCULAR CYLINDERS. What does the fossil record reveal regarding types of vascular patterns in ferns of the past? As in all groups of primitive vascular plants, the protostele is found in ferns from the Paleozoic and even from more recent eras. The protostelic condition was true of *Gleiche-*

nites, an extinct fern of the Triassic, which was undoubtedly related to the living genus *Gleichenia.* Of course the siphonostele is not without representation in the fossil record; it occurred in later ancient ferns.

Among living members of the Filicales the protostele has still persisted. For example, it occurs in the genera *Gleichenia* and *Lygodium,* and in the family Hymenophyllaceae. All living ferns having protosteles are considered the more primitive members of the Filicales, not only on the basis of this one feature but on others as well. Most species of *Gleichenia* are protostelic but of a rather special type (Fig. 13-20, A). The bulk of the central column is primary xylem in which tracheids of the metaxylem are interspersed

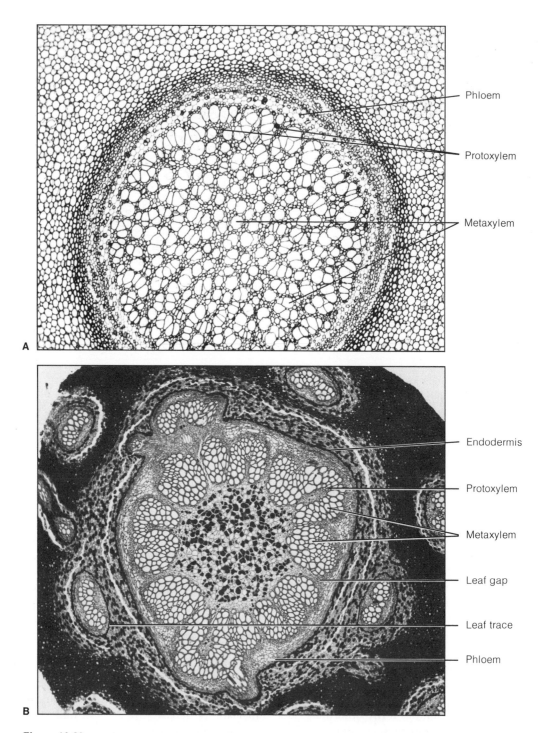

Figure 13-20
Transections of fern stems. **A**, *Gleichenia*, an example of a medullated or "vitalized" protostele; the metaxylem consists of tracheids intermingled with parenchyma; **B**, *Osmunda*, example of an ectophloic siphonostele.

with parenchyma cells. Protoxylem occupies definite loci near the periphery of the xylem. External to the xylem is a cylinder of phloem consisting of relatively large, thin-walled sieve cells and smaller parenchyma cells. The vascular cylinder is enclosed by a rather wide pericycle several cell layers in thickness which, in turn, is surrounded by the endodermis. Frequently the endodermis and particularly the Casparian strips are not well-defined. Topographically the endodermis is located at the boundary between thick-walled cortical cells and thin-walled pericyclic cells. The departure of a leaf trace does not materially affect the outline of the vascular cylinder of the stem and there is no leaf gap. The significance of a "vitalized" protostele in the evolution of the siphonostele has already been considered in Chapter 3.

The most commonly accepted viewpoint is that there are two types of siphonosteles: the *ectophloic siphonostele* (external cylinder of phloem only), and the *amphiphloic siphonostele* (internal and external cylinders of phloem) (see Chapter 3). *Osmunda* is one of the ferns commonly used to demonstrate an ectophloic siphonostele (Fig. 13-20, B). *Osmunda* has a short erect stem upon which is a crown of leaves. The stem is completely invested by leaf bases or stipes. A transverse section of an older shoot reveals the crowding of the leaves and the presence of a horseshoe-shaped vascular strand in each leaf. Centripetal to the free leaf bases is the outer cortex, consisting of thick-walled sclerenchyma with embedded leaf traces. The inner cortex is parenchymatous, the cells often packed with starch grains. The pith occupies a generous amount of the axis. Along the outer edge of the pith is a cylinder of more-or-less separate xylem strands, often U shaped in outline with the open side toward the pith. Protoxylem is in the sinus formed by the two arms of metaxylem. For this reason the xylem is mesarch in structure. Surrounding the xylem is a zone of paren-

chyma, several cells in thickness. External to this layer is the phloem with its relatively large sieve cells, often elongated in an oblique longitudinal direction. This is particularly true of the protophloem. Parenchyma cells also are interspersed among the sieve cells. External to the phloem is the *pericycle*, which, in turn, is surrounded by tannin-filled *endodermal* cells. The outline of the main vascular cylinder may be somewhat irregular, due to the divergence of leaf traces. Each leaf trace is confronted by a leaf gap which may extend longitudinally and obliquely for some distance in the xylem. The phloem, however, forms a continuous cylinder. In some species tracheids may be intermingled with parenchymatous pith cells, and in one species (*Osmunda cinnamomea*) an internal endodermis has been reported.

Ferns with vascular cylinders of the types just described have also been encountered as fossils. The stem anatomy of a late Paleozoic fern, *Thamnopteris*, closely resembled that of *Osmunda*. In *Thamnopteris*, however, the pith consisted of intermingled parenchyma and tracheids surrounded by a compact cylinder of xylem, without leaf gaps. The vascular cylinder was a "vitalized" protostele. Some species of the Mesozoic genus *Osmundacaulis* had stems that were even more similar in anatomy to those of *Osmunda* (see p. 341).

Without doubt many more ferns have the amphiphloic siphonostele than have the ectophloic type. Fundamentally there are two types of amphiphloic siphonosteles: the *solenostele*, and the *dictyostele*. The solenostele, without overlapping leaf gaps (Fig. 13-22, A), occurs in many creeping ferns that have conspicuous internodes (Fig. 13-21, A). Common examples are certain species of *Adiantum* and *Dennstaedtia* (hay-scented fern). A transverse section within an internodal region reveals a complete cylinder of vascular tissue surrounded by the cortex and enclosing a pith that is often sclerified (Fig. 13-21, A). Starting from the

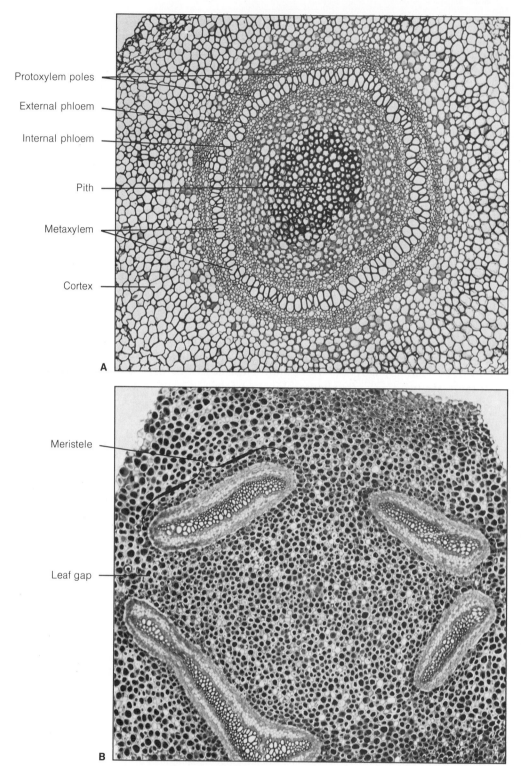

Protoxylem poles

External phloem

Internal phloem

Pith

Metaxylem

Cortex

A

Meristele

Leaf gap

B

Figure 13-21
Transections of fern steles. **A,** *Dennstaedtia,* solenostele, phloem, and endodermis occur on both sides of the xylem; section was made at a level between leaf gaps; **B,** *Phyllitis,* dictyostele.

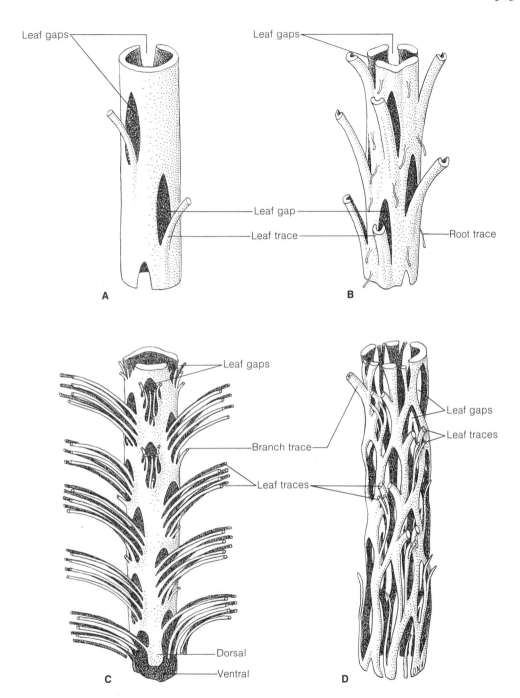

Figure 13-22
Types of stelar organization. **A,** solenostele, schematic; **B,** dictyostele (*Cheilanthes tenuifolia*);
C, dictyostele, rhizome has dorsiventral organization, leaves arising on dorsal side; some leaf traces are
attached along edge of a leaf gap (*Bolbitis diversifolia*); **D,** dictyostele (*Oleandra wallichii*). [B redrawn
from Nayar, *Jour. Linn. Soc. London Bot.* 58:449, 1963; C redrawn from Nayar and Kaur, *Jour. Linn. Soc.
London Bot.* 59:127, 1965; D redrawn from Nayar, Bajpai, and Chandra, *Jour. Linn. Soc. London Bot.*
60:265, 1968.]

outer edge of the vascular cylinder (stele), the outer endodermis can usually be distinguished by phenolic substances in its cells. Internal to the endodermis is the pericycle (one or more layers of cells in width) followed by the phloem (external phloem). As usual the ring of xylem is most conspicuous, being composed of scalariformly pitted tracheids. Internal phloem, pericycle, and inner endodermis follow, in that order, in progressing toward the pith. At a node the seemingly closed cylinder of vascular tissue is broken by the presence of a leaf gap. The leaf gap is not extensive, and the cylinder again appears closed within the next higher internode. This type of stele is not an entirely new innovation because species of *Adiantum* have been found from the Tertiary. The creeping water fern, *Marsilea*, is also solenostelic with a highly lacunate cortex typical of many hydrophytes (Fig. 13-61). Solenosteles may become highly elaborate, as in *Matonia* where there is a variable number of concentric, amphiphloic cylinders. At nodes each cylinder may have a leaf gap, and the entire leaf trace complex also may be a collection of concentric cylinders.

The most specialized type of siphonostele in ferns is the dictyostele. The dictyostele is an amphiphloic tube of vascular tissue in which parenchymatous leaf gaps overlap. A transverse section made at any level of a well-developed stem shows a ring of separate vascular bundles. These bundles vary considerably in size and shape, but each one consists of primary xylem enclosed by primary phloem. These strands are termed concentric, amphicribral vascular bundles, or *meristeles* (Fig. 13-21, B; also see Chapter 3, p. 61). The parenchymatous zones between meristeles are leaf gaps (Fig. 13-21, B). Most

of the Filicales are fundamentally dictyostelic and of the type just described (Fig. 13-22, B–D). This is true of upright as well as creeping species. If the leaf gaps are long and extensive and more than one leaf trace is associated with each gap, or if there are breaks in the cylinder not associated with leaves, the vascular cylinder becomes a tubular network. In one transverse section it is difficult if not impossible in these cases to distinguish the actual limits of the gap. We are describing the so-called dissected dictyostele. *Polypodium* is an example of a common fern with a *dissected dictyostele* (Fig. 13-23, A). An inner vascular cylinder may be present in some dictyostelic species as well as in some solenostelic species.

The rhizome of *Pteridium aquilinum* (bracken fern) has "medullary strands" within an outer vascular cylinder (Fig. 13-23, B). Depending upon the variety of *P. aquilinum*, the leaves may be borne on both long and short lateral shoots (O'Brien, 1963) or only on short shoots (Webster and Steeves, 1958). Therefore, the parenchymatous regions between meristeles of a stem which is devoid of leaves cannot be regarded as leaf gaps; their morphological significance is unknown. The meristeles of the outer ring tend to be smaller than the two strap-shaped medullary strands (Figs. 13-23, B; 13-24). Sclerenchyma may entirely or partially enclose the inner strands, or it may appear as two separate bands in transectional view. It should be emphasized that the cylinders of vascular tissue are not entirely independent of each other but are joined at a lower (older) level in the rhizome.

More highly complicated vascular structures have been described for ferns, namely, species with exceedingly long and large leaf

Figure 13-23 *(facing page)*
Transections of fern rhizomes. **A,** *Polypodium,* example of a dissected dictyostele;
B, *Pteridium,* a stele with outer and inner vascular cylinders and intervening sclerenchyma.

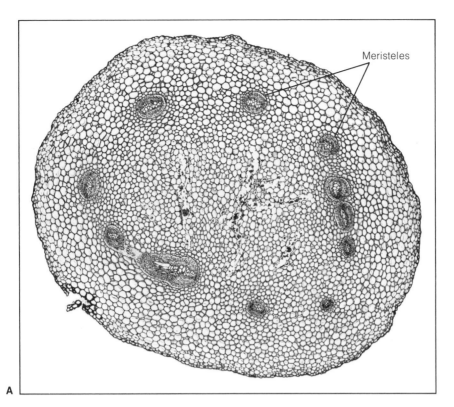

Meristeles

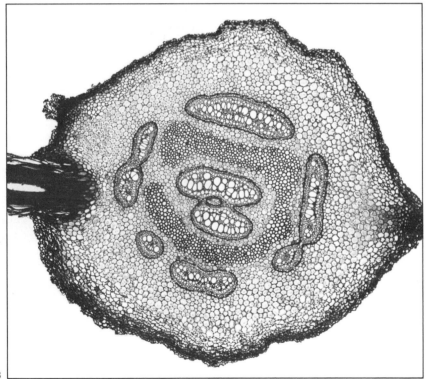

A

B

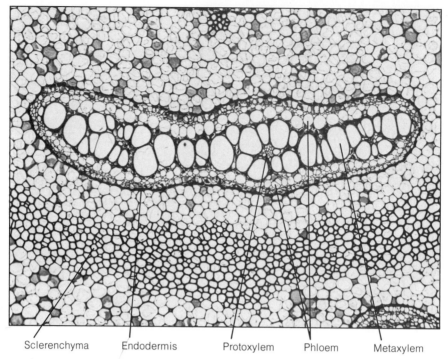

Sclerenchyma   Endodermis   Protoxylem   Phloem   Metaxylem

Figure 13-24
Transection of rhizome of *Pteridium* showing structural details of one meristele.

Figure 13-25
Trunk of tree fern showing conspicuous leaf scars and dense mat of roots forming a buttress at base of stem. Many roots have grown downward from points high on the trunk. [Courtesy Dr. T. E. Weier.]

gaps with numerous leaf traces attached to the sides of the gaps, and species with small innumerable meristeles, arranged in intricate patterns and often associated with leaf gaps in various ways. In concluding this section it should be pointed out that critical ontogenetic work, utilizing modern techniques, still remains to be done on fern vasculation before the results of earlier works can be fully accepted without reservation (Esau, 1954). Evolution of the stele in ferns is discussed in a later section on special problems in fern morphology.

The Root

Except for the first root or primary root of the embryo, all roots of a fern arise from the rhizome or upright stem, near leaf bases or below them. They are usually small in diameter, monopodially branched, and dark in color. The outer investment of fibrous material on the lower portion of the trunks of tree ferns is a mat of tangled, living, dead, and dying roots (Fig. 13-25). The actual stem axis may be only a relatively small cylinder within the covering of roots.

At the level of mature primary tissues in the root the epidermal cells are usually thin-walled (some cells may form root hairs). A generous portion of the inner cortex may be sclerenchymatous, or only the cells of a single layer around the endodermis may be thickened. The endodermis is always well-defined, and each cell has the usual Casparian strip; at maturity the cell walls vary in thickness according to the species. Fern roots are diarch to tetrarch with very few tracheids in the xylem. Xylem is exarch in maturation. No secondary growth occurs.

Roots commonly arise endogenously from cells of the stem endodermis. They force their way through the cortex to the outside. Early in their development a conspicuous apical cell makes its appearance and functions throughout the life of the root. In most investigated ferns the apical cell is pyramidal in shape, cutting off cells to the main body of the root on three sides and contributing segments to the root cap on the fourth side. The large apical cell is usually highly vacuolate and apparently divides only infrequently. The three primary meristems—protoderm, ground meristem, and procambium—have their origin from derivative cells of the apical initial, and they, in turn, give rise to the epidermis, cortex, and vascular cylinder, respectively. Lateral roots originate from cells of the endodermis, in contrast with higher plants in which the pericycle is usually the initiating layer (Ogura, 1972; Esau, 1965; Guttenberg, 1966).

The Gametophyte

The origin of the gametophyte is normally from a haploid cell, the spore. Members of the Filicales are homosporous and have exosporic gametophytes with the exception of *Platyzoma microphyllum*, which exhibits "incipient heterospory" and produces initially unisexual gametophytes (Tryon, 1964; Tryon and Vida, 1967). The "water ferns" are heterosporous and possess endosporic gametophytes. The latter type of gametophyte is described elsewhere (p. 365). With the renewed interest being shown in ferns our knowledge of the range in variability of the fern gametophyte has been greatly extended (Atkinson and Stokey, 1964; Nayar and Kaur, 1971).

Fern spores are either *tetrahedral* or *bilateral* in symmetry. If tetrahedral, the spore has a triradiate ridge on the proximal face (side in contact with other spores of a tetrad) and the spore is said to be trilete (Fig. 13-26, A). If bilateral, a distinctive portion of the exine is in the form of a line and the spore is said to be monolete (Fig. 13-26, B). The spore wall consists essentially of two parts; the inelastic outer layer, the *exine*, and the inner layer, the *intine*. In some species there may be an additional outer envelope termed the perispore or perine.

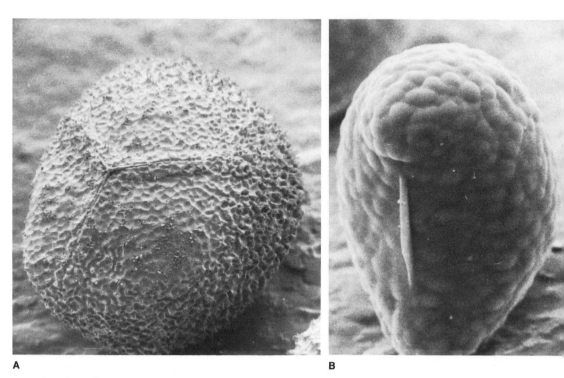

A **B**

Figure 13-26
Two representative types of fern spores. **A,** *Osmunda regalis* (trilete-type, × 1450); **B,** *Scyphularia pentaphylla* (monolete-type, × 1500). [Courtesy of Dr. R. H. Falk.]

SPORE GERMINATION. Spores of most ferns will germinate if provided with moisture and maintained at a pH of 4–8 and in the temperature range 15–30°C. Light in the visible spectrum is probably the most critical factor concerned with germination. In general, fern spores do not germinate in the dark, but there are a few exceptions. It has long been known that long and short wavelengths of light have different effects on fern-spore germination. Red light promotes germination while blue light is relatively ineffective or is inhibitory. Far-red light alone also exerts an inhibitory effect (Mohr, 1963). There are exceptions to the generalizations just made. For example, *Pteridium aquilinum* is indifferent to wavelength and germinates under light of all wavelengths as well as in total darkness. The spores of *Anemia phyllitidis* will not normally germinate in total

darkness, but they will if gibberellic acid is added to the culture medium (Schraudolf, 1962).

The difficulty of sectioning fern spores, due to the hard cell wall, has restricted histochemical studies. The results of one study, however, have shown that the spore of one species (Ostrich fern, *Matteuccia struthiopteris*) contained no starch, but it did have abundant storage protein granules. Starch did accumulate in chloroplasts when the spores were exposed to light during germination and the protein bodies disappeared as germination progressed (Gantt and Arnott, 1965).

There are variations in the sequence of wall formation as well as in polarity during spore germination (Fig. 13-27). In one type (*Osmunda*-type) the first division of the spore results in the formation of a small cell

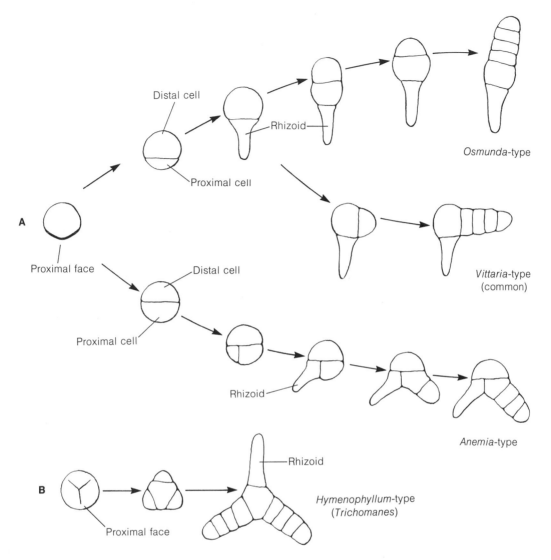

Figure 13-27
Schematic representation of certain types of spore germination in homosporous, leptosporangiate ferns. The proximal face (where spore was in contact with others in a tetrad) is shown schematically. (Consult text for details.) [Redrawn from Nayar and Kaur, *Bot. Rev.* 37:301, 1971.]

toward the proximal face and a larger one toward the distal face. The smaller cell will become a rhizoid and the larger one is the initial of the green prothallial filament. Additional transverse divisions in the initial and its derivatives produces a filament of several cells. A variation of this type and one that is common in the Filicales is shown in Fig. 13-27 (*Vittaria*-type). In this type the prothallial initial divides at right angles to the first

wall and one of the two cells proceeds to form a filament. In a third type (*Anemia*-type) the spore divides into two cells of equal size. The proximal cell divides unequally into a smaller rhizoidal cell and a larger prothallial initial; the distal cell does not divide (Fig. 13-27). Some ferns are "tripolar" (Fig. 13-27, B) in germination. (Additional examples can be found in the review article by Nayar and Kaur, 1971.)

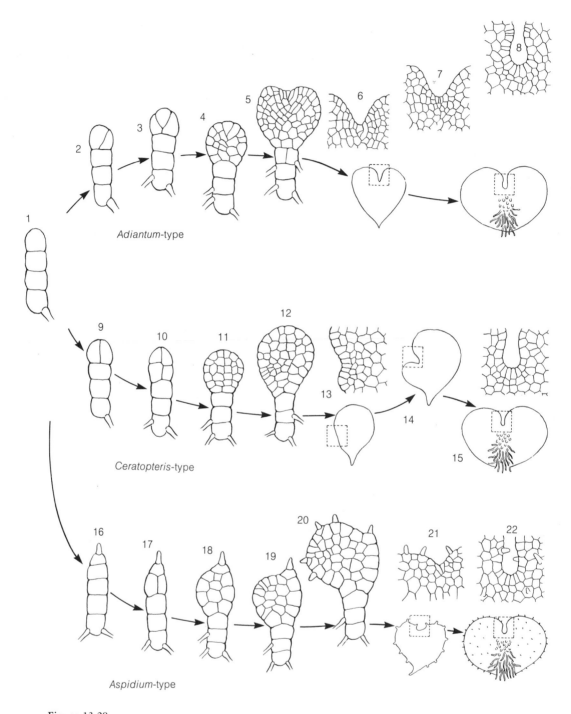

Adiantum-type

Ceratopteris-type

Aspidium-type

Figure 13-28
Schematic representation of three types of gametophyte development in homosporous, leptosporangiate ferns. The arrows indicate successive stages in each type of development as follows: numbers 1–8, *Adiantum*-type; 1, 9–15, *Ceratopteris*-type; 1, 16–22, *Aspidium*-type. [Redrawn from Nayar and Kaur, *Bot. Rev.* 37:304, 1971.]

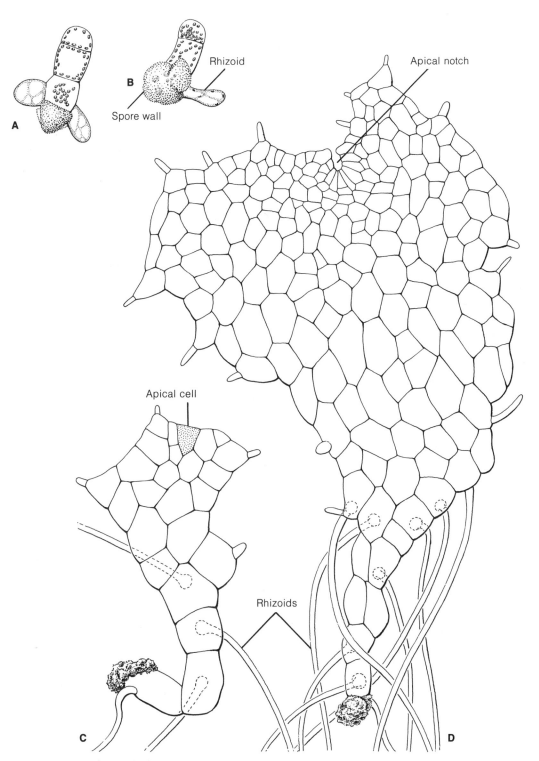

Figure 13-29
Germination of fern spores and development of the gametophyte. **A**, *Athyrium;*
B, *Pteridium;* **C**, **D**, *Dryopteris.*

DEVELOPMENT OF THE GAMETOPHYTE. Subsequent to spore germination each species tends to have a particular, established pattern of development; also, many larger groups of ferns share a common pattern (Nayar and Kaur, 1971). In general, seven different patterns are recognized. We will describe three patterns of development that lead to the common cordate type of gametophyte.

In *Adiantum*-type gametophytic development, divisions in the terminal cell of the prothallial filament result in the early establishment of a wedge-shaped apical cell. Cells are produced laterally from the apical cell and these in turn remain meristematic. The apical portion becomes notched and the apical cell then divides transversely and derivative cells continue to divide. Soon a thick midrib is formed by the cell divisions in the plane of the flat thallus. Ultimately the gametophyte becomes cordate with a midrib and thin lateral wings (Fig. 13-28, 1-8).

In *Ceratopteris*-type gametophytic development a broad plate of cells is formed, but without a definite apical cell. Meristematic activity is restricted to a group of cells along the margin of the gametophyte. A notch is produced laterally and one wing may become larger than the other, although the gametophyte may eventually become symmetrically cordate (Figs. 13-28, 9-15).

The early formation of a trichome (hair) from the terminal cell of the prothallial filament restricts apical growth in *Aspidium*-type development (Figs. 13-28, 16-22). A broad plate of cells is formed behind the anterior end of the developing gametophyte. An apical cell becomes established laterally but eventually the thallus becomes cordate (Fig. 13-29, D). Additional trichomes are formed from marginal and superficial cells.

The types of gametophyte development just described are considered representative of the Filicales, although development can vary, even within species of a section of a genus (e.g., *Pellaea*, Pray, 1968).

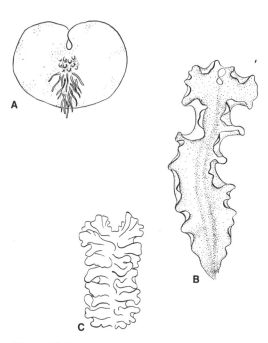

Figure 13-30
Representative types of gametophytes in the Filicales.
A, *Adiantum caudatum;* **B,** *Blechnum brasiliense;*
C, *Dipteris conjugata;* **D,** *Schizaea pusilla;* **E,** *Mecodium flabellatum;* **F,** *Lophidium (Schizaea) dichotoma.*
See text for explanation of types. [Redrawn from Nayar and Kaur, *Bot. Rev.* 37:295–396, 1971.]

FORM OF MATURE GAMETOPHYTES. As mentioned earlier the cordate-thallus type of gametophyte is the most common in the Filicales. The plant generally lies flat on the substrate, but in some species the wings are raised or have ruffled margins (Fig. 13-30).

Examples of other types of gametophytes, which are rare, are (1) strap or branched, ribbon-shaped (Fig. 13-30, B, E), (2) filamentous (Fig. 13-30, D), (3) tuberous and subterranean, and (4) more-or-less filamentous, with the diameters of some axes comprising several cells (Fig. 13-30, F).

GAMETANGIA. The gametophytes of the Filicales generally produce both antheridia and archegonia on the same thallus. They are bisexual. In a given population of gametophytes, antheridia generally appear

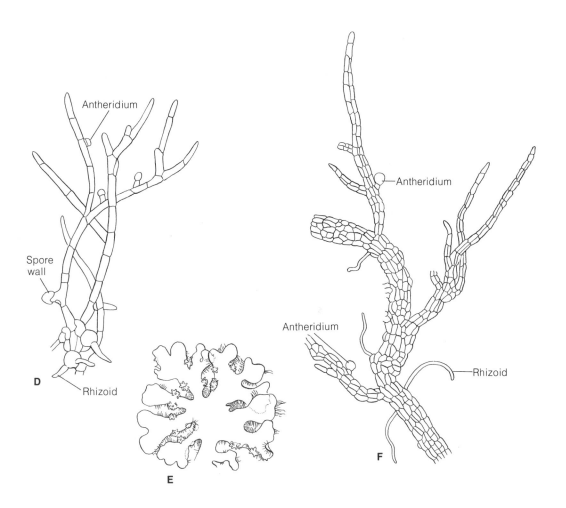

Antheridium

Spore
wall

D

Rhizoid

E

Antheridium

Antheridium

Rhizoid

F

earlier than archegonia. Antheridial formation may take place very early in development, even before the midrib is formed in cordate-types. Archegonia are formed later, on the midrib near the notch; antheridia generally are present toward the posterior end situated among rhizoids, but they may occur on the wings or near the notch. In thalloid, dorsiventral gametophytes, both types of sex organs are usually restricted to the ventral side. There are exceptions to these generalizations. From the evidence produced to date, it is fairly clear that antheridial initiation is controlled by naturally occurring hormones. The physiology of sexuality in ferns is discussed on page 327.

The mature antheridium of a specialized or derived homosporous leptosporangiate fern comprises three jacket cells—two ring cells and a cap cell—enclosing the spermatocytes or potential sperms (Fig. 13-31, I). The 16–32 sperms are eventually released by detachment of the cap cell or by the lifting-up of the cap cell as a hinged lid. Not all ferns in the Filicales have this type of antheridial organization. There are, for example, variations in number of jacket cells, and such variations are discussed in the section on fern systematics and phylogeny (p. 351).

The venter of the archegonium is normally embedded in gametophytic tissue and has a relatively short protruding neck comprising four rows of from five to seven tiers of cells enclosing a binucleate neck-canal cell, a ventral canal cell, and an egg cell (Fig. 13-31, F). (For a detailed account of antheridial and archegonial development see Chapter 5, pp. 86–91, and Fig. 13-31.)

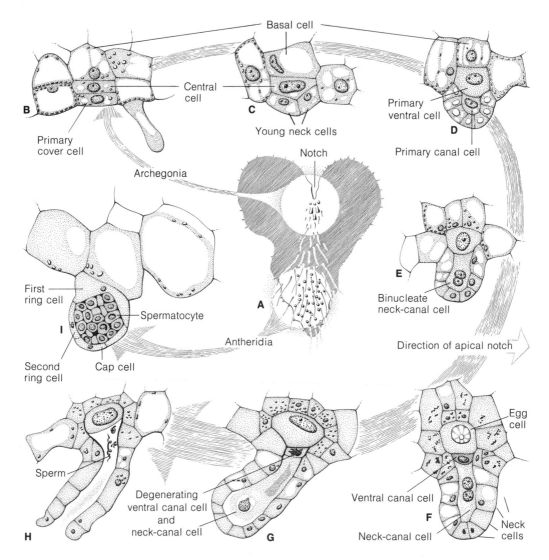

Figure 13-31
Gametophyte and gametangia of a fern in the Filicales. **A,** form of a sexually mature, heart-shaped gametophyte; **B–H,** stages in the development of an archegonium; **I,** a nearly mature antheridium. [From slides prepared by Dr. F. V. Ranzoni.]

Dynamics of Fertilization and Embryogeny

If sexually mature plants are flooded with water, antheridia may open within a few minutes. It has been shown that when flooded the antheridium shows signs of internal pressure followed by the loosening of the cap cell and the escape of the sperms, one by one. In a matter of seconds, or at the most a few minutes, the sperms lose their compact appearance and assume the loose, helical form of the swimming sperm (Ward, 1954). In some ferns a large vesicle of cytoplasm may remain attached to the multiflagellate nuclear band. There have been some interesting studies on sperm morphology in which the electron microscope and ultraviolet microscopy were employed. It has been shown by these studies that each flagellum actually consists of nine sets of peripheral tubules and one central pair (9 + 2 organization) (Manton, 1950; Manton and Clarke, 1951; Rice and Laetsch, 1967). It is a remarkable fact that the flagella of motile cells of many other plants and of animals exhibit this same type of organization.

In one study (Ward, 1954) mature archegonia are reported to open within one or two hours after immersion in water. The impending opening of an archegonium is heralded by enlargement of the distal end and separation of apical neck cells. This allows a tiny stream of mucilaginous material or slime to be released, which is followed by the forceful release of definable protoplasmic bodies — presumably they are entities of the axial row except for the basal egg cell. Following this phase, the upper tiers of neck cells split apart. Swimming sperms are at first not attracted to archegonia, but later, when the slime is diluted, they change direction chemotactically and swim toward an archegonium. In *Phlebodium aureum* from three to five sperms may swim into an open archegonium and occupy the ventral cavity at the same time. However, only one sperm effects syngamy.

The first division of the zygote is reported to take place anywhere from one hour to ten days after fertilization. In *Phlebodium aureum* it is five days (Ward, 1954). Even before the zygote divides, cells of the surrounding gametophyte divide and form a partially ensheathing calyptra. The first division wall of the zygote is generally parallel to the long axis of the archegonium (Fig. 13-32, A). This initial division separates an anterior cell or hemisphere that is directed toward the notch of the gametophyte and a posterior cell or hemisphere that is directed away from the notch. The former cell is in reality the apical pole or epibasal cell, and the latter is the basal pole or hypobasal cell. It can readily be

understood why the zygote in the Filicales is said to be "prone" in orientation. Each of the two cells divides and the new cell wall in each is perpendicular to the original or basal wall, resulting in the development of a four-celled embryo or "quadrant stage" of embryogeny (Fig. 13-32, B). Each of the four cells then undergoes (usually by synchronized divisions) more divisions (Fig. 13-32, C). According to the classical descriptions of fern embryogeny, the primary organs may be traced back to specific segments of the quadrant stage. The outer anterior quadrant or cell reportedly gives rise to the first leaf, the inner anterior quadrant to the future shoot apex. The primary root originates from the outer posterior quadrant, the foot from the inner posterior quadrant.

Some workers have questioned the preciseness of fern embryogeny (particularly of later stages) as presented in classical studies. In *Gymnogramme sulphurea* (Vladesco, 1935) the foot is reported to be derived from certain derivatives of the original inner anterior quadrant as well as from the inner posterior quadrant (Fig. 13-32, D). The setting aside of a shoot apical cell is delayed, and frequently it is a cell located very near the equatorial plane of the embryo and not a centrally located derivative of the inner anterior quadrant. According to another report (Ward, 1954) the inner two quadrants in *Phlebodium aureum* give rise to the foot, and the first leaf and root, appearing in that order, have their origin as described earlier. The origin of the future shoot apical meristem cannot be assigned to any definite quadrant, but the shoot apex arises from derivative cells of the inner and outer anterior quadrants midway between the organized foot and first leaf.

After whatever manner the organs are delimited, the first root and leaf begin to grow rapidly and pierce the calyptra (Fig. 13-33, B). The first leaf eventually grows forward and upward through the notch. After the first leaf has unfolded, more leaves are

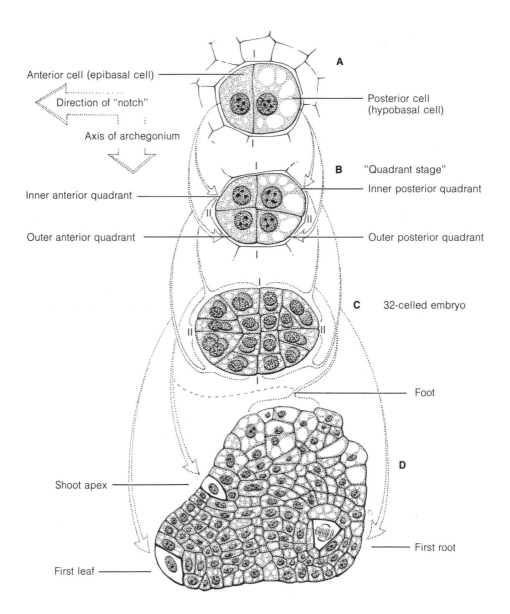

Anterior cell (epibasal cell)

Direction of "notch"

Axis of archegonium

A

Posterior cell
(hypobasal cell)

B "Quadrant stage"

Inner anterior quadrant

Inner posterior quadrant

Outer anterior quadrant

Outer posterior quadrant

C 32-celled embryo

Foot

D

Shoot apex

First root

First leaf

Figure 13-32
Developmental stages in the growth of a leptosporangiate fern embryo (*Gymnogramme sulphurea*).
[Redrawn from Vladesco, *Rev. Gen. Bot.* 47:513, 1935.]

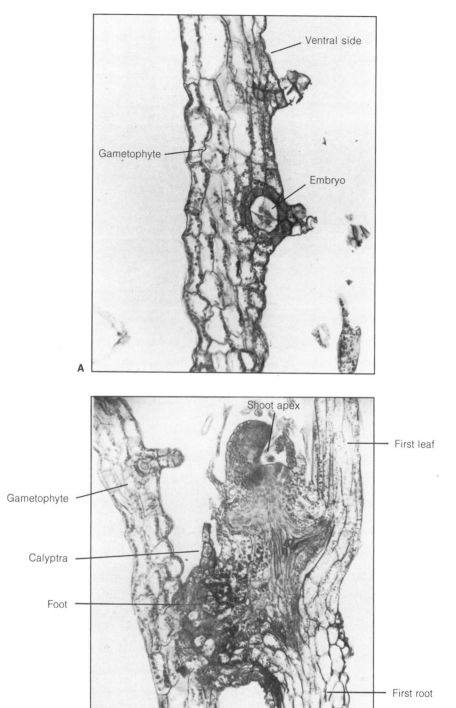

Figure 13-33
Vertical sections of gametophytes of *Adiantum* sp. showing attached eight-celled embryo (four cells visible) in **A**, and young developing sporophyte in **B**.

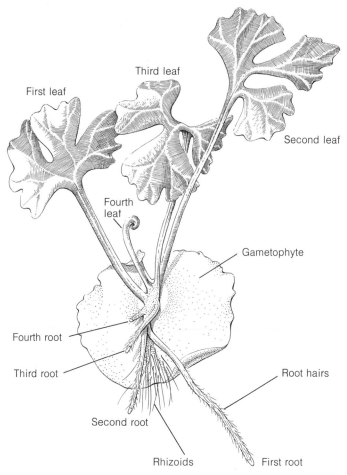

First leaf

Third leaf

Second leaf

Fourth leaf

Gametophyte

Fourth root

Third root

Root hairs

Second root

Rhizoids

First root

Figure 13-34
Ventral (lower) side of gametophyte of fern with attached young sporophyte.
Note circinate vernation of youngest (fourth) leaf.

formed by the shoot apical meristem, and roots are produced on the developing stem axis (Fig. 13-34). Sooner or later, the gametophyte degenerates and dies. Leaves of young sporophytes often do not resemble the leaves of adult plants.

We hope the reader has noted that the morphological features discussed thus far are those that Bower considered to be of utmost importance in fern phylogeny (see Chapter 11, p. 247). Furthermore, the student's attention is directed to Table 13-1 on p. 359, where he will find a comparison of presumed primitive and derived conditions

of certain fern characters. Attention should be given to Table 13-1 in reading the remainder of this chapter.

Special Problems in Fern Morphology

"PHYLETIC SLIDE" OF THE SORUS AND MODIFICATIONS OF THE INDUSIUM. The primitive position of sporangia is declared by most morphologists to be terminal if the leaf is essentially a branch system. Marginal sporangia of relatively narrow leaf segments or of larger laminae can also be thought of as occupying a primitive position. In the evolution of the fern leaf the processes of over-topping and webbing (see Chapter 3) would ultimately destroy the obvious terminal position of sporangia, but a study of the venation gives us a clue about their actual morphological position.

The location of sporangia on the lower or abaxial side of fronds is not solely a characteristic of living ferns because certain ferns in the Paleozoic, considered to be members of the Gleicheniaceae, had sori on the abaxial side of fronds. We can only assume that the transition has taken place several times and in different natural phylogenetic groups of ferns.

The question then is: How has this transition taken place and can we see it still in operation today? In the following discussion certain ferns are used illustratively in tracing the transition from the marginal position to the abaxial position. It must be kept in mind that not all of the examples are necessarily closely related. Their sori simply represent graphic examples of possible stages in the evolution of a single fern character—the position of the sorus.

The starting point for our story is a leaf with sporangia aggregated at the ends of veins. The sorus is clearly marginal, the indusium two-lipped and symmetrical. The receptacle is the continuation of the leaf margin, and the indusial flaps arise as adaxial and abaxial outgrowths. A vascular bundle ends at the base of the receptacle or traverses it if elongate in shape. Certain species of *Hymenophyllum* and *Trichomanes* (Figs. 13-7, 13-49) are illustrative of this organization. The sori in some species of *Davallia* and in *Dicksonia* appear to be marginal (and are in a general sense), but it can be demonstrated developmentally that the "adaxial flap" of the soral covering is actually a portion of the lamina; the "true indusium" arises abaxially in a submarginal position. The true indusium may be attached along its sides and base to the abaxial leaf surface (*Davallia*, Figs. 13-7, A; 13-51, A) or only at its base (*Dicksonia*, Figs. 13-8, B; 13-10, A). The submarginal (or intramarginal) and clearly abaxial positions of sori probably have had adaptive significance, and ontogenetic studies give us an insight into how the transition took place. In summary, the sorus gradually became shifted to the abaxial side, and during the countless years of evolution the original adaxial flap became transformed into a new leaf margin and was provided with vascular tissue (Fig. 13-35). Subsequently, the indusium could assume the various shapes (asymmetrical and symmetrical) described previously.

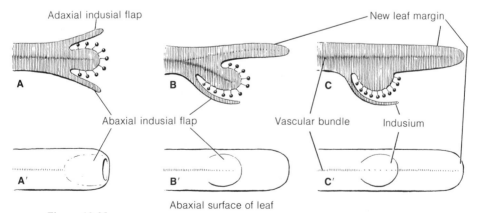

Figure 13-35
Possible evolutionary stages in the movement of the marginal sorus to the abaxial side of the fern leaf (the "phyletic slide"). **A–C**, vertical sections, hypothetical leaf lamina; **A′–C′**, abaxial surface views. (See text for details.)

It is further postulated that once the receptacle (therefore the sporangia also), assumed the abaxial position, the indusium disappeared so that today in certain genera the sporangia are protected only by the reflexed leaf margin. Associated with this transition was the fusion and loss of the individuality of sori. This transition is well represented by the genera *Pteridium* (Fig. 13-8, A) and *Pteris* (Fig. 13-12, C). The genera *Cheilanthes* and *Pellaea* may also be representative of this type of evolutionary development (Fig. 13-50).

Paleozoic ferns (such as those in the family Gleicheniaceae) had sori, without indusia, on their fronds (Figs. 13-7, E; 13-12, B; 13-52, B). Whether the processes of the "phyletic slide" were operative in the ancestry of the Gleicheniaceae is not clear. However, in the fern flora of today, a derivative family, the Polypodiaceae (Figs. 13-9, D; 13-11, B; 13-59), has sori that are without indusia. Likewise, in this general group of ferns (in, for example, *Platycerium* — the staghorn fern) there has been a breaking up of the soral organization in that sporangia may be scattered over the lower surface of the lamina.

EVOLUTIONARY AND ONTOGENETIC IMPLICATIONS OF STEM ANATOMY.   On the basis of comparative morphology of both living and extinct plants, the protostele, with little doubt, is the truly primitive type of vascular cylinder. This is true not only in ferns but also in other groups of vascular plants. Any departure from a simple protostele (haplostele) is commonly regarded as a derived condition.

The value of studies based on the comparative anatomy of mature stems should not be underestimated, but the student should be cognizant of the fact that most ferns, during their early ontogeny, have a vascular pattern quite different from that of the adult plant. For example, it has been shown by many investigators that the base of a fern stem (older portion) may be protostelic, whereas at a higher level of the same stem (younger portion) the vascular cylinder may be solenostelic or dictyostelic. Of course this change is ontogenetic, but it has been interpreted by some morphologists as representing an evolutionary recapitulation of stelar types. Rightfully, these same morphologists realize the importance of comparing the steles of plants of comparable age. Therefore the vascular cylinders of well-established adult plants can be compared, with some degree of assurance. Similarity of vascular cylinders, however, cannot be used in itself as an indication of close relationship between ferns. The attainment of similar levels of structural specialization may be reached in several evolutionary lines. We are dealing here with convergent evolution or homoplastic development (Bower, 1923, p. 159).

It is a well-known fact that the shoots of many plants, particularly herbaceous species, become obconical during growth, and, as a result, the stem at an upper level may be several times larger in diameter than the base. This phenomenon is particularly well displayed in ferns. Associated with this increase in size is generally a change in the form of the vascular cylinder. Utilizing Galileo's Principle of Similitude, Bower calculated that a mere increase in size or volume of a smooth cylinder of dead tracheids is not compensated for by a comparable increase in surface of the cylinder. The volume increases as the cube of the linear dimensions, and the surface increases as the square of the linear dimension, if the same shape is maintained. Bower believed that the proper relationship between dead conducting elements and living parenchyma must be maintained. This can be accomplished, then, in any one of several ways or by combinations of these methods: (1) formation of a central pith, (2) formation of flanges or irregular lobes on an otherwise smooth protostele, (3) formation of a highly intricate tubular vascular cylinder, and (4) presence of living cells among dead elements. In living, woody seed plants the necessary parenchyma cells are derived from the vascular cambium, constituting xylem parenchyma and wood rays. Whatever physiological advantages are achieved by these methods they appear to have been favorably acted upon through natural selection.

In recent years there has been a renewed interest in the problem of size and form by morphologists in the United States and Great Britain. In England, Wardlaw has been the leader of a group that has approached morphological problems from the experimental standpoint. Bower was impressed with the phylogenetic implications of stelar morphology, but he was aware also of the plasticity and variability of the vascular cylinder. Wardlaw (1949) has shown in *Dryopteris* that if the shoot apex is isolated from existing vascular tissue by vertical incisions, the vascular cylinder of the new "plug" of tissue will be solenostelic instead of dictyostelic for a short distance. Eventually the dictyostelic condition is achieved again. If the incisions are made very close to the apical cell, a prostostele is produced. Wardlaw interprets these results as an indication of the importance of nonhereditary factors in determining shoot organization. According to Wardlaw, the availability of nutrients (an extrinsic factor) perhaps has more of an influence on growth than do hereditary factors. Starvation experiments also support this general idea (Bower 1923, p. 179). An account of other experiments on ferns can be found in the following section.

EXPERIMENTAL MORPHOLOGY: SPORO-PHYTE. What are the methods or techniques open to the experimental morphologist? Changes in plants can be brought about by altering the culture medium, by applying chemicals directly to the intact surfaces that alter or provoke morphological changes, and through surgical manipulations. The techniques of tissue culture and the culturing of isolated organs should not be overlooked. The experimental morphologist, then, attempts to evoke changes in form and development of the normal plant grown under carefully controlled conditions. When changes are brought about he seeks to relate these changes to the conditions of the experiment, thereby hoping to arrive at a more complete understanding of the basic organization and development of the organism.

The ferns have proved to be particularly good experimental material. A complete discussion of experimental morphology is outside the scope of this book, but some of the interesting and stimulating research of Wardlaw, Cutter, and Wetmore on the shoot apical meristem of ferns will be described.

THE SHOOT APICAL MERISTEM. In review, the shoot apical meristem of a leptosporangiate fern is generally dominated by a definite apical cell situated at the tip of a conical protuberance or cone. In *Dryopteris* there are three different types of structures derived from the apical meristem: leaves, scales, and bud primordia. Leaves arise within the basiscopic margin of the apical cone; scales originate on the margin; buds have their origin on the broad subapical region (Fig. 13-36). It is a well-known fact that the shoot apex usually exerts physiological dominance over buds or bud primordia. That this is controlled by a substance (termed a growth hormone) diffusing basipetally has been demonstrated experimentally. For example, if the tip of a main shoot is removed, dormant buds or bud primordia will begin to grow. If, as in the ferns *Matteuccia* and *Onoclea,* the cut surface of the stem is smeared with indole-acetic acid in lanolin, the development of lateral buds does not take place (Wardlaw, 1946). The applied growth-regulating substance performs the same function as the natural growth hormone that is produced at the tip of the intact rhizome.

The shoot apical meristem of vascular plants has long been of interest to morphologists. It is in this region that new organs (e.g., leaves) and tissues are formed. We might ask if the apical meristem is a self-determining region or if its characteristics and activities are controlled by older, more mature tissues below it. Wardlaw (1949) demonstrated that if the conical apical meristem of *Dryopteris* is isolated from adjacent tissues by four deep longitudinal incisions, it is still capable of growing and giving rise to a leaf-bearing axis. The initial incisions isolated the apical meristem on a plug of parenchymatous tissue. The newly formed vascular tissue of the "isolated" shoot never became continuous with that below. Similar experiments have been performed with dicotyledons (Ball, 1952). It has been concluded that the shoot apex is a self-determining region, and, except for nutrients, is not dependent upon older tissues.

Additional evidence for the autonomous nature of the shoot apical meristem was provided by *in vitro* culture techniques. Wetmore (1950, 1954) was able to excise the shoot apical meristems of *Adiantum* and other lower vascular plants and grow normal plants from them on a mineral agar medium supplemented with 2 percent sucrose or dextrose. Nitrogen was supplied in the form of nitrate. Therefore, cells of the apex were capable of synthesizing complex nitrogenous substances from simple materials in the medium. The addition of auxin, yeast extract, or autoclaved coconut milk speeded up the rate of growth without changing the developmental pattern. Growing excised shoot apices (without attached leaf primordia) of angiosperms has been much more difficult, although they have

been cultured recently with success (Smith and Murashige, 1970).

The question was raised: Does the shoot apex (more specifically the apical cell) determine the position of leaves? In the test experiment great care was taken to puncture only the apical cell; existing primordia and loci of yet-uninitiated leaves were carefully avoided in order not to injure the tissues in any way. After the operation the existing primordia continued to grow, and new ones were formed in a normal pattern until all the space on the apical meristem was used. Apparently the apical cell does not determine leaf position, nor is it responsible for leaf initiation.

As a working hypothesis, Wardlaw assumes that the apical cell and all young leaf primordia constitute growth centers, each with its own physiological "field" (Fig. 13-36, A). This field restricts the initiation of new primordia to within a certain distance from its center. A new or presumptive leaf (I_1) arises in the first available space on the apical meristem—outside the fields of older adjacent leaves. Primordial meristems of lateral buds occupy positions some distance from the apical cell, in interfoliar positions. They, however, remain inhibited during normal growth.

To determine if leaf primordia actually influence each other in their development, a series of interesting experiments was performed. If a young primordium (e.g., L2) was isolated from adjacent primordia by deep longitudinal *radial* incisions, it grew very fast and soon became larger than older leaves (Fig. 13-36, B). If an incision was made in the position of a presumptive leaf (I_1), the site of origin of the next leaf in the phyllotactic pattern in that vicinity was shifted toward the incision. These experiments show that leaf primordia have inhibitory effects on each other.

In other experiments, if I_1 (or a L1 primordium and sometimes L2 or even L3) was isolated by a wide and deep longitudinal

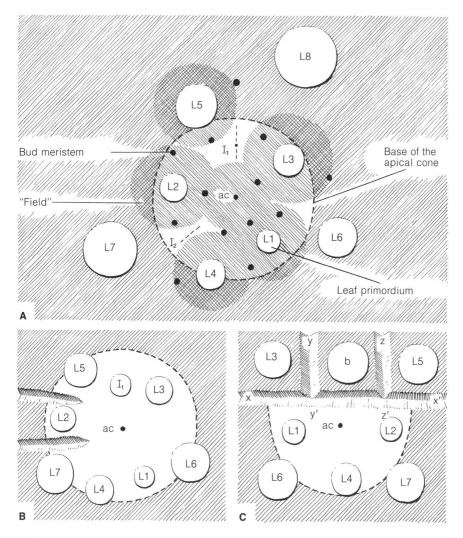

Figure 13-36
A, surface view of apical cone, shoot of *Dryopteris* (consult text for explanation);
B, isolation of young leaf primordium (L2) by longitudinal radial incisions; C, isolation
of I_1, by tangential incision (X–X′) and radial incisions (Y–Y′, Z–Z′), which resulted
in formation of a bud (b). ac, apical cell; b, bud; I_1, position of first presumptive
leaf; I_2, position of a second presumptive leaf. [Redrawn from *Phylogeny and
Morphogenesis* by C. W. Wardlaw. Macmillan, London, 1952.]

incision on the *adaxial* side of the primordium, a bud and not a leaf usually formed in
that position. The same effect was achieved
by making both radial and tangential incisions (Fig. 13-36, C). When, however,
shallow cuts were made, normal leaves
were formed. In the instances in which buds
developed from potential leaf primordia it
would appear that such primordia are undetermined in their early development.

According to Cutter (1954) the course of
development of a leaf primordium of *Dryopteris* can be changed unless its fate has
become determined by the appearance of
a large lenticular apical cell at the tip of the
leaf primordium (or by some other coordinated event). Also, all of these experiments
emphasize the importance of "prevascular
tissue" (a term used by Wardlaw) in early
development. *Deep* incisions would sever

the "prevascular" tissue but *shallow* incisions would not disrupt it. The controlling effects of the apical meristem are probably transmitted through prevascular tissue and a disruption of that tissue would affect the future development of primordia. (Additional information on these experiments and others can be found in the following selected publications: Wardlaw, 1952, 1965; Wardlaw and Cutter, 1954, 1956; Cutter, 1954, 1956, 1965).

In the surgical experiments just described for *Dryopteris,* leaf determination appears to be controlled initially by the shoot apex and then by the establishment of a distinctive lenticular apical cell at the tip of the young primordium. A different technique of investigating leaf determination was employed by T. A. Steeves and his associates. This technique involved growing excised primordia in sterile (aseptic) culture on nutrient agar. When excised primordia of *Osmunda cinnamomea* were cultured, they developed either as leaves or shoots depending on the age of the primordium. In these experiments the most recently initiated primordium (L1) developed into a shoot when excised and grown alone in culture. The tenth primordium (L10) always developed as a leaf. Between these extremes there was a range in response in which the probability of a primordium developing as a leaf was greater in older primordia (L7–L10). In contrast to *Dryopteris* the formation of a large apical cell at the tip of a primordium was delayed, in some cases as late as L6 or L7, although even these could develop into shoots. However, unlike *Dryopteris* the apical cells of shoots and leaves of *Osmunda* have the same form —they are both tetrahedral in shape. Steeves postulated that leaf determination in *Osmunda cinnamomea* is a gradual process and perhaps mediated by a leaf-forming substance.

Additional studies of excised primordia have shown that leaves influence each other during development. In these experiments excised leaves were cultured in pairs. When an L3 was paired with another L3, the excised primordia developed as leaves 35 percent of the time rather than as shoots. If L3 was paired with L10, the L3 primordium developed as a leaf 70 percent of the time. This effect was lost when L3 was paired with L14. These studies suggest that a diffusable leaf-determining substance is present in primordia of certain ages. However, more recent research supports the original tenet of Wardlaw and Cutter that the shoot apical meristem influences more directly the youngest primordium—but that older primordia can influence each other. (Additional information on culturing isolated primordia can be found in the following publications: Steeves, 1961, 1962; Kuehnert, 1967; Hicks and Steeves, 1969; White, 1971.)

FERTILITY OF THE SPOROPHYTE.   The responses of many angiosperms to daylength are fairly well known and have been documented rather thoroughly. For example, some angiosperms initiate flowers during short days (short light period and long dark period), others during long day conditions. Some are "day-neutral" in that flowering is dependent upon some factor other than daylength (e.g., vernalization). Much less is known about factors eliciting production of sporangia in ferns, but it is known that some ferns respond to daylength and others are day-neutral. Some examples of day-neutral ferns are certain species of *Pteris, Polypodium,* and *Ceratopteris* (Labouriau, 1958; Patterson and Freeman, 1963). In some other species fertility is independent of daylength but there is an annual periodicity in fertility.

Salvinia natans (heterosporous waterfern) is reported to be a short-day plant. *Salvinia rotundifolia* behaves as a short-day plant at temperatures above 20°C and as a long-day plant at 17°C. *Adiantum capillus-veneris* (maidenhair fern) becomes fertile on a long-day regime. It would appear that fern sporophytes are extremely variable in their response to environmental conditions as regards fertility.

EXPERIMENTAL MORPHOLOGY: CONTROL OF GAMETOPHYTIC GROWTH FORM. Spore germination and subsequent development of "typical" fern gametophytes have been described earlier (pp. 308–312), but the descriptive accounts of development did not include discussion of causal factors that might regulate growth.

During the last 10 years there has been a renewed interest in the fern gametophyte as an experimental object. The common type of fern gametophyte is autotrophic, relatively small, and can be grown in large numbers and manipulated with considerable ease. Light microscopy is sufficient for determining cell size, planes of cell division, etc.

Most ferns germinate to form a filament of cells as a result of transverse divisions (a new cell wall is perpendicular to the long axis of the filament). This is referred to as one-dimensional (1-D) growth. Sooner or later cells divide in another plane (longitudinal division) and a plate of cells (Fig. 13-37, A) is formed (two-dimensional, or 2-D, growth). What are the causal factors responsible for this change in growth pattern?

The transition from 1-D growth to 2-D growth in normal gametophyte development is in response to light—a photomorphogenic response. Light quality (wavelength) is the important factor, but light quantity also plays a role. Blue light (wavelengths less than 500 Å) is necessary for 2-D growth, and the quantity of it must exceed a certain amount. Under longer wavelengths, including red light, a gametophyte generally remains filamentous (Fig. 13-37, B) although there are exceptions (see p. 327). This all suggests that a photoreceptor-pigment is involved in the process. Most of the research on this problem has been done

in Germany by H. Mohr and his associates. Ohlenroth and Mohr (1964) suggested a working model in which blue light is absorbed by a photoreceptor-pigment, resulting in the production of a reaction product that enters the nucleus and evokes a gene response. A messenger RNA with new information is produced and then specific proteins (enzymes) concerned with 2-D growth are formed in the cytoplasm.

Other researchers have produced results that, according to them, support the concept that proteins and RNA have a specific effect on the transition to 2-D growth. Most of these investigations have utilized specific inhibitors of nucleic acid and protein synthesis. Some of these experiments have been criticized in that full growth data were not supplied (See Miller, 1968, for discussion). In one study, however, in which a wide range of nucleic acid and protein-synthesis inhibitors were used, their effect was to retard growth, but not to block the initiation of 2-D growth or antheridial differentiation (Schraudolf, 1967).

That RNA and proteins are involved in some way in the change from 1-D to 2-D growth nevertheless seems rather certain, although they may be produced as a consequence, rather than being the cause of 2-D growth. The results of experiments in which precursors of RNA were used have shown that there is a distinct increase in RNA 3–6 hours after plants previously cultured under red light were moved to blue light (Drumm and Mohr, 1967).

Another explanation for the change from 1-D to 2-D growth, which entails rates of cell division and cell elongation, has been suggested (Sobota and Partanen, 1966, 1967; Sobota, 1970). This hypothesis assumes that no single mechanism alone is responsible for the change. Sobota and Partanen cite Errera's law which states that a cell so divides in a plane that the new cell wall has minimal surface area. Exceptions do exist to the rule, but in the fern gametophyte the

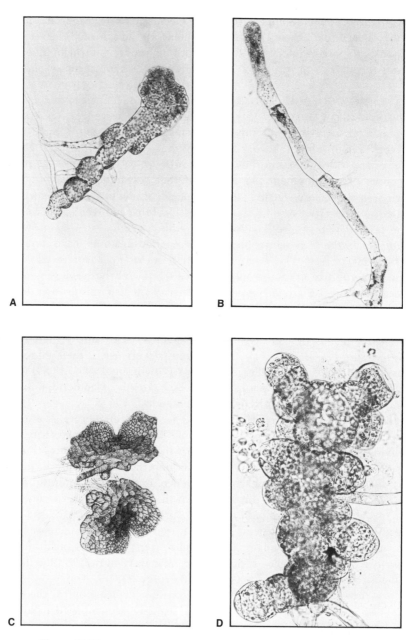

Figure 13-37
A, B, young gametophytes of *Pteridium aquilinum* (of same age) grown on
nutrient agar and illuminated with white light (**A**) and red light (**B**).
C,D, 14-day-old cultured gametophytes of *Onoclea sensibilis:* **C,** control
plants grown on nutrient agar without antheridial factor; **D,** gametophyte
grown from time of spore germination on medium containing antheridial
factor (*Pteridium*); note large number of antheridia, and the release of sperms
in one. (**C** and **D** are not at same magnification.) [Courtesy of
Dr. E. G. Cutter.]

plane of division would depend upon the length of the cell. If a cell of the germination filament were quite long when it divided, the new cell wall would be formed transverse to the long axis of the filament. If there were a division before the cell had elongated very much, the new cell wall could be either transverse or longitudinal and hence lead to 2-D growth.

Therefore, those external features which favor and promote a high sustained rate of cell division would influence the attainment and continuance of 2-D growth. Blue light favors 2-D growth but, as Sobota and Partanen (1966) have shown, 2-D growth can be maintained under red light if the intensity is high enough, contrary to the reports of others.

SEXUALITY AND GENETICS. The gametophytes of homosporous ferns at maturity are commonly bisexual (hermaphroditic) except in some obligate apogamous species. Spores grown on agar or on another substrate in order to provide gametophytes for teaching purposes often yield antheridial or archegonial plants. Antheridia are often found on archegonial plants. Antheridia are often found on gametophytes of any size, particularly on small ones.

Research in the last few years has shown that antheridial initiation is under the hormonal control of substances termed *antheridogens*. The following sequence of events that take place in a mass culture of gametophytes is based on a review of the subject by Voeller (1971).

In a population of gametophytes, some grow more rapidly than others and become two-dimensional (thalloid). These gametophytes produce a hormone (antheridogen) which diffuses into the substrate and which can induce the slower growing gametophytes to form many antheridia. The gametophytes producing the hormone form

archegonia, but they also apparently produce an inhibiting substance that blocks their capacity to form antheridia in response to the antheridogen they are producing. This mechanism promotes cross or intergametophytic fertilization.

If intergametophytic fertilization fails, antheridia form eventually on the archegoniate plants and self fertilization, or intragametophytic fertilization, can occur. It is thought that the basal parts of the older gametophyte are sufficiently distant from the apical meristem (where the inhibitor is produced) to escape the influence of the inhibitor. Also, the initially antheridial plants can become bisexual or hermaphroditic unless they are surrounded by high concentrations of antheridogen; antheridia are found near the base of the gametophyte and archegonia near the notch in a fashion usually illustrated for thalloid gametophytes (Fig. 13-31, A).

Antheridogen produced by one species of ferns is effective on others (Fig. 13-37, C, D). For example, the antheridogen produced by *Pteridium aquilinum* (bracken fern) is effective on certain species of twenty genera in four families. Members of certain groups, however, apparently have their own specific antheridogen—sometimes called antheridogen-B. One such group is the family Schizaeaceae. Gibberellic acid also can substitute for antheridogen-B (Fig. 13-38, A, B).

The general mechanism outlined above favors intergametophytic fertilization and hence heterozygosity. Self fertilization results in homozygosity with all of the bad features inherent in such a system. Klekowski (1971) has shown that single spore cultures (from same parent sporophyte) may reveal lethal genes in that sporophytes fail to form. However, placing portions of two "sporophyteless" hermaphroditic gametophytes of different origin together in culture results in the formation of one or more sporophytes. Klekowski has determined the

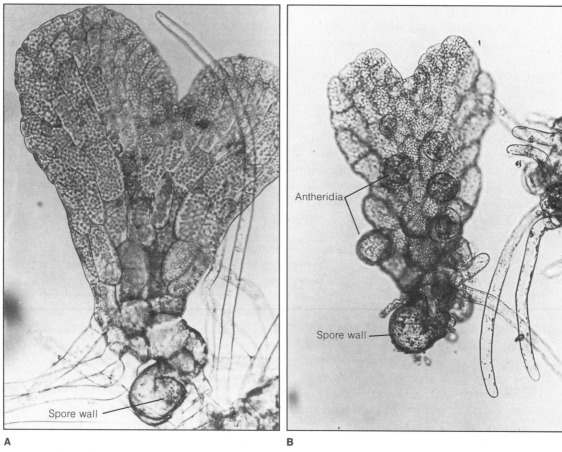

Figure 13-38
A,B, gametophytes of *Lygodium japonicum* fourteen days after spore germination; **A,** gametophyte grown on simple nutrient agar under white light without gibberellic acid added; **B,** grown as in **A,** but with gibberellic acid added to the culture medium. [Courtesy of Dr. E. G. Cutter.]

mating and breeding systems of many ferns and the interested student is referred to the following articles (Klekowski and Baker, 1966; Klekowski, 1969, 1971).

Homosporous ferns as well as other homosporous lower vascular plants have high chromosome numbers (96 percent have a basic chromosome number higher than 27). Klekowski and Baker (1966) think that in homosporous ferns this is important in creating and maintaining genetic variation in the face of the homozygotizing effects of possible habitual self fertilization in hermaphroditic gametophytes. To quote them, "Polyploidy increases gene redundancy and, if dominant alleles are present, any recessive alleles of the same genes elsewhere in the genotype are kept from immediate exposure to selection even when the plants are homozygous at each locus. Because polyploidy increases the dosage of genes beyond the disomic condition, genes can mutate and take on new functions without depriving the organism of essential processes which are maintained by their former homologs."

EVOLUTION OF THE GAMETOPHYTE. We are certain that there can be little doubt in the student's mind that the fern sporophyte has received more attention from botanists than has the gametophyte. Through the continued efforts of Alma G. Stokey and her associates, and more recently by B. K. Nayar, a considerable volume of information has been built up relative to the importance of the gametophyte generation in fern classification and phylogeny. The following account is based largely on their publications (Stokey, 1951; Atkinson and Stokey, 1964; Nayar and Kaur, 1971). Just as in the study of the sporophyte, certain structures and processes in gametophytic development can be used as bases for comparisons. We should consider: (1) type of spore germination, (2) early development and mature form of the gametophyte, (3) presence or absence of trichomes, and their morphology, (4) color, form, and distribution of rhizoids, and (5) morphology and position of sex organs. A primitive gametophyte of members of the Filicales is one that is slow to develop, perhaps requiring from six-to-eight weeks, to several months, or even a year or more to mature. Germination results in the formation of a plate of cells or a cylindrical mass or, rarely, a short filament which soon passes to a massive plate-like thallus. At maturity, primitive gametophytes may be subterranean, nonchlorophyllous, and tuberous-like (e.g., in *Actinostachys*) or green with a massive midrib and thick wings (e.g., in *Osmunda*). The antheridium is large with a layer of several jacket cells (from six to eight) enclosing several hundred sperms. Archegonia have relatively long, approximately straight necks, often with a wall between neck-canal nuclei.

There is an evolutionary trend toward shortening the time between germination of the spore and fertilization. There is a tendency to form a filamentous stage early in development. Frequently the filamentous form may be maintained (depending on environmental conditions) in older gametophytes, as is characteristic of certain species of *Schizaea* and *Trichomanes*. In other species the filamentous stage may be brief and the mature gametophyte may be cordate without a conspicuous midrib or it may be ribbon-like without a midrib. The size of the antheridium is reduced. The number of jacket cells is reduced to three, accompanied by a great reduction in sperm output. The archegonial neck is shorter and conspicuously curved away from the apical notch. (Refer to Fig. 13-30 for representative types of gametophytes.)

In summary, the gametophyte will undoubtedly play an ever increasing role in helping to answer questions about affinities. A summary of gametophytic features is presented in Table 13-1 at the end of this section.

EXPERIMENTAL EMBRYOLOGY. As the student is well aware by now, each phase of the fern life cycle — gametophytic and sporophytic — begins with a single-cell spore and zygote. Commonly a spore has a reduced number of chromosomes (haploid, n) and a zygote has a double set of chromosomes (diploid, $2n$) as a result of fertilization. It might be thought that the profound differences between gametophyte and sporophyte are related to chromosome numbers. However, the many well-documented cases of apogamy and apospory (see p. 332) have shown that the difference in chromosome numbers cannot fully account for the differences in morphology of the two generations. For example, in apogamously reproducing ferns both generations have the same chromosome number.

Many years ago the eminent British botanist, W. H. Lang postulated that physical and nutritional conditions must influence the early development of the spore and fertilized egg. He suggested that the restraints exerted upon the "encapsulated" zygote by the archegonium and surrounding gametophytic cells influence its characteristic growth; the spore, on the other hand, germinates in a "free" environment.

Since Lang's time some experiments have been undertaken in an effort to test his ideas. Incisions in the gametophyte around a fertilized archegonium of *Phlebodium* (*Polypodium*) *aureum* did effect a partial "release" of the young embryo. The course of embryogenesis was modified and an unusual sequence of stages occurred, but ultimately a normal sporophyte was produced (Ward and Wetmore, 1954).

In another set of experiments in which pure cultures of the fern *Thelypteris palustris* were used, departures were noted from the normal embryogenesis as a result of various surgical and isolation techniques (Jayasekera and Bell, 1959). By delicate and precise surgical techniques an archegonium and a small pad of gametophytic tissue were removed 3 or 5 days after fertilization and transferred to a mineral-agar culture medium. The course of embryo development was not altered materially except that there was a delay in the emergence of the embryo and in the appearance of the first root. In other experiments unfertilized archegonia were removed, inseminated in a suspension of sperms, and then transferred to the agar medium. If the small fragment of tissue contained the apical notch region of an actively growing gametophyte, embryogeny proceeded normally but at a slower pace (Fig. 13-39, A, B). If the fragment did not include an active apical notch region, development was delayed, especially the development of the root (Fig. 13-39, C, D). If indoleacetic acid was applied to a fragment that did not contain the apical notch region, the root was the first organ to emerge from the calyptra. Normally it is an embryonic leaf that emerges first. Removal of the archegonial neck of a fertilized archegonium induced aberrations in that the embryo emerged from the upper (dorsal) side of

the gametophyte and initially consisted of only leaves and shoot apices (Fig. 13-39, E). Roots appeared after 50 days. The investigators concluded that a supply of auxin, known to be produced by cells in the notch region (Albaum, 1938), may determine the plane of the first division and promote embryonic differentiation in general, especially of the root. The archegonial neck may place a restraint on the embryo and if it is removed, the orientation and future development of the embryo may be disturbed.

That the surrounding somatic tissue produces growth regulators which influence embryo development, especially of the root, was shown rather clearly by Rivières (1959). An isolated embryo of *Pteris longifolia* contained in an archegonium developed into a sporophyte, but the root did not develop.

DeMaggio (1963) has pursued the idea that physical restraint, environment, and nutrition are important in the embryogeny of ferns. He isolated twenty-day-old embryos of *Todea barbara* and grew them on a simple nutrient medium. Growth was delayed but the embryos developed normally. Zygotes isolated prior to the first division lost their globular form, assuming a two-dimensional form much like a gametophyte. Generally, growth ceased after 30 days. He concluded that restraint is important and that there is an essential nutritional interaction between the gametophyte and the embryo during the early stages of development.

It might be concluded that Lang's ideas about the importance of physical restraint and nutrition have been borne out by the experiments just described. Bell (1970) subscribes to the effects of physical restraint on the embryos of ferns, but, however, considers the egg (before fertilization) to be a rather specialized cell and quite different from a fern spore. On the basis of electron microscopy and autoradiography (i.e., use of radioactive substances), Bell has described what he considers to be significant features

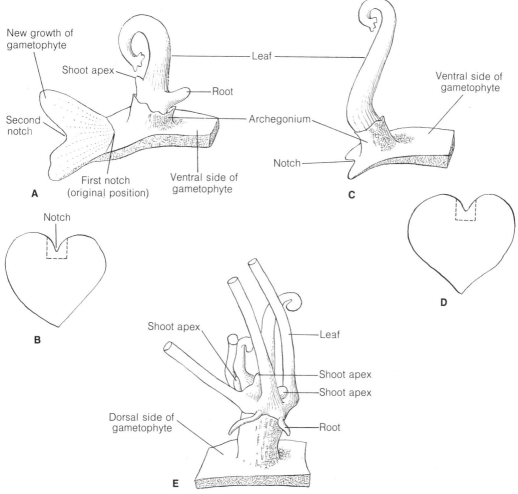

Figure 13-39
Experimental embryology in *Thelypteris palustris*. **A,** embryo 22 days old, the archegonium having been excised from the gametophyte prior to fertilization, but with the apical notch of the gametophyte attached and active; **B,** diagram showing portion of gametophyte excised. **C,** embryo 26 days old, the archegonium having been excised prior to fertilization, the apical notch of gametophyte attached but inactive; **D,** diagram showing portion of gametophyte excised. **E,** cylindrical body developing from zygote, 70 days after removal of archegonial neck on fifth day after fertilization. See text for details. [Redrawn from Jayasekera and Bell, *Planta* 54:1–14, 1959.]

of oögenesis (egg development) in *Pteridium* and *Dryopteris*. He has described a general but not complete degeneration of organelles shortly after the division of the central cell in archegonial development. The egg-cell nucleus becomes active in that invaginations of the nucleus penetrate into the cytoplasm. It is suggested that new organelles are formed from the invaginations and micronucleoli move into the invaginations where

strands of DNA may already be present. Just before fertilization the mature egg cell —in contrast to a spore—has many organelles, ribosomes, and is rich in RNA and basic proteins.

In summary, the "gametophytic template" in the egg is eliminated and a new "sporophytic scene" is established by gene activation even before fertilization occurs. However, this type of development does

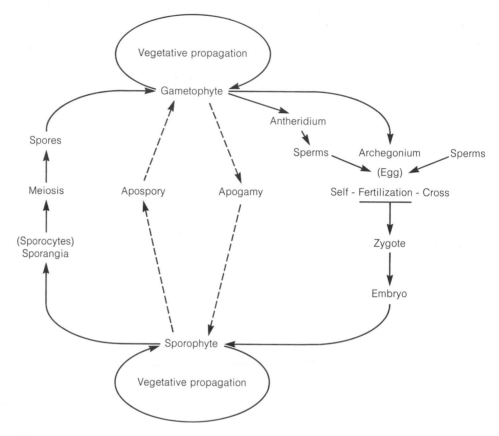

Figure 13-40
"Typical" generalized life cycle of a homosporous fern and possible deviations (apospory and apogamy) from the complete "sexual cycle."

not occur in the heterosporous fern *Marsilea* (Bell, 1970) or in the liverwort *Sphaerocarpus* (Diers, 1965, 1966).

Additional studies are needed before any final conclusions can be made about prefertilization stages and embryogeny in ferns and other lower archegoniate plants.

APOSPORY AND APOGAMY. In a "typical" sexual life cycle of an homosporous fern, the sporophyte forms sporangia that produce spores with a reduced chromosome number as a result of meiosis. These spores germinate, forming usually bisexual gametophytes. Antheridia and archegonia are formed on the gametophytes and self or cross fertilization occurs (Fig. 13-40). The embryo or new sporophyte is thus estab-

lished. As sporophyte and gametophyte often have a natural means of vegetative propagation or can be propagated experimentally, theoretically each generation could be perpetuated indefinitely.

There are some interesting and striking deviations from the "normal" life cycle in ferns, some occurring naturally, others being induced experimentally in the laboratory. One of the processes in the alteration of the life cycle is termed *apospory*. In the process of apospory a gametophyte is formed from a vegetative cell or cells of a sporophyte. Apospory has been known for many years and can result from the proliferation of leaf tissue in the formation of a gametophyte, taking place either naturally or being induced experimentally (Fig. 13-41, A, B). The gametophytes formed in this

Figure 13-41
Production of gametophytes directly from "juvenile" leaves of *Pteridium aquilinum.*
Leaves were placed on an agar medium that contained simple inorganic constituents.
This is an example of apospory.

Meiotic

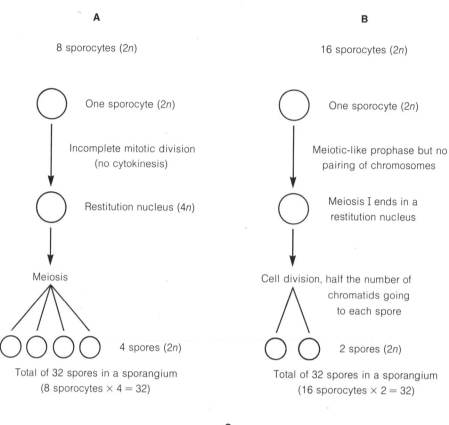

A

8 sporocytes (2n)

One sporocyte (2n)

Incomplete mitotic division
(no cytokinesis)

Restitution nucleus (4n)

Meiosis

4 spores (2n)

Total of 32 spores in a sporangium
(8 sporocytes × 4 = 32)

B

16 sporocytes (2n)

One sporocyte (2n)

Meiotic-like prophase but no
pairing of chromosomes

Meiosis I ends in a
restitution nucleus

Cell division, half the number of
chromatids going
to each spore

2 spores (2n)

Total of 32 spores in a sporangium
(16 sporocytes × 2 = 32)

C

Ameiotic

16 sporocytes (2n)

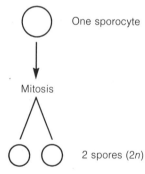

One sporocyte

Mitosis

2 spores (2n)

Total of 32 spores in a sporangium
(16 sporocytes × 2 = 32)

Figure 13-42
Irregularities in the process of sporogenesis in fern sporangia. Consult text for explanation.

manner have the same chromosome number as the sporophyte. For *Adiantum pedatum*, Morel (1963) has shown that juvenile leaves of young sporophytes will proliferate into gametophytes in culture, but that this capacity is lost in older leaves. Experimentally, gametophytes have been induced to form from the cut surface of fern petioles (Ward, 1963). Gametophytes as well as sporophytes will develop from the cut surface of a rhizome of *Phlebodium aureum* (Ward, 1963).

Apogamy, the development of a sporophyte from somatic tissue of a gametophyte, is a feature of even more significance than apospory in the life cycles of some ferns. In species in which apogamy occurs, antheridia and archegonia are commonly present, but the archegonia are nonfunctional. A cell or a group of cells of the gametophyte just behind the apical notch (sinus) undergoes cell divisions and from this mass of cells a new morphological sporophyte is formed. Since only mitotic divisions are involved, both gametophyte and sporophyte have the same chromosome number.

If spores are produced regularly by the sporophyte of a fern exhibiting apogamy on a regular basis (*obligate apogamy*) some type of compensating mechanism must be involved because both generations have the same chromosome number. Cytological studies of sporogenesis have revealed the answers. Three types of irregularities in spore production have been described. In two types meiosis is implicated but it is irregular. In the third, spores are produced without any evidence of meiosis taking place. In this type all cell divisions in the life cycle are mitotic.

MEIOTIC TYPES.   In some ferns there are eight sporocytes in a sporangium. The first division in each sporocyte is mitotic, but cytokinesis does not occur and the restitution nucleus has a double set of chromosomes. Meiosis then occurs in the restitution nuclei, resulting in the production of 32 spores with the same chromosome number as the original sporocytes (Fig. 13-42, A; Manton, 1950).

In a second type meiosis is impaired in that no pairing occurs in the 16 sporocytes and meiosis I ends in a restitution nucleus which then divides, half the number of chromatids going to each of two spores. A total of 32 spores is produced, each with the same number of chromosomes as in the original sporocytes (Fig. 13-42, B; Braithwaite, 1964).

AMEIOTIC TYPE.   Mitosis occurs in 16 sporocytes, which gives rise to 32 spores each with an unreduced chromosome number. (Fig. 13-42, C; Evans, 1964.)

A spore with an unreduced number of chromosomes, formed by one of the means outlined above, germinates and gives rise to a gametophyte. Then a sporophyte is formed apogamously and the cycle is complete without any change taking place in chromosome number of the two generations.

Apogamy can be induced experimentally in some ferns by altering the cultural conditions. For example, apogamy can be induced in *Pteridium aquilinum* (bracken fern) by the inclusion of glucose, sucrose, fructose, or maltose in the culture medium (Whittier and Steeves, 1960). Studies have indicated that induced apogamy is greater on a medium containing 5 percent sucrose than on the concentration of 2.5 percent glucose (Whittier, 1964a). Both naphthaleneacetic acid and gibberellic acid increased apogamy when supplied during the initiative phase of apogamous sporophytic development (Whittier, 1966). The apogamous response also was enhanced by high concentrations of phosphorous. The omission of any mineral elements from the medium inhibited apogamy (Treanor and Whittier, 1968) in contrast to reports that low mineral levels induce apogamy.

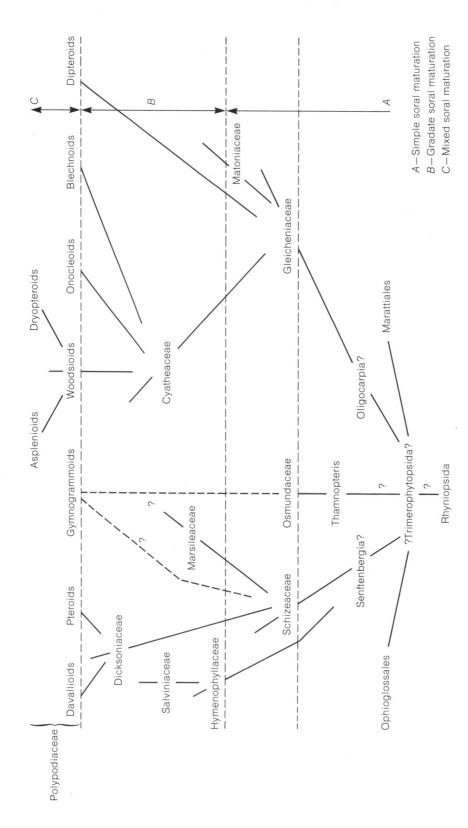

Figure 13-43

Possible phylogenetic interrelationships within the ferns. Consult text for discussion. [Based on a table from *Primitive Land Plants* by F. O. Bower. Macmillan, London, 1935. Extinct plants—Trimerophytopsida, *Senftenbergia, Thamnopteris*, etc.—added.]

As mentioned earlier some species of ferns are obligately apogamous and even in these sucrose hastened the onset of differentiation of apogamous sporophytes (Whittier, 1964b, 1965).

What is the significance of obligate apogamy? Obligate apogamy and some compensatory mechanism in spore production go hand in hand. The latter insures the production of spores for dissemination and eventual establishment of additional sporophytes. In the past it was postulated that obligate apogamy might be an adaptation to dry habitats. Water in the soil might be sufficient for growth, but insufficient for fertilization because sperms require free water in which to swim to the archegonia. However, the results of certain experiments on two obligate apogamous ferns, *Cheilanthes tomentosa* and *Cheilanthes alabamensis,* did not support the hypothesis (Whittier, 1965).

Some ferns have deviated from the "normal" life cycle to such an extent that they grow and reproduce only as gametophytes. The gametophytes reproduce vegetatively by multicellular propagules or gemmae. These gametophytes have been found in North Carolina and the Appalachian Mountains of the United States. The sporophyte has essentially been eliminated from the life cycle. In some instances small, dwarf sporophytes have been found in the vicinity of the gametophytes but they appear to be incapable of forming spores (Wagner and Sharp, 1963; Farrar, 1967; Farrar and Wagner, 1968).

For additional information on the value of ferns in an understanding of the alternation of generations see the review article by Whittier (1971).

Systematics and Phylogeny of the Filicales

It is entirely beyond the scope of this book to give extensive descriptions of fern families. We do hope, however, to give the student some of the criteria that are used in establishing certain families and arriving at reasonable phylogenies.

The classification scheme of F. O. Bower has been adopted by many botanists. Bower's concepts of fern classification and phylogeny were emphasized in the first edition of this textbook and are shown in modified form in Fig. 13-43. Certain basic families, recognized by Bower, are also accepted by pteridologists today, e.g., Osmundaceae, Schizaeaceae, Hymenophyllaceae, Gleicheniaceae, Matoniaceae. The Polypodiaceae, as constituted by Bower, is a large family comprising subgroups (e.g., davallioids, pteroids) that represent end points in descent from forms that had basically marginal sori. Bower was quite aware that the Polypodiaceae was polyphyletic. To qualify as a member of the Polypodiaceae a fern had to have certain morphological features: a highly specialized sporangium consisting of a long delicate stalk with transverse dehiscence and producing a small number of spores, mixed sorus, and a small specialized antheridium, generally consisting of three jacket cells and generally producing relatively few sperms. Genera combining these features automatically became members of the large family Polypodiaceae. The family is large as evidenced by the fact that $\frac{7}{8}$ of the fern genera are included in this highly artificial family.

Another general line of descent constituted Bower's "superficial" series in which sori, and especially the indusium, were said to have had a different origin than the margin of the leaf. The primitive family in this line is the Gleicheniaceae (devoid of an indusium) from which other families with cup-shaped indusia (e.g., Cyatheaceae), unilateral indusia (e.g., asplenioids), reniform and peltate indusia (e.g., dryopteroids) were derived. Some polypodiaceous ferns never developed indusia (e.g., dipteroids) in their line of descent.

In summary, the genera in the Polypodiaceae (in the sense of Bower) attained the same *evolutionary level,* although they were derived from different natural phylogenetic lines.

In more recent reviews and monographic treatments of the leptosporangiate ferns there is a tendency to establish more natural families out of the genera of the Polypodiaceae (*sensu* Bower). If this procedure is followed, the Polypodiaceae shrinks in number of genera. The davallioids become members of the Davalliaceae; the blechnoids become the Blechnaceae, and so on.

An example of a system of classification that reflects the recognition of several natural families within the originally large Polypodiaceae is that of Copeland (1947). Copeland attempted to make his taxonomic system phyletic with all members of common ancestry being included within a single family. He admitted that this approach often tends to make families large and difficult or impossible to define, although on the generic level it is possible to define limits. In his system, for example, the pteroids and gymnogrammoids of Bower's Polypodiaceae, as well as the Dicksoniaceae, become members of one family, the Pteridaceae. The dryopteroids, woodsioids, and onocleoids (of Bower's system) become united into a very large family, the Aspidiaceae. Copeland's system of classification is adopted in this textbook with full recognition of its shortcomings. The systems of Bower and Copeland are compared in Table 13-1.

Other classifications have been proposed since 1947. A notable feature of a system proposed by Holttum (1947, 1949) is the establishment of a very large central family, the Dennstaedtiaceae (with eleven subfamilies). This family includes among others, Bower's onocleoids, dryopteroids, asplenioids, and davallioids. The family is so large that taxonomic limits are difficult to define and there are several evolutionary lines of development within the family. Other systems have been proposed. In one instance 18 orders and 38 families are recognized (Pichi-Sermolli, 1958). In another, chromosome numbers (especially basic numbers) become the overriding diagnostic feature (Abraham, Ninan and Mathew, 1962).

In an even more recent system Wagner (1969) divides the leptosporangiate ferns into four major groups: (1) the phylogenetically isolated Osmundaceae, (2) the gymnogrammoid ferns, (3) the polypodioid ferns, and (4) the dennstaedtioid-aspidioid ferns (indusiate ferns). The gymnogrammoid ferns are composed largely of the Adiantaceae

Table 13-1
Fern classifications. A comparison of F. O. Bower's and E. B. Copeland's systems. See text for explanation

Bower	Copeland
FILICALES	FILICALES
OSMUNDACEAE	OSMUNDACEAE
SCHIZAEACEAE	SCHIZAEACEAE
HYMENOPHYLLACEAE	HYMENOPHYLLACEAE*
GLEICHENIACEAE	GLEICHENIACEAE
MATONIACEAE	MATONIACEAE
DICKSONIACEAE	
POLYPODIACEAE	
pteroids	PTERIDACEAE
gymnogrammoids	
davallioids	DAVALLIACEAE
asplenioids	ASPLENIACEAE
dryopteroids	
woodsioids	ASPIDIACEAE
onocleoids	
blechnoids	BLECHNACEAE
dipteroids	POLYPODIACEAE
CYATHEACEAE	CYATHEACEAE
MARSILEACEAE	MARSILEALES†
SALVINIACEAE	SALVINIALES†

*Copeland recognized 33 genera; others have limited the family to two genera.
†Recognized as families in the Filicales by Copeland, but treated as orders in this textbook.

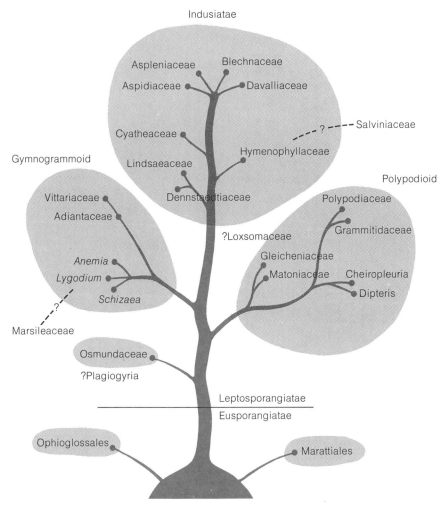

Figure 13-44
A phylogenetic scheme of the ferns. Consult text for details. [Based on Wagner, 1969; redrawn from Lloyd, *Univ. Calif. Publ. Bot.* 61:2, 1971.]

(a large part of Copeland's Pteridaceae, see p. 338), the members of which lack true indusia. In the gymnogrammoids, however, the modified leaf margin may serve functionally as an indusium. The polypodioid ferns comprise several families in which sori, devoid of indusia, are on the abaxial side of the leaves. The dennstaedtioid-aspidioid ferns, the largest and most advanced group of ferns, are characterized by the possession of true indusia (Fig. 13-44). The least specialized members of the latter group would be the Dennstaedtiaceae (certain genera in

Copeland's Pteridaceae). The primitive type of indusium is the cup-shaped structure found in the Dennstaedtiaceae. From this have evolved the various asymmetrical and modified indusia found, for example, in the Aspidiaceae, Davalliaceae, Aspleniaceae, and Blechnaceae (see pp. 347–354).

At first glance the three systems (Copeland's, Holttum's, Wagner's) appear to be quite different, but all three give recognition to the four general lines outlined by Wagner, even though the designation and size of families differ. Since a scheme of fern

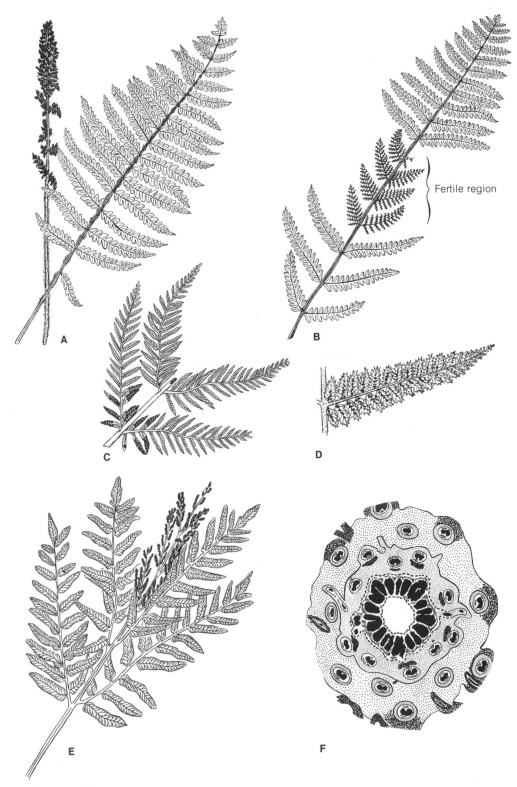

Figure 13-45
Osmundaceae. **A,** *Osmunda cinnamomea,* sterile and fertile fronds; **B,** *Osmunda claytoniana;* **C,** *Todea barbara;* **D,** *Leptopteris hymenophylloides;* **E,** *Osmunda regalis;* **F,** *Osmunda cinnamomea,* transverse section of rhizome. See text for details. [Redrawn from Hewitson, *Ann. Missouri Bot. Gard.* 49:57–93, 1962.]

Figure 13-46
Scanning electron micrograph of a cluster of mature sporangia of *Osmunda regalis;* dehiscence is longitudinal (× 40). [Courtesy of Dr. R. H. Falk.]

classification, generally acceptable to all fern specialists, has not been devised, the teacher of plant morphology must decide what best serves his purposes and most closely relates to his instructional goals. With the recently renewed interest in ferns, a system of classification, acceptable to most specialists, may become a reality within the next 10–15 years.

OSMUNDACEAE. The Osmundaceae, consisting of three living genera, *Osmunda, Todea,* and *Leptopteris*, has fossil relatives (*Thamnopteris*) in the rocks of the Permian. Other relatives (*Osmundacaulis*, formerly called *Osmundites*) were numerous in the Mesozoic. For more than 100 million years the family has displayed a remarkable constancy in characters. *Osmunda* itself is a genus that has probably existed for 70 million years and the family was probably more abundant in the past and more widely distributed than at present. Extant species are terrestrial, stand erect, and have simply uniseriate hairs. In *Osmunda cinnamomea* (cinnamon fern) there are two types of fronds—fertile and sterile (Fig. 13-45, A). In other species the frond consists of sterile and fertile regions (Fig. 13-45, B, C, E). In *Osmunda* large sporangia are attached along the margins of narrow leaf segments, but are not organized into definite sori. In *Todea* the sporangia occur on the abaxial surface of fertile regions of the lamina (Fig. 13-45, C). In *Osmunda* the origin of a sporangium is not always easily referable to a single superficial initial, and the mature sporangium has many eusporangiate features (Chapter 4). The sporangium is large with a lateral indusium, and at maturity the sporangium dehisces longitudinally (Fig. 13-46), liberating a large number of spores (up to 512). A basic chromosome number of $n = 22$ is uniform in the family. A transverse sec-

Figure 13-47
Schizaeaceae. *Anemia phyllitidis:* One leaf, showing
fertile and sterile portions.

formation of leaf gaps). The gametophyte
is a large, green thallus, often cordate, with
a conspicuous midrib. Long-lived gametophytes may reach 4–7 cm in length. Members of the Osmundaceae combine more
eusporangiate and leptosporangiate characteristics in their morphology than do members of any other family in the Filicales.
This is reflected in the establishment by
some pteridologists (e.g., Pichi-Sermolli,
1958) of a separate order, Osmundales, for
the group. Many morphologists consider
that no other group of ferns has been derived from the Osmundaceae, although
Bower considered that certain members of
the so-called gymnogrammoid ferns could
have been derived from the Osmundaceae.
Holttum (1949) was of the same opinion as
Bower. It is understandable that pteridologists would be divided in their views regarding the family. (Additional recent information
on the family, especially the fossil types,
can be found in the following references:
Hewitson, 1962; Miller, 1967; Tidwell and
Rushforth, 1970.)

SCHIZAEACEAE. This family is considered
to be an ancient one, and one from which
certain other fern families may have taken
their origin. The family is primarily tropical
or subtropical in its present distribution.
The rhizome is creeping or ascending in
growth habit. The leaf varies from simple to
dichotomously branched or even pinnately
compound. The leaf of *Lygodium* is distinctive in that it shows almost unlimited apical
growth and may attain the remarkable length
of 100 feet, and is adapted to climbing on
other vegetation (Fig. 13-4, F). Sporangia
are not grouped into sori, but occur singly,
and are unprotected except by the enrolling
of leaf segments or by marginal flaps of the
lamina (Figs. 13-7, D; 13-8, C). Sporangia
are regarded as marginal in origin, but are
often shifted to the abaxial side during
development. *Anemia phyllitidis* is another

tion of a stem reveals a large number of
closely packed leaf bases and the vascular
cylinder. The fossil genus *Thamnopteris*
was protostelic. Enough fossil forms are
known that an evolutionary sequence can
be traced from the protostelic type to that
found in *Osmunda* today, an ectophloic
siphonostele with overlapping leaf gaps
(Fig. 13-45, F). From all the evidence it
would appear that the siphonostelic organization in the family has been derived (in an
evolutionary sense) by medullation (intra-
stelar origin of the pith associated with the

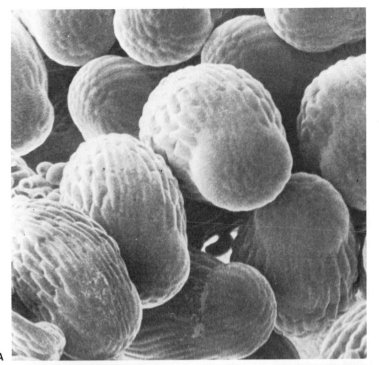

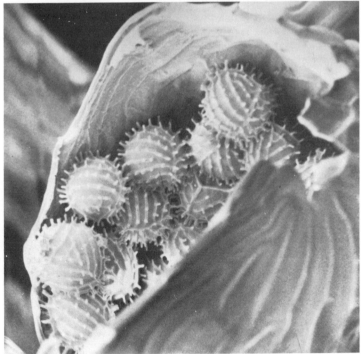

Figure 13-48
Schizaeaceae. *Anemia phyllitidis:* Scanning electron micrographs of
sporangia (**A**) and one sporangium which has dehisced, exposing the spores
(**B**). An annulus is apical in position, as shown by rounded tips in **A**.

representative species in which only the elongate basal pinnae of a leaf are fertile (Fig. 13-47). Sporangia are large, with short stalks, and they produce a large number of spores. Each sporangium has an apical, girdling annulus which brings about longitudinal dehiscence (Fig. 13-48, A, B). Some members are protostelic, others are siphonostelic. Research in recent years has resulted in the recognition of at least three different types of gametophytes in the family. The common green, cordate type is characteristic of *Lygodium* and *Anemia*. The species of one section of *Schizaea* have green, filamentous gametophytes and others (*Schizaea dichotoma*) are subterranean and filamentous, but the larger axes are multicellular. For *Actinostachys* the gametophytes are of the subterranean, axial, tuberous type and are associated with an endophytic fungus. Embryogeny is of the general leptosporangiate type except for *Actinostachys* in which the first division of the zygote is transverse, resulting in the formation of a shoot and foot, much like that of *Psilotum* and *Tmesipteris* (Psilopsida; p. 131). The genera described above apparently have had a common origin, but some morphologists, in an effort to point up the generic differences that do exist, separate the genera into two or three families, e.g., Schizaeaceae, Lygodiaceae, and Anemiaceae. (More information on the genera can be found in Bierhorst, 1971). Fertile leaves with sporangia (*Seftenbergia*) much like those of *Anemia* are known from the late Carboniferous and others are known from the Cretaceous and younger strata of the fossil record.

HYMENOPHYLLACEAE. The Hymenophyllaceae, or "filmy ferns," consists of two genera (*Hymenophyllum* and *Trichomanes*), or of many more (Copeland, 1947), depending upon the fern specialist's concept of taxonomic boundaries. The species of this group grow mainly in the tropics and many are epiphytic. Their leaves are often finely dissected, and the lamina (except at the vein) is usually one cell in thickness. Venation of the leaf is of the open dichotomous type (Fig. 13-49). The stems are protostelic. The receptacle of a sorus is marginal and may become elongated into a bristle and surrounded by a cup-shaped or two-lipped indusium (Fig. 13-49). The order of appearance of sporangia on the receptacle is generally basipetal, and hence of the gradate type. The sporangial stalks are short, and many spores are formed, which are liberated by oblique dehiscence of the spore case (Fig. 13-49, B, C).

PTERIDACEAE. Copeland (1947) considered the genera included in the Pteridaceae to constitute one large natural family, but he admitted that it might include more than one line of descent. As used here, it includes the pteroids, Dicksoniaceae, and gymnogrammoids of Bower (Fig. 13-43). Members of the family are typically terrestrial with a creeping or ascending rhizome; a few are arborescent. Fronds are pinnate in plan, but some are simple. The dermal appendages on rhizomes and fronds are either hairs or paleae. Rhizomes are either solenostelic or dictyostelic. The sori in some are marginal or near the margin and are protected by an indusium, or by a reflexed margin of the leaf, or borne on and protected by the reflexed leaf margin (*Adiantum*). In others, the sporangia are along veins or cover the entire surface of the fertile segment. A sporangium of some of the more primitive genera (e.g., *Dicksonia*) has an oblique annulus, whereas others have the longitudinal type. Most spores are tetrahedral and the gametophytes are of the green, cordate type.

The range in variation of organization and position of sori in the Pteridaceae can best be illustrated by discussing representative examples.

The *Dicksonia* sorus is intramarginal or submarginal and the indusium is bivalvate, the outer lip or flap being more lamina-like

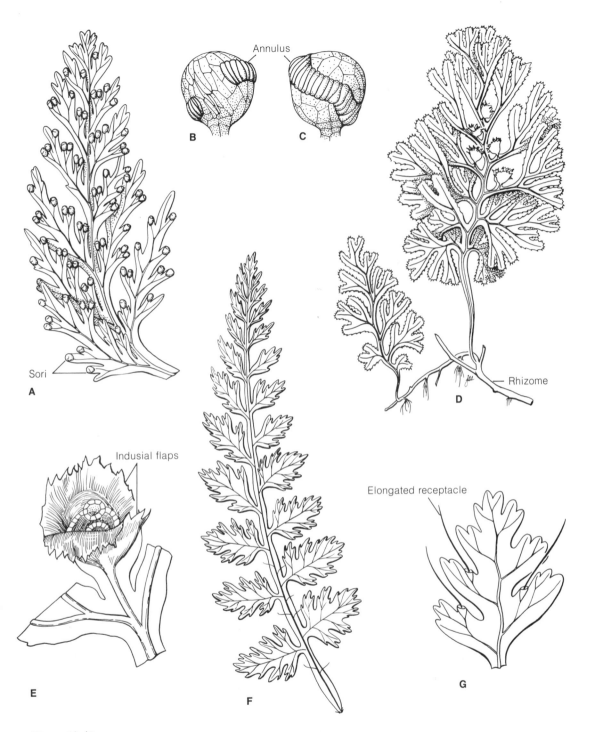

Figure 13-49
Hymenophyllaceae. **A–C,** *Hymenophyllum dilitatum:* portion of a frond (**A**), and sporangia (**B–C**) as seen from two sides; **D, E,** *Hymenophyllum tunbrigense:* portion of plant (**D**) and sorus, enlarged (**E**); **F, G,** *Trichomanes boschianum:* vegetative frond (**F**) and portion of fertile pinna (**G**). [A–C redrawn from *The Ferns,* Vol. II, by F. O. Bower. Cambridge University Press, London, 1926; D, E redrawn from *Welsh Ferns,* Ed. 5, by H. A. Hyde et al. Qualitex Printing Ltd., Cardiff, 1969; F, G redrawn from *Ferns of the Southeastern States* by J. K. Small. Hafner, New York, 1964.]

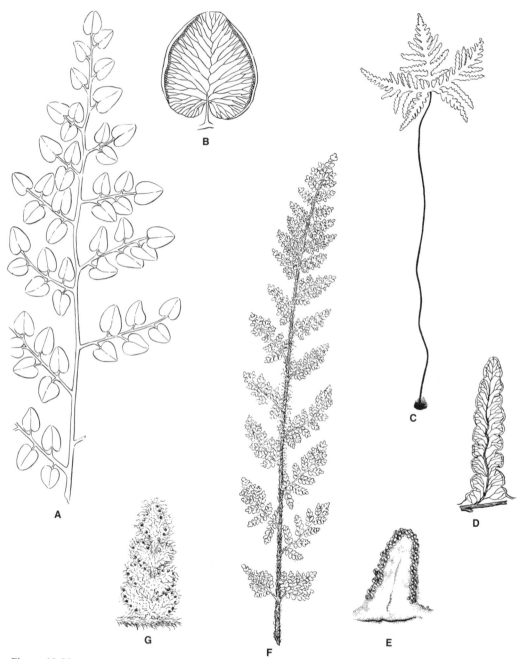

Figure 13-50
Pteridaceae. **A, B,** *Pellaea cardiomorpha:* portion of frond (**A**) and one pinnule enlarged (**B**) showing sporangia under reflexed leaf margin; **C–E,** *Notholaena standleyi:* a frond (**C**), sterile portion of lamina showing venation (**D**), and fertile portion (**E**); **F, G,** *Cheilanthes tomentosa:* frond (**F**) and fertile pinnule (**G**) showing sporangia and mat of hairs. [A,B,F,G redrawn from *Ferns and Fern Allies of Texas* by D. S. Correll. Texas Research Foundation, Renner, Texas, 1956; C–E redrawn from R. Tryon, *Contrib. Gray Herb. Harvard Univ.* No. 179:1–106, 1956.]

than the inner lip (an indusium in the more strict sense) (Fig. 13-10, A). Ferns comparable to *Dicksonia* have been reported from the Mesozoic.

The *Dennstaedtia* sorus is marginal in a sinus of the leaf, and is protected by a cup-shaped or slightly bivalvate structure, formed by the fusion of an inner indusial flap or lip and an outer flap (Fig. 13-10, B).

The *Pteridium* sorus (coenosorus) extends along the margin of a leaf segment and is protected by a strongly reflexed leaf margin and by an inner or "true" abaxial indusium (Fig. 13-8, A). In *Pteris* the continuous sorus is protected only by the structurally modified reflexed margin of the leaf.

The sori in *Pellaea* and other genera are also protected by reflexed, structurally modified leaf margins, but they are in a submarginal position along veins (Fig. 13-50, A, B). In still other genera, for example *Pityrogramma*, sporangia are borne along all veins and are not protected to any extent by the leaf margin. This is apparently a derived condition for the family, not a primitive one. *Adiantum* presents an interesting specialization in that the sori are formed on specialized reflex portions of the leaf margin (false indusium) (Fig. 13-9, B).

AN APPRAISAL OF THE PTERIDACEAE. In this textbook we have adopted Copeland's system of fern classification. We should hasten to point out that many fern specialists recognize certain phylogenetic lines within Copeland's Pteridaceae. For example, *Pteridium*, *Dennstaedtia*, and *Microlepia* are sometimes placed in the Dennstaedtiaceae. *Dicksonia* and its relatives are often placed in a separate family, the Dicksoniaceae, or aligned with members of the Cyatheaceae. This realignment is based upon chromosome numbers. Genera such as *Pellaea*, *Pityrogramma*, and *Adiantum* are often placed in a separate family—the Adiantaceae (Holttum, 1949; Wagner, 1969)—and these con-

stitute the gymnogrammoids of Bower's system (see Fig. 13-43). Cytological studies support the concept that the Adiantaceae should be recognized as a separate family (Manton, 1956).

DAVALLIACEAE. This family includes some of the ferns termed the davallioids in Bower's system (see Fig. 13-43). Ferns of this family have creeping rhizomes and are typically epiphytic. The stems are dictyostelic and dermal appendages are present in the form of paleae. The sori are near the margin or clearly on the abaxial side at the ends of veins. In *Davallia* and *Scyphularia* the indusium is attached to the leaf surface along its sides and base, forming a pouch-like structure (Fig. 13-51, A, B). In *Nephrolepsis* the sori may be located some distance from the leaf margin, but at the ends of veins. The indusium may be somewhat reniform. The annulus of the spore case in the family is longitudinal in orientation and the spores are bilateral. Basic chromosome numbers in the family are $n = 40$ and 41.

GLEICHENIACEAE. The family Gleicheniaceae, considered to be primitive, is a most natural group, according to Copeland (1947). It is primarily a tropical family of both hemispheres, often forming dense thickets at the edges of forests. The family occupies the same evolutionary level as does the Schizaeaceae in the other line of development in Bower's classification (Fig. 13-43). There were undisputed relatives in the Triassic, with soral impressions from the Carboniferous (*Oligocarpia*) resembling the sori of modern-day species.

Members of the Gleicheniaceae are terrestrial ferns with long creeping rhizomes, and the leaves may clamber on other vegetation. Fronds are once pinnate or pinnately compound (Fig. 13-52, A), but often they are pseudodichotomously branched by the abortion of terminal pinnae and the repetitive

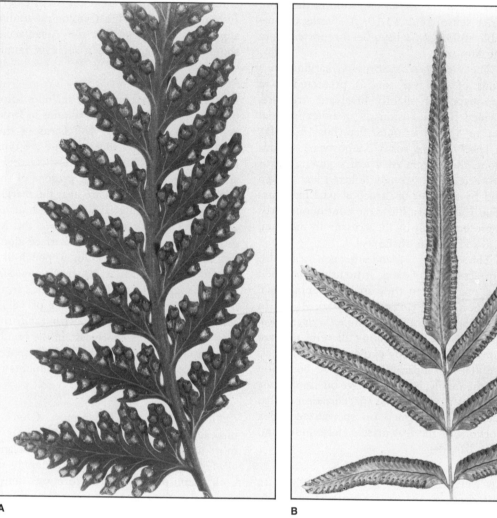

Figure 13-51
Davalliaceae. **A**, *Davallia divaricata:* portion of frond; **B**, *Scyphularia pentaphylla:* portion of frond. In both species, an indusium is pouch-like, united with the lamina, except where sporangia extend beyond open end of indusium.

development of lateral pinnae. The ultimate veins end freely. Stems are usually protostelic and of the "vitalized" type (Fig. 13-20, A). The sori of extinct and living members of this family are in two rows, one on each side of the leaf segment and are without indusia (Fig. 13-52, B). There are relatively few large sporangia in each sorus. Each sporangium produces a large number of spores (128–800), and dehiscence is brought about by the functioning of a transverse, to a transverse-oblique, girdling annulus (Fig. 13-52, C, D). The gametophyte is of the primitive type, namely, a large, green, dorsiventral plant with a conspicuous midrib. The antheridium is large and produces several hundred sperms. Considering all morphological features of this family, we can say that it obviously occupies a primitive position in the Filicales.

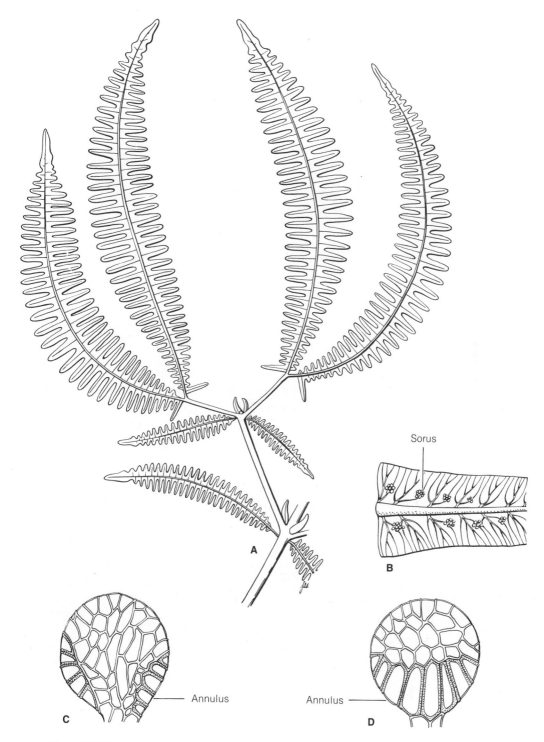

Figure 13-52
Gleicheniaceae. **A**, *Gleichenia linearis:* portion of a frond; **B–D**, *Gleichenia pectinata:* fertile portion
of frond (**B**) and sporangia as seen from two sides (**C**, **D**). [A redrawn from Pichi-Sermolli, *Webbia*
17:33–43, 1962; B–D redrawn from *Cryptogamic Botany,* Vol. II, Ed. 2., by G. M. Smith. McGraw-
Hill, New York, 1955.]

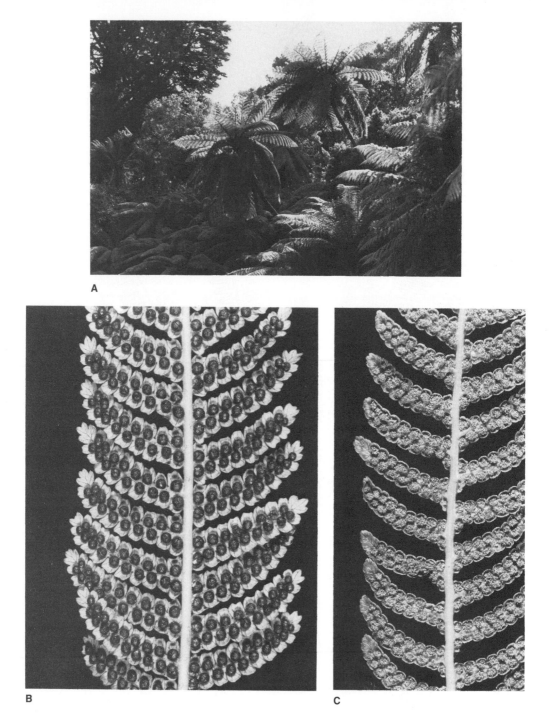

Figure 13-53
Cyatheaceae. A, tree ferns, some of which are members of the Cyatheaceae, growing in Golden Gate Park, San Francisco, California; B, *Cyathea medullaris:* portion of a pinnule showing membranous cup-shaped indusia; C, *Alsophila australis:* portion of a pinnule showing sori that are without indusia.

MATONIACEAE. The nearest affinity of this family is probably with the Gleicheniaceae, based on soral organization and structure of sporangia. Fossil members have been described from the Triassic and Cretaceous. The two extant genera in the family are from Indonesia, Borneo, and New Guinea. In *Matonia* the frond, above the petiole, is divided into two equal parts, each of which undergoes a series of unequal dichotomies. The leaf segments are pinnatifid, and solitary sori are borne at points near the midrib. Each sorus consists of a few sporangia arranged in a ring around the receptacle, which continues as the stalk of a peltate indusium. The sporangia of a sorus, which mature simultaneously, have very short stalks and oblique annuli. The stele of the stem is solenostelic with at least two concentric vascular cylinders and a central vascular cylinder. The spores are tetrahedral. The gametophyte is thalloid, green, and has a ruffled margin. Gametophyte morphology and the fact that the antheridial wall consists of several cells are suggestive of an affinity with the Gleicheniaceae.

CYATHEACEAE. In Bower's system the family Cyatheaceae occupies the same intermediate position in the "superficial series" as does the Dicksoniaceae in the "marginal series" (Fig. 13-43). It is truly the family of tree ferns, and there are species in all tropical lands south to Chile, New Zealand, and South Africa (Copeland, 1947). Individual trees may reach heights of 60–80 feet.

The typical tree fern has a trunk 4–12 inches, or more, in diameter, on which the leaf scars or leaf bases may be seen (Fig. 13-25). A crown of unusually large, pinnately compound leaves is present at the top of the trunk (Fig. 13-53, A). The sight of large unrolling fiddle-heads has evoked comment from many an observer. Older trunks are covered with masses of tangled roots; the base of the trunk is often enlarged, due to the presence of innumerable roots which have grown downward through the trunk from positions near leaf bases located high on the stem (Fig. 13-25). The stem itself is small in diameter. The vascular cylinders of all tree ferns are intricate dictyosteles of the dissected type (see p. 304).

Sori are located abaxially and are round without indusia, or with cup-shaped indusia that enclosed the sori during early development (Figs. 13-7, F; 13-53, B, C). Each sorus is of the gradate type, and each sporangium has an oblique annulus.

The gametophyte is the usual cordate type, although older specimens may have a conspicuous midrib.

ASPIDIACEAE. As constituted by Copeland (1947) this family is very large and difficult to define precisely. The name itself is illegitimate according to the rules of nomenclature, but it will continue to be used until taxonomists resolve the matter. The Aspidiaceae encompasses the dryopteroids, woodsioids, and onocleoids of Bower's system (Fig. 13-43). Most species are terrestrial, but some are epiphytic. Fronds are simple or pinnate and may be dimorphic. Dermal appendages are paleae and the stems are dictyostelic. Sori are on the abaxial side of fronds, being typically round or even elongate, or spread along veins. The indusium, when present, is attached around the base of the receptacle or only along one side, or it is sometimes a peltate structure. The annulus of the spore case is longitudinal and the spores are bilateral. Features of the gametophytes are "polypodiaceous" (p. 312). A chromosome number of $n = 41$ is common in the family.

Since it would be difficult to comment on all the various types of leaves and sori in the family, we will, instead, adopt the type method and describe representative genera.

352

Figure 13-54
Aspidiaceae. *Athyrium filix-foemina:* **A,** portion of a fertile frond; **B,** enlargement showing sori.

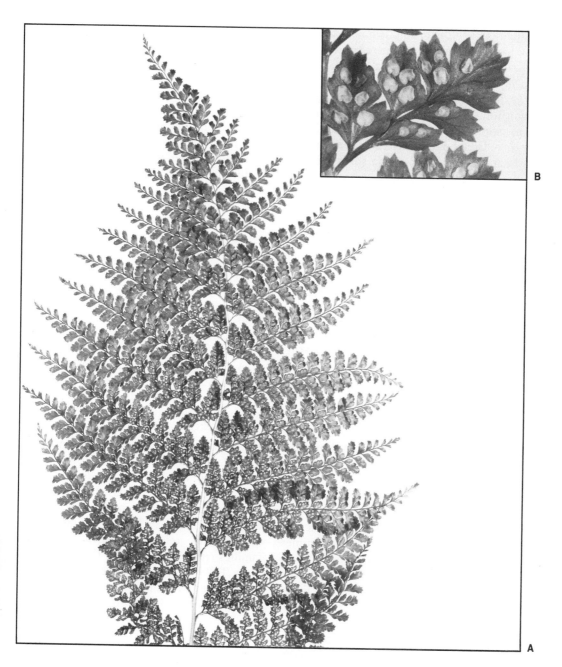

Figure 13-55
Aspidiaceae. *Aspidium* sp: **A,** portion of a fertile frond; **B,** enlargement showing sori.

Woodsia is a small, terrestrial genus with once-pinnate or bipinnate leaves. Sori are round, and the indusium is basal, consisting of hairs or scale-like structures which enclose the sporangia during early development, and then separate later (Fig. 13-7, G). The leaves of *Athyrium* may be large and pinnately compound or only once pinnate. The sorus is slightly elongate and lunate with a lateral indusium curved across the vein at the distal end of the sorus (Fig. 13-54). The indusium may be more or less reniform as in *Dryopteris,* and attached to the receptacle near the sinus (Fig. 13-55). The leaves are often large and bipinnatifid. In *Polystichum* and *Cyrtomium* the indusium is radially symmetrical, unlike *Dryopteris,* and it is peltate (Figs. 13-9, C; 13-11, A). The stalk of the indusium is attached at the top of the convex receptacle (Fig. 13-12, D).

Leaf dimorphism occurs in the Aspidiaceae and examples are *Onoclea sensibilis* (Fig. 13-56, A–D) and *Matteuccia struthiopteris* (Ostrich fern; Fig. 13-56, E–G). The rhizome of *Matteuccia* is ascending to erect, enclosed by leaf bases that are covered with chaffy scales. Both types of leaves, sterile and fertile, are once-pinnate and the sterile pinnae are pinnatifid, with open venation. The pinnae of fertile leaves are modified in that the segments are strongly revolute, enclosing the sori (Fig. 13-56, F, G). *Onoclea* is a swamp plant from the eastern United States and eastern Asia. The sterile leaves are pinnate, and the pinnae are sometimes pinnatifid, with a reticulate venation system. The pinnules are lobed on the fertile leaf, each lobe is recurved, and collectively the lobes form a globose structure enclosing a group of sori (Fig. 13-56, B, C). Each sorus is enclosed by a thin ephemeral indusium (Fig. 13-56, D). *Onoclea* and *Matteuccia* constitute the onocleoid ferns of Bower's classification (Fig. 13-43); the basis for their classification in the Aspidiaceae has been supported in a recent study made by Lloyd (1971).

AN APPRAISAL OF THE ASPIDIACEAE.   Some fern specialists have split the Aspidiaceae into two families. One family, the Athyriaceae, includes, for example, the genera *Woodsia, Athyrium, Diplazium, Cystopteris, Matteuccia,* and *Onoclea.* The genera *Dryopteris, Polystichum, Phanerophlebia,* and *Cyrtomium* are placed in the family Dryopteridaceae. If these "splinter" families are recognized, the family name Aspidiaceae is eliminated. Chromosome number in the Dryopteridaceae is uniformly $n = 41$.

ASPLENIACEAE.   This family is typified by *Asplenium* itself, which comprises about 700 species. Members of the family are terrestrial or epiphytic, with creeping or somewhat erect rhizomes, clothed with clathrate paleae. Fronds are simple to compound and range from very small to large; they are usually firm in texture. The elongate sori, which are bordered by indusia of equal length, are along the veins. The open or free margin of the indusium usually faces toward the apex of the pinna. The annulus of a sporangium is longitudinal and spores are bilateral in symmetry. Selected examples of the family are *Asplenium bulbiferum* (Fig. 13-8, D), *Asplenium nidus* (bird's-nest fern, Fig. 13-57, A, B), and *Phyllitis scolopendrium* (hart's-tongue fern; Fig. 13-57, D). Because of the shape of the indusium, *Athyrium* was placed in the asplenioids by Bower, but more modern systems relegate *Athyrium* to other families (the Aspidiaceae in Copeland's system). The base chromosome number in the Aspleniaceae is almost uniformly $n = 36$; that of *Athyrium* is $n = 40$.

BLECHNACEAE.   The fronds of species in this family are commonly large and coarse, usually pinnate or pinnatifid. The sporophytes are terrestrial and the rhizomes are dictyostelic and possess paleae that are nonclathrate. The sporangia are organized

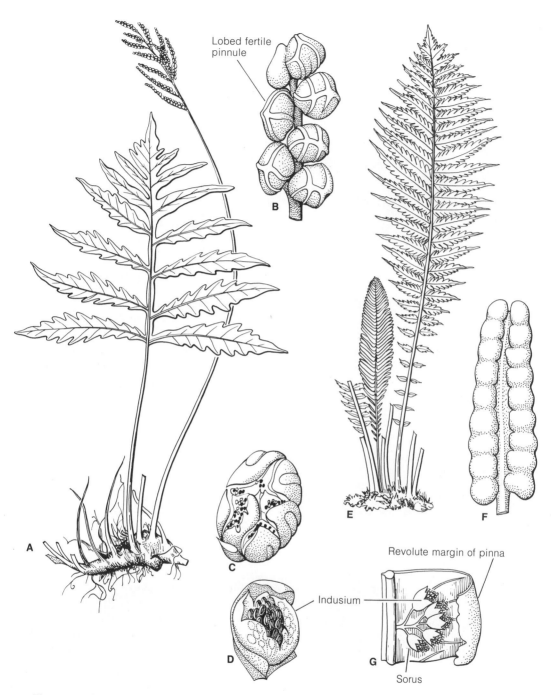

Lobed fertile
pinnule

Revolute margin of pinna

Indusium

Sorus

Figure 13-56
Aspidiaceae. **A–D,** *Onoclea sensibilis:* habit of plant, showing one fertile and one sterile leaf (**A**),
portion of fertile pinna (**B**), one fertile pinnule (**C**), and portion of fertile pinnule lobe removed to
show sorus (**D**); **E–G,** *Matteuccia struthiopteris:* habit of plant (**E**), fertile pinna (**F**), and position of sori
on pinna (**G**). [Redrawn from *The Ferns,* Vol. III, by F. O. Bower. Cambridge University Press,
London, 1928.]

Figure 13-57
Aspleniaceae. **A,** *Asplenium nidus;* **B,** portion of frond of *Asplenium nidus* showing numerous long sori; **C,** *Asplenium falcatum,* portion of frond showing unilateral indusia; **D,** leaf of *Phyllitis scolopendrium.*

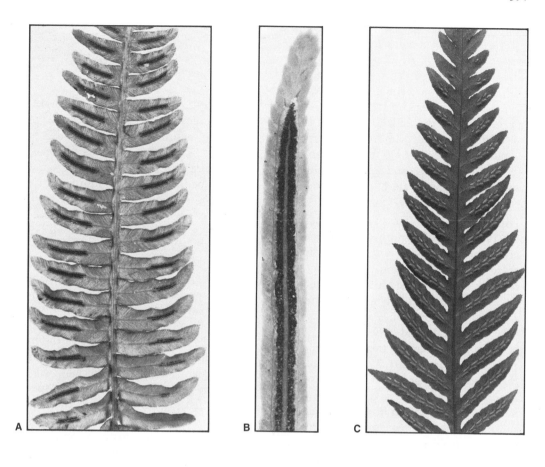

Figure 13-58
Blechnaceae. **A,** *Blechnum glandulosum,* portion of frond; **B,** *Blechnum* sp., pinna showing two coenosori, one on each side of midvein; **C,** portion of a pinnule of *Woodwardia;* **D,** one ultimate segment of *Woodwardia* frond, showing short coenosori, each facing the midrib; indusia attached on side away from midrib.

into long continuous coenosori on each side of the midrib of a pinna or segment (*Blechnum,* Fig. 13-58, A, B); or, the sori are shorter and interrupted, each one typically surrounded by a vein (e.g., *Woodwardia,* Fig. 13-58, C, D). The indusium opens on the side facing the midrib. *Blechnum* is considered to have the primitive sorus type and *Woodwardia* the derived type. Sporangia in the family are large and have a longitudinal annulus. The spores are of the bilateral type.

POLYPODIACEAE.   This family constitutes the dipteroids of Bower's system of classification (Fig. 13-43). Within the Polypodiaceae (Copeland's system) the genera (about 60) lack an indusium. Members of the family are primarily epiphytes of the tropics, although the common *Polypodium vulgare* (Fig. 13-59, A, B) and closely related species occur in Europe and in the north temperate zone. There are about 75 species of *Polypodium,* and they are most numerous in the American tropics. A dictyostelic rhizome is the prevailing type in the family and the rhizome is usually paleate. Leaves may be simple to once pinnate and many of them have a complex system of anastomosing veins in the lamina. Sori are round, sometimes elongate, or are spread over the entire surface, but they never have indusia. Sori may be crowded on the frond (Fig. 13-59, C), or in some genera, such as *Platycerium* (staghorn fern), sporangia are scattered without being organized into definite sori. The staghorn fern is interesting in that the leaves are dimorphic. Some large sterile clasping leaves (nest leaves) are appressed to the tree trunk or to the object on which the fern is growing, while sporangia occur in fertile regions on larger forked leaves (Fig. 13-59, D). The stalk of the sporangium in the family usually consists of three rows of cells, the annulus is longitudinal and the spores are predominately bilateral. Most pteridologists consider the family to have

been derived from the primitive Gleicheniaceae, although more directly, perhaps, from an extinct group related to the Matoniaceae.

AN APPRAISAL OF THE POLYPODIACEAE. It has become apparent that there are probably three natural series within the Polypodiaceae. Recognition of this fact had led to the establishment of separate families: the Polypodiaceae (in a more restricted sense), the Grammitidaceae, and the Dipteridaceae. Spores of the Dipteridaceae are trilete and contain chlorophyll. Monolete spores and lack of chlorophyll in the spores characterize the Polypodiaceae. The genera (*Cheiropleuria* and *Dipteris*) in the Dipteridaceae have dichotomously branched fronds, and stomatal development is of a simple type.

Table 13-2 summarizes some of the more important primitive and derived forms of certain characters in the Filicales. We are well aware of the fact that some fern specialists will disagree with some of the conclusions and it is their prerogative to do so. In the absence of definitive evidence for a feature, any conclusion, at best, is subjective.

Marsileales: *Marsilea, Regnellidium,* and *Pilularia*

This order is one of the two groups of "water ferns," both of which are heterosporous. In some earlier systems of classification this group constituted a family in the Filicales (the other group of water ferns constituted, in those systems, the family Salviniaceae). In other systems, the families have been placed in the order Hydropteridales—to emphasize the fact that they are aquatic as well as heterosporous. More recently the two orders Marsileales and Salviniales have been established to emphasize their structural differences.

Table 13-2
Primitive and derived forms of certain characters in the Filicales

Character	Primitive	Derived
	SPOROPHYTE	
Habitat	Terrestrial	Aquatic; epiphytic
Growth form	Ascending to erect	Strongly rhizomatous; dendroid
Leaf	Pinnate, determinate growth	Modified pinnate; simple; indeterminate growth
	Open dichotomous venation	Modified dichotomous to reticulate venation
	Monomorphic	Dimorphic
	Simple trichomes	Complex trichomes to scales
	Single, simple arc of vascular tissue in petiole	Dissected cylinder of tissue
Stem	Dichotomous branching	Types of branching other than dichotomous
	Protostelic	Solenostelic to dictyostelic
	Single leaf trace per leaf gap	Two to several leaf traces per leaf gap
	Simple trichomes	Complex trichomes to scales
Sorus	Sporangia single	Few to many sporangia in a sorus
	Marginal position	Superficial (abaxial) position
	Indusium absent	Present to secondarily absent
	Simple (simultaneous) maturation	Gradate to mixed maturation
Sporangium	Large, short stalks	Delicate, long stalks
	Massive, irregular tapetum	Two-layered tapetum
	Apical or subapical annulus	Oblique to vertical annulus
	Longitudinal dehiscence	Oblique to transverse dehiscence
	Numerous spores (e.g., 128, 256)	Few spores (e.g., 32, 64)
Spores	Spores tetrahedral (trilete)	Spores bilateral (monolete)
	Without perispore	With perispore
	GAMETOPHYTE	
Form and growth	Three-dimensional mass formed early	Early filamentous stage
	When mature, tuberous or elongate with conspicuous midrib	Cordate; ribbon-shaped without midrib; filamentous
	Without trichomes	With trichomes
	Maturing slowly	Maturing rapidly
Antheridium	Large, several jacket cells	Small, jacket of 3 cells
	Several hundred sperms	Commonly 14, 16, or 32 sperms
Archegonium	Relatively long, straight neck	Shorter, strongly curved neck
	A cell wall between neck-canal nuclei	No cell wall between neck-canal nuclei
Embryo	Early divisions irregular	More precise cleavages

A

B

C

D

Figure 13-59

Polypodiaceae. **A,** portion, frond of *Polypodium vulgare,* showing sori; **B,** subdivision of leaf in **A** enlarged, showing exindusiate sori; **C,** frond of *Pyrrosia angustifolia,* showing crowded exindusiate sori; **D,** *Platycerium willinckii* growing in greenhouse; large, clasping, "nest leaves" can be seen at center of photograph; sporangia (not visible) are scattered over lower surface of lamina near tips of forked leaves. [D courtesy of Dr. T. E. Weier.]

There are three genera in the Marsileales—(family Marsileaceae): *Marsilea, Regnellidium,* and *Pilularia* (Figs. 13-60; 13-62, A, B). All are aquatic plants in that they are generally rooted in mud or damp soil, often submerged in shallow water with just the leaf blades (*Marsilea* and *Regnellidium*) floating on the surface of the water. They are rhizomatous with roots arising from the lower side of the rhizome, usually at the nodes. *Marsilea* is generally found in warmer climates—in particular, a greater abundance of species is found in Africa and Australia. *Regnellidium* is a South American plant and *Pilularia* is more widespread, being found in North America, Africa, Europe, and Australia. Very little is known from the fossil record about the past history of this line of ferns.

Marsilea

The lamina of *Marsilea* is quadrifid (Fig. 13-60) and looks like a four-leaf clover and has, incidentally, been used on occasion as a substitute for the true four-leaf clover. Each pinna has basically dichotomous venation, but there are cross connections that unite the major vein system into a closed reticulum. A young leaf is circinate in vernation and the pinnae are folded together until late in development. At maturity the pinnae are extended perpendicular to the petiole. Leaves that remain totally submerged ("water form") are different morphologically from the "land form" and the two types have been studied experimentally (see p. 369). Reproductive structures, termed sporocarps, are found attached to or near petioles (Fig. 13-60).

ANATOMY OF THE RHIZOME. The prostrate rhizome of *Marsilea* is solenostelic. The inner portion of the cortex usually consists of a compact tissue; the outer portion is lacunate with large air spaces around radiating rows of parenchyma. The rhizomes

of submerged plants generally have a thin-walled parenchymatous pith, whereas those growing on mud or damp soil have a sclerotic pith (Fig. 13-61).

ROOTS. Roots arise at the bases of leaves. Internally a root is much like that of other ferns, but vessels occur in the xylem. As mentioned previously (p. 298), vessels in the rhizome, petiole, and root were discovered and described many years ago (Bliss, 1939) in the bracken fern (*Pteridium aquilinum*). More recently vessels have been found *only in the roots* of the heterosporous ferns, *Marsilea quadrifolia, Marsilea drummondii,* and *Marsilea hirsuta* (White, 1961). The end walls of individual vessel members vary from steeply oblique with scalariform perforation plates to transverse with simple perforation plates. Definitive proof of the occurrence of vessels was obtained by immersing plants with decapitated roots in

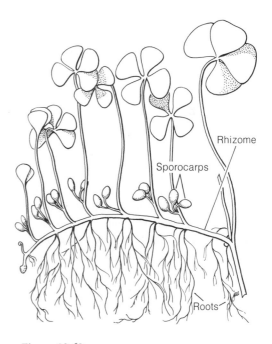

Figure 13-60
Sporophyte of *Marsilea quadrifolia*. [Redrawn from *Morphology of Vascular Plants: Lower Groups,* by A. J. Eames. McGraw-Hill, New York, 1936.]

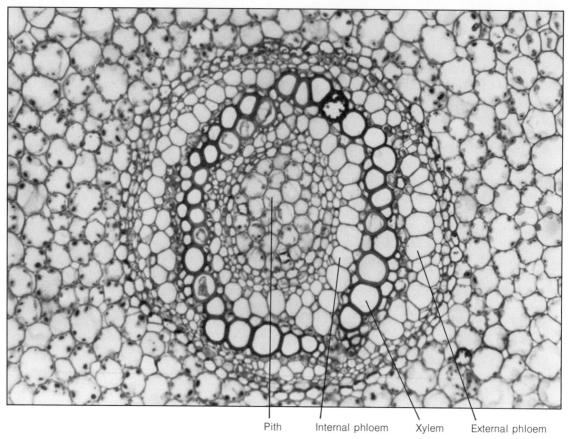

Pith   Internal phloem   Xylem   External phloem

Figure 13-61
Transection of the rhizome of *Marsilea* sp. showing details of the solenostele.

ink; the particles passed throughout the length of the roots. If actual perforations were not present, the particles could not have passed from cell to cell because ink particles are too large to pass through pit membranes. The ink particles, however, did not enter the stems. *Marsilea* and *Pteridium* are rather far apart taxonomically and the presence of vessels in these two genera indicates that there were at least two independent origins of vessels in ferns.

Regnellidium and *Pilularia*

Regnellidium (Fig. 13-62, A), a monotypic genus, differs from *Marsilea* in having leaves that are bifid; venation is of the dichotomous type (Pray, 1962) although the entire sys-

tem of veins is united by a marginal vein. Also, *Regnellidium* is the only known non-flowering plant that produces latex. It is formed in unbranched laticiferous ducts in the cortex of the rhizome, petiole, and in the lamina (Labouriau, 1952). *Pilularia,* a relatively small plant, has a filiform leaf that consists of only a petiole; no lamina is formed (Fig. 13-62, B). The petiole, just as is true of the leaves of *Marsilea* and *Regnellidium,* is circinate during development. The internal anatomy of the rhizomes of *Pilularia* is essentially the same as that of *Marsilea*. The leaf of *Pilularia* is a simple structure, but it probably became simple through evolutionary processes of reduction. In the course of time there has been a loss of the lamina.

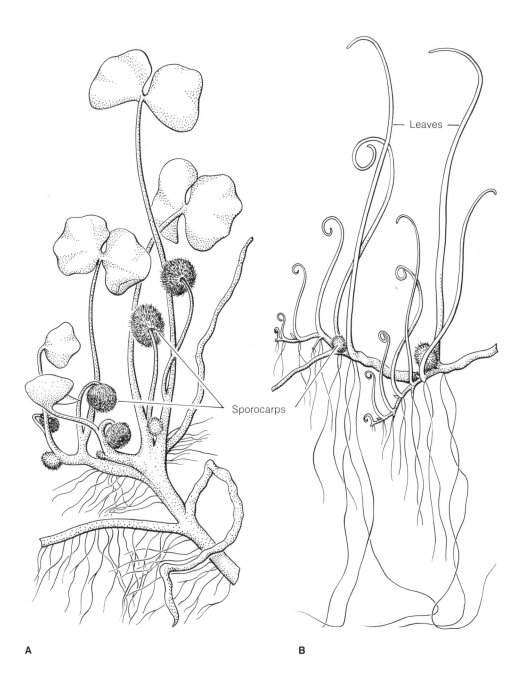

Figure 13-62
Sporophytes of *Regnellidium diphyllum* (**A**) and *Pilularia globulifera* (**B**). Natural size. [Redrawn from *Morphology of Vascular Plants: Lower Groups,* by A. J. Eames. McGraw-Hill, New York, 1936.]

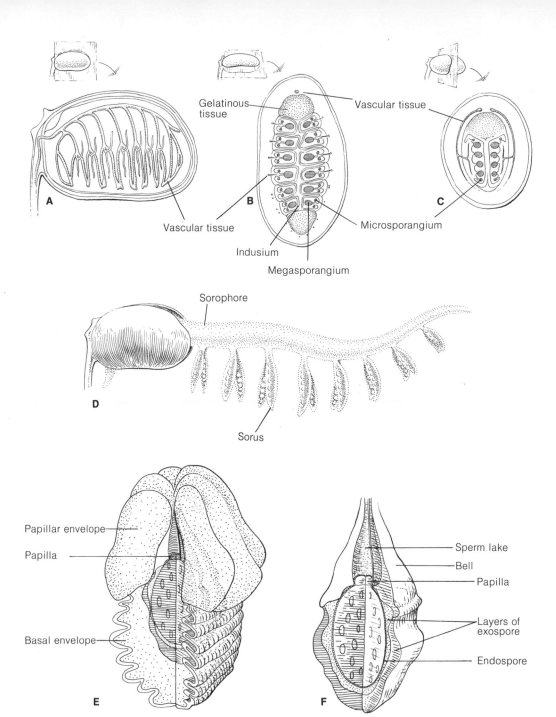

Figure 13-63

A–C, Structure of sporocarps of *Marsilea*. **A**, *Marsilea minuta*, longitudinal section in the plane of the stalk, showing the vasculature of one-half of the sporocarp as viewed from the inside; **B**, *Marsilea* sp., longitudinal section at right angles to **A**, showing soral organization. **C**, *Marsilea* sp., transverse section; **D**, *Marsilea quadrifolia*, extrusion of gelatinous cylinder (sorophore) to which sori are attached; **E,F**, *Marsilea vestita*, megaspores shortly after hydration, the papillar and basal envelopes removed in **F**. (See text for details.) [A redrawn from Puri and Garg, *Phytomorphology* 3:190, 1953; B–D adapted from *Morphology of Vascular Plants: Lower Groups*, by A. J. Eames. McGraw-Hill, New York, 1936; E,F redrawn from Machlis and Rawitscher-Kunkel, *Amer. Jour. Bot.* 54:692, 1967.]

Reproductive Cycle of *Marsilea*

Species of *Marsilea* are generally available for laboratory studies and hence we will adopt the "type method" and describe the reproductive cycle of the genus.

Reproductive structures, termed *sporocarps* and containing sporangia, are hard, bean-shaped bodies. In most species they occur singly, but there are exceptions (Fig. 13-60). Sporocarps withstand desiccation and are reported to be viable even after 20–35 years. These specialized structures are thought to be laminar in evolutionary origin. The question does arise, however, of whether a sporocarp represents a simple, folded basal pinna, or whether in evolution it has arisen from an entire pinnate leaf (for discussions see Eames, 1936; Smith, 1955; Puri and Garg, 1953). There is one main vein in the sporocarp, with lateral side branches that dichotomize (Fig. 13-63, A). Sori are on the inner side of each half of the sporocarp, oriented parallel to the lateral veins. Each sorus consists of a receptacle to which are attached megasporangia along its crest and microsporangia along the flanks (Fig. 13-63, B, C). An indusium covers each sorus and is a hood-like structure which extends to the margin of the sporocarp (Fig. 13-63, B, C). Developmentally each sorus is generally of the gradate type, with megasporangia being initiated first, followed by microsporangia. Sporangial development is of the leptosporangiate type. Only one spore in each megasporangium is functional after meiosis has taken place. The sporangium wall shows no sign of cellular specialization such as the formation of an annulus—hence there is no mechanism for dehiscence. Since the overwhelming majority of leptosporangiate ferns are homosporous, the heterosporous condition in *Marsilea* and other water ferns is especially noteworthy.

OPENING OF THE SPOROCARP AND [DE]VELOPMENT OF GAMETOPHYTES. [As men]tioned earlier a sporocarp is a hard s[tructure] and can withstand desiccation. The [spores] are capable of germinating as soon a[s the] sporocarp is mature, but do not d[o so] until the sporocarp opens. In nature [the] sporocarps do not open until two or three years after their formation. This delay is probably due to the imperviousness of the hard sporocarp wall. Opening of the sporocarp can be hastened by cracking or scoring the stony covering. When the sporocarp is then placed in water the tissues within begin to imbibe water and swell. As the sporocarp swells it splits open and within a few minutes a long, worm-like gelatinous structure (*sorophore*) emerges, attached to which are the sori (Fig. 13-63, D). The sorophore, a gelatinized portion of the sporocarp wall, swells and emerges from the sporocarp, carrying the sori to the outside. The sorophore may become 10–15 times longer than the sporocarp. The indusial and sporangial walls are intact at first, but with continued imbibition there is a general breakdown of the tissues and the spores are released. In the laboratory the separation of spores from surrounding tissues can be hastened by stirring the water or agitating the dish containing the sporocarps.

MALE GAMETOPHYTE. The numerous *microspores* begin to germinate almost immediately (by cell divisions within the spore wall). After several divisions two groups of 16 spermatocytes are formed, surrounded by jacket cells (Fig. 13-64, A–H; Sharp, 1914). Spermatocyte development and discharge of sperms are temperature dependent. For example, sperm discharge can be expected to occur in about six hours at 25°C. Prior to actual liberation of the sperms, the outer layer of the microspore,

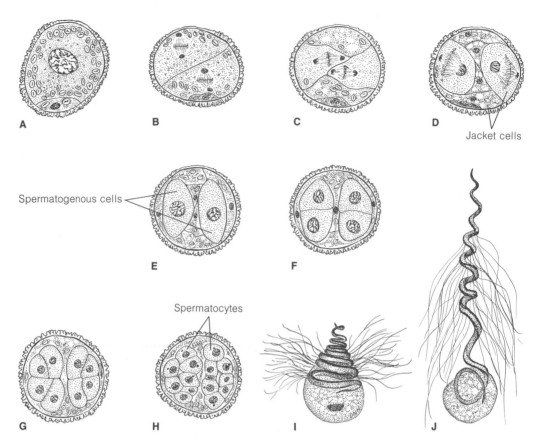

Jacket cells

Spermatogenous cells

Spermatocytes

Figure 13-64
Marsilea quadrifolia. **A–H**, stages in development of male gametophyte; **I**, sperm at time of liberation from microspore wall; **J**, sperm entangled in gelatinous material around megaspore. [After Sharp, redrawn from *Morphology of Vascular Plants: Lower Groups,* by A. J. Eames. McGraw-Hill, New York, 1936.]

Figure 13-65 (*facing page*)
A–D, *Marsilea quadrifolia.* **A**, longitudinal section of uninucleate megaspore, outer wall layers (exospore) removed; **B**, mature archegonium and portion of basal or nutritive cell; **C**, two-celled embryo; **D**, embryo enclosed by developing calyptra; **E**, *Marsilea vestita,* megaspore with enclosed mega-gametophyte surrounded by gelatinous sheath in which sperms are embedded; **F**, *Marsilea vestita,* external appearance of young sporophyte attached to megaspore and enclosed megagametophyte. [A–D redrawn from *Plant Morphology* by A. W. Haupt. McGraw-Hill, New York, 1953.]

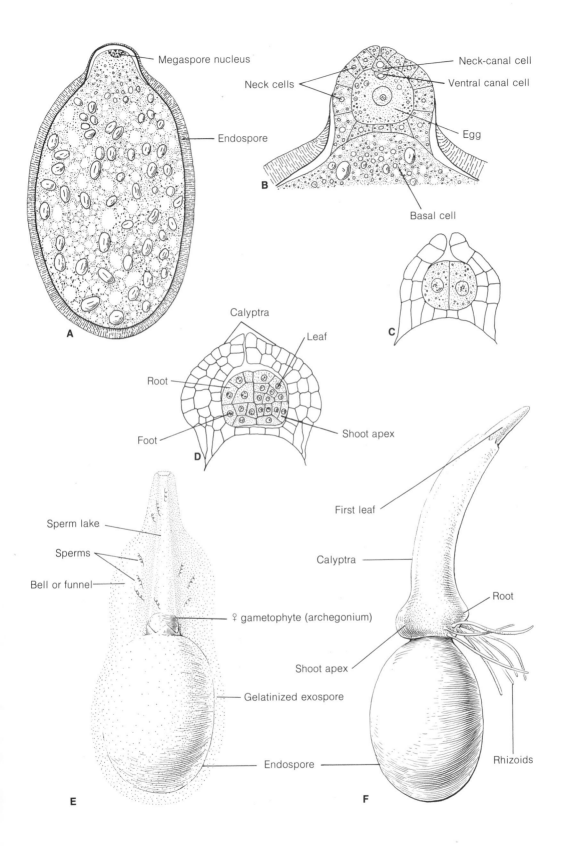

Megaspore nucleus

Endospore

A

Neck cells

Neck-canal cell

Ventral canal cell

Egg

Basal cell

B

C

Calyptra

Leaf

Root

Foot

Shoot apex

D

Sperm lake

Sperms

Bell or funnel

♀ gametophyte (archegonium)

Gelatinized exospore

Endospore

E

First leaf

Calyptra

Root

Shoot apex

Endospore

Rhizoids

F

termed the *exospore,* ruptures along the tri-radiate ridge. The sperms, enclosed by the inner layer of the spore wall, the *endospore,* emerge. The endospore then breaks, liberating the two groups of spermatocytes. Each motile sperm is then liberated by dissolution of the membrane of the spermatocyte.

Each mature swimming sperm consists of a large posterior cytoplasmic vesicle, a nucleus and a flagella-bearing coil at the anterior end (Figs. 13-64, I; 13-66). A sperm may be active for about one hour after liberation at 22–25°C. An individual flagellum has the usual $9 + 2$ arrangement of fibrils, an organization common to motile cells of plants and animals (Rice and Laetsch, 1967). When sperms are released in the vicinity of megaspores they become trapped in the gel around a megaspore. After motility is lost, a sperm uncoils and the cytoplasmic vesicle is lost (Figs. 13-64, J; 13-65, E).

FEMALE GAMETOPHYTE. As stated earlier there is one functional *megaspore* in each megasporangium. Approximately five minutes after a dry megaspore is moistened, the layers of the spore wall imbibe water and expand to form a gelatinous mass around the megaspore. At the apical end (where the endospore wall layer expands to form a small papilla) the gelatinous mass is characterized by longitudinal folds; the comparable structure around the basal end has horizontal folds (Fig. 13-63, E). During this process the sporangial wall is shed. If a large concentration of sperms is present, these envelopes are rapidly destroyed (Machlis and Rawit-scher-Kunkel, 1967). Within the envelopes just described is a thick gelatinous structure shaped like a bell whose interior is liquid. These are referred to as the "bell" and "sperm lake" (Fig. 13-63, F). The former constitutes the "funnel"—a term often used in the literature on *Marsilea.* If sperms are in the vicinity they become trapped in the gel or can be seen swimming in the sperm lake. The bell and subjacent layers of the exospore merge into a single gelatinous matrix a few hours after hydration. This is the stage generally illustrated in textbooks (Fig. 13-65, E).

The first cell division of the large megaspore occurs very soon after hydration—reported to be within $2\frac{1}{2}$ hours in one instance (Demalsy-Feller, 1957). There is apparently some variation in the events that follow the first division, but very soon a group of cells is formed at the apical end (within the papilla) and a cell wall separates the group of cells from a very large nutritive basal cell. An archegonial initial is formed and from it an archegonium is initiated consisting of an egg cell, a ventral canal cell, a neck-canal cell and neck cells (Fig. 13-65, A, B). Growth of the archegonial complex ruptures the thin endospore-wall layer at the apical end and a conspicuous papilla or protuberance is formed. With a slight separation of the neck cells and disintegration of the ventral canal and the neck-canal cell, a passage is created for a sperm to enter from the sperm lake.

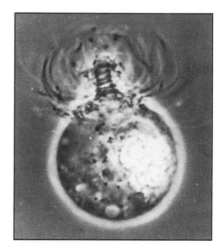

Figure 13-66
Swimming sperm of *Marsilea vestita* as seen with phase-contrast microscopy. Note the posterior cytoplasmic vesicle and the anterior nucleus- and flagella-bearing coil ($\times$ 1650). [Courtesy of Dr. W. M. Laetsch.]

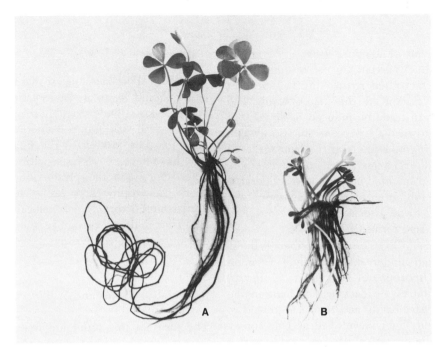

Figure 13-67
Sporophytes of *Marsilea* sp. **A**, grown in an inorganic nutrient solution with 4 percent sucrose added; **B**, grown as in **A**, but with the addition of 1 percent sucrose. The "land form" is exhibited in **A**, the "water form" in **B**.

EMBRYO DEVELOPMENT.   Following fertilization, which may occur as early as ten hours after hydration of the sporocarp, the first division of the zygote is longitudinal (in reference to the neck of the archegonium). Each cell of the two-celled embryo (Fig. 13-65, C) then divides transversely, resulting in the quadrant stage of the embryo. There are, however, reports that the first division may be transverse (Smith, 1955). Subsequent development is much like that of members of the Filicales in that the first root, leaf, foot, and shoot apex can be related to specific quadrants of the four-celled embryo (Fig. 13-65, D). Within four or five days the young sporophyte has a well-developed primary root and leaf (Fig. 13-65, F). If a mineral nutrient solution is provided, the sporophyte will continue to grow in a water environment, producing several "juvenile" type leaves before the quadrifid "adult" leaves develop.

Sterile cultures (free of bacteria, etc.) of sporophytes can be established rather easily using special methods that have been developed (Laetsch, 1967).

Land and Water Leaf Forms

Many species of plants exhibit differences in leaf form depending upon whether they are growing submerged in water or on land.

Marsilea is useful in a study of this phenomenon. Adult land and water forms of this genus are easily distinguished from each other. A "water form" has elongated internodes, short petioles, and leaves with a divided lamina, expanded in the plane of the petiole. A "land form" has short, thick internodes and long petioles, with the lamina quadrifid and expanded at right angles to the petiole (Fig. 13-67, A, B).

Allsopp (1963) has related these differences primarily to nutrition—specifically to the internal sugar concentration. Young sporophytes of *Marsilea drummondii* grown *in vitro* on 5 percent glucose resembled the land form; others grown on 1–2 percent glucose resembled the water form. The water form could be induced to develop in plants grown on 4 percent glucose — which normally would result in the land form — if gibberellic acid was added to the medium. This supplemented medium also accelerated growth. Inhibition of protein synthesis was reported to be associated with the development of land forms (White, 1966).

Gaudet (1963, 1965) described leaf development under different light regimes. Young sporophytes of *Marsilea vestita,* grown initially in darkness or light in the far-red wavelength, and then transferred to continuous light, developed as land forms. Even sporophytes cultured in a medium that normally permits only the water form to develop responded to a transfer to continuous light in this way. Light of far-red wavelength was more effective than darkness in causing this conversion. Long wavelengths of far-red light do not penetrate as far as blue into water. Gaudet suggested that, therefore, the water form would develop under natural conditions.

The light effect just described would appear to be a plausible explanation of the phenomenon, but the results of some recent experiments tend to confound the final solution of the problem. By growing young plants of *Marsilea* on a solid substrate in an atmosphere with increased partial pressure of CO_2, water forms were produced instead of land forms. The increased concentrations of CO_2 were effective whether the sporophytes were grown in darkness or light (Bristow and Looi, 1968). These studies throw some doubt on the role of internal physiological differences of the types described earlier. It is clear, as White (1971) has stated, that the possible causes are not as simple as first thought — i.e., nutrition and light quality.

Phylogeny

Some fern specialists suggest that the Marsileales should be treated as a family in the Filicales (Fig. 13-43), possibly related to the Schizaeaceae. The group is undoubtedly related in some fashion to the Filicales, but they have become very specialized through heterospory and the production of endosporic gametophytes. It seems reasonable, until contradictory paleobotanical evidence is forthcoming to recognize the group as constituting a separate order.

Salviniales: *Azolla* and *Salvinia*

The ferns in this order are referred to as "water ferns" and the descriptive term is probably more justified for them than for the Marsileales. Members of the Salviniales are small and float free in ponds and lakes. There are two genera — *Azolla* and *Salvinia* (Figs. 13-68; 13-69). Both genera are heterosporous, and bear sporangia in sporocarps which, in contrast to those in the Marsileales, are modified sori. The gametophytes are endosporic in development as in the Marsileales.

Salvinia, known to have been present in the Cretaceous, is now primarily a genus of tropical regions, with one species growing in the United States. *Azolla,* with six extant species, is worldwide in distribution, with two species native to the United States. Species of *Azolla* are known from the Cretaceous and were widespread during the Tertiary (Hall and Swanson, 1968; Hall, 1969).

The leaves of *Salvinia* are in whorls of three on the floating rhizome. There are no roots, but one leaf of each whorl is submerged and is composed of dissected, hairlike filaments (Fig. 13–68, B). These filiform structures may serve the function of roots. The two floating leaves are concave and bear

A

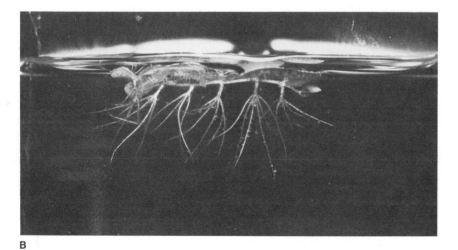

B

Figure 13-68
Salvinia sp. **A,** group of floating plants as viewed from above (about natural size); note that two of the three leaves at each node can be seen; **B,** a floating plant viewed from the side showing the branched, filiform third leaf at each node, which extends into the water.

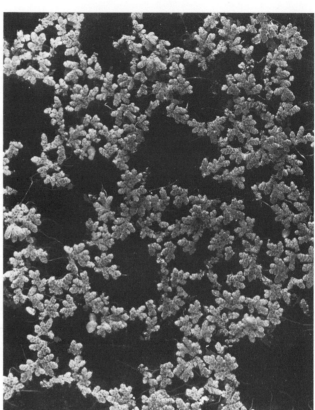

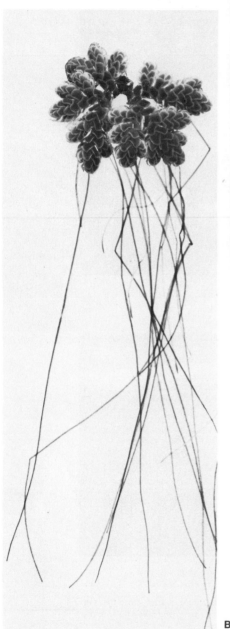

Figure 13-69
Azolla filiculoides. **A,** a group of floating plants (about natural size); **B,** view showing overlapping leaves of the floating shoots and their slender pendulous roots.

A

B

numerous waxy trichomes which can trap air if the plants become immersed.

The thin, horizontal rhizome of *Azolla* bears leaves and roots (Fig. 13-69, B). The crowded leaves are small, each being bilobed. The upper lobe is photosynthetic, while the lower lobe is colorless and more delicate, perhaps functioning in the absorption of water. Roots generally arise at the junctions of the branches. At the base of each leaf there is a cavity that contains the nitrogen-fixing, blue-green alga, *Anabaena azollae.* When growing in mass, the plants may conceal the water below, and they may develop a distinctive reddish color. There are methods for obtaining aseptic clones of *Azolla* for experimental purposes (Posner, 1967).

Reproductive Cycle

Space does not permit a detailed account of reproduction, but we will describe in a general way that of *Azolla* and make general comments on the life history of *Salvinia.*

Mature sporocarps of *Azolla* are of two sizes. The larger one contains the microsporangia surrounded by an indusium (Fig. 13-70, A). At maturity the megasporocarp contains one functional megasporangium. The differentiation of sporocarps occurs rather late in development, by the survival of either one megasporangium or many microsporangia. A sorus is gradate in development, the potential megasporangium being formed first. A mature megasporangium contains one functional megaspore,

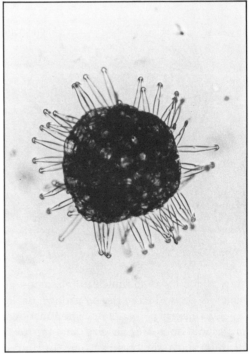

A B

Figure 13-70
Azolla filiculoides. A, ventral side of a sporophyte showing four "sporocarps."
B, one microsporic massula showing barbed glochidia (greatly enlarged).

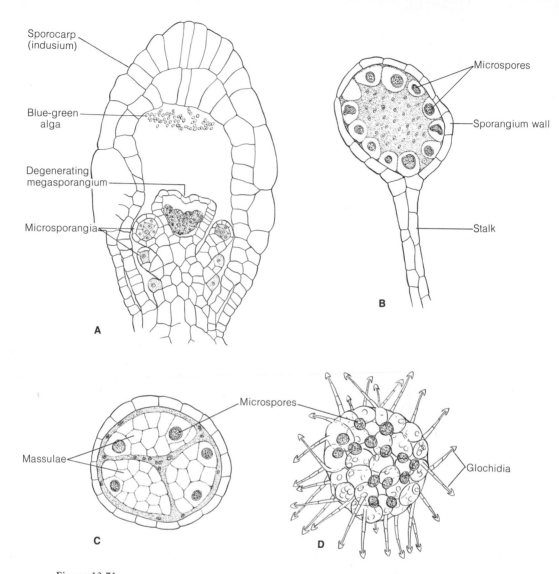

Figure 13-71
Azolla filiculoides. **A,** longitudinal section of young microsporangial sporocarp; **B,C,** stages in development of massulae within microsporangia; **D,** microsporic massula after liberation from microsporangium. [A–C redrawn from *Cryptogamic Botany,* Vol. II, Ed. 2, by G. M. Smith. McGraw-Hill, New York, 1955; D redrawn from *Morphology of Vascular Plants: Lower Groups,* by A. J. Eames. McGraw-Hill, New York, 1936.]

surrounded by a multinucleate plasmodium, which is derived from the breakdown of the tapetal cells (Fig. 13-72, A). The functional megaspore increases in size and the plasmodium above the spore becomes organized into large masses, called *massulae* or "floats" (Fig. 13-72, B). The embedded megaspore soon forms a group of cells (megagametophyte) at its apical end (Fig. 13-72, C). With continued growth the spore wall breaks and

the megagametophytic tissue is exposed. Archegonia are then formed at the surface. During the growth processes, the original sporangial wall and sporocarp wall (indusium) are ruptured (Fig. 13-72, D), liberating the megaspore and its surrounding massulae.

All of the microspores within a microsporangium are functional (Fig. 13-71, A). During meiosis there is a breakdown of the tapetum into a multinucleate plasmodium,

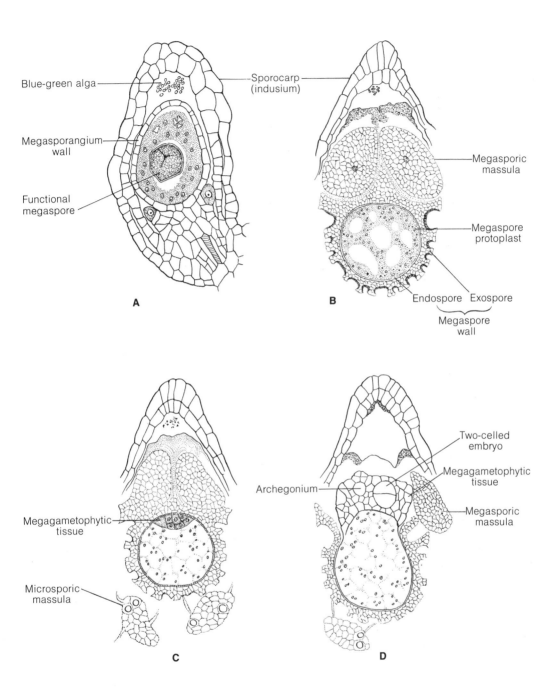

Figure 13-72
Azolla filiculoides. **A,** megasporangium and surrounding sporocarp wall (indusium); **B–D,** stages in development of megagametophyte and mussulae. (See text for details.) [Redrawn from *Cryptogamic Botany,* Vol. II, Ed. 2, by G. M. Smith. McGraw-Hill, New York, 1955.]

just as in a megasporangium. The spores move to the periphery of the sporangium and become encased in three or more massulae (Fig. 13-71, B, C). In some species of *Azolla,* elongate, hooked processes (*glochidia*) are formed on the massulae (Figs. 13-70, B; 13-71, D). These serve the purpose of attaching the "microsporic massulae" to the "megasporic massulae" (or megaspore wall), thereby bringing the male gametophytes into the vicinity of a megametophyte and better assuring that fertilization will take place (Fig. 13-72, C). The microspores in the "microsporic massulae" germinate by undergoing several cell divisions in the production of sperms. The sperms then swim to the archegonia. The first division of the zygote is reported to be transverse (Fig. 13-72, D) to oblique followed by a longitudinal division in each of the two cells. The four portions of the embryo—first leaf, root, shoot apex, and foot—can be related to the four quadrants of the embryo much like those of other ferns.

As many of the details have been omitted from the account of this extraordinary sexual cycle in ferns, the interested student should consult the pertinent literature, much of it quite old (See Smith, 1955, and Bonnet, 1957, for references).

The reproductive cycle in *Salvinia* is much the same as in *Azolla* except that the sporocarps are similar in size and a megasporocarp contains more than one functional megasporangium. There are also some differences in the germination of microspores and in the development of microgametophytes and megagametophytes.

References

Abraham, A., C. A. Ninan, and P. M. Mathew
    1962. Studies on the cytology and phylogeny of the pteridophytes. VII. Observations on 100 spp. of South Indian ferns. *Jour. Indian Bot. Soc.* 41:339–421.

Albaum, H. G.
    1938. Inhibitions due to growth hormones in fern prothallia and sporophytes. *Amer. Jour. Bot.* 25:124–133.

Allsopp, A.
    1963. Morphogenesis in *Marsilea. Jour. Linn. Soc. London Bot.* 58:417–427.

Atkinson, L. R., and A. G. Stokey
    1964. Comparative morphology of the gametophyte of the homosporous ferns. *Phytomorphology* 14:51–70.

Ball, E.
    1952. Morphogenesis of shoots after isolation of the shoot apex of *Lupinus albus. Amer. Jour. Bot.* 39:167–191.

Bell, P. R.
    1970. The archegoniate revolution. *Sci. Progr.* (England) 58:27–45.

Bierhorst, D. W.
    1971. *Morphology of Vascular Plants.* Macmillan, New York.

Blake, S. F.
    1942. The ostrich fern as an edible plant. *Amer. Fern Jour.* 32:61–68.

Bliss, M. C.
    1939. The tracheal elements in the ferns. *Amer. Jour. Bot.* 26:620–624.

Bonnet, A. L.-M.
1957. Contribution à l'étude des Hydropté-ridées. III. Recherches sur *Azolla fili-culoides* Lamk. *Rev. Cytol. Biol. Veg.* 18:1–88.

Bower, F. O.
1923. *The Ferns,* Vol. I. Cambridge University Press, London.
1935. *Primitive Land Plants.* Macmillan, London.

Braithwaite, A. F.
1964. A new type of apogamy in ferns. *New Phytol.* 63:293–305.

Briggs, W. R., and T. A. Steeves
1959. Morphogenetic studies on *Osmunda cinnamomea* L. The mechanism of crozier uncoiling. *Phytomorphology* 9:134–147.

Bristow, J. M., and A. Looi
1968. Effects of carbon dioxide on the growth and morphogenesis of *Marsilea. Amer. Jour. Bot.* 55:884–889.

Copeland, E. B.
1939. Antarctica as the source of existing ferns. *Pac. Sci. Congr. Proc.* 4:625–627.
1942. Edible ferns. *Amer. Fern Jour.* 32: 121–126.
1947. *Genera Filicum* (The Genera of Ferns). Chronica Botanica Co., Waltham, Mass.

Coquillat, M.
1950. Au sujet du "pain de fougère" en Mâconnais. *Bull. Mens. Soc. Linn. Lyon* 19:173–175.

Cutter, E. G.
1954. Experimental induction of buds from fern leaf primordia. *Nature* 173: 440–441.
1956. Experimental and analytical studies of Pteridophytes. XXXIII. The experimental induction of buds from leaf primordia in *Dryopteris aristata* Druce. *Ann. Bot.* n.s. 20:143–165.
1965. Recent experimental studies of the shoot apex and shoot morphogenesis. *Bot. Rev.* 31:7–113.

De Maggio, A. E.
1963. Morphogenetic factors influencing the development of fern embryos. *Jour. Linn. Soc. London Bot.* 58:361–376.

Demalsy-Feller, M.-J.
1957. Etudes sur les Hydropteridales. V. Gamétophytes et gamétogenèse dans le genre *Marsilea. Cellule* 58:169–207.

Diers, L.
1965. Elektronmikroscopische Untersuchungen über die Eizellbildung und Eizellreifung des Lebermooses *Sphaerocarpus donnellii* Aust. *Zeit. Naturforsch.* 20(b): 795–801.
1966. On the plastids, mitochondria, and other cell constituents during oogenesis of a plant. *Jour. Cell Biol.* 28:527–543.

Drumm, H., and H. Mohr
1967. Die Regulation der RNS-Synthese in Farngametophyten durch Licht. *Planta* 72:232–246.

Eames, A. J.
1936. *Morphology of Vascular Plants. Lower Groups.* McGraw-Hill, New York.

Esau, K.
1954. Primary vascular differentiation in plants. *Biol. Rev.* 29:46–86.
1965. *Plant Anatomy,* Ed. 2. Wiley, New York.

Evans, A. M.
1964. Ameiotic alternation of generations: A new life cycle in the ferns. *Science* 143: 261–263.

Farrar, D. R.
1967. Gametophytes of four tropical fern genera reproducing independently of their sporophytes in the southern Appalachians. *Science* 155:1266–1267.

Farrar, D. R., and W. H. Wagner, Jr.
1968. The gametophyte of *Trichomanes holopterum* Kunze. *Bot. Gaz.* 129:210–219.

Gantt, E., and H. J. Arnott
1965. Spore germination and development of the young gametophyte of the ostrich fern (*Matteuccia struthiopteris*). *Amer. Jour. Bot.* 52:82–94.

Gaudet, J. J.
1963. *Marsilea vestita:* Conversion of the water form to the land form by darkness and by far-red light. *Science* 140:975–976.
1965. The effect of various environmental factors on the leaf form of the aquatic fern *Marsilea vestita. Physiol. Plant.* 18:674–686.

Guttenberg, H. von
  1966. Histogenese der Pteridophyten. *In* Zimmermann, W. (ed.), *Handbuch der Pflanzenanatomie*, Band VII, Teil 2. Gebrüder Borntraeger, Berlin.

Hall, J. W.
  1969. Studies on fossil *Azolla*: primitive types of megaspores and massulae from the Cretaceous. *Amer. Jour. Bot.* 56:1173–1180.

Hall, J. W., and N. P. Swanson
  1968. Studies on fossil *Azolla*: *Azolla montana*, a Cretaceous megaspore with many small floats. *Amer. Jour. Bot.* 55:1055–1061.

Hewitson, W.
  1962. Comparative morphology of the Osmundaceae. *Ann. Missouri Bot. Gard.* 49: 57–93.

Hicks, G. S., and T. A. Steeves
  1969. In vitro morphogenesis in *Osmunda cinnamomea*. The role of the shoot apex in early leaf development. *Can. Jour. Bot.* 47:575–580.

Holttum, R.
  1947. A revised classification of the leptosporangiate ferns. *Jour. Linn. Soc. London Bot.* 53:123–158.
  1949. The classification of ferns. *Biol. Rev.* 24: 267–296.

Jayasekera, R. D. E., and P. R. Bell
  1959. The effect of various experimental treatments on the development of the embryo of the fern *Thelypteris palustris*. *Planta* 54:1–14.

Klekowski, E. J., Jr.
  1969. Reproductive biology of the Pteridophyta. II. Theoretical considerations. *Jour. Linn. Soc. London Bot.* 62:347–359.
  1971. Ferns and genetics. *BioScience* 21:317–322.

Klekowski, E. J., Jr., and H. G. Baker
  1966. Evolutionary significance of polyploidy in the Pteridophyta. *Science* 153:305–307.

Kuehnert, C. C.
  1967. Developmental potentialities of leaf primordia of *Osmunda cinnamomea*. The influence of determined leaf primordia on undetermined leaf primordia. *Can. Jour. Bot.* 45:2109–2113.

Labouriau, L. G.
  1952. On the latex of *Regnellidium diphyllum* Lindm. *Phyton* 2:57–74.
  1958. Studies on the initiation of sporangia in ferns. *Arq. Mus. Nac. Rio de Janeiro* 46: 119–201.

Laetsch, W. M.
  1967. Ferns. Pp. 319–328 *in* Wilt, F. H., and N. K. Wessels (eds.), *Methods in Developmental Biology*. Crowell, New York.

Lloyd, R. M.
  1971. Systematics of the onocleoid ferns. *Univ. Calif. Publ. Bot.* 61:1–86.

Machlis, L., and E. Rawitscher-Kunkel
  1967. The hydrated megaspore of *Marsilea vestita*. *Amer. Jour. Bot.* 54:689–694.

Manton, I.
  1950. Demonstration of compound cilia in a fern spermatozoid by means of the ultraviolet microscope. *Jour. Exp. Bot.* 1:68–70.
  1956. Chromosomes and fern phylogeny with special reference to "Pteridaceae". *Jour. Linn. Soc. London Bot.* 56:73–92.

Manton, I., and B. Clarke
  1951. Demonstration of compound cilia in a fern spermatozoid with the electron microscope. *Jour. Exp. Bot.* 2:125–128.

Miller, C. N., Jr.
  1967. Evolution of the fern genus *Osmunda*. *Contr. Mus. Paleontol. Univ. Mich.* 21: 139–203.

Miller, J. H.
  1968. Fern gametophytes as experimental material. *Bot. Rev.* 34:361–440.

Mohr, H.
  1963. The influence of visible radiation on the germination of archegoniate spores and the growth of the fern protonema. *Jour. Linn. Soc. London Bot.* 58:287–296.

Morel, G.
  1963. Leaf regeneration in *Adiantum pedatum*. *Jour. Linn. Soc. London Bot.* 58:381–383.

Nayar, B. K., and S. Kaur
  1971. Gametophytes of homosporous ferns. *Bot. Rev.* 37:295–396.

O'Brien, T. P.
  1963. The morphology and growth of *Pteridium aquilnum* var. *esculentum* (Forst.) Kuhn. *Ann. Bot.* n.s. 27:253–267.

Ogura, Y.

1972. Comparative anatomy of vegetative organs of the Pteridophytes. *In* Zimmerman, W. (ed.), *Handbuch der Pflanzenanatomie*, Ed. 2, Band VII, Teil 3. Gebrüder Borntraeger, Berlin.

Ohlenroth, K., and H. Mohr

1964. Die Steuerung der Protein-synthese durch Blaulicht und Hellrot in den Vorkeimen von *Dryopteris filix-mas* (L.) Schott. *Planta* 62:160–170.

Patterson, P. M., and A. S. Freeman

1963. The effect of photoperiod on certain ferns. *Amer. Fern Jour.* 53:126–128.

Pichi-Sermolli, R. E. G.

1958. The higher taxa of the Pteridophyta and their classification. Pp. 70–90 *in* Hedberg, O., (ed.), *Systematics of Today*, Almquist and Wiksells, Uppsala.

Posner, H. B.

1967. Aquatic vascular plants. Pp. 301–317 *in* Wilt, F. H., and N. K. Wessells (eds.), *Methods in Developmental Biology*. Crowell, New York.

Pray, T. R.

1962. Ontogeny of the closed dichotomous venation of *Regnellidium*. *Amer. Jour. Bot.* 49:464–472.

1968. The gametophytes of *Pellaea* section *Pellaea*: dark-stiped series. *Phytomorphology* 18:113–143.

Puri, V., and M. L. Garg

1953. A contribution to the anatomy of the sporocarp of *Marsilea minuta* L. with a discussion of the nature of sporocarp in the Marsileaceae. *Phytomorphology* 3: 190–209.

Rice, H. V., and W. M. Laetsch

1967. Observations on the morphology and physiology of *Marsilea* sperm. *Amer. Jour. Bot.* 54:856–866.

Rivières, R.

1959. Sur la culture in vitro d'embryons isolés de Polypodiacées. *Compt. Rend. Acad. Sci.* (Paris) 248:1004–1007.

Schraudolf, H.

1962. Die Wirkung von Phytohormonen auf Keimung und Entwicklung von Farnprothallien. I. Auslösung der Antheridienbildung und Dunkelkeimung bei Schizaeaceen durch Gibberellinsäure. *Biol. Zentralbl.* 81:731–740.

1967. Wirkung von Hemmstoffen der DNS-, RNS- und Proteinsynthese auf Wachstum und Antheridienbildung in Prothallien von *Anemia phyllitidis* L. *Planta* 74:123–147.

Sharp, L. W.

1914. Spermatogenesis in *Marsilea*. *Bot. Gaz.* 58:419–431.

Smith, G. M.

1955. *Cryptogamic Botany*, Vol. II, Ed. 2. McGraw-Hill, New York.

Smith, R. H., and T. Murashige

1970. In vitro development of the isolated shoot apical meristems of angiosperms. *Amer. Jour. Bot.* 57:562–568.

Sobota, A. E.

1970. Interaction of red light and an inhibitor produced in the meristem of the fern gametophyte. *Amer. Jour. Bot.* 57:530–534.

Sobota, A. E., and C. R. Partanen

1966. The growth and division of cells in relation to morphogenesis in fern gametophytes. I. Photomorphogenetic studies in *Pteridium aquilinum*. *Can. Jour. Bot.* 44:497–506.

1967. The growth and division of cells in relation to morphogenesis in fern gametophytes. II. The effect of biochemical agents on the growth and development of *Pteridium aquilinum*. *Can. Jour. Bot.* 45:595–603.

Steeves, T. A.

1961. A study of the developmental potentialities of excised leaf primordia in sterile culture. *Phytomorphology* 11:346–359.

1962. Morphogenesis in isolated fern leaves. *Symp. Soc. Study Develop. Growth* 20: 117–151.

Steeves, T. A., and R. H. Wetmore

1953. Morphogenetic studies on *Osmunda cinnamomea* L.: some aspects of the general morphology. *Phytomorphology* 3:339–354.

Stokey, A. G.

1951. The contribution by the gametophyte to classification of the homosporous ferns. *Phytomorphology* 1:39–58.

Tidwell, W. D., and S. R. Rushforth
   1970. *Osmundacaulis wadei,* a new osmunda-
   ceous species from the Morrison Formation (Jurassic) of Utah. *Bull. Torrey
   Bot. Club* 97:137–144.
Treanor, L. L., and D. P. Whittier
   1968. The effect of mineral nutrition on apogamy in *Pteridium. Can. Jour. Bot.* 47:
   773–777.
Tryon, A. F.
   1964. *Platyzoma* – A Queensland fern with incipient heterospory. *Amer. Jour. Bot.* 51:
   939–942.
Tryon, A. F., and G. Vida
   1967. *Platyzoma*: a new look at an old link in
   ferns. *Science* 156:1109–1110.
Tryon, R. M.
   1964. Evolution in the leaf of living ferns.
   Mem. Torrey Bot. Club 21:73–85.
Vladesco, M. A.
   1935. Recherches morphologiques et expéri-
   mentales sur l'embryogénie et l'organo-
   génie des fougères leptosporangiées.
   Rev. Gen. Bot. 47:513–528; 564–588.
Voeller, B.
   1971. Developmental physiology of fern gametophytes: relevance for biology. *Bio-
   Science* 21:266–270.
Wagner, W. H., Jr.
   1969. The construction of a classification. Pp.
   67–103 in *Systematic Biology.* (Publ.
   1692.) Nat. Acad. Sci., Washington, D.C.
Wagner, W. H., Jr., and A. J. Sharp
   1963. A remarkably reduced vascular plant in
   the United States. *Science* 142:1483–
   1484.
Ward, M.
   1954. Fertilization in *Phlebodium aureum* J. Sm.
   Phytomorphology 4:1–17.
   1963. Developmental patterns of adventitious
   sporophytes in *Phlebodium aureum* J. Sm.
   Jour. Linn. Soc. London Bot. 58:377–380.
Ward, M., and R. H. Wetmore
   1954. Experimental control of development in
   the embryo of the fern, *Phlebodium
   aureum. Amer. Jour. Bot.* 41:428–434.
Wardlaw, C. W.
   1946. Experimental and analytical studies of
   pteridophytes. VIII. Further observations on bud development in *Matteuccia*

struthiopteris, Onoclea sensibilis, and species of *Dryopteris. Ann. Bot.* n.s. 10:117–
   132.
   1949. Further experimental observations on
   the shoot apex of *Dryopteris aristata*
   Druce. *Phil. Trans. Roy. Soc. London*
   223(B):415–451.
   1952. *Phylogeny and Morphogenesis.* Macmillan,
   London.
   1958. Reflections on the unity of the embryonic tissues in ferns. *Phytomorphology*
   8:323–327.
   1962. The sporogenous meristems of ferns: A
   morphogenetic commentary. *Phytomorphology* 12:394–408.
   1965. The organization of the shoot apex. Pp.
   966–1076 in Ruhland, W. (ed.), *Encylo-
   pedia of Plant Physiology,* Vol. XV.
   Springer-Verlag, New York.
Wardlaw, C. W., and E. G. Cutter
   1954. Effect of deep and shallow incisions on
   organogenesis at the fern apex. *Nature*
   174:734–735.
   1956. Experimental and analytical studies of
   pteridophytes. XXXI. The effect of shallow incisions on organogenesis in *Dryopteris aristata* Druce. *Ann. Bot.* n.s. 20:
   39–56.
Wardlaw, C. W., and D. N. Sharma
   1961. Experimental and analytical studies of
   pteridophytes. XXXIX. Morphogenetic
   investigations of sori in leptosporangiate
   ferns. *Ann. Bot.* n.s. 25:477–490.
Webster, B. D., and T. A. Steeves
   1958. Morphogenesis in *Pteridium aquilinum*
   (L.) Kuhn. General morphology and
   growth habit. *Phytomorphology* 8:30–41.
Wetmore, R. H.
   1950. Tissue and organ culture as a tool for
   studies in development. *Rep. Proc. 7th
   Int. Bot. Congr.* (Stockholm) p. 369.
   1954. The use of "in vitro" cultures in the investigation of growth and differentiation
   in vascular plants. *Brookhaven Symp. Biol.*
   6:22–40.
White, R. A.
   1961. Vessels in roots of *Marsilea. Science* 133:
   1073–1074.
   1963a. Tracheary elements of the ferns. I. Factors which influence tracheid length;

correlation of length with evolutionary divergence. *Amer. Jour. Bot.* 50:447–455.

1963b. Tracheary elements of the ferns. II. Morphology of tracheary elements; conclusions. *Amer. Jour. Bot.* 50:514–522.

1966. The morphological effects of protein synthesis inhibition in *Marsilea. Amer. Jour. Bot.* 53:158–165.

1971. Experimental studies of the sporophytes of ferns. *BioScience* 21:271–275.

Whittier, D. P.

1964a. The influence of cultural conditions on the induction of apogamy in *Pteridium* gametophytes. *Amer. Jour. Bot.* 51:730–736.

1964b. The effect of sucrose on apogamy in *Cyrtomium falcatum* Presl. *Amer. Fern Jour.* 54:20–25.

1965. Obligate apogamy in *Cheilanthes tomentosa* and C. *alabamensis. Bot. Gaz.* 126:275–281.

1966. The influence of growth substances on the induction of apogamy in *Pteridium* gametophytes. *Amer. Jour. Bot.* 53:882–886.

1971. The value of ferns in an understanding of the alternation of generations. *BioScience* 21:225–227.

Whittier, D. P., and T. A. Steeves.

1960. The induction of apogamy in the bracken fern. *Can. Jour. Bot.* 38:925–930.

14

General Morphology
of Gymnosperms

The word "gymnosperm" literally means "naked seed" and serves to designate an important character of those groups of seed-bearing plants in which the ovules and seeds are not enclosed within a carpel—as in angiosperms—but are borne in an exposed position on sporophylls, scales, or comparable structures. As we will show later in this chapter, there is an increasing body of paleobotanical evidence to support the idea that naked seeds probably evolved independently in several Paleozoic lines of plants. This concept has led to the widespread rejection of Gymnospermae as a natural taxon and the recognition, in its place, of several parallel groups of naked-seeded plants. We have adopted this viewpoint in the present book and have

classified the extinct and living gymnosperms under three major taxa: Cycadopsida, Coniferopsida, and Gnetopsida.

The gymnosperms include extremely ancient lines of seed-bearing plants; their long evolutionary history—extending back at least two or three hundred millions of years—contains many examples of organisms which flourished for a time and then became extinct. One of the most interesting and phylogenetically important of the extinct gymnosperms was an assemblage of plants decidedly fernlike in foliage and general appearance which possessed a primitive type of seed. Indeed, the pinnatifid leaves of some of these plants for a long time were classified as the foliar organs of ferns, and it was not until seeds were found attached to the fronds that the unique nature of these organisms was fully appreciated. This group was well named the "seed ferns" or "pteridosperms" and their existence in the Carboniferous Period was regarded as evidence of the origin of the gymnosperms from an ancient stock of fern-like plants (Fig. 15-2).

In addition to the seed ferns, two other ancient, extinct orders of gymnosperms are well known from the fossil records: the Cordaitales and the Cycadeoidales. The Cordaitales, according to Chamberlain (1935) "formed the world's first great forests." Many members of this essentially Paleozoic order reached considerable height and differed from the seed ferns and other cycadophytes by their simple leaves, which in some were as long as one meter (Fig. 14-1). The widely distributed genus *Callixylon* — notable for the beautiful preservation of the wood of the trunk and branches — was considered for many years a member of the Cordaitales until Beck (1960b) demonstrated that it represented the stem of a group of plants to which he gave the name Progymnospermopsida (see p. 393 for a detailed discussion of Beck's reconstruction of the progymnosperm *Archaeopteris*).

The Cycadeoidales, which resembled certain modern cycads in their short or columnar, weakly branched trunks and large pinnately compound leaves, flourished during the Jurassic period (Fig. 14-1). The cycadeoids were contemporaneous with the large herbivorous dinosaurs and the abundance of cycadeoid plants in the Mesozoic period is responsible for the designation of this era as the "Age of Cycads."

The remaining members of the gymnosperms include both extinct and living forms and are conveniently grouped in four orders: Cycadales, Ginkgoales, Coniferales, and "Gnetales." A detailed account of the morphology, anatomy, and reproduction of selected representatives of these orders will be presented in Chapters 15–18. At this point, however, it is desirable to describe briefly the general nature and geographical distribution of these surviving gymnosperms. The largest group, and the one containing the most familiar gymnosperms, is the Coniferales. Examples of this order are pines, spruces, firs, cedars, and junipers. Some conifers are not only among the largest plants on earth, but are also (for example, *Sequoiadendron giganteum* and *Pinus aristata*) organisms with a life-span reaching or even exceeding 4,000 years. The Coniferales are worldwide in distribution and many of them form extensive forests in various parts of both the northern and southern hemispheres. In contrast, the living representatives of the Ginkgoales and Cycadales are veritable "living fossils" in the modern world of plants. *Ginkgo biloba*, for example, is the sole living member of the Ginkgoales and exists in the wild state only in certain mountains in southeastern China. The living cycads are also relicts from a past age and at present are confined to limited areas in the subtropics and tropics. The cycads only rarely form continuous stands or represent abundant components of a given flora. The final living order of the gymnosperms is the "Gnetales," a very puzzling group of plants from an evolutionary point of view. *Welwitschia* is a monotypic genus and is restricted to certain desert regions in southwest Africa. The other two genera, *Ephedra* and *Gnetum*, are more diversified taxonomically and have a wider range of distribution. *Ephedra* — usually a shrub or small tree — occurs in tropical and temperate Asia and in certain temperate regions of North and South America. *Gnetum*, which is most commonly a vine, is found in tropical areas in Asia, Africa, and South America. As will be discussed in Chapter 18, the "Gnetales" have frequently been regarded as a group bearing certain relations to the angiosperms, but because at present they are only represented in the fossil record by pollen remains, they constitute a

Table 14-1
Summary of principal distinctions between cycadophytic and coniferophytic
gymnosperms

Character	Cycadophytes	Coniferophytes
Habit	Weakly branched to unbranched with short conical or columnar trunks	Profusely branched, often with tall excurrent trunks
Leaves	Relatively large, pinnately compound	Relatively small, simple and needle-like, scale-like or laminate in form
Stem anatomy	Pith and cortex large; little secondary wood, manoxylic in type	Pith and cortex small; copious secondary wood, pycnoxylic in type

very isolated and poorly understood group of gymnospermous plants from an evolutionary point of view.

In concluding this section, it is necessary to discuss briefly the very broad, *organographic scheme* of classifying gymnosperms which apparently first appeared in the textbooks written by Coulter and Chamberlain (1917) and Chamberlain (1935). According to these authors, there are two *major* evolutionary lines within the gymnosperms as a whole, the "cycadophytes" and the "coniferophytes" (Table 14-1). The cycadophytes include all gymnosperms with fern-like pinnatifid leaves, columnar or weakly branched trunks, and large and conspicuously developed pith and cortical zones in the stems. The secondary xylem of typical cycadophytes is relatively small in amount and is composed of tracheids mixed with relatively abundant parenchyma; wood of this type has been termed *manoxylic*. The cycadophytes are essentially a fossil group and comprise the extinct seed ferns and the cycadeoids; the only surviving representatives of this line in

gymnosperm evolution are the modern cycads. In contrast, the coniferophytes are distinguished by their more profusely branched trunks, their simple leaves (needle-like, scale-like, or laminate in form) and the relatively small pith and cortical zones in the stem. The secondary xylem of typical coniferophytes is less parenchymatous than that of cycadophytes; wood of this type is termed *pycnoxylic*. The coniferophytes include the Cordaitales, the Paleozoic and living conifers, and the Ginkgoales. Because of their uncertain phylogeny, the "Gnetales" are usually not included in either the cycadophytes or coniferophytes but are regarded as an isolated group of unknown affinity (Chapter 18).

Although the fossil record has convinced many morphologists and paleobotanists that the cycadophytes and coniferophytes very probably represent distinct, parallel lines of gymnosperm evolution, it is still uncertain whether the two lines arose from a common ancestor or originated independently. As we will show later in this chapter, the answer to the problem perhaps lies in the continued

Figure 14-1 (*facing page*)
Reconstructions showing the habit and general organography of extinct types of gymnosperms. Left, *Dorycordaites,* a member of the Cordaitales; center, *Cycadeoidea,* right, *Williamsonia sewardiana,* both members of the Cycadeoidales. Although *Dorycordaites* was not coexistant with other genera shown, it has been included in the figure to emphasize contrasts in growth habits. [*Dorycordaites* redrawn from *Studies in Fossil Botany* by D. H. Scott. Adam and Charles Black, London, 1920; *Williamsonia* adapted from *An Introduction to Paleobotany* by C. A. Arnold. McGraw-Hill, New York, 1947.]

study of the progymnosperms—a plexus of Devonian and Carboniferous plants which combined, in a remarkable way, free-sporing reproduction with a gymnospermous type of wood anatomy.

Ontogeny and Structure of the Seed

The precursor of a gymnosperm seed is the *ovule*, a structure which has often been characterized as "an integumented megasporangium." From a phylogenetic standpoint, however, it is not yet certain whether primitive types of ovules originated from the sporangia of homosporous or heterosporous plants (for a discussion of this problem, see Doyle, 1953). Furthermore, delineating the morphology and phyletic origin of the integument are still elusive problems, although considerable progress has been made with reference to the evolutionary history of the integument in the seeds of the pteridosperms (see Chapter 15).

Despite the present inadequate state of knowledge of the phylogeny of seeds, considerable information exists regarding the ontogeny and structure of the seeds of living gymnosperms. In this portion of the chapter, we present a critical analysis and summary of the processes that occur during the development of the ovule into a seed. To promote clarity and continuity many of the details of gametophyte morphology and embryogenesis have been omitted and will be reserved for the more extended treatments in Chapters 15–18.

In simplest ontogenetic terms, a gymnospermous seed is initiated as the result of the fertilization of the egg cell of the female gametophyte contained within the ovule. The diploid zygote produces the embryo, which remains embedded in the nutritive tissue of the female gametophyte, and the integument gives rise to the seed coat. The entire composite structure becomes detached from the parent sporophyte and ultimately germinates

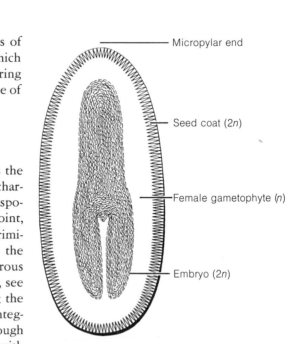

Figure 14-2
Structure of a gymnospermous type of seed in longisectional view.

to produce a new plant. From the standpoint of alternation of generations, the seed is thus a remarkable combination of two sporophytic generations and one gametophytic generation (Fig. 14-2). The seed coat is diploid and represents a part of the previous sporophyte; the nutritive tissue is the haploid female gametophyte, and the embryo is the new diploid sporophyte generation.

With this general orientation in mind, it is now appropriate to discuss the main stages in the development of an ovule into a seed. Frequent reference to the generalized diagram of the life cycle of an hypothetical gymnosperm, given in Fig. 14-3, will greatly aid in understanding the following discussion.

The Ovule and Megasporogenesis

One of the essential prerequisites for the development of a seed is the production of two different spore types, i.e., microspores and megaspores. Unfortunately, these names,

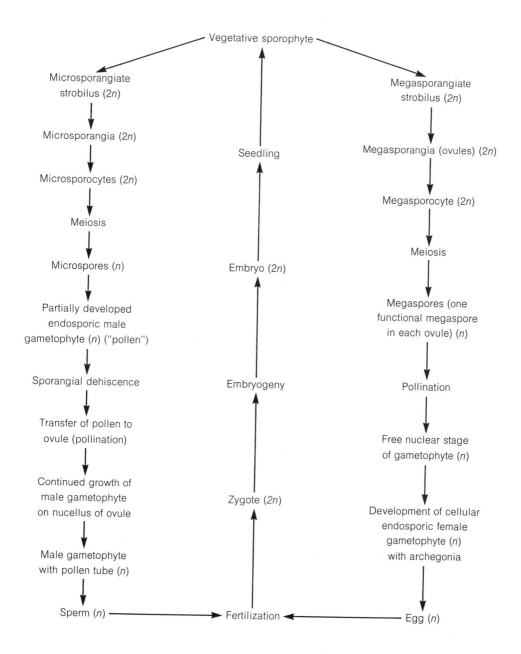

Figure 14-3
Generalized representation of the reproductive cycle in gymnosperms.

when used to designate the two kinds of spores in seed plants, suggest a direct comparision, or homology, with heterosporous vascular plants that have free spores, such as *Selaginella* and *Isoetes*, in which the megaspore is a relatively enormous cell as compared with the much smaller microspore. Several investigations however, have revealed that the size relations between microspores and megaspores in the gymnosperms as a whole tend to be the *reverse* of those characteristic of lower heterosporous plants. Very commonly the microspore is the larger of the two spore types, or else the two kinds of spores are approximately equal in size. On the basis of their studies on spore dimensions, both Thomson (1927, 1934) and Doyle (1953) maintain that gymnosperms may have had an homosporous rather than an heterosporous ancestry, and that pending more decisive evidence, noncommittal terms should be used. Thompson originally proposed "pollen spores" and "seed spores" as substitutes, respectively, for microspores and megaspores, whereas Doyle suggested, as alternate terms, "androspores" and "gynospores." Although it can be argued that there is considerable merit in these proposals, the terms microspore and megaspore are so widely used in morphological literature that it seems preferable to retain them in this textbook. But it must be emphasized that this is done with the understanding that these terms do not necessarily imply a *direct phylogenetic connection* between *modern* gymnosperms and any group of living, heterosporous vascular plants that have free spores.

Much more fundamental than a mere distinction in size is the marked difference in the *behavior* of the two spore types in the gymnosperms. The numerous microspores develop in distinct microsporangia of the eusporangiate type, and the endosporic male gametophytes have already begun their development at the time of dehiscence of the sporangium (Fig. 17-24). In contrast, the usu-ally solitary functional megaspore arises from deep within the tissue of the ovule and is never released by dehiscence from this type of sporangium. On the contrary, the megaspore enlarges and gives rise to an endosporic female gametophyte which is nourished and sheltered by the enveloping tissues of the ovule. This characteristic *retention of the megaspore* and its growth and development into a female gametophyte within its own sporangium represent fundamental prerequisites for seed development. For this reason it will now be necessary to examine briefly the structure of the gymnospermous ovule, which differs in a number of respects from a typical megasporangium.

Figure 14-4, A, depicts somewhat diagrammatically an ovule in median longisectional view. The main body of the ovule is known as the nucellus and consists of parenchymatous tissue. Deeply embedded within the nucellus there is evident a linear tetrad of four well-defined haploid megaspores which have originated by meiosis from a single megasporocyte. The lowermost of these megaspores (i.e., the one farthest away from the micropyle) is destined to enlarge and to give rise to the female gametophyte, whereas the remaining spores above it will ultimately degenerate (Fig. 14-4, B). Surrounding the upper free end of the nucellus is a collar or rim-like integument which is pierced by a small central opening or *micropyle.* As will be described more fully later, the micropyle in gymnospermous ovules provides the means of entrance to the interior of the ovule for the endosporic male gametophytes. Except for the Gnetopsida, which are frequently said to develop ovules with two integuments, all other living gymnosperms consistently form a single integument which is free from the nucellus only near the upper or micropylar end of the ovule. The lower portion of the ovule, where integument and nucellus are firmly joined, is termed the *chalaza.*

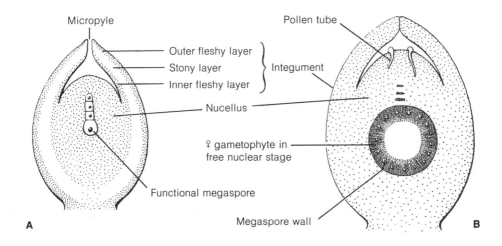

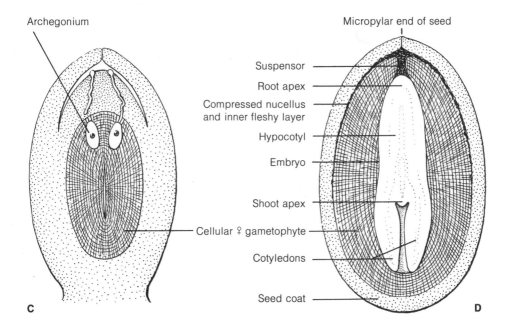

Figure 14-4
The processes and structures concerned in the development of a seed in gymnosperms.
A, longisection of ovule, showing linear tetrad of megaspores, the lowermost of which will develop into the female gametophyte; **B,** female gametophyte in free nuclear stage of development (note early phases in growth of pollen tubes in nucellus); **C,** cellular female gametophyte with two archegonia (note that a pollen tube has reached the archegonium at the right); **D,** mature seed, consisting of seed coat, remains of the nucellus, female gametophytic tissue, and endoscopically oriented embryo.

Although the nucellus of the ovule clearly appears to be *functionally equivalent* to a megasporangium, the integument represents an accessory structure which is not duplicated in the megasporangia of lower heterosporous vascular plants. It has been conjectured that the integument represents phylogenetically an indusium-like structure and that the nucellus may represent the only surviving sporangium of a hypothetical fern sorus. But as Schnarf (1937) has emphasized, this theory and other hypotheses regarding the origin of the integument have little evidence in their favor, and at present the evolutionary history of the integument is an unsolved problem (see Chapter 15 for more recent ideas).

From an anatomical standpoint, however, the integument has been rather thoroughly studied, and certain facts deserve brief mention at this point. Throughout the gymnosperms the integument is histologically differentiated into three zones or layers: an outer fleshy layer, a middle "stony" or sclerenchymatous layer, and an inner fleshy layer (Fig. 14-4, A). The degree of development of each of these layers during seed ontogeny varies within the different groups of the gymnosperms; in some (e.g., many conifers) the outer fleshy layer is rudimentary, whereas in others (e.g., the cycads and *Ginkgo*) this layer is thick, and may be conspicuously pigmented in the mature seed. In all, however, the inner fleshy layer tends to collapse and in the mature seed appears as a papery layer lining the inner surface of the stony layer. Further indication of the histological complexity of the integument is shown in many gymnosperms by the development of a vascular system (Fig. 15-21, D). In the cycads vascular strands traverse *both* the outer and inner fleshy layers, but in *Ginkgo* only the inner bundle system is developed. In members of the pine family both sets of vascular bundles have been eliminated. These varying conditions are of interest when compared with the structure of certain types of Paleozoic seeds in which the nucellus itself, as well as the outer fleshy layer of the integument, contained vascular tissue (Fig. 15-6). Therefore it seems reasonable to conclude that there has been a general tendency in seed evolution to eliminate vascular strands first from the nucellus and then from one or both fleshy layers of the integument.

Very commonly reference is made to the process of megasporogenesis in *Selaginella* as an approach to one of the processes underlying the seed habit. In two reported instances the development of the female gametophyte and the embryo may occur within the sporangial cavity (see Chapter 9). However, the megasporangium of *Selaginella* lacks an integument; in this respect and in the absence of any vascular system it differs markedly from a gymnospermous ovule. Furthermore, the growth of the female gametophyte and the embryo in *Selaginella* is at the expense of food previously stored in the megaspore itself, whereas the gametophyte and embryo of a seed depend for nutrition on the food material supplied to them by the closely adjacent tissue of the ovule.

The Female Gametophyte

A detailed description of the structure and development of the female gametophyte of various gymnosperms will be found in the next four chapters. At this point, discussion will be restricted to the more general morphological aspects of the female gametophyte with special reference to seed development.

The first phase of development of the gymnospermous female gametophyte is characterized by an extensive series of free-nuclear divisions (Fig. 14-4, B). Ultimately wall formation begins at the periphery and proceeds centripetally until the entire gametophyte consists of cells which later become richly stored with reserve food materials (Fig. 14-4, C). Throughout the ontogeny of the gametophyte this structure is encased by the well-defined *megaspore wall*, which appears in some species to increase in thickness during the development of the ovule. It may attain a thickness of 9 or 10 microns in the adult seeds of some of the cycads. As Schnarf (1937, p. 18) has emphasized, the presence of a conspicuous megaspore wall surrounding the female gametophyte is one of the most important and definitive features common to gymnospermous seeds; it may serve to indicate the phylogenetic connection between the megagametophytes of free-sporing plants and those of primitive gymnosperms.

During or following the process of wall formation, certain individual superficial cells of the gametophyte—usually those near its micropylar end—give rise to archegonia (Fig. 14-4, C). The number of archegonia varies considerably, ranging from usually two in *Ginkgo*, to many, as in some of the conifers. All living gymnosperms, with the exception of *Gnetum* and *Welwitschia*, are thus archegoniate plants, despite the fact that the female gametophyte is no longer a freely exposed sexual plant. (See Chapter 5 for a general discussion of the Archegoniatae.)

Pollination, the Formation of Pollen Tubes, and Fertilization

Since the ovules of gymnosperms are exposed, the process of pollination consists in the transferral (usually by wind) of the partly developed endosporic male gametophyte (pollen grain) to the micropylar end of the ovule. In many gymnosperms, the pollen is said to adhere to a drop of fluid (the so-called "pollination drop") which exudes from the micropyle. According to Chamberlain (1935), the retraction of the pollen drop causes the pollen grains to be drawn through the micropylar canal to the nucellus. This explanation, however, is not supported by the detailed studies of Doyle and O'Leary (1935) on a number of conifers. In *Pinus*, for example, they found that the pollen *at first* adheres to the projecting arms of the integument, which thus function somewhat like a stigma. Subsequently, a small amount of liquid is secreted by the ovule, filling the micropylar canal and spreading as a film on the projecting arms. Pollen grains float in this liquid and then, since the ovules of *Pinus* are inverted, the grains begin to *rise upwardly in the liquid*, transversing the micropylar canal and finally reaching the nucellus. In Douglas fir (*Pseudotsuga*), however, no fluid is secreted by the ovule and the pollen grains, "trapped" within the stigmatic lobes of the integument, germinate and form pollen tubes which elongate and finally reach the surface of the nucellus (see Allen and Owens, 1972).

In some of the gymnosperms, a well-defined recess, termed the *pollen chamber*, is developed at the free apical end of the nucellus (Figs. 15-21, D; 16-7; 18-14, B). Pollen chambers were highly developed in the ovules of Paleozoic gymnosperms and are believed to have served as liquid-containing cavities within which the flagellated sperms were liberated by the dehiscence of the wall of the pollen grain. Among living gymnosperms, pollen chambers occur only in *Ginkgo*, the cycads, and *Ephedra*.

One of the distinctive morphological characters of *living* gymnosperms is the production of a more-or-less tubular outgrowth of the male gametophyte known as the pollen tube (Fig. 14-4, B, C). In *Ginkgo* and the cycads the pollen tube is largely haustorial in function, and grows, often for several months, like a fungal hypha *laterally* into the tissue of the nucellus; its function seems to be the absorption of food materials which are used by the gametophytic cells at the lower end of the tube. At the time of fertilization the basal end of the pollen tube bursts, liberating the two large flagellated sperms together with some liquid into the cavity directly above the female gametophyte (Fig. 15-21, C). One or both sperms may then enter an archegonium, and one of them fertilizes the nucleus of the large egg cell. In contrast, the pollen tubes of the conifers are sperm carriers, and after growing downward through the intervening nucellar tissue they convey the nonflagellated male gametes directly to the archegonium (Fig. 14-4, C).

The term "siphonogamous" has been used collectively to designate plants in which the sperms are directly conveyed to the egg by means of a pollen tube. Lower vascular plants, by contrast, have been designated as "zooidogamous" because the motile flagellated sperms are freely liberated from the antheridium into water through which they must move (often for some considerable distance) to reach and fertilize the eggs. The evolutionary steps in the transition from zooidogamy to siphonogamy are by no means entirely clear, but the development of sperm-carrying pollen tubes, typical of higher gymnosperms and of all angiosperms, was surely a significant achievement. By means of the pollen tube the considerable chances and hazards of the aquatic zooidogamous method were eliminated, and much greater assurance of fertilization was made possible. In the three extinct orders of gymnosperms—Pteridospermales, Cordaitales, and Cycadeoidales—there is no evidence of the existence of pollen tubes. It is possible that in these groups the motile sperms were liberated directly into the pollen chambers by the rupture of the spore wall. Among living types the cycads and *Ginkgo* appear to have a rather primitive type of pollen tube which serves primarily as a haustorium. It seems significant in this connection that these are the only known living seed plants

that have retained the flagellated type of sperm characteristic of zooidogamous tracheophytes.

Embryogeny and the Maturation of the Seed

One of the most distinctive features of seed development in the majority of living gymnosperms is a period of free nuclear divisions at the beginning of embryogeny. The only known exceptions are found in *Welwitschia*, *Gnetum*, and *Sequoia sempervirens* in which, as in angiosperms (aside from *Paeonia*) and all lower vascular plants, the first division of the zygote is followed directly by the development of a wall separating the two cells. In most gymnosperms, beginning with the first mitotic division of the nucleus of the fertilized egg, there is a more-or-less protracted phase of nuclear multiplication unaccompanied by the formation of walls. This initial phase of embryogeny thus resembles the early stage of development of the female gametophyte (Fig. 14-4). The number of free diploid nuclei formed in the young embryo of gymnosperms varies widely. In some of the cycads (e.g., *Dioon edule*), about 1,000 free nuclei are formed before walls begin to appear, whereas in *Pinus* the number is four.

The phylogenetic significance of the period of free nuclear divisions in gymnosperm embryogenesis presents a difficult problem. With reference to the cycads, Chamberlain (1935, p. 142) maintained that "a free nuclear period arose as a consequence of the enlarging egg. The mass of protoplasm became so large that the early mitotic figures could not segment it." From both a phylogenetic and morphogenetic viewpoint, however, it is not clear (1) why there is so much variation in the *extent* of the free nuclear period among various gymnosperm taxa or (2) why in *Sequoiadendron giganteum* (giant sequoia), the proembryo is initiated by free nuclear divisions while in the related *Sequoia sempervirens* (coast redwood) the first division of the zygote is accompanied by wall formation (Buchholz, 1939).

Following the free nuclear phase, the embryo in gymnosperms becomes cellular and gradually differentiates into a suspensor, shoot apex, cotyledons, hypocotyl, and radicle. From the standpoint of polarity, the embryo of gymnosperms is strictly endoscopic with the shoot end directed away from the micropyle (Fig. 14-4, D).

In most of the gymnosperms there is a marked tendency toward the condition of polyembryony; i.e., the formation of several embryos in a single gametophyte. This is possible because more than one archegonium is commonly fertilized, and hence several zygotes may be produced. But more remarkable is the process of cleavage polyembryony, which is characteristic of many conifers. In this process, which will be described in more detail in Chapter 17, certain cells of the young embryo become separated from one another and give rise to a system of four or more distinct embryos. In some of the conifers both types of polyembryony may occur in the same developing seed. Physiological competition between the various embryos usually results in the elimination of all but one, which continues its differentiation and becomes the dominant embryo in the fully developed seed.

During the last phases of embryogeny the nucellar tissue of the ovule becomes disorganized and frequently persists only as a paper-like cap of dry tissue at the micropylar end of the seed (Fig. 14-4, D). Further histological maturation of the various layers of the integument continues, and the stony layer becomes an extremely hard, resistent shell which effectively encloses and mechanically protects the female gametophyte and the embryo. Except for the cycads and *Ginkgo*, the detached seed of gymnosperms remains

dormant for some time. Under favorable conditions the embryo resumes growth and, rupturing the seed coat, develops into a new sporophyte plant (Chapter 17).

Archaeopteris and the Origin of Gymnosperms

Because the gymnosperms include extremely ancient lines of seed-bearing plants, the ultimate solution of the problem of the type—or types—of vascular plants from which they may have originated is primarily dependent on the analysis of the fragmentary fossil record. The formidable nature of the task which confronts paleobotanists is aggravated by the fact that many gymnosperm fossils—like the remains of other, more primitive tracheophytes—are isolated portions of leaves, strobili, sporangia, spores, stems, roots, etc. Many of these separate parts have been provisionally assigned generic names, but efforts to reconstruct a "complete plant" from the synthesis of such "organ genera" are often highly theoretical and subject to drastic revision as new paleobotanical discoveries are made. The most significant progress in understanding, however, is made when two or more "organ genera" can be demonstrated to be physically interconnected. One of the most remarkable discoveries of this sort in recent years was made by Beck (1960a, 1960b) who found two well-known organ genera, *Archaeopteris* (the "frond" of a presumed fern) and *Callixylon* (the stem of a gymnospermous tree) in organic connection (Fig. 14-5). The plant represented by the interconnection of *Archaeopteris* and *Callixylon* was placed by Beck (1960b) in a new class of tracheophytes which he named the Progymnospermopsida. As one of the relatively advanced representatives of this class, *Archaeopteris* was shown by Beck to combine the anatomy typical of

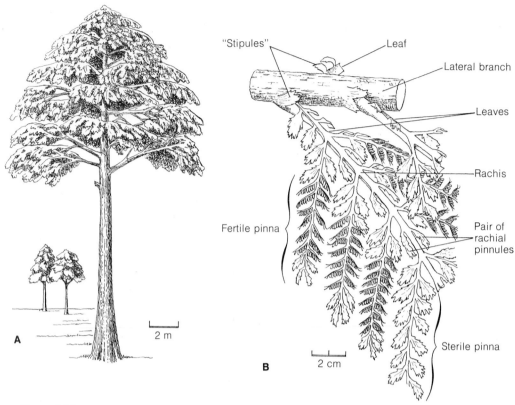

Figure 14-5
Restorations of *Archaeopteris*. **A**, habit; **B**, branch with basal parts of three "leaves." [Redrawn from Beck, *Amer. Jour. Bot.* 49:373, 1962.]

gymnosperms (i.e., secondary xylem tracheids with circular bordered pits) with "fern-like" fronds and so-called "pteridophytic reproduction" (i.e., reproduction by free spores, rather than by seeds). According to Beck's (1970, p. 379) interpretation, "the progymnosperms are of great significance because they seem to be the immediate ancestors of seed plants."

Following the original reconstruction of *Archaeopteris*, Beck (1962, 1964, 1970, 1971) and others (Carluccio et al., 1966; Banks, 1968) have made further studies on this remarkable progymnosperm and have considerably advanced our knowledge of its morphology, anatomy, and phylogenetic relationships. To recount in detail the "con-

tinuing story" of *Archaeopteris* is far beyond the scope or intention of the present text. But it will be useful (1) to note the salient features of Beck's discovery, because it represents an important milestone in paleobotany; (2) to review the changing concepts with reference to the morphology and vasculature of the "frond"; and (3) to consider briefly recent ideas of the evolution of the stele in progymnosperms and gymnosperms.

Historical Résumé

Prior to Beck's (1960b) investigations, *Callixyon* and *Archaeopteris* had been recognized by paleobotanists as two of the most widespread "organ genera" of the Late Devonian Period. The generic name *Archaeopteris* was given to compressions of large fern-like "leaves" which were considered by Arnold (1947) to represent "the most widely distributed fern in the Upper Devonian." According to Beck (1960a), *Archaeopteris* is a common fossil in eastern North America and has been collected in various localities from Gaspé Peninsula, Quebec, to southwestern Pennsylvania; the specimen used by Beck in his original reconstruction was found by him in beds of Late Devonian age near Sidney, New York. *Callixylon*, on the other hand, is a name originally applied to petrifications of certain Devonian trees with gymnospermous secondary xylem. Stumps of *Callixylon* five feet in diameter have been reported and logs 20–28 feet long or more have been uncovered as fossils in certain areas of Kentucky and Texas.

The specimen, which revealed the organic connection between the frond of *Archaeopteris* and the stem of *Callixylon*, consisted of a small axis "to which are attached several fragments of bipinnately compound leaves, one of which bears both fertile and vegetative parts" (Beck 1960b, p. 352). In accordance with the rule of priority, established by the International Code of Nomenclature, Beck used the earlier name, *Archaeopteris*, for the entire plant represented by the two organ genera; he considered that the closest affinity of his specimen was with *Archaeopteris macilenta* Lesquereux.

Obviously, the critical point in Beck's discovery was his demonstration that the axis of his specimen represents a portion of the stem of *Callixylon*. This was shown by the presence of well-defined *groups* of circular bordered pits on the radial walls of the tracheids of the secondary xylem, a feature which had long been considered distinctive of the wood of *Callixylon* (see Beck 1960b, 1970 for a detailed analysis of the secondary xylem of *Callixylon*). Beck has contributed

an informal and highly readable account of his discovery of *Archaeopteris* to the recent textbook by Jensen and Salisbury (1972, pp. 493–495).

Using data based on previous descriptions of the habit of *Callixylon* by Arnold (1931), *Archaeopteris* was considered by Beck (1962) to have been a large tree—attaining a height of 60 feet or more—with a crown of branches bearing large bipinnately compound leaves (Fig. 14-5). The "fertile pinnules," according to his interpretation, bore either one or two rows of sporangia. Although Beck found that the spores of *Archaeopteris macilenta* varied from 44–68 μ in diameter, it is uncertain whether the genus as a whole was homosporous or whether it comprised *both* heterosporous and homosporous species; according to Arnold (1947, p. 177), *Archaeopteris latifolia* was heterosporous. Fig. 14-6 represents the reconstruction made by Carluccio et al. (1966) and shows that the sporangia were borne in two rows on the adaxial side of the fertile pinnule. These investigators described the structure of the epidermis of the sporangium wall and its longitudinal method of dehiscence but provided no further information on spore morphology in *Archaeopteris*.

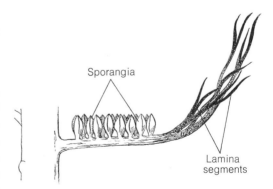

Sporangia

Lamina segments

Figure 14-6
Reconstruction of a fertile "pinnule" of *Archaeopteris*. Note the two rows of adaxial sporangia. [Redrawn from Carluccio et al., *Amer. Jour. Bot.* 53:719, 1966.]

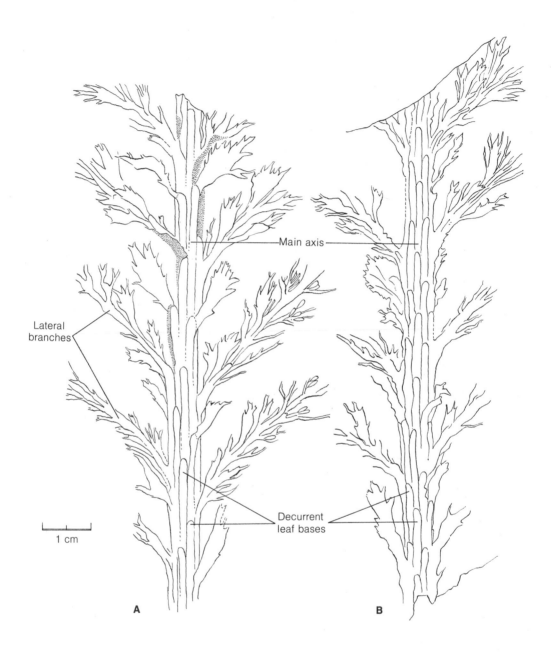

Main axis

Lateral
branches

Decurrent
leaf bases

1 cm

A

B

Figure 14-7
Camera lucida drawings of "part" and "counterpart" of a compression specimen ("375-Y") of *Archae-opteris,* showing helical phyllotaxis of the decurrent leaf bases on main axis and distichous arrangement of lateral branches. See text for further explanation. [Redrawn from Beck, *Amer. Jour. Bot.* 58:758, 1971.]

Morphology of the "Frond"

One of the basic questions posed by Beck's original interpretation concerns the nature of the "fronds" which he found attached to the stem of *Callixyon*. In his 1962 paper, he interpreted the frond as a bipinnately compound leaf which consists of a "rachis" bearing pairs of "rachial pinnules" between the pairs of opposite or subopposite sterile and fertile "pinnae" (Fig. 14-5, B). In the light of this interpretation, the frond of *Archaeopteris* could be regarded as a megaphyll, comparable in its general morphology to the leaf of ferns and certain pteridosperms.

A very different interpretation of the frond of *Archaeopteris* was subsequently proposed by Carluccio et al. (1966). They found that the so-called "rachis" has a radially symmetrical vascular system like a stem and that the axes of the lateral pinnae are characterized by a similar type of vasculature. Using these facts, Carluccio et al. concluded that the frond of *Archaeopteris* is not a pinnately compound leaf but rather should be regarded as a leafy "branch system." According to this interpretation, the actual leaves are helically arranged and comprise (1) the pairs of "rachial pinnules" and (2) the fertile and sterile appendages (pinnules) borne on the lateral axes which had previously been regarded as pinnae. Beck (1970) later adopted this revised interpretation and further noted that the frond appears to be subtended by a leaf, which he had earlier (1962) designated as a stipule (Fig. 14-5, B). This latter term, which is more properly applied to one of the paired basal appendages of the leaves of certain dicotyledons, seems an extremely inappropriate word to use with reference to the morphology of such an archaic plant as *Archaeopteris* (see Chapter 19, pp. 554–556 for a detailed account of stipule morphology in the dicotyledons.)

In his most recent paper, Beck (1971) reported the first major breakthrough in the interpretation of the *external morphology* of compression specimens of *Archaeopteris macilenta*. By removing the carbonaceous film covering the branch system, Beck discovered that in almost every specimen in his collections, a remarkable pattern of closely approximated *decurrent leaf bases* was revealed on the main axis (see Beck's new reconstruction shown in Fig. 14-8). The number of decurrent leaf bases usually differs on the two sides of the axis. In other words, it was essential to examine both "part" and "counterpart" of a compressed specimen. When this precaution was observed, it became evident that the lateral branches of the main axis are arranged distichously (i.e., in two opposite or subopposite series) while the phyllotaxis of

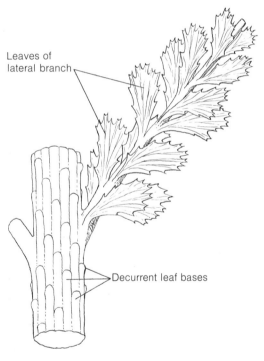

Leaves of lateral branch

Decurrent leaf bases

Figure 14-8
Archaeopteris macilenta. Restoration of the external morphology of part of a lateral branch system, showing the spirally arranged decurrent leaf bases on main axis and part of a lateral branch with its leaves. [Redrawn from Beck, *Amer. Jour. Bot.* 58:758, 1971.]

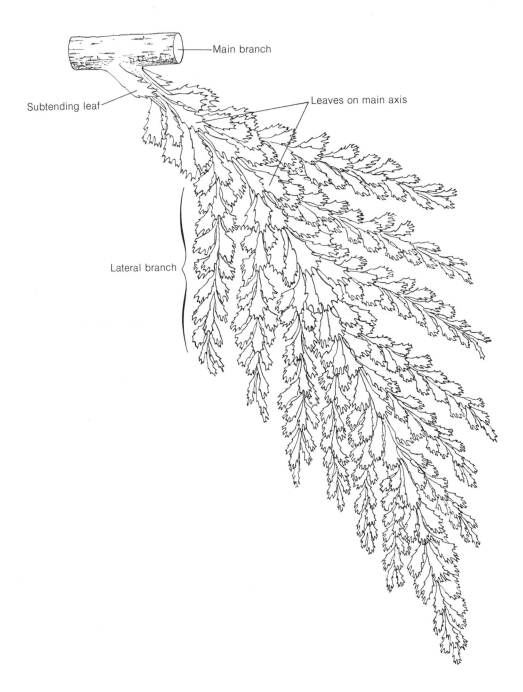

Figure 14-9
Archaeopteris macilenta. Restoration of a complete, vegetative, lateral branch system attached to a larger axis. [Redrawn from Beck, *Amer. Jour. Bot.* 58:758, 1971.]

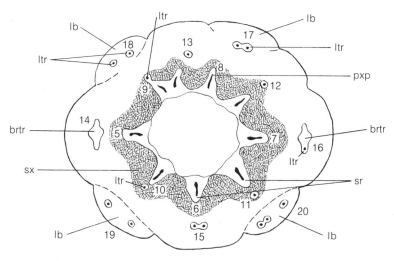

Figure 14-10
Restoration of the vascular anatomy of the main axis of *Archaeopteris macilenta* (Beck's specimen "375-II"). The stele consists of nine "ribs" (*sr*) and is enclosed by a cylinder of secondary xylem (*sx*) indicated by hatching. The helical sequence in the divergence of branch traces (*br tr*) and leaf traces (*l tr*) is indicated by numbers 5–20; higher numbers indicate older traces, lower numbers younger traces. l b, leaf base; px p, protoxylem pole. See text for further explanation. [Redrawn from Beck, *Amer. Jour. Bot.* 58:758, 1971.]

the leaf bases on the main axis is helical (Fig. 14-8). The greater density of leaves is postulated by Beck to occur on the adaxial side of a given specimen, although in some, there may be equal numbers of leaf bases on either side. These facts, according to Beck (1971, p. 775), make it certain that the earlier descriptions by many investigators of opposite or subopposite "rachial pinnules do not reflect the total morphology" of the frond-like branch system of *Archaeopteris*. Although Carluccio et al. (1966, p. 728) stated that the sterile leaves are spirally arranged on the lateral branches, Beck's new observation supported his earlier conclusion that the phyllotaxis is decussate rather than helical (see Beck, 1971, Figs. 20 and 40).

Beck (1971) synthesized the results of his new studies by means of two new restorations, one (Fig. 14-8) representing the external morphology of a segment of the main axis with its decurrent leaf bases and a lateral branch, the other (Fig. 14-9) portraying a complete, vegetative, lateral branch system.

Beck emphasized that "these restorations should supersede one previously published by me (Beck, 1962, Fig. 1), which is hereby shown to be an incomplete and thus inaccurate representation of a lateral branch system."

Vascular Anatomy of the "Frond"

The conclusions reached by Beck (1971) from his study of the external morphology of compression specimens were strongly supported by his investigations on the anatomy of the stele and the mode of divergence of the leaf traces and branch traces. The most complete information was provided by a study of transections of specimen "375-II." A reconstruction of the vasculature of the main axis of this specimen is shown in Fig. 14-10. According to Beck (1971, p. 770), the internal structure of specimen "375-II" correlates exactly with the external morphology of specimen "375-Y," which is shown in Fig. 14-7.

As shown in Fig. 14-10, the stele of specimen "375-II" is a cylindrical system of primary xylem consisting of 9 more-or-less completely interconnected radiating ribs. Branch traces and leaf traces diverge from these ribs as follows: 2 of the 9 stelar ribs give rise to traces which vascularize the members of the 2 *distichous* rows of lateral branches (Fig. 14-10, br tr). The remaining 7 ribs—3 on one side of the stele and 4 on the opposite side—are the points of origin of the traces which extend into the helically arranged leaves of the main axis (Fig. 14-10, ltr). Transections at successive levels reveal that *each* stelar rib, associated with a leaf trace, consists of 2 radially aligned *protoxylem poles* connected by a sheet of protoxylem tracheids (Fig. 14-10, pxp 8). The outer protoxylem pole, together with some metaxylem, separates as a leaf trace; within a longitudinal distance of 2–3 mm, the outer pole is reconstituted and *at a higher level,* gives rise to another leaf trace (Fig. 14-10, l tr 9). As indicated by the series of numbers in Fig. 14-10, leaf and branch traces originate from *alternate ribs* of the stele in helical sequence. Each leaf trace follows a steep, obliquely radial course through the cortex of the axis and then becomes divided into 2 strands as it enters the base of a leaf (Fig. 14-10, l b 17 and 18). The further branching of these 2 strands produces the dichotomous venation characteristic of the leaves of *Archaeopteris* (see Beck, 1962, p. 375).

Phylogenetic Significance of *Archaeopteris*

Beck's (1971) description and analysis of the external morphology and vascular anatomy of *Archaeopteris* supports the idea that the so-called "frond" of this progymnosperm represents a complex "branch system." The fact that leaf and branch traces of the main axis arise along a continuous ontogenetic spiral emphasizes that the lateral branches and leaves of this branch system are not sharply defined organ categories. If this is a correct interpretation, it seems questionable that the branch system of *Archaeopteris* typifies a stage in the evolution of the megaphyllous type of leaf characteristic of either the ferns or of such ancient gymnosperms as the pteridosperms. In his most recent paper, Beck (1971, p. 783) concludes that "the accumulative evidence suggests persuasively that *Archaeopteris* should be either considered, with related progymnosperms, as a major taxon, or be included in the Coniferophyta." In his view, there is a striking resemblance between the leafy branch systems of *Archaeopteris* and the frond-like branch systems of such living conifers as *Metasequoia* and *Taxodium*. He also notes that some of the Permian conifers, such as *Buriadia* and *Carpentieria* "closely resembled *Archaeopteris* even in details of leaf morphology."

The Progymnospermopsida, as originally conceived by Beck (1960b, p. 364), include not only *Archaeopteris* but also other Devonian and early Carboniferous genera which were characterized by secondary xylem and pteridophytic reproduction. Figure 14-11 shows a number of the genera included by Banks (1968) under the Progymnospermopsida. Among the primitive members—placed in the order Aneurophytales—may be mentioned *Aneurophyton*, in which the branching was spiral and *Tetraxylopteris,* distinguished by its opposite and decussate pattern of branching. In both genera, the branch systems have been regarded "as primitive fronds" (for detailed information on the morphology and anatomy of *Tetraxylopteris*, see Beck, 1957). With reference to Fig. 14-11, it is particularly interesting to note that although some morphologists regard *Protopteridium* as a primitive fern, this genus is classified as a progymnosperm by Banks. With reference to such problems in classification, it should be realized that the *degree* of preservation of the remains of many

Devonian plants is variable; some fossils are only represented by compressions of the branch systems, while for others there are well-preserved fragments of the stele and secondary xylem. For this reason, present concepts and classifications of progymnospermous plants must be regarded as tentative and subject to change as new paleobotanical studies continue.

If *Archaeopteris* is neither a primitive fern nor the progenitor of the ferns, as Beck (1962, 1964, 1970, 1971) has maintained, what evidence may be cited to support the possible evolution of coniferophytic gymnosperms from the progymnosperms? The investigations of Namboodiri and Beck (1968a,b,c) suggests that a possible solution of this problem lies in the correct interpretation of the structure and evolution of the stele in progymnosperms and gymnosperms. If the "eustele" in these two groups is indeed devoid of leaf gaps—and thus *different* from the steles of most ferns—the widespread concept of the fern-origin of gymnosperms would be dealt a severe blow. The presence of a leaf gap, associated with the divergence of a leaf trace from the *primary vascular cylinder*, was one of the definitive characters used by Jeffrey (1917) in establishing the "Pteropsida," a major taxon under which he grouped ferns, gymnosperms, and angiosperms. In connection with his assumption that the gymnosperms evolved from ferns, he stated (Jeffrey, 1917, p. 291) that it is now clear "that the lower gymnosperms have come from the Filicales as the result of the simplification and reduction of the primary structure of the stele on the one hand,

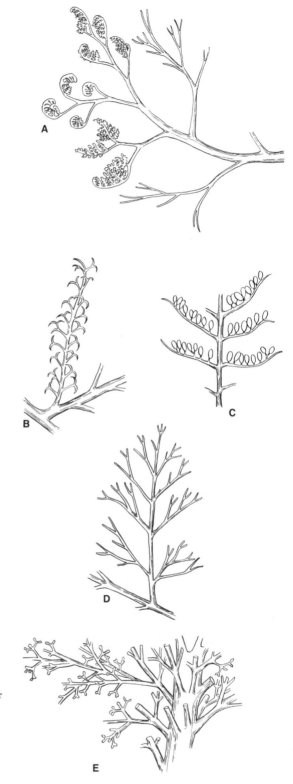

Figure 14-11
Examples of genera classified by Banks (1968) under the Progymnospermopsida. **A,** *Protopteridium hostimense;* **B,** *Aneurophyton;* **C–D,** *Svalbardia;* **E,** *Tetraxylopteris.* [Redrawn from Banks, pp. 73–107 *in* E. T. Drake (Ed.) *Evolution and Environment.* Yale University Press, New Haven, 1968: after Kraüsel and Weyland, 1933; Høeg, 1942; Beck, 1957.]

accompanied by the marked development of secondary fibrovascular tissues on the other." Beck (1964, p. 30, footnote) has expressed his dissent from such a viewpoint by remarking "that there have never been leaf gaps in the conifers, some other gymnosperms and their probable precursors, and that stelar evolution in ferns and gymnosperms comprises two entirely separate lines." Let us now examine the nature of the evidence derived from the studies of Namboodiri and Beck which appears to support this conclusion.

Figure 14-12 represents diagrammatically the main steps in the evolution of the primary vascular system of progymnosperms and gymnosperms. The most primitive stage was a protostele, the ribs of which represent the points of origin of the radially diverging traces to the "appendages" (Fig. 14-12, A). This kind of stele is illustrated by certain primitive progymnosperms such as *Aneurophyton* and *Tetraxylopteris*. The next stage in stelar evolution was characterized by the gradual dissection of the protostele into "longitudinal columns," a process accompanied by "medullation," i.e., the origin of a central pith region (Fig. 14-12, B). Further evolutionary specialization led to a more definitive system of discrete *sympodial bundles*, each of which, at the nodal level of the appendages, divided *tangentially* into two *radially aligned* strands, an outer leaf trace and an inner "reparatory strand;" the latter continued its upward course and a leaf gap was not formed (Fig. 14-12, C). This is the type of primary vascular system which Beck (1971) found in the main axis of the "frond" of *Archaeopteris* (see Fig. 14-10). The next step in evolution is characterized by a change from a tangential to a radial plane of division of the sympodial bundles and resulted in the orientation of the reparatory strand and its associated leaf trace along the same tangential plane (Fig. 14-12, D). This mode of separation of leaf traces likewise occurred

without the formation of leaf gaps and is illustrated by the Carboniferous seed fern, *Lyginopteris*. According to Beck (1970, pp. 387–388) the stelar system of *Lyginopteris oldhamia*, "in all of its major features, including the clear absence of a leaf gap, is identical with that of the conifers."

Figure 14-12, E, depicts the eustele typical of living conifers with helically arranged leaves. As in progymnosperms and *Lyginopteris*, leaf gaps are absent and in addition, the sympodial bundles at various points become closely approximated but remain unfused throughout their course in the stem. For more detailed information on the organization of the primary vascular system of conifers (including genera with decussate and whorled phyllotaxis) the student should consult the papers by Namboodiri and Beck (1968a,b).

The salient feature of the theory of stelar evolution proposed by Namboodiri and Beck (1968c) concerns the apparent similarity between the general mode of divergence of leaf traces in progymnosperms and conifers. In both groups, leaf traces are said to arise *directly* from the sympodial bundles of the eustele without the formation of leaf gaps (see Fig. 14-12, C–E). Under this interpretation, the major steps in stelar evolution of gymnosperms did not include a stage comparable to the siphonosteles of ferns in which leaf gaps subtend the divergence of the foliar traces from the primary vascular cylinder. Quite apart from phylogenetic theories, however, it is interesting to note that the delimitation of leaf gaps turns out to be a difficult problem for the comparative anatomist who is concerned with the description of the structure and ontogeny of the eustele in living seed plants. In stems in which the primary vascular system is composed of anastomosing strands, the recognition of leaf gaps, according to Esau (1965, p. 364), "is rather uncertain because the parenchyma that occurs above the diverging leaf

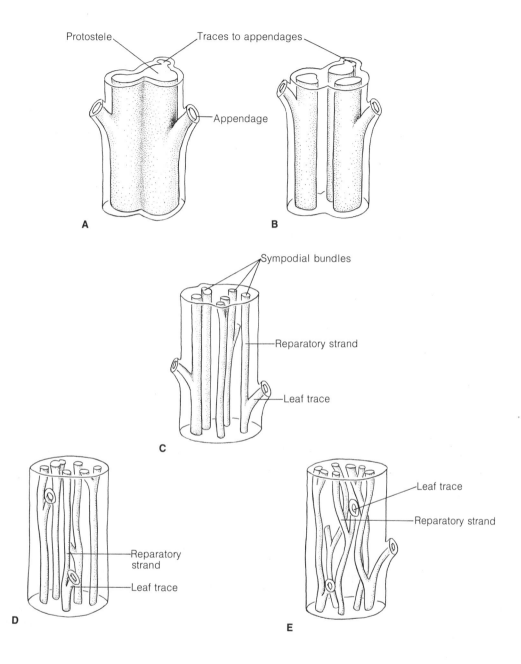

Figure 14-12
Diagrams showing the probable evolutionary development of the primary vascular systems of progymnosperms and gymnosperms: **A,** the primitive protostelic condition; **B,** division of protostele into longitudinal columns; **C,** vascular system composed of discrete sympodial bundles, each of which, at a nodal level, divides tangentially into a reparatory strand and a leaf trace, as illustrated in *Archaeopteris;* **D,** vascular system in which the reparatory strand and leaf trace are formed by the radial division of the sympodial bundle, as illustrated in *Lyginopteris,* a seed fern; **E,** primary vascular system characteristic of living conifers with helical phyllotaxis. See text for further explanation. [Redrawn from Namboodiri and Beck, *Amer. Jour. Bot.* 55:464, 1968.]

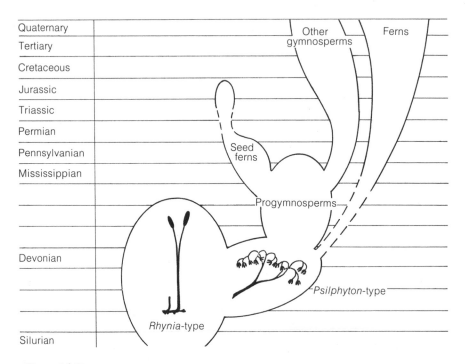

Figure 14-13
The possible evolutionary history of ferns, progymnosperms and gymnosperms. [Redrawn from *Evolution and Plants of the Past* by H. P. Banks. Wadsworth, Belmont, California, 1970.]

trace is confluent with the interfascicular regions" of the stele. But after secondary growth has occurred *in the interfascicular areas*, foliar gaps are definable in the *secondary xylem* and, in transectional view, are strictly correlated *in position* with the divergence of the leaf traces. It is this aspect of anatomy which has made possible the recognition of the various types of nodal anatomy characteristic of woody dicotyledons and conifers (for a description of the patterns of nodal anatomy in dicotyledons, see Chapter 19, pp. 570–577).

In the light of Beck's comprehensive studies on *Archaeopteris* and the theory of stelar evolution proposed by Namboodiri and Beck, the origin of gymnospermous plants from ferns—as postulated by Jeffrey (1917)—no longer appears to be a useful hypothesis. For this reason, we have given up the "Pteropsida" as a taxon and adopted, with some modification, the classification of the Tracheophyta devised by Banks (see Chapter 2). In his recent treatise on vascular plant evolution, Banks (1970) proposes that both seed ferns and other gymnosperms (e.g., conifers) may have evolved directly from the progymnosperms, and that the ferns possibly had an independent origin from some group of ancient plants of the type of *Psilophyton* (Fig. 14-13).

References

Allen, G. S., and J. N. Owens
   1972. *The Life History of Douglas Fir.* Information Canada, Ottawa.
Arnold, C. A.
   1931. On *Callixylon newberryi* (Dawson) Elkins et Wieland. *Contr. Mus. Paleontol. Univ. Mich.* 3:207–232.
   1947. *An Introduction to Paleobotany.* McGraw-Hill, New York.
Banks, H. P.
   1968. The early history of land plants. *In* Drake, E. T. (ed.) *Evolution and Environment.* Yale University Press, New Haven.
   1970. *Evolution and Plants of the Past.* Wadsworth, Belmont, Calif.
Beck, C. B.
   1957. *Tetraxylopteris schmidtii* gen. et sp. nov., a probable pteridosperm precursor from the Devonian of New York. *Amer. Jour. Bot.* 44:350–367.
   1960a. Connection between *Archaeopteris* and *Callixylon. Science* 131:1524–1525.
   1960b. The identity of *Archaeopteris* and *Callixylon. Brittonia* 12:351–368.
   1962. Reconstructions of *Archaeopteris* and further consideration of its phylogenetic position. *Amer. Jour. Bot.* 49:373–382.
   1964. The woody, fern-like trees of the Devonian. *Mem. Torrey Bot. Club* 21:26–37.
   1970. The appearance of gymnospermous structure. *Biol. Rev.* 45:379–400.
   1971. On the anatomy and morphology of lateral branch systems of *Archaeopteris. Amer. Jour. Bot.* 58:758–784.
Buchholz, J. T.
   1939. The embryogeny of *Sequoia sempervirens* with a comparison of the Sequoias. *Amer. Jour. Bot.* 26:248–257.
Carluccio, L. M., F. M. Heuber, and H. P. Banks
   1966. *Archaeopteris macilenta,* anatomy and morphology of its frond. *Amer. Jour. Bot.* 53:719–730.

Chamberlain, C. J.
   1935. *Gymnosperms. Structure and Evolution.* University of Chicago Press, Chicago.
Coulter, J. M., and C. J. Chamberlain
   1917. *Morphology of Gymnosperms.* University of Chicago Press, Chicago.
Doyle, J.
   1953. Gynospore or megaspore—a restatement. *Ann. Bot.* n.s. 17:465–476.
Doyle, J., and M. O'Leary
   1935. Pollination in *Pinus.* Sci. Proc. Roy. Dublin Soc. 21:181–190.
Esau, K.
   1965. *Plant Anatomy.* Ed. 2. Wiley, New York.
Jeffrey, E. C.
   1917. *The Anatomy of Woody Plants.* University of Chicago Press, Chicago.
Jensen, W. A., and F. B. Salisbury
   1972. *Botany: An Ecological Approach.* Wadsworth, Belmont, Calif.
Namboodiri, K. K., and C. B. Beck
   1968a. A comparative study of the primary vascular system of conifers. I. Genera with helical phyllotaxis. *Amer. Jour. Bot.* 55:447–457.
   1968b. A comparative study of the primary vascular system of conifers. II. Genera with opposite and whorled phyllotaxis. *Amer. Jour. Bot.* 55:458–463.
   1968c. A comparative study of the primary vascular system of conifers. III. Stelar evolution in gymnosperms. *Amer. Jour. Bot.* 55:464–472.
Schnarf, K.
   1937. *Anatomie der Gymnospermen—Samen.* In Linsbauer, K. (ed.), *Handbuch der Pflanzenanatomie.* Band X, Teil 1. Gebrüder Borntraeger, Berlin.
Thomson, R. B.
   1927. Evolution of the seed habit in plants. *Proc. Trans. Roy. Soc. Canada* 21:229–272.
   1934. Heterothally and the seed habit *versus* heterospory. New Phytol. 33:41–44.

15

The Cycadopsida

The class Cycadopsida includes the living and extinct representatives of the cycadophyte line of gymnosperm evolution. As we have pointed out in Chapter 14, cycadophytes as a whole are distinguished by their large, pinnately compound leaves, columnar or sparingly branched trunks, and manoxylic secondary xylem. The only surviving cycadophytes are the modern cycads (order Cycadales), which will receive the major attention in this chapter. But we will first briefly consider two extinct orders of the Cycadopsida, the Pteridospermales, or "seed ferns," and the Cycadeoidales, or "fossil cycads." Both of these orders are of exceptional evolutionary interest: the seed ferns provide important clues about the nature of the seed in the earliest gymnosperms; the Cycadeoidales, although sharing some features in habit and anatomy with modern cycads, appear to represent a separate "blind end" in cycadophyte evolution.

Pteridospermales

The Paleozoic pteridosperms were a group of plants of the Carboniferous Period which combined, in a most remarkable way, the general habit and foliage of ferns with the formation of gymnospermous seeds (Figs. 15-1, 15-6). Since the Carboniferous Period had often been characterized as the Age of Ferns, the demonstration that many of the presumed "ferns" were in reality seed-bearing plants must be regarded as an outstanding achievement in the history of paleobotany.

A. Fronds with differentiated pinnules
   B. Secondary veins forming a network
      C. Pinnules attached by whole breadth of base
          Lonchopteris ——————

      C'. Pinnules attached by a single point
          Linopteris ——————

   B'. Secondary veins not forming a network
      D. Pinnules with a distinct midrib
         E. Pinnules small, attached by whole base
            Pecopteris ——————

         E'. Pinnules large
           F. Decurrent at base
            Alethopteris ——————

           F'. Constricted at base
            Taeniopteris ——————

      D'. Pinnules with an indistinct midrib
         G. Pinnules attached by whole base
            Odontopteris ——————

         G'. Pinnules attached by a point
            Neuropteris ——————

A'. Fronds with pinnules lobed, or dissected; rarely with rounded pinnules.
         H. Pinnules small, veins radiate
            fan-like from base of pinnule
            Sphenopteris ——————

         H'. Pinnules large, basal pinnule decurrent and bifid
            Mariopteris ——————

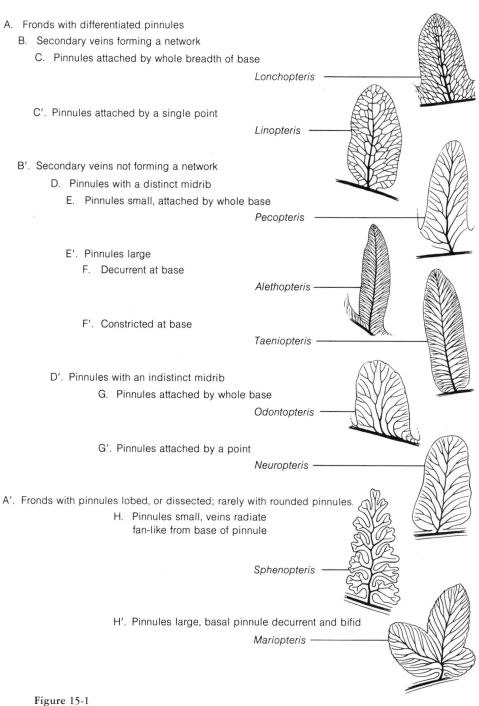

Figure 15-1
Key based on venation and mode of attachment of pinnules of Paleozoic seed ferns. [Based on key in *Textbook of Paleobotany* by W. C. Darrah. Copyright D. Appleton-Century Co., New York, 1939; by permission of Appleton-Century-Crofts, Inc.]

Figure 15-2
Reconstruction of a specimen (12–15 feet high) of *Medullosa noei*. [Redrawn from W. N. Stewart and T. Delevoryas, *Bot. Rev.* 22:45, 1956.]

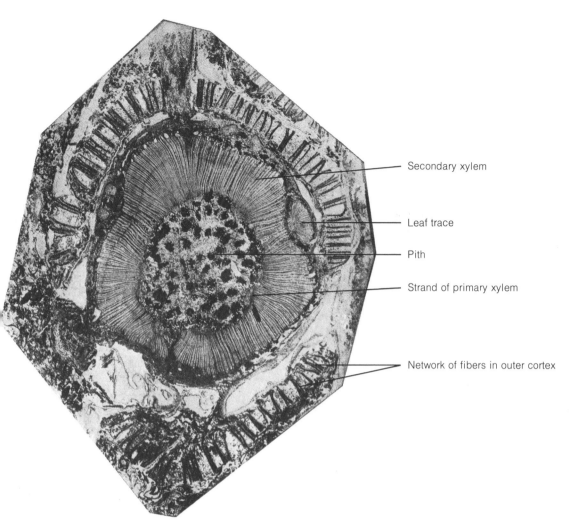

Secondary xylem

Leaf trace

Pith

Strand of primary xylem

Network of fibers in outer cortex

Figure 15-3
Transection of "monostelic" stem of *Lyginopteris oldhamia,* showing primary vascular strands, leaf traces, and the well-developed cylinder of secondary xylem. Note the characteristic network of fibrous strands in the outer cortex. [From *An Introduction to the Study of Fossil Plants* by J. Walton. Adam and Charles Black, London, 1953.]

The classical investigations which led to the recognition of the seed ferns as a separate group of plants were made at the turn of this century by Oliver and Scott (1905). Since that time, much study has been devoted to the comparative morphology and systematics of the pteridosperms.

The two principal families of Paleozoic seed ferns are the Lyginopteridaceae and the Medullosaceae. The members of the Lyginopteridaceae are distinguished by *"monostelic"* stem anatomy, i.e., stems with the vascular tissue arranged in one cylinder (Fig. 15-3), and by seeds in which the integument is joined, nearly throughout its length, with the nucellus (Fig. 15-6, E). In contrast, the stems of members of the Medullosaceae are *"polystelic,"* i.e., the vascular system consists of three independent steles (e.g., *Medullosa noei*) or of many (e.g. *Medullosa primaeva*), in each of which the primary xylem is surrounded by secondary xylem (Fig. 15-4). The seeds of the Medullosaceae are larger than those of the Lyginopteridaceae and, as is clearly illustrated

410

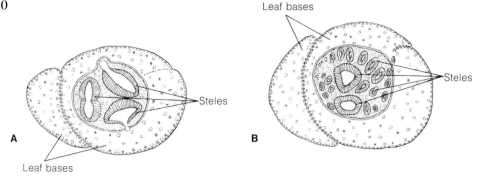

Figure 15-4
Diagrammatic transections showing the "polystelic" anatomy typical of stems of the Medullosaceae.
A, *Medullosa noei,* with three steles; **B,** *Medullosa primaeva,* with many steles. [Redrawn from
Morphology and Evolution of Fossil Plants by T. Delevoryas. Holt, Rinehart, and Winston,
New York, 1962.]

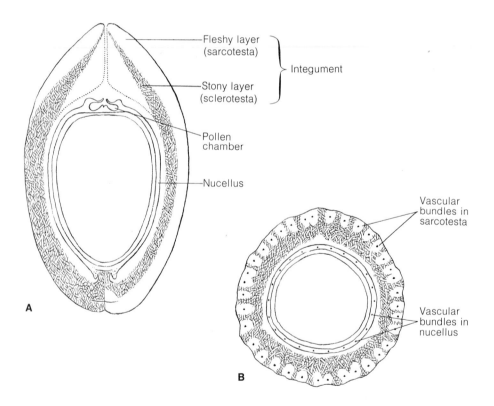

Figure 15-5
Reconstruction of the seed of *Pachytesta illinoensis* showing the absence of union between the integument and
the nucellus. **A,** longitudinal view; **B,** transectional view. Note numerous vascular bundles in both the integument and the nucellus. [Redrawn from Taylor, *Palaeontographica* 117(B):1, 1965.]

by *Pachytesta*, are provided with an integument which is completely free from the nucellus. As shown in Fig. 15-5, a vascular system is present in both the integument and the outer region of the nucellus of the seed of *Pachytesta* (for detailed descriptions of seed structure in *Pachytesta*, see Taylor, 1965.).

No attempt will be made in this book to review the voluminous literature on seed ferns or to consider the involved problems of their nomenclature and classification. Information on these matters can be found in standard paleobotanical textbooks (see, for example, Arnold, 1947; Walton, 1953; Andrews, 1961; Delevoryas, 1962). But we will briefly indicate some of the salient morphological characters of *Lyginopteris oldhamia* (=*Calymmatotheca hoeninghausi*) because this is the fossil plant upon which Oliver and Scott (1905) founded their original concept of the class "Pteridospermae."

The great merit of the pioneering work of Oliver and Scott was that the stem genus *Lyginodendron* (now called *Lyginopteris*) and the seed genus *Lagenostoma lomaxi*, were shown to be parts of the same plant. In transectional view, the primary vascular system of the stem of *Lyginopteris* is a "typical" gymnospermous eustele, which consists of a series of mesarch bundles arranged around the periphery of the pith (Fig. 15-3). External to this primary vascular system is a well-developed cylinder of secondary xylem, consisting of tracheids provided with bordered pits on their radial walls. The seed of *Lagenostoma*, although gymnospermous in structure, was partially enclosed within a cupule, the pedicel and lobes of which bore distinctive stalked glands (Fig. 15-6). Entirely similar glands were also found on the stems of *Lyginopteris* and on the leaf fragments (presumably of the form genus *Sphenopteris*) attached to such stems. These distinctive glands were considered to provide compelling evidence of the original connection between the seed (*Lagenostoma*) and the vegetative structures of *Lyginopteris*. The rachis of the compound leaf of *Lyginopteris* is believed to have branched dichotomously above the lowest pinnae, thus resembling the reconstruction of the leaf of certain medullosan seed ferns (Fig. 15-2). The relatively slender stem of *Lyginopteris* suggests that the plant was a climber (Zimmermann, 1959) or else leaned against neighboring plants for support (Arnold, 1947).

It has already been pointed out in Chapter 14 that the nature and origin of the integument of the seed in gymnosperms represent difficult morphological problems. Perhaps a partial answer is provided by recent studies on the seeds of certain pteridosperms of the Carboniferous Period. According to Andrews (1963), "the most primitive fossil that may be called a seed is *Genomosperma kidstoni*" (Fig. 15-7, A). The nucellus of this seed, in place of being enclosed by and partially joined with an integument, is surrounded basally by a whorl of eight free filamentous processes. A more advanced condition towards the formation of an integument is represented by the seed of *Genomosperma latens,* in which the eight integumentary "lobes" are apically appressed—forming what Andrews regards as a "rudimentary micropyle"—and joined with each other for about a third of their length (Fig. 15-7, B). Fusion of the integumentary lobes is almost complete in the seed of *Eurystoma* and complete union is shown by the integument of the seed of *Stamnostoma* (Fig. 15-7, C–D). Much remains to be learned about Paleozoic seeds, including the nature and origin of cupulate types. A most interesting example of a cupulate seed has recently been described by Pettitt and Beck (1968) and given the name *Archaeosperma arnoldii*. The particular interest of *Archaeosperma* is that it represents the first record of the existence of seed-bearing gymnosperms in the Devonian Period.

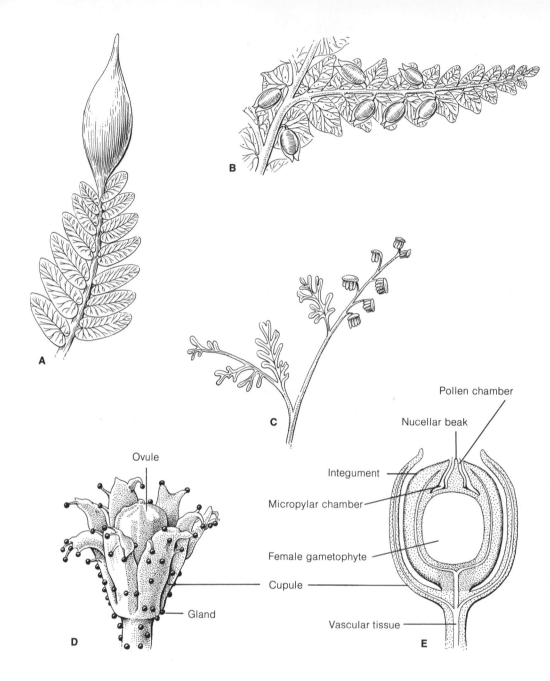

Pollen chamber

Nucellar beak

Integument

Micropylar chamber

Female gametophyte

Cupule

Vascular tissue

Ovule

Gland

Figure 15-6
Reproductive structures in seed ferns. **A,** pinna with terminal seed of *Neuropteris heterophylla;*
B, *Emplectopteris triangularis,* seeds attached to subdivisions of leaf; **C,** portion of fertile leaf of
Crossotheca bearing clusters of pendant microsporangia; **D,** reconstruction of the ovule and cupule of
Lagenostoma lomaxi; **E,** longisection of ovule of *Lagenostoma* showing details of internal structure.
[**A** redrawn from *Textbook of Paleobotany* by W. C. Darrah. Copyright D. Appleton-Century Co.,
New York, 1939, by permission Appleton-Century-Crofts, Inc; **B** redrawn from *Ancient Plants and
the World They Lived In* by H. N. Andrews. Comstock, New York, 1947; **C** and **D** from *Les Plantes
Fossiles dans leurs Rapports avec les Végétaux Vivants* by L. Emberger. Masson et Cie, Paris, 1944;
E, adapted from *An Introduction to Paleobotany* by C. A. Arnold. McGraw-Hill, New York, 1947.]

Cycadeoidales

The order Cycadeoidales (= Bennettitales of some authors) includes a remarkable group of cycad-like plants that flourished during most of the Mesozoic Era and became extinct at the end of the Cretaceous Period. Two principal families of cycadeoids are commonly recognized by paleobotanists: (1) the Williamsoniaceae, the members of which had more-or-less columnar, sparingly branched trunks bearing crowns of pinnately compound leaves (Fig. 14-1); one genus in this family, *Williamsoniella*, was a plant of more spreading form than *Williamsonia* that branched dichotomously and produced simple rather than compound leaves; and (2) the Cycadeoidaceae represented by the genus *Cycadeoidea*. In contrast to *Williamsonia*, the trunk of *Cycadeoidea* was relatively short, massive, and unbranched (Fig. 14-1). Because of their distinctive conical or ovoid form, the silicified trunks of *Cycadeoidea* have attracted much popular attention and, before such specimens had been studied by paleobotanists, they were frequently thought to represent fossil beehives, wasps' nests, or corals! (See Fig. 15-8.)

Remains of the Cycadeoidales have been found in England, Scotland, and the continent of Europe and India. In North America, the most famous and productive localities for "fossil cycads" were certain areas in the Black Hills of South Dakota. The silicified and beautifully preserved trunks of hundreds of specimens of *Cycadeoidea* collected from the Black Hills deposits provided the critical material upon which G. R. Wieland (1906, 1916) based his classical morphological and taxonomic studies on American fossil cycads.

In this textbook it will not be possible to consider comparatively the complex and puzzling aspects of the vegetative and reproductive structures of the cycadeoids. But

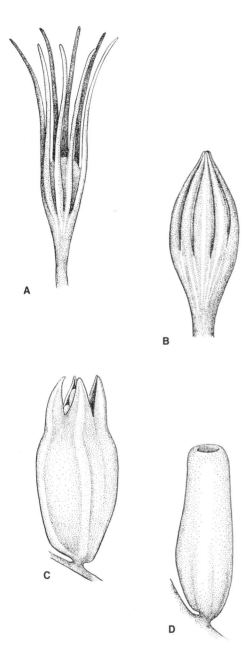

Figure 15-7
Types of pteridosperm seeds, illustrating the theoretical origin and phylogenetic development of the integument. **A**, *Genomosperma kidstoni;* **B**, *Genomosperma latens;* **C**, *Eurystoma angulare;* **D**, *Stamnostoma huttonense.* [Redrawn from Andrews, *Science* 142:925, 1963. Copyright © 1963 by the American Association for the Advancement of Science.]

Figure 15-8
Trunk of *Cycadeoidea marylandica,* the earliest described American specimen of a fossil cycad. The specimen, according to Wieland, was discovered in 1860, in iron-ore beds of the Potomac formation of Maryland, and is represented here about one-fourth its natural size. Note the numerous strobili distributed among the persistent leaf bases which constitute the armor of the trunk. [From *American Fossil Cycads,* Vol. I. by G. R. Wieland. Carnegie Institution, Washington, D.C., 1906.]

it is important and interesting to comment briefly on the morphology of the strobili of *Cycadeoidea,* which have attracted great interest and speculation among paleobotanists and morphologists.

As is shown in Fig. 15-8, the strobili of *Cycadeoidea* were born laterally among the persistent leaf bases which formed the armor of the trunk. According to Wieland's (1906) reconstruction, the strobilus was *bisporan-* *giate* and consisted of a peduncle bearing a series of spirally arranged bracts and terminating in a conical receptacle, on the surface of which were very numerous tightly packed ovules and "interseminal scales" (Fig. 15-9). At the base of the receptacle of *expanded cones* there was a *whorl* of basally fused reflexed *pinnate microsporophylls,* each pinna bearing two rows of *synangia,* i.e., structures containing numerous microsporangia.

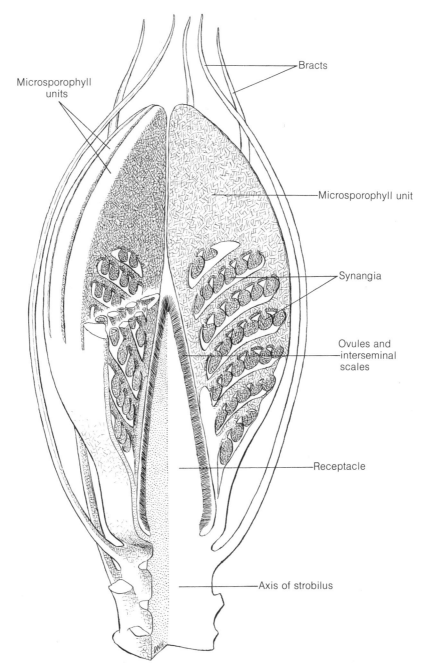

Microsporophyll units

Bracts

Microsporophyll unit

Synangia

Ovules and interseminal scales

Receptacle

Axis of strobilus

Figure 15-9
Idealized reconstruction of the bisporangiate cone of *Cycadeoidea*. At the right, one of the microsporophyll "units" is shown in median radial sectional view; note trabecular structures that bear the synangia. At the left, a portion of one microsporophyll unit is cut transversely. The prominent conical receptacle bears numerous small ovules and interseminal scales all over its surface. [Redrawn from Delevoryas, *Palaeontographica* 121(B):122, 1968.]

In the light of comparative studies, it is noteworthy that Wieland's general description of the ovuliferous receptacle in the cone of *Cycadeoidea* is still valid. According to Delevoryas (1968b, p. 141), "in all known members of the Cycadeoidales that have ovulate structures preserved, ovules are borne on a fleshy receptacle. It may be elongated and conical, or short and almost globose, but the essential features are common throughout." The morphology of the pollen-forming organs of *Cycadeoidea,* as reinterpreted by Delevoryas (1963, 1968a), differs considerably however from Wieland's classical reconstruction. According to Delevoryas, the microsporangiate structures were not free, reflexed appendages but, as is shown in Fig. 15-9, the "units" were joined for most of their length and only distally show any trace of "free" sporophyll-like structure. The structures which bear the sporangia are not the "pinnae" reconstructed by Wieland, but rather are rods of tissue (trabeculae) which connect the outer and inner parts of each sporophyll unit (Fig. 15-9, sectional view at left). The highly peculiar microsporangial "organ" of the cone of *Cycadeoidea* is regarded by Delevoryas as an advanced type of structure, probably derived phylogenetically from the fusion of once separate microsporophylls. If the microsporangiate "organ" in the cone of *Cycadeoidea* is regarded as a structure which did not open at the time of pollination, it becomes necessary to attempt to explain how the pollen grains in fact were disseminated (see also Crepet, 1972). Delevoryas (1963, p. 51) believes that the entire microsporangiate organ may have become detached from the base of the receptacle and "upon drying, split, to allow the pollen to blow away." Many cones of *Cycadeoidea* appear to be devoid of microsporangiate organs, possibly as the result of their precocious abscission.

It seems to be rather generally agreed that the cycadeoids represent a terminal line in gymnosperm evolution. Cycadeoids differ so markedly in cone structure from any of the living cycads that it would be impossible to imagine any phylogenetic connection between the two groups. Furthermore, the stomata of the Cycadeoidales are *syndetocheilic* in contrast to the *haplocheilic* stomata typical of the leaves of the Cycadales. (For a description of the ontogeny of syndetocheilic and haplocheilic stomata, see Chapter 17, p. 479.) This additional distinction between the two orders has proved extremely helpful in the correct identification of isolated cycadophyte fronds of the Mesozoic Era. The evolutionary origin of the Cycadeoidales is still an unsettled question although the most likely ancestral forms would seem to be the pteridosperms (Delevoryas, 1968b).

Cycadales

According to paleobotanical studies, modern cycads appear to represent the surviving members of a former larger cycad flora that extended over much of the earth during the Mesozoic Era. Arnold (1953) stated that the cycads as a whole have existed over a period of at least 200 million years. As fossils, they have been found in deposits that range in age from the Late Triassic to the Early Cretaceous; cycad remains have been discovered in such widely separated regions of the world as Siberia, Manchuria, Oregon, Alaska, several islands in the Arctic Ocean, Greenland, Sweden, England, central Europe, India, Australia, and the Antarctic continent. The distribution of the cycads was probably at its maximum during the Jurassic, coinciding with the age of the giant dinosaurs.

The most comprehensive account of the cycads—based on studies in both the field and laboratory—was given by Chamberlain

(1919, 1935), and his comparative studies have profoundly influenced our present understanding of their reproductive morphology and relationships. Chamberlain (1919) recognized nine genera, which he placed in the single family Cycadaceae under the order Cycadales. More recently Johnson (1959) added the genus *Lepidozamia* and recognized three separate families of cycads: the Cycadaceae and Stangeriaceae, each represented by a single genus, and the Zamiaceae, which includes the remaining eight genera.

The pattern of distribution of the ten genera of cycads is peculiar and difficult to interpret from a phytogeographical point of view. In the Western Hemisphere there are four genera: *Microcycas* (a monotypic genus restricted to western Cuba), *Cerato-zamia* and *Dioon* (endemic to Mexico), and *Zamia,* which occurs in southern Florida (the only genus of cycads native to the United States), the West Indies, Mexico, Central America, and northern and western South America. In the Eastern Hemisphere, the important cycad areas are: Australia, where the endemic genera *Bowenia, Macro-zamia,* and *Lepidozamia,* as well as *Cycas,* are found, and Africa, which is the home of *Stangeria* and *Encephalartos.* The relatively large genus *Cycas* occurs not only in Australia but is also represented by species in India, China, the southern islands of Japan, Madagascar, East Africa, and certain islands of the south Pacific Ocean. If we attempt, speculatively, to reconstruct the paths of past migration which led to the present geographical distribution of the living cycads,

Figure 15-10
A microsporangiate plant of *Microcycas calocoma* growing in the colony at Cayo Ramones, Province of Pinar del Río, Cuba. The trunk of this large specimen measured 28 feet in height and was 44 inches in circumference near the base. [From Foster and San Pedro, *Mem. Soc. Cubana Hist. Natur.* 16:105, 1942.]

Figure 15-11
Habit and general organography of cycads.
A, specimen of *Cycas revoluta* growing in
cultivation in Pasadena, California (note
the crowded zones of persistent leaf bases
which comprise the armor of the trunk);
B, specimen of *Dioon spinulosum* growing in
the Conservatory at Golden Gate Park,
San Francisco, California (note the
enormous size of the pendulous megasporangiate cone and the conspicuous
armor of leaf bases). [**B** courtesy
Dr. T. E. Weier.]

it becomes very difficult to explain the present isolation of such monotypic genera as *Stangeria* (Africa) and *Microcycas* (Cuba). Such isolated but morphologically well-defined genera are perhaps "leftovers," as Arnold (1953) suggested, of a former, more-extensive cycad population.

Because of their large, attractive palm-like leaves, many of the cycads are widely grown as ornamentals. The commonest species in cultivation throughout most of the Northern Hemisphere is the so-called Sago Palm (*Cycas revoluta*). This form requires greenhouse protection in regions of cold winters but is quite hardy outdoors in Southern California, parts of Louisiana, and Florida (Fig. 15-11, A). The most extensive outdoor collections of cycads in the United States are found in the Coconut Grove Palmetum, Coral Gables, Florida, and the Huntington Botanical Gardens, San Marino, California. The growth and general aspects of the species found at the latter institution are interestingly described and photographically illustrated in a book by Hertrich (1951).

Vegetative Organography and Anatomy

The general habit of cycads ranges from types in which the stem is tuberous and partly or wholly subterranean (species of *Zamia, Macrozamia,* and *Encephalartos*; and the genera *Stangeria* and *Bowenia*) to relatively tall plants with the general aspect of tree ferns or palms. The latter category is illustrated by the Cuban cycad *Microcycas calocoma* (Fig. 15-10). The columnar trunk of this inappropriately named genus was found by Foster and San Pedro (1942) to attain a height of about 30 feet as measured from the ground to the base of the terminal crown of leaves. According to Chamberlain (1919), the tallest of the cycads is *Macrozamia hopei,* a native of Queensland, Australia, which may reach a height of 60 feet.

The same author records specimens of the Mexican cycad *Dioon spinulosum,* which measured 50 feet in height. Since cycads in general are conspicuously sluggish in their rate of growth, large arborescent types probably are extremely old. Estimates of the age of large specimens have been based on a study of the persistent leaf bases which constitute the characteristic armor of the trunk (Fig. 15-11). If the average number of leaves produced each year can be determined, this number divided into the total number of leaf bases on the entire trunk yields some approximation of the age of the plant. By this method Chamberlain concluded that a plant of *Dioon edule* with a trunk only six feet in height was at least 1,000 years old. Age determinations based on the study of the leaf armor, however, are probably conservative because the sluggish period of development of the seedling and the prolonged periods of dormancy of older plants are not taken into consideration.

Cycads as a whole, particularly the arborescent types, are unbranched, but there are exceptions — in species having the tuberous habit of growth as well as those of the columnar habit (Fig. 15-12) (Swamy, 1948). In *Cycas revoluta,* for example, lateral buds commonly develop from various areas of the trunk, and their irregular expansion into shoots results in the production of irregularly or grotesquely branched individuals. These lateral buds, which apparently originate from the living tissue of the persistent leaf bases, are capable of rooting and producing new plants if detached from the trunk. But in other instances branching is clearly the result of the injury or destruction of the main terminal bud. Striking examples of this were discovered by Foster and San Pedro (1942) in *Microcycas.* In this genus adventitious buds arise following injury (often as a result of wind) to the upper part of the trunk, and may give rise to large candelabra-like branches.

Figure 15-12
A, large specimen of *Encephalartos transvenosus,* a cycad native
to Africa. Note huge ovulate cone at tip of the single lateral
branch. **B,** ovulate cones of *Encephalartos umbeluziensis.*
[From Dyer, *Cactus Succulent Jour.* 44:209, 1972.]

B

A

STEM. As viewed in transverse section, the stem of cycads is distinguished by several anatomical characteristics. First, the cortex and pith are exceptionally large, as compared with the relatively narrow diameter of the vascular cylinder (Fig. 15-13). For example, the diameter of the pith in large trunks of *Cycas revoluta* may measure as much as 6 inches. Secondly, the course of the vascular strands or traces which supply each of the leaves is complex. Since the various strands entering a given leaf extend horizontally through the cortex before entering the leaf base, a girdling arrangement is seen in thick or cleared sections of the stem. Although secondary growth by means of a vascular cambium occurs in the stems of many gen-era, the amount of secondary vascular tissues produced is relatively small, and the xylem is traversed by numerous broad parenchymatous rays. Seward (1917, p. 7) designated the loose-textured wood of cycads as "manoxylic," in contrast with the more compact and dense xylem of conifers, which he termed "pycnoxylic." Moreover, although concentric and definable growth rings may be seen in certain genera, they do not appear to represent true annual increments or annual rings. According to Chamberlain's study of *Dioon spinulosum,* the rings correspond to those periods in development when new crowns of leaves were produced by the terminal bud.

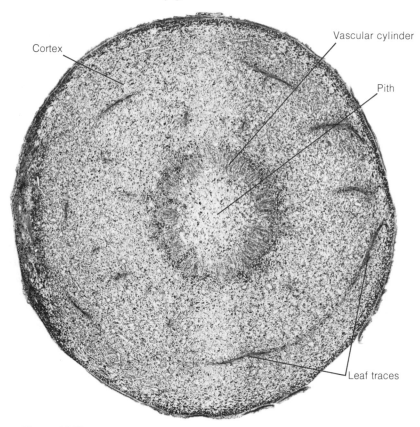

Figure 15-13
Transection of stem of *Zamia sp.* demonstrating the large pith and cortex and the characteristic girdling leaf traces (×4).

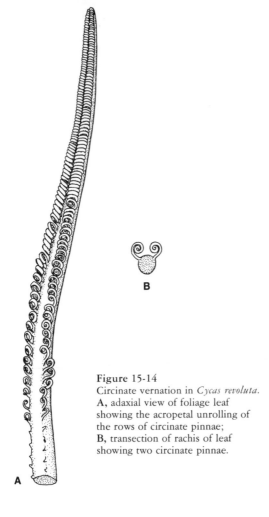

Figure 15-14
Circinate vernation in *Cycas revoluta*.
A, adaxial view of foliage leaf
showing the acropetal unrolling of
the rows of circinate pinnae;
B, transection of rachis of leaf
showing two circinate pinnae.

like the bases of the foliage leaves, persist for many years on the trunk and thus further contribute to the characteristic leaf armor. In many cycads the new leaves are erect, and the leaflets are flat during the period of expansion of the crown. But in the genus *Cycas,* the vernation of the young leaflets is distinctively *circinate* as in many ferns. Figure 15-14, A shows that the unrolling of the successive coiled pinnae of each leaf proceeds from the base upwardly and is thus properly designated as acropetal.

There are various types of venation patterns in the leaflets or pinnae of cycad leaves and some of these patterns provide consistent characters useful in the recognition of genera. The genus *Cycas* is unique because each pinna is traversed by only a single prominent midvein and lateral veins are absent. In contrast, the leaflets of *Stangeria* develop a pinnate type of venation consisting of a midvein from which radiate lateral veins that are closely spaced and dichotomously branched. Many of the veinlets in the pinnae of *Stangeria* anastomose near the margin, forming conspicuous closed loops as is shown in Fig. 15-15, B. In the remaining eight genera, the pinnae are devoid of a distinct midvein and the venation consists of a series of longitudinal veins which branch dichotomously at various levels in their course through the pinnae (Fig. 15-15, A). Although Troll (1938) called attention to the vein anastomoses in the pinnae of *Ceratozamia mexicana* and *Stangeria,* this aspect of leaf venation in the cycads has received much less attention than it deserves. The genus *Zamia,* for example, provides striking examples of the union between the branches derived from the dichotomy of two adjacent veins (Fig. 15-16). This distinctive type of anastomosis is evidently a very primitive kind of vein union as it occurs not only in many ferns but also in the dichotomously veined lamina of *Ginkgo biloba* (Arnott, 1959). It is in-

LEAVES. Cycads are distinguished from all other living gymnosperms by their pinnate leaves, which, in certain genera, may reach a length of 4–5 feet. The genus *Bowenia* is exceptional in having large bipinnate leaves; in all other living cycads the leaf develops a single series of leaflets (Fig. 15-14, A). Except during the juvenile phases of development, the leaves in most of the genera are produced in crowded clusters, or crowns, which thus impart a palm-like aspect to many of the arborescent types (Figs. 15-10; 15-11, A). Prior to the expansion of a crown of new foliage leaves the latter are overlaid and protected by an extensive imbricated series of tough bud scales. These structures,

teresting, from a taxonomic point of view, that, although Johnson (1959) placed considerable emphasis on venation patterns in his characterization of the three families of cycads, he failed to mention the occurrence in some genera of vein anastomoses. It seems entirely possible that more detailed studies will reveal that the presence or absence of anastomoses has value in the separation of cycad taxa at both the genera and specific levels (see Brashier, 1968).

Certain interesting aspects of leaf histology may be briefly noted at this point. In *Cycas,* as previously mentioned, each

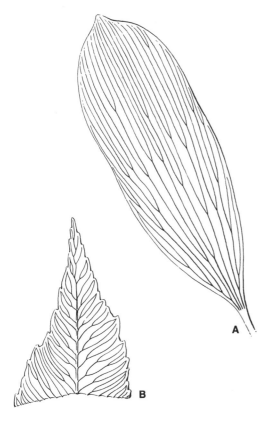

Figure 15-15
Dichotomous venation in leaflets of cycads. **A,** *Zamia wallisii;* **B,** *Stangeria schizodon,* tip of leaflet (note midvein, dichotomously branched lateral veins, and union of certain vein endings). [From *Vergleich. Morphologie der höheren Pflanzen* by W. Troll. Gebrüder Borntraeger, Berlin, 1938.]

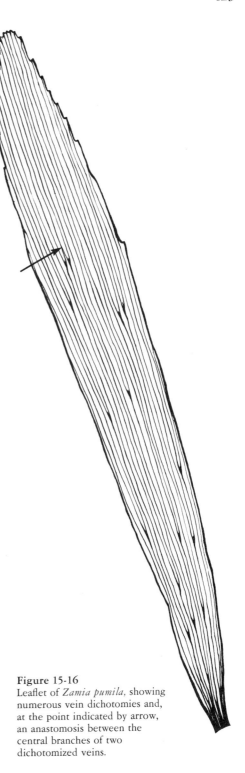

Figure 15-16
Leaflet of *Zamia pumila,* showing numerous vein dichotomies and, at the point indicated by arrow, an anastomosis between the central branches of two dichotomized veins.

pinna is vascularized by a single bundle which extends through the midrib region. Although lateral veins are absent, a sheet of so-called "accessory transfusion tissue" extends from each side of the midrib to the margins of the pinna. Some of the cells in this transfusion tissue resemble tracheids in their elongated form and wall structure and perhaps facilitate the translocation of material between the midvein and the mesophyll of the pinna. (See Brashier, 1968 for further details regarding transfusion tissue in the pinnae of cycad leaves.) Of additional interest, with reference to *Cycas,* is the fact that there are *both* endarch and exarch xylem in the midrib bundle of the pinnae as well as in the vascular strands in the leaf rachis.

APOGEOTROPIC ROOTS. According to Chamberlain (1935), the sporophytes of all cycads form *apogeotropic roots* which grow upwardly and, near the soil surface, branch dichotomously, forming peculiar coralloid masses of rootlets. Coralloid rootlets may form in seedlings but are particularly conspicuous near the base of the trunk of well-developed cultivated specimens (e.g., in *Cycas revoluta*), especially if the plants are grown in tubs. One of the distinctive features of coralloid roots is the presence in the root cortex of a well-defined cylindrical zone of cells inhabited by a blue-green alga, *Anabaena cycadeae* Reinke. In a recent study on *Cycas revoluta,* Storey (1968) found that "somatic reduction" occurs in many of the cells of the developing "algal zone" and results in the formation of numerous small cells with greatly reduced chromosome number interspersed among larger diploid cells. The small cells form the "slime" that apparently is the medium essential for the establishment of the colonies of blue-green algae. The biological role of the algae — if there is any — has as yet not been determined.

The Reproductive Cycle

STROBILI. All carefully investigated species of cycads are strictly dioecious in the sense that microsporangiate and megasporangiate strobili are borne on separate plants. Chamberlain stated emphatically that "in thirty years of study in the field and in greenhouses I have never seen anything to indicate that the cycads are not absolutely dioecious." In all genera, with the exception of *Cycas,* both types of strobili are compact cone-like structures with determinate growth; in position they are apparently terminal in such genera as *Dioon,* although in certain species of *Macrozamia, Bowenia,* and *Encephalartos* they are axillary and lateral. According to Chamberlain (1935), the *first* cone produced by *Dioon* is terminal, and a new vegetative meristem, developing *laterally* at the base of this cone, gives rise to the next crown of leaves and a new cone. The possible occurrence of this type of "sympodial growth" in other cycad genera deserves study.

The unique feature of *Cycas* consists in the fact that the megasporophylls are produced, like the foliage leaves, in a relatively loose crown which surrounds the shoot apex of the terminal bud (Fig. 15-17, B). After seed maturation, a new crown of foliage leaves expands above the cluster of megasporophylls; the latter persist as dead or dying structures below this new foliage. The microsporangiate strobilus of *Cycas,* however, is a terminal, compact, determinate cone as in the other genera (Fig. 15-17, A).

According to Chamberlain, "the largest cones that have ever existed are found in the living cycads." Apparently the largest dimensions and greatest weight are characteristic of the megasporangiate strobili of certain species of *Encephalartos, Dioon,* and *Macrozamia* (Figs. 15-11, B; 15-12). Weights up to 90 pounds have been recorded for *Encephalartos,* and Chamberlain describes

Figure 15-17
Strobili of *Cycas revoluta*. **A,** microsporangiate strobilus (cultivated specimen, Pasadena, California); **B,** megasporangiate strobilus (note pinnatifid form of megasporophylls). [**B,** courtesy Dr. T. E. Weier.]

the seed cones of *Macrozamia denisoni* as measuring two feet in length, nearly a foot in diameter at the base, and with a weight of 50–70 pounds. In comparison with such gigantic cones, the cones of the more familiar gymnosperms, such as pine, indeed seem insignificant in size.

SPOROPHYLLS AND SPORANGIA. The microsporophylls in the cycads, although varying in size and form, all are rather thick, scale-like structures that bear the microsporangia on their lower or abaxial surfaces (Fig. 15-18, B). This resemblance to the abaxial position of sporangia typical of many ferns is strengthened by the fact that the microsporangia are arranged in somewhat definite soral clusters. The number of microsporangia borne by a single sporophyll varies from more than a thousand in *Cycas media* to several dozen in *Zamia floridana*. Although much remains to be done in the study of the origin and early ontogeny of the microsporangium, it is known that its structure at maturity is typically eusporangiate with a wall several cell layers in thickness, a tapetum, and numerous small microspores (Fig. 15-18, A). The surface cells of the microsporangium are large and very thick walled (except at the point where dehiscence will occur) and have collectively been regarded by Jeffrey (1917, Fig. 160) as an annulus.

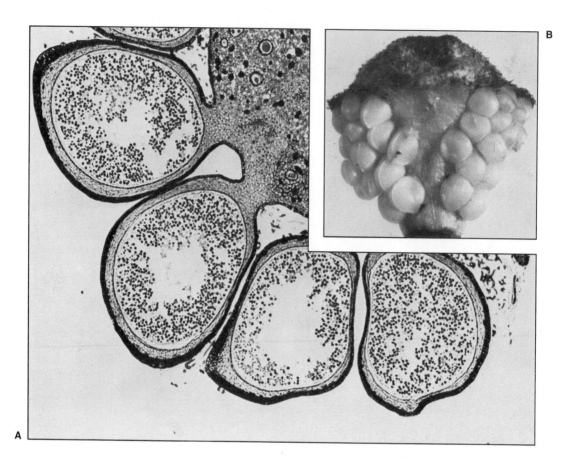

Figure 15-18
Zamia sp. **A**, transection of microsporophyll; **B**, abaxial view of microsporophyll, showing soral clusters of microsporangia. Note thick walls of the stalked microsporangia and the abundant microspores produced in each of them.

Figure 15-19 (*facing page*)
Megasporophylls in Cycads. **A**, *Cycas revoluta*, pinnatifid type of megasporophyll with developing seeds; **B**, *Cycas circinalis*, note very rudimentary pinnae [redrawn from *Die natür. Pflanzenfamilien* by Engler and Prantl. Leipzig: Wilhelm Englemann, 1926]; **C**, *Dioon edule*, megasporophyll with expanded lamina tip and two ovules; **D**, *Ceratozamia*, peltate type of megasporophyll bearing two ovules (the two spines at the top of the sporophyll are characteristic for this genus); **E**, *Zamia*, peltate type of megasporophyll with two ovules. [D redrawn from *Syllabus der Pflanzenfamilien* by A. Engler and E. Gilg. Gebrüder Borntraeger, Berlin, 1924.]

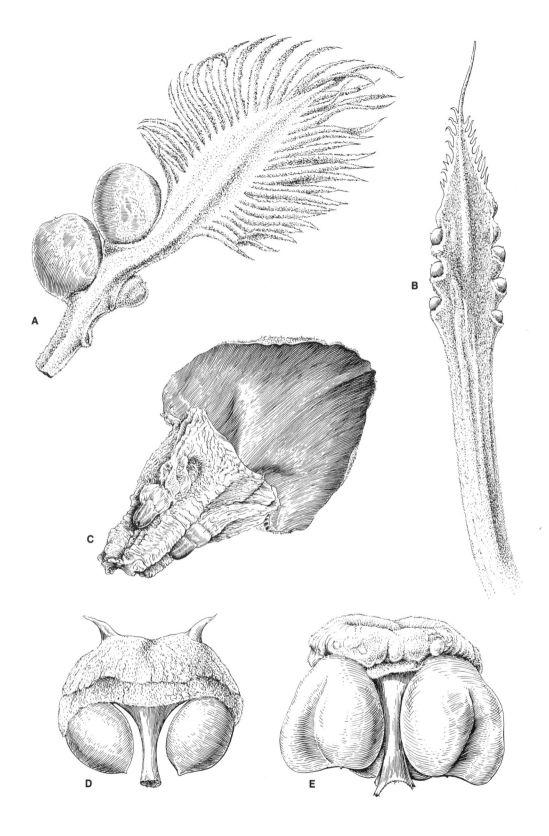

428

The megasporophylls vary considerably in size and form; in many cases their shape is of great systematic value in the characterization of genera or even species. Two extreme types occur. In *Zamia, Microcycas,* and *Ceratozamia,* for example, the megasporophylls are peltate, scalelike organs, each bearing two ovules (Fig. 15-19). In marked contrast, the megasporophylls of *Cycas revoluta* are conspicuously pinnatifid, leaf-like structures that bear 6–8 ovules laterally arranged on the sporophyll axis below the terminal group of rudimentary pinnae (Fig. 15-19). In other genera, such as *Dioon* and *Encephalartos,* the sporophylls, although essentially scale-like in form, may show extended tips or marginal serrations suggestive of reduced leaf blades. The common interpretation of this range in sporophyll form and ovule number is to regard the condition in *Cycas revoluta* as primitive and the other conditions as being the result of varying degrees of suppression and ultimate elimination of a definable lamina in the megasporophyll (Fig. 15-19).

The fossil record appears to give some support to this proposed evolutionary series and to emphasize the primitive character of the leaf-like megasporophyll of the modern *Cycas.* Delevoryas (1962), for example, concluded that the megasporophyll of *Cycas revoluta* "represents a reduction from a seed-fern type of fertile leaf that was pinnately divided with ovules taking the place of pinnules." More recently, Mamay (1969) reported his discovery of new types of cycad megasporophylls in the Lower Permian strata of the southwestern United States. In one of his types, the megasporophyll consists of a sterile *undivided* distal blade and 4–6 pairs of basal ovules partially enclosed by a "lamina" which he compares to the marginal indusium of a fern. Beginning with this precursory type of megasporophyll, Mamay proposed a hypothetical reduction series, which resulted

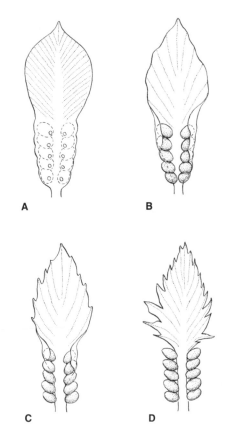

Figure 15-20
Semidiagrammatic reconstruction of the phylogeny of the *Cycas*-type of megasporophyll. **A,** megasporophyll of new Permian cycad genus, showing pairs of ovules below the sterile, terminal lamina; **B–C,** hypothetical intermediate forms; **D,** "modern" type of *Cycas* megasporophyll with pinnatifid lamina. [Redrawn from Mamay, *Science* 164:295, 1969. Copyright © 1969 by the American Association for the Advancement of Science.]

in the elimination of the basal lamina and the origin of "an essentially modern cycadean megasporophyll" (Fig. 15-20).

The ovules of cycads are erect in orientation, the micropyle lying opposite the point of attachment of the ovule to the megasporophyll. In the genus *Cycas* the micropyles of the ovules are turned obliquely outward and because of the loose arrangement of the pinnatifid megasporophylls are a striking example of veritable naked ovules (Fig. 15-19). In the other genera the micropyles of the paired ovules of each sporophyll are directed inward toward the axis of the strobilus and, except during the brief period when pollination occurs, are not directly exposed to the air (Fig. 15-19).

On the whole, cycad ovules are large as compared with the ovules of other gymnosperms. According to Chamberlain, the ovules of *Cycas circinalis* and *Macrozamia denisonii* may reach 6 cm in length; in other genera the ovules are smaller, the most diminutive being those of *Zamia pumila*, which measure only 5–7 mm in length. Structurally the cycad ovule consists of the integument laterally joined with the massive nucellus except near the micropylar end of the ovule (Fig. 15-21, D). During its ontogeny, the integument becomes histologically differentiated into three layers: an outer and an inner fleshy layer and a middle stony layer, which is sclerified and hard in texture. A very interesting feature of the ovule in cycads is the development of two separate vascular systems. The outer system extends as a series of unbranched veins through the outer fleshy layer, whereas the inner system, consisting of numerous dichotomously branched strands, vascularizes the inner fleshy layer (Fig. 15-21, D, E). It has been reported that in several genera, some of the branches of the inner vascular system may terminate in the base of the free part of the nucellus (Kershaw, 1912; Reynolds, 1924; Shapiro, 1951). The inner layer of the integument usually breaks down during seed development, but the outer fleshy layer persists and, in the ripe seed, may be bright red (*Encephalartos, Zamia*), salmon pink (*Microcycas*), or nearly white (*Dioon, Ceratozamia*).

GAMETOPHYTES. Our knowledge of the gametophytes, fertilization, and embryogeny in cycads is still somewhat fragmentary, and the brief discussion of these topics here necessarily represents a composite picture derived from the available data. In many respects the investigations of Chamberlain, especially on *Dioon,* are classical and have provided the firm basis for most of the subsequent investigations on other species.

MALE GAMETOPHYTE. The development of the endosporic male gametophyte of cycads proceeds through the early stages before the pollen grains are shed from the microsporangium. The mitotic division of the microspore nucleus forms two cells: a small basal *prothallial cell* and a larger meristematic initial. The latter functions as an *antheridial initial* and its division yields a *generative cell,* which lies next to the prothallial cell, and a *tube cell* comprising a nucleus and its associated cytoplasm (Fig. 15-21, A). At this three-celled stage, the partially developed male gametophyte, enclosed within the original spore wall, is released from the microsporangium as a pollen grain.

The manner in which the pollen of such dioecious plants as the cycads reaches the megasporangiate cones needs more careful study in the field. Although Chamberlain (1935) maintained that all cycads are probably wind pollinated, there have been reports of insect pollination in such genera as *Encephalartos* and *Macrozamia* (see Rattray, 1913, and Baird, 1939). Whatever may be the exact mechanism of pollen transfer, the pollen grains, on reaching the ovules, are said by Chamberlain to be drawn into the pollen chamber by the "retraction" of the pollination droplet that exudes from the micropylar end of the ovule. Within the pollen chamber, pollen tubes begin to form and the male gametophyte completes its development in the manner represented diagrammatically in Fig. 15-22.

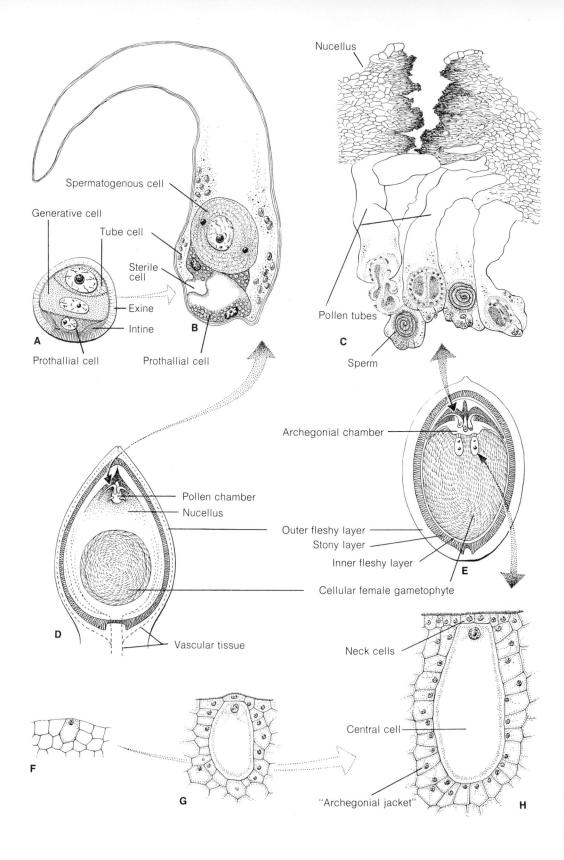

Spermatogenous cell

Generative cell

Tube cell

Sterile cell

Exine

Intine

Prothallial cell

A

Prothallial cell

B

Nucellus

Pollen tubes

C

Sperm

Pollen chamber

Nucellus

D

Vascular tissue

Archegonial chamber

Outer fleshy layer

Stony layer

Inner fleshy layer

E

Cellular female gametophyte

Neck cells

Central cell

F

G

"Archegonial jacket"

H

Figure 15-21 *(facing page)*
The reproductive cycle in cycads. **A–C**, development of male gametophyte in *Cycas*. **A**, the three-celled stage at time of shedding; **B**, the generative cell has divided forming the spermatogenous cell and the sterile cell. This stage occurs after pollination and during the invasion of the nucellus by the pollen tube; **C**, reconstruction of adjacent serial sections showing five pollen tubes with sperms in various stages of development, hanging from the nucellus in the archegonial chamber (note the spirally arranged band of flagella on two of the mature sperms); **D**, diagram of longisection of ovule of *Dioon edule* showing early stage in development of pollen tubes and the cellular stage of the female gametophyte; **E**, diagram of longisection of ovule of *Dioon edule* showing pollen tubes hanging downward in the archegonial chamber and the position of the archegonia in the female gametophyte; **F–H**, ontogeny of archegonium in *Dioon edule*. [**A–C** redrawn from Swamy, *Amer. Jour. Bot.* 35:77, 1948; **D–H**, redrawn from *Gymnosperms. Structure and Evolution* by C. J. Chamberlain, The University of Chicago Press, Chicago, 1935.]

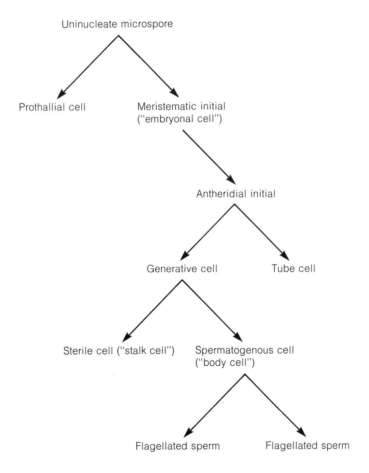

Figure 15-22
Diagram, based on Sterling's (1963) description, showing the successive stages in development of the male gametophyte in the cycads. See text for further explanation.

According to Chamberlain's studies, the interval between pollination and fertilization varies from four to six months. During this interval, a pollen tube, produced by the tubular extension of the upper portion of the pollen-grain wall (i.e., the end *opposite* the basal prothallial cell), invades the tissue of the nucellus (Fig. 15-21, D, E). In some cycads, this haustorial portion of the pollen tube may become irregularly branched. Concomitant with the haustorial growth of the upper part of the tube, the lower swollen prothallial end of the tube grows downward and the pollen chamber becomes enlarged by the continued breakdown of the adjacent nucellar tissue (Fig. 15-21, D). Finally the pollen tubes become freely suspended in the archegonial chamber above the exposed female gametophyte (Fig. 15-21, E).

During the growth of the pollen tube, the generative cell divides, producing a *sterile cell* located next to the prothallial cell and a terminal *spermatogenous cell* (Fig. 15-21, B). These terms, proposed by Sterling (1963) seem preferable to the older terminology in which the two cells derived from the division of the generative cell were designated as the "stalk cell" and the "body cell" (see Fig. 15-22). The term stalk cell suggests an unproved homology with the stalk of the antheridium of bryophytes and certain leptosporangiate ferns. Furthermore, the so-called stalk cell in the majority of cycads is nonfunctional—and hence "sterile"—although in *Microcycas,* according to Downie (1928), the stalk cell divides repeatedly, forming a linear series of body cells each of which forms two sperms. It is also evident that the term "body cell" should be replaced by the more explicit term "spermatogenous cell," which clearly defines the basic role of this cell, i.e., the production of a pair of gametes. Figure 15-21, C, based on the reconstruction of several serial sections, depicts a group of suspended pollen tubes and their contained male gametophytes just prior to the release of the sperms. A more detailed description of the structure and behavior of the remarkable flagellated sperms will be given in the discussion of fertilization in cycads.

FEMALE GAMETOPHYTE.   As stated in the previous chapter, the female gametophyte in gymnosperms arises by the enlargement and division of a single functional megaspore which is deeply situated within the nucellar tissue of the ovule (Fig. 14-4, A). However, because of the many technical difficulties of securing young ovules at the critical stages, our present knowledge of the details of megasporogenesis in the cycads is both meager and contradictory. One of the most thorough investigations was made by Smith (1910), who found that in *Zamia floridana* a linear tetrad of megaspores is produced from the megasporocyte; the three upper megaspores degenerate, and the lowermost megaspore enlarges and gives rise to the female gametophyte. In certain other cycad genera, it has been reported that a row of three rather than four cells is produced during megasporogenesis (Maheshwari and Singh, 1967). In this case the uppermost cell in the series is an *undivided dyad cell* and the two cells below it are megaspores derived from the division of the lower dyad cell. As in *Zamia,* however, only the lower of the two megaspores is functional and the cells above it degenerate (DeSloover, 1961).

Following the enlargement of the functional megaspore, the female gametophyte is initiated by a series of synchronized free nuclear divisions. The extremely numerous nuclei that are formed during this stage in development become arranged in the peripheral layer of cytoplasm which is found between the megaspore wall and the large central vacuole of the coenocytic gametophyte. Because of the very delicate structure of the gametophyte, it is extremely difficult to estimate accurately the number of free nuclei. Chamberlain (1935) stated that about 1,000 free nuclei are present at the end of the coenocytic period of gametophyte development in *Dioon edule*. Much larger numbers of nuclei—some greater than 3,000—were reported by DeSloover (1964) in his study of the female gametophyte of *Encephalartos poggei*. During its free nuclear phase, the female gametophyte of cycads becomes surrounded by one or two layers of nucellar cells, which constitute the so-called "endosperm jacket." This jacket apparently functions somewhat like a tapetum and is thought to convey soluble food material from the nucellus to the enlarging gametophyte.

The exact manner in which the young coenocytic female gametophyte becomes converted into a cellular structure, bearing archegonia, has been reviewed for the gymnosperms as a whole by Maheshwari and Singh (1967). At the end of the free nuclear period, each of the peripheral nuclei becomes connected by means of "secondary spindles" with six adjacent nuclei. Anticlinal walls then begin to develop centripetally, resulting in the formation of cells which are termed "alveoli." Each alveolus is hexagonal in transection and its nucleus lies at its base or inner end, which is open next to the vacuole. The female gametophyte at this stage has been likened to a honeycomb, each cavity representing a single alveolus. As the tubular alveoli extend towards the center of the gametophyte, many of them become "closed," at various distances from the center, by nuclear division followed by the development of more-or-less obliquely oriented walls. The small triangular cells which result are termed "precociously closed alveoli" by Maheshwari and Singh. The triangular cells, as well as the other alveoli, by means of repeated periclinal divisions (with reference to the gametophytic surface), form rows of radially arranged cells; but later, as the result of more divisions in irregular planes, this arrangement may become lost.

Very little information is available regarding the process of alveolation in cycads and it is not yet clear whether all the genera conform to the general plan of wall formation which we have just briefly outlined. The only recent detailed account is that given for *Encephalartos* by DeSloover (1964). According to his description, two stages in the partitioning of the coenocytic gametophyte can be recognized: a first stage characterized by the formation of alveoli and "pyramidal cells" (= "triangular cells"), followed by the stage of "repartitioning" of the alveoli, which begins at the periphery of the gametophyte. These two stages overlap to some extent because in *Encephalartos* the process of repartitioning may begin before the original alveoli have extended to the center of the gametophyte.

After the female gametophyte has become cellular, certain of the surface cells at the micropylar end become differentiated as *archegonial initials,* the number fluctuating between different genera or even within the same species. In most cycads, the number of archegonia which reach maturity is usually 2–6, according to Maheshwari and Singh (1967). *Microcycas* is apparently unique because large numbers of archegonia, arranged in groups, are initiated over the sides as well as the apex of the gametophyte. However, it is only the archegonia in the micropylar group that attain functional maturity.

The following account of the development of the archegonium in the cycads is based on Chamberlain's (1919, 1935) study on *Dioon edule*. The first step in ontogeny consists in the periclinal division of the archegonial initial into two unequal cells, a small outer *primary neck cell* and a much larger *central cell* (Fig. 15-21, F–G). Then the primary neck cell, by means of an anticlinal division, forms the two neck cells which,

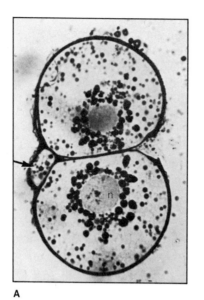
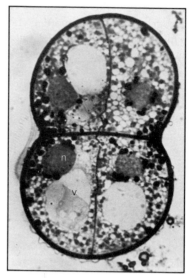

A

B

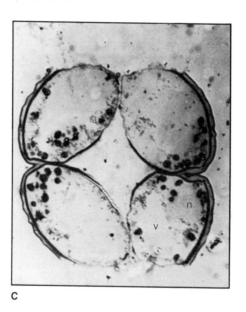

C

Figure 15-23
Transections showing successive stages in development of the four-celled neck
of the archegonium in *Zamia integrifolia.* **A,** pair of "secondary neck cells."
The arrow indicates an adjacent cell of the female gametophyte. **B,** later stage
following the division of each of the secondary neck cells; n, nucleus, v, vacuole.
C, mature archegonial neck at the time of fertilization, showing the opening
formed by the partial separation of the four neck cells. [Courtesy of
Dr. Kurt Norstog.]

according to Chamberlain, are characteristic of the mature archegonia of all cycads. Following the formation of the neck, the central cell enlarges and becomes highly vacuolated. At this time the gametophytic cells that surround the central cell become defined as an "archegonial jacket," which possibly functions in the translocation of food materials to the archegonium (Fig. 15-21, H). A short while before fertilization —only a few days in *Dioon*—the nucleus of the central cell divides, forming the small *ventral-canal-cell nucleus* and the much larger nucleus of the *egg cell*. No wall is formed between these nuclei, and the ventral-canal-cell nucleus remains near the neck of the archegonium and soon degenerates, while the egg nucleus moves towards the center of the egg and becomes relatively enormous in size. Chamberlain states that the egg nucleus in *Dioon* may reach a diameter of 500μ and Maheshwari and Singh (1967) remark that the egg in the cycads is perhaps the largest in the plant kingdom.

In concluding our brief description of the structure and ontogeny of the cycad archegonium, it should be noted that several recent studies fail to support Chamberlain's generalization that the neck of the archegonium consists of only two cells. Norstog and Overstreet (1965), for example, found that the archegonial neck in *Zamia integrifolia* consists of a tier of four cells. In a more detailed study of neck development in this same species, Norstog (1972) found that the young neck of the archegonium, prior to the stage at which fertilization can occur, consists of two "secondary neck cells" (Fig. 15-23, A). Just before fertilization, however, each of the secondary neck cells becomes divided by a very delicate wall; at this stage the neck is thus composed of a quartet of cells (as is shown in Fig. 15-23, B). According to Norstog, increasing turgor pressure in each of the four cells results in their enlargement and partial separation, creating a space or channel between them (Fig. 15-23, C). Although he did not actually observe sperms passing through the open archegonial neck, flagella were seen beating within one of the archegonia. Norstog believes that "distortion of the sperm must occur in fertilization since the neck opening is on the order of 50–70 μ in diameter, while the sperm is about 200 μ in diameter."

In addition to his intensive study on *Zamia*, Norstog found that the mature archegonial necks in *Encephalartos, Ceratozamia, Cycas,* and *Macrozamia* are likewise composed of four cells. On the basis of his studies, he reached the conclusion "that a four-celled neck apparatus is typical of cycads rather than the two-celled condition thought to be prevalent." (For a review on the archegonium in gymnosperms, see Maheshwari and Sanwal, 1963.)

FERTILIZATION.   Although many aspects of fertilization in the cycads remain to be studied, the large multiflagellated sperms have attracted much attention ever since they were first observed in *Cycas revoluta* in 1896 by Ikeno. His discovery represented a milestone in the history of plant morphology because the motile sperms of cycads, like those of *Ginkgo,* are the only known authentic examples of flagellated spermatozoids in the living seed plants. (See Ogura, 1967, for an interesting essay on the discovery of motile sperms in *Ginkgo* and *Cycas revoluta.*) Following the pioneering work of Ikeno, Webber (1901) and Chamberlain (1935) presented detailed accounts of the morphology and behavior of cycad sperms, and very recently, the ultrastructure of the spermatozoids of *Zamia* has been described by Norstog (1967).

The sperms of cycads are huge cells, those of *Zamia integrifolia* measuring about 180μ in diameter while still in the pollen tube (Norstog, 1967); the spermatozoids of *Dioon edule,* according to Chamberlain (1935), attain a diameter of 230μ and a length of 300μ after their release from the pollen tube. A fully mature sperm is top-shaped and the very numerous flagella are attached in ranks along the 5–6 turns of a spiral band which occupies the distal half of the cell (Fig. 15-24). Norstog (1967) estimated that the spermatozoid of *Zamia* develops about 20,000 flagella! His investigations with the electron microscope revealed that each flagellum has the structure typical of most animal and plant flagella, i.e., nine pairs of peripheral tubular fibers and a pair of central tubular fibers enclosed in an outer membrane.

Both Webber and Chamberlain observed that the two spermatozoids move freely for some time within the intact pollen tube. Webber remarked that "it is an interesting sight to see the two giant spermatozoids moving around vigorously in the pollen tube, bumping against each other and the wall of the tube in their reckless haste." When the prothallial end of the pollen tube ruptures, the two spermatozoids, together with some liquid, are discharged into the archegonial chamber. According to Swamy (1948), the entrance of the sperms into the archegonium appears to be aided by their pulsating and amoeboid movements. The exact physical mechanism, however, deserves study because the sperm is much larger than the opening created by the separation of the neck cells of the archegonium (Norstog, 1972). Once inside the cytoplasm of the egg, the flagellated band of the functional sperm becomes detached and remains plainly visible near the neck of the archegonium. The nucleus of the spermatozoid sinks down into the egg cytoplasm and finally comes into direct contact with the egg nucleus. This is fertilization. Although several sperms may enter an archegonium, only one male nucleus fuses with the nucleus of the egg; the other spermatozoids remain near the top of the egg and finally disintegrate.

EMBRYOGENY AND SEED DEVELOPMENT. In the previous chapter it was emphasized that the embryogeny of most gymnosperms begins with a period of free nuclear divisions. The cycads as a whole are distinguished by the relatively large number of free nuclei produced before wall formation starts.

Following the division of the zygotic nucleus, successive divisions for some time are definitely synchronized. But after about eight successive divisions, which result in 256 nuclei, irregularities appear, some of the nuclei either failing to divide or at least not keeping pace with the rate of division of the remainder. Hence the ultimate number of free nuclei is likely to be less than the theoretical expectation. In *Dioon edule* Chamberlain found about 1,000 free nuclei before the initiation of cell walls. Other cycads have smaller numbers; the lowest recorded is 64 in the genus *Bowenia*.

Wall formation in the embryo begins at the lower end of the archegonium and progresses toward the neck end. In some genera this segmentation process extends completely throughout the entire mass of multinucleate protoplasm. But in others only the inward-facing portion of the embryo develops cell walls—the remainder retains the free nuclear condition and ultimately disintegrates. This is the situation in *Zamia*, and the salient features of its early embryogeny will now be outlined on the basis of the study by Bryan (1952).

At the final phase of the free nuclear period a large number of nuclei are aggregated at the inward-directed end of the embryo (Fig. 15-25, A). Two successive, simultaneous divisions of these nuclei occur, and during these divisions cell walls form progressively in an upward direction (Fig. 15-25, B, C). As a result, the embryo now consists of two well-defined regions: a tissue of walled cells, and a region of vacuolated cytoplasm with scattered nuclei. This

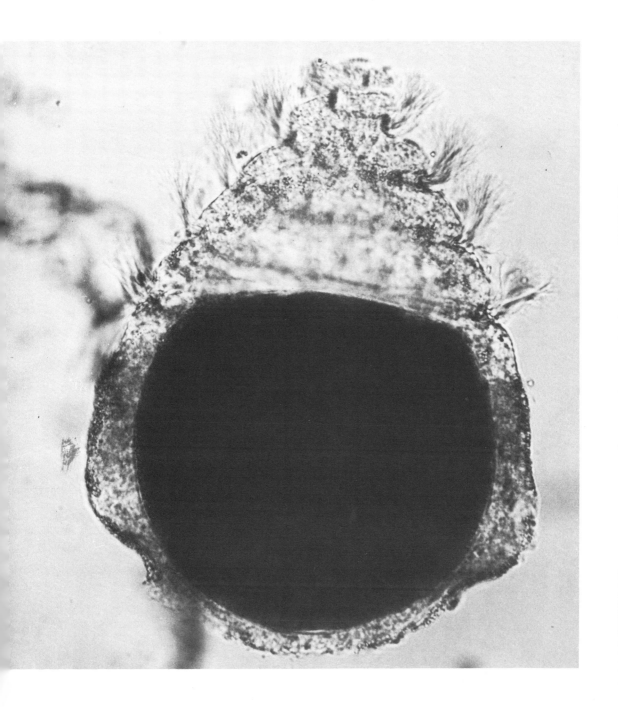

Figure 15-24
Photomicrograph of a living spermatozoid of *Zamia integrifolia*, showing the flagellated spiral band and the large central nucleus (dark hemisphere) (× 625). [From Norstog and Overstreet, *Phytomorphology* 15:46, 1965.]

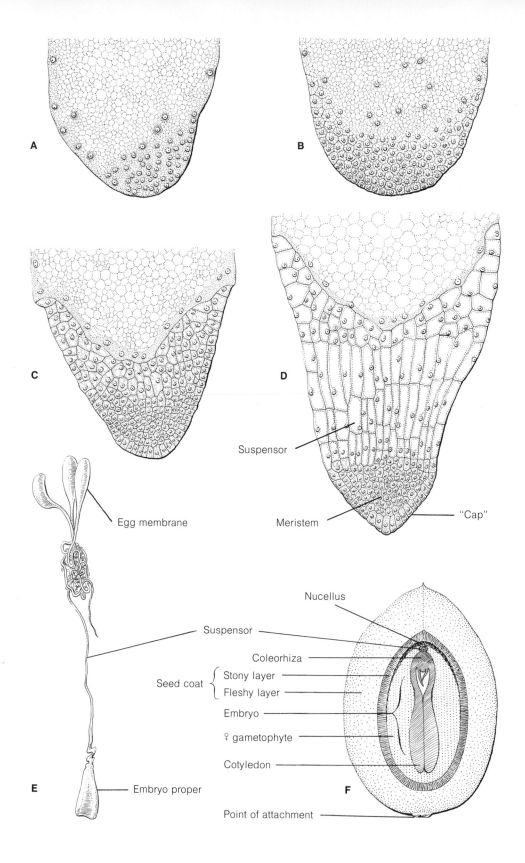

A

B

C

D

Suspensor

Meristem

"Cap"

Egg membrane

Suspensor

Embryo proper

E

Nucellus

Coleorhiza

Seed coat { Stony layer
            { Fleshy layer

Embryo

♀ gametophyte

Cotyledon

Point of attachment

F

latter portion of the embryo apparently serves a nutritive function for a time but finally disintegrates. Active cell division in the cellular part of the embryo results in the gradual differentiation of (1) a conspicuously meristematic zone, from which the main organs of the embryo ultimately arise, and (2) a posterior region of elongating cells arranged in vertical series, which mark the origin of the massive suspensor typical of cycad embryos (Fig. 15-25, D). The cells lying above the suspensor region are smaller and constitute a zone that possibly serves as a buffer, which may, to some degree, direct the downward extension of the suspensor. A distinctive feature of the young embryo of *Zamia* is the layer of discrete superficial cells that constitutes a cap over the meristematic zone (Fig. 15-25, D). Bryan found that the outer walls of the cap cells are extremely thick. Gradually the cap cells distintegrate, and the outermost cells of the meristematic zone become organized as a new surface layer. The functional significance, if any, of this "deciduous" cap in the embryo of *Zamia* is problematical. It may constitute a protective layer during the early period when the meristematic zone is pushed deeply into the gametophytic tissue by the elongating suspensor. Comparative embryological studies are highly desirable in order to determine whether caps occur in the young embryos of other cycad genera.

In *Zamia,* as well as in other genera, the elongation of the suspensor forces the apex of the embryo through the tough membrane of the original egg into contact with the nutritive tissue of the large gametophyte. During this invasion the suspensor usually becomes coiled or twisted and often reaches a considerable length. Simple polyembryony (i.e., the development of the zygotes of several archegonia in the same ovule) is extremely common in cycads. According to Chamberlain's investigations, the coiled suspensors of several young embryos may intertwine to form a compound structure (Fig. 15-25, E). Typically, only one embryo normally survives and is functional.

As stated above, the main organs of the embryo develop from the meristematic, inward-facing end of the embryo which is thus of the endoscopic type. (See Chapter 6 for a general discussion of endoscopic polarity in embryos.) Usually two cotyledons, which flank the shoot apex, are soon differentiated, although *Ceratozamia* is characterized by having a single cotyledon. The posterior end of the embryo in direct contact with the suspensor develops into the radicle. In some cycads the shoot apex forms several bud-scale primordia and possibly the primordium of the first foliage leaf before seed development is completed. The structure of a ripe cycad seed is shown in longisectional view in Fig. 15-25, F.

The Seedling

There is no fixed period of seed dormancy in cycads, and germination may begin as soon as the seeds fall on the ground. The micropylar end of the stony layer of the seed coat is first ruptured by the elongation of

Figure 15-25 (*facing page*)
A–D, early embryogeny in *Zamia umbrosa.* A, free nuclear period; B, formation of walls between nuclei in lower end of embryo; C, beginning of cellular differentiation (note lower region of small actively dividing cells above which are the first elongating cells of the future suspensor); D, later stage showing clear differentiation of surface layer or "cap," meristem, suspensor, and "buffer" region (note that upper portion of embryo has remained in the free nuclear condition); E, young embryo of *Cycas* dissected from seed (note long, slender suspensor and, above, the persistent membranes of several egg cells); F, diagram of a longitudinal section of a cycad seed. [A–D redrawn from Bryan, *Amer. Jour. Bot.* 39:433, 1952; E redrawn from Swamy, *Amer. Jour. Bot.* 35:77, 1948.]

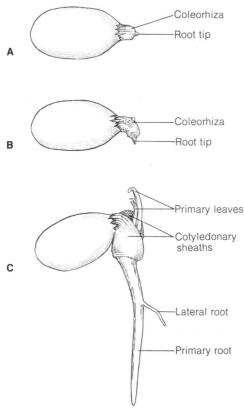

Figure 15-26
Seedling development in *Dioon edule*. **A, B,** early
stages in seed germination, showing the emergence
of the root tip from the coleorhiza; **C,** young seedling
with several primary leaves enclosed by the sheaths of
the two cotyledons. [Redrawn from *Gymnosperms
Structure and Evolution* by C. J. Chamberlain. The
University of Chicago Press, Chicago, 1935.]

the *coleorhiza,* a sheath of tissue which encloses the root tip. The root tip then protrudes through the coleorhiza and begins to
grow down into the soil (Fig. 15-26, A–B).
A distinctive feature of the subsequent
development of the seedling is that the
blades of the two cotyledons remain per-

manently within the seed, apparently functioning as haustorial organs, whereas the
cotyledonary sheaths and the first primary
leaves soon emerge from the ruptured seed
coat (Fig. 15-26, C). In seedlings two or
more years old, the seed may still remain
attached by the withered cotyledons to the
axis of the young plant. For several years
after germination, the shoot apex of the
seedling produces scale leaves (*cataphylls*)
and, at intervals, *single* foliage leaves. As
development of the young plant continues,
the number of foliage leaves increases until
well-defined crowns of foliage leaves,
alternating with zones of scale leaves, become the established pattern of growth.

During the transition from seedling to
adult sporophyte, the shoot apex increases
greatly in size and in *Cycas revoluta* may
attain a diameter of about 3.5 mm, a dimension greatly exceeding that typical of the
apices of other vascular plants. The complex
zonal structure of the large shoot apices
of cycads is strikingly illustrated in *Microcycas* (Fig. 15-27). One of the most distinctive features of the apex of this cycad is the
convergence of well-defined files of cells
from the lateral part of the initiation zone
towards the inner cells of the zone of central mother cells. This fan-like arrangement
of cells, so strikingly displayed in *Microcycas,* presents a difficult problem in any
effort to interpret the patterns of cell
division and cell polarity in the shoot apex;
perhaps additional examples of the *Microcycas* type of zonal structure will appear as
comparative studies continue. (For further
details on the shoot apex of *Microcycas* and
other cycad genera, see Foster 1939, 1940,
1941, 1943; and Johnson 1944, 1951.)

Figure 15-27 (*facing page*)
Median longisection of the shoot apex of *Microcycas calocoma*. Note particularly the fan-like arrangement of
the tiers of cells that constitute the initiation zone, and the deeply situated zone of central mother cells, which
gives rise to the shallow rib meristem zone and the inner portion of the peripheral zone. [From Foster, *Amer.
Jour. Bot.* 30:56, 1943.]

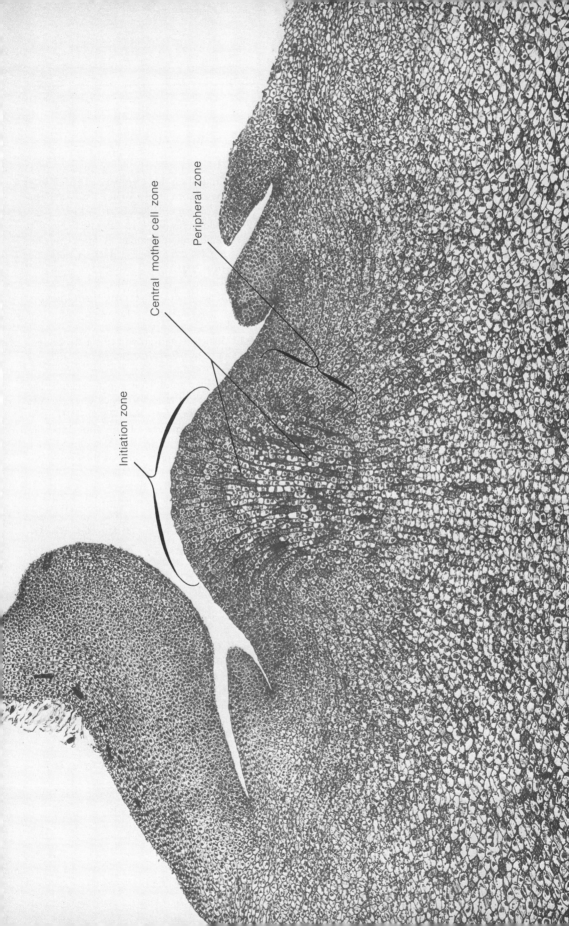

Initiation zone

Central mother cell zone

Peripheral zone

References

Andrews, H. N., Jr.
  1961. *Studies in Paleobotany.* Wiley, New York.
  1963. Early seed plants. *Science* 142:925–931.
Arnold, C. A.
  1953. Origin and relationships of the cycads. *Phytomorphology* 3:51–65.
  1947. *An Introduction to Paleobotany.* McGraw-Hill, New York.
Arnott, H. J.
  1959. Anastomoses in the venation of *Ginkgo biloba. Amer. Jour. Bot.* 46:405–411.
Baird, A. M.
  1939. A contribution to the life history of *Macrozamia reidlei. Jour. Roy. Soc. West. Aust.* 25:153–175.
Brashier, C. K.
  1968. Vascularization of cycad leaflets. *Phytomorphology* 18:35–43.
Bryan, G. S.
  1952. The cellular proembryo of *Zamia* and its cap cells. *Amer. Jour. Bot.* 39:433–443.
Chamberlain, C. J.
  1919. *The Living Cycads.* University of Chicago Press, Chicago.
  1935. *Gymnosperms. Structure and Evolution.* University of Chicago Press, Chicago.
Crepet, W. L.
  1972. Investigations of North American Cycadeoids: Pollination mechanisms in *Cycadeoidea. Amer. Jour. Bot.* 59:1048–1056.
Delevoryas, T.
  1962. *Morphology and Evolution of Fossil Plants.* Holt, Rinehart and Winston, New York.
  1963. Investigations of North American Cycadeoids: cones of *Cycadeoidea. Amer. Jour. Bot.* 50:45–52.
  1968a. Investigations of North American cycadeoids: Structure, ontogeny, and phylogenetic considerations of cones of *Cycadeoidea. Palaeontographica* 121(B):122–133.
  1968b. Some aspects of cycadeoid evolution. *Jour. Linn. Soc. London Bot.* 61:137–146.

DeSloover, J.-L.
  1961. Études sur les Cycadales. I. Méiose et mégasporogenèse chez *Encephalartos poggei* Asch. *Cellule* 62:105–116.
  1964. Études sur les Cycadales. III. Nucelle, gamétophyte femelle et embryon chez *Encephalartos poggei* Asch. *Cellule* 64:149–200.
Downie, D. G.
  1928. Male gametophyte of *Microcycas calocoma. Bot. Gaz.* 85:437–450.
Foster, A. S.
  1939. Structure and growth of the shoot apex of *Cycas revoluta. Amer. Jour. Bot.* 26:372–385.
  1940. Further studies on zonal structure and growth of the shoot apex of *Cycas revoluta. Amer. Jour. Bot.* 27:487–501.
  1941. Zonal structure of the shoot apex of *Dioon edule. Amer. Jour. Bot.* 28:557–564.
  1943. Zonal structure and growth of the shoot apex in *Microcycas calocoma.* (Mig.) A.D.C. *Amer. Jour. Bot.* 30:56–73.
Foster, A.S., and M. R. San Pedro
  1942. Field studies on *Microcycas calocoma. Mem. Soc. Cubana Hist. Natur.* 16:105–121.
Hertrich, W.
  1951. *Palms and Cycads.* Privately printed, San Marino, California.
Ikeno, S.
  1896. Spermatozoiden von *Cycas revoluta. Bot. Mag.* (Tokyo) 10:367–368.
Jeffrey, E. C.
  1917. *The Anatomy of Woody Plants.* University of Chicago Press, Chicago.
Johnson, L. A. S.
  1959. The families of cycads and the Zamiaceae of Australia. *Proc. Linn. Soc. New South Wales* 84:64–117.
Johnson, M. A.
  1944. On the shoot apex of the cycads. *Torreya* 44:52–58.
  1951. The shoot apex in gymnosperms. *Phytomorphology* 1:188–204.

Kershaw, E. M.
1912. Structure and development of the ovule of *Bowenia spectabilis. Ann. Bot.* 26:625–646.

Maheshwari, P., and M. Sanwal
1963. The archegonium in gymnosperms: A review. *Mem. Indian Bot. Soc.* No. 4: 103–119.

Maheshwari, P., and H. Singh
1967. The female gametophyte of gymnosperms. *Biol. Rev.* 42:88–130.

Mamay, S. H.
1969. Cycads: Fossil evidence of late Paleozoic origin. *Science* 164:295–296.

Norstog, K.
1967. Fine structure of the spermatozoid of *Zamia* with special reference to the flagellar apparatus. *Amer. Jour. Bot.* 54:831–840.
1972. Role of archegonial neck cells of *Zamia* and other cycads. (in press).

Norstog, K., and R. Overstreet
1965. Some observations on the gametophytes of *Zamia integrifolia. Phytomorphology* 15:46–49.

Ogura, Y.
1967. History of discovery of spermatozoids in *Ginkgo biloba* and *Cycas revoluta. Phytomorphology* 17:109–114.

Oliver, F. W., and D. H. Scott
1905. On the structure of the Paleozoic seed *Lagenostoma lomaxi,* with a statement of the evidence upon which it is referred to *Lyginodendron. Phil. Trans. Roy. Soc. London* 197(B):193–247.

Pettitt, J. M., and C. B. Beck
1968. *Archaeosperma arnoldii* — a cupulate seed from the Upper Devonian of North America, *Contr. Mus. Paleontol. Univ. Mich.* 22:139–154.

Rattray, G.
1913. Notes on the pollination of some South African cycads. *Trans. Roy. Soc. S. Africa* 3:259–270.

Reynolds, L. G.
1924. Female gametophyte of *Microcycas. Bot. Gaz.* 77:391–403.

Seward, A. C.
1917. *Fossil Plants,* Vol. III. Cambridge University Press, London.

Shapiro, S.
1951. Stomata on the ovules of *Zamia floridana. Amer. Jour. Bot.* 38:47–53.

Smith, F. G.
1910. Development of the ovulate strobilus and young ovule of *Zamia floridana. Bot. Gaz.* 50:128–141.

Sterling, C.
1963. Structure of the male gametophyte in gymnosperms. *Biol. Rev.* 38:167–203.

Storey, W. B.
1968. Somatic reduction in cycads. *Science* 159:648–650.

Swamy, B. G. L.
1948. Contributions to the life history of a *Cycas* from Mysore (India). *Amer. Jour. Bot.* 35:77–88.

Taylor, T. N.
1965. Paleozoic seed studies: a monograph of the American species of *Pachytesta. Palaeontographica* 117(B):1–46.

Troll, W.
1938. *Vergleichende Morphologie der höheren Pflanzen.* Bd. l. Zweiter Teil, Lieferung l.

Walton, J.
1953. *An Introduction to the Study of Fossil Plants.* Adam and Charles Black, London.

Webber, H. J.
1901. *Spermatogenesis and Fecundation of Zamia.* U.S. Dep. Agr. Bur. Plant Ind. Bull. No. 2:1–100.

Wieland, G. R.
1906. *American Fossil Cycads.* (Publication 34, 1.) Carnegie Institution, Washington, D.C.
1916. *American Fossil Cycads. Taxonomy.* (Publication 34, 2.) Carnegie Institution, Washington, D.C.

Zimmermann, W.
1959. *Die Phylogenie der Pflanzen.* Gustav Fischer, Stuttgart.

16

The Coniferopsida: *Ginkgo biloba*

The first European botanist to study the maidenhair tree was Kaempfer, who observed it in cultivation in Japan in 1690 and later published a botanical description of the tree. He proposed for it the name *Ginkgo,* and Linnaeus adopted this generic appellation in 1771, adding the descriptive specific epithet *biloba* in reference to the deep notching that characterizes the lamina of the leaf on many specimens. Although there has been considerable difference of opinion regarding the etymology of the word "Ginkgo," there is, according to Li (1956), no justification for rejecting the name on the basis that it is a distortion of the word "Ginkyo." Li maintains that the maidenhair tree has been correctly designated as *Ginkgo* for nearly 250 years "and certainly will be known as such forever."

During the latter part of the nineteenth century, *Ginkgo biloba* was placed in the family Taxaceae under the order Coniferales because of the resemblance of its large fleshy seeds to the seeds of *Torreya* and *Cephalotaxus.* The discovery by Hirase (1896) of motile flagellated spermatozoids in *Ginkgo* led to a complete reappraisal of the systematic position of *Ginkgo biloba,* which today is regarded as the only surviving representative of the Ginkgoales (Engler, 1954).

Like the cycads, *Ginkgo biloba* is a veritable "living fossil," and perhaps, as Arnold (1947) suggested, "may indeed be the oldest living genus of seed plants." Leaves very similar in form and venation to the leaves of the modern *Ginkgo* have been found as fossils in rocks deposited during the Mesozoic Era, when ginkgo-like plants were apparently worldwide in distribution. As an example we may mention *Ginkgo digitata,* which was a Jurassic species that grew in such widely separated areas as western North America, Alaska, Australia, Japan, and England. Other fossil ginkgophytes are *Ginkgoites* and *Baiera*. The latter genus is particularly notable because its multilobed leaves resemble the deeply incised leaf blades often observed in seedlings and in coppice shoots of the living *Ginkgo*. According to Arnold (1947), the leaves of the form-genus *Ginkgoites* so closely resemble *Ginkgo* as to be indistinguishable from those of the living species.

Following the Mesozoic Era, *Ginkgo* declined progressively in its distribution and, according to Seward (1936), "there can be no doubt that China was the last, if it is not the present, natural home of the Maidenhair tree."

Whether *Ginkgo* still exists in the wild state in the more remote and poorly explored forests of China has been regarded an unsettled question by many writers. Some evidence has been produced in favor of the existence of native stands of *Ginkgo* trees, though many botanists contend that such trees may represent the offspring of cultivated specimens. In a thorough discussion, however, Li (1956) presents evidence that *Ginkgo* still exists in the wild state in southeastern China, and that the last refuge of this living fossil is a mountainous area "along the northwestern border of Chekiang and southeastern Anhwei." From a broad evolutionary viewpoint, *Ginkgo* is to be regarded "as one of the wonders of the world; it has persisted with little change until the present through a long succession of ages when the earth was inhabited by animals and plants for the most part far removed, in kind as in time, from their living descendants. *Ginkgo* is one of a small company of living plants which illustrates continuity and exceptional power of endurance in a changing world" (Seward, 1938, p. 424).

Ginkgo biloba is widely cultivated as a park specimen, or street tree in many temperate areas of the world (Figs. 16-1, 16-2). In accordance with the vitality which has enabled it to survive as a distinct organism for millions of years, the modern *Ginkgo* appears exceptionally resistant to the attacks of insects and fungi, and grows successfully amidst the smoke and gasoline fumes of modern cities (Fig. 16-1). The outer fleshy coat of the seed emits an odor like rancid butter, and for this reason the male or microsporangiate trees are preferable to female trees for park or street planting.

Although there is no known way of separating male from female trees on the basis of external morphological characters prior to the formation of reproductive structures, the work of Lee (1954) and Pollock (1957) suggests that it is possible to differentiate between chromosomal complements in the two sexes. If a study of chromosome morphology by means of stem-tip and leaf-primordia smears proves practical in distinguishing the sex of young *Ginkgo* plants, it would then be possible to eliminate potential female trees from street plantings.

The kernel of the seed (i.e., the female gametophytic tissue and the embryo) of *Ginkgo* is highly nutritious and is used as food in China and Japan. *Ginkgo* seeds are imported into this country, and we, the authors, have secured seeds for class study from the various markets in San Francisco's Chinatown.

Vegetative Organography and Anatomy

Habit

Young *Ginkgo* trees have a pronounced excurrent habit of growth, resembling that of many conifers (Fig. 16-1). With increasing age, the crown becomes broad and irregular and the pattern of branching variable (Fig. 16-2). Exceptionally robust trees, such as those found near certain temples and shrines in China and Japan, may attain a height of 100 feet.

Figure 16-1 (*facing page*)
Ginkgo biloba. Young specimen illustrating dense foliage and
characteristic excurrent habit of growth. This tree is vigorous
and healthy although it is growing in an urban environment
which might not be expected to be favorable for it.

Figure 16-2
A large, microsporangiate specimen of *Ginkgo biloba* in leafless condition, University of
California Campus, Berkeley. Note very irregular pattern of branching.

Burls

In very old cultivated specimens of *Ginkgo,* stalactite-like *burls,* called "chichi" (nipples) by the Japanese, hang downward from the lower sides of many of the larger branches. These peculiar burls may occur either singly or in clusters and according to Fujii (1895), may attain a length of 2.2m and a diameter of 30 cm. If one of these strange growths reaches the ground, it may take root and form leaves. Fujii's anatomical study revealed that a chichi, near its point of attachment to the parent branch, contains a central, deeply imbedded, spur shoot together with its associated buds. These buds keep pace with the secondary growth of the burl and appear as small protuberances on the outer surface of the thick cylinder of xylem. Although Fujii concluded that the chichi of *Ginkgo* represent a "pathological" formation, he provided no explanation of the "causal factors" responsible for their origin and unusual mode of development.

Shoot Dimorphism

During the development of a *Ginkgo* tree, a marked distinction becomes increasingly evident between two types of shoots: *long shoots,* distinguished by their widely separated nodes and numerous leaves, and the more slowly growing *spur shoots,* characterized by short, crowded internodes and the annual expansion of only a few leaves. Spur shoots begin their development as buds, which arise in the leaf axils of long shoots, and may continue their sluggish pattern of vegetative growth for many years. Anatomical studies have shown that the zonal structure of the apical meristem is similar in the terminal buds of both long and spur shoots; the histogenetic difference between the two shoot types is the result of the longer duration of cell division and cell elongation in the primary stem tissues

derived from the terminal meristem of the long shoots (Foster, 1938).

It is particularly interesting, from a physiological viewpoint, that the pattern of growth in the two types of shoots is reversible: a spur shoot may abruptly proliferate into a long shoot and conversely, the terminal growth of a long shoot may be greatly retarded for several seasons, thus simulating the growth pattern of a lateral spur shoot. The physiological basis for shoot dimorphism in *Ginkgo* has not been fully explained, although there is experimental evidence that auxin—produced in an elongating long shoot—inhibits the expansion of the axillary buds into long shoots but does not prevent the formation of spur shoots (Gunckel, Thimann, and Wetmore, 1949).

Leaves

One of the most distinctive morphological characters of *Ginkgo* is the foliage leaf, which consists of a petiole and a fan-shaped dichotomously veined lamina (Fig. 16-3). Although Linnaeus' specific epithet *biloba* correctly describes the form of the lamina of many *Ginkgo* leaves, there is an enormous range of variation, with respect to the *degree* of lobing and dissection, among the leaves of a single tree. Critchfield (1970) has recently studied "heterophylly" in *Ginkgo* and has shown that the form of the lamina is correlated with the position of a given leaf or leaf series in the shoot system of the tree. On spur shoots and the basal region of long shoots, the leaf blades are either entire or divided by a distal notch into two lobes. Leaves of this type are present in a partly developed stage in the winter buds and complete their growth during bud expansion the following spring. In contrast, most of the leaves found on the upper part of long shoots are initiated and complete their development during the same season;

Figure 16-3
Cleared leaf illustrating the dichotomous pattern of venation. The arrows indicate vein anastomoses. See text for further explanation [From Arnott, *Amer. Jour. Bot.* 46:405, 1959.]

the lamina of such leaves is always divided by a very deep sinus into two major lobes, each of which in turn is further dissected into segments. Critchfield suggested that auxin may control the extended period of growth of long shoots and the correlated formation of multilobed leaves.

The regular dichotomous pattern of venation in the lamina is one of the striking morphological characters of the leaf of *Ginkgo*. Two leaf traces, derived from separate bundles of the stele, extend through the petiole and give rise to two systems of dichotomously branched veinlets which vascularize the two halves of the lamina (Fig. 16-3). This distinctive type of venation was also present in the leaves of many of the extinct members of the Ginkgoales (Florin, 1936; Arnold, 1947).

Throughout morphological literature, the dichotomous venation of *Ginkgo* is de-

scribed as "open" in type, i.e., devoid of vein unions. Arnott's (1959) detailed survey of more than 1,000 leaves of *Ginkgo biloba* revealed for the first time that there are various types of anastomoses. Approximately 10 percent of the leaves that he studied showed one or more anastomoses; the largest number of vein unions observed in a single lamina was five (see Fig. 16-3).

The most common type of anastomosis — representing 41.1 percent of the total anastomoses found — consists in the union, near the lamina margin, of the two branches derived from a single vein dichotomy (indicated in Fig. 16-3 by the arrows at the edge of the leaf). Another common form of anastomosis — representing 31.3 percent of the anastomoses — is produced by the union of the adjacent branches produced by two separate vein dichotomies (indicated in Fig. 16-3 by the arrow placed in the right-

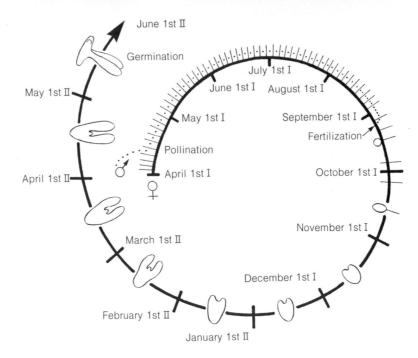

Figure 16-4
Diagrammatic representation of the cycle of reproduction in *Ginkgo biloba*. See text for explanation. [Adapted from Favre-Duchartre, *Phytomorphology* 8:377, 1958.]

hand lobe of the leaf). From a comparative standpoint, it is interesting that both of these types of anastomoses occur in the leaf pinnae of certain genera in the living cycads (Chapter 15, Fig. 15-16).

Ginkgo is deciduous and before the leaves are shed in the autumn, they turn a beautiful golden-yellow. During the leafless period (see Fig. 16-2), the dormant buds of the spur and long shoots are protected by a series of tightly imbricated bud scales.

Stem

Compared with the cycads, the pith and cortex of young stems of *Ginkgo* are relatively small. Chamberlain (1935) noted, however, that the pith and cortical zones of spur shoots are larger in diameter than the corresponding tissue areas in the stem of long shoots. As in typical coniferophytes, cambial activity in *Ginkgo* is vigorous and sustained and produces a pycnoxylic type of secondary xylem with well-defined annual rings.

The Reproductive Cycle

Phenology

The series of events that culminates in the production of ripe seeds in *Ginkgo* takes approximately 14 months, according to the detailed investigations made by Favre-Duchartre (1956) in the area of Paris, France. As is shown diagrammatically in Fig. 16-4, pollination, the maturation of the male and female gametophytes, and fertilization occur in one season (April–September), and embryogeny is not completed until the spring of the following year (November–May). In Fig. 16-4, the upper part of the solid black spiral line represents the period during which the development of the female gametophyte (♀) takes place and the dotted line depicts the period of association between the developing male gametophyte (♂) and the megagametophyte within the ovule. Frequent reference to Fig. 16-4 should be made in connection with the detailed descriptions of sporogenesis, gametogenesis, and embryogeny that follow.

A

B

Figure 16-5
Ginkgo biloba. **A,** spur shoot with expanding
leaves and microsporangiate strobili;
B, spur shoot with young leaves and
pairs of ovules borne on slender stalks.

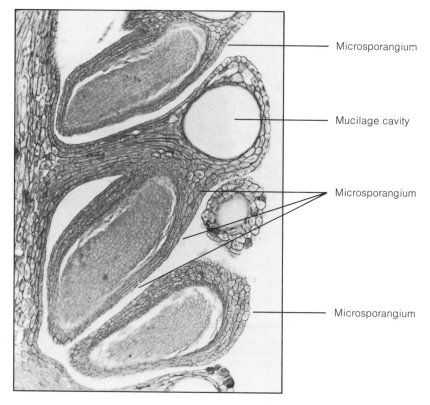

Microsporangium

Mucilage cavity

Microsporangium

Microsporangium

Figure 16-6
Longisection showing a microsporophyll, with one of its two pendant microsporangia, attached to the axis of a microsporangiate strobilus; note conspicuous mucilage cavity in the apical "hump" of this microsporophyll. The compact tissue within the thick walls of the microsporangia shown in this section is composed of the primary sporogenous cells.

Microsporangiate and Megasporangiate Structures

Ginkgo is dioecious, and when the reproductive stage has been reached, the pollen-forming organs and the ovules are produced on separate trees. A distinctive feature is the restriction of the microsporangiate and ovuliferous structures to the spur shoots, where they are evident in the spring in the axils of the inner bud scales and the foliage leaves (Fig. 16-5).

THE MICROSPORANGIATE STROBILUS. The microsporangiate strobilus is a loose, pendulous catkin-like structure, consisting of a main axis to which are attached numerous appendages each of which bears two (sometimes three or four) pendant microsporangia at its tip (Fig. 16-5, A). Such terms as "sporangiophore" and even "stamen" have been applied to the fertile appendages of the microsporangiate strobilus, thus indicating the uncertainty that prevails about their homologies (Fig. 16-6). Although it is not yet clear whether the microsporangia originate from superficial or hypodermal initials, the general plan of development is eusporangiate as in the cycads. A nearly mature microsporangium has a wall of several layers of cells (including the tapetum) that encloses a central group of microsporocytes.

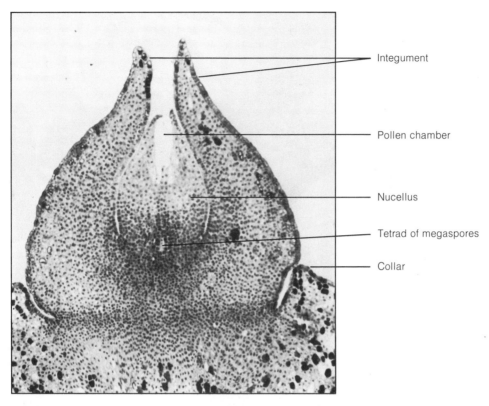

Integument

Pollen chamber

Nucellus

Tetrad of megaspores

Collar

Figure 16-7
Median longisection of an ovule of *Ginkgo,* showing nucellus with conspicuous pollen chamber, enclosed by the single integument. Megasporogenesis has occurred and a linear tetrad of megaspores is present in the basal region of the nucellus. Note sectional view of the "collar" below the point of divergence of the ovular integument.

These microsporocytes, by meiotic division, produce numerous tetrads of haploid microspores.

THE OVULIFEROUS STRUCTURE. In striking contrast to the megasporophylls of the cycads, each of the ovuliferous organs of *Ginkgo* arises in the axil of a leaf of the spur shoot and consists of a stalk or "peduncle" that bears at its tip two (occasionally three or more) erect ovules (Figs. 16-5, B; 16-8). Each ovule is subtended, below the point of divergence of the integument from the nucellus, by a rim-like outgrowth that was termed the "collar" by Chamberlain (1935).

The so-called collar has been interpreted as a vestigial sporophyll and the entire ovuliferous structure as a strobilus. The recent histogenetic studies of Pankow and Sothman (1967) provide no support for this speculative interpretation. They found that the collar originates as a rim of tissue *after* the integument of the ovule is well advanced in its ontogeny. This fact—coupled with the absence of any vascular tissue in the collar—led them to conclude that the ovules of *Ginkgo* are cauline and terminal on lateral axes, and that the assumed "foliar" nature of the collar is highly questionable, at least from an ontogenetic point of view (Figs. 16-7; 16-8).

Figure 16-8
Portion of a branch of an ovulate tree of *Ginkgo biloba*, showing the attachment of the
peduncles of the ripening ovules to the tips of the spur shoot.

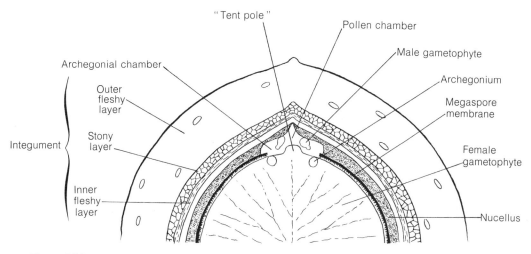

Figure 16-9
Longitudinal section (diagrammatic) showing the structure of the upper part of a mature ovule of *Ginkgo*. See text for further explanations. [Redrawn from Favre-Duchartre, *Rev. Cytol. Biol. Veg.* 17:1, 1956.]

The relatively extensive free portion of the nucellus of the ovule of *Ginkgo* is enclosed by a single integument which becomes anatomically differentiated into a thick outer fleshy layer, a more compact and thinner inner fleshy layer and a hard stony middle layer; the latter constitutes the "shell" of the mature seed (Fig. 16-9). In contrast to the ovules of cycads, the vascular system of the ovule of *Ginkgo* is weakly developed and consists of two bundles which are restricted to the inner fleshy layer of the integument (Favre-Duchartre, 1956).

The Male Gametophyte

The early ontogeny of the male gametophyte of *Ginkgo* resembles that typical for the cycads except that two prothallial cells—rather than one—are developed. After the second prothallial cell has been formed, the meristematic initial—functioning as an "antheridial initial"—divides and produces the generative cell and the tube cell (see Fig. 17-24). At this four-celled stage, the young male gametophyte, enclosed within the wall of the pollen grain, is released into the air by the dehiscence of the microsporangium wall.

The pollen is carried to the megasporangiate tree by wind currents and adheres to the mucilaginous pollination droplet which exudes from the micropyles of the ovules. Retraction of this droplet brings the pollen into the pollen chamber where the formation of the haustorial pollen tube and the final stages in the development of the male gametophyte take place (see Fig. 16-4 for the approximate date of pollination).

Megasporogenesis

According to Favre-Duchartre (1956), megasporogenesis in *Ginkgo* occurs towards the end of April. The single megasporocyte is a cell deeply situated in the ovule approximately at the level of separation of the integument from the nucellus (Fig. 16-7). Meiosis occurs and produces a row of four (or sometimes three) cells, the lowermost of which develops into the female gametophyte. In their study of the ultrastructure of the megaspore mother cell in *Ginkgo,* Stewart and Gifford (1967) made the interesting observation that following meiosis I, *all* the plastids and mitochondria present at the chalazal end of the sporocyte become segregated in the lower dyad cell that is destined to produce the functional megaspore.

The investigations of Favre-Duchartre (1956) and DeSloover-Colinet (1963) have shown that megasporogenesis and the formation of the *pollen chamber* in the upper part of the nucellus are synchronized events. A group of internal cells, at the micropylar end of the nucellus, enlarge and then begin to degenerate, creating a cavity that is the beginning of the pollen chamber. The breakdown of nucellar tissue continues, and finally the nucellar epidermis, which lies over the cavity, ruptures and a large open pollen chamber is formed (Fig. 16-7).

The Female Gametophyte

As in the cycads, the development of the female gametophyte in *Ginkgo* begins with the *coenocytic* stage, characterized by extensive free nuclear divisions, and is followed by the *cellular stage,* during which the coenocyte becomes converted by wall formation into a cellular gametophyte bearing archegonia at its micropylar end. These two phases in the development of the megagametophyte may now be considered in more detail.

COENOCYTIC STAGE. This phase in development results from the enlargement of the functional megaspore accompanied by a succession of free nuclear divisions which occur in the peripheral cytoplasm situated between the megaspore membrane and the large central vacuole. Favre-Duchartre (1958) reported that as the result of thirteen successive mitotic divisions, approximately 8,000 free nuclei are produced. He found that the free nuclear divisions are not synchronized but rather proceed from the chalazal to the micropylar end of the coenocyte. During the coenocytic phase, the megaspore membrane progressively thickens and its outer surface can be resolved into a series of short vertical "filaments" which are pressed against each other forming the outer coherent surface layer of the membrane.

CELLULAR STAGE. At the close of the period of free nuclear divisions, anticlinal walls begin to develop from the periphery of the coenocyte towards the central vacuole. The *alveoli* that are produced are long tubular uninucleate cells that for some time are "open," i.e., devoid of internal end walls (Fig. 16-10, B). Some of the centripetally developing alveoli meet in the center of the gametophyte and, according to Favre-Duchartre (1956), a membrane is formed across their open ends, transforming them into cells. After the first pyramidal cells have formed, the closed alveoli divide by formation of periclinally oriented walls creating files of cells that are oriented radially with respect to the megaspore membrane. As the young cellular gametophyte increases in volume, anticlinal divisions also occur and the previous regular alignment of the files of cells is lost (Fig. 16-10, B, C).

ARCHEGONIA. Although the number of archegonia varies from one to five, according to Favre-Duchartre (1956), the usual number is two (Fig. 16-9). Each archegonial initial is a superficial cell at the micropylar end of the gametophyte and by means of a periclinal division, forms a large *central cell* and a smaller *primary neck cell.* The latter soon divides anticlinally, forming a pair of neck cells. The archegonial neck remains in this two-celled stage for a very long time during which the central cell becomes greatly enlarged. Coinciding with the division of the nucleus of the central cell,

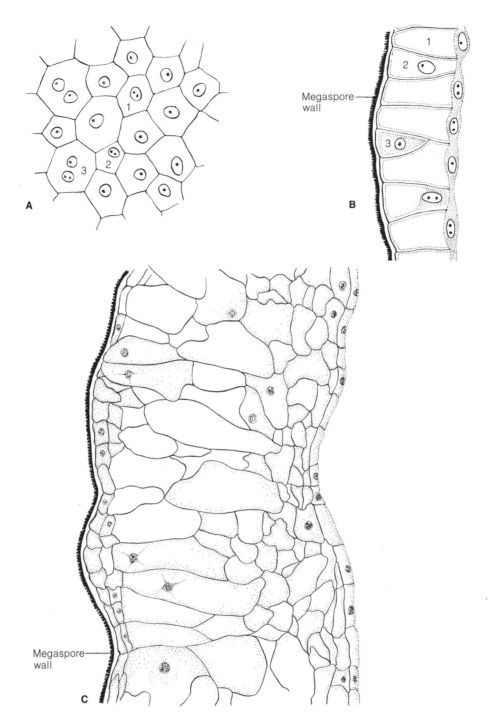

Figure 16-10
Cellular development of the female gametophyte in *Ginkgo*. **A, B,** young alveoli as seen in surface (**A**) and sectional (**B**) views. Various types of alveoli are represented by corresponding numbers as follows: 1, normal uninucleate alveoli; 2, alveoli preparatory to the initiation of pyramidal cells; 3, alveoli which have formed pyramidal cells. **C,** sectional view of cellular female gametophyte. See text for further explanation. [Redrawn from Favre-Duchartre, *Rev. Cytol. Biol. Veg.* 17:1, 1956.]

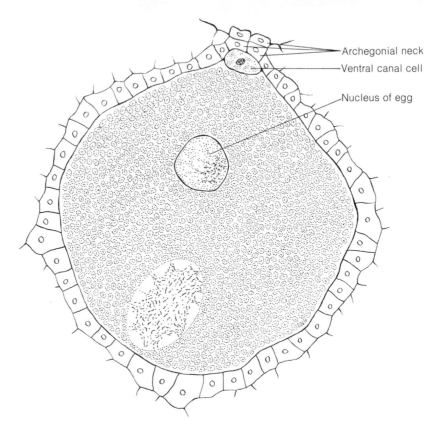

Archegonial neck
Ventral canal cell
Nucleus of egg

Figure 16-11
Sectional view (diagrammatic) of a fully developed archegonium of *Ginkgo* showing the
four-celled neck, the ventral canal cell, and the large egg. The numerous small bodies
within the egg cytoplasm are interpreted by Favre-Duchartre as "protein-lipid granules."
[Redrawn from Favre-Duchartre, *Rev. Cytol. Biol. Veg.* 17:1, 1956.]

which initiates the ventral canal cell and
the egg nucleus, each of the two neck cells
divides thus producing the four-celled neck
typical of the mature archegonium of *Ginkgo*
(Fig. 16-11).

After the archegonia have completed
their development, a peculiar column of
gametophytic tissue—known as the "tent
pole"—becomes elevated between them.
This column at first extends towards the
part of the nucellus lying below the pollen
chamber. In late August, the nucellar
tissue and the adjacent region of the megaspore membrane become destroyed. This
creates an *archegonial chamber* which has
the form of a circular crevice surrounding
the central tent pole (Fig. 16-9).

Fertilization

The final stages in development of the male
gametophyte, prior to fertilization, take
place within the ovule. The pollen grain,
lodged in the pollen chamber, germinates
and the haustorial end of the pollen tube
attaches itself to the wall of the pollen
chamber where it ramifies, sending numerous delicate rhizoid-like processes
between the cells of the nucellus (Favre-
Duchartre, 1956). The prothallial end of
the pollen tube elongates and is freely
suspended within the archegonial chamber.
Soon the generative cell divides *anticlinally*,
giving rise to the sterile cell and the spermatogenous cell (Fig. 16-12).

Fertilization in *Ginkgo* occurs during September (Fig. 16-4). The nucleus of the spermatogenous cell divides without forming an intervening wall, and two spermatozoids are formed which, aside from their smaller size, resemble the motile sperms of the cycads. A mature spermatozoid of *Ginkgo* is top-shaped and the numerous flagella are attached along the three turns of a spiral band located at one end of the cell (Lee, 1955). When the wall of the spermatogenous cell ruptures, the two spermatozoids are released into the pendulous end of the pollen tube. An opening then develops at the tip of the tube and the spermatozoids are set free into the liquid which is present at this time in the archegonial chamber. Flagellar movements propel the sperms for some time in this liquid medium before they enter an archegonium. During its passage between the swollen and reflexed cells of the archegonial neck, a spermatozoid becomes greatly stretched but the entire gamete, including the flagellated band, enters the cytoplasm of the upper part of the egg (Shimamura, 1937). According to Lee (1955), usually only one of the two spermatozoids produced by a pollen tube enters an archegonium; the other sperm, if it enters the same archegonium, soon degenerates and is eventually "absorbed." The nucleus of the functional spermatozoid separates from the flagellated sheath and moves downward until it contacts the egg nucleus, forming a diploid zygote, which gives rise to the future embryo.

Eames (1955) confirmed earlier reports that fertilization and embryogenesis in *Ginkgo* may occur either on the tree or after the ovule has fallen to the ground. The latter possibility is of phylogenetic interest with reference to some of the Paleozoic gymnosperms (e.g., the Cordaitales) in which the embryo, like that of *Ginkgo,* may have only begun to develop after the ovules had been shed from the parent tree.

Embryogeny

The early phase of embryogeny in *Ginkgo* is characterized by numerous free nuclear divisions, as in the cycads. After a series of about eight successive divisions (256 nuclei), centripetal wall formation begins and the young embryo becomes cellular throughout. In contrast with cycad embryogeny, no well-defined suspensor is formed. The lower end of the embryo, by means of active cell divisions, becomes a meristem from which the shoot apex and cotyledons are developed (Fig. 16-13); the cells immediately behind this portion ultimately differentiate into the primary root or radicle (Ball, 1956a, b). Usually there are two cotyledons, but occasionally three are developed. In addition to the cotyledons, the embryo of ripe seeds commonly contains the primordia of several additional foliar structures which, together with the shoot apex, constitute the first terminal bud of the plant.

The germination of the seed closely resembles that typical of cycads. The primary shoot and root emerge by the rupture of the micropylar end of the seed, but the tips of the cotyledons remain within the nutritive tissue of the female gametophyte. The original seed may still cling to the base of a seedling a year or more in age.

The apical meristem of both the long and spur shoots of *Ginkgo* has a characteristic zonal structure. As shown in Fig. 16-14, the subsurface zone of the shoot apex consists of a conspicuous group of enlarged, highly vacuolated, central mother cells from which the more actively dividing and smaller cells of the peripheral and rib-meristem zones take their origin. The type of zonation in the apical meristem of *Ginkgo* has been very helpful in the interpretation of the structure and growth of the shoot apex in the cycads and in certain genera of the coniferales. (See Foster, 1938; Johnson, 1951; Esau, 1965; Gifford and Corson, 1971.)

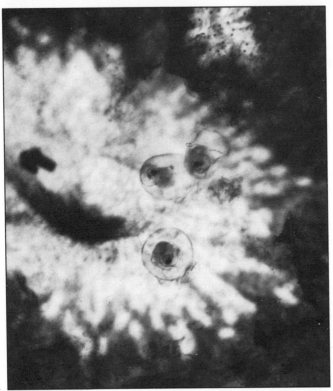

A

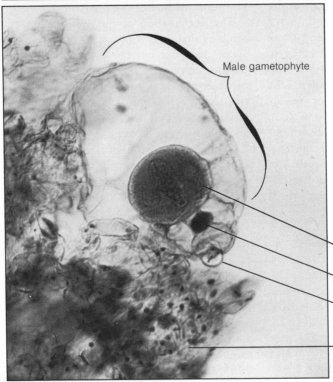

B

Male gametophyte

Spermatogenous cell

Nuclei of prothallial and sterile cell

Remains of pollen grain

Nucellus

Figure 16-12 (*facing page*)
Ginkgo biloba. **A,** whole mount of portion of nucellus showing
three attached male gametophytes as they might appear if
viewed by looking upwardly from the archegonial chamber;
B, male gametophyte (attached to surface of nucellus) just
prior to the division of the spermatogenous cell.

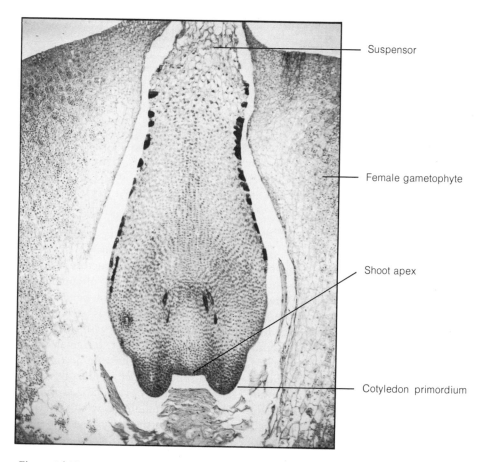

Suspensor

Female gametophyte

Shoot apex

Cotyledon primordium

Figure 16-13
Longitudinal section of an advanced stage in the embryogeny of *Ginkgo biloba.* The future
shoot apex region is flanked by a pair of developing cotyledon primordia. Note the poorly defined
suspensor of the embryo and the adjacent food-storing tissue of the female gametophyte.

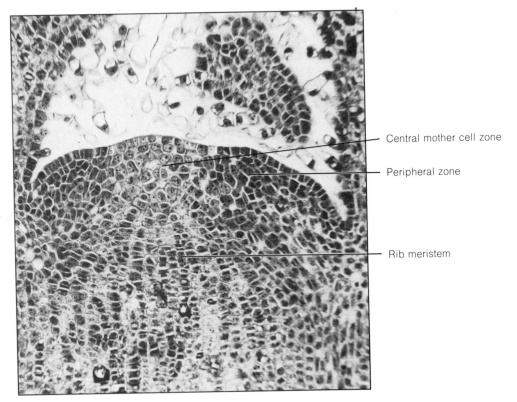

— Central mother cell zone

— Peripheral zone

— Rib meristem

Figure 16-14
Ginkgo biloba. Median longisection of shoot apex showing typical zonal structure.

Summary and Conclusions

Ginkgo biloba is a large profusely branched tree, with deciduous leaves and clearly defined *spur shoots* and *long shoots.* One of the most distinctive characters of the vegetative sporophyte is the petiolate foliage leaf, with its entire or variously lobed, dichotomously veined lamina. Vein unions (anastomoses), although few in number, are found in various parts of the lamina and closely resemble in type the anastomoses in the pinnae of certain cycads. As in typical coniferophytes, the pith and cortical regions of the young stems are relatively small. Cambial activity is vigorous and sustained and the secondary xylem is pycnoxylic.

Ginkgo is dioecious and the microsporangiate strobili and ovuliferous structures arise in the axils of the foliar organs of the spur shoots. The microsporangiate strobilus is a lax, catkin-like structure, the main axis of which bears a series of appendages, each terminating in a pair of pendant microsporangia. The ovuliferous structure consists of a stalk or peduncle with two (occasionally three or more) erect, terminal ovules. The integument of each ovule becomes differentiated into a thick, outer fleshy layer, a thinner, weakly vascularized inner fleshy layer, and a middle stony layer that becomes the "shell" of the mature seed.

With reference to sporogenesis, gametogenesis, and embryogeny, there are many points of close similarity between *Ginkgo* and the living cycads. Among the most important of these resemblances may be listed the following: (1) the development of haustorial pollen tubes and large multi-flagellated spermatozoids; (2) the occurrence of a coenocytic phase, followed by the formation of *alveoli* in the development of the female gametophyte; (3) the development of large archegonia, each with four neck cells; and (4) the extensive period of free nuclear divisions during early embryogeny. The ripe seed of *Ginkgo,* like that of the cycads, consists of an endoscopically oriented embryo, usually with two cotyledons, imbedded in the tissue of the female gametophyte and surrounded by a thick seed coat. When the seed germinates, the laminar portions of the cotyledons remain as haustorial structures within the gametophytic tissue and the seed may remain attached for some time to the base of the seedling.

What may be concluded regarding the evolutionary relationship between *Ginkgo* and the cycads? Both are veritable "living fossils" but the similarities between their gametophytes and seeds probably represent the retention of ancient patterns of reproduction that were shared by a number of Paleozoic and Mesozoic gymnosperms. If this view is correct, *Ginkgo* and the cycads should be regarded as extremely primitive, surviving members of two independent lines of gymnosperm evolution. The living cycads are the only extant representatives of the cycadophyte branch of evolution, whereas *Ginkgo,* on the basis of its general organography and wood anatomy, is a typical coniferophyte.

References

Arnold, C. A.
　1947. *An Introduction to Paleobotany.* McGraw-Hill, New York.
Arnott, H. J.
　1959. Anastomoses in the venation of *Ginkgo biloba. Amer. Jour. Bot.* 46:405–411.
Ball, E.
　1956a. Growth of the embryo of *Ginkgo biloba* under experimental conditions I. Origin of the first root of the seedling *in vitro. Amer. Jour. Bot.* 43:488–495.
　1956b. Growth of the embryo of *Ginkgo biloba* under experimental conditions. II. Effects of a longitudinal split in the tip of the hypocotyl. *Amer. Jour. Bot.* 43:802–810.
Chamberlain, C. J.
　1935. *Gymnosperms. Structure and Evolution.* University of Chicago Press, Chicago.

Critchfield, W. B.
　1970. Shoot growth and heterophylly in *Ginkgo biloba. Bot. Gaz.* 131:150–162.
De Sloover-Colinet, A.
　1963. Chambre pollinique et gamétophyte mâle chez *Ginkgo biloba. Cellule* 64:129–145.
Eames, A. J.
　1955. The seed and *Ginkgo. Jour. Arnold Arboretum* 36:165–170.
Engler, A.
　1954. *Syllabus der Pflanzenfamilien,* Ed. 12, Band 1. Gebrüder Bortraeger, Berlin.
Esau, K.
　1965. *Plant Anatomy,* Ed. 2. Wiley, New York.
Favre-Duchartre, M.
　1956. Contribution à l'étude de la reproduction chez le *Ginkgo biloba. Rev. Cytol. Biol. Veg.* 17:1–218.
　1958. *Ginkgo,* an oviparous plant. *Phytomorphology* 8:377–390.

Florin, R.

    1936. Die fossilen Ginkgophyten von Franz-Joseph-Land nebst Erörterungen über vermeintliche Cordaitales mesozoischen Alters. *Palaeontographica* 81(B):71–173.

Foster, A. S.

    1938. Structure and growth of the shoot apex in *Ginkgo biloba*. *Bull. Torrey Bot. Club* 65:531–556.

Fujii, K.

    1895. On the nature and origin of so-called "Chichi" (nipple) of *Ginkgo biloba* L. *Bot. Mag.* (Tokyo) 9:444–450.

Gifford, E. M., Jr., and G. E. Corson, Jr.

    1971. The shoot apex in seed plants. *Bot. Rev.* 37:143–229.

Gunkel, J.E., K.V. Thimann, and R.H. Wetmore

    1949. Studies of development in long shoots and short shoots in *Ginkgo biloba* L. IV. Growth habit, shoot expression, and the mechanism of its control. *Amer. Jour. Bot.* 36:309–316.

Hirase, S.

    1896. Spermatozoid of *Ginkgo biloba*. *Bot. Mag.* (Tokyo) 10:171 (in Japanese).

Johnson, M. A.

    1951. The shoot apex in gymnosperms. *Phytomorphology* 1:188–204.

Lee, C. L.

    1954. Sex chromosomes in *Ginkgo biloba*. *Amer. Jour. Bot.* 41:545–549.

    1955. Fertilization in *Ginkgo biloba*. *Bot. Gaz.* 117:79–100.

Li, Hui-Lin

    1956. A horticultural and botanical history of *Ginkgo*. *Bull. Morris Arboretum* 7:3–12.

Pankow, H., and E. Sothmann

    1967. Histogenetische Untersuchungen an den weiblichen Blüten von *Ginkgo biloba* L. *Ber. Deutsch. Bot. Ges.* 80:265–272.

Pollock, E. G.

    1957. The sex chromosomes of the Maidenhair tree. *Jour. Hered.* 48:290–294.

Seward, A. C.

    1938. The story of the Maidenhair tree. *Sci. Progr.* (England) 32:420–440.

Shimamura, T.

    1937. On the spermatozoid of *Ginkgo biloba*. *Fujii Jubilaei Volume* pp. 416–423.

Stewart, K. D., and E. M. Gifford, Jr.

    1967. Ultrastructure of the developing megaspore mother cell of *Ginkgo biloba*. *Amer. Jour. Bot.* 54:375–383.

17

The Coniferopsida: Coniferales

The most dominant and conspicuous gymnosperms of the modern world belong to the order Coniferales. Included in this order are such familiar and widely cultivated trees as pine, spruce, fir, cedar, yew, and redwood (Fig. 17-1). Many conifer genera are of great economic importance to man as sources of lumber, wood pulp for the manufacture of paper, and turpentine. The evolutionary history of the conifers is as long or probably longer than that of *Ginkgo* and the cycads, and extends from the Carboniferous and Permian Periods up to the present. According to Florin (1963), *recent* conifer families can be traced back to the Mesozoic Era; most genera appear to have had representatives in all periods from the Jurassic to the present and several genera may have existed as early as the Triassic. Students interested in the past migrations of the conifers and in a detailed analysis of their present geographical ranges should consult Florin's (1963) definitive monograph, which is at present the most scholarly and comprehensive treatment of these complex subjects.

Unlike the living cycads, which consist of only a few relict genera, modern conifers are represented by 51 genera and approximately 548 species, and are widespread in both the Northern and Southern Hemispheres. Western North America and eastern Asia are regions characterized by an exceptional diversity of conifers, with some of the genera forming extensive forests. Parts of Australia and New Zealand are

Figure 17-1
Cedrus deodara, growing on campus of University of California, Berkeley.
Note large erect seed cones on specimen at right.

A

likewise notable for the abundance and diversity of their conifers. The conifers are plants of the more temperate regions of the world; in contrast to the cycads, only a comparatively small number of conifers are found in tropical areas.

Systematics and Geographical Distribution

Many efforts have been made to define genera and to group these genera into tribes and families. Probably all of the proposed schemes of classification are artificial to some degree, because of the large amount of parallel evolution of the morphological structures that are employed as the basis for separating genera and families. As is true for other groups of vascular plants, an ideal classification of conifers should rest on the *totality* of morphological features, including the morphology and anatomy of the entire sporophyte as well as the gametophyte and embryo. This goal may ultimately

B

Figure 17-2
A, Branch of *Cedrus deodara* with two ripe pollen cones, each borne on a spur shoot;
B, two ovulate cones of *Pseudotsuga* (douglas fir) showing the prominent exserted tridentate bracts characteristic of this genus.

be attained but at present the systematic treatments emphasize characters of (1) the form, phyllotaxy, and anatomy of the leaves, (2) the organization of the strobili, especially the megasporangiate cones, and (3) to a lesser degree the mature pollen grains.

The various living genera of the Coniferales have been arranged by Pilger (1926) under seven families. This scheme of classification has found wide acceptance by many systematists and morphologists. Although the great majority of conifers produce their seeds in well-defined cones — as is illustrated by the familiar seed cones of pines, firs, and spruces — the family Taxaceae provides a notable exception to this generalization. In this family, each seed is terminal on a short lateral shoot and is partially (*Taxus*) or completely (*Torreya*) enclosed by an outer fleshy envelope known as the *aril* (Fig. 17-6). Because of the absence of a clearly defined seed cone, the family Taxaceae has been removed, by some botanists, from the Coniferales and placed in the separate order Taxales (Sporne, 1965). Florin (1951), the great student of the conifers, states that there is no evidence from the fossil record that the distinctive ovuliferous structure of the Taxaceae is the result of phylogenetic reduction. On the contrary, he favors the idea "that the taxads have always had isolated female flowers in the axils of foliage or scale leaves."

In the following conspectus we present some of the distinguishing characters of each of the seven families of living conifers. The total number of genera in each family is shown by the number in parenthesis, but for the sake of brevity, only a few of the representative genera of the larger families are listed.

PINACEAE (9): *Pinus, Pseudotsuga, Abies, Picea, Cedrus.* Monoecious trees with spirally arranged linear or needle-like leaves; each microsporophyll has 2 microsporangia; ovules borne in pairs on woody cone scales subtended by more-or-less free bracts; pollen grains usually have 2 wings. (Fig. 17-2)

TAXODIACEAE (10): *Taxodium, Metasequoia, Sequoia, Sequoiadendron, Cryptomeria.* Monoecious trees with spirally arranged or opposite (*Metasequoia*) needle-like or linear leaves; microsporophylls with 2–9 microsporangia; ovuliferous scale joined with its bract and bearing 2–9 ovules; pollen grains devoid of wings. (Fig. 17-3)

CUPRESSACEAE (18): *Cupressus, Juniperus, Thuja, Calocedrus, Libocedrus.* Monoecious or dioecious trees or shrubs; leaves opposite or whorled, usually scale-like in form; microsporophylls with 3–6 (or more) microsporangia; ovulate cones woody or somewhat fleshy at maturity; ovuliferous scale united with its bract and bearing 2 to many ovules; pollen grains devoid of wings. (Fig. 17-4)

ARAUCARIACEAE (2): *Araucaria, Agathis.* Dioecious or monoecious trees; leaves linear-broad, opposite or spirally arranged; microsporophylls with 5–20 pendant microsporangia; ovuliferous scale fused with its bract and bearing a single median ovule; pollen grains devoid of wings. (Fig. 17-19, I–K)

PODOCARPACEAE (7): *Podocarpus, Dacrydium, Phyllocladus.* Trees or shrubs, most of which are dioecious with scale-like, linear or broad spirally arranged leaves; microsporophylls with 2 microsporangia; megasporangiate strobili cone-like or greatly modified in structure (in which case they consist of a rudimentary axis, a reduced scale or scales, each with a single ovule, and a subtending bract); pollen grains usually with one to several wings. (Fig. 17-5)

CEPHALOTAXACEAE (1): *Cephalotaxus.* Trees or shrubs, most of which are dioecious;

leaves needle-like and opposite in arrangement; microsporophylls with 3–8 microsporangia; ovulate cones with decussately arranged bracts, each bract subtending 2 ovules; pollen grains devoid of wings.

TAXACEAE (4): *Taxus, Torreya.* Dioecious trees or shrubs; leaves linear or needle-like, spirally arranged; microsporophylls peltate with 2–8 microsporangia; ovuliferous branch with a single terminal ovule partly or entirely enclosed at maturity by a fleshy aril; pollen grains devoid of wings. (Fig. 17-6; Fig. 17-19, L)

Certain facts about the geographical distribution of the families and genera of conifers deserve brief mention. The majority of the members of the Pinaceae are found in the Northern Hemisphere. The largest genus is *Pinus,* which includes about 90 species. In contrast, the Araucariaceae and most of the Podocarpaceae are restricted to the Southern Hemisphere, the largest genus in these families being *Podocarpus,* comprising about 105 species. *Podocarpus* and *Araucaria* (e.g., the "monkey puzzle tree") are widely cultivated in the warmer parts of the world. The Cupressaceae and Taxodiaceae are large families that have representatives in both hemispheres. The family Taxodiaceae is of particular botanical and general interest for several reasons. First, it includes the coast redwood (*Sequoia sempervirens*), limited to a narrow coastal fog belt of northern California and southern Oregon, and the giant redwood, or "big tree" (*Sequoiadendron giganteum*), found only in a few groves on the western slopes of the Sierra Nevada in California. Both of the sequoias include individual trees—fortunately protected by the U.S. Park Service—which are among the largest trees in existence. The Taxodiaceae is also notable because this family now includes, as one of its most interesting species, the "dawn redwood" (*Metasequoia glyptostroboides*). Because *Metasequoia* had long been known only from its fossil record, uncovered in various parts of western North America and Asia, the discovery in 1944 by Chinese botanists of living specimens of the genus in Szechuan Province, China, was of exceptional scientific importance. The morphology and systematic affinities of the living species with other extant genera of the Taxodiaceae have been discussed in detail by Sterling (1949). According to Chaney (1950), who has investigated the paleobotanical aspects of fossil *Sequoia* and *Taxodium, Metasequoia* "was the most abundant and widely distributed genus of the Taxodiaceae in North America from Upper Cretaceous to Miocene time. There is no known record of its occurrence on this continent in rocks younger than Miocene; it survived into the Pliocene epoch in Japan, and a few hundred trees are still living in the remote interior of China." Thus, *Metasequoia,* somewhat like *Ginkgo biloba* and the living cycads, is truly a living fossil, and further studies of its structure and reproductive cycle should result in discoveries of considerable evolutionary interest.

Habit and Longevity

Many of the conifers are trees that may attain an enormous size—e.g., *Sequoia, Sequoiadendron, Pseudotsuga,* certain species of *Pinus, Picea,* and *Abies,* and in the Southern Hemisphere, the genus *Agathis.* Very commonly the habit is prominently excurrent with a persistent central trunk and a tiered or whorled arrangement of branches. Some species, however, may exhibit a more diffuse or deliquescent pattern of growth as is true of the "digger pine" (*Pinus sabiniana*) and the Monterey cypress (*Cupressus macrocarpa*). In New Caledonia, *Podocarpus ustus* has been reported to grow as a parasite on certain other conifers, a relationship which is apparently unique among all known gymnosperms (Laubenfels, 1959).

Figure 17-3
Seed cones of two genera in the Taxo-diaceae. **A,** branch of *Sequoia sempervirens* (California coast redwood) with five larger seed cones; **B,** branch of *Cryptomeria japonica* showing a group of seed cones; note that the ovuliferous scales terminate in four or five free tips, a distinctive character for this genus.

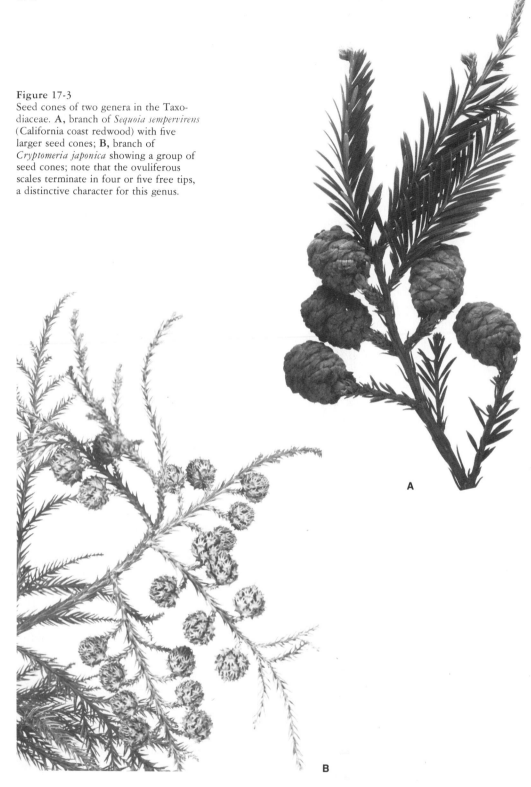

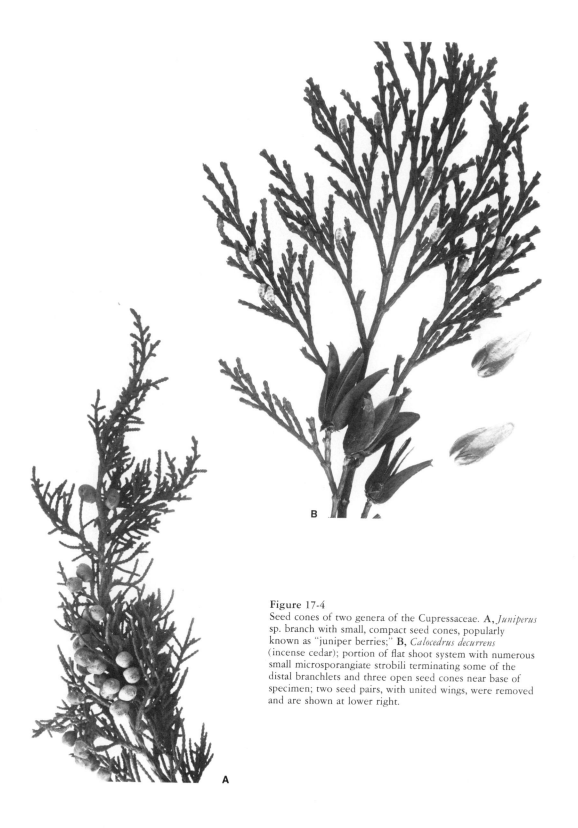

Figure 17-4
Seed cones of two genera of the Cupressaceae. **A,** *Juniperus*
sp. branch with small, compact seed cones, popularly
known as "juniper berries;" **B,** *Calocedrus decurrens*
(incense cedar); portion of flat shoot system with numerous
small microsporangiate strobili terminating some of the
distal branchlets and three open seed cones near base of
specimen; two seed pairs, with united wings, were removed
and are shown at lower right.

Figure 17-5
Podocarpus sp. Mature megasporangiate "cone," consisting of a single seed (s) borne at the apex of a fleshy receptacle (r). [Photo courtesy of Dr. T. E. Weier.]

Except for a very few deciduous genera (*Larix, Pseudolarix, Taxodium,* and *Metasequoia*), the Coniferales are "evergreen"; i.e., the foliage leaves function for more than one season. In *Araucaria* sp. the dead foliage leaves may remain attached for many years to the branches. Usually the leaves of conifers are individually shed, but in *Pinus, Sequoia sempervirens, Taxodium,* and *Metasequoia,* entire shoots are abscised from the older portions of the branch system.

It is of great interest to note the amazing longevity of certain genera in the Coniferales. Individual specimens of *Sequoiadendron giganteum* may reach an age of more than 3,000 years. Molisch (1938) states "that some trees may attain an age of about 4,000 years and are thus the longest-lived organisms on our planet today." During the past decade, much attention has been directed to the even greater longevity of the bristlecone pine (*Pinus aristata* Englm.). The pi-

Figure 17-6
Ovulate structures in two genera of the Taxaceae. **A,** ripe seeds of *Taxus* (yew), each borne on a short axillary branch and partially enclosed by a cup-shaped fleshy aril; **B,** *Torreya californica* (California nutmeg), branch with a cluster of ripe seeds. Each seed is completely covered by a fleshy aril.

oneering work on this species was done by Edmund Schulman, who concentrated most of his studies on the ancient trees growing —under the protection of the U.S. Forest Service—in the White Mountains of southeastern California. (For detailed information see Ferguson, 1968, and Lindsay, 1969.) By means of an increment borer, narrow cores of wood are removed from the trunks and careful determinations made of the number of growth layers or annual rings. Figure 17-7 shows the grotesque form of "Pine Alpha"—4,300 years old. Other trees in the same grove were determined to be older than 4,000 years. One of them— appropriately named "Methuselah" by Dr. Schulman—had reached an age of 4,600 years. How these trees have been able to survive through the millenia under the harsh conditions of their alpine environ-

ment is still a challenging problem in tree physiology. One possible factor responsible for their survival may be the retention of functional needles for 20–30 years. According to Ferguson (1968) "this insures a somewhat stable photosynthetic capacity that can carry a tree over several years of stress." As is shown in Fig. 17-7, "Pine Alpha" is largely a twisted, contorted "slab" of dead wood; only a narrow strip of living bark—and presumably some functional wood—connects its single leafy branch with the root system.

The oldest bristlecone pine thus far discovered grew in a grove of trees on Wheeler Peak in eastern Nevada (Currey, 1965). A horizontal section, made near the base of the trunk, revealed that this specimen was approximately 4,900 years old. The nearly 5,000-year life-span of such organisms as

the bristlecone pines can only be fully appreciated when viewed in comparison with events in man's own history. To quote from Lindsay's (1969) article, "Pine Alpha was growing when pyramids were being built in Egypt, and it was a very old tree when Christ was born."

Vegetative Organography and Anatomy

Foliage Leaves

MORPHOLOGY AND VENATION. Although the foliage leaves of all *living* conifers are simple, they vary considerably in both size and form. In many conifers, such as spruce (*Picea*), fir (*Abies*), and pine (*Pinus*), the leaves are commonly described as "needles" because of their narrow attenuated form. According to Chamberlain (1935), the slender needle-leaves of *Pinus palustris* may reach a length of 40 cm and are the longest leaves of any of the living coniferophytes. At the opposite extreme, in *Thuja* ("arbor vitae"), *Juniperus* (juniper), *Calocedrus* (incense cedar), and cypress (*Cupressus*), the leaves are very small and scale-like, superficially resembling the microphylls of the shoots of *Lycopodium* and *Selaginella* (Fig. 17-4). From an anatomical standpoint, however, scale-leaves in the conifers are not strictly comparable with microphylls; the vascular supply of microphylls is derived from a protostele, whereas the traces of coniferous scale leaves arise from the sympodial bundles of a eustele, as we have explained in Chapter 14, page 402. Lastly, it must be noted that certain species of *Agathis, Araucaria*, and *Podocarpus*, native to the Southern Hemisphere, are distinctive because of their broad leaves, which present a striking contrast to the usual type of foliage in the conifers. Chamberlain (1935) states that the leaves of *Podocarpus wallichianus* reach a length of 12.5 cm and a width of 3.5 cm.

In most conifers the leaves develop on long shoots, and the phyllotaxy is spiral and alternate. A conspicuous exception is the Cupressaceae, all members of which have decussate or whorled phyllotaxy. The pattern of branching with reference to phyllotaxy is complex and variable. In such genera as *Pinus, Picea, Abies*, and *Pseudotsuga* a large proportion of the needle-like leaves produced during a growing season are devoid of axillary buds; in these plants branching proceeds from a few axillary buds located in a pseudowhorl just below the terminal bud. The ramification of members of the Cupressaceae, by contrast, is often very profuse, and commonly results in the development of flattened spray-like branch systems (Fig. 17-4, B).

A few genera (*Larix, Cedrus, Pseudolarix*) produce their foliage leaves on spur as well as long shoots, thus recalling the marked dimorphism of shoots characteristic of *Ginkgo biloba* (see Chapter 16). With reference to spur shoots, the situation in *Pinus* is unique among all living conifers (Fig. 17-8). In this genus the photosynthetic needle leaves of well-developed sporophytes are restricted to lateral spur shoots which are ultimately shed as units from the tree; following the seedling stage, the foliar appendages produced on the primary axes are exclusively nonphotosynthetic scale leaves. The spur shoots are often referred to as "leaf fascicles," a misleading term which tends to obscure the real morphology of these structures. A spur shoot in pine arises from a bud developed in the axil of a scale-leaf produced on the main axis. The young spur consists of a very short stem, a series of

Figure 17-7 (*facing page*)
"Pine Alpha," a 4,300-year-old specimen of bristlecone pine (*Pinus aristata*) growing in Schulman Grove in the White Mountains of California. [Infrared photograph courtesy of Mr. Lloyd Ullberg, California Academy of Sciences.]

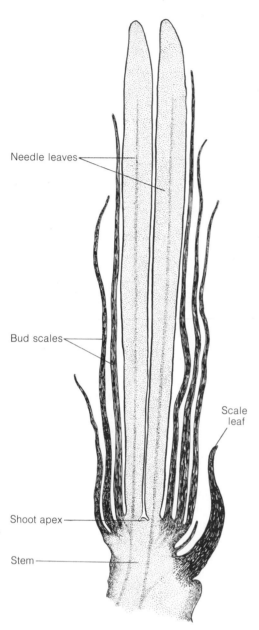

Needle leaves

Bud scales

Scale leaf

Shoot apex

Stem

Figure 17-8
Longisection of a spur shoot of *Pinus laricio* showing the diminutive shoot apex between the bases of the two needle leaves. See text for further explanation. [Redrawn from *Gymnosperms: Structure and Evolution* by C. J. Chamberlain. The University of Chicago Press, Chicago, 1935.]

imbricated membranous bud scales and, depending on the species, from one to five foliage leaves. (See Sacher, 1955a, for a detailed study of bud scale ontogeny in *Pinus*.) The diminutive shoot apex is usually recognizable in a longisection of a mature spur, but the cells composing it are vacuolate and inactive in appearance, and the surface and subsurface cells may ultimately die, become desiccated, and collapse (Sacher, 1955b). Under conditions of unusual stimulation, fully mature spur shoots may proliferate into long shoots through the reactivation of their shoot apices. The peculiar segregation of the foliage leaves to spur shoots can be followed in the progressive development of the pine seedling. When a pine seed germinates, the leaves that develop after the whorled cotyledons are green and needle-like, but are borne in a spiral series on the *main axis* of the seedling (Fig. 17-33). The first spurs arise as buds in the axils of certain of these juvenile leaves; all subsequently formed spurs, however, arise from buds subtended by nonphotosynthetic scale leaves. The deciduous spur shoots of *Pinus* not only are an interesting example of the vegetative specialization in the Coniferales but have figured prominently as an aid in the interpretation of the ovuliferous scales of the megasporangiate strobili in the conifers (Doak, 1935).

In a number of living conifers there is a marked difference in form between the seedling or juvenile foliage leaves and the leaves produced during the subsequent growth of the sporophyte. This heterophylly is particularly striking in certain genera of the Cupressaceae in which the juvenile leaves are needle-like and are ultimately followed by the appressed scale-like leaves characteristic of the adult plant. By means of cuttings it is possible to propagate the juvenile phase and to produce well-developed trees or shrubs bearing only (or largely) juvenile foliage. The various horticultural forms produced in this way are popular

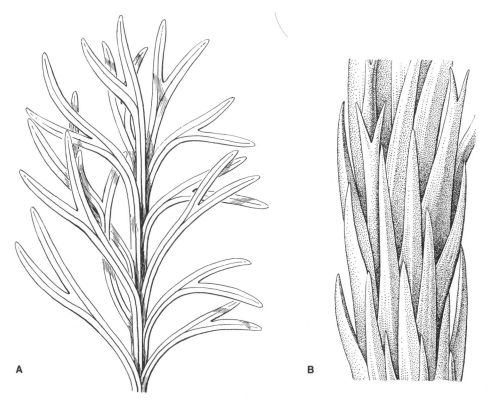

Figure 17-9
Foliage leaves of extinct conifers. **A,** dichotomously forked leaves of *Carpentieria frondosa;*
B, simple and forked leaves at base of an ovuliferous cone of *Lebachia piniformis.* [Redrawn
from Florin, *Palaeontographica* Abt. B, Bd. 85, 1944.]

ornamentals and are often collectively designated Retinospora.

It should be evident from the preceding paragraphs that the external form of the foliage leaf not only varies between genera or families but may even change during the ontogeny of the individual. For this reason it is difficult to trace the ultimate phylogenetic origin of the leaf of modern members of the Coniferales. Laubenfels (1953), on the basis of a comprehensive survey, has grouped the leaves of living conifers into the following major categories: type 1 includes needle-like leaves which are univeined and tetragonal in transection. This type is regarded by Laubenfels as the commonest form in fossil conifers, and it is widely distributed among the Pinaceae, Araucariaceae, Podocarpaceae, and Taxodiaceae; type 2 comprises leaves which also are univeined, but which are linear or lanceolate in contour and bifacially flat-

tened. This type is regarded as the most common among living conifers, and members of all families "have type II in some genus at some period in ontogeny"; type 3 comprises all scale-like forms of leaves, e.g., the adult foliage leaves of the Cupressaceae; type 4 includes the broad, multiveined leaves of *Agathis,* and of species of *Araucaria* and *Podocarpus.* From a phylogenetic standpoint it would be tempting to regard the leaves of type 4 as the primitive form from which the smaller and more simply veined types of leaves have been derived. However, the evidence from paleobotanical research does not support this conjecture (see Laubenfels' discussion).

Florin (1950, 1951) has shown that in Paleozoic conifers two distinct types of leaves (sometimes connected by intermediate forms on the same individual) were present (Fig. 17-9). In the probably more primitive of the two types, the leaf, although vascularized by a single trace, was dichotomously lobed or bifurcated, and was dichotomously veined. There were bifurcated leaves on the main axis and principal branches of *Lebachia* and *Ernestiodendron*, and they were followed in branches of the ultimate order by a second type which were needle-like or scale-like and very probably unilined. The occurrence of heterophylly in these very ancient conifers is of considerable interest, and led Florin to conclude that the simple leaf arose by reduction. In certain genera, such as *Carpentieria*, even the leaves of small lateral shoots were forked (Fig. 17-9, A). This condition is interpreted by Florin as a slightly modified persistence of the juvenile type of foliage. From this brief discussion it may be concluded that (1) the needle-like type of leaf probably arose early in the phylogeny of the Coniferales from a dichotomously branched appendage, and (2) this simple leaf type, modified to varying degrees, has persisted to the present time as the characteristic foliar appendage of most living conifers.

Napp-Zinn (1966) in his comprehensive monograph on gymnosperm leaf anatomy, states that the leaves of most conifer genera are vascularized by a single median bundle. Examples of this common type of venation are found in the leaves of *Abies, Picea, Pseudotsuga*, all members of the Cupressaceae, and certain species of *Pinus* (Fig. 17-12). As might be expected, the venation pattern developed in the broad leaves of *Araucaria, Agathis*, and *Podocarpus* is more complex. It consists of a longitudinal series of dichotomously branched veins, resembling in this respect the venation of the pinnae of the leaf in certain cycads (see Chapter 15, Fig. 15-15, A). An outstanding example of the systematic value of leaf venation is provided by the genus *Pinus*. As is well known (Shaw, 1914), in the "soft pines" (e.g., *Pinus strobus, Pinus monophylla*, and *Pinus lambertiana*), the vasculature of the mature needle consists of a single median vein; in the "hard pines" (e.g., *Pinus ponderosa* and *Pinus laricio*), the mature leaf has two more-or-less widely separated veins (Figs. 17-10; 17-12). Many years ago, Koehne (1893) utilized this difference in vein number as a basis for dividing the genus *Pinus* into two subgenera, namely, *Haploxylon* (species with single-veined leaves) and *Diploxylon* (species with two-veined leaves). Although some morphologists in the past have interpreted the two-veined form as a "reversion" to an ancient pattern of dichotomous venation, this view is not supported by histogenetic studies. In pine leaves that at maturity are two-veined, the young developing needle *at first* is traversed by a single median vein with protophloem and protoxylem. As differentiation continues, a median ray-like wedge of parenchyma is formed that destroys the median portions of the protophloem and protoxylem and finally causes a division of the original bundle into two strands. In some instances, one or two additional rays are formed, dividing the original vein into three or four separated bundles. Thus, from an ontogenetic standpoint, the mature two-veined form seems to be a specialization of the basic univeined form that is so widespread in the leaves of conifers.

HISTOLOGY. Numerous comparative studies have shown that the histology of relatively small simple leaves of conifers is extremely complex. The epidermal, fundamental, and fascicular tissue systems (see Chapter 3) are characteristically well-defined, but the cellular structure of these systems, particularly the last two, varies widely even among genera or species. Although our present knowledge of the comparative leaf anatomy of the conifers as a whole is incomplete, two parts of leaf histology — the stomata and the transfusion tissue — have been widely studied and discussed in the literature, and these deserve brief consideration at this point.

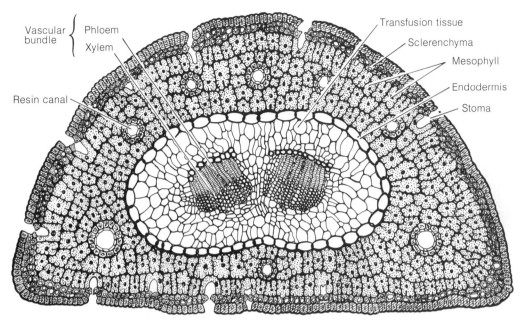

Vascular bundle { Phloem / Xylem

Resin canal

Transfusion tissue

Sclerenchyma

Mesophyll

Endodermis

Stoma

Figure 17-10
Transection of needle leaf of *Pinus laricio,* showing the two vascular bundles characteristic of the leaves of the "hard pines." [Redrawn from *Gymnosperms. Structure and Evolution* by C. J. Chamberlain. The University of Chicago Press, Chicago, 1935.]

STOMATA. In many of the living conifers, the stomata are on both surfaces of the leaf, but in certain taxa, stomata are restricted to either the abaxial or adaxial epidermis. The leaf of *Pinus,* for example, is termed *amphistomatic* because there are stomata on all epidermal surfaces of the needle (Fig. 17-12, A). A typical example of a *hypostomatic* leaf (having the stomata confined to the lower or abaxial epidermis) is provided by *Pseudotsuga.* Regardless of their topography, the stomata in the leaves of most genera are arranged in rather well-defined longitudinal rows.

Without question, the most comprehensive survey of the structure, ontogeny, and systematic value of stomata in gymnosperms was made by Florin (1931, 1951). The impetus for Florin's very detailed investigations was the extraordinarily well-preserved cuticular pattern of the epidermis of fossil gymnospermous leaves, which made it possible to compare the stomatal morphology of extinct and living taxa. In addition, Florin studied the ontogeny of the stomata in all of the major groups of living gymnosperms (i.e., cycads, *Ginkgo biloba,* Coniferales, and "Gnetales"). On the basis of his detailed investigations, he proposed a classification of gymnospermous stomata into two major types. In the *haplocheilic,* or "simple-lipped," type, the two guard cells of the stoma arise *directly* by the division of a mother cell (Fig. 17-11, A). In this type, the epidermal cells surrounding the stoma may function as subsidiary cells or each of them may divide into an inner subsidiary cell and one or more radially arranged "encircling cells"; in the second, or *"syndetocheilic,"* type (compound lipped), the mother cell of the stoma divides into three cells, of which the median cell functions as the mother cell of the two guard cells while the two lateral cells become differentiated as subsidiary cells. The distinctive feature of this type of stomatal development is the origin of the guard-cell mother cell and the

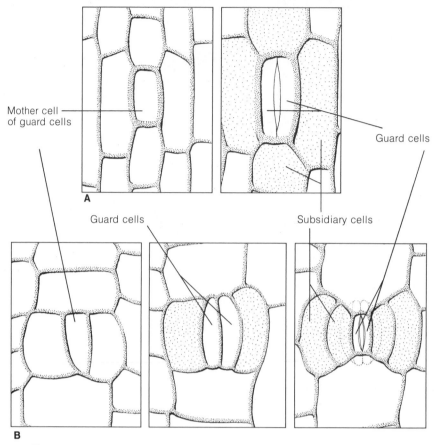

Mother cell
of guard cells

Guard cells

A

Guard cells

Subsidiary cells

B

Figure 17-11
The two main types of stomata in gymnosperms. **A**, haplocheilic; **B**, syndetocheilic.
See text for discussion. [Redrawn from Florin, *Act Horti Bergiani* 15:285, 1951.]

two lateral subsidiary cells from the same "parent" cell (Fig. 17-11, B). One or both of the subsidiary cells may subsequently divide and produce an outer pair of "encircling cells" as is shown in Fig. 17-11, B, right).

It is particularly significant, from an evolutionary viewpoint, that both of Florin's types of stomatal apparatus apparently can be traced through the fossil record and that they serve so consistently to define major gymnosperm taxa. The haplocheilic type is characteristic of (1) living and extinct conifers (e.g., *Lebachia, Ernestiodendron*), (2) the Pteridospermales, (3) Cordaitales, (4) Cycadales, (5) Ginkgoales, and (6) the genus *Ephedra* in the Ephedrales. The apparently advanced, syndetocheilic type is more limited in occurrence and is found in the extinct Cycadeoi-

dales and in the living genera, *Welwitschia* and *Gnetum*. For more extensive descriptions of the ontogeny of the stomata in gymnosperms, the student is referred to the monographs of Florin (1931, 1951) and the more recent accounts by Arnold (1953), Maheshwari and Vasil (1961), Pant and Mehra (1964), and Martens (1971).

TRANSFUSION TISSUE. The term transfusion tissue is applied to the tracheids and associated parenchyma cells, which are at the periphery of the leaf veins in all living representatives of the gymnosperms (Worsdell, 1897; Abbema, 1934; Griffith, 1957, 1971). We will use the leaf of *Pinus* as a basis for a preliminary discussion of transfusion tissue in the leaves of conifers.

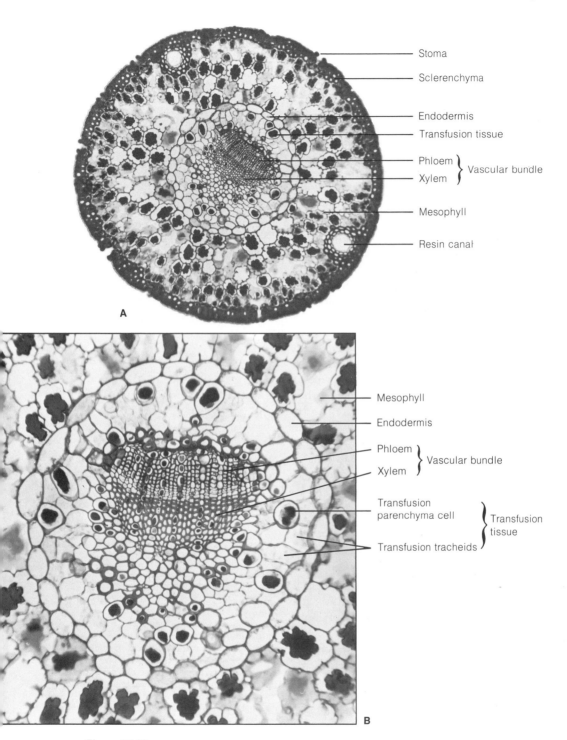

Figure 17-12
Histology of the leaf of *Pinus monophylla*. **A**, transection of leaf; **B**, details of endodermis, transfusion tissue, and vascular bundle.

As seen in transectional view, the transfusion tissue in a pine needle consists of several layers of compactly arranged cells which surround the vascular bundle or bundles and which are separated from the inner region of the mesophyll by the endodermis (Fig. 17-12). Sacher (1953), who has made a detailed ontogenetic study of the leaf of *Pinus lambertiana*, found that the transfusion tissue arises from the outer region of the procambium in the young needle, and that its maturation proceeds basipetally and centripetally. The transfusion tissue in *Pinus* consists of two principal cell types: *transfusion tracheids*, which have thick, lignified secondary walls with conspicuous circular bordered pits; and *transfusion parenchyma* cells, which retain their protoplasts and are further characterized by the development of tannin-like substances (Fig. 17-12, B). These two cell types form a complex interconnected system; both types occur in direct contact with the cells of the endodermis. The physiological role of the transfusion tissue in *Pinus* and other gymnosperms is generally considered to be conduction of materials between the vascular bundle and the mesophyll.

The spatial arrangement of foliar transfusion tissue in certain other conifer genera is quite different from that of *Pinus*. In *Podocarpus macrophyllus*, for example, two wing-like extensions of transfusion tissue occur at either side of the single vein (Fig. 17-13, A). Immediately adjacent to the xylem, the transfusion tissue consists only of short tracheids, whereas the outer parts of each wing—termed the accessory transfusion tissue by Griffith (1957)—consist of both parenchyma cells and tracheids, and extend between the palisade and spongy mesophyll nearly to the leaf margin. In *Cephalotaxus*, according to Griffith (1971), the spatial relation of the transfusion tissue varies at different levels in the same leaf. Near the leaf apex, transfusion tissue forms an arc adjacent to the xylem (Fig. 17-13, B); in the

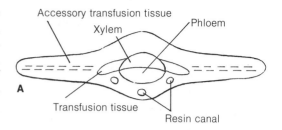

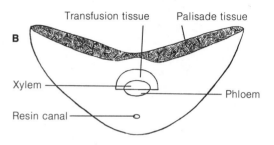

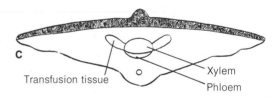

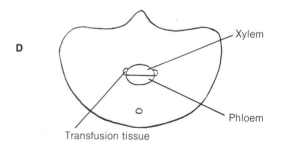

Figure 17-13
Diagrams showing arrangement of transfusion tissue in the leaves of conifers. **A,** *Podocarpus macrophyllus,* transection of leaf showing areas of transfusion tissue at each side of xylem of midvein and wings of accessory transfusion tissue extending to each margin of lamina; **B–D,** *Cephalotaxus harringtonia,* transections through terminal (**B**), middle (**C**), and basal (**D**) portions of same leaf. See text for further explanation. [**A** from Griffith, *Amer. Jour. Bot.* 44:705, 1957; **B–D** from Griffith, *Phytomorphology* 21:86, 1971.]

middle region of the lamina, the xylem of the vein is bordered by two short wings of transfusion tissue (Fig. 17-13, C); and at the base of the leaf, only a few transfusion tracheids occur next to the xylem (Fig. 17-13, D). In both *Podocarpus* and *Cephalotaxus*, the transfusion tissue first appears in the apical region of the young leaf primordium and subsequently develops in a basipetal direction.

The association of transfusion tissue with the veins of the leaves in the Coniferales and other gymnosperms poses interesting but difficult morphological problems. According to one widely expressed view, transfusion tissue phylogenetically originated from the primitive, centripetal, primary xylem that is characteristic of the leaves of certain Paleozoic gymnosperms (Worsdell, 1897; Jeffrey, 1917). A contrary view holds that transfusion tissue represents the modified development of the parenchyma surrounding or flanking the vascular bundle. Both of these interpretations appear somewhat conjectural, and a better understanding of the morphology and history of foliar transfusion tissue will depend on further studies of its occurrence in gymnosperm leaves as well as an intensive ontogenetic investigation of this tissue in a wide range of living gymnosperms. In this connection, it is interesting that in the leaf of such a "living fossil" as *Metasequoia* (dawn redwood), the vein is flanked on both sides by well-defined transfusion tracheids (Shobe and Lersten, 1967).

Stem Anatomy

It will be possible only to review here certain of the main anatomical features of the stem in the Coniferales. The transection of the young stem of *Pinus* will be used to illustrate the general topography and cellular structure of the various tissues (Fig. 17-14, A). Unlike stems in the cycads, this stem is characterized by a prominent secondary vascular system. At the developmental stage illustrated in Fig. 17-14, A, this system is composed of secondary phloem and secondary xylem, both developed from the persistent type of cambial activity characteristic of all conifers. Aside from the resin ducts, the secondary xylem consists of tracheids, which are elongate cells arranged in regular radial rows, and xylem rays, which consist of living parenchyma cells arranged in radially directed sheets. Figure 17-14, B, shows the form and relationship of the tracheids and rays in longitudinal section and the conspicuous circular bordered pits and pit-pairs of the tracheids. The wood of *Pinus*, like that of all gymnosperms except members of the Gnetopsida, is characterized by the absence of vessels, and in this respect is more homogeneous and primitive than the xylem of the majority of angiosperms. As shown in Fig. 17-14, the secondary phloem resembles the xylem in its relatively simple structure, and consists of elongated sieve cells, phloem parenchyma, and rays; in many conifers, fibers and sclereids may also develop in the secondary phloem. The pith and cortex of the pine stem are composed largely of parenchyma tissue, but the cortex is particularly distinguished by the presence of numerous, large resin canals. The epidermis has thickened outer walls overlaid by a massive cuticle, and during later stages in the increase in circumference of the stem it is sloughed away by the development of a cylinder of periderm.

Because of the continued activity of the vascular cambium for hundreds or even thousands of years (e.g., *Sequoiadendron*), the stems of many conifers attain enormous diameters. Usually, as shown in Fig. 17-14, A, the so-called annual rings are clearly defined, each ring representing the increment of secondary xylem produced annually by the cambium. Not infrequently false annual rings are produced as the result of abnormal cambial activity; they, of course, must be

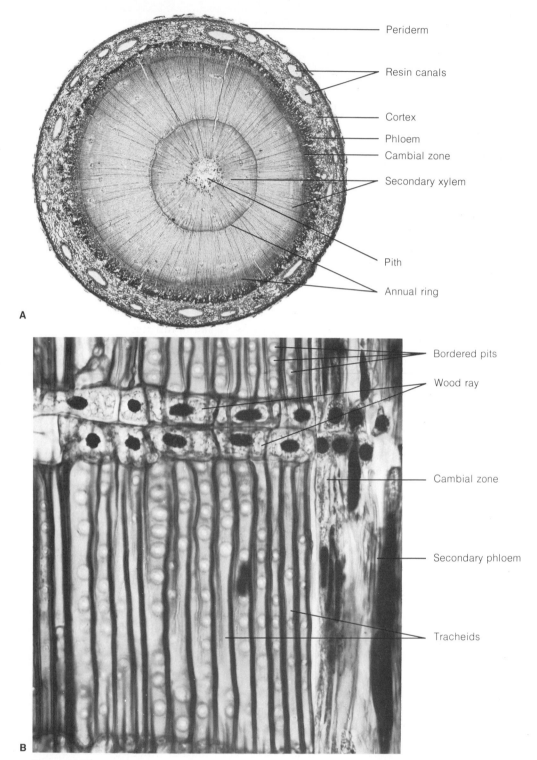

Periderm

Resin canals

Cortex

Phloem

Cambial zone

Secondary xylem

Pith

Annual ring

A

Bordered pits

Wood ray

Cambial zone

Secondary phloem

Tracheids

B

Figure 17-14
Stem anatomy in *Pinus*. **A,** transection of two-year-old stem; **B,** radial longisection of secondary
vascular tissues and cambial zone. Note especially the conspicuous circular bordered pits of the
tracheids and the living cells of the wood ray.

A

B

Figure 17-15
Pinus. A, terminal portion of branch showing clusters of microsporangiate strobili below the expanding spur shoots; B, tip of branch showing large terminal bud, and below it, three young megasporangiate strobili. [Courtesy of Dr. T. E. Weier.]

considered when attempting to estimate the age of a given stem.

The scientific interest in conifers and their importance as sources of lumber have been responsible for the numerous studies on cambial activity and wood histology in these plants. For more detailed treatments of these topics and other aspects of stem anatomy, the student should consult Esau (1965), and Brown, Panshin, and Forsaith (1949).

Strobili and Sporangia

One of the most uniform characteristics of the Coniferales is the monosporangiate nature of their cones or strobili. This means that in all normal instances two distinct types of strobili are formed: the microsporangiate, or pollen-bearing, cone, and the megaspo-

rangiate, or seed, cone (Fig. 17-15). The latter is the larger, and is exemplified by the familiar cones of pines, firs, and spruces (Figs. 17-1, 17-2, B; 17-3; 17-4). Bisporangiate cones (structures that develop both microsporangia and ovules) have been observed in many genera of living conifers, but they clearly represent abnormalities and do not provide evidence about the phylogenetic history of the two cone types in living forms. According to Florin, no bisporangiate cones have been encountered in the fossil remains of either Paleozoic or Mesozoic conifers—he concludes that monosporangiate cones represent a primary and fundamental condition in the evolutionary history of conifers.

As stated earlier in the chapter, the living conifers are predominantly monoecious, both kinds of cones occurring on the same individual. This was very probably the pre-vailing situation in the extinct conifers also. The dioecious condition, in which the two kinds of cones are produced by separate individuals of a species, is found in the Taxaceae, in a majority of the Araucariaceae, in *Podocarpus*, and in a number of genera in the Cupressaceae.

The Microsporangiate Strobilus (Pollen Cone)

Compared with those of the cycads, the microsporangiate strobili of most conifers are relatively small, commonly measuring only a few centimeters or less in length. The longest that have been recorded are those of *Araucaria bidwilli*, which may reach a length of 10–12 cm. The position of the microsporangiate strobili varies considerably within the Coniferales. In *Pinus* these strobili arise in the axils of scale leaves and are produced in subterminal clusters (Fig. 17-15, A), whereas in *Cedrus* solitary microsporangiate cones develop at the tips of certain of the spur shoots (Fig. 17-2, A). In other families, e.g., the Cupressaceae, the strobili are developed terminally on certain specialized lateral shoots.

The microsporophylls vary in form from flattened leaf-like appendages with expanded tips to peltate organs. In all species, the microsporangia develop on the lower surface of the sporophylls, but the number of sporangia is not constant throughout the order (Fig. 17-16). In the Pinaceae the number is consistently two, whereas in the other families there may be from two to seven sporangia on each sporophyll. Certain species of *Araucaria* and *Agathis* are notable because thirteen to fifteen sporangia may be produced by a single sporophyll.

Throughout much of the literature it is stated that the microsporangium of conifers, although eusporangiate in its general pattern of development, *originates* from the peri-clinal division of a series of hypodermal cells (Chamberlain, 1935; Campbell, 1940; Haupt, 1953). In other words, in marked contrast with the *superficial* position of sporangial initials in the lower vascular plants, the parent cells of the microsporangia of conifers are asserted to lie *below* the surface or epidermal layer of the microsporophyll.

Several investigations reveal, on the contrary, that the method of origin of the microsporangium is not standardized but that there are two different patterns of initiation. In two members of the family Pinaceae— *Pseudotsuga menziesii* (Allen, 1946b; Allen and Owens, 1972) and *Cedrus deodara* (Erspamer, 1952)—the sporangial initials consist of a group of *superficial cells*. As in the lower vascular plants (e.g., *Lycopodium*), these initials divide periclinally, forming an outer series of cell layers (1–4) that comprise

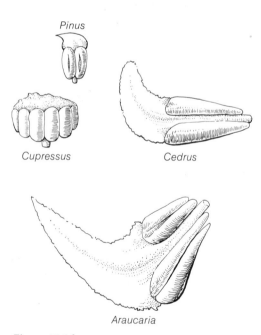

Figure 17-16
Microsporophylls in the conifers showing variations in their form and in the number of microsporangia. [Redrawn from *Gymnosperms. Structure and Evolution* by C. J. Chamberlain. The University of Chicago Press, Chicago, 1935.]

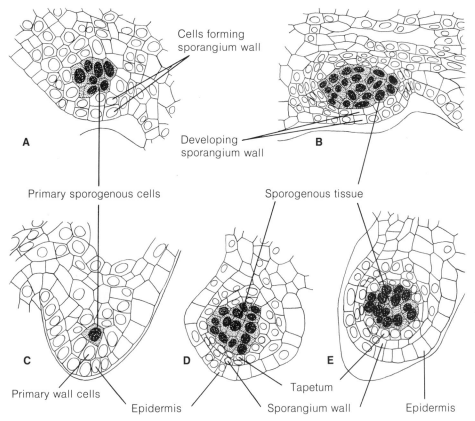

Cells forming sporangium wall

A

Developing sporangium wall

B

Primary sporogenous cells

Sporogenous tissue

C

D

E

Primary wall cells

Tapetum

Epidermis

Sporangium wall

Epidermis

Figure 17-17
Initiation and early development of the microsporangium in conifers. **A, B,** *Cedrus deodara,* showing origin of microsporangium from superficial initials; **C–E,** *Chamaecyparis lawsoniana,* showing origin of microsporangium from hypodermal initials (note that epidermis is a distinct cell layer throughout development). [Redrawn from J. L. Erspamer, Ph.D. dissertation, University of California, Berkeley, 1952.]

the *sporangium wall* and, in most species, the tapetum and an inner group of *primary sporogenous cells* (Fig. 17-17, A, B). According to Fagerlind (1961), the microsporangia of *Pinus, Picea,* and *Larix* likewise originate from superficial initials. In all these plants, therefore, the superficial cells of the mature microsporangium are ontogenetically a part of the sporangium wall and collectively represent the sporangial epidermis.

The mode of microsporangial initiation is different in *Cryptomeria japonica* (Erspamer, 1952; Singh and Chatterjee, 1963), *Chamaecyparis lawsoniana* (Erspamer, 1952), *Cupressus arizonica* (Owens and Pharis, 1967), *Taxus baccata* (Erspamer, 1952), and *Cepha-*

lotaxus drupacea (Singh, 1961). In all these conifers, the sporangial initials are strictly *hypodermal* in position and after the usual eusporangiate development has been completed, the mature microsporangium is externally bounded by an epidermal layer that has had no role in the formation of the embedded sporangium (Fig. 17-17, C–E).

The studies that we have just reviewed emphasize the great need for a more comprehensive survey of the method of initiation and early ontogeny of the microsporangium throughout the Coniferales as well as in other groups of the gymnosperms. Furthermore, as was pointed out in Chapter 4, all studies on sporangial development should

488

be made in the light of detailed information on (1) the structure and growth of the vegetative and reproductive apices and (2) the method of initiation of foliage leaves and microsporophylls. In members of the Pinaceae, the superficial position of the microsporangial initials is matched by the active contribution of superficial cells of the apical meristem to the initiation of leaf and sporophyll primordia. In contrast, the hypodermal position of microsporangial initials in *Cryptomeria, Cupressus arizonica*, and *Taxus* is closely correlated with the presence in the shoot apex of a well-defined surface layer, which does not contribute to the *inner tissue* of either the foliage leaf or the microsporophyll. Despite these apparent correlations, Fagerlind (1961) believes that the two patterns of microsporangial initiation may eventually prove to be connected by "transitions." He cites, in support of his opinion, the occasional periclinal division of epidermal cells in *Taxus*, by means of which cells may be contributed to the development of the underlying sporangium. In his view, future studies may show other instances of this sort and thus reveal a rather continuous "typological series."

At maturity the wall of the microsporangium consists of one layer or, as in *Cedrus*, several layers of cells. Commonly many or all of the internal layers of the wall become

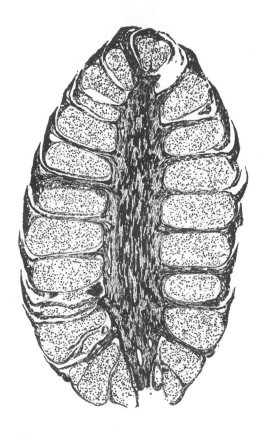

Figure 17-18
Longitudinal section of mature microsporangiate strobilus in *Pinus*.

Figure 17-19 (*facing page*)
Morphology of ovuliferous structures in various conifers. **A**, *Pinus banksiana*, longisection of young seed cone showing relation of developing ovuliferous scales and their associated bracts; **B**, *Pinus banksiana*, longisection of vegetative bud showing the general structure and axillary position of spurshoot primordia; **C**, *Pinus maritima*, longisection of immature seed cone showing several examples of the orientation of phloem (white) and xylem (black) in the vascular traces of the ovuliferous scale and its associated bract; **D**, *Pinus maritima*, longisection showing separate origin of bract trace and ovuliferous scale traces; **E**, *Pinus maritima*, transection showing vascular system of bract and ovuliferous scale (note inverted orientation of scale bundles); **F, G**, *Abies balsamea*, abaxial and adaxial views, respectively, of ovuliferous scale; **H**, *Chamaecyparis lawsoniana*, longitudinal section of seed cone showing union between ovuliferous scale and bract; **I, J**, *Araucaria bidwillii*. Adaxial and end views, respectively, of ovuliferous scale joined with its bract; **K**, *Araucaria balansi*, longisection showing vascular supply of bract and scale (note single basal ovule); **L**, *Taxus canadensis*, median longitudinal section of vegetative shoot and of secondary lateral fertile shoot which terminates in a single ovule. [Redrawn from following sources: **A, B** from *Gymnosperms. Structure and Evolution* by C. J. Chamberlain. University of Chicago Press, Chicago, 1935; **C, D, E, H, K** from Aase, *Bot. Gaz.* 60:277, 1915; **F, G, I, J, L** from Florin, *Acta Horti Bergiani* 15:285, 1951.]

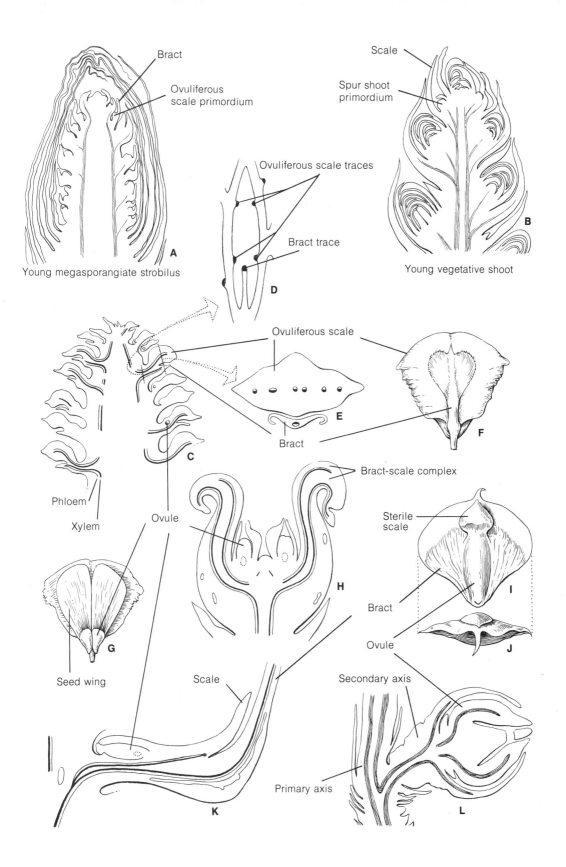

Bract

Ovuliferous scale primordium

Young megasporangiate strobilus

A

Scale

Spur shoot primordium

Young vegetative shoot

B

Ovuliferous scale traces

Bract trace

D

Ovuliferous scale

Bract

E

F

C

Phloem

Xylem

Ovule

Bract-scale complex

Sterile scale

Bract

Ovule

I

J

Seed wing

G

H

Scale

Secondary axis

Primary axis

Ovule

K

L

crushed or obliterated. The outermost layer consists very often of cells with reticulate, helical, or annular thickenings, which closely resemble the patterns of the secondary wall of tracheary elements. This surface layer is concerned with the mechanical rupturing of the sporangium at the period of release of the pollen grains. Although a tapetum is characteristically developed, its origin is quite variable (Fig. 17-17, D, E), since it may be the innermost layer of the sporangial wall (*Chamaecyparis*), or else be a derivative of the sporogenous tissue (*Taxus*). In most genera, particularly in the Pinaceae and Araucariaceae, a large number of microspores are produced in each sporangium (Fig. 17-18). A very characteristic feature of the microspores of most genera in the Pinaceae is the development of two air-filled lateral bladders or wings (Fig. 17-24). Winged microspores are present also in members of the Podocarpaceae.

The Megasporangiate Strobilus (Seed Cone)

In this section we are concerned with the morphology of the megasporangiate strobilus, or seed cone, which is characteristically developed in the majority of the families of living conifers. It would not be an exaggeration to state that the interpretation of the structure and evolution of the seed cone represents one of the most difficult and controversial problems in plant morphology. Rudolf Florin (e.g., 1944, 1950, 1951) has made highly significant contributions to a better understanding of the structure and evolutionary history of the megasporangiate cone in the Coniferales. The necessarily brief discussion given here is based largely on his discoveries and conclusions, and we hope that from it the student will gain a fairly clear understanding of the problems and the kinds of evidence that have been used to solve them.

The essence of the morphological problem posed by the seed cone of conifers is clearly illustrated in the family Pinaceae. If a longisection of a young seed cone of *Pinus* is examined, it becomes evident that each of the scales is associated with a small bract which appears adnate to the abaxial basal region of the scale (Fig. 17-19, C). Dissection of a young cone further reveals that each scale bears a pair of ovules which are attached basally to its adaxial surface (Fig. 17-20, B). The ovules are *inverted* in orientation; i.e., the micropyles are directed towards the base of the scale. As each ovuliferous scale is subtended by a bract, the seed cone is thus a *compound strobilus* and has often been likened to an inflorescence. In contrast, the pollen cone represents a *simple strobilus* because the microsporangia are directly attached to the primary appendages or microsporophylls (compare Figs. 17-18 and 17-19, C).

As shown in Fig. 17-19, C, the bracts of the seed cone of *Pinus* are much shorter than their associated ovuliferous scales, whereas in *Pseudotsuga,* the bracts are tridentate structures, that are very conspicuous in the mature seed cone (Fig. 17-2, B). Prominent exerted bracts are also characteristic of the seed cones of certain species of *Larix* and *Abies*.

Additional evidence that the seed cone in conifers is morphologically a compound strobilus is provided by its ontogeny; the relatively few studies using this approach to the problem have been reviewed in detail by Gifford and Corson (1971). In contrast to the voluminous literature examining the early differentiation of inflorescences and flowers in angiosperms, the "transition" from the type of structure and growth characteristic of a vegetative shoot apex to the pattern of growth typical of either the microsporangiate or megasporangiate strobili of conifers has received very little attention. As an example of the salient features of early development of the ovuliferous cone we have selected *Pinus ponderosa* (Fig. 17-21).

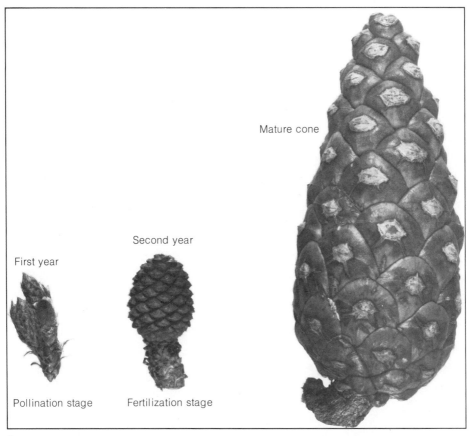

First year

Pollination stage

Second year

Fertilization stage

Mature cone

A

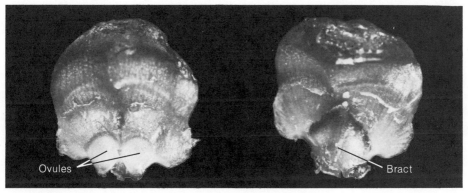

Ovules

Bract

B

Figure 17-20
Pinus halapensis. **A,** developing seed cones showing numerous overlapping ovuliferous scales
(cone scales); **B,** ovuliferous scale dissected from cone; adaxial view at left, showing the
attachment of the two ovules; abaxial view at right, showing the short basally adnate bract.

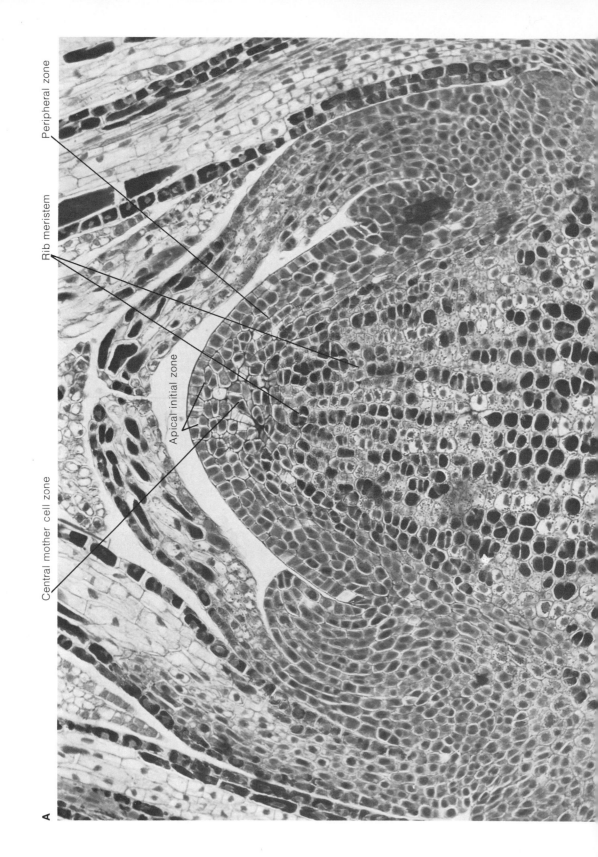

Peripheral zone

Rib meristem

Apical initial zone

Central mother cell zone

A

493

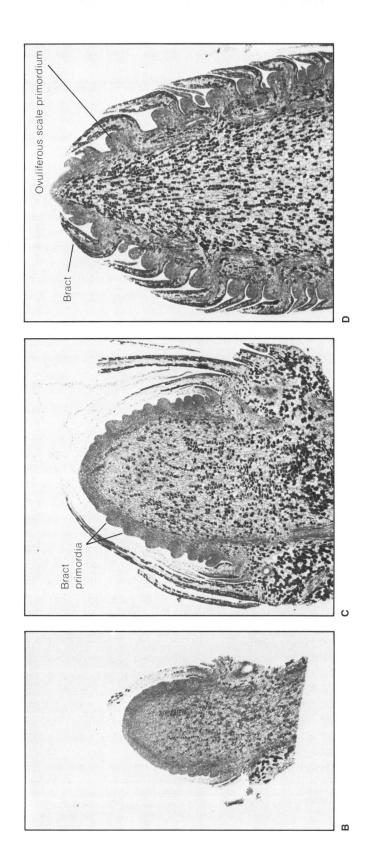

Figure 17-21

Pinus ponderosa. A, median longisection showing zonal structure of the vegetative shoot apex; B–D, successively older stages in the differentiation of the ovulate cone. B, young cone dissected from terminal bud November 29, 1957, showing the initiation of the first bract primordia; C, later stage (March 31, 1958), showing acropetal development of additional bract primordia; D, structure of cone on May 1, 1958, showing well-defined, ovuliferous, scale primordia developing in the axils of the bracts. (Illustrations not at same magnification.) [A from Sacher, *Amer. Jour. Bot.* 41:749, 1954; B–D from Gifford and Mirov, *Forest Sci.* 6:19, 1960.]

According to Sacher (1954), the vegetative shoot apex of this species is histologically demarcated into four tissue zones: an *apical initial zone*, a *peripheral zone* from which foliar appendages originate, a *central-mother-cell zone,* and a *zone of rib meristem,* which forms the pith of the shoot axis. Gifford and Corson (1971) recommend the use of the term apical zone to include those cells at the summit of the apex and their immediate basal derivatives, i.e., the cells of the central-mother-cell zone (Fig. 17-21, A). Gifford and Mirov (1960) found that young undifferentiated megasporangiate cone primordia, each subtended by a scale, are present by mid-September in the terminal buds of *Pinus ponderosa* trees growing at Placerville, California. The zonal structure of the apex of the cone primordium at this period is essentially similar to that of the vegetative shoot apex. In the latter part of November, the peripheral tissue zone of the cone apex appears as a mantle of small densely staining cells and the pith region of the cone axis has noticeably increased in size. At this stage, the first *bract primordia* have been initiated from the mantle at the base of the young cone (Fig. 17-21, B). Gifford and Mirov's study revealed the interesting fact that the young seed cone continues to grow through the winter and early spring months in the vicinity of Placerville. By March of the year following cone initiation, the acropetal development of bracts is very evident and procambial strands have differentiated into some of the lower bract primordia (Fig. 17-21, C). During April, the bracts attain nearly their full size, and the primordia of ovuliferous scales are developing in their axils. By May 1, well-defined primordia of the ovuliferous scales, each subtended by a maturing bract, are very conspicuous (Fig. 17-21, D). At this period of development, the young ovulate cone is structurally comparable to a young vegetative shoot with a bract subtending each of the developing spur shoots (compare Figs. 17-21, D, and 17-19, B).

Although the evidence from organography and ontogeny supports the view that the ovulate cone in the conifers is a compound strobilus, the *phylogenetic* origin of the ovuliferous scale remains to be explained. The scale is sporophyll-like in that it bears ovules, but its axillary position with reference to a bract represents a puzzling situation, from an evolutionary point of view. Of all the conflicting and involved theories that have been proposed during the past century,[*] the most plausible and best-supported view is that advanced by Florin who originated the idea that, phylogenetically, the ovuliferous scale is a highly condensed and modified *fertile shoot* and hence is not a sporophyll. In other words, the scale evolved from an originally leafy, ovule-bearing, dwarf shoot and its present "simple" appearance is the result of the fusion and specialization of both the sterile and fertile components of such a composite ancestral structure.

One of the earliest precursors of the modern ovuliferous scale occurred in *Lebachia* and *Ernestiodendron,* two genera of Paleozoic conifers which have been studied in great detail by Florin. In *Lebachia* the megasporangiate cone consisted of a series of spirally arranged, bifid bracts, in the axils of which developed short, uniovulate, radial, leafy shoots. Usually all but one of the scale-like leaves were sterile; the fertile appendage or megasporophyll, situated near the base or middle region of the dwarf shoot, faced the main axis of the cone and bore a single erect ovule (Fig. 17-22, A). The structure of the axillary fertile shoots in the cone of *Ernestiodendron* was somewhat similar to that in *Lebachia* except for the larger number of ovuliferous appendages (Fig. 17-22, B). Florin has termed the fertile shoots in these and other conifers "seed-scale complexes,"

---

[*]For a critical discussion of many of the older interpretations of the ovuliferous scale, the student is referred to Worsdell (1900), Coulter and Chamberlain (1917), and Chamberlain (1935).

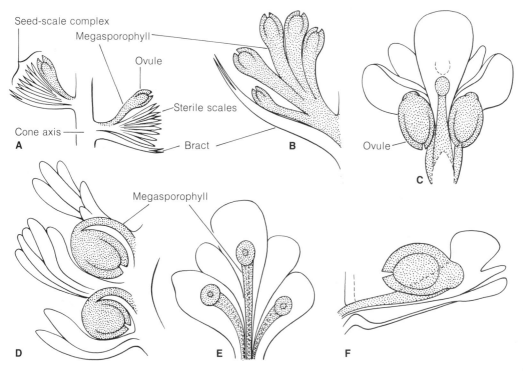

Figure 17-22
Various types of ovuliferous structures in extinct conifers. **A,** *Lebachia goeppertiana;* **B,** *Ernestiodendron filiciforme;* **C, D,** adaxial and side views, respectively, in *Pseudovoltzia liebeana;* **E, F,** adaxial and side views, respectively, in *Voltzia* sp. [Redrawn from Florin, Palaeontographica Abt. B, Bd. 85, 1944.]

and believes they demonstrate that the compound type of megasporangiate strobilus is a primary form in the Coniferales, with the exception of the Taxaceae. He regards the Taxaceae as having a separate evolutionary history because they lack definable seed cones. Very little is known about the internal structure of the ovule in such ancient conifers as *Lebachia* or *Ernestiodendron.* A single integument was present, and it is thought that the female gametophyte produced two archegonia at its micropylar end.

In certain conifers of the Late Permian the fertile ovuliferous dwarf shoots were characterized by a significant series of structural modifications. The cone of *Pseudovoltzia,* for example, consisted of spirally arranged entire bracts. Each axillary seed-scale complex was more-or-less flattened and was composed of a short axis bearing five sterile scales and two or three stalked megasporophylls. Each sporophyll bore a single inverted ovule (Fig. 17-22, C, D). Florin concluded that this type of fertile shoot "corresponds to the ovuliferous scale of a recent pine cone with its two inverted ovules." Additional reduction and modification of the seed-scale complex occurred in the Triassic genus *Voltzia.* The significant features here are that the five sterile scales are basally fused to form a flat sterile component, and that the three megasporophylls were adnate for most of their length to the inner surface of these united scales (Fig. 17-22, E–F).

The further evolutionary development of the seed-scale complex in the Coniferales is still imperfectly understood, and reference to Florin's various papers will show the many gaps in the fossil record. Essentially, however, the general trend seems to have been toward (1) the elimination of all but a few sterile scales, which became fused into the so-called ovuliferous scale, and (2) the ultimate suppression of the sporophylls, which finally became incorporated with the lower part of the ovuliferous scale.

In modern conifers, evolutionary advancement in the ovulate cone is shown by the various degrees of fusion between the ovuliferous scale and its subtending bract. Throughout the Pinaceae, the bract is only basally adnate to the scale (Fig. 17-19, C); whereas these structures are more-or-less completely united in members of the Taxodiaceae, Cupressaceae, and Araucariaceae (Fig. 17-19, H–K). Comparative studies on the vascular anatomy of the ovulate cone in the Pinaceae reveal that the traces which enter the ovuliferous scale diverge from the stele in fundamentally the same manner as the branch traces of an axillary vegetative shoot (Fig. 17-19, D). In *Pinus maritima,* for example, Aase (1915) found that a single vascular strand, with the xylem oriented toward the adaxial surface, as in a vegetative leaf, extends as a separate trace into the bract (Fig. 17-19, D). In contrast, the branch-like vascular supply of the ovuliferous scale consists of three or four traces which diverge at a higher level from the stele; these strands branch dichotomously in the scale and form a series of veins with their xylem oriented towards the lower surface of the scale (Fig. 17-19, E). According to Florin (1951), the so-called "inversion" of the vascular bundles in the ovuliferous scale in the Pinaceae indicates its phylogenetically original radial symmetry. Each of the two ovules borne on the ovuliferous scale is vascularized by a single strand derived as a branch from an adjacent lateral bundle (for a detailed account of the profuse dichotomous venation characteristic of the ovuliferous scales of *Abies, Pseudotsuga, Pinus,* and *Cedrus,* see Tison, 1913).

Recently Lemoine-Sebastian (1968, 1969) has studied the vasculature of the bracts and ovuliferous scales of the ovulate cones of various genera in the Taxodiaceae and Cupressaceae. The bract is vascularized by a single trace that only rarely branches in its course through the appendage. But the pattern of vasculature of the ovuliferous scale is usually very complex and varies according to the genus. In some genera, only a single system of inversely oriented strands is formed; in other genera, radial branching of the strands forms two vascular arcs: an adaxial series of strands with inverted orientation of the xylem and an abaxial system comprising normally oriented bundles (see Lemoine-Sebastian, 1968, Pl. XI, and 1969, Fig. 76). These varied patterns of vasculature of the ovuliferous scales in the Taxodiaceae and Cupressaceae have considerable morphological and systematic interest, but their phylogenetic significance is an open question which depends for its solution upon further analysis and broadened comparative studies.

The Reproductive Cycle in *Pinus*

Numerous comparative studies have demonstrated the wide variation in the details of sporogenesis, gametophytic development, and embryogeny within the living conifers. To choose one genus from the large number of investigated plants as being "typical" of the Coniferales as a whole is doubtless impossible. Most conifers, on the contrary, prove to have a blend of gametophytic and embryological characters, some advanced, some primitive, and still others shared by even the cycads and *Ginkgo.* For these reasons there is probably no living genus which can serve alone as the measure of phylogenetic trends within the order.

The selection of *Pinus* as the basis for the following discussion of reproduction in the conifers is admittedly arbitrary, but yet not without some justification. Various species of pine are widely cultivated in many areas

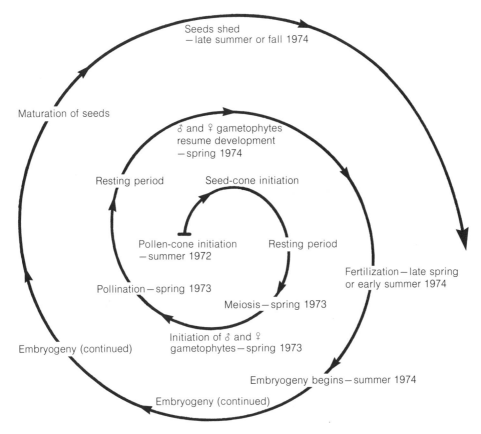

Seeds shed
—late summer or fall 1974

Maturation of seeds

♂ and ♀ gametophytes
resume development
—spring 1974

Resting period   Seed-cone initiation

Pollen-cone initiation   Resting period
—summer 1972

Fertilization—late spring
or early summer 1974

Pollination—spring 1973

Meiosis—spring 1973

Embryogeny (continued)

Initiation of ♂ and ♀
gametophytes—spring 1973

Embryogeny begins—summer 1974

Embryogeny (continued)

Figure 17-23
Diagrammatic representation of a complete reproductive cycle in *Pinus*.
See text for detailed explanation.

of North America and Europe and provide easily accessible material for class study. Furthermore, prior to the publication of Allen and Owen's (1972) comprehensive monograph on the life history of the Douglas Fir (*Pseudotsuga menziesii*), the reproductive cycle of *Pinus* had been studied more comprehensively than that of any other conifer (Ferguson, 1904; Buchholz, 1918; Haupt, 1941; Johansen, 1950). As a result, it will be possible to present a rather complete and connected account of its cycle of reproduction. For the student, the advantages of a coherent description of the processes and structures that result in the formation of a seed in such a familiar type of conifer as *Pinus* are obvious. However, to insure an

understanding of the variations in the details of reproduction in the Coniferales as a whole, comparisons between *Pinus*, *Pseudotsuga*, and several other conifer genera will be made later in the chapter.

Phenology

Many conifers, such as *Pseudotsuga* (Allen and Owens, 1972, p. 83), are *both* pollinated and fertilized during the same season. The phenology of *Pinus*, however, is atypical among conifers in that a period of about 12–14 months intervenes between the pollination of the ovules and the actual fertilization of the archegonia. If the period of initiation and early development of the

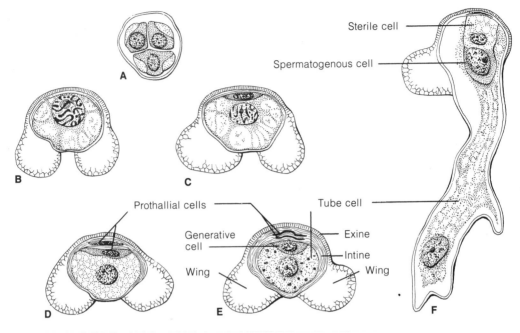

Figure 17-24
Development of the male gametophyte in *Pinus laricio*. **A,** tetrad of young microspores enclosed within wall of microsporocyte (note developing wings of microspores); **B,** prophase of first division of microspore nucleus; **C,** the first prothallial cell has been formed and the microspore nucleus is preparing to divide again; **D,** a second prothallial cell has been formed; **E,** the microspore nucleus ("antheridial initial") has divided, producing the tube and generative cells; at this four-celled stage the male gametophyte is liberated from the microsporangium; **F,** young pollen tube as it would appear in the nucellus of ovule after pollination (note that generative cell has divided to form the sterile and spermatogenous cells.) [Redrawn from *Gymnosperms. Structure and Evolution* by C. J. Chamberlain. University of Chicago Press, Chicago, 1935.]

strobili is included, the *complete* life cycle of *Pinus* extends over a period of three years, rather than two years, the period in many other conifers.

The main events in the life cycle of pine are shown by the diagram in Fig. 17-23. In order to be explicit, the cycle is depicted as beginning with cone initiation in 1972 and culminating with the maturation of the ripe seeds in 1974. We have deliberately avoided indicating specific months at which the various processes take place because it is evident that there is considerable fluctuation and variation, depending on the latitude, altitude, weather, etc., which vary in different parts of the geographical range of such a large genus as *Pinus*. This is illustrated clearly with respect to the exact period at which mega-

sporangiate cones are initiated in different species of *Pinus*. Gifford and Mirov (1960) found that young ovulate cones could be identified within the terminal buds of *Pinus ponderosa* in mid-September in the vicinity of Placerville, California. In contrast, Konar and Ramchandani (1958), in their study of the Himalayan species, *Pinus wallichiana*, reported that the ovulate cones are initiated in January. There are also interspecific differences in the exact timing of other events, which, of course, are not shown in the generalized type of life-cycle diagram illustrated in Fig. 17-23. The value of this diagram consists in aiding the student to follow the morphological details of the reproductive cycle in proper chronological sequence.

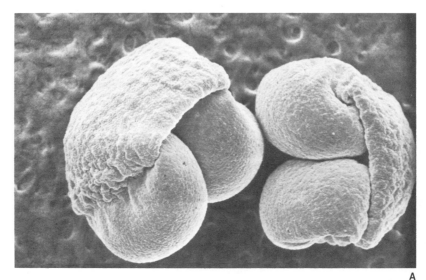

A

B

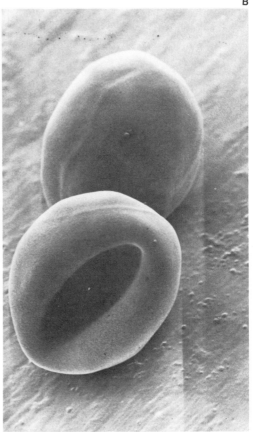

Sporogenesis

MICROSPOROGENESIS.   The entire developmental history of the microsporangiate
strobilus in certain species of pine extends
over a period of approximately a year
(Ferguson, 1904). The strobili are initiated
in the axils of scale leaves in the spring or
early summer, and by winter the microsporangia contain well-defined sporogenous
tissue. Meiosis and the formation of the
characteristic winged microspores, however,
are processes that do not occur until the
following spring (see Fig. 17-23). Each functional microsporocyte gives rise to four
haploid microspores, which are enclosed
within the wall of the microsporocyte for
some time (Fig. 17-24, A). According to
Ferguson, the two bladders or wings of each
spore are formed by the separation of the
outer (exine) and inner (intine) layers of the
microspore wall; this takes place while the
members of a spore tetrad are still surrounded by the common mother-cell wall.
The outer surface of the mature microspore,
especially the wings, exhibits a reticulate
sculpture (Fig. 17-25, A). The lower end of
the spore wall, between the wings, is relatively thin and smooth; this region is the
point at which the pollen tube later emerges
(Fig. 17-24, B–F).

500

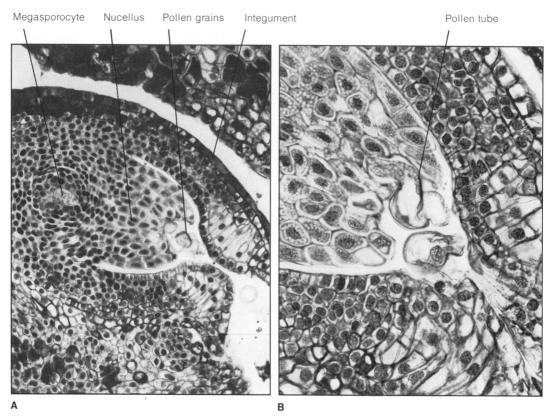

Megasporocyte   Nucellus   Pollen grains   Integument   Pollen tube

A   B

Figure 17-26
Pinus sp. **A,** median longisection of ovule showing a single deeply embedded megasporocyte (note large radially elongated cells of the middle layer of the integument and the pollen grains); **B,** longisection of ovule showing young pollen tube penetrating the nucellus.

MEGASPOROGENESIS. Before describing the process of megasporogenesis, it is essential to review briefly the ontogeny of the ovuliferous scale in *Pinus*. Figure 17-19, A, represents somewhat diagrammatically a longisection of a young megasporangiate strobilus enclosed within a series of overlapping bud scales. In *Pinus ponderosa*, this stage in development is reached during the month of May (Gifford and Mirov, 1960). The young strobilus at this time consists of an axis bearing a series of bracts, in the axil of each of which is the primordium of an ovuliferous scale (Fig. 17-21, D). As in the microsporangiate strobilus, however, the differentiation of the young ovulate cone is interrupted by the onset of

winter. According to Ferguson (1904), the first indications of ovule development in certain pines cannot be discerned until April or late May of the year following the initiation of the cone. Prior to megasporogenesis, the pine ovule consists of a nucellus and a single integument; the position of the ovule is inverted and the conspicuous micropyle points inward, toward the cone axis (Fig. 17-26).

In the species of *Pinus* that have been intensely studied, the single megasporocyte of each ovule just prior to meiosis is located deep within the nucellus (Fig. 17-26, A). According to Konar's (1960) study on *Pinus roxburghii*, the deeply situated position of the megasporocyte arises as follows. A hypodermal initial cell, at the apex of the nucellus, divides periclinally into an outer *primary parietal cell* and the *megaspore mother cell.* As a result of the numerous divisions of the primary parietal cell, the megasporocyte in this species ultimately becomes deeply embedded in the nucellus.

Theoretically the meiotic division of the megasporocyte should yield a tetrad of four haploid megaspores. Ferguson's (1904) detailed investigations, however, showed variation in the number of cells produced by the megasporocyte, not only among different species of pine but even within the same species. In *Pinus austriaca* a linear tetrad of four megaspores results from the division of *each* of the first two cells produced by the division of the sporocyte. In *Pinus strobus* and *Pinus rigida*, however, only three cells are formed. This is the result of the degeneration of the upper dyad cell of the two formed by the division of the sporocyte; only the lower dyad cell (i.e., the cell farthest from the micropyle) of the two divides. Regardless of these variations, the lowest cell of the series produced by the sporocyte becomes the functional haploid megaspore, with the two or three cells above it soon disintegrating.

The Male Gametophyte and Pollination

As in the cycads and *Ginkgo,* the early development of the endosporic male gametophyte takes place before the dehiscence of the microsporangium. As a result of three successive nuclear divisions, the young male gametophyte consists of two prothallial cells (which soon become flattened and dead), a generative cell, and a tube cell. This is the usual stage in *Pinus* for the liberation of the pollen grains into the air (Fig. 17-24, E).

The ovules of *Pinus* are wind pollinated. During the late spring or early summer, the axis of the young megasporangiate cone elongates and the ovuliferous scales become separated. The abundant pollen grains, borne by the wind, sift down between the scales and adhere to the pollination drops, which exude from the open ends of the *inverted* ovules. According to Doyle and O'Leary (1935), a pollen grain is very buoyant; at first it floats on the surface of the pollination drop and then begins to rise *upwardly* in the liquid, which fills the micropylar canal of the ovule. Doyle emphasizes that the function of the wings is to so orient the pollen grain that the end *away from* the prothallial cells is ultimately brought into direct contact with the exposed surface of the nucellus. In this position, the pollen grain germinates and forms a pollen tube.

Ferguson (1904) found that the free end of the nucellus is slightly depressed in *Pinus*, and that the pollen grains lie in this shallow cavity. According to her observations, the radial elongation of cells of the middle layer of the integument, a short distance above the apex of the nucellus, forms a rim-like outgrowth that seals off the outer part of the micropylar canal (Fig. 17-26, A). After pollination, the ovuliferous scales become drawn together and remain closely appressed until the seeds are released from the mature cones. In the so-called "closed-cone pines" (e.g., *Pinus attenuata*), the heat produced by a forest fire is usually what is required to separate the cone scales and liberate the seeds.

As we have already emphasized, a salient feature of the reproductive cycle in *Pinus* is the relatively long interval between pollination and fertilization. In many pines, this interval is about twelve months (Fig. 17-23). At the time of pollination, the male gametophyte consists of four cells and megasporogenesis has just begun in the ovule. Following pollination there is a slow period of development of the male gametophyte; this comprises the emergence of the pollen tube and the division of the generative cell to form a sterile cell and a spermatogenous cell (see diagrams of male-gametophyte development shown in Figs. 17-27 and 17-24, F).

Concomitantly the single functional megaspore within the nucellus of the ovule enlarges, and there is a series of free nuclear divisions. According to Ferguson (1904), about five successive nuclear divisions, yielding 32 nuclei, take place before the period of dormancy is reached. Thus, both the male and female gametophytes are in a comparatively early stage of development throughout the winter, and male gametes and archegonia do not form until the following spring.

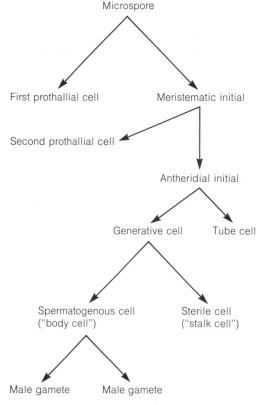

Figure 17-27
Diagram showing the successive stages in development of the male gametophyte in *Pinus*. Male-gametophyte development is similar in *Ginkgo biloba*. [Adapted from Sterling, *Biol. Rev.* 38:167, 1963.]

The Female Gametophyte

The development of the female gametophyte of *Pinus* resembles in many respects that which has already been described for the cycads and *Ginkgo*. At the close of the first period in its development, the young female gametophyte of *Pinus* is a more-or-less spherical sac, bounded by the megaspore wall, and containing about 32 free nuclei, embedded in a parietal layer of cytoplasm that surrounds a large central vacuole. When development resumes in the spring, there is a very rapid formation of additional free nuclei. Ferguson estimated that about 2,000 free nuclei are present in the female gametophyte of *Pinus strobus* at the time when cell walls begin to form.

Unfortunately, since the publication of Ferguson's (1904) study on *Pinus*, which has long been considered a classic, very little modern investigation has been devoted to the manner in which the coenocytic megagametophyte becomes converted, by wall formation, into a cellular gametophyte. Chamberlain's (1935) brief account suggests that the centripetal formation of *alveoli* and the subsequent development of radial rows of cells takes place very much as we have indicated in describing the ontogeny of the female gametophyte in *Ginkgo biloba* (see Chapter 16, Fig. 16-10). The recent description of the cellular phase in development of the female gametophyte in *Pseudotsuga* (a member of the family Pinaceae) very possibly may be shown to parallel that of *Pinus* (Allen and Owens, 1972).

The development of archegonia may be deferred until the female gametophyte is completely cellular, but not infrequently archegonia are initiated before wall formation has terminated. According to Ferguson, archegonia are detectable about two weeks before fertilization. The ontogeny of the archegonium is markedly like that of a cycad (Fig. 15-21, F–H). The initial, which is a superficial cell differentiated at the micropylar end of the gametophyte, divides periclinally into a small outer primary neck cell and a larger inner central cell. By means of two successive anticlinal divisions, the primary neck cell forms four neck cells, a condition apparently rather common in *Pinus strobus*. In other species, however (e.g., *Pinus austriaca, Pinus rigida*, and *Pinus resinosa*), these four neck cells may divide periclinally, forming an eight-celled neck with the cells arranged in two tiers. Apparently the number of neck cells varies somewhat even in the same species. The central cell enlarges, but its nucleus remains close to the neck of the archegonium. Ultimately the nucleus of the central cell divides, forming a small ventral canal cell which is separated by a definite wall from the large egg cell. The ventral canal cell eventually degenerates, and the nucleus of the egg cell becomes enlarged and descends to a central position in the egg. When mature, the egg of the archegonium is jacketed by a distinct layer of cells, and numerous proteinaceous bodies (so-called protein vacuoles) become evident in the cytoplasm; these bodies superficially resemble nuclei (Fig. 17-30, A).

The number of archegonia produced by a single gametophyte varies considerably, depending somewhat on the species. According to Johansen (1950) there is usually a single archegonium in the Monterey pine (*Pinus radiata*), whereas in the majority of the other species investigated the number of archegonia ranges from two to six (Fig. 17-28).

Fertilization

During the resumption of growth of the female gametophyte in the spring (Fig. 17-24, F), the sterile cell and spermatogenous cell of the male gametophyte move down toward the lower end of the pollen tube. Approximately a week before fertilization, the nucleus of the spermatogenous cell divides to form two male gametes, which are unequal in size. In marked contrast with the large motile

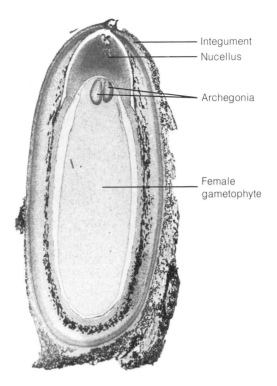

Figure 17-28
Pinus sp. Median longisection of ovule showing female gametophyte with two archegonia.

spermatozoids found in *Ginkgo* and the cycads, the male gametes in *Pinus* are not highly individualized cells, and they are entirely devoid of flagella. Following the formation of the two male gametes, the pollen tube grows actively down through the nucellus (Fig. 17-29, A). Since several pollen grains may reach the apex of the nucellus, a corresponding number of tubes may be formed; usually only two or three tubes, however, reach the female gametophyte. The tip of a pollen tube forces itself between the neck cells of the archegonium and then ruptures, discharging the two male gametes, the tube nucleus, and the sterile cell into the cytoplasm of the egg. Fertilization consists in the entrance of the larger of the two male gametes into the nucleus of the egg cell (Fig. 17-30, A, B). Usually the smaller male gamete, the tube nucleus, and the sterile cell

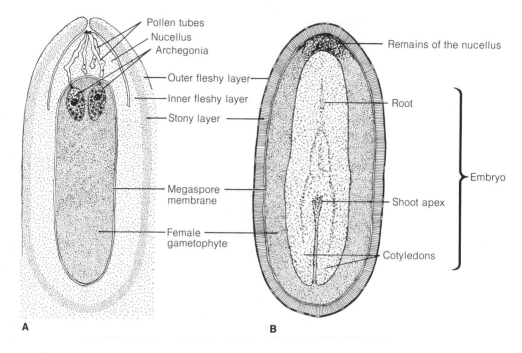

Figure 17-29
Pinus laricio. **A**, longisection of ovule at time of fertilization; **B**, longisection of mature seed. [Redrawn from *Gymnosperms. Structure and Evolution,* by C. J. Chamberlain. University of Chicago Press, Chicago, 1935.]

can be seen at the top of the egg after sexual fusion has been accomplished. Eventually these three nuclei degenerate.

Embryogeny and Seed Development

The early embryogeny of *Pinus*, like that of the majority of the members of the Coniferales, is characterized by an extremely short period of free nuclear divisions. Following syngamy the nucleus of the fertilized egg contains two distinct haploid groups of chromosomes, one paternal, the other maternal (Fig. 17-30, B). These two sets of chromosomes soon become arranged at the equatorial region of a common spindle, and at this time Ferguson (1904) and Haupt (1941) were able to count 24 chromosomes, apparently the typical diploid number for various species of *Pinus* (Fig. 17-30, C). The division of the zygote nucleus yields two nuclei, each of which promptly divides,

forming a total of four free nuclei located near the middle of the cytoplasm of the egg. These four nuclei represent the extent of the free nuclear period in *Pinus*. Soon after their formation, they move to the lower end of the archegonium where a third division accompanied by wall formation occurs (Fig. 17-30, D, E). The first wall formed is transverse to the long axis of the archegonium and separates the eight nuclei into two tiers, then vertical walls appear separating the nuclei in each tier; the proembryo now consists of a lower tier of four cells, completely bounded by walls and an upper tier devoid of walls adjacent to the cytoplasm of the egg (Fig. 17-30, E). The next division usually occurs in the cells of the upper tier and is followed by a similar transverse division in the lower tier. The *proembryo*, as it is termed, now consists of sixteen cells arranged in four superposed tiers (Fig. 17-30, F, G). This tiered arrangement of the cells of the proembryo is found

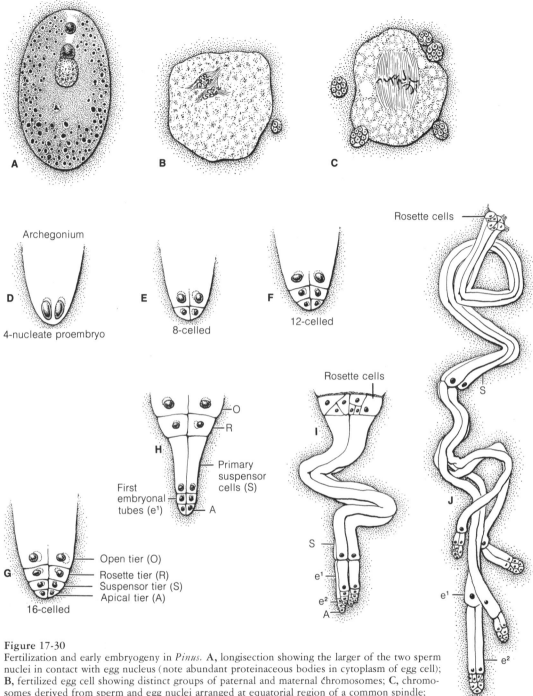

Figure 17-30
Fertilization and early embryogeny in *Pinus*. **A,** longisection showing the larger of the two sperm nuclei in contact with egg nucleus (note abundant proteinaceous bodies in cytoplasm of egg cell); **B,** fertilized egg cell showing distinct groups of paternal and maternal chromosomes; **C,** chromosomes derived from sperm and egg nuclei arranged at equatorial region of a common spindle; **D,** free nuclear phase in embryogeny with four nuclei; **E–G,** development of the four superposed cell tiers of the proembryo; **H,** elongation of primary suspensor cells and formation of first embryonal tubes (e^1); **I,** the lower end of the proembryo has begun to separate (cleavage polyembryony) into four vertical series of cells (note the formation of a second series of embryonal tubes (e^2) and the divisions which have occurred in the rosette cells); **J,** series of four competing embryos formed by cleavage (note conspicuous apical cell in each embryo). [**A–C** redrawn from Haupt, *Bot. Gaz.* 102:482, 1941; **D–J,** redrawn from Buchholz, *Trans. Illinois Acad. Sci.* 23:117, 1931.]

in many genera of the Coniferales, and the precision with which it originates offers a marked contrast with the comparatively unstratified cell arrangement characteristic of the early embryogeny of the cycads and *Ginkgo* (see Chapter 16).

One of the outstanding features of the later phases of embryogeny in *Pinus* (and in many other conifers) is the development, from the walled cells of the proembryo, of from four to eight separate competing embryos. This early formation of multiple embryos becomes particularly remarkable when it is recalled that embryogeny in pine — as in all tracheophytes — begins with a single fertilized egg cell.

To understand clearly the process of polyembryony it is essential to begin with the structure and further development of the cell tiers of the sixteen-celled proembryo shown in Fig. 17-30, G. The uppermost cell tier, sometimes termed the open tier, is in open communication with the egg cytoplasm and possibly serves for a short time to transmit reserve food materials to the lower portion of the proembryo; soon, however, the cells of this tier disintegrate. The next tier comprises the so-called rosette cells which earlier investigators of pine embryogeny considered functionless. Buchholz (1918, 1931), however, has clearly demonstrated that several, or all four, rosette cells may give rise during early embryogeny to small embryos, which often become twisted or curved in orientation (Fig. 17-30, I). The development of the rosette embryos, however, is not extensive — they soon abort and do not include more than a dozen or so cells.

The dominant aspect of polyembryony in *Pinus* is the result of the ultimate separation of the lower cell tiers of the proembryo into four filamentous embryos (Fig. 17-30, I). This process has been termed *cleavage polyembryony*, and its salient features may now be discussed. Shortly before the proliferation of the rosette cells into embryos, the cells in the tier below them elongate markedly

(Fig. 17-30, H, I). These are the *primary suspensor cells*, and their vigorous extension forces them and the apical tier of cells through the membrane of the original egg into the female gametophyte (Fig. 17-31). Buchholz (1918) has described in detail the breakdown of the starch-containing cells in the upper part of the female gametophyte and the formation of a "corrosion cavity" into which the growing system of embryos intrudes. Because of the limited confines of the corrosion cavity, the primary suspensors soon become coiled and buckled (Fig. 17-30, I, J). During the early elongation of the primary suspensors the apical tier of embryonal cells gives rise, by a series of transverse divisions, to several additional cell tiers (e^1, e^2, etc. in Fig. 17-30, H, I). These cells quickly elongate, like the primary suspensor cells behind them, and are termed *embryonal tubes*; their extension serves to push the embryo system farther into the corrosion cavity of the gametophyte. Following the formation of the first series of embryonal tubes the lower end of the embryo system separates or cleaves into four distinct vertical series of cells. Each series consists of an apical cell, two or more embryonal tubes, and a primary suspensor cell, and represents an independently developing embryo (Fig. 17-30, I, J). Prior to elongation any of the successively formed embryonal tubes may divide, forming vertical walls; as a consequence, each unit embryo may be attached to a complex and collaterally arranged series of embryonal tubes. In contrast, the primary suspensor cells never divide but eventually collapse and die later in embryogeny.

Within the system of embryos derived by the cleavage of one proembryo, there is apparently intense competition. One of the four embryos, usually the lowest and most aggressive member of the group, continues to develop and becomes the differentiated embryo of the pine seed; the other embryos abort and cannot be detected in the mature seed.

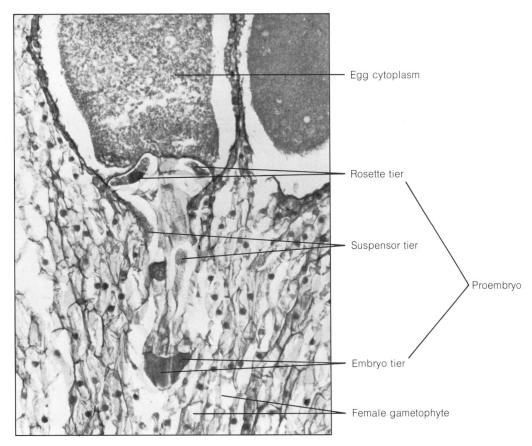

Egg cytoplasm

Rosette tier

Suspensor tier

Proembryo

Embryo tier

Female gametophyte

Figure 17-31
Longitudinal section of the proembryo of *Pinus* sp. Note that the elongation of the primary suspensor cells has forced the apical tier deep into the tissue of the female gametophyte.

From the foregoing discussion it should be clear that each proembryo in *Pinus* is theoretically capable of forming eight embryos, four derived from the rosette tier and four from the cleavage process just outlined. If all the archegonia of a single gametophyte are fertilized, an extraordinary number—as many as 48—separate embryos might begin development. Buchholz (1918) actually found as many as four separate embryo systems, each the product of a fertilized archegonium, in some species of pine. These embryo systems grow in competition for variable periods of time, but only a single embryo among them normally reaches a fully developed condition in the ripe seed.

The later stages of development of the successful embryo are complex and can be only considered very briefly in this chapter. According to Buchholz the terminal cell of the embryo soon takes on the character of a pyramidal apical cell which forms derivative cells or segments very much like the apical cell of *Equisetum* or a leptosporangiate fern (Fig. 17-30, J). Buchholz (1931) and Chamberlain (1935) regard the presence of an apical cell in conifer embryogeny as a "primitive character." After the main body of the pine embryo consists of several hundred cells a definitive apical cell is no longer apparent, and the extreme apex of the embryo is occupied by a group of equivalent apical

A

B

Figure 17-32
A, *Pinus monophylla* (piñon pine); portion of
an open cone, showing pairs of wingless seeds
attached to the adaxial surface of the reflexed
ovuliferous scales; **B,** *Pinus sabiniana* (digger
pine); seeds with prominent wings.

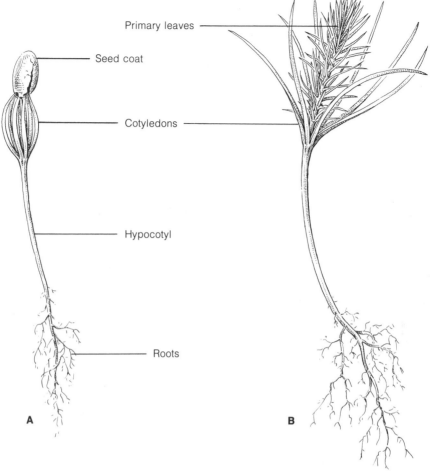

Primary leaves

Seed coat

Cotyledons

Hypocotyl

Roots

A

B

Figure 17-33
Pinus edulis. **A,** young seedling showing whorl of cotyledons and remains of seed coat;
B, older seedling showing cotyledons and spirally arranged primary leaves. [Redrawn from
Gymnosperms. Structure and Evolution by C. J. Chamberlain. University of Chicago Press,
Chicago, 1935.]

initials. The further histogenesis of the embryo includes the differentiation, at the suspensor end, of the initial cells of the root apex, and the ultimate formation of a series of cotyledon primordia from the shoot apex region. For a detailed description of tissue and organ formation during the later stages of embryogeny the student is referred to the work of Spurr (1949) on *Pinus strobus*.

The fully developed embryo of *Pinus* consists of a whorl of cotyledons (the average number in the genus, according to Butts and Buchholz, 1940, is 8.1) that surrounds the shoot apex, a short hypocotyl, and a primary root or radicle. The embryo is embedded in the tissue of the massive female gametophyte which, in turn, is surrounded by the seed coat (Fig. 17-29, B). The seed coat consists primarily of a hard outer coat, which is derived from the stony layer of the integument. The inner fleshy layer of the integument degenerates during seed development, and in the ripe seed is reduced to a thin papery membrane. Usually the remains of the nucellus can be seen at the micropylar end of the female gametophyte.

Although some kinds of pine (e.g., piñon pine) produce seeds devoid of wings, in most species the seed, at the time it is shed from the cone, is attached to a thin membranous wing which apparently aids in its dispersal (Fig. 17-32). The seed wing, according to Sporne (1965), is formed by the separation of a portion of the adaxial surface of the cone scale adjacent to the ovule, and hence, morphologically, is not part of the seed itself.

When the pine seed germinates, the entire embryo emerges from the ruptured seed coat, which may adhere for a short time to the tips of the cotyledons (Fig. 17-33, A). The primary shoot formed by the terminal bud of the seedling at first bears only a spiral series of needle-like leaves (Fig. 17-33, B). Later, as we have described on page 476, the first spur shoots arise in the axils of some of the primary leaves.

Sacher (1954) has described in detail the zonal structure of the shoot apex in *Pinus lambertiana* and *Pinus ponderosa*. The structure of the apex of the latter species has already been considered in our account of the ontogeny of the ovuliferous cone (see p. 494 and Fig. 17-21, A).

Comparisons Between *Pinus* and Other Conifers

In this book it is not possible to present a detailed comparison between the gametophytic and embryological features of *Pinus* and those of other conifers. For broad comparative treatments of the gametophytes and embryos of a very wide range of conifer genera, the student is referred to Chamberlain's (1935) review of the older literature and to the more recent reviews of Sterling (1963), Maheshwari and Singh (1967), and Doyle and Brennan (1972). The purpose of the brief account that follows is to emphasize some of the most outstanding differences between *Pinus* and other conifers with reference to (1) the development and structure of the male gametophyte, (2) the types of pollination mechanisms, and (3) the development and behavior of the proembryo.

The Male Gametophyte

Although the formation of two prothallial cells appears to be a consistent feature of the male gametophyte of *Pinus, Pseudotsuga* (Fig. 17-34, A), and other members of the Pinaceae, prothallial cells are absent from the pollen grains of the Taxodiaceae, Cupressaceae, Cephalotaxaceae, and Taxaceae. In these families, the first division of the microspore nucleus gives rise directly to the generative cell and the tube cell (Sterling, 1963). At the opposite extreme, the very numerous prothallial cells, typical of the Araucariaceae, have attracted study and aroused speculation about their phylogenetic significance (Burlingame, 1913; Eames, 1913; Chamberlain, 1935). In *Araucaria*, for example, the first two prothallial cells, instead of becoming senescent, as in members of the Pinaceae, undergo active divisions and form at first several tiers of cells. Then the cell walls break down and the free prothallial nuclei are liberated into the general cytoplasm of the pollen grain, as is illustrated in the sectional view of the young male gametophyte shown in Fig. 17-34, B. It is said that there are as many as 40 prothallial cells (or nuclei) in the male gametophyte of *Agathis*, which is regarded by Zimmermann (1959, p. 454) as the largest male gametophyte in living seed plants. The evolutionary significance of the large number of prothallial cells in the male gametophyte of the Araucariaceae is an unsettled question. Chamberlain (1935) regarded this as a primitive feature, whereas Eames (1913), with reference to *Agathis*, believed that numerous prothallial cells represent a phylogenetically derived condition.

In many conifers, the male gametes are represented by two nuclei (which may be associated with cytoplasm), as in *Pinus, Pseudotsuga*, and other members of the Pinaceae. But in certain genera of the Taxodiaceae and Cupressaceae, the male gametes have been described as well-defined cells, separated by a wall after their formation in

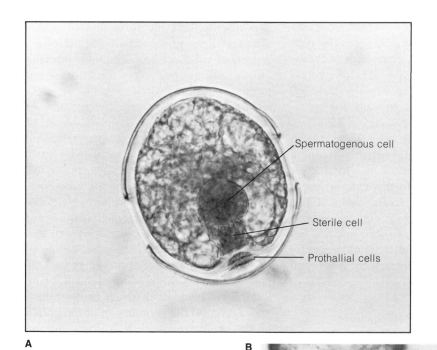

Spermatogenous cell

Sterile cell

Prothallial cells

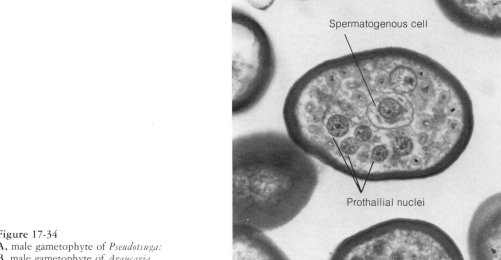

Spermatogenous cell

Prothallial nuclei

Figure 17-34
A, male gametophyte of *Pseudotsuga;*
B, male gametophyte of *Araucaria.*
See text for explanation.

the spermatogenous cell. Aside from the absence of flagella, male gametes of this type resemble the young spermatozoids of cycads and *Ginkgo*. As the pollen tubes of conifers function primarily to convey the male gametes directly to the archegonia, the apparently highly organized gametes of *Taxodium* and related plants may conceivably represent gametes that only recently — from a phylogenetic standpoint — have "lost" their locomotor apparatus. In this connection, it should be noted that Christiansen (1969) reported that he observed "ciliated" sperms in *Pseudotsuga* (Douglas fir) but this was not confirmed by the careful study of spermatogenesis in the same species by Allen and Owens (1972).

Types of Pollination Mechanisms

The comprehensive investigations of Doyle (1945) revealed a surprising range in the types of pollination mechanisms in the living conifers. According to his interpretation, the exudation of pollination fluid, i.e., the pollination drop, and the formation of pollen grains with wings are salient and correlated aspects of the mechanism of pollination, not only in various members of the Pinaceae but in the conifers as a whole (Fig. 17-25, A). As we have previously noted, the function of the wings or bladders of the pollen grains of *Pinus* is to orient the grains as they rise through the fluid in the micropylar canal towards the nucellus of the inverted ovule.

According to Doyle, various "modifications" of the basal type, represented in *Pinus*, have evolved in the Coniferales as the result of (1) the loss of the pollination exudate, (2) the assumption of a "stigmatic" or receptive function by the free tips of the ovular integument, (3) the reduction and ultimate loss of the bladders or wings of the pollen grains, and (4) the germination of the pollen on the ovuliferous scale. *Pseudotsuga* may be used to illustrate one of the modified types of pollination mechanisms in the Pinaceae (see Fig. 17-35).

In *Pseudotsuga*, the pollen grains as shown in Fig. 17-25, B, are devoid of wings and

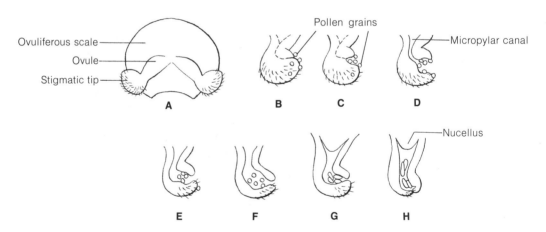

Figure 17-35
The pollination mechanism in Douglas fir (*Pseudotsuga menziesii*). **A,** adaxial view of ovuliferous scale showing the stigmatic tips of the two ovules at the time of pollination; **B,** adherence of pollen to hairs on stigmatic tips; **C–F,** formation of depression in lower lip into which pollen grains sink; **G,** approximation of integumentary lips and enclosure of pollen grains in micropylar canal; **H,** elongation of pollen grains prior to the formation of pollen tubes. [Redrawn from *The Life History of Douglas Fir* by Allen and Owens. Information Canada, Ottawa, 1972.]

there is no exudate of liquid from the ovule (Allen, 1963; Allen and Owens, 1972). The pollen grains, falling inwardly along the upper surface of the bracts of the seed cone, land upon the sticky hairs of the unequal "lips" of the integument of the ovule (Fig. 17-35, B). At this time, a depression begins to form on the upper surface of the larger lip, apparently as the result of the collapse of surface and subsurface cells (Fig. 17-35, C-F). Some of the pollen grains sink into this shallow cavity and, as a result of the inward growth and approximation of the two lips, become enclosed within the upper end of the micropylar canal of the ovule (Fig. 17-35, G, H). In marked contrast to *Pinus*, the pollen grains of *Pseudotsuga* germinate within the micropylar canal rather than on the surface of the nucellus. A pollen tube does not form in *Pseudotsuga* until the young gametophyte has ruptured the exine of the spore wall and has become a much elongated structure. The pollen tube emerges only after the elongating gametophyte has come into contact with the nucellus (for further details and photographic illustrations, see Allen and Owens, 1972, p. 96).

One of the most remarkable deviations from the *Pinus* or *Pseudotsuga* type of pollination mechanism occurs in the Araucariaceae. In this family, the pollen germinates on the ovuliferous scale—or in its axil—at a point far removed from the ovule. Eames (1913) describes in great detail how the pollen tubes of *Agathis* at first branch and grow into the tissue of the ovuliferous scale; some of the branches, in their haustorial growth, even penetrate the phloem and xylem tissues of the cone axis. Although it has been suggested that the germination of the pollen on the ovuliferous scale in araucarians may represent a primitive pattern in conifers, this idea is rejected by Doyle (1945), who maintains that the Carboniferous and Permian ancestors of the modern conifers "all had direct ovular reception of the pollen," and

that hence the araucarian type of pollination mechanism should be regarded as a derived and advanced character.

The Development of the Proembryo

As compared with *Ginkgo* and the cycads, the period of free nuclear divisions in most conifers is brief, and very commonly, as in *Pinus*, terminates after four nuclei have been formed. In some genera, such as *Taxus*, walls do not form until 16–32 free nuclei have developed (Sterling, 1948). The early embryogeny of the Araucariaceae, however, is unique. The free nuclear period in this group is relatively protracted, and 32–64 nuclei are present before walls are formed. Unlike most conifers, the free nuclei do not migrate to the base of the egg but remain in the center. A massive suspensor develops from the upper cells of the proembryo, whereas the outer cells of its lower portion form a protective cap, recalling the analogous structure formed during the early embryogeny of *Zamia* (see Chapter 15, p. 439). The free end of the araucarian embryo, after the cap cells have degenerated, gives rise to the main body of the embryo (Johansen, 1950).

As far as is known, *Sequoia sempervirens* is the only conifer devoid of a free nuclear period in embryogeny. After the first division of the zygote there is a transverse wall, yielding a bicellular embryo. This is a puzzling problem, especially because in *Sequoiadendron giganteum* embryogeny begins with the formation of eight free nuclei (Buchholz, 1939).

Polyembryony is widespread among members of the Coniferales, and, according to the extensive investigations of Buchholz, cleavage polyembryony occurs in 30 of the 37 genera that have been intensively studied. The occurrence of cleavage polyembryony, however, is variable even among genera of the same family. In the Pinaceae, for example, *Pinus, Cedrus, Tsuga,* and *Pseudolarix* exhibit cleavage polyembryony, whereas *simple polyembryony* seems to prevail in *Abies, Picea*, and *Pseudotsuga.*

The expression *simple polyembryony* designates the condition in which each of the fertilized eggs of one gametophyte produce separate but undivided embryos. In other words, each proembryo produces a single embryo and the four cells of the apical tier function as a unit without cleaving. Simple polyembryony is widespread among conifers and a good example is provided by *Pseudotsuga* (Allen, 1946a; Allen and Owens, 1972). During the early embryogeny of *Pseudotsuga*, rosette cells are not formed and the twelve-celled proembryo consists of three (rather than four) tiers of cells: an open tier in contact with the egg cytoplasm, a middle tier of suspensor cells, and an apical tier. Although the four cells of the apical tier may contribute more-or-less equally to the later development of the embryo, Allen and Owens state that in the most common pattern of development, two of the cells of the apical tier become dominant and "overtop" the other two by their more active growth and division. Cells of the dominant pair divide in two planes and give rise, *without cleavage*, to the main body of the embryo.

Although Buchholz (1926) regarded cleavage polyembryony as the primitive type of embryogeny in the conifers as a whole, this interpretation has been severely criticized by Doyle and Brennan (1972) on the basis of their wide survey of embryogenesis. They concluded that "the absence of cleavage was the primitive condition in conifers and taxads." From this standpoint, Doyle and Brennan maintain that cleavage polyembryony has arisen separately in many lines of conifers, and that the early embryogeny of *Pinus* "is not a prototype for conifers in general but shows a specialized derived condition." (For a detailed review on embryogeny of conifers see also Chowdhury, 1962.)

Summary and Conclusions

In this chapter we have attempted to present certain of the important morphological features of the leaf, stem, and strobili of the Coniferales as a whole. In addition, we have given a rather full and connected description of the reproductive cycle in *Pinus*, a genus that seems to illustrate some of the widespread characteristics of the gametophytes and embryo that are common to many conifers.

Although the leaves of conifers are simple and commonly univeined, these organs are not microphylls in a morphological sense because their traces are associated with definable "leaf gaps." In certain of the Paleozoic conifers the foliage leaf was bifid or dichotomously lobed; from this primitive foliar type the simple leaf of modern conifers may have evolved.

With reference to leaf histology, emphasis has been placed on the structure and method of development of stomata because of their systematic and phylogenetic importance in the study of ancient as well as living conifers. Attention has been directed also to the characteristic transfusion tissue which accompanies the vascular system in conifer leaves. The need for comprehensive ontogenetic and comparative studies on transfusion tissue was emphasized.

The histology of the stem of *Pinus* was briefly described and the absence of vessels in the secondary xylem was shown to be a general feature of all conifers.

Throughout the Coniferales the microsporangiate strobilus is "simple," consisting of an axis bearing a series of microsporophylls. The microsporangia are borne on the lower surface of the sporophylls and vary from two to seven, depending upon the genus. In the light of recent study we have shown that there is no standardized method of initiation of the microsporangium. In

Pseudotsuga and *Cedrus* the sporangial initials are superficial cells, whereas in a series of other genera these parent cells are hypodermal in position. In each type, however, the pattern of development of the microsporangium is of the eusporangiate type.

With the exception of the Taxaceae and related forms, the megasporangiate strobilus of the Coniferales is "compound" and consists of an axis bearing a series of bracts, each of which subtends or is more-or-less fused with an ovuliferous scale which develops one or more ovules. According to Florin's extensive studies on the megasporangiate strobili of Paleozoic and Mesozoic conifers, the so-called ovuliferous scale of modern conifers is the evolutionary modification of a primitive, radially organized, leafy, ovule-bearing lateral shoot.

The reproductive cycle of *Pinus* exhibits many features shared by the majority of gymnosperms including the cycads and *Ginkgo*, namely: endosporic and reduced male gametophytes; a protracted period of free nuclear division followed by centripetal wall formation in the female gametophyte;

highly specialized archegonia with neck cells, an evanescent ventral canal cell, and a large egg cell; and a period of free nuclear divisions in embryogeny. The absence of flagellated sperms and the development of pollen tubes which convey the male gametes to the archegonia are important features distinguishing *Pinus* and other conifers from both *Ginkgo* and the cycads. Additional characters that *Pinus* shares with many other conifers include highly evolved nonflagellated male gametes; a very restricted period of free nuclear divisions in embryogeny; and the tiered arrangement of the cells of the proembryo. This structure in *Pinus* and in many other genera typically produces, by cleavage of the two lowermost cell tiers, four separate competitive embryos; the rosette cells in *Pinus* may also proliferate into small and abortive embryos.

The chapter concludes with a comparison between *Pinus* and other conifers with reference to (a) the male gametophyte, (b) types of pollination mechanisms, (c) the early development of the proembryo, and (d) simple and cleavage polyembryony.

References

Aase, H. C.
  1915. Vascular anatomy of the megasporophylls of conifers. *Bot. Gaz.* 60:277–313.
Abbema, T. van.
  1934. Das Transfusiongewebe in den Blättern der Cycadinae, Ginkgoinae und Coniferen. *Trav. Bot. Neerl.* 31:310–390.
Allen, G. S.
  1946a. Embryogeny and development of the apical meristems of *Pseudotsuga*. I. Fertilization and early embryogeny. *Amer. Jour. Bot.* 33:666–677.

  1946b. The origin of the microsporangium of *Pseudotsuga*. *Bull. Torrey Bot. Club* 73: 547–556.
  1963. Origin and development of the ovule in Douglas Fir. *Forest Sci.* 9:386–393.
Allen, G. S., and J. N. Owens
  1972. *The Life History of Douglas Fir.* Information Canada, Ottawa, Canada.
Arnold, C. A.
  1953. Origin and relationships of the cycads. *Phytomorphology* 3:51–65.

Brown, H. P., A. J. Panshin, and C. C. Forsaith
 1949. *Textbook of Wood Technology*, Vol. I. *Structure, Identification, Defects, and Uses of the Commercial Woods of the United States.* McGraw-Hill, New York.

Buchholz, J. T.
 1918. Suspensor and early embryo of *Pinus. Bot. Gaz.* 66:185–228.
 1926. Origin of cleavage polyembryony in conifers. *Bot. Gaz.* 81:55–71.
 1931. The pine embryo and the embryos of related genera. *Trans. Illinois Acad. Sci.* 23:117–125.
 1939. The embryogeny of *Sequoia sempervirens*, with a comparison of the Sequoias. *Amer. Jour. Bot.* 26:248–257.

Burlingame, L. L.
 1913. The morphology of *Araucaria brasiliensis*. I. The staminate cone and male gametophyte. *Bot. Gaz.* 55:97–114.

Butts, D. and J. T. Buchholz
 1940. Cotyledon number in conifers. *Trans. Illinois Acad. Sci.* 33:58–62.

Campbell, D. H.
 1940. *The Evolution of the Land Plants.* Stanford University Press, Stanford, California.

Chamberlain, C. J.
 1935. *Gymnosperms. Structure and Evolution.* University of Chicago Press, Chicago.

Chaney, R. W.
 1950. A revision of fossil *Sequoia* and *Taxodium* in western North America based on the recent discovery of *Metasequoia. Trans. Amer. Phil. Soc.* 40:172–239.

Chowdhury, C. R.
 1962. The embryogeny of conifers: a review. *Phytomorphology* 12:313–338.

Christiansen, H.
 1969. On the pollen grain and the fertilization mechanism of *Pseudotsuga menziesii* (Mirbel) Franco. var. *viridis* Schwer. *Silvae Genet.* 18:97–104.

Coulter, J. M., and C. J. Chamberlain
 1917. *Morphology of Gymnosperms.* University of Chicago Press, Chicago.

Currey, D. R.
 1965. An ancient bristlecone pine stand in eastern Nevada. *Ecology* 46:564–566.

Doak, C. C.
 1935. Evolution of foliar types, dwarf shoots, and cone scales in *Pinus. Illinois Biol. Monogr.* 13:1–106.

Doyle, J.
 1945. Developmental lines in pollination mechanisms in the Coniferales. *Sci. Proc. Roy. Dublin Soc.* 24:43–62.

Doyle, J., and M. Brennan, S. J.
 1972. Cleavage polyembryony in conifers and taxads—A Survey. II. Cupressaceae, Pinaceae, and conclusions. *Sci. Proc. Roy. Dublin Soc.* 4(A):137–158.

Doyle, J., and M. O'Leary
 1935. Pollination in *Pinus. Sci. Proc. Roy. Dublin Soc.* 21:181–190.

Eames, A. J.
 1913. The morphology of *Agathis australis. Ann. Bot.* 27:1–38.

Erspamer, J. L.
 1952. Ontogeny and morphology of the microsporangium in certain genera of the Coniferales. Ph.D. Diss., University of California, Berkeley.

Esau, K.
 1965. *Plant Anatomy,* Ed. 2. Wiley, New York.

Fagerlind, F.
 1961. The initiation and early development of the sporangium in vascular plants. *Svensk Bot. Tidskr.* 55:299–312.

Ferguson, C. W.
 1968. Bristlecone pine: Science and Esthetics. *Science* 159:839–846.

Ferguson, M. C.
 1904. Contributions to the knowledge of the life history of *Pinus* with special reference to sporogenesis, the development of the gametophytes, and fertilization. *Proc. Wash. Acad. Sci.* 6:1–202.

Florin, R.
 1931. Untersuchungen zur Stammesgeschichte der Coniferales und Cordaitales. *Svenska Vetensk. Akad. Handl.* Ser. 5. 10:1–588.
 1944. Die Koniferon des Oberkarbons und des unteren Perms. *Paleontographica* 85(B):365–654.
 1950. Upper Carboniferous and Lower Permian Conifers. *Bot. Rev.* 16:258–282.

1951. Evolution in Cordaites and Conifers. *Acta Horti Bergiani* 15:285–388.

1963. The distribution of conifer and taxad genera in time and space. *Acta Horti Bergiani* 20:121–312.

Gifford, E. M., Jr., and N. T. Mirov

1960. Initiation and ontogeny of the ovulate strobilus in Ponderosa Pine. *Forest Sci.* 6:19–25.

Gifford, E. M., Jr., and G. E. Corson, Jr.

1971. The shoot apex in seed plants. *Bot. Rev.* 37:143–229.

Griffith, M. M.

1957. Foliar ontogeny in *Podocarpus macrophyllus*, with special reference to transfusion tissue. *Amer. Jour. Bot.* 44:705–715.

1971. Transfusion tissue in leaves of *Cephalotaxus*. *Phytomorphology* 21:86–92.

Haupt, A. W.

1941. Oogenesis and fertilization in *Pinus lambertiana* and *P. monophylla*. *Bot. Gaz.* 102:482–498.

1953. *Plant Morphology*. McGraw-Hill, New York.

Jeffrey, E. C.

1917. *The Anatomy of Woody Plants*. University of Chicago Press, Chicago.

Johansen, D. A.

1950. *Plant Embryology*. Chronica Botanica, Waltham, Mass.

Koehne, E.

1893. *Deutsche Dendrologie. Kurze Beschreibung der in Deutschland im Freien aushaltenden Nadel—und Laubholzgewächse*. F. Enke, Stuttgart.

Konar, R. N.

1960. The morphology and embryology of *Pinus roxburghii* Sar. with a comparison with *Pinus wallichiana* Jack. *Phytomorphology* 10:305–319.

Konar, R. N., and S. Ramchandani

1958. The morphology and embryology of *Pinus wallichiana*. *Phytomorphology* 8:328–346.

Laubenfels, D. J. de

1953. The external morphology of coniferous leaves. *Phytomorphology* 3:1–20.

1959. Parasitic conifer found in New Caledonia. *Science* 130:97.

Lemoine-Sebastian, C.

1968. La vascularisation du complexe bractée-écaille chez les Taxodiacées. *Trav. Lab. Forest. Toulouse* 7. Article 1, 1968.

1969. Vascularisation du complexe bractée-écaille dans le cone femelle des Cupressacées. *Bot. Rhedonica* Ser. 7. pp. 3–27.

Lindsay, G.

1969. The ancient bristlecone pines. *Pac. Discovery* 22:1–8.

Maheshwari, P., and H. Singh

1967. The female gametophyte of gymnosperms. *Biol. Rev.* 42:88–130.

Maheshwari, P., and V. Vasil

1961. The stomata of *Gnetum*. *Ann. Bot.* n.s. 25:313–319.

Martens, P.

1971. *Les Gnétophytes*. (Handbuch d. Pflanzenanatomie, Band 12, Teil 2.) Gebrüder Borntraeger, Berlin-Nikolassee.

Molisch, H.

1938. *The Longevity of Plants*. English Translation by E. H. Fulling. New York.

Napp-Zinn, K.

1966. *Anatomie des Blattes. I. Blattanatomie der Gymnospermen*. (Handbuch d. Pflanzenanatomie, Band 8, Teil 1.) Gebrüder Borntraeger, Berlin-Nikolassee.

Owens, J. N., and R. P. Pharis

1967. Initiation and ontogeny of the microsporangiate cone in *Cupressus arizonica* in response to gibberellin. *Amer. Jour. Bot.* 54:1260–1272.

Pant, D. D., and B. Mehra

1964. Development of stomata in leaves of three species of *Cycas* and *Ginkgo biloba*. *Jour. Linn. Soc. London Bot.* 58:491–497.

Pilger, R.

1926. Coniferae. Pp. 164–166 *in* Engler and Prantl (eds.), *Die natürliche Pflanzenfamilien*, Ed. 2, Vol. 13.

Sacher, J. A.

1953. Structure and histogenesis of the buds of *Pinus lambertiana*. Ph.D. Diss., University of California, Berkeley.

1954. Structure and seasonal activity of the shoot apices of *Pinus lambertiana* and *Pinus ponderosa*. *Amer. Jour. Bot.* 41:749–759.

Sacher, J. A. (*continued*)
1955a. Cataphyll ontogeny in *Pinus lambertiana. Amer. Jour. Bot.* 42:82–91.
1955b. Dwarf shoot ontogeny in *Pinus lambertiana. Amer. Jour. Bot.* 42:784–792.
Shaw, G. R.
1914. *The Genus Pinus.* (Arnold Arboretum Publication No. 5.) Riverside Press, Cambridge, Mass.
Shobe, W. R., and N. R. Lersten
1967. A technique for clearing and staining gymnosperm leaves. *Bot. Gaz.* 128: 150–152.
Singh, H.
1961. The life history and systematic position of *Cephalotaxus drupacea* Sieb. et Zucc. *Phytomorphology* 11:153–197.
Singh, H., and J. Chatterjee
1963. A contribution to the life history of *Cryptomeria japonica* D. Don. *Phytomorphology* 13:429–445.
Sporne, K. R.
1965. *The Morphology of Gymnosperms.* Hutchinson University Library, London.
Spurr, A. R.
1949. Histogenesis and organization of the embryo in *Pinus strobus. Amer. Jour. Bot.* 36:629–641.

Sterling, C.
1948. Proembryo and early embryogeny in *Taxus cuspidata. Bull. Torrey Bot. Club* 75:469–485.
1949. Some features in the morphology of *Metasequoia. Amer. Jour. Bot.* 36:461–471.
1963. Structure of the male gametophyte in gymnosperms. *Biol. Rev.* 38:167–203.

Tison, A.
1913. Sur la persistance de la nervation dichotomique chez les Conifères. *Bull. Soc. Linn. Normandie* 6e Ser. 4:31–46.

Worsdell, W. C.
1897. On "transfusion tissue:" its origin and function in the leaves of gymnospermous plants. *Trans. Linn. Soc. London Bot.* Ser. II. 5:301–319.
1900. The structure of the female "flower" in Coniferae. *Ann. Bot.* 14:39–82.

Zimmermann, W.
1959. *Die Phylogenie der Pflanzen,* Ed. 2. Gustav Fischer, Stuttgart.

18

The Gnetopsida

The subject of this chapter is the morphology of a small group of seed plants represented by three living genera, *Ephedra, Welwitschia,* and *Gnetum.* The distinctive organography and anatomy of the sporophyte and the many peculiar features of the reproductive cycle of these genera have attracted the attention of morphologists for more than a century and have resulted in an extremely voluminous literature. Very recently, Martens (1971) has brought together in a single volume monographic treatments of each of the three genera of the gnetophytes and has examined critically the many conflicting theories that have been advanced to explain their interrelationships and their affinities with other groups of living seed plants. Martens' comprehensive treatise has

provided an invaluable source of information and orientation in the preparation of the necessarily brief account given here.

Throughout much of the literature on the morphology of the gnetophytes, repeated efforts have been made to demonstrate that these plants form a "connecting link" between gymnosperms and angiosperms (Arber and Parkin, 1908). Among the "angiospermic" features usually mentioned in defense of this idea are (1) the compound nature of *both* the microsporangiate and megasporangiate strobili, which have been interpreted as "inflorescences" by many botanists, and (2) the presence of vessels in the xylem, a feature that has been regarded as a major divergence from the vessel-less wood characteristic of the cycads,

Ginkgo biloba, and all members of the Coniferales. The presumed isolation of the
gnetophytes from all living and extinct
gymnosperms is clearly emphasized in
Pulle's (1938) classification of the major
groups of seed plants, in which *Ephedra,
Welwitschia,* and *Gnetum* are segregated
under the subdivision "Chlamydospermae,"
which is placed *between* the "Gymnospermae" and "Angiospermae." The word
Chlamydospermae (from the Greek words
meaning seeds with an envelope or "cloak")
refers to the so-called "outer integument"
of the ovule, which has been interpreted as
a pair of fused appendages or "bracteoles"
(see Fig. 18-14, B, outer envelope). In this
connection, Martens (1971, p. 259) astutely
remarks that an ovule provided with an
accessory envelope is not "naked" in the
sense employed to characterize the ovules
of gymnospermous plants. Some authors
have even compared the external envelope
of the ovule of the gnetophytes to the
"ovary" of the angiosperms!

Aside from the remains of pollen grains,
there is no evidence at present from the
fossil record about the age or phylogenetic
history of the gnetophytes. Therefore, all
classifications are, to a large degree, artificial and simply reflect our present ignorance of the true affinities of living gnetalean
plants (see Nemejc, 1967). In the past,
Ephedra, Welwitschia, and *Gnetum* were
placed in the family Gnetaceae under the
order *Gnetales.* The tendency now is to
split this old order into three orders, namely,
Ephedrales, Welwitschiales, and Gnetales,
each consisting of a single family and a single
genus (see Eames, 1952, and Martens, 1971).
We have adopted this systematic treatment
in the present text and have grouped the
three orders, for convenience, under a
single class, the Gnetopsida.

Because of the considerable botanical
interest of the gnetophytes, brief descriptions of their habit, geographical distribution, vegetative organography, and anatomy
will first be given. This will be followed by a
concise account of the main steps in the
reproductive cycle of *Ephedra.* This genus
was selected because it is quite conifer-like
in many details of its morphology and because its gametophytes and embryogeny are
now fairly well understood. Furthermore,
several species of *Ephedra* are frequently
grown in the northern hemisphere as cultivated plants in greenhouses and botanical
gardens, and thus provide living material
for classroom study. *Gnetum* — and particularly *Welwitschia* — are much less readily
available and many aspects of their reproductive cycles are either imperfectly understood
or subject to varied interpretations. At the
close of the chapter an effort will be made
to point out certain of the most striking
differences between these genera and
Ephedra.

Geographical Distribution and Habit

The genus *Ephedra,* consisting of about 35
species, is confined to cool, usually arid
regions in both the Eastern and Western
Hemispheres (Fig. 18-1). In the New World,
Ephedra is restricted to western North
America, parts of Mexico, and a wide area
in South America. According to Cutler's
(1939) monograph, the 16 species found
in the United States occur in dry or desert
areas of California, Nevada, Utah, Arizona,
and New Mexico. Most species of *Ephedra*
are profusely branched shrubs (Fig. 18-2),
although a few are scandent, and one species, *Ephedra triandra,* native to Brazil,
Uruguay, and Argentina, is a small tree.

In marked contrast, *Gnetum* inhabits tropical rain forests in parts of Asia, northern
South America, Africa, and certain Pacific
islands between Australia and Asia (Fig.
18-1). Most of the 30 or more described
species of *Gnetum* are lianas that climb high

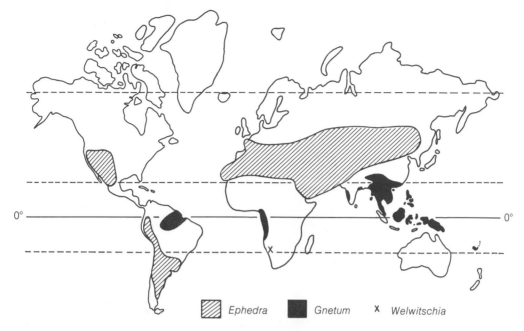

Figure 18-1
Present distribution of the three genera of the gnetophytes. See text for further explanation. [Based on Hutchinson, *Kew Bulletin of Miscellaneous Information*, No. 2, 1924. By permission of Her Majesty's Stationery Office.]

Figure 18-2
Ephedra sp. photographed near Monitor Pass, California. Note the large size of the basal stems of this shrubby species. [Courtesy Dr. T. E. Weier.]

Figure 18-3
Welwitschia mirabilis, young sporophyte showing the two permanent foliage leaves. The specimen has been grown in a section of pipe to provide space for the development of the long tap root. [Courtesy of Dr. T. E. Weier.]

Figure 18-4
Welwitschia mirabilis. Old microsporangiate specimen growing in a desert area near Brandberg in southwestern Africa. Each of the two huge leaves has become longitudinally split into a series of contorted and buckled "segments." Note the numerous clusters of strobili above the leaf axils at the periphery of the woody "crown." [Courtesy of Dr. R. J. Rodin.]

into the crowns of various trees in the rain forest; one species, *Gnetum gnemon,* becomes a small tree (Maheshwari and Vasil, 1961).

The most bizarre and geographically restricted gnetophyte is the African genus *Welwitschia,* which consists of the single species, *Welwitschia mirabilis* (Fig. 18-1). The specific epithet "mirabilis" is quite appropriate because the adult sporophyte, in habit and general organography, is unlike that of any known plant on the earth (Figs. 18-3, 18-4). The exposed portion of the sporophyte consists of a massive, woody, concave "disc" or "crown" that bears two huge strap-shaped leaves. These are the only *permanent* photosynthetic organs of the plant and they become split and frayed in old plants and extend in a twisted and contorted manner along the surface of the ground. The leaves of old specimens may reach a length of 9 feet and a width of 3 feet! Martens (1971), the leading student of the genus, characterizes *Welwitschia* as "an extraordinary plant monster" and remarks that the age of certain individual plants may exceed 1,000 years.

Welwitschia is dioecious, and the microsporangiate and megasporangiate cones are borne terminally on ramified branch systems. As Martens (1971, p. 172) emphasizes, these strobiliferous shoots represent the *only authentic branches* of this amazing plant (Fig. 18-4). According to Pearson (1929), the earliest fertile branches arise from a crescentic ridge situated above the axil of each of the foliage leaves. As the crown continues to expand in diameter, a succession of new ridges, each potentially capable of forming fertile shoots, arises from the center outward between the first ridge and the leaf bases.

Welwitschia was discovered in 1860 in Angola by Dr. Frederic Welwitsch, in whose honor the plant was named by the British botanist J. D. Hooker (1863). Since then, the range of *Welwitschia* has been discovered

to be of considerable extent. Rodin (1953a) found the plant growing in southwestern Africa including the Kaokoveld, a semidesert region about 100 miles inland from the coast. In most of the desert areas in which *Welwitschia* grows, the rainfall is meager and perhaps its principal source of water is fog or heavy dew, which may be absorbed directly by the leaves. Also, the very long taproot characteristic of old plants may be able to absorb subsurface water. In any event, the survival of *Welwitschia* under extremely arid conditions poses an interesting problem in plant physiology that needs more study.

Vegetative Organography and Anatomy

In this section our discussion will be limited to only a few of the unique and well-investigated aspects of the vegetative sporophyte of the gnetophytes. Despite the voluminous literature, which has been reviewed by Pearson (1929), Chamberlain (1935), and Martens (1971), it is evident that much still remains to be done before an adequate treatment of the sporophytes of gnetalean plants as a whole can be accomplished. *Gnetum* and *Welwitschia* are both extremely complex anatomically and require further ontogenetic study.

The Leaf

The foliage leaves of *Ephedra, Gnetum,* and *Welwitschia* are strikingly different in form and venation and provide morphological characters that are definitive for each of the three genera.

In *Ephedra,* the phyllotaxis varies from decussate to whorled and the leaves are basally joined by a membranous commissure to form a more-or-less conspicuous sheath at each node (Fig. 18-5, B). Throughout most of the literature on *Ephedra,* the leaves are characterized as "reduced" or "scale-

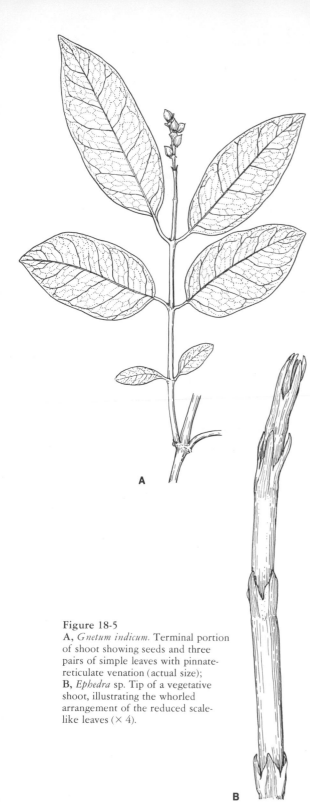

Figure 18-5
A, *Gnetum indicum.* Terminal portion of shoot showing seeds and three pairs of simple leaves with pinnate-reticulate venation (actual size); B, *Ephedra* sp. Tip of a vegetative shoot, illustrating the whorled arrangement of the reduced scale-like leaves (× 4).

like" in form. This generalization, however, needs to be qualified because in several species (e.g., *Ephedra foliata* and *Ephedra altissima*) the lamina is slender and needle-like, and, according to Stapf (1889), may attain a length of 3 cm and a width of 1–1.5 mm (Fig. 18-6, A). In this connection, it should be emphasized that there is considerable variation in the proportional development of the sheath and lamina in the successive leaves of a single shoot. In *Ephedra foliata,* for example, the leaves in the middle or upper part of a shoot may have well-developed laminae whereas the basal leaves of the same shoot tend to be much smaller, with laminae only 3–4 mm long (compare Fig. 18-6, A and B). In this and other species, there is obviously no sharp boundary between "well-developed" and "reduced" scale-like leaves.

It is widely stated in the literature that the leaf in *Ephedra* is vascularized by a pair of traces that neither branch nor anastomose in their course through the sheath and lamina. Although this *bifasciculate pattern of venation* seems to be consistent for a number of species (Fig. 18-6, A, B), an interesting divergence has recently been described in detail with reference to *Ephedra chilensis* (Foster, 1972). The distinctive feature of the foliar venation in this species is the formation of a *midvein* in certain leaves (Fig. 18-7). One of the most puzzling aspects of this *trifasciculate type of venation* is its sporadic and unpredictable occurrence in the leaves of a single shoot. Most commonly, the members of a leaf pair or whorl are two-veined but numerous examples were found in which one member of a leaf pair is two-veined while the other leaf is vascularized by three veins (Fig. 18-8). Comparable variation in venation pattern may also occur among the members of a three-leaved whorl (Fig. 18-9). In a relatively small number of instances, *both members* of a leaf pair are three-veined. Three-veined leaves have also been observed in *Ephedra fragilis,* where they seem to be even more common than in *E. chilensis.* The morphological and phylogenetic significance of trifasciculate venation in *Ephedra* is an open question at present and requires for its solution a comprehensive survey of the venation patterns in a wide range of species.

The foliage leaf of *Gnetum* presents a complete morphological contrast to the leaf of *Ephedra.* In *Gnetum* the foliage leaves are arranged in basally connate pairs and in their form and pinnate-reticulate venation bear a striking resemblance to the simple leaves of many dicotyledons (Fig. 18-5, A). Very little comparative information is available regarding the number of leaf traces that vascularize the leaf in the various species of *Gnetum.* Rodin and Paliwal (1970) found that usually an *odd number of traces* (5–7, or sometimes 8) extend into the leaf of *Gnetum ula,* but whether this type of nodal anatomy characterizes the genus as a whole is apparently unknown. According to Rodin (1967), the 5–7 leaf traces in *Gnetum gnemon* extend into the midrib of the young leaf as a series of longitudinal bundles that, at successive levels, dichotomize and form *pairs of secondary* veins that curve toward the margins of the lamina. In the submarginal region, each secondary vein bifurcates and the derivative branches unite, creating a series of coarse "meshes" at each side of the midrib. Within each mesh, a more delicate reticulum of veinlets is later differentiated (for additional information and illustrations of venation patterns in several other species of *Gnetum,* see Rodin, 1966).

The vegetative organography of *Welwitschia,* as we have already noted, is without parallel among all living vascular plants. Except for the two short-lived cotyledons, produced during the seedling phase, the *permanent* photosynthetic organs are represented by a single pair of enormous strap-shaped leaves, which continue to grow

indefinitely in length through the activity of an intercalary meristem located at the base of each appendage (Fig. 18-3). According to Martens (1971), the young leaf of *Welwitschia* is at first vascularized by two and then four strands, and additional lateral bundles continue to differentiate as the leaf increases in width basally. A well-developed leaf in *Welwitschia* is traversed by numerous "parallel" longitudinal veins that become

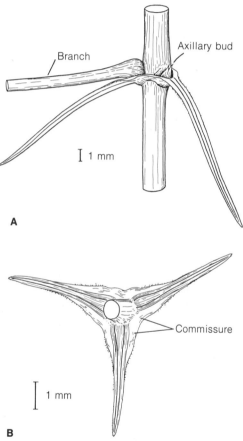

Figure 18-6
Variation in size and morphology of the leaves in *Ephedra foliata.* **A,** leaves with well-developed laminae and weakly connate sheaths; the leaf at the right is traversed by two freely terminating veins; **B,** whorl of three scale-leaves, detached from shoot and drawn from lower surface. Each leaf is vascularized by a pair of veins and is basally connected, by a prominent commissure, with adjacent leaves of the whorl. [From Foster, *Jour. Arnold Arboretum* 53:364, 1972.]

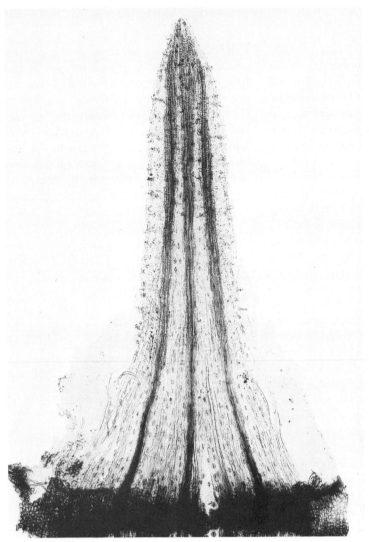

Figure 18-7
The three veined pattern of leaf vasculature in *Ephedra chilensis*.

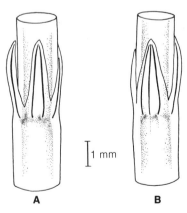

Figure 18-9
A, B, *Ephedra chilensis*. Variation in venation pattern among leaves of the same whorl. One leaf (B) is three-veined while each of the other leaves is two-veined (A). [From Foster, *Jour. Arnold Arboretum* 53:364, 1972.]

1 mm

A   B

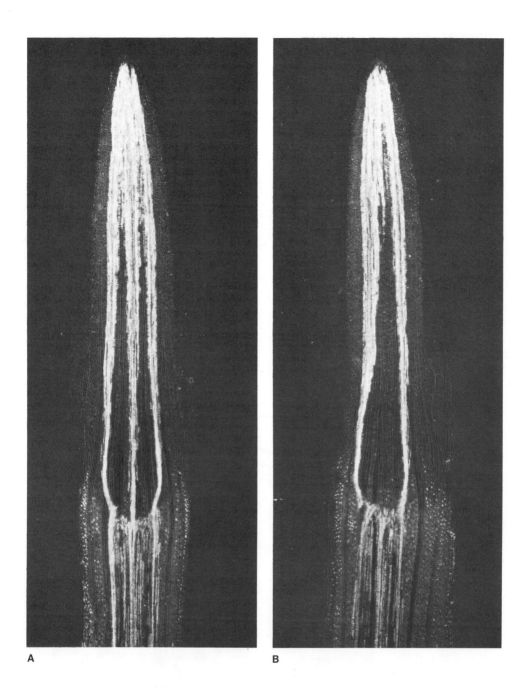

A

B

Figure 18-8
A pair of cleared stained leaves of *Ephedra chilensis.* One leaf (A) is three-veined; the other (B), two-veined. In each leaf, the vein tips are embedded in a mass of transfusion tracheids. [From Foster, *Jour. Arnold Arboretum* 53:364, 1972.]

interconnected by smaller obliquely oriented veins. These may anastomose in various ways to form irregular areoles (i.e., meshes) or may fuse in pairs and terminate blindly in the mesophyll. This peculiar and highly distinctive type of venation also occurs in the blades of the cotyledons and seems unique as compared with the venation of *Ephedra, Gnetum,* and other gymnosperms (see Rodin, 1953b, 1958a, 1958b).

The Shoot Apex

Comparative studies have revealed a remarkable similarity in the *basic structure* of the shoot apex of the gnetophytes: in all three genera a more-or-less discrete surface layer, or *tunica,* is present, the continuity of which is interrupted only by occasional periclinal divisions. The investigations of Gifford (1943) on *Ephedra altissima* and Seeliger's (1954) study on *Ephedra fragilis* var. *campylopoda* emphasized the infrequency of periclinal divisions in the surface layer of the apex and its tunica-like structure and growth (Fig. 18-10). Johnson (1950), in a study of 85 apices of *Gnetum gnemon* found "no evidence of periclinal divisions in the tunica clothing the summit of the dome." The aberrant behavior of the shoot apex in *Welwitschia* presents a striking contrast to the "open system" of growth characteristic of *Ephedra* and *Gnetum.* According to the detailed study made by Martens and Waterkeyn (1963), the shoot apex of the seedling of *Welwitschia* first initiates a pair of primordia which later develop into the two permanent leaves of the sporophyte. But contrary to the classical interpretation, an *additional pair* of appendages — the so-called

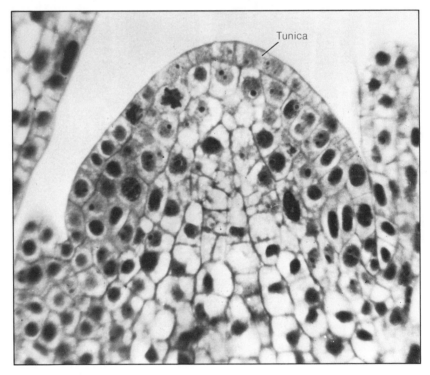

Figure 18-10
Median longisection of the shoot apex of *Ephedra altissima.* Note the single, clearly defined tunica.

"scaly bodies"—are formed by the apex in a plane at right angles to the plane of the foliage-leaf primordia. Martens and Waterkeyn reject the prevalent view according to which the "scaly bodies" are cotyledonary buds, and interpret them as a *third pair of foliar appendages*. Soon after the formation of the scaly bodies, the tissue of the shoot apex loses its meristematic character and finally degenerates. This precocious death of the terminal meristem of the *young sporophyte* of *Welwitschia* is unique among vascular plants and is an additional example of the bizarre morphology of this extraordinary plant.

The phylogenetic significance of the angiospermic organization of the shoot apex of the gnetophytes is, of course, problematical and very possibly is the result of evolutionary convergence. Martens (1971, p. 252) believes that the structural similarity of the apices of the three genera of gnetophytes supports the evolutionary position of this group at a level intermediate between those of the conifers and angiosperms.

Vessels

As stated in the beginning of this chapter, the presence of vessels in the Gnetopsida distinguishes them from other living gymnosperms and has often been used as an argument for their presumed evolutionary relationship to the angiosperms. But it has now been clearly established that the *method* of origin of the perforations in the vessel members of the gnetophytes differs from that of all other vascular plants including the angiosperms. In the angiosperms, as well as in certain species of *Pteridium* and *Selaginella,* the initial step in vessel evolution was the loss of the membranes from the *transversely elongated* bordered pits situated at each sloping end of a tracheid-like cell. Further elaboration of these slit-like perforations led to the development of vessel members with well-defined scalariform perforation plates. The most advanced vessel members in angiosperms possess large circular or oval simple perforations which originated by the elimination (phylogenetically and ontogenetically) of the bars between the slit-like openings. In contrast, as Thompson (1918) has shown, the initial step in vessel evolution in the Gnetopsida began with the loss of membranes from a series of *circular* bordered pits located near the ends of long tracheid-like cells. In *Ephedra,* there are transitional conditions between intact bordered pits and *bordered foraminate perforations,* i.e., there are clear transition forms between typical tracheids and vessel members (Fig. 18-11). The vessel members of *Gnetum* commonly possess large circular or elliptical simple perforations and thus markedly resemble the specialized vessel members of many angiosperms. Thompson (1918) found, however, that the *Gnetum* type of vessel perforation is the result of the further enlargement of a series of circular perforations and the elimination of the portions of the end wall between them.

Thus in the initial steps of their origin as well as in their subsequent specialization, the vessels of the gnetophytes have evolved differently and independently from those in the angiosperms. As Bailey (1953) has remarked in this connection, "although the highly evolved vessels of *Gnetum* resemble those of comparably specialized vessels of angiosperms, the similarity cannot be used as an indication of close relationship, but provides a very significant illustration of convergent evolution in plants."

The Reproductive Cycle in *Ephedra*

By way of introduction it will be essential to comment briefly on the confusing "angiosperm-centered" terminology commonly

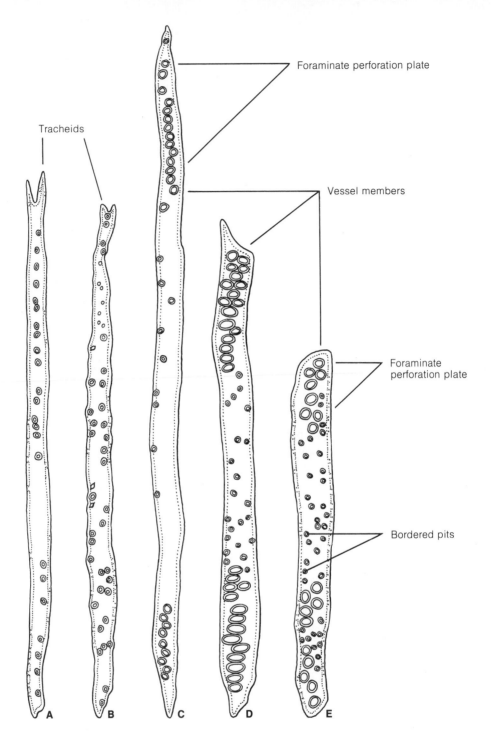

Tracheids

Foraminate perforation plate

Vessel members

Foraminate perforation plate

Bordered pits

A B C D E

Figure 18-11
Tracheary elements from the secondary xylem of *Ephedra californica.* **A, B,** tracheids with numerous circular bordered pits; **C–E,** vessel members with foraminate perforation plates on end walls. [From *Plant Anatomy* by K. Esau. Wiley, New York, 1953.]

used in describing the reproductive structures of the gnetophytes. To many writers the strobili of *Ephedra, Gnetum,* and *Welwitschia* are directly comparable with angiospermic inflorescences, and the parts of the gnetalean "flower" are very frequently designated by such terms as "perianth," "stamen," "anther," "column," etc. This kind of nomenclature is particularly confusing and misleading for *Ephedra,* which is notable for

the conifer-like morphology of its gametophytes and embryo; these structures are commonly described in the literature with the aid of "gymnosperm-centered" terminology! To eliminate the implication of homology between sporogenous structures of the Gnetopsida and those of angiosperms, the terms and general interpretations proposed by Eames (1952) will be adopted in the following résumé of the life cycle of *Ephedra.*

Figure 18-12
Ephedra chilensis. Microsporangiate (left) and megasporangiate (right) strobili. Note protrusion of tubular integument of the ovules of the megasporangiate strobili.

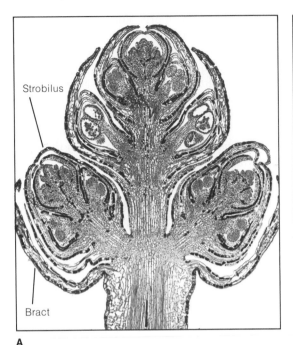
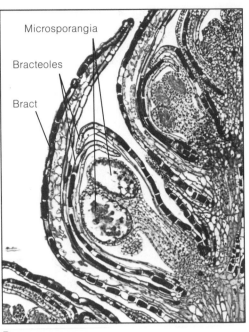

A

B

Figure 18-13
Structure of microsporangiate strobili in *Ephedra chilensis*. **A,** longisection of tip of reproductive shoot (each of the microsporangiate strobili is situated in the axil of a bract); **B,** enlargement of portion of **A** showing a microsporangiate shoot which consists of an axis with apical microsporangia enclosed by a pair of fused bracteoles.

The Strobili

Most species of *Ephedra* are strictly dioecious and both the microsporangiate and ovulate cones are compound in structure; i.e., the cone axis bears pairs of bracts which subtend either microsporangiate or ovuliferous structures (Fig. 18-12). In certain monoecious species (e.g., *Ephedra campylopoda*) some of the strobili are *bisporangiate* with the microsporangiate structures developed in the axils of the lower bracts and the ovules located in the terminal part of the same cone.

The microsporangiate strobilus consists of a number of pairs of bracts; the lowest pairs are usually sterile, whereas each of the other pairs subtend microsporangiate shoots (Fig. 18-13, A–B). The microsporangiate shoot, or "microsporangiophore," consists of an axis bearing a pair of fused bracteoles at its base, which enclose a terminal group of microsporangia, the number varying with the species (see Fig. 18-13, B). According to Eames (1952) the evidence from ontogeny and vascular anatomy indicates that the microsporangiophore of *Ephedra* is the result of the phylogenetic fusion of a pair of microsporophylls; in certain species, recognized as primitive by taxonomists, the two sporophylls are free and each bears a terminal cluster of four microsporangia.

The ovulate cone of *Ephedra* also consists of an axis bearing decussately arranged pairs of bracts. However, most of the bracts are sterile, and the cones of many species contain only two ovules, one in the axil of each of the upper bracts (Fig. 18-14, A). In some

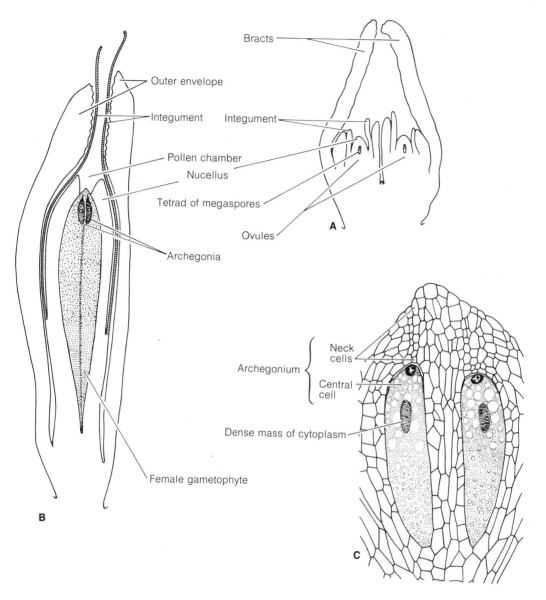

Figure 18-14
The ovule and mature female gametophyte of *Ephedra foliata*. **A**, longisection of megasporangiate strobilus showing two young ovules, each with a linear tetrad of megaspores; **B**, median longisection of an ovule showing the female gametophyte, the conspicuous pollen chamber, and elongated integument; **C**, details of micropylar region of female gametophyte showing structure of archegonia. [From Maheshwari, *Proc. Indian Acad. Sci.* 1:586, 1935.]

species the cones are *uniovulate* and this condition is commonly the result of the abortion of one ovule and the crowding of the other into a false "terminal" position (see Eames, 1952, p. 83 and Fig. 1, E–I).

The nucellus of the ovule of *Ephedra* is enclosed by two envelopes, *each* of which has been regarded as an ovular integument. Particularly notable is the marked elongation of the upper region of the integument, which protrudes from the tip of the ovule as a delicate open tube (Fig. 18-14, B). This *micropylar tube,* as it is frequently called, functions as a receptive organ for the pollen and represents one of the salient characters shared by the ovules of *all* gnetophytes. Description of the morphology of the outer envelope of the ovule of *Ephedra,* however, is highly controversial; Martens (1971, p. 41) has summarized the conflicting opinions that have been advanced with reference to this question. We have adopted the viewpoint that the so-called "outer integument" of the ovule of *Ephedra* represents a pair of connate bracteoles, comparable to the pair of bracteoles of the microsporangiate shoot, whereas the inner envelope represents the only true integument.

Microsporogenesis and the Male Gametophyte

Microsporogenesis has been carefully investigated in several species of *Ephedra* (Land, 1904; Maheshwari, 1935; Singh and Maheshwari, 1962). The sporangial initials are hypodermal and divide periclinally, forming an outer layer of *primary wall cells.* The periclinal division of the layer of primary wall cells yields a single wall layer, which eventually becomes crushed, and the tapetum. During the meiotic division of the microsporocytes, the tapetal cells become multinucleate and finally degenerate. At maturity, the microsporangium contains tetrads of microspores enclosed within a thick-walled epidermis.

The early steps in the ontogeny of the endosporic male gametophyte of *Ephedra* closely parallel the mode of development of the male gametophyte of *Pinus* (see Chapter 17). The first two mitotic divisions yield two lens-shaped *prothallial cells* which begin to degenerate soon after their formation. Then the nucleus of the meristematic, or antheridial, initial divides again, forming the generative and tube nuclei (Fig. 18-15, A–D). The division of the nucleus of the generative cell produces a sterile cell and a spermatogenous cell (Fig. 18-15, E, F). In this five-celled stage in development, the pollen grain is shed from the microsporangium.

The mature pollen grains of *Ephedra* are ellipsoidal in form and the exine is characteristically sculptured into a series of ridges —extending from pole to pole—separated by longitudinal furrows (Fig. 18-16). Because of these definitive structural characters, it has been possible to identify *Ephedra* pollen in the fossil record and to draw structural comparisons between it and the pollen of living species (see the detailed study made by Steeves and Barghoorn, 1959).

Megasporogenesis and the Female Gametophyte

The megasporocyte, prior to meiosis, is a large, conspicuous cell located rather deeply within the nucellus of the ovule (Fig. 18-17, A). According to some investigators (Maheshwari, 1935; Seeliger, 1954; Singh and Maheshwari, 1962), the sunken position of the megasporocyte is due (1) to the formation above it of several layers of parietal cells, derived from the early division of a hypodermal "archesporial cell," and (2) the very active periclinal division of the cells of the nucellar epidermis (Fig. 18-17, B). The meiotic divisions of the megasporocyte result in the formation of four megaspores (Fig. 18-17, C), although in several instances Maheshwari observed a row of three cells. This probably was the result of the failure

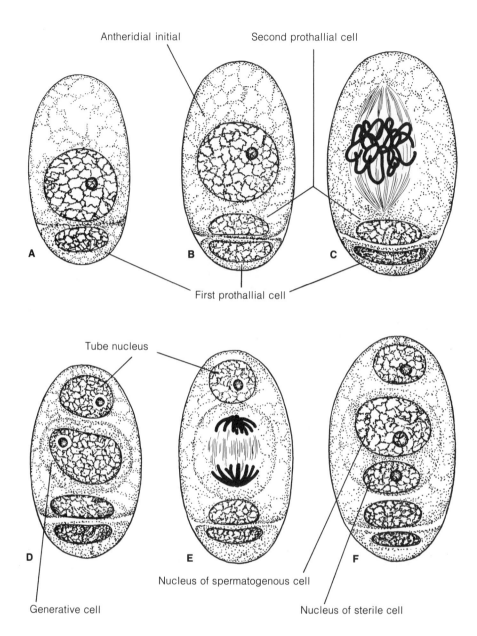

Figure 18-15
Development of male gametophyte of *Ephedra trifurca*. **A**, formation of first prothallial cell;
B, gametophyte with two prothallial cells; **C, D**, formation of tube and generative cells;
E, F, division of nucleus of generative cell and formation of nuclei of sterile and sperma-
togenous cells. [Redrawn from Land, *Bot. Gaz.* 38:1, 1904.]

Figure 18-16
Scanning electron micrograph of a group of pollen grains of *Ephedra* sp. Note the prominent longitudinal (meridional) ridges characteristic of the pollen of this genus; × 1100.

of the upper dyad cell to divide, as is not uncommon in *Pinus* and other gymnosperms. In either case, the lowermost cell in the series enlarges and a series of *free-nuclear divisions* occur in the peripheral cytoplasm which surrounds the large central vacuole (Fig. 18-17, D). Land (1907) found 256 free nuclei in the coenocytic female gametophyte of *Ephedra trifurca* but higher numbers have been reported on other species, 500 in *Ephedra foliata* and about 1,000 in *Ephedra distachya*.

Following the period of free nuclear divisions, anticlinal walls begin to develop centripetally, forming long, tubular uninucleate *alveoli* which remain "open" (i.e., devoid of inner walls) for some time next to the central vacuole. Before the centripetally elongating alveoli meet in the center, they begin to divide by forming periclinally oriented walls (Lehmann-Baerts, 1967). Active cell divisions continue throughout the young gametophyte which soon becomes histologically differentiated into two regions: a lower (chalazal) zone of small, compact, frequently dividing cells, and an upper (micropylar) zone of longer, thinner-walled cells; certain of the superficial cells of the upper zone later function as *archegonial* initials. The number of archegonia formed varies from two to eleven, according to Martens (1971).

The archegonial initial divides periclinally into an outer *primary neck cell* and an inner *central cell*. The most distinctive feature of

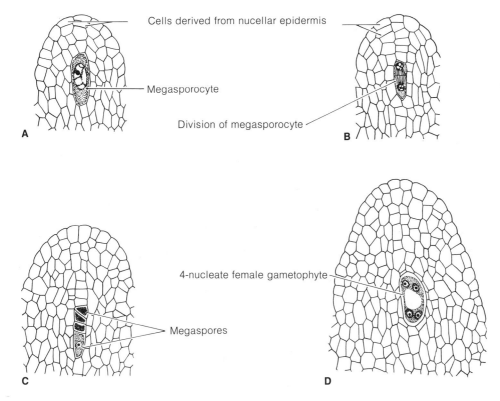

Cells derived from nucellar epidermis

Megasporocyte

Division of megasporocyte

A

B

4-nucleate female gametophyte

Megaspores

C

D

Figure 18-17
Megasporogenesis and the young female gametophyte of *Ephedra foliata*. **A–C**, development of linear tetrad of megaspores; **D**, four-nucleate stage of female gametophyte. [Redrawn from Maheshwari, *Proc. Indian Acad. Sci.* 1:586, 1935.]

archegonial development in *Ephedra* is the formation of a relatively massive neck which is produced by the repeated periclinal and anticlinal divisions of the derivatives of the primary neck cell. At first, the divisions are so regular that 3–5 tiers of cells are produced, each tier composed of a quartet of cells. Subsequent divisions are less regular and the cells of the archegonial neck merge with the adjacent cells of the gametophyte (Fig. 18-14, C). The mature archegonial neck in *Ephedra trifurca* consists of about 30–40 cells and Land (1904) remarks: "of all gymnosperms, *Ephedra* has the longest-necked archegonium." The enlargement of the central cell is followed by the division of its nucleus to form the *ventral-canal-cell nucleus*

and the *egg nucleus*. Sometimes the ventral-canal-cell nucleus appears to degenerate soon after its formation, but in some species it persists and remains intact near the upper part of the archegonium (Lehmann-Baerts, 1967).

Following the first division of the archegonial initial, the contiguous cells of the gametophyte divide, producing a "nutritive jacket" around the entire developing archegonium.

Pollination

Prior to pollination, the nucellar tissue lying directly above the archegonia of the gametophyte begins to break down creating a very

Figure 18-18
Ephedra sp. Basal cluster of three megasporangiate strobili. Note conspicuous "pollination drop" at tip of two of the strobili. See text for further discussion. [Courtesy of Dr. E. G. Cutter.]

conspicuous funnel-shaped *pollen chamber* (Fig. 18-14, B). During pollen-chamber formation, the disintegrating cells, at the summit of the nucellus, produce a liquid, rich in sugar, which fills the micropylar canal of the ovule and, at the time of pollination, exudes as a "pollination drop" from the open end of the exserted integument (Fig. 18-18).

With the exception of *Ephedra campylopoda,* which Porsch (1910, 1916) believed is pollinated by various types of insects that are attracted by the nectar-like pollination drops, the pollen grains of *Ephedra* are normally carried by wind currents to the ovulate strobili. The grains adhere to and "float" on the sticky surface of the pollination drop. As it begins to evaporate, the column of liquid

in the micropylar canal shortens and the pollen grains are pulled inward, into the pollen chamber (Pearson, 1929, p. 106).

Fertilization

According to Land (1907), the interval between pollination and fertilization in *Ephedra trifurca* may be as short as ten hours, which is a remarkable contrast with the more extended interval typical of *Pinus* and a number of other conifers. A unique morphological feature of *Ephedra* is the fact that at the time of pollination, the archegonial end of the female gametophyte is freely exposed at the base of the deep pollen chamber (Fig. 18-14, B). As a result, when the pollen grain germinates, the pollen tube penetrates *gametophytic tissue,* i.e., the tissue of the archegonial neck. This represents a striking contrast to the growth of the pollen tube through the nucellar—i.e., *sporophytic*—tissue of the ovule, characteristic of the conifers.

Following the emergence of the pollen tube, the exine of the pollen grain is shed and the spermatogenous cell divides, forming two male gametes. After reaching the egg, the tip of the pollen tube ruptures and the two sperms, together with the sterile cell and the tube cell, are discharged into the egg cytoplasm.

Although one of the male gametes unites with the egg nucleus and forms a diploid zygote, the behavior of the other sperm is extraordinary in certain species. Khan (1943) working on *Ephedra foliata,* reported that the second male gamete may fuse with the nucleus of the ventral canal cell. Although an embryo does not result from this fusion, Khan observed, in one specimen, two nuclei that he considered were formed by the division of the fertilized ventral canal cell. The union of both sperms in *Ephedra* with nuclei in the archegonium recalls the process of "double fertilization" typical of the reproductive cycle of the angiosperms (see Chapter 20). But Khan concluded that "the type of double fertilization seen in *Ephedra* may simply be the natural outcome of a tendency towards fusion between any two nuclei of opposite sexual potencies that happen to lie free in a common chamber."

Embryogeny

There is, in both *Ephedra trifurca* and *Ephedra foliata,* a process of free nuclear division, beginning with the division of the zygote. Each of the eight free diploid nuclei that are produced first becomes surrounded by a densely staining sheath of cytoplasm; later, each cell forms an external cellulosic wall

(Fig. 18-19, A). As each of these cells may *independently* develop into an embryo, *Ephedra* exhibits a distinctive and precocious type of cleavage polyembryony. In a functional sense, the free proembryonal cells in *Ephedra* are comparable to the separate development of the four cells of the apical tier of the proembryo of *Pinus* (Chapter 17). From an ontogenetic standpoint, polyembryony in *Ephedra* has been "pushed back" to the free nuclear stage in embryogeny. According to Land (1907), it is usually the lower proembryonal cells that successfully begin to develop into embryos. Khan observed in one of his preparations six embryos in different stages of development (Fig. 18-19, B).

In *E. foliata* a proembryonal cell first produces a tubular projection, termed the *suspensor tube,* into which the nucleus descends, where it then divides; the two nuclei become separated by the formation of a transverse wall and the proembryo then consists of a terminal embryonic initial and an elongating suspensor (Fig. 18-19, C). A similar pattern is found in the proembryo of *E. trifurca* except that the nucleus divides before the suspensor tube is initiated and there is a long interval between this nuclear division and the formation of a transverse wall. A transverse division of the embryonic initial, followed by a longitudinal division of the terminal cell, yields a group of three cells from which, by further cell divisions, the embryo proper originates (Fig. 18-19, D). A multicellular *secondary suspensor* differentiates next, forming the suspensor cell, and the distal end of the embryo produces two cotyledons and the shoot apex. In connection with the description of the vegetative shoot apex of *Ephedra* given earlier in this chapter, it is interesting to note that a well-defined *tunica* originates in the apex of the embryo even before the cotyledons develop. Of the several embryos which competitively develop in a single ovule, only one reaches

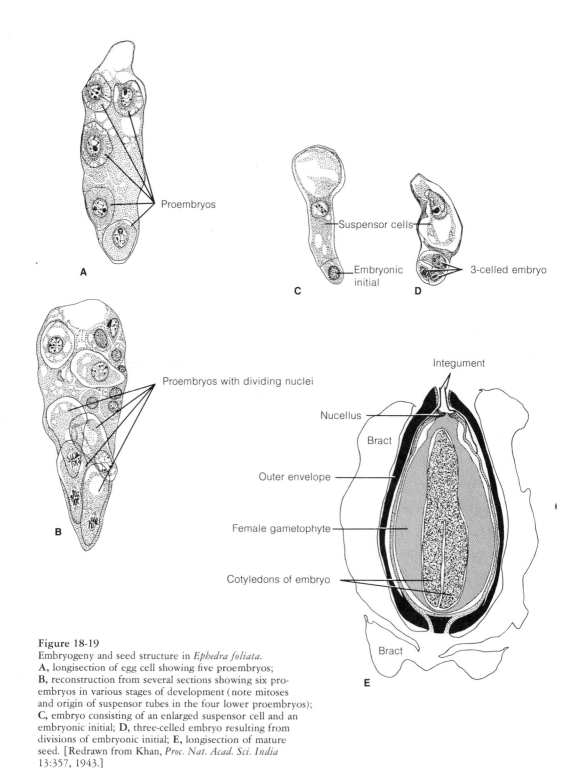

Figure 18-19
Embryogeny and seed structure in *Ephedra foliata*.
A, longisection of egg cell showing five proembryos;
B, reconstruction from several sections showing six proembryos in various stages of development (note mitoses
and origin of suspensor tubes in the four lower proembryos);
C, embryo consisting of an enlarged suspensor cell and an
embryonic initial; **D,** three-celled embryo resulting from
divisions of embryonic initial; **E,** longisection of mature
seed. [Redrawn from Khan, *Proc. Nat. Acad. Sci. India*
13:357, 1943.]

Figure 18-20
Ephedra chilensis. Mature ovulate strobili, showing
thick fleshy bracts.

a fully developed stage in the seed. Khan observed in one ovule 18–19 separate embryos. He interpreted this large number as the result of the combination of simple and cleavage polyembryony.

The general structure of a ripe seed in *E. foliata* is shown in Fig. 18-19, E. The conspicuous embryo, with its two large cotyledons, is embedded within the tissue of the female gametophyte, and the remains of the nucellus are evident as a disorganized sheath of cells. At the micropylar end of the seed the remains of the "true" integument are evident and the entire seed is externally enclosed by the so-called outer integument (= outer envelope), many cells of which develop thick hard walls. As the seed matures

in *E. foliata*, and many other species, the adjacent subtending bracts of the ovulate strobilus become thick and fleshy, forming an additional investment (Figs. 18-19, E, and 18-20). According to Land's (1907) study on *E. trifurca*, there appears to be no resting, or "dormant," period for the seed, which may even begin to germinate within the parent strobilus.

Morphological Comparisons Between *Ephedra, Gnetum,* and *Welwitschia*

The three genera of gnetophytes share the following morphological characters: compound microsporangiate and megasporangiate strobili, an extended micropylar tube formed by the integument of the ovule, vessels in the secondary xylem, shoot apices having a well-defined surface layer (tunica), decussate phyllotaxis, and embryos with two cotyledons. The phylogenetic and taxonomic significance of these points of resemblance, however, must be judged in the light of equally impressive *differences* between the genera with reference to (1) the development and structure of both the male and female gametophytes, (2) the methods of fertilization, and (3) the types of embryogenesis. Let us briefly summarize these differences.

The male gametophyte of *Ephedra* is very much like that of certain conifers (e.g., *Pinus, Pseudotsuga*) and, when mature, consists of two ephemeral prothallial cells, a sterile and tube cell, and two male gametes formed by the division of the spermatogenous cell. As compared with *Ephedra*, the male gametophytes in *Gnetum* and *Welwitschia* appear greatly "reduced" in structure and more difficult to interpret. In both of these genera, only a *single* prothallial cell, a tube cell, and a generative cell are present at the time the pollen is liberated from the microsporangia. When the pollen grain germinates in the pollen chamber of the ovule, a sterile and spermatogenous cell are not formed, and the generative cell *directly* produces, by its division, the two male gametes.

The difference between the development and structure of the female gametophyte in the three genera are extraordinary and have led to many contradictory interpretations. *Ephedra* has a typical *gymnospermous* cellular gametophyte with well-defined archegonia. Moreover, its gametophyte is *monosporic* in method of origin since it begins its development from a *single* haploid megaspore. In contrast, the female gametophyte of both *Gnetum* and *Welwitschia* is devoid of archegonia and is *tetrasporic* in its mode of origin. In each of these two genera, the meiotic divisions in the *megasporocyte* yield *four free haploid nuclei, all* of which contribute, by free nuclear divisions, to the formation of the female gametophyte. The only other known examples of seed plants with tetrasporic female gametophytes are found in certain angiosperms (see Chapter 20).

Although the first phase in the development of the female gametophyte of *Gnetum* and *Welwitschia* is characterized by numermous free nuclear divisions, the subsequent development of the female gametophytes is distinctive for each genus and quite unlike that of *Ephedra*. In *Welwitschia*, groups of nuclei become isolated into "compartments" by wall formation, whereas in *Gnetum* the upper (i.e., micropylar) end of the female gametophyte *remains* in the free nuclear phase and lies above a basal cellular zone.

Correlated with these striking differences, there are equally remarkable differences between the methods of fertilization in the three genera. Fertilization in *Ephedra* occurs as in many gymnosperms: the pollen tube, after penetrating the neck of the archegonium, releases two gametes into the egg cell where one of them unites with the egg-cell nucleus and forms the zygote. In *Gnetum*, one or more of the free nuclei at the micro-

pylar end of the female gametophyte function as *female gametes*. Maheshwari and Vasil (1961, p. 85) state that "both the male cells from a pollen tube can function provided two eggs are present close to the pollen tube." Unlike *Ephedra*, there is apparently no free nuclear period in embryogeny in *Gnetum*.

The method of fertilization in *Welwitschia* is apparently without a parallel among either gymnosperms or angiosperms. According to the classical work of Pearson, a number of the multinucleate cells at the micropylar end of the female gametophyte develop long tubes — termed embryo-sac tubes — which grow *upward* through the nucellus. Many of these tubes meet the downward-growing pollen tubes. At the point of contact between a pollen tube and an embryo-sac tube, the walls of both tubes break down and several female nuclei *enter the pollen tube*, where one of them is fertilized by a male gamete. Thus, unlike any other known seed plant, syngamy takes place in the pollen tube itself rather than in an archegonium or an embryo sac. When the zygote of *Welwitschia* divides, a two-celled proembryo is formed, and as in *Gnetum,* there is no free nuclear phase in embryogenesis.

In conclusion it must be emphasized that the gnetophytes represent, morphologically and phylogenetically, a very paradoxical group of seed plants. On the one hand, as Chamberlain (1935) maintained, the *combination* of certain characters — such as the compound nature of the microsporangiate strobili and the presence of vessels in the xylem — "is sufficient to keep the three genera together." But in opposition to this viewpoint, the wide divergence in the morphology of the vegetative sporophyte of *Ephedra* from that of the two other genera and particularly the remarkable differences between the female gametophytes and methods of fertilization in *Ephedra, Gnetum,* and *Welwitschia*, all suggest that the gnetophytes are a very *heterogeneous assemblage* of plants. This is the viewpoint of Martens (1971), who remarks that he is more struck by the contrasts between the three genera than by the traits they have in common. Many years ago, Pearson (1929) expressed the opinion that "the relationship between the three genera of the Gnetales are perhaps as obscure now as they have been at any time within the last decade, and proof that they are of near affinity is lacking." This conclusion seems equally valid today, despite the great advances in our knowledge of the comparative morphology of the gnetophytes so admirably presented in Marten's (1971) monograph.

References

Arber, E. A., and J. Parkin
  1908. Studies on the evolution of the angiosperms. The relationship of the angiosperms to the Gnetales. *Ann. Bot.* 22: 489–515.
Bailey, I. W.
  1953. Evolution of the tracheary tissue of land plants. *Amer. Jour. Bot.* 40:4–8.
Chamberlain, C. J.
  1935. *Gymnosperms. Structure and Evolution.* University of Chicago Press, Chicago.
Cutler, H. C.
  1939. Monograph of the North American species of the genus *Ephedra. Ann. Missouri Bot. Gard.* 26:373–428.
Eames, A. J.
  1952. Relationships of the Ephedrales. *Phytomorphology* 2:79–100.
Foster, A. S.
  1972. Venation patterns in the leaves of *Ephedra. Jour. Arnold Arboretum* 53:364–378.
Gifford, E. M., Jr.
  1943. The structure and development of the shoot apex of *Ephedra altissima* Desf. *Bull. Torrey Bot. Club* 70:15–25.
Hooker, J. D.
  1863. On *Welwitschia*, a new genus of Gnetaceae. *Trans. Linn. Soc. London* 24:1–48.

Johnson, M. A.
  1950. Growth and development of the shoot of *Gnetum gnemon* L. I. The shoot apex and pith. *Bull. Torrey Bot. Club* 77:354–367.
Khan, R.
  1943. Contributions to the morphology of *Ephedra foliata* Boiss. II. Fertilization and embryogeny. *Proc. Nat. Acad. Sci. India* 13:357–375.
Land, W. J. G.
  1904. Spermatogenesis and öogenesis in *Ephedra trifurca. Bot. Gaz.* 38:1–18.
  1907. Fertilization and embryogeny in *Ephedra trifurca. Bot. Gaz.* 44:273–292.
Lehmann-Baerts, M.
  1967. Ovule, gamétophyte femelle et embryogenèse chez *Ephedra distachya* L. Cellule 67:51–87.
Maheshwari, K.
  1935. Contributions to the morphology of *Ephedra foliata* Boiss. I. The development of the male and female gametophytes. *Proc. Indian Acad. Sci.* 1:586–606.
Maheshwari, P., and V. Vasil
  1961. *Gnetum.* Council of Scientific and Industrial Research, New Delhi.
Martens, P.
  1971. *Les Gnétophytes.* (Handbuch d. Pflanzenanatomie, Band 12, Teil 2.) Gebrüder Borntraeger, Berlin.

Martens, P., and L. Waterkeyn

1963. The shoot apical meristem of *Welwitschia mirabilis* Hooker. *Phytomorphology* 13:359–363.

Nemejc, F.

1967. Notes on the origin and the past of the Chlamydospermophyta. *Preslia* 39:345–351.

Pearson, H. H. W.

1929. *Gnetales.* Cambridge University Press.

Porsch, O.

1910. *Ephedra campylopoda*, eine entomophile Gymnosperme. *Ber. Deutsch Bot. Ges.* 28:404–412.

1916. Die Nektartropfen von *Ephedra campylopoda. Ber. Deutsch. Bot. Ges.* 34:202–212.

Pulle, A.

1938. The classification of the spermatophytes. *Chron. Bot.* 4:109–113.

Rodin, R. J.

1953a. Distribution of *Welwitschia mirabilis, Amer. Jour. Bot.* 40:280–285.

1953b. Seedling morphology of *Welwitschia. Amer. Jour. Bot.* 40:371–378.

1958a. Leaf anatomy of *Welwitschia.* I. Early development of the leaf. *Amer. Jour. Bot.* 45:90–95.

1958b. Leaf anatomy of *Welwitschia.* II. A study of mature leaves. *Amer. Jour. Bot.* 45:96–103.

1966. Leaf structure and evolution in American species of *Gnetum. Phytomorphology* 16:56–68.

1967. Ontogeny of foliage leaves in *Gnetum. Phytomorphology* 17:118–128.

Rodin, R. J., and G. S. Paliwal

1970. Nodal anatomy of *Gnetum ula. Phytomorphology* 20:103–111.

Seeliger, I.

1954. Studien am Sprossvegetationskegel von *Ephedra fragilis var. campylopoda* (C. A. Mey.) Stapf. *Flora* 141:114–162.

Singh, H., and P. Maheshwari

1962. A contribution to the embryology of *Ephedra gerardiana. Phytomorphology* 12:361–372.

Stapf, I.

1889. Die Arten der Gattung *Ephedra. Denkschr. Akad. Wien. Math.-Naturw.* Cl. II. 56:1–112.

Steeves, M. W., and E. S. Barghoorn

1959. The pollen of *Ephedra. Jour. Arnold Arboretum* 40:221–255.

Thompson, W. P.

1918. Independent evolution of vessels in Gnetales and angiosperms. *Bot. Gaz.* 65:83–90.

19

General Morphology and Evolution of the Angiosperms

The angiosperms, or flowering plants, constitute the dominant and most ubiquitous vascular plants of modern floras on the earth. The term angiosperm (literally, a vessel seed) was devised to designate one of the most definitive characteristics of flowering plants, namely the enclosure of the ovules or potential seeds within a hollow ovary. In this respect angiosperms are considered to be advanced, as compared with the naked seeded gymnosperms.

Angiosperms far exceed in number and diversity of form and structure all other major groups of living plants, and more than 200,000 different species have been named and classified. Flowering plants occupy a very wide range of ecological habitats (including both salt and fresh water), and extend far toward the polar extremities of the earth. Although scientific study of the classification, morphology, and geographical distribution of angiosperms has been pursued for more than 200 years, we still do not possess even a reasonably complete census of living flowering plants. Modern scientific botany had its origin in the studies of the north temperate angiosperms of England and Europe, and our knowledge of the richly diversified floras of tropical areas is still extremely incomplete. Such regions of the earth as New Guinea, New Caledonia, tropical Asia and Africa, and the vast Amazon rain forest still await comprehensive botanical exploration. A fuller knowledge of the angiosperms in these regions may ultimately modify many of our present concepts, which, it must be emphasized, are based largely on the intensive study of north temperate plants.

Aside from bacteria and pathogenic fungi, angiosperms are the plants that most obviously affect the existence of man on earth. The basic food supply of the world is derived from the seeds and fruits of angiosperms (rice, wheat, corn, are outstanding examples), and fibers, wood, drugs, and other products of great economic value come from flowering plants.

Thus far the known fossil record of the angiosperms has failed to provide any reliable clues about their origin and evolutionary development. The fragmentary evidence provided by fossilized wood, and especially by impressions of leaf form and venation, shows that by the Middle Cretaceous Period angiosperms had reached a high stage of morphological specialization. Even if due allowance is made for errors in identifying genera by means of leaf impressions, it seems clear that by the latter part of the Mesozoic Era many modern families were clearly defined. How should this apparently sudden appearance of highly specialized angiosperms in the Cretaceous Period be interpreted? Does it indicate that angiosperms evolved at a much faster rate than was true of the gymnosperms and lower vascular plants? Or is this abrupt appearance of angiosperms in the fossil record a clear demonstration of a long pre-Cretaceous period in their evolutionary development?

Because of the very meager fossil record, it is not possible to answer these questions in any satisfactory manner. Fragmentary evidence, however, is accumulating in support of the inference that, like other groups of vascular plants, the angiosperms have probably developed over an extremely long period of time. Their so-called abrupt rise to dominance during the Mesozoic may thus be an illusion. Axelrod (1952) has discussed the paleobotanical evidence suggesting the existence of angiosperms during the Triassic and Jurassic periods—they may even, as he postulates, have been in existence at the end of the Paleozoic Era. An interesting, frequently cited example of the existence of angiosperms as early as the Triassic Period is provided by impressions of the pleated lamina of a palm-like leaf that Brown (1956) named *Sanmiguelia lewisi.* Unfortunately, the cellular structure of this leaf genus is not preserved and its presumed systematic affinity with the palms must for the present be considered an open question.

A most discouraging aspect of the paleobotanical study of angiosperms is the possibility that they initially developed in upland areas, i.e., in regions most unlikely for the preservation of fruits, flowers, leaves, wood, and pollen (Axelrod, 1952, 1961, 1970). If they did, we may search in vain for tangible fossil records of truly ancient and primitive angiosperms. (For further discussion of this problem, see Delevoryas, 1962, and Takhtajan, 1969.)

Because the fossil record is sparse and equivocal, theories regarding the ancestral stock or stocks from which the angiosperms originated are extremely speculative and contradictory. (For critical reviews of this problem, see Eames, 1961; Cronquist, 1968; Takhtajan, 1969; and Sporne, 1971.) Nearly every group in the Tracheophyta—including non-seed-bearing plants such as lycopods, sphenopsids, and ferns—has been regarded at one time or another as the possible "precursor" of the angiosperms. At present one of the most widely held views, as expressed, for example, by Takhtajan (1969), is that the "ancestors of the angiosperms must be sought amongst the gymnosperms." In connection with this viewpoint, the *origin of the ovule* (particularly its integument) is a matter of basic concern, and we have already briefly discussed this problem in Chapters 14 and 15. If we adopt the common postulate that the primitive type of angiosperm flower was bisporangiate, i.e., that it contained both stamens and carpels, it becomes evident that none of the living gymnosperms (cycads, *Ginkgo*, conifers, gnetophytes) could be regarded as "ancestral" to the angiosperms because their strobili are typically monosporangiate (see Chapters 15–18). Among extinct groups of gymnosperms, the Cycadeoidales might at first seem, because of their bisporangiate strobili, to provide the key to the mystery of angiosperm origin (Arber and Parkin, 1907). But the naked ovules and the peculiar synangial pollen-forming organs of the Cycadeoidales would seem to preclude the possibility that this group of ancient gymnosperms gave rise to the flowering plants (Delevoryas, 1962).

It has also been suggested that the angiosperms may have been derived from the seed ferns (Andrews, 1947; Thomas, 1955). However, the differences between their reproductive structures and wood anatomy and those of *living* primitive angiosperms raise formidable difficulties to the theory of a pteridosperm ancestry of flowering plants (see Eames, 1961, and Delevoryas, 1962).

The "abominable mystery" of the origin of the angiosperms still remains to be solved —perhaps through future discoveries by paleobotanists of truly ancient "missing links." Meanwhile, as will be pointed out later in this chapter, a reliable body of comparative data derived from the study of surviving types of primitive angiosperms has accumulated and holds considerable promise in orienting and guiding paleobotanists in their continued search for fossil evidence of ancient angiosperms.

The studies of John Ray in the seventeenth century on the structure of seeds eventually led to the recognition of two major groups within the living angiosperms: Dicotyledoneae and Monocotyledoneae (see Engler, 1926, 1964). In the classification adopted in the present book, these groups represent the two subclasses of the Angiospermopsida. According to Cronquist's (1968) rough estimate, approximately 165,-000 species of dicotyledons and about 55,000 species of mocotyledons have been described.

The difference in the number of cotyledons present in the embryo—one in the monocotyledons and two in dicotyledons—provides the most familiar distinction between the two subclasses of the angiosperms, although as we will show in Chapter 20, this classical distinction is by no means without exceptions. Further differences between monocotyledons and dicotyledons are based on (1) the *numerical plan of the flower* (3 in monocots, 4 or 5 in dicots); (2) the *venation pattern* ("parallel" in monocots, reticulate in dicots); (3) the *vascular bundle arrangement* in the stem (scattered in monocots, arranged in a cylinder in dicots); and (4) the *presence* (dicots) *or absence* (monocots) *of a vascular cambium* that produces secondary phloem and secondary xylem.

As with cotyledon number, these additional distinctions between monocotyledons and dicotyledons overlap or have exceptions, as we will point out at appropriate points in our presentation of the vegetative and reproductive morphology of the angiosperms (Chapters 19–21). Despite the absence of a sharp demarcation between monocotyledons and dicotyledons, the opinion is widely held that the monocotyledons arose during the early evolution of the angiosperms from some type of dicotyledonous ancestors. Cronquist (1968) and Takhtajan (1969) share the opinion that the "Liliatae" (their taxonomic term for monocotyledons) were derived from the "Magnoliatae" (their taxonomic term for dicotyledons). The dicotyledonous order Nymphaeales (which includes "water lilies") seems to these authors to fit the theoretical morphological and ecological requirements for the kind of ancestral stock from which monocotyledons may have taken their origin.

Organography and Anatomy

In Chapter 3 an effort was made to characterize the fundamental aspects of organography and anatomy illustrated by vascular plants. That chapter should be carefully studied as an essential background to the necessarily brief and condensed treatment of the vegetative sporophyte of angiosperms now to be presented.

The extensive descriptions and interpretations of organography given by Goebel (1905) and Troll (1935, 1937, 1938, 1939, 1949, 1964, 1967, 1969), and the comprehensive accounts of comparative anatomy found in the treatises by Solereder (1908), Metcalfe and Chalk (1950), Bailey (1954), and Esau (1965), ably demonstrate the structural complexity and diversity of living angiosperms. But it should be emphasized that many of our past concepts and generalizations, especially of comparative anatomy, were based to a very large extent on the study of the dicotyledons rather than on a comprehensive and balanced knowledge of *both* of the classes of the angiosperms. Fortunately, broad surveys of the systematic anatomy of the vegetative organs of the monocotyledons were initiated more than a decade ago, at the Jodrell Laboratory, Royal Botanic Gardens, Kew, England. The result of these and later studies have been published, under the able editorship of Dr. C. R. Metcalfe, as a series of volumes bearing the general title *Anatomy of the Monocotyledons* (see Metcalfe, 1960, 1971; Tomlinson, 1961, 1969; Cutler, 1969; Ayensu, 1972). These monographs are not only highly important contributions to the comparative anatomy and systematics of the monocotyledons, but they are reminders of the great caution that must be observed when generalizations are attempted regarding the trends of anatomical specialization in this group of flowering plants.

General Morphology of Foliage Leaves

The truly enormous range in form and organization of the foliage leaves of angiosperms precludes an adequate "review" or condensation of the subject in the present text. Our account, therefore, will be necessarily brief and limited to the general morphology of leaves and the unsolved problems of their evolutionary history. Without question, Wilhelm Troll's (1935a, 1935b, 1937, 1938, 1939a, 1939b) comprehensive analyses of leaf morphology are outstanding in quality and students are referred to his publications for detailed information and for references to the voluminous literature.

Although the foliage leaf in some angiosperms appears to consist only of a leaf blade —and for this reason is described as *sessile*— more commonly the leaf consists of three parts that are more-or-less well defined: (1) the *leaf base*, developed, in many species, as a sheath or provided with a pair of stipules; (2) the *petiole*, and (3) the leaf blade, or *lamina*. The form, proportions, and structure of each of these parts, but particularly of the lamina, varies widely, not only between the leaves of different taxa but even among the succession of leaf types produced during the seedling and post-seedling phases of development of a single species. In so-called "simple leaves," the lamina is an undivided unit structure, highly variable in form and with entire, dentate, serrate, or crenate margins. Leaves in which the lamina is more-or-less conspicuously divided into pinnately or palmately arranged *lobes* are also common, especially in various genera or species of dicotyledons (Fig. 19-9, H).

The most complex type of lamina organization occurs in "compound leaves," characterized by the formation of separate pinnae or leaflets, which are attached in various ways to the portion of the leaf axis known as the *rachis*. In palmately compound leaves, the

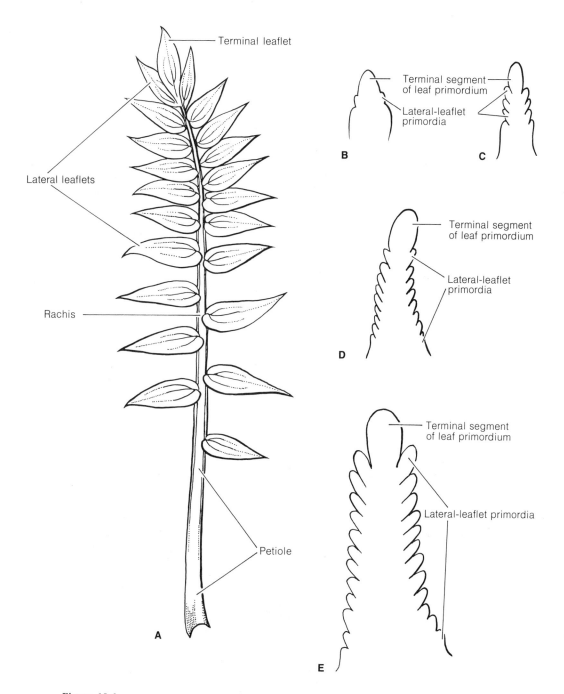

Figure 19-1
Morphology and development of the leaf of *Polemonium caeruleum*. **A**, adult leaf; **B–E**, successive stages in the basipetal formation of lateral leaflet primordia. [Redrawn from Troll, *Nova Acta Leopoldina* n.f. 2:315, 1935.]

individual leaflets radiate in a digitate pattern from a very short rachis; in pedately compound leaves, e.g., those of *Kingdonia,* the lamina is composed of a series of lobed "segments" attached to a transversely expanded rachis (Fig. 19-15). In the most familiar and widespread type of compound leaf, the rachis is elongated and bears two rows of simple or divided leaflets; the leaflets may be arranged alternately or in pairs along the rachis (Fig. 19-1). There is a superficial resemblance between a shoot with simple leaves and a pinnately compound leaf, but the latter usually subtends an axillary bud and ontogenetically, unlike a shoot, it is a strictly "determinate" organ—i.e., a structure in which all apical growth ceases at an early stage in development.

One of the most interesting features of pinnately compound leaves in the dicotyledons is the variable patterns in which lateral leaflets are formed during the early phases of leaf ontogeny. Leaves that at maturity appear very similar in general morphology may prove strikingly different with respect to the order of formation of lateral-leaflet primor-

dia. Three general developmental sequences of leaflet initiation are recognized, namely: *basipetal, acropetal,* and *divergent* (Fig. 19-2).

The basipetal order of formation of lateral leaflets is the most common pattern, according to Troll, and a clear example is provided by leaf development in *Polemonium caeruleum* (Fig. 19-1). At a very early stage, the young lamina consists of a prominent terminal segment—destined to develop into the terminal leaflet—and a pair of lateral-leaflet primordia (Fig. 19-1, B). As development continues, two rows of lateral-leaflet primordia develop basipetally (i.e., downwardly) toward the base of the lamina (Fig. 19-1, C-E). Lateral-leaflet initiation is accompanied by the intercalary elongation of the rachis and its continued extension finally results in the separation of the leaflets along the mature rachis. As shown in Fig. 19-1, A, the lowest leaflets may become alternately arranged while the upper leaflets are more crowded and are disposed in well-defined pairs.

The acropetal order of leaflet initiation represents the reverse of the pattern illustrated in *Polemonium,* in that the two rows of

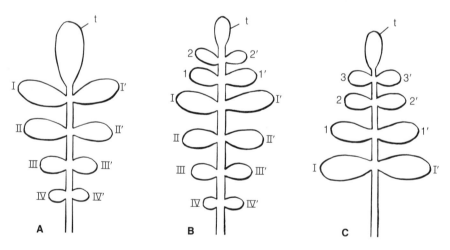

Figure 19-2

Diagrams showing the basipetal (A), divergent (B), and acropetal (C) orders of lateral-leaflet formation in pinnately compound leaves. In each diagram, the basipetal series is indicated by Roman numerals (I–I', II–II', etc.), the acropetal series by Arabic numerals (1–1', 2–2', etc.). In each figure, t designates the terminal leaf segment. [Redrawn from Troll, *Nova Acta Leopoldina* n.f. 2:315, 1935.]

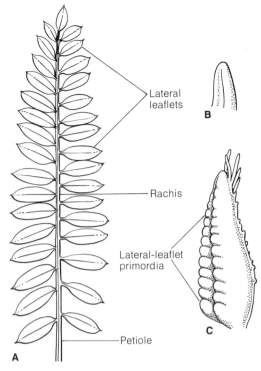

Figure 19-3
Morphology and development of the leaf of *Astragalus cicer*. **A,** adult leaf; **B,** adaxial view of leaf primordium showing the two marginal ridges from which the lateral-leaflet primordia will develop; **C,** later stage in leaf development, showing acropetal order of leaflet formation. [Redrawn from Troll, *Nova Acta Leopoldina* n.f. 2:315, 1935.]

lateral-leaflet primordia develop upwardly, from the base of the lamina towards its apex (Fig. 19-3, B, C). According to Troll, the acropetal formation of leaflets occurs only in leaves that exhibit pronounced apical growth for a relatively long period in leaf ontogeny. In this connection it is interesting to note that in *bipinnate leaves*, the first order of lateral-leaflet primordia commonly develops basipetally and the second order of leaflets, or "pinnules," arising from them form in an acropetal sequence.

In the divergent type of leaflet development, lateral-leaflet initiation begins near the middle of the young lamina and proceeds *both* basipetally and acropetally (Fig. 19-4). This type, therefore, combines the basipetal and acropetal orders of leaflet initiation. In extreme examples, illustrated diagrammatically in Fig. 19-2, C, all but the lowest pair of leaflets belong to the acropetally developed

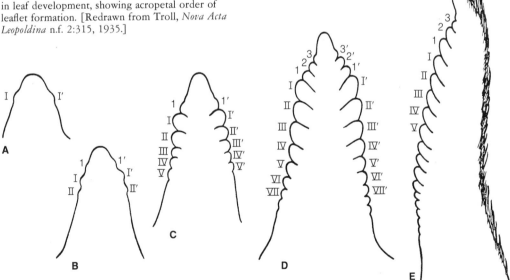

Figure 19-4
The divergent pattern of lateral-leaflet formation in *Achillea millefolium*. **A–D,** adaxial views of successive stages in development, showing the basipetal series (I-I'–VII-VII') and the acropetal series (1-1'–3-3') of lateral-leaflet primordia; **E,** side view of later stage in leaf development. [Redrawn from *Vergleichende Morphologie der höheren Pflanzen* by W. Troll. Gebrüder Borntraeger, Berlin, 1937.]

series; in *Achillea*, there is a marked emphasis on the formation of a larger number of basipetally formed pinnae (see Fig. 19-4).

Despite the considerable morphogenetic interest of Troll's studies, they do not resolve the difficult question of the phylogenetic relationship between "simple" and "compound" leaves in the dicotyledons. Which is the primitive type? In members of certain families, such as the Quiinaceae, the juvenile leaves are pinnately compound or pinnately lobed and "merge" gradually into the simple leaves characteristic of the adult tree (Foster, 1950a, 1951). This type of *heterophylly* might be used to support the idea that a pinnately compound leaf is the ancestral type *in this family*. More commonly, however, the seedling or juvenile leaves in angiosperms tend to be simpler than the adult type of foliage and this fact might encourage the speculation that simple leaves, rather than compound ones, are the primitive form. Thus far, the fossil record sheds no convincing light on the problem, although a number of morphologists regard the simple, pinnately veined leaf as the primitive leaf type in modern angiosperms (Eames, 1961; Cronquist, 1968).

Since our discussion thus far has been centered on the leaves of dicotyledons, it is essential, for comparative purposes, to comment briefly on certain problems which have arisen in the interpretation of leaf morphology in monocotyledons. In general, the leaves of monocotyledons develop a simple lamina and compound leaves are extremely uncommon. The most striking examples of compound leaves in monocotyledons are found in palms, in certain genera of which

the leaf is said to reach a length of 30–50 feet and thus to be the largest leaves among seed plants. However, the leaflets or subdivisions of the lamina in both fan and pinnate-leaf types of palms do not arise from separate leaflet primordia as do the compound leaves of dicotyledons. On the contrary, the young lamina of a palm leaf first develops a series of compressed "folds" which, later in ontogeny, split along either the adaxial or abaxial ridges of the lamina into pinnately or palmately arranged leaflets. The histogenetic aspects of this extraordinary method of leaflet formation are very complex and far too involved for consideration here (for details, see Eames, 1953, and Corner, 1966). But it must be noted that the remarkable type of leaf ontogeny, typical of palms, seems to emphasize the isolated systematic position of these angiosperms and poses the difficult question of the evolutionary relationship between the leaves of monocotyledons and dicotyledons (see Corner, 1966, Chapter 3).

Another problem in leaf morphology among monocotyledons is the interpretation of the so-called "unifacial" type of leaf of certain members of the Liliaceae, Araceae, Iridaceae, Amaryllidaceae, and Juncaceae. A unifacial leaf differs from a "typical" bifacial monocotyledonous leaf in the following ways: (1) although it generally possesses a sheathing dorsiventral leaf base, the remainder of the leaf is not differentiated into petiole and lamina; (2) it is flattened in a median rather than in a transverse plane, and (3) the collateral vascular bundles are arranged in radially rather than in the crescentic pattern characteristic of a bifacial lamina.

Figure 19-5 (*facing page*)
Diagrams contrasting the morphology and vascular anatomy of the unifacial leaf of *Acorus calamus* with the type of foliage leaf characteristic of most Araceae. **A**, adaxial view of leaf of *Acorus*; **B–E**, transections at the levels indicated by broken lines in **A**. **F**, adaxial view of leaf typical of other members of the Araceae; **G–I**, transections at the levels indicated by the broken lines in **F**. Phloem of vascular bundles represented in white, xylem in black. Photosynthetic tissue indicated by hatching. dm, dorsal median bundle; L1–L1′, L2–L2′, etc., lateral vascular strands; vm, ventral median bundle. [Redrawn from Kaplan, *Amer. Jour. Bot.* 57:331, 1970.]

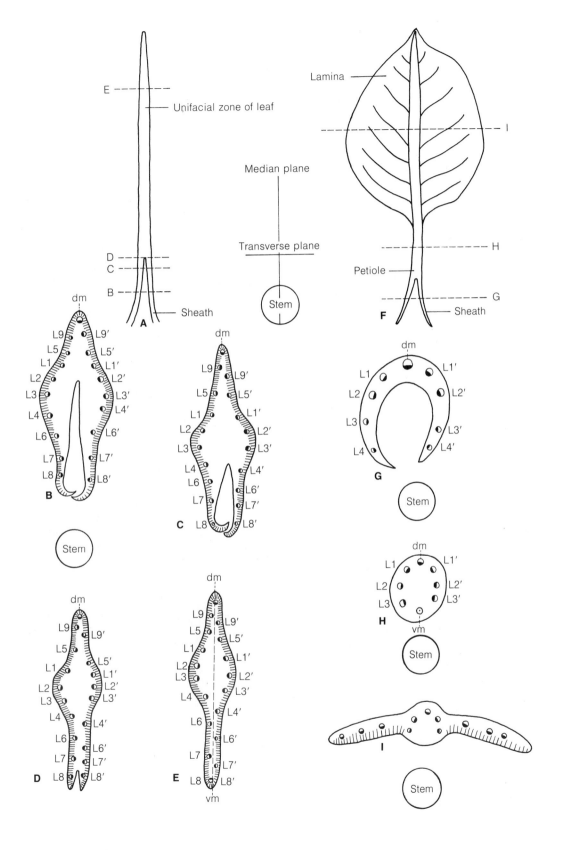

E - - - - - - Unifacial zone of leaf

Median plane

Transverse plane

Stem

D - - - - -
C - - - - -

B - - - - - Sheath

A

Lamina —— I

Petiole

H

G

Sheath

F

dm

L9 L9'
L5 L5'
L1 L1'
L2 L2'
L3 L3'
L4 L4'
L6 L6'
L7 L7'
L8 L8'

B

Stem

dm

L9 L9'
L5 L5'
L1 L1'
L2 L2'
L3 L3'
L4 L4'
L6 L6'
L7 L7'
L8 L8'

C

dm

L1 L1'
L2 L2'
L3 L3'
L4 L4'

G

Stem

dm

L9 L9'
L5 L5'
L1 L1'
L2 L2'
L3 L3'
L4 L4'
L6 L6'
L7 L7'
L8 L8'

D

dm

L9 L9'
L5 L5'
L1 L1'
L2 L2'
L3 L3'
L4 L4'
L6 L6'
L7 L7'
L8 L8'
vm

E

dm

L1 L1'
L2 L2'
L3 L3'
vm

H

Stem

I

Stem

These features are diagrammatically represented in Fig. 19-5, which is reproduced from Kaplan's (1970) study on leaf histogenesis in *Acorus calamus* (Araceae). His investigation revealed that in this species apical growth ceases at an early stage and that marginal growth in the upper part of the developing leaf is suppressed in favor of *radial development*, which occurs by means of an active meristem situated on the *adaxial side* of the leaf primordium. A very similar pattern of histogenesis was described earlier by Boke (1940) in his study of the *Acacia* phyllode, a foliar structure which he interpreted as equivalent to the petiole-rachis regions of a pinnate foliage leaf. Kaplan concluded that on the basis of their ontogenetic similarity, the unifacial leaf of *Acorus* and the phyllode of *Acacia* are homologous structures. Kaplan's detailed study of *Acorus* emphasizes the fundamental need for ontogenetic evidence in the interpretation of leaf morphology, not only with respect to the Araceae but with reference to other examples of unifacial leaves among monocotyledons as a whole.

Stipules

An enigmatic character of the foliage leaf of many dicotyledons is the presence of a pair of small appendages, known as stipules, near the base of the leaf (Fig. 19-6). The expression "free lateral stipules" is often used to describe such stipules that *appear* to be "independent" organs. The impression that they are independent is strengthened by the early abscission of lateral stipules during leaf expansion in many trees and shrubs; in such plants (e.g., *Quercus, Fagus, Populus, Ulmus*) the deciduous stipules leave separate "stipule scars" on the stem at each side of the leaf. But in many other plants stipules are apparently fused with the leaf base and these are termed "adnate stipules" (Figs. 19-7, A; 19-8, A). In plants with decussate phyllotaxis, *interpetiolar stipules* may develop

between the two leaves at each node. Interpetiolar stipules are dual structures, i.e., they represent the partial or complete union of the adjacent stipules at each side of the node. In *Magnolia*, the margins of the two stipules of a leaf are loosely joined to form a conical "stipular sheath" which encloses the younger leaves of the terminal bud.

The above résumé gives only a very incomplete sketch of some of the diverse forms of stipules in the dicotyledons and reference should be made to Glück (1919) and Troll (1939a, pp. 1248–1290) for detailed descriptions of stipule morphology in both monocotyledons and dicotyledons.

Regardless of the great variability in the size, form, and methods of attachment of stipules, it is evident that they are integral parts of a leaf and not "accessory" or "independent" appendages of the shoot. Support for this statement is provided first of all from ontogeny, which shows that stipules develop from the leaf-base region of a foliage leaf. (Fig. 19-7). Additional evidence about

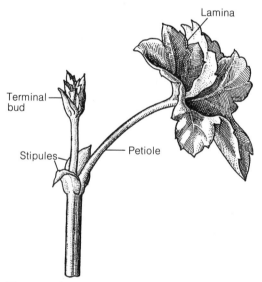

Figure 19-6
Shoot tip of *Pelargonium peltatum*, showing the pair of free lateral stipules associated with a recently expanded leaf. [Redrawn from *Vergleichende Morphologie der höheren Pflanzen* by W. Troll. Berlin, Gebrüder Borntraeger, 1939.]

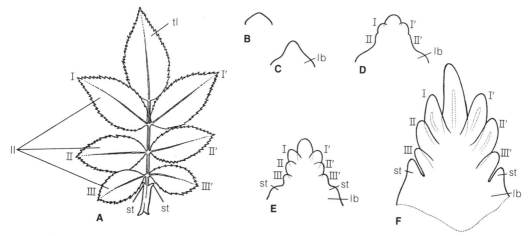

Figure 19-7
Leaf development in *Rosa*. **A**, adult leaf with adnate stipules (st) and pinnately compound lamina, composed of terminal leaflet (tl) and three pairs of lateral leaflets (ll). **B–F**, successively later stages in leaf development, showing origin of stipules (st) from leaf base (lb) and basipetal order of formation of lateral-leaflet primordia (I-I'–III-III'). [Redrawn from *Vergleichende Morphologie der höheren Pflanzen* by W. Troll. Gebrüder Borntraeger, Berlin, 1938.]

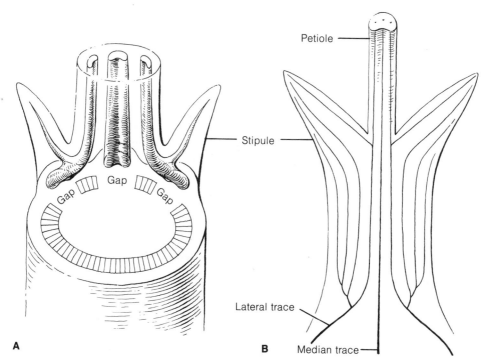

Figure 19-8
Vasculation of stipules. **A**, diagram of a trilacunar node showing that the vascular supply of each stipule arises as a branch from the adjacent lateral leaf trace; **B**, *Trifolium medium,* diagram showing that the veins entering the paired adnate stipules represent branches of the corresponding lateral leaf traces. [**A** redrawn from Sinnott and Bailey, *Amer. Jour. Bot.* 1:441, 1914; **B** redrawn from *Vergleichende Morphologie der höheren Pflanzen* by W. Troll. Gebrüder Borntraeger, Berlin, 1939.]

the intimate relation between stipules and the leaf comes from the fact that the vascular supply of each member of a pair of stipules is derived from the corresponding lateral leaf trace (Fig. 19-8). Sinnott and Bailey (1914), in their classical study of stipules, found that the presence or absence of stipules is correlated with the type of nodal anatomy. In the majority of stipulate-leaved families which they surveyed, the node is either tri- or multilacunar; stipules are usually absent from plants whose node is unilacunar. As the trilacunar node was previously assumed by Sinnott (1914) to be primitive, Sinnott and Bailey (1914) concluded that a leaf with two lateral stipules represents the primitive form in angiosperms. Two trends of evolutionary specialization were recognized: (1) the formation, in some taxa, of a sheathing leaf base as a result of the fusion of the stipules with the petiole and (2) the phylogenetic loss of stipules, which in turn is correlated with the loss of the lateral leaf traces. However, as we will discuss later (see pp. 573–574), recent studies support the view that a unilacunar two-trace node, rather than a trilacunar node, may be primitive. If this conclusion is valid, the problem of the morphology and evolution of stipules will require complete reexamination. As Takhtajan (1969) remarks, "the question of whether or not the earliest angiosperms had stipulate leaves must therefore remain open."

Cataphylls

The development of the shoot of many angiosperms is accompanied by the periodic formation of scale leaves, or cataphylls. The term "cataphylls" (derived from the German word *Niederblätter,* meaning "lower leaves") collectively designates the scales of the rhizomes of monocotyledons and dicotyledons as well as the protective scales of the winter buds of shrubs and trees. Cataphylls, like foliage leaves, originate from primordia

formed by the shoot apex and for this reason the two types of appendages may be regarded as *serially homologous,* despite their adult divergence in form, anatomy, and function (see Chapter 1 for a discussion of serial homology). In many dicotyledons, the homology between the bud scale and the foliage leaf is clearly demonstrated by the presence of a *rudimentary lamina* at the apex of each of the sheath-like scales (Fig. 19-9, A–D), and by the occurrence of "transitional forms" between typical scales and foliage leaves (Fig. 19-9, E, F). This type of bud scale morphology is found in many taxa and has been studied ontogenetically in *Aesculus* (Foster, 1929) and *Paeonia* (Foster and Barkley, 1933). In other genera, however, such as *Carya,* the *morphological divergence* between bud scale and foliage leaf takes place at such an early stage in ontogeny that a *direct comparison* between the mature unsegmented scale and corresponding parts of the pinnately compound foliage leaf cannot be drawn (see Foster, 1931, 1932, 1935). The bud scales of some woody dicotyledons (e.g., *Betula, Alnus, Fagus*) represent modified stipules, and each pair of scales is associated with a rudimentary lamina. Stipular bud scales in some genera (e.g., *Malus, Prunus*) are tridentate in form, with the median pointlet representing the rudimentary lamina and the two lateral pointlets corresponding to a pair of adnate stipules. (For a detailed review of the history of investigations on the ontogeny, anatomy, and morphology of bud scales, see Foster, 1928.)

Venation

A striking and definitive morphological character of modern angiosperms is found in the complex and diversified venation patterns of their leaves. Even casual inspection of the gross venation of the foliar lamina of common dicotyledons reveals a more-or-less prominent *midvein* that terminates at the

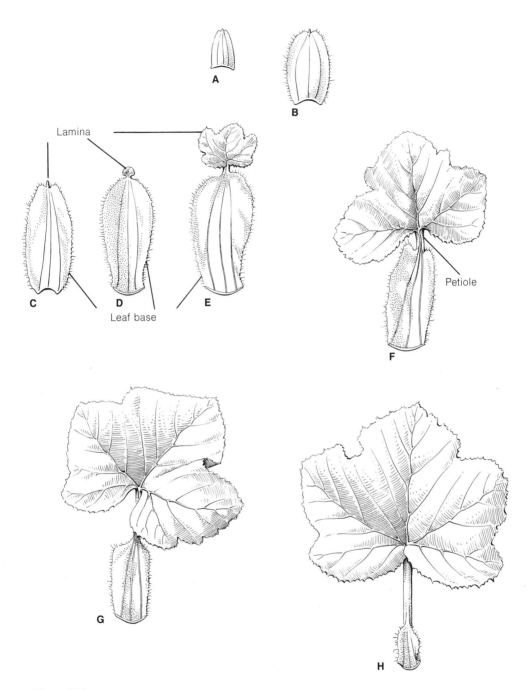

Figure 19-9

Ribes sanguineum. **A–D,** outer scales of a bud showing enlarged leaf base and rudimentary lamina;
E–F, transitional forms between bud scales and foliage leaves (note enlarged lamina and short petiole
of the appendages); **G–H,** foliage leaves each with a well-developed palmately lobed lamina, an
elongated petiole, and a sheathing leaf base. [Redrawn from *Vergleichende. Morphologie der höheren
Pflanzen* by W. Troll. Gebrüder Borntraeger, Berlin, 1939.]

558

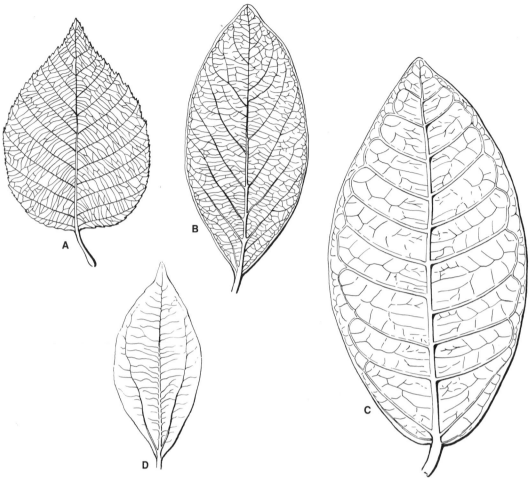

Figure 19-10

Some of the types of pinnate-reticulate venation found in the leaves of dicotyledons. Only a part of the minor intercostal venation is represented in these drawings. **A,** craspedodromous venation of *Betula medwedewii;* **B,** brochidodromous venation of *Laurus canariensis;* **C,** camptodromous venation of *Ficus hombroniana;* **D,** palmate (acrodromous) venation of *Cinnamomum glabrum.* [**A, B** redrawn from *Die Evolution der Angiospermen* by A. Takhtajan. Gustav Fischer, Jena, 1959; **C** redrawn from E. J. H. Corner, *Phil. Trans. Roy. Soc.* B 253:23, 1967; **D** redrawn from *Die Blatt-Skelete der Dikotyledon* by C. R. von Ettinghausen. Kais. Kon. Hof und Staatsdruckerei, Wien, 1861.]

apex of the lamina, and a series of pinnately or palmately arranged secondary veins (Fig. 19-11). The "course," i.e., the direction and mode of termination, of the main secondary veins in pinnate venation fluctuates widely between genera (or even between species) and was used more than a century ago by Von Ettinghausen (1861) to establish a series of "venation types." Some of his terms for describing the course of the main veins

in angiosperm leaves have proved useful in describing venation patterns and examples are given in Fig. 19-10. (See also Hickey's 1973 paper on leaf "architecture.")

The vasculature of the intercostal areas between the major veins in angiosperm leaves consists of a complex network of anastomosed "minor" veins and veinlets and for this reason, typical angiosperm venation is described as *reticulate.* When the minor

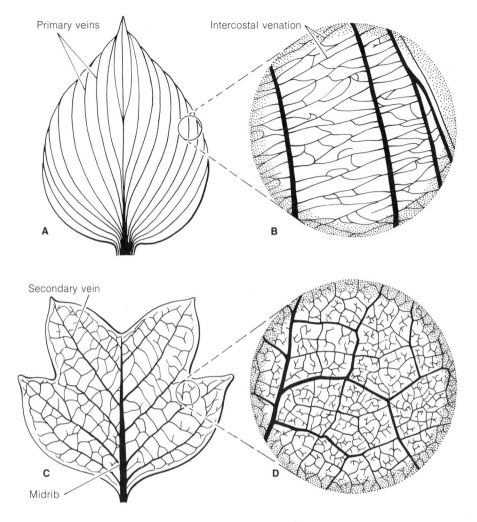

Figure 19-11
Venation patterns in angiosperms. **A,** the campylodromous venation of *Hosta caerulea* (note
the arcuate course and the progressive union, from base of lamina upwards, of the primary
veins); **B,** details of network of veinlets from area indicated in the circle of **A** (note the freely
terminating vein endings); **C,** major venation in the leaf of *Liriodendron tulipifera;* **D,** details of
area indicated in the circle of **C,** showing complex reticulum of veinlets and numerous branched
vein endings. [**A, B** redrawn from Pray, *Amer. Jour. Bot.* 42:611, 1955; **C, D** redrawn from Pray,
Amer. Jour. Bot. 41:663, 1954.]

venation is carefully studied at high magnification, it becomes evident that the ultimate
(i.e., the smallest) meshes, or areoles, contain solitary or often branched vein endings
which terminate freely in the mesophyll
(Fig. 19-11, D). In recent years, increased
attention has been given to the ontogenetic
and histological aspects of minor venation in
the leaves of both monocotyledons and dicotyledons (Pray, 1963; Lersten, 1965; Esau,

1967). Comparative studies emphasize the
need for the consideration of the *total vasculature* (i.e., both the major and minor venation) in any effort to describe or to interpret
the foliar venation of an angiosperm taxon.
Outstanding examples of the systematic
value of minor venation at the generic level
are provided by the bizarre and highly distinctive venation patterns characteristic of
members of the Quiinaceae, a small family

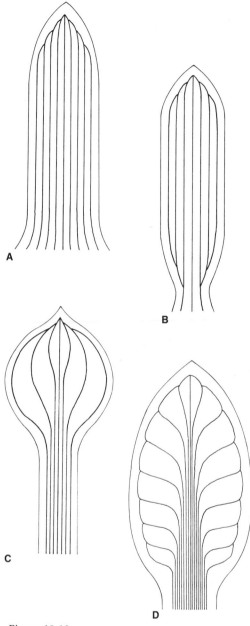

Figure 19-12
Diagrams showing various types of striate venation in the leaves of monocotyledons. **A, B,** longitudinally striate venation typical of the leaves of the Gramineae; note apical convergence and fusion between successive primary veins; **C,** arcuate-striate venation pattern; **D,** pinnate-striate venation pattern. [Adapted from *Vergleichende Morphologie der höheren Pflanzen* by W. Troll. Gebrüder Borntraeger, Berlin, 1938.]

of tropical dicotyledons (Foster 1950a, 1950b, 1951, 1952, 1953).

A detailed description and comparison of the extremely varied types of angiosperm venation patterns cannot be undertaken in the present book (for comprehensive accounts of venation patterns, see Von Ettinghausen, 1861; Kerner and Oliver, 1895; and Troll, 1938). Instead, we will examine two basic morphological problems in the interpretation of venation, the ultimate solution of which would contribute significantly to an understanding of the origin and trends of phylogenetic specialization of flowering plants.

Venation Patterns in Monocotyledons and Dicotyledons

The first problem concerns the validity of the common idea, held by many morphologists and taxonomists, that monocotyledons are characterized by "parallel venation" in contrast to the reticulate pattern of venation typical of dicotyledons. Although it is true that reticulate venation is highly developed in dicotyledons, the concept of so-called parallel venation should be rejected because it is both inaccurate and misleading. In grasses, for example, the leaf superficially appears to be traversed by a closely spaced series of independent, longitudinal, parallel veins. But careful examination, using cleared* and stained leaves, reveals that the main veins do not extend equidistant throughout their course but, on the contrary, *converge* and progressively anastomose toward the apex of the lamina. These features are clearly analyzed by Troll (1938) who terms this type of closed venation striate rather than parallel (Fig. 19-12, A, B). It should be further emphasized that the longi-

---

*For information on the techniques used to clear and stain leaves see Foster, 1953, 1955, and Shobe and Lersten, 1967.

tudinally striate type of venation found in grasses is by no means representative of the monocotyledons as a whole. In members of such families as the Liliaceae, Marantaceae, and Araceae, the diverging lateral veins form an *arcuate* pattern in which many of the lateral veins do not reach the apex of the lamina but join the adjacent inner veins as they converge in the upper region of the leaf (Pray, 1955). This distinctive venation pattern was termed campylodromous by von Ettinghausen (1861) and is well illustrated in the liliaceous genus *Hosta* (Fig. 19-11, A). As shown in Fig. 19-12, C, campylodromous venation is somewhat intermediate in pattern between the longitudinal and pinnate forms of striate venation.

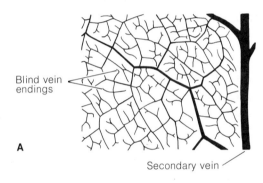

A

Blind vein endings

Secondary vein

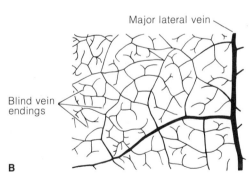

Major lateral vein

Blind vein endings

B

Figure 19-13
Illustrations of reticulate venation patterns in angiosperm leaves. A, in *Liriodendron,* a dicotyledon; B, in *Smilax,* a monocotyledon. Note remarkable similarity in overall venation pattern including the presence of blind vein endings. [Redrawn from Pray, *Phytomorphology* 13:60, 1963.]

Accompanying the continued inaccurate use of the term parallel venation there has been a conspicuous neglect of the minor venation which is usually present between the main longitudinal veins in the leaves of monocotyledons. In grass leaves, the minor venation is relatively simple and consists of delicate transverse or oblique *commissural veinlets* which interconnect the primary veins (Blackman, 1971). Commissural veinlets are also well defined in the leaf of *Maranta* (Troll, 1938, p. 1075, Fig. 857) and are so regularly spaced in the leaf of *Bambusa* (bamboo) that a mosaic-like reticulum is produced. More complex types of intercostal minor venation are found in members of the Araceae and Orchidaceae and also are present in the leaf of *Hosta* (Fig. 19-11, B). One of the most interesting types of intercostal minor venation may be observed in the leaf of *Smilax.* According to Pray (1963) the minor venation "is composed of a reticulum of polygonal areoles . . . and is remarkably similar in pattern to that of *Liriodendron* and many other dicotyledons" (Fig. 19-13). The example of *Smilax* clearly demonstrates that with our present knowledge, there is no single type of foliar venation which distinguishes the leaves of monocotyledons from those of dicotyledons. This fact must be carefully considered in future efforts to reconstruct the phylogenetic origin and systematic relationship of the monocotyledons.

Phylogeny of Venation Patterns in Angiosperms

The second problem that arises in the interpretation of venation patterns concerns the nature of the foliar vasculature in the remote ancestors of present-day angiosperms. It must be emphasized that the meager fossil record of the flowering plants sheds little light on this problem. On the contrary, the abundant leaf impressions of angiosperms

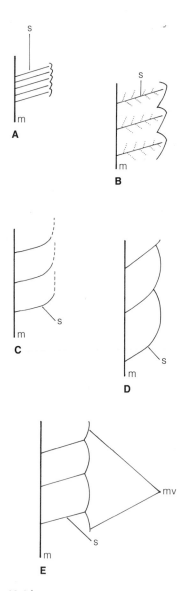

Figure 19-14
Diagrams showing the midvein (m) and the course
of the secondary veins (s) in various types of pinnate
venation in the leaves of dicotyledons. **A, B,** early
stages, respectively, of craspedodromous venation,
each secondary vein terminating in a tooth at the
lamina margin; **C, D,** two forms of camptodromous
venation, characterized by the distal curvature (**C**)
and fusion (**D**) of the successive secondary veins;
E, pinnate venation in which the united tips of
the secondary veins produce an apparent marginal
vein (mv). [Redrawn from *Vergleichende Morphologie
der höheren Pflanzen* by W. Troll. Gebrüder
Borntraeger, Berlin, 1938.]

from Cretaceous and Tertiary strata are
remarkably "modern" in contour and gross
venation and do not seem to represent examples of truly ancestral leaves. (For a different interpretation, see Rüffle, 1968, 1969.)
Because the present evidence from paleobotany is equivocal, many morphologists
have attempted to reconstruct the primitive
type of angiosperm venation on the basis
of the form and venation of the foliage leaf
in presumably primitive living families and
genera (Cronquist, 1968; Takhtajan, 1959,
1969).

Many years ago Sinnott and Bailey (1915)
postulated that the primitive angiospermic
leaf probably was three-lobed and palmately
veined. More recently, this view has been
supplanted by the idea that primitive angiosperm leaves were simple, entire and provided with pinnate-reticulate venation (see
Fig. 19-10). One of the main reasons advanced in support of this hypothesis is the
occurrence of simple pinnately veined leaves
in many members of the Ranales (e.g., Magnoliaceae, Chloranthaceae, Winteraceae).
Leaves with this type of venation are commonly associated with unilacunar two-trace
nodes (e.g., in *Austrobaileya, Ascarina*), and
they also are found in a number of families
that have primitive vessel-less xylem (e.g.,
Winteraceae, Trochodendraceae, Amborel-
laceae). If the idea that pinnate venation is
primitive is accepted, it follows, according to
Takhtajan (1959), that palmate venation, so-
called parallel venation, and advanced forms
of pinnate venation are derivative types of
venation, from a phylogenetic standpoint.
Takhtajan maintains that the most primitive
form of pinnate venation is von Ettinghausen's (1861) craspedodromous type,
which is characterized by the termination of
all the main secondary veins at the lamina
margin (Fig. 19-10, A). A possible sequence
in the phylogenetic development of various
"derived" types of pinnate venation is represented diagrammatically in Fig. 19-14, A–E.

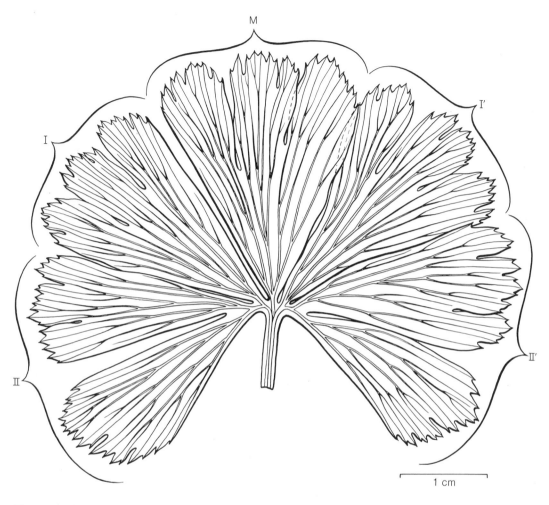

Figure 19-15
Mature foliage leaf of *Kingdonia uniflora* showing the open dichotomous venation of the five major segments (M, I–I′, II–II′) of the pedately divided lamina. [Redrawn from Foster and Arnott, *Amer. Jour. Bot.* 47:684, 1960.]

Although pinnate-reticulate venation undoubtedly is predominant in the leaves of ranalean plants, it still remains an open question whether this form of vasculature prevailed in the extinct ancestors of these and other surviving types of primitive angiosperms. Recent investigations have revealed that there are authentic examples of *dichotomously veined* leaves (with few vein anastomoses) in two living ranalean herbs, *Kingdonia uniflora* and *Circaester agrestis* (Figs.

19-15, 19-16). The systematic affinities of these two plants are obscure and, for the present, it seems desirable to treat these genera as the sole representatives, respectively, of the families Kingdoniaceae and Circaeasteraceae (Foster, 1961a, 1963; Willis, 1966). Both *Kingdonia* and *Circaester* are montane plants; *Kingdonia*, an herbaceous perennial, is endemic to western China and the range of *Circaester*, a small annual, extends from northwestern India and Nepal

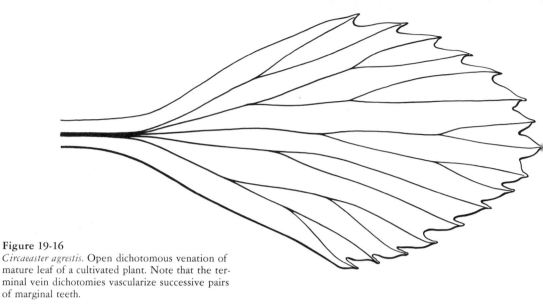

Figure 19-16
Circaeaster agrestis. Open dichotomous venation of
mature leaf of a cultivated plant. Note that the terminal vein dichotomies vascularize successive pairs
of marginal teeth.

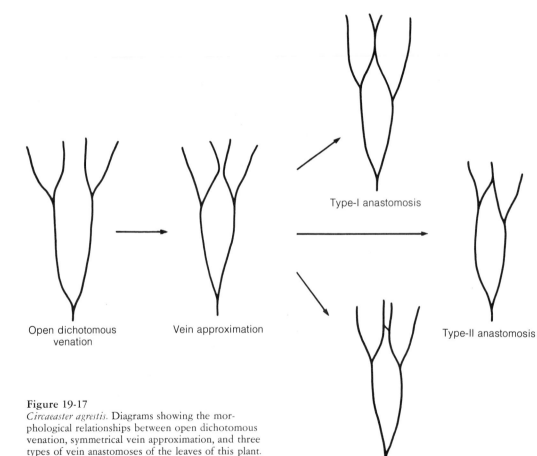

Type-I anastomosis

Open dichotomous
venation

Vein approximation

Type-II anastomosis

Type-III anastomosis

Figure 19-17
Circaeaster agrestis. Diagrams showing the morphological relationships between open dichotomous
venation, symmetrical vein approximation, and three
types of vein anastomoses of the leaves of this plant.
[Redrawn from Foster, *Jour. Arnold Arboretum*
49:52, 1968.]

Figure 19-18
Circaeaster agrestis. Tracing of the dichotomous venation pattern of a cleared and stained leaf of an herbarium specimen of a plant collected in Szechuan Province, China. The arrow indicates a prominent blind vein ending. (Specimen obtained courtesy of Botanical Museum and Herbarium, Lund, Sweden.)

to Tibet and northwestern China. (Foster, 1959, 1963; Foster and Arnott, 1960; Hu, Li, and Lee, 1964.)

What is the phylogenetic significance of dichotomous venation in *Kingdonia* and *Circaeaster*? Do these plants illustrate the *persistence* of a very ancient (ancestral?) form of angiospermic venation, or should the dichotomous venation be interpreted as the result of *"reversion"* to an ancient type of foliar vasculature? In this connection, it is interesting —and probably phylogenetically significant —that various types of anastomoses occur in both genera and have been intensively studied, particularly in *Circaeaster* (Foster, 1966, 1968). The vein unions are sporadic and apparently comparable to certain of the types of anastomoses in the leaves of *Ginkgo* and certain cycads (Chapters 15, 16). Although the occasional anastomoses which develop in the dichotomous venation of *Kingdonia* and *Circaeaster* might be interpreted as vestiges of some form of ancestral reticulate venation, there is no evidence at present to support this conjecture (Foster, 1966; Foster and Arnott, 1960). But it seems equally possible that anastomoses, as well as the common blind vein endings (Fig. 19-18) may simply represent modifications or evolutionary advances, which have been superimposed on a fundamentally primitive pattern of dichotomous venation (Foster, 1970, 1971). The combination, in *Kingdonia* and *Circaeaster*, of dichotomous venation with such obviously specialized characters as an *herbaceous* habit, small or reduced flowers, and xylem with vessels seems just as significant as the frequent association of primitive and advanced characters in *woody* members of the Ranales. Clearly, however, the phylogenetic significance of dichotomous venation in angiospermic leaves cannot be fully resolved until paleobotanical research uncovers the type of evidence needed to solve the riddle of the ancestry of modern flowering plants (Foster, 1961b).

Histology of Leaves

In Chapter 3 the general application of Sach's scheme of tissue systems to the description of leaf anatomy was briefly discussed. Figure 19-19 shows transections of portions of the leaf blades of *Pyrus* (a dicotyledon) and *Lilium* (a monocotyledon), and will serve to illustrate the necessarily brief account of foliar anatomy presented here. The abaxial and adaxial surface layers collectively represent the dermal system or epidermis, and consist of tightly joined epidermal cells and stomata. The latter are often restricted to the lower epidermis but may be present in both epidermal layers. Stomata represent the ports of exit and entrance of air and water vapor and are thus highly important to the normal functioning of the leaf. A great variety of trichomes (i.e., hairs, glands, scales) may develop from the leaf epidermis, and often provide valuable diagnostic characters in the taxonomic definition of species and genera.

The internal tissue systems of the lamina are the mesophyll, or *fundamental system*, and the veins and veinlets, which comprise the fascicular or *conducting system* (Fig. 19-19). The mesophyll is composed of living, thin-walled cells rich in chloroplasts, and in many dicotyledons it is clearly differentiated into one or more relatively compact layers of palisade parenchyma and a region of more loosely arranged spongy parenchyma. Very commonly the palisade parenchyma is situated directly below the upper epidermis (dorsiventral type of anatomy), but is also often found below both epidermal layers (isolateral type of anatomy). Palisade cells are usually columnar in form, but in some angiosperms these cells are armed or lobed (Fig. 19-19, B). The mesophyll functionally represents the chief photosynthetic tissue of the plant.

The histology of the veins and veinlets in the lamina depends on their degree of de-

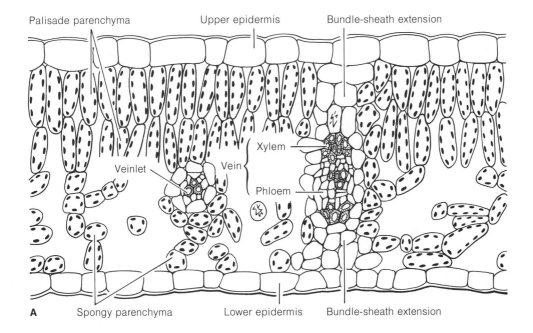

Palisade parenchyma   Upper epidermis   Bundle-sheath extension

Xylem

Veinlet   Vein

Phloem

A   Spongy parenchyma   Lower epidermis   Bundle-sheath extension

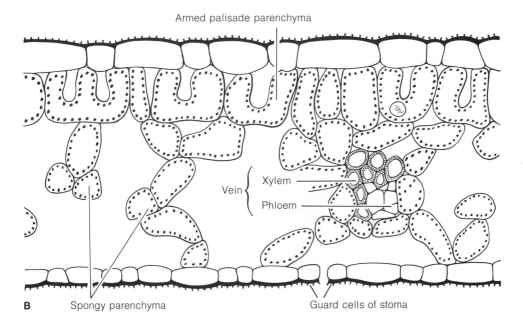

Armed palisade parenchyma

Vein

Xylem

Phloem

B   Spongy parenchyma   Guard cells of stoma

Figure 19-19
Transections illustrating the histology of the lamina of leaves in angiosperms. **A,** *Pyrus:* **B,** *Lilium* (see text for discussion of this figure). [Redrawn from *Plant Anatomy* by K. Esau. Wiley, New York, 1953.]

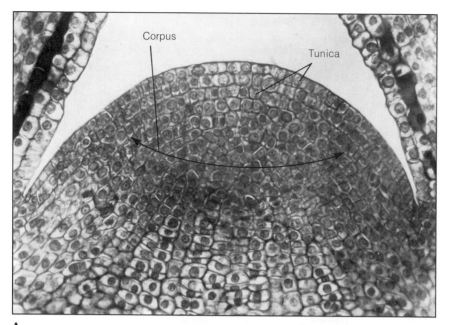

Figure 19-20
Median longitudinal sections of the vegetative shoot apices, showing tunica-corpus organization. **A**, *Trochodendron aralioides;* **B**, *Chenopodium album.* (See also diagrammatic representation in Fig. 19-21.)

velopment, and also varies widely according to the species or genus of plant. Each of the larger veins and veinlets consists of a collateral strand of phloem and xylem enclosed within a bundle sheath. In many woody dicotyledons the bundle sheaths are connected with one or both epidermal layers by plates of cells known as bundle-sheath extensions (Fig. 19-19, A). These structures are believed to participate in the conduction of liquids between the veins and the epidermis. The vascular tissues of the smaller veinlets and veinlet ends in the leaf are derived from the procambium, and are thus exclusively primary. Cambial activity, if it occurs in a given leaf, is therefore restricted to the coarser

veins in the lamina and to the vascular system of the petiole. Generally speaking, there is relatively little secondary xylem and phloem in the fascicular system of the leaf.

This account of leaf histology reflects very little of the wide range in the details of structure of the dermal, fundamental, and fascicular tissue systems of the leaf. Students interested in more extensive discussions of the structure and histogenesis of angiospermous leaves should consult Solereder (1908), Metcalfe and Chalk (1950), Esau (1965), von Guttenberg (1960), and the recent monographs on the comparative anatomy of monocotyledons cited on p. 548 of this chapter.

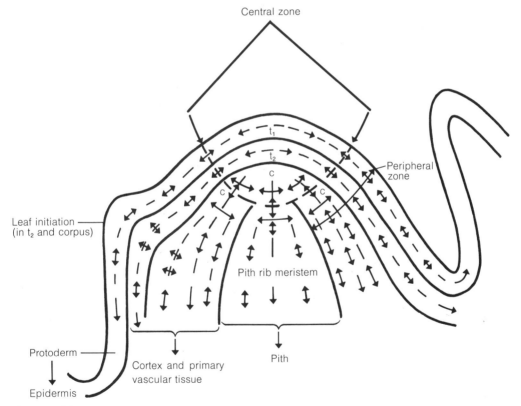

Figure 19-21
Diagrammatic representation of zonation and growth in the vegetative shoot apex of a hypothetical dicotyledon possessing a two-layered tunica. The distance between a pair of arrowheads indicates the degree of mitotic activity; the shorter the distance, the greater the mitotic activity. Contributions of various zones of the shoot apex to primary tissue layers and regions of the stem are indicated. t_1, first tunica layer; t_2, second tunica layer; c, corpus.

570

The Shoot Apex

A prominent feature of the shoot apices in angiosperms is the presence of a *tunica-corpus* type of cellular zonation. (Figs. 19-20, 19-21). This kind of zonal structure is not restricted to angiosperms, but is also found in the shoot apices of such conifers as *Araucaria* and *Agathis*, as well as in the gnetophytes *Ephedra* and *Gnetum* (Chapter 18). The *tunica* is represented by the layer or layers of cells in which the anticlinal plane of cell division predominates except at the sites of leaf initiation, where cells of the second and deeper tunica layers divide periclinally. In contrast to the prevailingly anticlinal plane of division in cells of the tunica, the *corpus* is a zone in which cells divide in varied planes. The boundary between tunica and corpus zones may fluctuate even in the same species because (1) at certain phases of growth, the outer cells of the corpus may be arranged in a tunica-like layer, and (2) periclinal division may occur, in certain monocotyledons, in the summital cells of the outermost tunica layer.

In addition to the topographical delimitation of tunica and corpus, most angiosperm shoot apices also exhibit a *cytohistological* pattern of zonation (Fig. 19-21). A *central zone* is apparent which includes some cells of both the tunica and corpus. Cells of the central zone are frequently larger, more vacuolate, and often exhibit a lower rate of mitotic division than the cells in the other regions of the shoot apex. In contrast, the *peripheral zone*, located laterally to the central zone, consists of cells that are relatively smaller and with a higher mitotic rate than those of the central zone. Cells of the *pith rib meristem* zone, which is found at the base of the central zone, divide predominantly in a transverse plane (with reference to the longitudinal axis of the stem) and form more-or-less well-defined files of cells which differentiate into the pith.

Additional information on the structure and growth of the shoot apex in angiosperms can be found in the review articles by Wardlaw (1965), Cutter (1965), and Gifford and Corson (1971).

Stems: General Structure

The comparative anatomy of the stem in angiosperms is a subject of great complexity and can be treated only in very brief outline here. In addition to the wide fluctuation in the structure of the primary tissue systems, the stems of all woody and of many herbaceous dicotyledons develop secondary vascular tissues from a cambium and usually also form cork and phelloderm tissues from the phellogen or cork cambium. Extensive secondary growth results in the crushing and ultimate elimination of the epidermis, cortex, and primary phloem areas of a stem. It is generally held that herbaceous angiosperms phylogenetically have arisen from woody types—from this point of view, the narrow vascular cylinder or the ring of vascular bundles commonly developed in herbaceous dicotyledons represent highly reduced and specialized types of vascular systems (Fig. 19-22, A). In the stems of very many monocotyledons the vascular bundles are scattered throughout the fundamental tissue system, and there is no definable boundary between cortex and pith (Fig. 19-22, B). Monocotyledons as a class lack the cambial activity commonly present in dicotyledons, but in a few genera in the Liliaceae (e.g., *Yucca, Dracaena, Cordyline*) a peculiar type of secondary growth occurs which adds new vascular bundles and additional parenchyma to the primary body (Zimmermann and Tomlinson, 1970).

Nodal Anatomy

A salient feature of stem anatomy that has received considerable attention in recent years is the vascular anatomy of the node. In

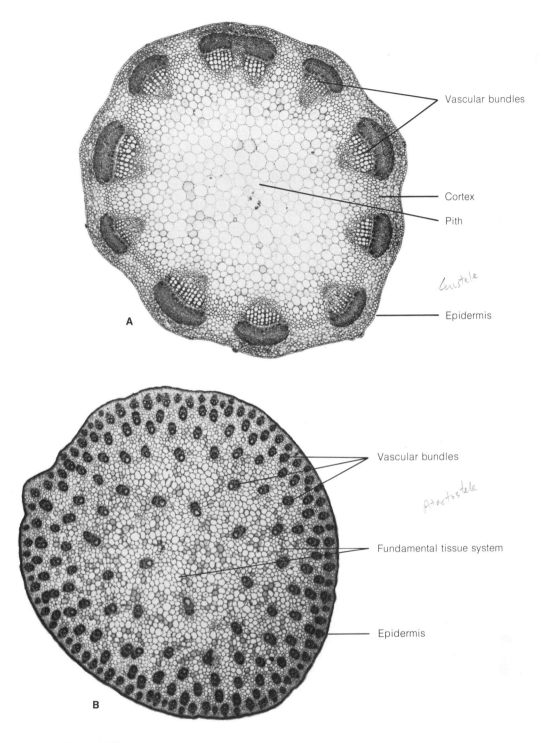

Vascular bundles

Cortex

Pith

Cnstele

Epidermis

A

Vascular bundles

Atactostele

Fundamental tissue system

Epidermis

B

Figure 19-22
Types of stem anatomy in angiosperms. **A,** transection of stem of *Trifolium* (a dicotyledon);
B, transection of stem of *Zea mays* (a monocotyledon) (note scattered arrangement of vascular
bundles and the absence of definable cortex and pith).

dicotyledons, for example, the vascular supply of a foliar appendage consists of one or more *leaf traces*, which diverge at the node into the leaf base and are associated with corresponding parenchymatous areas in the vascular cylinder known as *leaf gaps* (Fig. 19-23). We have already discussed, in Chapter 14, the difficulties which arise in attempting to define precisely what is meant by a "leaf gap" with reference to the primary vascular system. But in young stems of dicotyledons which have experienced secondary growth, serial transections of the node reveal well-defined parenchymatous areas — i.e., "gaps" in the secondary vascular cylinder — each related to the divergence of one or more leaf traces. Leaf gaps, in this descriptive sense, are characteristic features of the nodal anatomy of dicotyledons.

The significant investigations of Sinnott (1914) revealed that there are three principal forms of nodal anatomy in the dicotyledons as a whole: the *unilacunar node*, in which the leaf trace or traces are related to a single gap (Fig. 19-23, A, D); the *trilacunar node*, characterized by the association of three gaps with the diverging leaf traces (Fig. 19-23, B, C); and the *multilacunar node*, in which numerous leaf traces are related to a corresponding number of leaf gaps (Fig. 19-23, E). Sinnott's comprehensive survey led him to conclude that the trilacunar node is the primitive type in the angiosperms and that the unilacunar node arose phylogenetically either by the elimination of the two lateral gaps and their traces or as the result of the approximation of the two lateral traces, which, together with the median trace, formed a tripartite leaf trace related to a single gap in the vascular cylinder (Fig. 19-23, D). According to Sinnott's interpretation, the multilacunar node is also a derivative type which evolved from the trilacunar node by the formation of additional pairs of leaf traces (Fig. 19-23, E).

Sinnott's conclusions on the evolution of nodal structure and the subsequent investi-

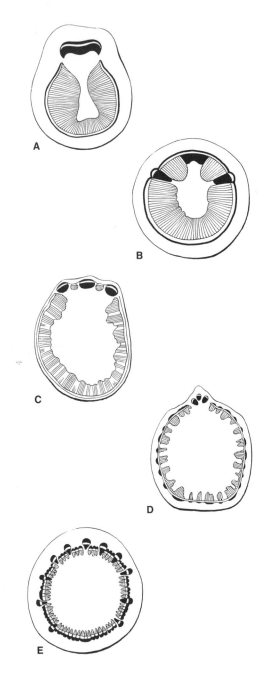

Figure 19-23
Transections illustrating some of the main types of nodal anatomy in dicotyledons. **A**, unilacunar node of *Eucalyptus* sp., **B**, trilacunar node of *Ilex opaca;* **C**, trilacunar node of *Sisymbrium leiocarpum* (Cruciferae); **D**, unilacunar node of *Barbarea* (Cruciferae); **E**, multilacunar node of *Acanthopanax.* [Redrawn from Sinnott, *Amer. Jour. Bot.* 1:303, 1914.]

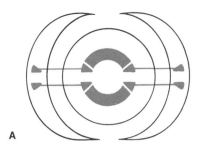

A

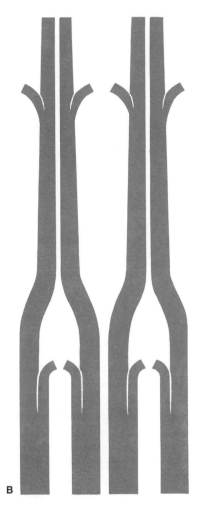

B

Figure 19-24
Diagrams illustrating the nodal anatomy of *Ascarina*.
A, transection showing that each member of a pair
of opposite leaves is vascularized by two traces,
related to a single gap; **B,** vascular system of stem
spread out in one plane to demonstrate the *independent attachment,* at the subnodal level, of the two
strands of each leaf trace. [Redrawn from Swamy,
Jour. Arnold Arboretum 34:375, 1953.]

gations of Sinnott and Bailey (1914, 1915)
seemed to indicate that the primitive leaf in
angiosperms was probably simple, palmately
veined and three-lobed, and associated with
a trilacunar node. More recent studies have
confirmed their conclusions that in many
genera and even families, the node is remarkably stable and hence of considerable
diagnostic value in systematic analyses. But
it is also evident, on the other hand, that
considerable variation in nodal anatomy may
be found, not only between the genera in
one family but even within a single genus.
Good examples of variable nodal anatomy
are provided by Post's (1958) investigations
on *Frasera* and *Swertia* (Gentianaceae),
Dickison's (1969) survey of the family Dilleniaceae and Philipson and Philipson's (1968)
very detailed study of nodal anatomy in
Rhododendron (Ericaceae) in which five different nodal types are described and used as a
basis for the systematic analysis of the genus.

The assumption that the trilacunar type of
node represents the ancient and primitive
form in the dicotyledons was later challenged by Canright (1955), Marsden and
Bailey (1955), and Bailey (1956). As Bailey
(1956) remarked, "it is essential to determine how reliable a working hypothesis is
provided by nodal anatomy."

First of all, Bailey questioned whether
all forms of unilacunar nodes are strictly
homologous — i.e., comparable from a
morphological-phylogenetic point of view.
In a number of seed plants with unilacunar
nodes, such as *Ginkgo*, certain species of
Ephedra, and several ranalean dicotyledons
with decussately arranged leaves, the foliage
leaf is vascularized by *two traces* which, at
subnodal levels, do not unite to form a single
vascular strand but are connected to separate
"bundle systems" of the stele (Figs. 19-24,
19-27). A good example of this distinctive
kind of vasculature among the angiosperms
is found in *Ascarina*, a member of the family
Chloranthaceae, as is shown in Fig. 19-24. In
Ascarina, as well as in the monotypic genus

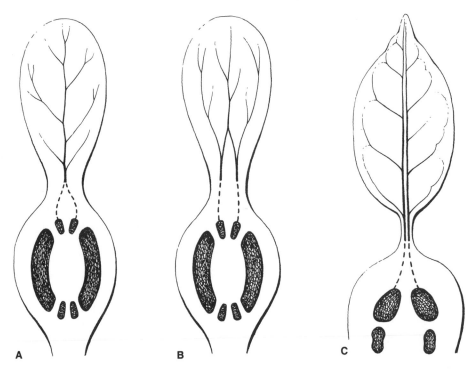

Figure 19-25
A, B, unilacunar nodal anatomy of cotyledons in dicotyledonous seedlings (note that in A the cotyledonary traces unite directly to form a midvein, whereas in B each trace dichotomizes, the two central strands joining to form a midvein); C, unilacunar nodal anatomy and vascularization of the foliage leaf in *Austrobaileya* (note that the two bundles remain separate in their course through the petiole and lamina). [A, B redrawn from Bailey, *Jour. Arnold Arboretum* 37:269, 1956; C redrawn from Bailey and Swamy, *Jour. Arnold Arboretum* 30:211, 1949.]

Austrobaileya (Austrobaileyaceae), the two leaf traces extend as separate strands through the petiole and may remain unconnected throughout the midrib or else unite to form a midvein in the middle or upper part of the lamina (Fig. 19-25, C). The unilacunar, two-trace type of node just described raises an important phylogenetic question: Is this form of nodal structure, rather than the trilacunar node, the truly primitive form that has been retained during the evolution of certain angiosperm genera and families? The commonness of this kind of nodal architecture in ferns and certain gymnosperms (e.g., *Ephedra* and *Ginkgo*), is cited by Bailey (1956, p. 284) in giving an affirmative answer to this question. But additional support

for the concept of the primitiveness of the two-trace unilacunar node was found by Bailey in his study of the nodal anatomy and venation of angiosperm cotyledons. In his survey of 99 families of dicotyledons, he discovered that in 77 percent of them the cotyledons have an *even number* of traces (i.e., 2 or 4) at the nodal level and that 60 percent of them have two *independent* traces associated with a single gap. Very commonly, as is shown in Fig. 19-25, B, the two traces dichotomize and the two derivative central strands then unite to form the midvein of the lamina. In other patterns, however, the midvein is formed by the precocious fusion (without preliminary dichotomy) of the two cotyledonary traces (Fig. 19-25, A). In other words,

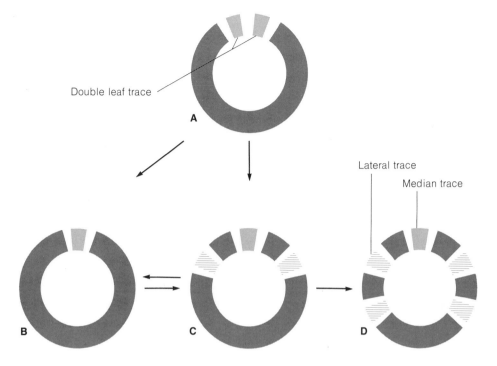

Figure 19-26
Phylogenetic origin of the uni-, tri-, and multilacunar types of nodal anatomy in dicotyledons.
A, primitive type of unilacunar node with double leaf trace; **B,** unilacunar node derived from **A** by
fusion of two traces; **C,** trilacunar node derived from either **A** or **B** by addition of a pair of lateral
traces and gaps (arrow pointing left indicates possible origin of **B** from **C** by secondary elimination
of lateral gaps and traces); **D,** multilacunar node derived from **C** by addition of a pair of lateral
traces and gaps. [Redrawn from Marsden and Bailey, *Jour. Arnold Arboretum* 36:1, 1955.]

the vasculature of many cotyledons represents a "transition" from an *even* number of
traces at the nodal level to an *odd* number
(1 or 3) of primary veins in the base of the
lamina. Comparable transitions from an even
to an odd number of strands also occur in
the *foliage leaves* of certain genera of the
Monimiaceae (Money, Bailey and Swamy,
1950) and the Chloranthaceae (Swamy,
1953). These facts led Bailey (1956) to conclude that differences between the vascularization of cotyledons and foliage leaves, at
the nodal level, "are quantitative rather than
qualitative, an *even* number of vascular
strands being commoner in the case of cotyledons and an *odd* number in the case of
leaves."

Once established in evolution, the primitive two-trace type or the derived one-trace
type of unilacunar node is believed to have
evolved first into the trilacunar and then into
the multilacunar node by the formation of
additional pairs of lateral leaf traces. Fig.
19-26 depicts this possible line of evolution
of nodal types and also suggests that in some
instances, a unilacunar node may *secondarily*
arise from a trilacunar node by the loss of the
two lateral traces. According to Bailey (1956,
p. 275), such families as the Leguminosae
and Anacardiaceae provide illustrations of
reversible changes in evolution.

Bailey's conclusion that the unilacunar,
two-trace node is the primitive form in the
angiosperms has served to stimulate a

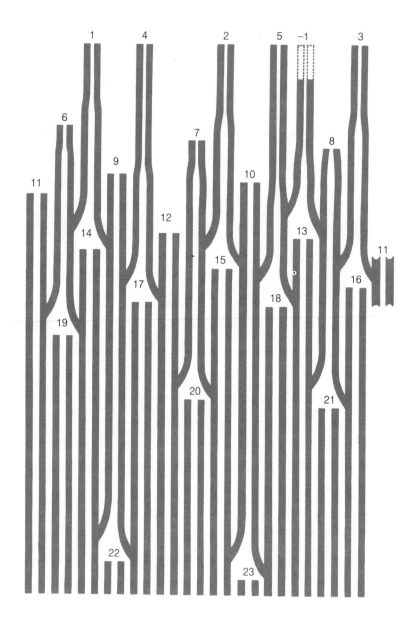

Figure 19-27
Diagram showing the interconnected strands of the vascular cylinder of
the stem of *Ginkgo biloba* spread out in one plane. Each numeral (1–23)
in the diagram indicates the position of a leaf and its double trace.
Note particularly that the two strands composing *each* leaf trace do not
become joined at subnodal levels but are attached to separate bundles
of the stele. [Redrawn from Gunckel and Wetmore, *Amer. Jour. Bot.*
33:532, 1946.]

needed revival of interest in the phylo- genetic and systematic implications of nodal anatomy. Although his general viewpoint has been accepted by a number of morphol- ogists (e.g., Eames, 1961; Carlquist, 1961; Philipson and Philipson, 1968), several in- vestigators have raised objections to his theory. In particular, doubt has been cast on the validity of using evidence from the cotyledonary node in the solution of phylo- genetic problems. Benzing (1967), for exam- ple, completely rejects Bailey's conclusions on the evolution of nodal types and main- tains that "the anatomy of cotyledonary nodes does not necessarily reflect ancestral conditions in the mature stem," an opinion which is also shared by Conde and Stone (1970). Benzing's argument rests on his belief that *alternate leaf arrangement* is primi- tive and that hence the paired arrangement of cotyledons and the decussate phyllotaxis of foliage leaves (e.g., the leaves in members of such ranalean families as the Chlorantha- ceae, Monimiaceae, and Austrobaileyaceae) both are "derived" patterns. In this connec- tion, however, it should be emphasized that at present there is no compelling evidence to support the conclusion that decussate phyllotaxis is a "derivative" pattern of leaf arrangement *throughout* the angiosperms. Cronquist (1968, p. 66) for example, in his discussion of phyllotaxis, notes that in such an advanced family as the Compositae, "it is perfectly clear that opposite leaves are primi- tive and alternate leaves are advanced." It should also be emphasized, in fairness to Bailey's broad treatment of nodal anatomy in the Tracheophyta, that one of the most outstanding examples of a unilacunar, two- trace type of node is that of *Ginkgo biloba*, a primitive living gymnosperm in which the phyllotaxis of the dichotomously veined foliage leaves is *alternate* and not decussate (Fig. 19-27).

In conclusion, it is fortunate for the prog- ress of our knowledge that the phylogeny

and classification of nodal types in the angio- sperms are not "closed issues" at the present time. Takhtajan (1969, p. 52), for example, holds the view that "the basic type of angio- sperm node was one with three or more gaps" and was more likely to have been multilacunar than trilacunar. He is also con- vinced that in the primitive trilacunar node, the *median* leaf trace was double. His theory, in contrast to Bailey's view, thus postulates that the unilacunar two-trace node arose secondarily by the *loss* of the two lateral leaf traces (see Takhtajan, 1969, p. 51, Fig. 3). Apart from the need for continued explora- tions of nodal anatomy among living angio- sperms (see, for example, Howard, 1970), more detailed information is also required about how common the unilacunar two-trace type of node is in fossil groups, such as the seed ferns, progymnosperms, and gymno- sperms. For a critical view of this aspect of the problem of nodal evolution the student is referred to the paper by Pant and Mehra (1964).

Wood Anatomy

One of the most productive approaches of modern comparative investigations on the systematics and phylogeny of the angio- sperms is the study of the structure and development of secondary xylem ("wood") that constitutes the major *permanent* tissue of the stems and roots of woody plants (see Metcalfe and Chalk, 1950; and Cheadle, 1956). From an histological point of view, secondary xylem is a highly complex tissue which may be resolved, in its basic organi- zation, into two integrated systems. These are clearly distinguished by the type and orientation of the component cells. In the *axial system*, which is composed of tracheary elements (tracheids, vessel members), wood fibers, and axial strands of living parenchyma, the cells are arranged with their long axes parallel to the longitudinal axis of the stem and root. The *ray system*, in contrast, comprises the xylem rays, which are sheets (one or more cells in thickness) of parenchyma cells, oriented radially with reference to the cells of the axial system. The cells of the two systems, interconnected by means of pit pairs, originate from the vascular cambium. The *fusiform initials* of the cambium, by means of tangential divisions, add new cells to the axial system; the *ray initials*, by similar divisions, contribute additional cells to the existing rays. (Fig. 19-28). As growth in diameter of the xylem cylinder continues, new ray initials originate within the cambium and produce new xylem rays.

To understand the complex structure of any wood, it is essential to examine the tissue in transverse, radial, and tangential planes of section. By way of illustration, a small portion of the secondary xylem of *Liriodendron* (Magnoliaceae) is shown in Fig. 19-28 in the form of a three-dimensional "block diagram." This type of representation is particularly helpful to the student in his effort to visualize properly the structure and spatial extent of the wood rays as seen in the three planes of section. In transverse section, the wood rays appear as radially oriented series of living parenchyma cells which extend from the vascular cambium for varying distances into the wood. Radial sections reveal that each xylem ray consists of several tiers of "procumbent cells" flanked marginally by a series of "upright cells" that are conspicuously elongated in the vertical direction (Fig. 19-28). A ray of this type is termed *heterocellular*, in contrast to the *homocellular* rays found in the woods of certain other dicotyledons, which consist of only procumbent or only upright cells. Tangential sections reveal the rays in transverse view; in this plane of section, each ray is characteristically lenticular or elliptical (Fig. 19-28). The combined use of transverse, radial, and tangential sections is also essential to an understanding of the axial system of the wood, particularly with reference to (1) the

distribution and structure of vessel members and other cell types; (2) the location and types of perforation plates in the vessel members; and (3) the nature of the pit pairs on the lateral walls of contiguous cells. As shown in Fig. 19-28 for example, the scalariform perforation plates, characteristic of the vessel member of *Liriodendron*, are displayed in face and sectional views, respectively, in the radial and tangential sections of the wood.

In recent years, increasing attention has been focused on the problems of the origin, structure, and systematic distribution of *vessels* in angiosperm woods. Vessels seem to provide some of the most reliable type of evidence regarding the general phylogeny of the xylem, not only in the angiosperms but in the vascular plants as a whole (Bailey, 1944, 1953). The salient feature of a vessel is that it consists of a series of cells which are interconnected by *perforations*—i.e., open areas located most commonly on the end walls of the cells (Fig. 19-29, F–I). The cells of a vessel are termed *vessel members* or *vessel elements*. In contrast to vessel members, *tracheids* are imperforate cells which are interconnected by bordered pit pairs and which do not form well-defined longitudinal cell series.

Comprehensive surveys on the comparative anatomy of the xylem of extinct and living tracheophytes have revealed that in all probability vessel members originated phylogenetically from tracheids, as the result of the dissolution of the membranes of bordered pit pairs located at the terminal points of contact of the members of a vertical series of cells. In the Gnetopsida, as we have already pointed out (Chapter 18), the *initial steps* in the formation of vessel elements resulted from the loss of the membranes between groups of circular bordered pit pairs. But in *Selaginella*, in certain ferns (*Pteridium, Marsilea*), and in the angiosperms, the initial step in vessel evolution was the loss of the membranes of a series of

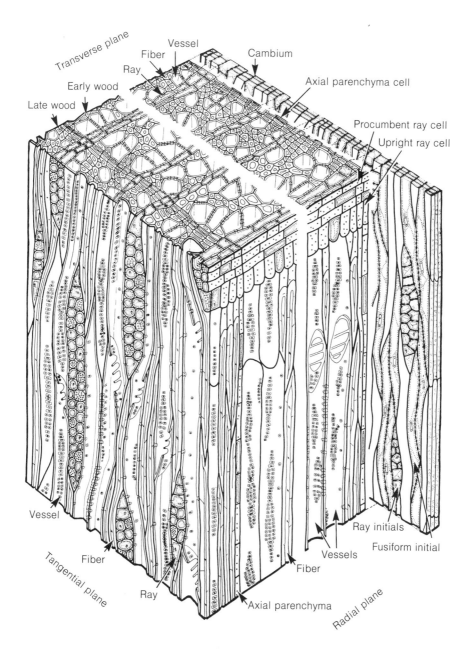

Figure 19-28
Block diagram showing the cells of the axial and ray systems of the secondary xylem of
Liriodendron tulipifera in transverse, radial, and tangential planes of section. A small
portion of the vascular cambium (separated from the wood for the sake of clarity) is
shown at the right of the diagram. See text for further explanation. [From *Plant Anatomy*,
Ed. 2, by K. Esau. Wiley, New York, 1965.]

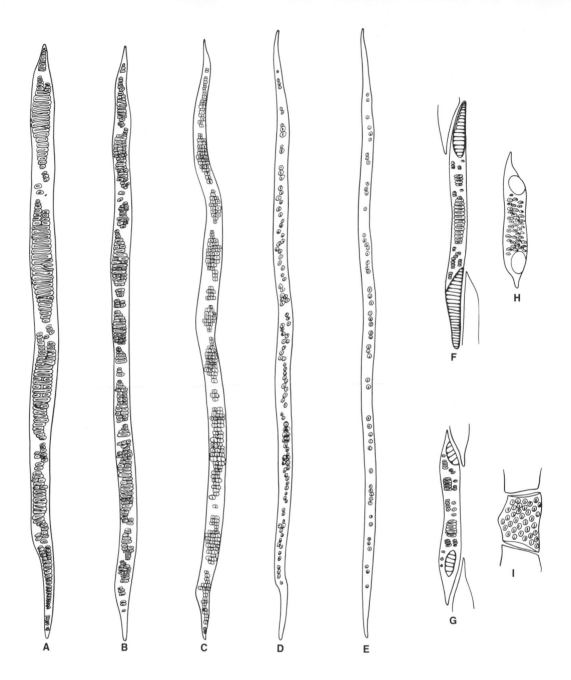

Figure 19-29
Various types of tracheary elements of the secondary xylem of dicotyledons. **A–E,** tracheids characteristic of the primitively vessel-less wood of *Tetracentron, Trochodendron,* and *Drimys,* showing scalariform bordered pits (**A**), circular bordered pits (**D, E**) and transitions between scalariform and opposite pitting (**B, C**). **F–I,** various types of vessel members formed in more specialized woods: **F, G,** vessel members with scalariform perforation plates and lateral walls with scalariform-opposite pitting; **H, I,** vessel members with simple perforations and lateral walls with alternately arranged bordered pits. In **F, G,** and **I,** portions of the vessel members lying above and below the cell shown in detail are represented in outline only. See text for further discussion.
[Redrawn from Bailey and Thompson, *Ann. Bot.* 32:503, 1918.]

transversely elongated pit pairs located near the ends of long tracheid-like cells. These facts all point to a very important general conclusion: vessels originated *independently* in a number of the major taxa of vascular plants and hence the presence of vessels is not by itself a valid proof of systematic affinity, at least for the major groups in the Tracheophyta (Bailey, 1944, 1953; Cheadle, 1953; Cheadle and Tucker, 1961).

Within the angiosperms—both mono-cotyledons and dicotyledons—the conclusion seems inescapable that the most primitive type of vessel member closely resembles a long scalariformly pitted tracheid from which it differs only in the presence of a series of transverse slit-like openings which collectively constitute the scalariform perforation plates at each end of the cell (Fig. 19-29, F). The prototype of tracheid, from which such a primitive vessel member originated, fortunately is preserved in the woods of a series of living vessel-less ranalean plants which have received considerable attention from both anatomists and systematists (Bailey and Thompson, 1918).

At the beginning of the present century, vessel-less wood had been found in *Trochodendron, Tetracentron,* and various genera in the family Winteraceae. Van Tieghem (1900), the celebrated French anatomist, was so impressed with this unique type of xylem structure that he grouped all these vessel-less dicotyledons into a special order which he named the "Homoxylées." This taxonomic segregation was, however, rejected by later anatomists as it became increasingly evident that, although *Trochodendron* and *Tetracentron* show some degree of relationship, they clearly have no direct affinity with the Winteraceae when morphological characters other than wood structure are given due consideration (Bailey and Nast, 1945a, b). Moreover, two additional genera of vessel-less dicotyledons have been discovered and

studied: *Amborella,* a monotypic genus endemic to New Caledonia (Bailey and Swamy, 1948; Bailey, 1957), and *Sarcandra,* an Asiatic genus in the Chloranthaceae (Swamy and Bailey, 1950). It is thus evident that the living, primitively vessel-less dicotyledons can no longer by regarded as a small "anomalous" group of angiosperms. On the contrary, the ten known genera are morphologically diversified and in habit range from shrubs (*Sarcandra, Amborella*) to large, long-lived trees (*Trochodendron, Tetracentron,* and certain representatives of the Winteraceae). It seems likely that, as comparative anatomical studies on the angiosperms continue, other examples of vessel-less dicotyledons will be discovered.

Because of the phylogenetic importance of vessel-less angiosperm wood it will be now desirable to describe briefly a representative example. We have selected for this purpose the wood of *Tetracentron sinense,* a large tree native to the montane forests of central and western China. As seen in transverse section, the wood of *Tetracentron* consists of well-defined growth layers (i.e., "annual rings") in which the tracheids of both the early and late wood are arranged in very regular radial rows, much like the orderly series of tracheids in the wood of a gymnosperm (Fig. 19-30, A). The rays, as seen in a transection of the wood, are relatively numerous and include both uniseriate and multiseriate types. Tangential sections (Fig. 19-30, B) emphasize the structural difference between the two kinds of rays and also reveal the very long tracheids with their characteristically overlapping ends. One of the most significant aspects of tracheid structure in *Tetracentron* is the character of the pitting on the lateral walls. As seen in radial section (Fig. 19-30, C) the walls of the tracheids of the early wood have typical scalariform pitting; i.e., the closely spaced, transversely elongated pits are arranged like the rungs of a ladder, hence

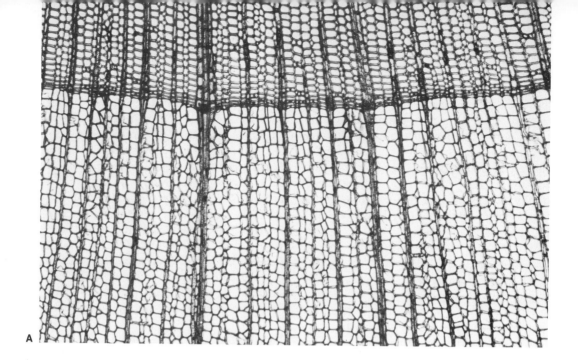

A

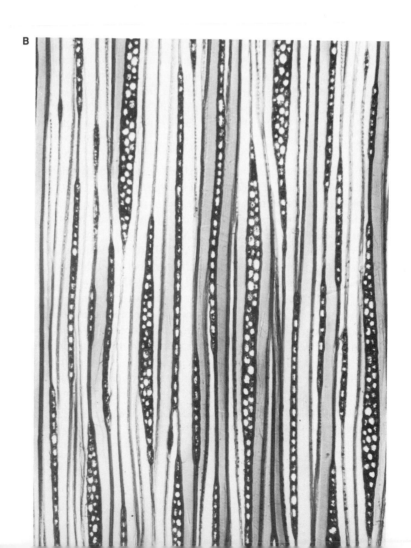

B

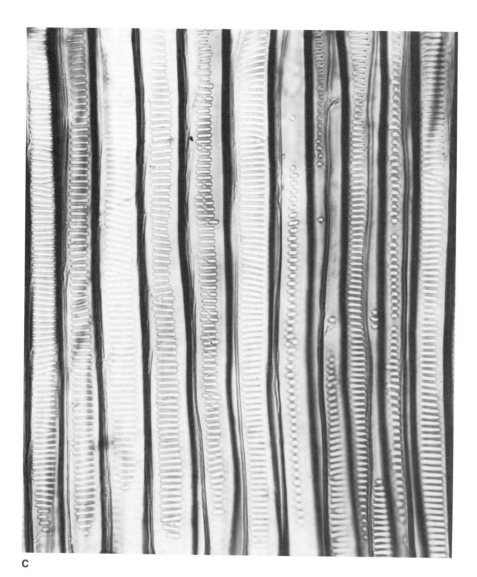

C

Figure 19-30
Structure of the primitively vessel-less wood of the stem of *Tetracentron sinense*. **A**, transection, showing part of a growth layer ("annual ring") composed of tracheids arranged in very regular radial rows and the characteristic uniseriate and multiseriate rays. Note the sharp demarcation between the early, or spring, wood (large, thin-walled tracheids) and the late, or summer, wood (smaller, thicker-walled tracheids); **B**, tangential section showing the characteristic elongated tracheids with overlapping ends, and the structure of the uni- and multiseriate rays; **C**, radial section passing through the early wood, showing the conspicuous scalariform pitting on the lateral walls of the tracheids. See text for further comments. **A**, × 92; **B**, **C**, × 124. [Photomicrographs made by A. A. Blaker, from sections prepared by Charles Quibell.]

the term "scalariform." (See also the isolated cell shown in Fig. 19-29, A, and the diagram in Fig. 19-31, A). In contrast, the thicker-walled tracheids of the late wood of *Tetracentron* — as in the late wood of *Trochodendron* and *Drimys* — are provided with circular bordered pits on their radial facets (Fig. 19-29, D, E). Frequently, as is shown in Fig. 19-29, B, certain parts of the wall of a tracheid may show transitions between typical scalariform and opposite pitting (i.e., the formation of several oval pits in a transverse

row) as well as transitions between opposite and alternately arranged pits (see Fig. 19-31 for diagrammatic representations of the various patterns of bordered pits of tracheary elements).

It is significant that in dicotyledonous woods which form vessels, scalariform pitting is frequently, although by no means invariably, associated with vessel members that develop long scalariform perforation plates (Fig. 19-29, F). Such primitive vessel members are thus closely similar in structure to the longer scalariformly pitted tracheids of *Tetracentron, Trochodendron,* and other vessel-less dicotyledons. The various steps in the further evolutionary development of vessel members from tracheid-like cells are extensively preserved in the secondary xylem of more advanced dicotyledons and have been analyzed in detail by Bailey (1944). The major trends of specialization include (1) the marked *decrease* in the length of the vessel members, a change which reflects a corresponding decrease in length of the fusiform cambial initials, (2) an *increase* in the lateral expansion of vessel members during their ontogeny, culminating, in the most specialized examples, in short drum-shaped elements with truncated ends, (3) a gradual transition, by a reduction in the number of perforations, from scalariform to simple perforations, and (4) the replacement of scalariform intervessel pitting by circular bordered pits, arranged first in opposite and finally in alternate seriation. These four major trends in the phylogeny of vessel members are represented diagrammatically in Fig. 19-29, F–I.

Thus far our discussion of vessel evolution in the angiosperms has been restricted to the dicotyledons, in which the evidence is clear that vessels first appeared in the secondary xylem and subsequently "worked backward," in their ontogenetic appearance, through the primary xylem (Bailey, 1944). A different and *independent mode of origin* of vessels,

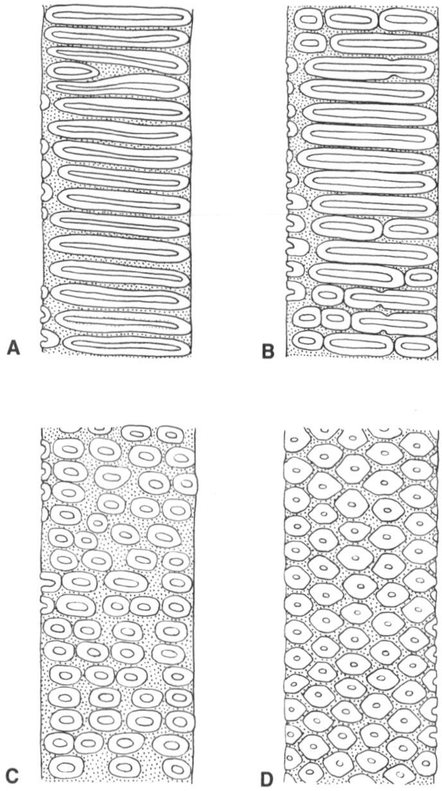

Figure 19-31
Form and arrangement of bordered pits on the lateral walls of tracheary elements. **A,** transversely elongated pits arranged in a scalariform series; **B,** scalariform opposite pitting; **C,** pits arranged in horizontal rows (opposite pitting); **D,** alternately arranged, crowded, bordered pits. [Redrawn from *Plant Anatomy* by A. Fahn. Pergamon, Oxford, 1967.]

however, has occurred in the monocotyledons. According to the voluminous data assembled through the researches of Cheadle (1953, 1955), vessels in the monocotyledons first appeared in the *metaxylem* (i.e., the last formed part of the primary xylem) *of the roots.* Indeed, in many monocotyledonous taxa, vessels are entirely restricted to the roots and the aerial parts of such plants are vessel-less! Because of this remarkable fact, Cheadle, in his detailed studies, has been able to trace the progressive development of vessels from an organographic standpoint (see also Cheadle and Tucker, 1961). A low level of specialization, for example, would be represented by those taxa that have vessels only in the roots, an intermediate level by taxa that have vessels in both roots and stems, and the highest level of specialization by those that have vessels throughout all parts (stems, inflorescence axes, leaves) of the plant. It is highly significant that Cheadle's investigations also show that the *structural evolution* of vessel members in monocotyledons closely parallels that characteristic of the dicotyledons. Vessel members in monocotyledons were derived from scalariformly pitted tracheids; subsequent specialization, as in dicotyledons, involved a decrease in length and an increase in diameter of vessel members accompanied by a transition from scalariform to simple perforations.

In conclusion, the broad systematic implications of the independent origin and parallel trends of specialization of vessels in dicotyledons and monocotyledons need consideration. Bailey (1944) remarked that "if the angiosperms are monophyletic, the monocotyledons must have diverged from the dicotyledons before the acquisition of vessels by their common ancestors. This renders untenable all suggestions for deriving monocotyledons from vessel-bearing dicotyledons or *vice versa.*" In Cheadle's (1953) view, however, the most likely

hypothesis is to assume "that the monocotyledons were derived (by reduction of vascular cambium activity, or by modification of this activity, as illustrated in the Agavales and elsewhere) from undiscovered fossil or living woody dicotyledons that lacked vessels throughout the plant." (For further critical discussions of the significance of vessels with reference to the origin of monocotyledons, see Kosakai, Moseley, and Cheadle, 1970.)

Roots

Although roots perform very important functions, such as anchorage, absorption and conduction of water and mineral solutes, and food storage, these essential organs of vascular land plants have never been studied morphologically or anatomically as thoroughly as leaves and stems. Some of the reasons for the neglect of broad comparative studies on roots are: (1) the technical problems of dealing satisfactorily with the growth and structure of subterranean organs, (2) the wide divergence in anatomy between "typical" subterranean roots and the highly modified roots of tropical epiphytes, mangroves, and climbing plants, and (3) the absence of convincing paleobotanical evidence regarding the phylogenetic origin of the root and the trends in its evolutionary specialization.

Among the most comprehensive of modern publications on roots may be cited the exhaustive monographic work by Troll (1967) and the detailed histological treatments of the primary structure of the root in gymnosperms and angiosperms provided by von Guttenberg (1941, 1968). Since it is obviously impossible in the present text to consider in any detail the development and structure of roots, the following account is simply a brief résumé of the "highlights" of a very complex and inadequately explored subject.

GENERAL MORPHOLOGY. It was pointed out briefly in Chapter 6 that the terms *homorhizic* and *allorhizic* were introduced by Goebel (1930) to designate the two main patterns of origin of the first root in the embryo of vascular plants. In allorhizic plants, represented by the living gymnosperms and angiosperms, the embryo is bipolar—i.e., root and shoot apices lie at opposite ends or poles of the embryonic axis (Chapter 20). In homorhizic plants, illustrated by all lower tracheophytes, the first root is lateral with reference to the embryonic axis (Chapter 6, Fig. 6-2).

Allorhizic plants or "allorhizophytes" as Troll (1949) designates them, may form a well-defined root system consisting of the main or primary root (derived from the radicle of the embryo) and its lateral branches (Fig. 19-32, A). On the other hand, in typical "homorhizophytes," such as *Lycopodium* and ferns, aside from the first root in the embryo, all subsequent roots arise from the stem of the sporophyte (Chapter 3, Fig. 3-10). According to Troll's (1967, p. 2126–2127) viewpoint, plants characterized by "primary homorhizy" (i.e., the lower vascular plants) should not be considered to have a definable root system because *all* roots are shoot borne. An analogous kind of organization, termed "secondary homorhizy" by Troll (1949) is characteristic of many angiosperms and is virtually the predominant type of root organization throughout monocotyledons. In angiosperms that exhibit secondary homorhizy, the primary root of the seedling is short-lived or else weakly developed, and lateral roots form mainly from various regions of the stem (Fig. 19-32, B).

Despite the often consistent pattern of origin of shoot-borne roots from specific nodal or internodal regions of the stem, such roots are very commonly termed "adventitious." We believe this term is inappropriate, because the shoot-borne "prop roots" of

Zea, and the aërial roots of English Ivy, to cite striking examples, are as much a part of the "normal" morphology of these plants as their leaves or flowers! It therefore seems more in keeping with the facts to restrict the expression "adventitious roots" to those roots which arise *de novo* as a result of regenerative processes in callus tissue of stem cuttings or from the "dedifferentiation" of cells in detached leaves of such plants as *Begonia* and African violet (*Saintpaulia ionantha*). In these instances, as well as in laboratory cultures of callus tissue, the formation of root primordia is truly an adventive phenomenon and not part of the normal developmental morphology of the plant as a whole.

ANATOMY. Some of the remarkable anatomical and ontogenetic differences between the primary structure of roots and shoots deserve attention in our consideration of the vegetative morphology of the angiosperms and will be treated briefly at this point. For more detailed accounts of the comparative and developmental anatomy of the root the student should consult Troll (1967), Esau (1965), and von Guttenberg (1968).

THE ROOT APEX. The apices of shoots and roots in vascular plants differ strikingly in their growth and structure. In the shoot, the apical meristem is *superficial* and is overarched by the exogenous foliar organs which it has produced while the corresponding meristem in the root tip is, strictly speaking, *subterminal* because it is covered by a protective root cap (Chapter 3, Fig. 3-1).

Expressed in ontogenetic terms, cell differentiation in the root tip is bidirectional, in that the outwardly produced cells become part of the root cap while the inwardly derived cells are added to the body of the root. The precision of these opposed patterns of cell formation is vividly illustrated in certain lower plants (e.g., leptosporangiate

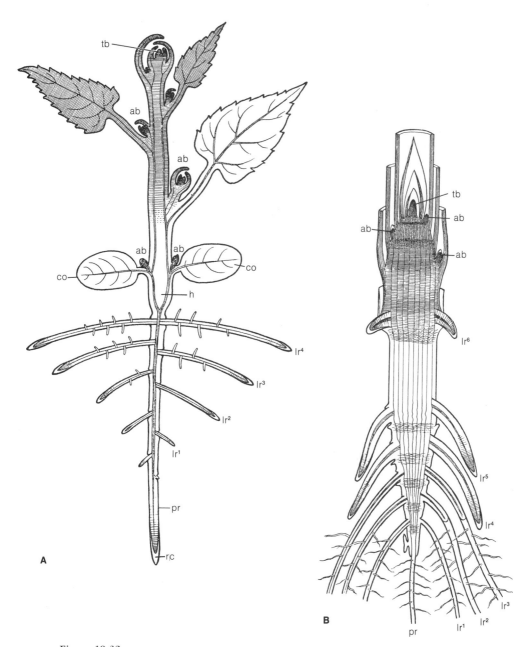

Figure 19-32
A, diagrammatic representation of a typical allorhizic dicotyledonous plant, as seen in longisectional view. Note acropetal sequence in formation of the endogenous lateral roots from
primary root; **B,** diagram of a young homorhizic monocotyledonous plant (*Zea mays*), as
seen in longitudinal section, to illustrate development of lateral roots (lr^1–lr^6) from lower
internodes of stem. Note weak development of primary root and its branches. ab, axillary
bud; co, cotyledon; h, hypocotyl; lr, lateral root; pr, primary root; rc, root cap; tb, terminal
bud. [Redrawn from *Vergleichende Morphologie der höheren Pflanzen* by W. Troll. Gebrüder
Borntraeger, Berlin, 1935.]

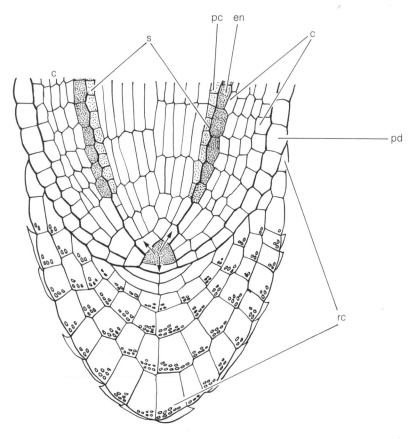

Figure 19-33
Diagrammatic representation of median longisection through the root tip of *Equisetum arvense,* showing conspicuous apical cell (demarcated by stippling). The arrows indicate that cell lineages of root cap as well as root body originate from segments derived from the apical cell. c, cortex; en, endodermis; pc, pericycle; pd, protoderm; rc, root cap; s, stele. [Redrawn from *Grundzüge der Pflanzenanatomie* by B. Huber. Springer-Verlag, Berlin, 1961.]

ferns such as *Pteris,* and *Equisetum*) in which *all* tissue at the root apex is derived from the segments of a single conspicuous four-sided apical cell (Fig. 19-33). Divisions parallel to the *outer face* of the apical cell contribute exclusively to the root cap; divisions parallel to its three *inner faces* yield additions to the root body. An apical cell is also present in the shoot apex of the same plants but there are no divisions parallel to the outer face of the apical cell and hence a cap is not produced. In angiosperms the root cap originates in various ways: in many monocotyledons from a separate multi-cellular "histogen" termed the *calyptrogen*;

in a wide range of dicotyledons from a *dermatocalyptrogen* which forms *both* the root cap and the protoderm layer of the root body (see Esau, 1965, pp. 120–123).

THE VASCULAR CYLINDER. One of the most interesting aspects of the anatomy of vascular plants is the difference in structure between the primary vascular systems of root and stem. In typical stems of gymnosperms and dicotyledons, the primary phloem and xylem are *collateral* in arrangement and in the form either of a "ring" of discrete vascular bundles or a more-or-less

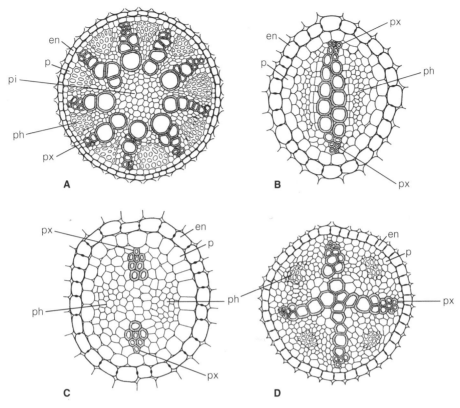

Figure 19-34
Transections of the vascular cylinders of the roots of various angiosperms. **A,** polyarch root of *Acorus calamus;* **B,** diarch root of *Nigella damascena;* **C,** diarch root of *Fumaria grandiflora;* **D,** tetrarch root of *Ranunculus acris.* en, endodermis; p, pericycle; ph, phloem; pi, pith; px, protoxylem. [Redrawn from *Vergleichende Morphologie der höheren Pflanzen* by W. Troll. Gebrüder Borntraeger, Berlin, 1967.]

continuous cylinder which encloses the central pith (Fig. 19-22, A). In sharp contrast, the primary phloem and primary xylem in roots are arranged in an alternate and radial pattern—i.e., phloem and xylem do not lie on the same radius but are relatively discrete, alternating strands (Fig. 19-34). The center of the vascular cylinder, especially in monocotyledons, may be occupied by a pith surrounded by the separate strands of phloem and xylem (Fig. 19-34, A). But very commonly in the roots of dicotyledons, the originally separate xylem strands or "rays" become joined in the center,

forming a ridged or fluted core of primary xylem whose pattern resembles the actinostele characteristic of both the stem and root of such lower vascular plants as *Lycopodium* (Fig. 19-34, D).

It must be strongly emphasized that the direction of radial maturation of the primary xylem in roots is centripetal, with the protoxylem as the outer pole of each strand or ray and the metaxylem in a more central position (Fig. 19-34). Thus the root is characterized by *exarch xylem* in contrast to the prevalent *endarch xylem* of angiospermous stems (for further information on patterns of primary xylem differentiation, see Chapter 3, Fig. 3-20).

On the basis of the number of protoxylem poles the vascular structure of roots is designated as diarch (2), triarch (3), tetrarch (4), etc.; for convenience, roots with a large number of protoxylem poles are termed polyarch (Fig. 19-34, A). Polyarch roots are very frequently encountered in the monocotyledons—in, for example, certain palms, the roots may have 100 or more alternating strands of xylem and phloem (Tomlinson, 1961, p. 51). Dicotyledonous roots very commonly are diarch, triarch, or tetrarch, as illustrated in Fig. 19-34, B–D). According to von Guttenberg (1968), the number of xylem rays may be consistent within taxa and even may be of systematic value in characterizing certain families. Diarch roots, for example, seem to predominate in the Cruciferae, Caryophyllaceae, Umbelliferae, Chenopodiaceae, Compositae, and Valerianaceae; the roots of many members of the Euphorbiaceae and Cucurbitaceae are tetrarch. But frequently the number of protoxylem poles may fluctuate, not only between the different roots of one plant but even at different levels in the same root. Apparently, such variation is correlated in many instances with the diameter of the vascular cylinder. In roots with slender cylinders, the number of xylem poles is likely to be less than in roots with wider vascular cylinders.

Despite the considerable morphological and physiological interest of the vascular system in roots, surveys of a wide range of both monocotyledons and dicotyledons are essential in order to determine the value of root anatomy in elucidating taxonomic problems. From available published descriptions on root anatomy in monocotyledons, it is evident that in certain genera and families there are many bizarre and unusual structural features which should be considered by the systematist. In grasses, for example, one

or more large vessels develop in the inner or central region of the vascular cylinder in addition to the typical peripheral series of alternately arranged phloem and xylem strands (Esau 1965, Pl. 83 and 84). Perhaps the most complex and puzzling type of root anatomy is illustrated in certain members of the Strelitziaceae and Musaceae (see Tomlinson, 1969, and Riopel and Steeves, 1964) in which the center of the root is traversed by numerous scattered vessel strands and separate strands of internal phloem tissue

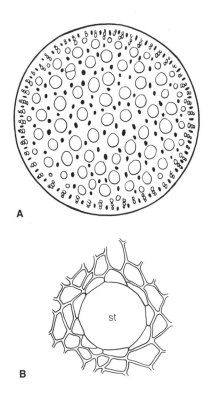

Figure 19-35
A, structure of the vascular cylinder of the root of *Ravenala madagascariensis* as seen in transection. In addition to a large number of alternating xylem and phloem strands at the periphery of the vascular cylinder, the ground tissue (white) is traversed by numerous scattered strands of phloem (black dots) and xylem (circles). B, details of a phloem strand, showing the single large sieve tube (st). [Redrawn from *Anatomy of Monocotyledons* by P. B. Tomlinson. Clarendon Press, Oxford, 1969.]

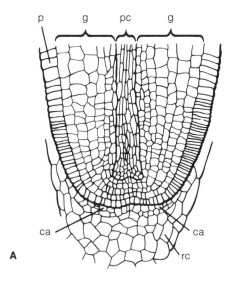

p g pc g

ca ca

A rc

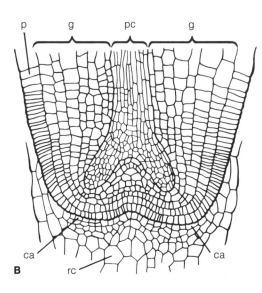

p g pc g

ca ca

B rc

Figure 19-36
Median longisection of the dichotomizing root tip
of *Lycopodium inundatum*. **A,** early phase in dichotomous branching, indicated at the apex of the procambial cylinder as well as in the calyptrogen, which
has begun to form two new root caps; **B,** later stage,
showing forked procambial cylinder and the corresponding formation of two separate groups of
calyptrogen initials. ca, calyptrogen initials of the
new root apices; g, ground meristem; p, protoderm;
pc, procambial cylinder; rc, root cap of "mother
root." [Redrawn from *Vergleichende Morphologie der
höheren Pflanzen* by W. Troll. Gebrüder Borntraeger,
Berlin, 1967.]

(Fig. 19-35). It seems reasonable to suppose
that "anomalous" root anatomy of this sort
may prove very useful in the future study
of systematic relationships within the
monocotyledons.

ORIGIN OF LATERAL ROOTS. In a few of
the lower vascular plants, such as *Lycopodium,
Selaginella,* and *Isoetes,* the roots branch dichotomously (Fig. 19-36). Aside from these
exceptions, the method of origin of lateral
roots is uniformly *endogenous* in sharp contrast to the *exogenous* development of
branches from buds in the shoot system
(Chapter 3, Fig. 3-1, B).

The primordia of lateral roots in seed
plants typically arise by the localized division of cells of the pericycle, a cylinder of
potentially meristematic cells representing
the outer boundary of the vascular cylinder
(Fig. 19-37, lr'). Usually the first evidence of
lateral-root initiation is seen at some distance from the apex of the parent root and
the normal sequence of formation of root
primordia is strictly acropetal. During its
differentiation and enlargement, a lateral-
root primordium gradually pushes its way
through the cortical parenchyma of the
parent root. Before it enters the soil as a
functional organ, the young lateral root has
a definable root cap, a subterminal meristem
and an embryonic axis (Fig. 19-37, lr'');
usually it has also established vascular connections with the adjacent phloem and
xylem strands of the mother root (Fig.
19-32, A).

The *rhizotaxy,* or position of lateral roots,
is not random but is clearly related to the
radial arrangement of the xylem and phloem
strands in the parent root (Mallory et al.,
1970). In triarch, tetrarch, and pentarch
roots, for example, lateral roots are initiated
in those parts of the pericycle which lie
opposite the protoxylem poles (Fig. 19-37).
Thus, unless modified by injuries, the number of vertical rows of lateral roots very

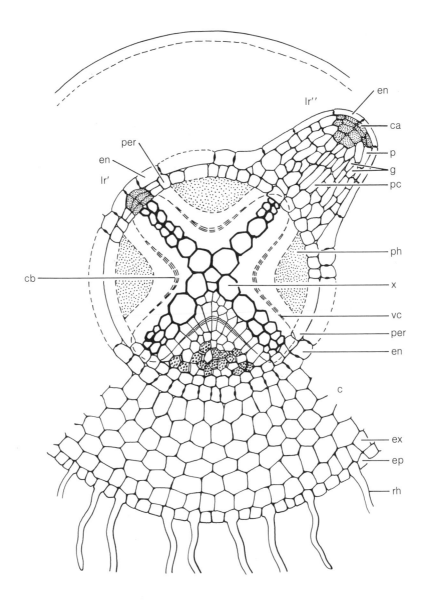

Figure 19-37

Diagrammatic transection of a tetrarch root to illustrate the endogenous initiation of a lateral-root primordium (lr′) and the early histogenetic differentiation of a young lateral root during its growth through the cortex of the parent root (lr″). The broken outline (cb) indicates schematically the position of the vascular cambium of the parent root. c, cortex; ca, calyptrogen; en, endodermis; ep, epidermis; ex, exodermis; g, ground meristem; lr′, primordium of lateral root; lr″, young lateral root with root cap; per, pericycle; pc, procambium; p, protoderm; ph, phloem; rh, root hair; vc, vascular cambium; x, xylem. [Redrawn from *Grundzüge der Pflanzenanatomie* by B. Huber. Springer-Verlag, Berlin, 1961.]

commonly equals the number of xylem poles in the main root. There are exceptions to this widespread type of rhizotaxy in the diarch roots of many dicotyledons, in which lateral-root primordia arise either *between* the xylem and phloem strands, yielding four vertical rows of branch roots, or else *opposite* each phloem strand, yielding only two vertical rows of lateral roots. In the polyarch roots of certain monocotyledons (e.g., members of the Gramineae), the pericycle is interrupted by the outermost tracheary elements of each xylem ray and the lateral-root primordia are initiated in those portions of the pericycle which lie opposite the phloem strands.

On the basis of its apical ontogeny, type of primary vascular system, and method of branching, the root must be regarded as a very distinct type of organ not simply in angiosperms, but in vascular plants as a whole. Doubtless the fundamental similarity in structure between the roots of all the major groups of tracheophytes, the absence of convincing "transitions" between roots and shoots, and the very incomplete fossil record of the root have all tended to emphasize the abiding mystery of root origin and evolution. From a purely "typological" point of view, Troll (1967, pp. 2007–2010) rejects the idea of roots originating from shoot-like axes and maintains that the root is a fundamental organ in the morphological organization of all vascular plants. On the other hand, it is quite possible that continued studies on the fossil remains of ancient vascular plants may someday reveal the steps in evolution which occasioned the present morphological differences between roots and shoots in vascular plants.

The Flower

As angiosperms are commonly designated the flowering plants it might be assumed that there is rather general agreement about the scientific concept of a flower. Unfortunately this is not the case, and the literature on floral organography, ontogeny, and structure displays widely divergent viewpoints of the fundamental nature of the flower as well as on the interpretation of its component organs (sepals, petals, stamens, and carpels) (Fig. 19-38).

One of the basic difficulties lies in our complete ignorance of the evolutionary history of the flower. Floral organs are generally fragile, ephemeral structures and are rarely preserved in the fossil record. Thus it becomes largely a matter of conjecture whether it is justifiable to draw comparisons between modern angiospermous flowers and the spore-producing structures of other tracheophytes. If such comparisons are attempted, it is quite possible to reach either a very broad or a very restricted concept or definition of a flower.

As an example of the excessively broad concept, favored particularly by certain German morphologists, we cite the definition of Goebel (1905, p. 469): "I understand here by the term 'flower' a shoot beset with sporophylls, that is to say, leaves bearing sporangia." From this standpoint, and with the additional qualification that the flower is a shoot of limited growth, there are "flowers," according to Goebel, in all the strobilus-bearing lower vascular plants as well as in the gymnosperms. From a similar viewpoint, Engler (1926) refers to the microsporangiate and megasporangiate strobili of the conifers as "male" and "female" flowers, respectively. We have also pointed out in Chapter 18 that the strobili of the gnetophytes have been likened to inflorescences which bear flowers with a vestigial perianth (see also Fagerlind, 1946).

On the other hand, there has been a strong tendency on the part of many morphologists and taxonomists to limit the concept of the flower to the angiosperms. From this *restricted* viewpoint, a flower is a *particular type of determinate sporogenous shoot*, and one of its

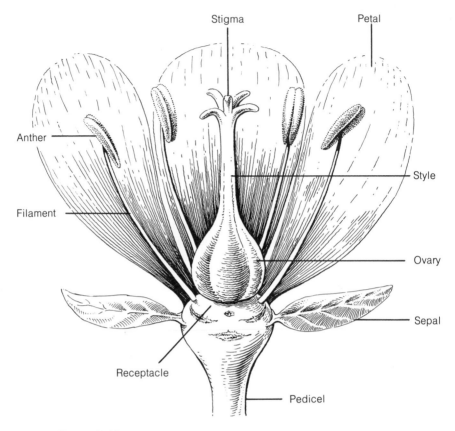

Figure 19-38
Organography of a flower. [Redrawn from *Lehrbuch der allgemeinen Botanik* by von Guttenberg, Ed. 2. Akademie-Verlag, Berlin, 1952.]

most definitive organs is the carpel. The carpel, which resembles a megasporophyll in general *function,* is regarded as morphologically distinctive because the ovules (i.e., megasporangia) are usually enclosed within a hollow basal portion designated as the ovary. Furthermore, in contrast with the megasporophylls of gymnosperms, most carpels terminate in a stigma, which serves as a receptive structure for the pollen.

In view of the absence of paleobotanical evidence about the primitive angiosperms, the evolutionary history of the carpel as well as that of the stamen, petal, and sepal remain unknown. Fortunately, however, there still exist angiosperms with carpels

and stamens which may well typify early steps in the evolutionary history of these debatable organs from some more elemental type of sporophyll. Later in this chapter a brief description of the floral morphology of these relict angiosperms will be given.

Modern work on the flowers of angiosperms has utilized evidence from general organography, vascular anatomy, and ontogeny in attempting to define morphologically a flower and its appendages. It will not be possible in this book to review the major results of enquiry along these various lines. Students interested in detailed discussions of these important areas of morphological research should consult the review articles and special treatises by Eames (1931, 1951), Eames and McDaniels (1947), Puri (1951), Wilson and Just (1939), and Melville (1962, 1963), to learn where the pertinent literature is to be found.

In this book it is necessary to emphasize, however, that despite the still controversial status of many aspects of floral morphology, the so-called classical theory of the angiosperm flower has much evidence in its favor (Mason, 1957). According to this theory, which derives from the early writings of Goethe (1790) on metamorphosis in plants and which was developed in detail by De Candolle (1844), the flower is a shoot and the appendages (i.e., floral organs) are the morphological equivalents, or homologues, of leaves (Arber, 1937, 1946). In the light of modern ideas concerning the origin of the megaphyllous leaf from a branch system, the question of the presumed homology between *foliage leaves* and floral appendages is indeed a controversial matter, at least from the standpoint of phylogeny (Wilson, 1953). But marked *resemblances* can be demonstrated between vegetative leaves and floral appendages, in initiation, early ontogeny, and basic plan of vasculature. It will therefore be desirable to present concisely some of the evidence from vascular anatomy and ontogeny on which the present widespread adherence to the classical theory of the flower depends.

Vascular Anatomy

Since the classical investigations of Van Tieghem (1875) on the anatomy of the pistil and fruit, much labor and considerable enthusiasm have been devoted to the study of the vasculation of the flower (Moseley, 1967).

In addition to providing data of importance in the morphological interpretation of fusion, adnation, and the inferior ovary, the anatomical method has, in general, furnished

strong support for the classical theory of the flower. A. J. Eames (1931), an outstanding leader in the study of floral anatomy, maintains that "flowers, in their vascular skeletons, differ in no essential way from leafy stems." He emphasizes, however, that because of the determinate pattern of growth of a flower and the crowding and fusion of the numerous appendages on the short floral axis, anatomical interpretation is often difficult, and, to be successful, requires broad comparative knowledge of the vasculature of both flowers and vegetative shoots. Puri (1951), who has reviewed in detail the very extensive literature on floral anatomy, concludes that evidence derived from the anatomical method has made a significant contribution to a better understanding of the angiosperm flower. But he emphasizes the necessity of regarding vasculation as only *one* of the important sources of morphological ideas — the evidence from organography and floral ontogeny also needs full consideration.

Numerous comparative studies on the vasculation of angiosperm flowers make it possible to select relatively simple types of flowers for brief consideration. In many respects, the flower of *Aquilegia,* a genus in the Ranunculaceae, is ideal because the successive floral appendages are free from one another and are borne on a well-defined stem-like axis or receptacle. According to the detailed investigations of Tepfer (1953) the flower of *Aquilegia formosa* var. *truncata* commonly consists of an axis bearing 70 appendages as follows: a calyx of 5 sepals, a corolla of 5 petals, an androecium of 45 stamens and 10 staminodia (petaloid organs devoid of anthers) and a gynoecium of 5 carpels, free at first but becoming basally concrescent later. Serial transections of the flower of *Aquilegia* reveal that the vascular system of the pedicel and receptacle is stem-like, consisting of a dissected cylinder of phloem and xylem from which are derived

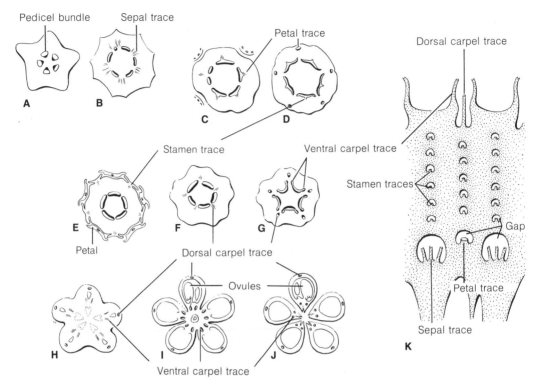

Figure 19-39
A–J, transections illustrating vascular anatomy of the flower of *Aquilegia*. **A**, transection of pedicel;
B, C, departure of sepal and petal traces; **D, E,** departure of stamen traces; **F,** departure of dorsal
traces of each carpel; **G,** departure of ventral carpel traces; **H,** transection of basal region of united
carpels; **I,** the five carpels nearly free; **J,** the carpels at higher level, each with a dorsal and two
ventral bundles; **K,** pattern of the secondary vascular cylinder of a mature flower of *Aquilegia*
split longitudinally and spread out in one plane (this diagram shows only a portion of the floral
traces and gaps depicted in A–G). [A–J redrawn from Eames, *Amer. Jour. Bot.* 18:147, 1931;
K redrawn from Tepfer, *Univ. Calif. Publ. Bot.* 25:513, 1953.]

the traces to the successive floral appendages
(Fig. 19-39, A-J). These floral traces, like
leaf traces, are associated with definable
gaps *after* secondary growth has occurred
(Fig. 19-39, K).

As is true of a great many flower types
which have been studied anatomically, there
are remarkable differences in *Aquilegia* in
the number of traces and the venation characteristic of the successive sets of floral
appendages. Figure 19-39, K, illustrates the
basic plan of vasculation in the flower, and
Fig. 19-40 depicts the venation patterns
typical of the sepal, petal, stamen, staminode,

and carpel. At the nodal level of the calyx,
each of the five sepals receives three traces
which typically are associated with a single
gap (Fig. 19-39, K). The three traces become more-or-less joined as a single bundle
in the base of the sepal, and then separate
into a palmate series at the base of the lamina, the venation of which consists of a series
of irregularly dichotomizing and interconnected bundles (Fig. 19-40, A). The five
petals, and each of the numerous stamens
and staminodia, in contrast, are vasculated
by single traces, each trace related to a distinct and separate gap (Fig. 19-39, C-E, K).

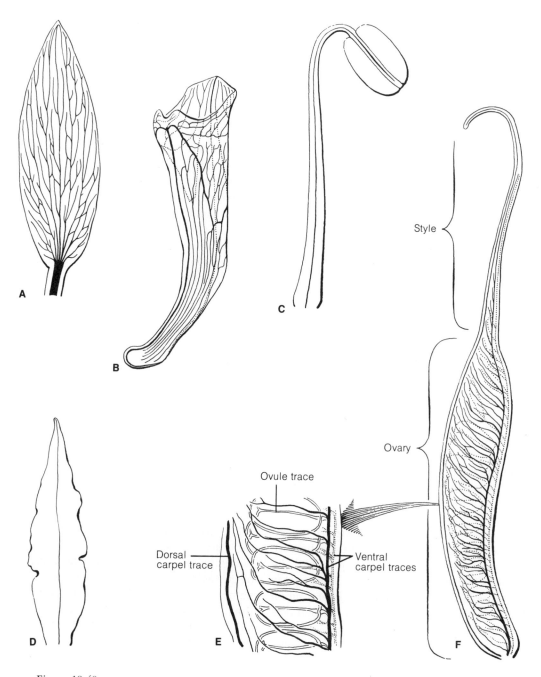

Figure 19-40
Venation patterns in the floral organs of *Aquilegia formosa* var. *truncata*. **A**, sepal; **B**, petal; **C**, stamen; **D**, staminodium; **E**, details of vasculature of portion of ovary (note the lateral, dichotomizing branches of the ventral traces and the derivation of the ovule traces); **F**, total venation of a carpel (the ovules, ovule traces, and wall venation of lower side are shown by dotted lines). [Redrawn from Tepfer, *Univ. Calif. Publ. Bot.* 25:513, 1953.]

Like the sepals, petals* of the flower exhibit a complex pattern of venation that differs markedly from the simple unbranched vascular strand which traverses each of the stamens and the staminodia (Fig. 19-40, B-D). Each of the five carpels receives three traces (associated with a single gap) — a dorsal trace which extends up the abaxial side of the carpel into the terminal region of the style, and two ventral traces which traverse the fused adaxial margins of the ovary and end in the style a short distance below the dorsal bundle (Fig. 19-39, F-K). As shown in Fig. 19-40, E, F, these ventral carpellary bundles give rise to lateral dichotomizing veins, one series of which enters the stalks of the ovules, the other series constituting the venation of the two walls of the ovary. From this description it is evident that the basic vasculation of the flower of *Aquilegia* closely resembles that typical of a vegetative shoot. In both the shoot and the flower the successive appendages are vascularized by traces associated with gaps in the secondary vascular cylinder of the axis (i.e., stem or receptacle). Indeed, the vascular plan of the flower in *Aquilegia* appears to lend strong support to the classical interpretation of the flower, and serves as an example of a simple and possibly primitive type of floral vasculature in the angiosperms.

Many flowers, however, are more complex because of the reduction in size or phylogenetic loss of certain organs, or because of the very common tendency toward the fusion between floral organs. Fusion may consist in partial or complete lateral union or cohesion between the adjacent members of the calyx, corolla, androecium, or gynoecium of a flower (Fig. 19-41, A, K-M); or it

may consist in adnation between members of successive floral whorls, for example, the union of stamens to petals or to the tube of a sympetalous corolla (Fig. 19-41, B-F). These organographic specializations are, in turn, reflected in corresponding modifications of the vasculature of complex types of flowers.

It will not be possible in this book to discuss fully the application of the anatomical method to the interpretation of the diversified levels of specialization in angiosperm flowers, but a few examples, selected from anatomical studies on the gynoecium, will serve at least to illustrate the effects of cohesion. In a free carpel the margins may become closely appressed without fusion, or, as in *Aquilegia,* may fuse during early phases in carpel ontogeny (Tepfer, 1953). In either case, the two ventral bundles are separate and distinct throughout the carpel (Figs. 19-40, F; 19-41, G). In instances in which the carpellary margins fuse earlier in ontogeny, the two ventral bundles may be joined as a double or an apparently single bundle at various levels in the carpel (Fig. 19-41, I, J). Syncarpous gynoecia reflect in their vasculature the varied degrees of lateral cohesion between adjacent carpels. In the simplest examples, the paired ventral bundles remain distinct and constitute the vascular system of the axile placenta (Fig. 19-41, K). When the carpels are closely united to form a tri- or multiloculate ovary, the placental region is often vascularized by one-half the expected number of bundles. This is interpreted as the result of fusion either between the ventrals belonging to each carpel or between the ventral bundles of laterally adjacent carpels (Fig. 19-41, L, M). Students interested in the further complications in the vasculature of gynoecia, stamens, sympetalous corollas, and synsepalous calyces of specialized flowers should consult Eames (1931), Puri (1951), Eames and MacDaniels (1947), and Kaplan (1967).

---

*The venation patterns of petals in other taxa in the Ranunculaceae range from open dichotomous in *Adonis* (Hiepko, 1965) to dichotomous-reticulate in *Ranunculus* (Arnott and Tucker, 1963, 1964). For further information on the vasculature of sepals and petals in other angiosperms see Glück, 1919; Chrtek, 1962, 1963; and Hiepko, 1965.

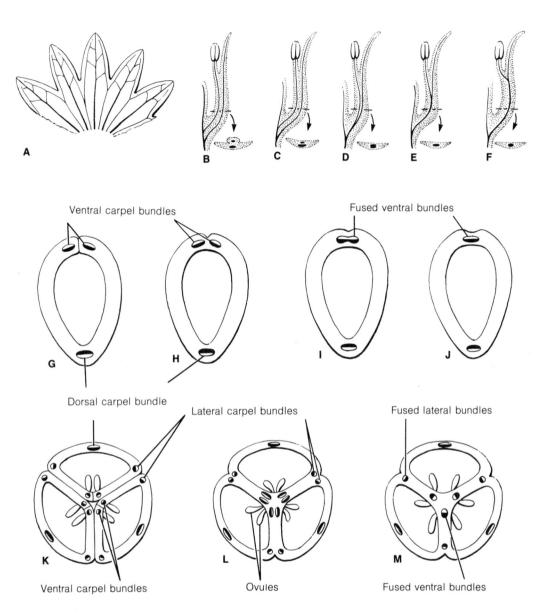

Figure 19-41

The effects of cohesion and adnation on the vasculature of floral organs. **A,** the calyx of *Ajuga reptans,* showing the five basally fused sepals (the adjacent lateral veins of each sepal are united below the sinus); **B–F,** corresponding longitudinal and transectional diagrams showing the results of successive degrees of adnation between stamen and petal upon the pattern of vasculature; **B, C,** weak adnation, with traces of stamen and petal separate; **D–F,** progressive steps in the radial fusion between stamen and petal bundles; **G–J,** transectional diagrams showing effect of union of carpel margins on ventral bundles; **G, H,** the ventral bundles are distinct; **I,** a double ventral bundle; **J,** fusion of ventrals to a single bundle; **K–M,** transections showing effects of cohesion between three carpels; **K,** carpels in close contact but with distinct margins and paired ventral bundles; **L,** carpels fused, the ventral and lateral bundles of adjacent carpels arranged in pairs; **M,** the pairs of ventral and lateral bundles united. [Adapted from *An Introduction to Plant Anatomy,* Ed. 2, by Eames and MacDaniels. McGraw-Hill, New York, 1947.]

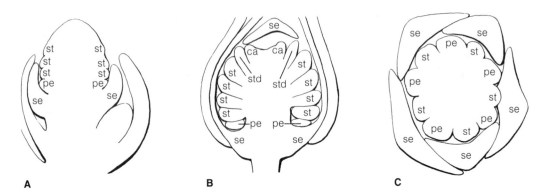

Figure 19-42

Organogeny of the flower of *Aquilegia formosa* var. *truncata*. **A, B,** outlines of longisections of the floral apex, illustrating acropetal development of sepals (se), petals (pe), stamens (st), staminodes (std), and carpels (ca); **C,** outline of a transection of a developing flower showing the five sepals and the primordia of petals and stamens. [Redrawn from Tepfer, *Univ. Calif. Publ. Bot.* 25:513, 1953.]

Ontogeny

The recent revival of interest, throughout the world, in problems of growth and differentiation in plants has resulted in a rich literature, particularly on the subject of apical ontogeny in vegetative shoots, inflorescences, and flowers (Gifford and Corson, 1971). In addition to numerous studies on the structure of the shoot apex and the development of leaves and stem tissues, much attention has been given to the ontogeny of the flower. A few ontogenetic investigations have yielded conclusions radically opposed to the classical view that a flower is a special or modified type of shoot (see Thompson, 1934; Grègoire, 1938; Buvat, 1952). But, in general, developmental studies appear to strengthen the classical interpretation and to complement the conclusions reached through the study of the vasculature of adult flowers. As an example, a brief discussion will now be presented of Tepfer's (1953) detailed study of floral ontogeny in *Aquilegia* and *Ranunculus.*

In both of these genera the sepals, petals, stamens, and carpels are produced in an acropetal sequence from the meristematic floral apex (Fig. 19-42). During the period of initiation of the appendages, however, there are rather striking changes in the dimensions and cellular structure of the floral apex. These changes involve a transition from a relatively small apex of uniformly and densely staining cells to a broad dome-like apical meristem with a conspicuously vacuolated parenchymatous core overlaid by several layers of meristematic cells. The latter structure is particularly characteristic of the apex of *Ranunculus* during the formation of the carpel primordia. The massive parenchymatous core of the floral apices of *Aquilegia* and *Ranunculus,* however, is correlated with the elongated floral receptacle in these genera, and therefore does not demonstrate a fundamental difference between vegetative and floral meristems as was postulated by Grègoire (1938). In other angiosperms, for example, *Vinca, Umbellularia, Laurus,* and *Frasera* (Fig. 19-43), the structure and growth of the floral apex during the formation of appendages resemble closely the organization of a vegetative shoot apex (McCoy, 1940; Boke, 1947; Kasapligil, 1951).

Tepfer's investigations further demonstrate that the method of initiation and the early phases of cellular differentiation of

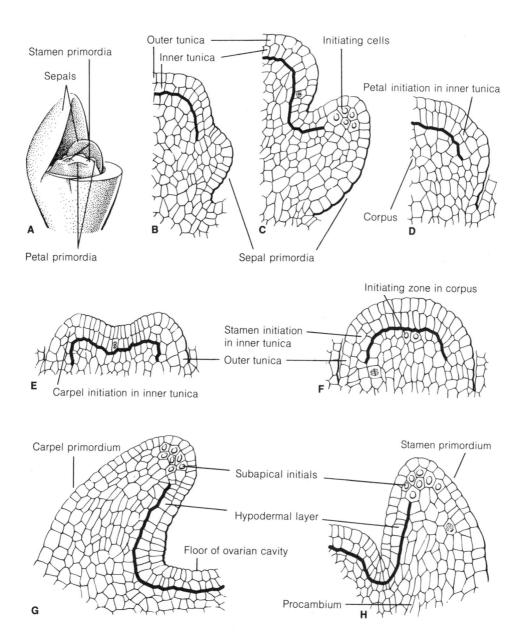

Figure 19-43

Initiation of floral appendages in *Frasera carolinensis*. **A,** flower bud with two of the sepals removed showing primordia of petals and stamens. **B–H,** longisections showing that all floral organs are initiated by periclinal divisions in the second tunica layer of the floral apex (a heavy line demarcates the tunica from the corpus in all the figures); **B, C,** initiation and early development of the sepal primordium; **D,** initiation of petal; **E,** initiation of the two carpel primordia; **F,** initiation of stamen primordium; **G,** longisection of young carpel showing group of subapical initials; **H,** longisection showing apical growth of stamen primordium. [Redrawn from McCoy, *Amer. Jour. Bot.* 27:600, 1940.]

sepal, petal, stamen, and carpel primordia are fundamentally similar and closely agree with the early ontogeny of bracts and foliage leaves. The striking resemblances include the method of development of the procambium, which is produced acropetally in *both* vegetative and floral apices. The early ontogeny of the carpels in *Aquilegia* is of particular morphological interest. Soon after their initiation from the floral meristem, the carpel primordia appear horseshoe-shaped in transectional view (Fig. 19-44, D). Well-defined marginal growth—resembling the marginal growth of a vegetative leaf—results ultimately in the *actual* ontogenetic fusion of the margins of each carpel and the formation of the ovarian region. Ovules are initiated from hypodermal cells situated about midway between the fused edges of the carpel and the inner surface of the ovarian cavity (Fig. 19-44, A). Because of this, Tepfer regards the placentation of the ovules as laminal rather than marginal. During the early phases of development of the gynoecium, the five carpel primordia are free from one another (Fig. 19-44, B, D). Later the bases of adjacent carpels become fused by the cohesion of their meristematic epidermal layers (Fig. 19-44, C). At maturity the original, separate epidermal layers of the joined carpels can no longer be distinguished (Fig. 19-44, E). Thus in *Aquilegia* the union of the carpel margins and the fusion between the bases of adjacent carpels are truly ontogenetic processes. Many examples of such postgenital carpel fusions are found in the angiosperms (Baum, 1948a, 1948b).

From a physiological and ontogenetic standpoint, flowering in angiosperms is expressed in diverse and varied ways. The production of a single terminal flower in the life of a plant characterizes some species. In the majority of angiosperms, inflorescences, with or without terminal flowers, are formed. In other species, flowers are

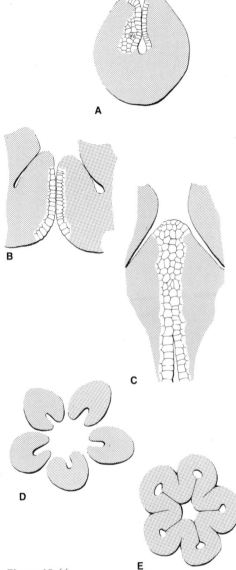

Figure 19-44
Carpel primordia and fusion of adjacent carpels in *Aquilegia formosa* var. *truncata*. **A,** transection of carpel showing initiation (two hypodermal cells with nuclei) of ovule distal to margin; **B, C,** stages in the ontogenetic fusion of the edges of two adjacent carpels; **B,** carpel edges in contact with distinct epidermal layers; **C,** edges of carpels fusing, boundary of epidermal layers becoming indistinct; **D,** transection showing typical horseshoe shape of the free carpel primordia; **E,** transection showing postgenital fusion of carpel primordia. [Redrawn from Tepfer, *Univ. Calif. Publ. Bot.* 25:513, 1953.]

formed in the axils of typical foliage leaves (see pages 620–627 for a more complete discussion of this topic). The times of initiation and maturation of flowers are exceedingly variable. For annuals, the time interval between seed germination, flowering, and the production of fruits is a single growing season. Other species are biennials—seed germination and vegetative growth take place during the first growing season and flowering occurs in the second year. Woody perennials of temperate regions present other complications. For example, inflorescences or single flowers may be formed terminally or they may be initiated in the axils of leaves either in the spring of the year in which they bloom or during the previous summer or fall. The physiological factors controlling flowering in the perennial plants are complex and not very well understood at present.

Many plants exhibit a *photoperiodic* response—that is, they respond to the length of the daylight hours. Hence, there are two types of photoperiodic flowering responses —long-day and short-day responses. A long-day plant flowers when the light period *exceeds* a certain value—generally 12–14 hours. A short-day plant flowers when the day length is *less* than a certain critical value. Actually, it is the length of the night (dark period) that is critical for these plants. Still other plants are day neutral in that flowering is regulated by other factors, e.g., number of leaves produced beforehand or some other morphological or physiological condition or combination of conditions.

SHORT-DAY PLANTS.   Several short-day species have been investigated rather thoroughly. In general the sequence of events in flowering is as follows: photo-perception and induction in the leaves result in the production of a flowering stimulus (presumably a hormone) which moves to the vegetative shoot apex by way of the petiole and stem, where floral initiation takes place. During floral initiation RNA and protein molecules essential to the flowering process are synthesized. This is rapidly followed by the mitotic and morphological phases which often result in marked changes in the architecture of the shoot apex, before floral appendages are formed.

Some short-day species are extremely sensitive to photoperiod. For example, *Pharbitis nil* (Japanese morning glory) seedlings can be induced to flower even at the cotyledon stage if given one dark period of 14–16 hours. If the dark period is interrupted in the middle and the seedlings are exposed to red light, even for a few minutes, the plants will not flower. In contrast, morphogenesis is rapid in plants that do not receive the red light interruption and within 5 days stamen primordia are evident. Other species that are especially sensitive to the short-day–long-night regime include certain species of *Chenopodium* (lambsquarters) and *Xanthium* (cocklebur). Literature pertinent to short-day plants can be found in Salisbury (1963), Nougarède (1967), and Gifford and Corson (1971).

LONG-DAY PLANTS.   *Sinapis alba* (=*Brassica alba*), white mustard, remains vegetative if grown under a short-day regime. If a plant is grown under short-day conditions for 60 days and then transferred to long-days, the shoot apex begins to change as early as 20 hours after the beginning of the long day. This example of sensitivity is remarkable in that induction, the production of the floral stimulus, and translocation of the stimulus to the reactive site, the shoot apex, take place in such a short period (Bernier, 1962; Bernier, Kinet, and Bronchart, 1967).

It is impossible in this textbook to describe the many well-documented descriptions of *transition from vegetative growth to flowering.* We will, however, describe the ontogeny of the flower in *Liriodendron tulipifera* (tulip tree), which provides some support for the classical theory of the angiosperm flower (that is, the flower is a modified shoot). The flower of the tulip tree embodies several presumedly primitive features of organization: (1) the presence of numerous and helically arranged parts that are not fused and (2) the determinate nature of the floral apex is expressed only after an extended period of morphogenesis. Morphologically, the flower consists of 3 sepals, 6 petals, 30–50 stamens, and about 100 carpels on an elongate receptacle (Fig. 19-49).

From a topographical standpoint the vegetative shoot apex is a low dome (Fig. 19-45, A) and possesses a tunica and corpus (see page 570 for discussion of tunica-corpus organization). Cells of the tunica divide anticlinally except during the initiation of leaf primordia, where cells of T_2 divide periclinally. Cells of the corpus divide in various planes. During August or September (depending upon the latitude) the growth of some buds on a tree changes from vegetative to reproductive. There is little structural change in the vegetative apex as the meristem enters the period of floral development. A biseriate tunica remains as distinct as it was in the vegetative apex. The receptacle begins to elongate at the time of sepal initiation and there is an increase in mitotic activity in all regions of the apex. Petals are initiated and then the floral meristem enlarges greatly during stamen initiation (Fig. 19-45, C). Carpels are initiated acropetally, and throughout the stamen and carpel development, a rib meristem is active in the formation of the greatly elongated receptacle (Fig. 19-45, B–D). This account is based upon a study by G. L. Vertrees (1962).

The scanning electron microscope has provided the morphologist with a powerful new tool for the study of plant form and development. For example, in his study of floral development in *Adonis aestivalis,* a member of the Ranunculaceae, Joseph Lin was able to obtain excellent three-dimensional photographs of living floral buds. In the examples included here (Fig. 19-46, A, B), the sepals and petals were removed before the young flowers were placed in the vacuum chamber of the scanning electron microscope. The numerous developing stamens, with conspicuously lobed anthers, surround the young carpels in the central region of the flower bud. Carpels, as can be seen in Fig. 19-46, B, are initiated from the floral apical meristem. It will be noted in Fig. 19-46, B that a carpel, early in its development, becomes a concave, hooded structure, open along a slit-like crease on the adaxial side (facing the floral axis). The cleft, which is the future "suture" of the carpel, becomes deeper due to marginal growth along the edge of the opening. A short style will later develop as an extension of the upper end of the carpel (not yet evident in these young carpels).

Figure 19-47, A, B, presents views of carpels as seen in a longitudinal section of a developing flower. A median longitudinal section of a carpel (upper one) is shown in Fig. 19-47, A. In this plane it would appear that the carpel is open, adaxially and laterally. However, sections which are slightly off median (through the marginal meristems) reveal that the carpel is only concave in form as is readily apparent in Fig. 19-47, B. In transection a carpel primordium appears to be horse-shoe shaped if the section is made at a high level (cf. Fig. 19-47, A, C).

Figure 19-45 (*facing page*)
Liriodendron tulipifera. **A,** vegetative shoot tip. **B–D,** stages in development of the flower from a vegetative shoot apex. Note the great enlargement of the young floral apex (**B, C**) and the elongate form of the receptacle (**D**) near the end of the phase of carpel initiation. See text for more complete explanation.

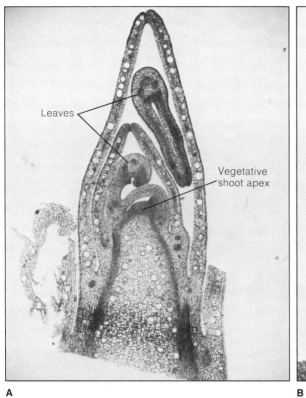

A

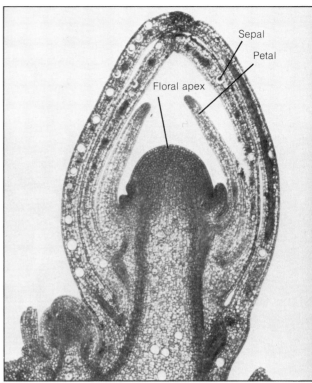

B

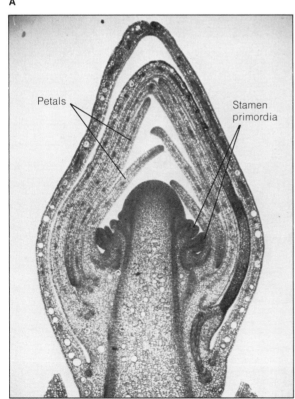

C

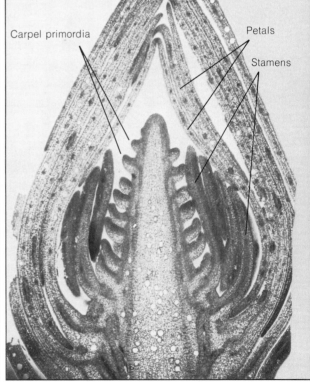

D

A

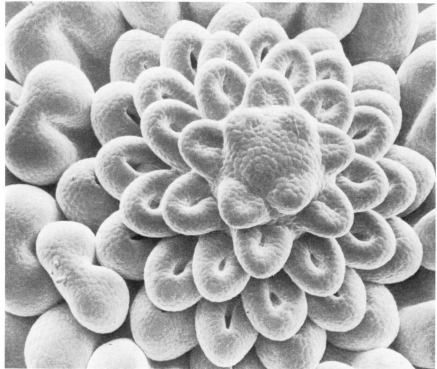

B

Figure 19-46
Scanning electron micrographs of developing flowers of *Adonis aestivalis* (sepals and petals not included). **A,** lower magnification than **B,** showing numerous stamens surrounding the many developing carpels. **B,** enlargement of the floral apex showing stages in the initiation and growth of carpels. [Courtesy Josèph Lin.]

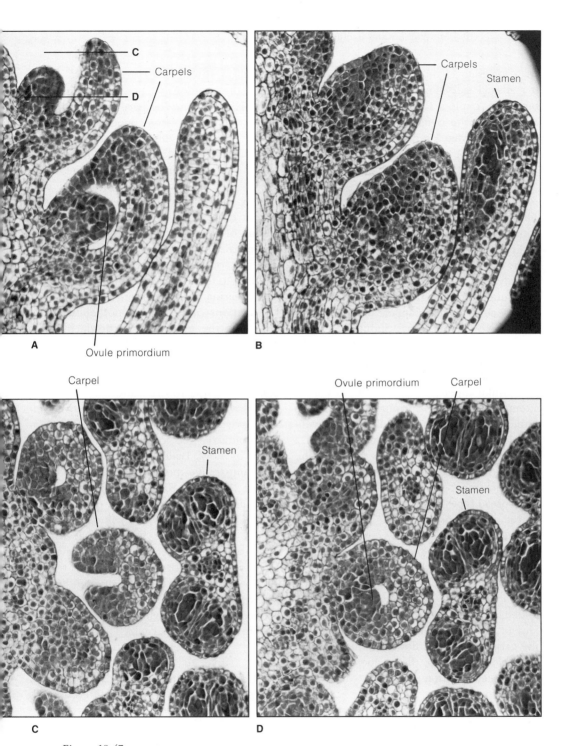

Figure 19-47
Adonis aestivalis. **A, B,** longitudinal sections, portions of young flowers. The upper carpel in **A** has been sectioned in the median longitudinal plane; the lower one in **A** is slightly off median (tangential). **C, D,** transections, portions of young flowers. The carpels are about the same age as those in **A** and **B**; **C,** carpel cut at level C in **A**; **D,** carpel cut at level D in **A**. See text for a more complete explanation.

At a lower level the carpel is not open on the adaxial side (Fig. 19-47, D). Only one functional ovule is formed in each carpel of *Adonis*, and the ovule primordium is positioned somewhat laterally in the lower part of the carpel (Fig. 19-47, A, D).

Phylogeny of Stamens and Carpels

The subject of floral morphology may be appropriately concluded by a consideration of certain modern views on the evolution of the stamen and carpel in the angiosperms. Usually these sporangium-bearing organs are quite different from foliage leaves in their general form and appearance, and there are still conflicting views about their morphology and evolution (Parkin, 1951). In the light of the evidence already presented in this chapter, stamens and carpels seem to be leaf-like appendages. But how should we attempt to visualize a primitive stamen or a primitive carpel? The common or conventional type of modern stamen consists of a delicate filament, which bears at its tip the anther containing usually four embedded microsporangia (see Fig. 20-1, A). Is this widespread type of stamen highly specialized or relatively conservative in character? As for the carpel, the classical view regards this organ fundamentally as a foliar organ in which the fused involute margins serve as the point of attachment of the enclosed ovules. Does this concept of an involutely folded, leaf-like structure provide a reliable idea of the nature and method of origin of the carpel in ancient flowering plants? Provisional answers to these questions have been recently attempted, and they merit careful consideration.

On the basis of a series of investigations on the vasculation of stamens, Wilson (1937, 1942) proposed that these organs be interpreted with the aid of the Telome Theory. This theory, discussed earlier in Chapter 3, attempts to derive vegetative megaphylls as well as sporogenous organs from primitive, dichotomously branched axes. In Wilson's view the modern angiosperm stamen, with its slender filament and compact anther, represents the end product of reduction and fusion of a dichotomous branch system with terminal sporangia. According to his idea the number of sporangia ultimately was reduced to four, and the modern anther morphologically is a synangium consisting of four microsporangia. Wilson concluded also that the carpel originated from a series of fertile telomes which became webbed to form ultimately a foliar-like structure bearing marginal ovules. Infolding of the margins of such an hypothetical structure was the final step in producing the modern ovary which encloses the ovules.

If the telome interpretation of the stamen and carpel were accepted, it would appear somewhat futile to draw any *direct* morphological comparisons between vegetative and floral organs in present-day angiosperms. On the contrary, the comprehensive surveys made by I. W. Bailey and his associates on the comparative morphology of many families in the order Ranales have produced a new and hopeful line of attack on the problem of stamen and carpel evolution (Bailey, 1954). These surveys indicate that within *living* woody members of the Ranales there have persisted not only primitive trends of wood specialization—including primitively vessel-less xylem—but also types of stamens and carpels which appear relatively primitive and unspecialized in character.

Degeneria, a recently investigated ranalean genus native to Fiji, provides a striking illustration of primitive stamens and carpels. The stamen of *Degeneria* is *not* differentiated into filament, anther, and connective but is a broad, foliaceous, three-veined sporophyll which develops four, slender, elongated microsporangia deeply embedded in its abaxial surface (Fig. 19-48, C, E). It should be noted that the paired sporangia are

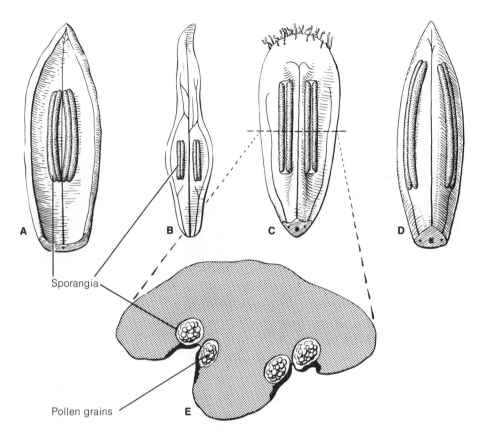

Sporangia

Pollen grains

Figure 19-48

Primitive types of stamens in the Ranales. **A,** *Austrobaileya maculata,* adaxial surface showing paired sporangia at each side of midvein; **B,** *Himantandra baccata,* abaxial surface showing sporangia and relatively complex venation; **C,** *Degeneria vitiensis,* abaxial view showing pairs of sporangia between lateral veins and midvein; **D,** *Magnolia maingayi,* adaxial view showing venation and paired sporangia; **E,** diagrammatic view of transection of stamen of *Degeneria* showing the four sporangia embedded in abaxial surface. [Redrawn from Canright, *Amer. Jour. Bot.* 39:484, 1952.]

laminal rather than marginal in position, and that they lie between the lateral veins and the midvein of the stamen. There are closely similar types of broad microsporophylls in other ranalean genera (Figs. 19-48, 19-49) such as *Austrobaileya, Himantandra* and certain members of the Magnoliaceae (Canright, 1952; Ozenda, 1952). In the Magnoliaceae Canright was able to arrange the various stamen types in an evolutionary series, beginning with broad, three-veined microsporophylls and culminating in types with marginal sporangia and with a definable differentiation into anther and filament. He regards the stamen of *Degeneria,* however, as

"the closest of all *known* types to a primitive angiosperm stamen."

The megasporophyll of *Degeneria* differs in many fundamental ways from the typical angiosperm carpel with its differentiation into a closed ovary, style, and stigma. When studied in transectional view during early ontogeny, the carpel of *Degeneria* is a conduplicately folded structure, the closely adjacent margins of which flare outwardly but remain unfused for a considerable time (Fig. 19-50). Thus, during this stage in development, the open carpellary margins form a narrow cleft extending from the exterior to the inner cavity or locule.

Figure 19-49
Flower of *Liriodendron tulipifera* showing external perianth, numerous stamens with abaxially embedded sporangia, and central group of closely crowded carpels. [From *Botany. An Introduction to Plant Science.* Ed. 2, by Robbins, Weier, and Stocking. Wiley, New York, 1957.]

Swamy's (1949) investigation shows that, beginning with the flared carpellary margins, an extensive development of epidermal hairs proceeds inwardly toward the locule, ultimately extending beyond the points of attachment of the ovules (Fig. 19-50). The space between the closely adjacent carpellary margins becomes filled with interlocking hairs, whereas the trichomes on the inner surface of the carpel are short and papillate. Collectively, all these hairs represent a stigmatic surface. Pollination requires the deposition of pollen grains on the adaxial hairy divergent margins of the open carpel. Swamy observed that the developing pollen tubes grow into the cavity of the carpel between the hairs and along the papillate surface — in no instance do the tubes penetrate the tissue of the carpel. The position and source of vascular supply of the ovules in *Degeneria* are remarkable also. The two rows of ovules are remote during initiation and development from the true margins of the carpel, and some are vascularized from branches of the ventral bundles, some from branches of the dorsal bundle, and still others by strands derived from both the main ventral and dorsal bundles. Following pollination and the fertilization of the ovules, the adjacent adaxial surfaces of the carpel become concrescent, and the recurved portions of the margins persist as parallel corky ridges on the mature fruit.

Additional examples of conduplicate carpels with extensive marginal stigmatic surfaces are found in the Winteraceae, another highly interesting ranalean family exhibiting many primitive morphological characters including the total absence of vessels in the xylem. According to Bailey and Swamy (1951) "the least modified form of surviving carpel" occurs in *Drimys piperita* and allied species of this genus. The young carpel is stalked, but, like the carpel of *Degeneria*, consists of a conduplicately folded lamina enclosing a series of ovules attached to the adaxial inner surface, distal to the carpellary margins (Fig. 19-51, B, D). This carpel, during the period of anthesis, can readily be unfolded and cleared, revealing that the ovules are attached to areas between the dorsal and two lateral veins (Fig. 19-51, C). As in *Degeneria*, the ovules of *Drimys piperita* are vascularized by extensions from either the dorsal or ventral carpellary bundles or by bundles that arise from strands derived from both the median and lateral systems. Pollen grains adhere to the external stigmatic marginal hairs, and, as in *Degeneria*, the pollen tubes reach the ovules by growing through the mat of hairs which extend inwardly to the surface of the locule (Fig. 19-51, B, D). The varied and complex trends of carpel specialization that have presumably arisen from the primitive condition exemplified in *Degeneria* and *Drimys piperita* are described in detail by Bailey and Swamy (1951) but will not be discussed in this book.

In summary, the recent investigations on ranalean families strongly support the interpretation of stamens and carpels (and also sepals and petals) as phyllomes, i.e., modified leaf-like appendages. Periasamy and Swamy (1956) have emphasized the remarkable similarity in the method of marginal growth of the conduplicate carpel of *Cananga* (Annonaceae) to that typical of many simple foliage leaves. Swamy states that the carpel of this species "is a true homologue of a nearly mature, adaxially conduplicate foliar appendage, the entire inner surface of the cleft corresponding to the adaxial surface of the foliar lamina."

Further insight regarding the nature and origin of the stamen and carpel in the angiosperms must await the results of more comprehensive morphological surveys of a wide range of angiosperms. It is only by such laborious procedures that we shall be able to test the importance of recent investigations on ranalean families. Bailey has continually emphasized the great caution which

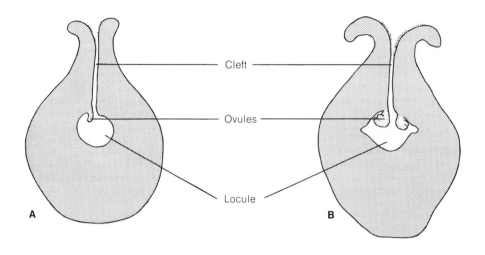

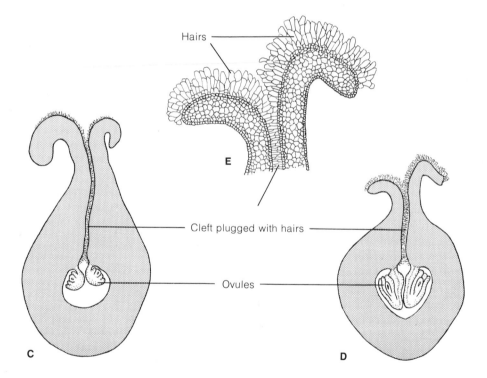

Figure 19-50
Transections of conduplicate carpel of *Degeneria* showing position of ovules and gradual occlusion of cleft between margins by development of hairs. **A,** young stage with open glabrous cleft; **B,** initiation of hairs on adaxial surface of flared ventral halves of the carpel; **C, D,** progressive inward development of hairs and occlusion of the cleft; **E,** detailed structure of dense mat of hairs within the cleft and on the adaxial flared ventral edges of the carpel. [Redrawn by Mrs. Emily E. Reid from Swamy, *Jour. Arnold Arboretum* 30:10, 1949.]

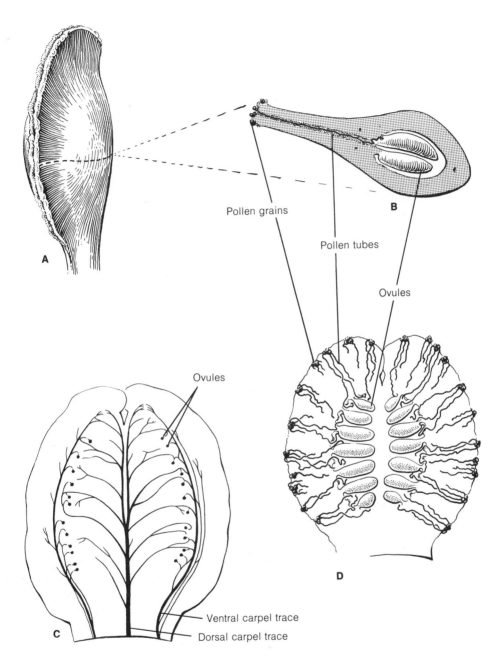

Figure 19-51
The primitive conduplicate carpel of *Drimys piperita*. **A,** side view showing the paired stigmatic crests; **B,** transection showing attachment of ovules and growth of a pollen tube through the mass of hairs which lies between the ventral surfaces of the carpel; **C,** cleared unfolded carpel showing vasculature and variations in derivation of ovule traces; **D,** unfolded carpel showing the course of the pollen tubes and the laminal placentation of the ovules. [Redrawn from Bailey and Swamy, *Amer. Jour. Bot.* 38:373, 1951.]

must be observed in attempting to reconstruct the phylogeny of tissues and organs in vascular plants. Evolution has evidently proceeded at different rates, not only among different organs or tissues within the plant but with respect to characters of a single organ or structure. It is thus essential to avoid overemphasis on certain selected end products of evolution and to consider whenever possible the *totality* of evidence derived from a study of the ontogeny and morphology of a wide range of structural features. From this standpoint, the sound development of phylogenetic ideas must be based upon a full consideration of the *successively modified ontogenies* of a given tissue or organ. Bailey remarks that with our present knowledge "it is not possible to derive the angiospermic carpel from any known group of the gymnosperms without the interpolation of fundamental morphological changes that are excessively speculative." In his view, the most promising solution to the abiding mystery of the origin of the angiosperms would appear to be in the paleobotanical study of plant remains in the Austro-Malayan and Indo-Malayan regions.

Pollen

A comparatively new but rapidly expanding branch of modern botany is the field of *palynology,* which deals with the varied and complex aspects of spores and pollen grains. The outer layer of the wall of these structures contains *sporopollenin,* one of the most resistant substances produced by plants and the one that is responsible for the well-preserved condition of spores and pollen grains in the fossil record. As Erdtman (1969) has emphasized, palynology has many contacts with other scientific disciplines, including paleobotany, taxonomy, economic geology, and medicine (e.g., the study of pollen as a cause of certain types of allergies, such as hay fever).

It would be pretentious, in this text, to attempt a résumé of the results of palynological research on the highly diversified types of pollen grains in the angiosperms. There is no field in comparative plant morphology in which there is such a rich and constantly changing terminology as that associated with the descriptions and interpretations of pollen-grain structure. Heslop-Harrison (1968) refers to the "intimidating terminology" that has developed with reference to the wall structure of pollen grains and Erdtman (1969), a recognized leader in this field, confessed that "this fascinating subject is far from easy, and the intricate details of the terminology involved might dishearten the beginner." But pollen morphology, aside from its own inherent attractions, has increasingly become a "tool" in systematic and phylogenetic studies and for this reason deserves some attention in our book. As far as possible, every effort has been made to define, as simply and clearly as possible, those *basic terms* that are indispensable in describing the general morphology of angiosperm pollen. Students who wish to pursue the subject more intensively, especially from a taxonomic viewpoint, should consult the texts written by Wodehouse (1935), Faegri and Iverson (1950), Erdtman (1969), and Kapp (1969).

In the following discussion, we will restrict ourselves to a brief presentation of two of the most salient features of pollen grains: (1) the morphology and positions of *apertures,* i.e., the thin areas in the wall that are the points of exit for the emerging pollen tubes and (2) the major layers of the outer wall or *exine* of the grain. Before doing this, however, it is essential to discuss the *polarity* of a pollen grain.

The process of microsporogenesis yields tetrads of haploid microspores each of which, prior to the dehiscence of the anther, gives rise to a two- or three-celled endosporic male gametophyte (see Chapter 20).

Figure 19-52
Scanning electron micrograph of a tetrad of pollen grains of *Drimys winteri* var. *chilensis.*
This is the shedding condition of pollen in *Drimys* and other members of the Winteraceae.
Note the large pore evident in the distal face of one of the grains in the tetrad and the
coarsely reticulate sculpture of the exine. [Courtesy Dr. R. H. Falk.]

Very commonly, the pollen grains of each tetrad become separated while still in the ripe anther and are shed as solitary grains known as *monads*. In this event, the position of the aperture or apertures with reference to the two "poles" of the grain, is difficult or indeed impossible to ascertain. But in some dicotyledonous taxa (e.g., Winteraceae, and certain genera of Annonaceae), the pollen is shed in the form of coherent tetrads; pollen units of this type permit an accurate analysis of the polarity of the grains (Fig. 19-52). They are so arranged in the tetrad that each grain is bisected by an imaginary "polar axis" which passes from the *distal* (i.e., the outer) face to the *proximal* (i.e., the inner) face of the pollen grain. The "equator" is the zone of the pollen grain which is located approximately midway between its two poles and the term "meridional" designates those surface features of the grain which are oriented at right angles to the equator. The relevance of these terms may now be examined with reference to the position and the form of the apertures.

APERTURES. As we have already mentioned, an aperture is a specific area in a pollen grain through which a pollen tube emerges. Two general types of apertures, distinguished by their form, are generally recognized: *furrows,* which are slit-like depressions in the wall, and *pores,* which are more-or-less isodiametric in shape. The number (N), position (P), and character (C) of

Polar view   Lateral view

Monocolpate (monosulcate)
A

Monoporate
B

Tricolpate
C

Tricolporate
D

Stephanocolpate (polycolpate)
E

Stephanocolporate
(polycolporate)
F

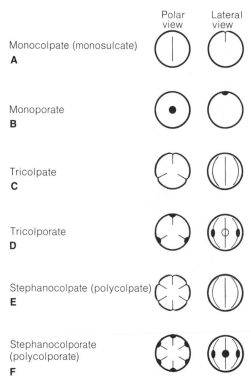

Figure 19-53
Diagrammatic representations of various types of apertures found in the pollen grains of angiosperms. [Redrawn from *Textbook of Modern Pollen Analysis* by K. Faegri and J. Iversen. Ejnar Munksgaard, Copenhagen, 1950.]

apertures are seemingly "endless" in variability, and form the basis for Erdtman's (1963, 1969) classification of pollen grains. Terminologies reach a peak of complexity in his "NPC-System" and cannot be considered in this brief treatment. But it is necessary to examine carefully the meaning and application of certain of the most widely employed terms used to describe pollen-grain apertures.

According to Walker (1971), the most primitive form of aperture in angiosperm pollen is the *sulcus,* a groove or furrow that is located at the distal pole of a pollen grain (Figs. 19-53, A, 19-54, A). He terms pollen grains with a single distally placed furrow

anasulcate in distinction to the *catasulcate* type of grain in which the furrow is located at the proximal pole. From a broad phylogenetic standpoint, it is interesting to note that anasulcate pollen is found not only in certain living as well as extinct types of gymnosperms but also is characteristic of the pollen grains of many monocotyledons and of certain ranalean genera such as *Degeneria* and all members of the Magnoliaceae (Canright, 1953; Dahl and Rawley, 1965). The evolutionary significance of this taxonomic pattern of distribution of anasulcate pollen, particularly its prevalence among certain taxa of the Ranales, represents a difficult problem which awaits further comparative studies for its solution. In this general connection it is noteworthy that Walker (1971) found that in the large ranalean family Annonaceae, only the pollen of *Pseudoxandra* has retained the primitive anasulcate condition; in the other genera the furrow is on the proximal pole, a circumstance which he believes is significant in understanding the phylogenetic specialization of the Annonaceae.

In contrast to the sulcate type of aperture, the pollen grains of a great many dicotyledons form three or more meridionally placed furrows, each of which is termed a *colpus* (Fig. 19-53). Apparently the *tricolpate* type of pollen grain is restricted to the dicotyledons and probably represents one of the basic forms of grains in this class of the flowering plants (Figs. 19-53, C, 19-54, B). Phylogenetic specialization, beginning with the tricolpate type of grains, has given rise presumably to pollen grains with more than three furrows (polycolpate or stephanocolpate grains) and to colpi each of which may have an included pore (tricolporate and stephanocolporate grains). Examples of these types are seen in Fig. 19-53, D–F.

Porate pollen grains, with one or more pores in place of furrows, are distinctive of certain taxa in both the monocotyledons

A

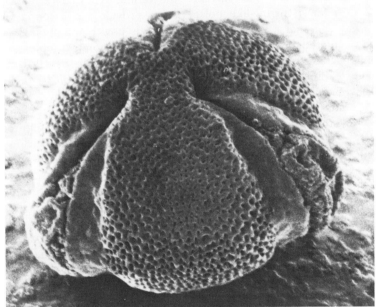

B

Figure 19-54
Scanning electron micrographs illustrating two important types of apertures in pollen grains. **A,** *Magnolia* sp; cluster of monosulcate pollen grains, one of which shows the conspicuous furrow or *sulcus* on the distal pole of the grain; × 1110; **B,** *Euphorbia bupleurifolia:* single *tricolpate* pollen grain; note finely reticulate sculpture of the exine. [B courtesy Dr. G. L. Webster.]

and dicotyledons (see Faegri and Iversen 1950, Pl. VIII). The monoporate condition, illustrated by the pollen of grasses, is regarded by Eames (1961) as the result of the phylogenetic reduction of the distal furrow to a small circular pore (Fig. 19-53, B). The pollen grains of members of the dicotyledonous family Winteraceae are also provided with single distal pores as shown in Fig. 19-52. Bailey and Nast (1943), however, expressed considerable doubt about whether an homology should be drawn between this type of aperture and the distally located furrow characteristic of the pollen of the Magnoliaceae and other ranalean families.

In concluding this description of pollen-grain apertures, a brief comment should be made regarding the current application of the terms "sulcus" and "colpus." Walker (1971, p. 11) maintains that the correct palynological term for a furrow-like polar aperture is *sulcus* and that a *colpus,* which is meridionally centered on the equatorial zone of a pollen grain, represents "a new and fundamentally different type of aperture." In this connection it is interesting to discover that Erdtman (1945, p. 190), at one phase in his studies, conceded that if tricolpate and monocolpate apertures are not homologous structures, "grains of the typical monocotyledonous monocolpate type may be termed "monosulcate." But in his 1969 treatise, the term sulcus has apparently been entirely replaced in favor of colpus. The problem we are discussing is, however, much more than a mere choice of terms. At present we have virtually no fully convincing evidence about the phylogenetic origin of the tricolpate form in the pollen of angiosperms, although Wilson (1964) postulated that the peculiar three-parted ("trichotomosulcate") distal polar aperture, characteristic of some of the pollen of members of the Canellaceae, represents an intermediate evolutionary stage between the monosulcate and the tricolpate condition.

SPORODERM. The term sporoderm is now widely used to designate collectively the two *major* wall layers of a mature pollen grain. The inner of these two layers is the *intine* while the outer layer is termed the *exine.* The exine alone survives the drastic chemical treatment, known as "acetolysis" used to prepare pollen for microscopical study; for this reason, the emphasis has inevitably centered on the structure and external sculpturing of the exine (for information on the acetolysis method, see Erdtman, 1969, pp. 213–216). When the surface of the exine is studied at high magnification, particularly with the use of the scanning electron microscope, exquisite patterns of sculpturing, in the form of reticula, warts, spines, striations, etc. are revealed (Figs. 19-52, 19-54). Surface features of pollen grains, together with the various types and arrangement of apertures, provide a rich source of information which is now being applied in the elucidation of taxonomic problems in the angiosperms (see Walker's 1971 comprehensive survey of the pollen of the Annonaceae).

From the standpoint of its internal structure, the exine can usually be resolved into two principal layers and different terms have been proposed to designate them. Faegri and Iversen (1950) for example, term the outer and inner layers, respectively, the "ektexine" and the "endexine." Although these terms have been adopted by some palynologists (e.g., Felix, 1961; Wilson, 1964), Erdtman (1969) considers them unwieldy and proposes the substitute terms *sexine* for the outer layer and *nexine* for the inner layer of the exine. His terminology, which seems linguistically more acceptable, is now in wide use by students of pollen-grain structure (see Heslop-Harrison, 1968; Walker, 1971). The ultrastructure of the sexine is extremely complex and the investigation of it has, in turn, given rise to a confusing and inconsistent terminology, which is evident in Erdtman's recent treatise (see

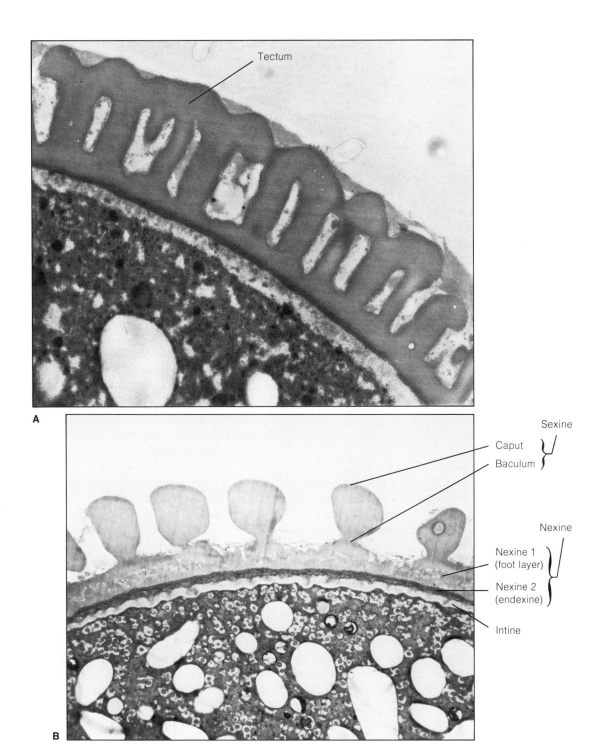

Figure 19-55
Electron micrographs of small portions of the sporoderm of pollen grains. **A**, sporoderm of *Euphorbia esculenta* in which the fusion of the bacula of the sexine forms a tectum. **B**, sporoderm of *Cnidoscolus rotundifolia*, illustrating an intectate type of sporoderm. [Courtesy Mr. James Tanno.]

Erdtman, 1969, pp. 39–46). Suffice it to state that the sexine appears to consist of capitate rods (bacula); if these elements are mostly free from each other, the exine is termed intectate; if a union between the heads (capita) of the rods forms a roof or *tectum,* the sexine is termed tectate (Fig. 19-55).

In conclusion, it should be emphasized that the morphological study of pollen grains, despite the formidable and ever changing terminology, has already contributed significantly to the elucidation of taxonomic problems and in the future—if intensive studies on apertures and wall structure are pursued within a broad frame of reference—may be expected to shed important light on the persistent mystery of the origin of the angiosperms and their salient trends of specialization.

Inflorescences

In some angiosperm taxa, the flowers are solitary and are borne in a terminal position on the axis. Although this position has been regarded by Parkin (1914) and others as primitive in the angiosperms, the phylogenetic significance of large terminal flowers in such genera as *Paeonia, Magnolia, Liriodendron, Calycanthus,* and *Eupomatia* is an open question primarily because of our present ignorance of the structure and position of the flowers in the remote ancestors of modern flowering plants. Among living angiosperms, most solitary terminal flowers probably represent the surviving members of either terminal or lateral clusters of flowers (Eames, 1961). In the Winteraceae, for example—a family distinguished by several primitive features—there are numerous transitions between complex branched inflorescences and solitary axillary or terminal flowers. In one genus of this family, *Zygogynum,* the solitary terminal flower

represents "the end of a reduction series" (Bailey and Nast, 1945a). According to Troll's (1969, pp. 330–452) comprehensive survey, the inflorescence has been reduced to a solitary terminal flower in a great many dicotyledonous genera, some of which are strictly uniflorus, whereas in others, only certain of the species have solitary terminal flowers.

Most commonly the flowers of angiosperms develop in more-or-less well-demarcated clusters known as inflorescences. During the long history of taxonomy, many efforts have been made to characterize and to classify the bewildering "types" of angiosperm inflorescences. As might be expected, a correspondingly rich terminology has inevitably arisen and we are indebted to Rickett (1944, 1955) for his scholarly review of the subject and for an effort to redefine the confusing terms that are still used to classify inflorescences. Without question, the most comprehensive modern account of inflorescences has been given by Troll (1964, 1969), who has published the first two volumes of his projected four-volume treatise on inflorescence morphology. Later in this section, we will review briefly Troll's synthetic treatment, which has culminated in his "typology" of inflorescences. But first it will be essential to examine certain rather general organographic and ontogenetic aspects of inflorescences.

Organography of Inflorescences

From a broad organographic standpoint, an inflorescence "is a flower-bearing branch or system of branches" (Rickett, 1944, p. 224). When inflorescences are characterized in this way, no sharp boundary can be drawn between terminal bracteate clusters of flowers and floriferous leafy shoots in which the lateral flowers arise in the axils of foliage leaves. Diagrammatic examples of well-demarcated inflorescences are shown in

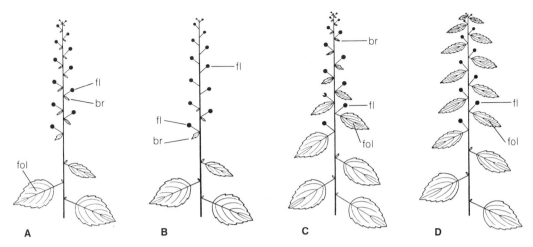

Figure 19-56
Diagrams showing variations in the types of foliar organs that subtend the lateral flowers (fl) of inflorescences. **A,** well-defined terminal raceme with each flower subtended by a diminutive bract (br); **B,** similar type of inflorescence illustrating progressive reduction and ultimate suppression of bracts; **C,** transitional form, the lower flowers arising in the axils of small foliage leaves (fol); **D,** leafy or "frondose" inflorescence with all flowers subtended by small petiolate foliage leaves. [From *Die Infloreszenzen* by W. Troll. Gustav Fischer, Stuttgart, 1964.]

Fig. 19-56, A, B. Each flower may be subtended by a minute — and often ephemeral — bract, or a progressive reduction in the size of the bracts may culminate in the formation of bract-less flowers in the upper region of the inflorescence. According to Troll (1964, p. 6), many floral bracts are merely small, greatly simplified leaf blades or else correspond morphologically to the basal sheath region of foliage leaves. In the type of organization which Troll designates as a "frondose inflorescence" (Fig. 19-56, D), each flower is subtended by a slightly reduced foliage leaf. As is shown in Fig. 19-56, C, there are transitional forms between typical "bracteate" and "frondose" types of inflorescences.

Ontogeny of Inflorescences

Inflorescences have long been classified, on a developmental basis, into two major categories: "closed" or determinate and "open" or indeterminate types (Fig. 19-57).

Troll (1964) has strongly emphasized this traditional method of treatment, particularly with reference to his typological classification of inflorescences.

In closed inflorescence development (Fig. 19-57, A), the meristematic apex of the developing inflorescence ceases producing bracts and gives rise to the primordia of the outer appendages of a terminal flower (Fig. 19-58, A, B). The initiation of the terminal flower is thus precocious, i.e., it occurs when the uppermost lateral flowers are still in a very early stage in their ontogeny (Fig. 19-58, B). Correlated with this fact, the terminal flower, as a rule, opens before the expansion of the lateral flowers. From a comparative standpoint, it is interesting that lateral flowers, in a closed type of inflorescence, may open acropetally, basipetally, or in a divergent pattern with respect to the terminal flower. In the divergent sequence, expansion of the lateral flowers begins in the middle region of the inflorescence and continues both acropetally and basipetally (Troll, 1964, p. 13, Abb. 8).

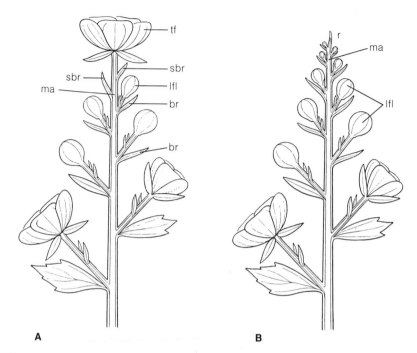

Figure 19-57

Diagrams contrasting the morphology of "closed" and "open" types of inflorescences. **A,** closed type, the main axis (ma) of the inflorescence ending in a terminal flower (tf). Note the two sterile bracts (sbr) intervening between the terminal and uppermost lateral flower (lfl); **B,** open type, the main axis (ma) ending blindly in the form of a small terminal "rudimentary" tip (r). See text for further explanations. [From *Die Infloreszenzen* by W. Troll. Gustav Fischer, Stuttgart, 1964.]

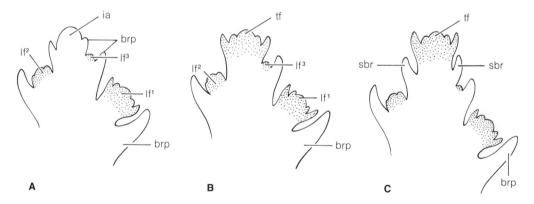

Figure 19-58

Diagrammatic longisections showing early stages in the development of the terminal and lateral flowers of a closed type of inflorescence. **A,** stage prior to the initiation of the terminal flower; the inflorescence apex (ia) has formed bract primordia (brp) that subtend an acropetal series of developing lateral flowers (lf^1–lf^3); **B,** later stage illustrating origin of terminal flower (tf) from the inflorescence-apex; **C,** stage similar to **B** but with a pair of sterile bracts (sbr) at the base of the terminal flower. [From *Die Infloreszenzen* by W. Troll. Gustav Fischer, Stuttgart, 1964.]

In the open inflorescence, the apex of the young inflorescence, in place of initiating a terminal flower, continues in an "indefinite" manner to form bracts (with lateral flowers) but ultimately ceases growth and ends in the form of a rudimentary stub (Fig. 19-57, B, r). Troll describes in detail the considerable variation in morphology of the abortive tip in open inflorescences. In some instances, *all* the lateral flowers complete their development and the true apex of the inflorescence may be minute or even unrecognizable. In contrast, the entire terminal region of the inflorescence may consist of the apical meristem together with a few undeveloped bracts and lateral flower primordia.

Classification of Inflorescences

Many of the terms which are still used to characterize the common types of inflorescences, such as those presented together with diagrams in Fig. 19-59, were used as early as 1751 by Linnaeus in his *Philosophia Botanica.* During the long history of morphology and taxonomy, the meaning of some of these terms changed very little or not at all, while other terms became more-or-less drastically modified. In addition, new terms were introduced during the nineteenth century. As a result, there now exists in botanical literature a confused labyrinth of terminologies which has been analyzed in great detail in Rickett's (1944, 1955) papers.

Several major factors appear to have contributed to the present rather chaotic state of inflorescence classification. In the first place, we are still uninformed about the evolutionary history of angiosperms, i.e., what constitutes the basic or "primitive" type of inflorescence? Does such an apparently "simple" type as the *raceme,* which consists of an axis bearing pedicillate lateral flowers subtended by bracts, represent a "prototype" of the other forms of inflorescences, such as the *spike, panicle, corymb,* and *umbel?* (See Fig. 19-59 for illustrations of these types.) Or should we begin our phylogenetic series with complex branched inflorescences — such as the panicle — and postulate that from them were derived, for example, the *umbel* and *head* in which the flowers are congested at the end of the inflorescence apex? (See Fig. 19-59 for illustrations of these types.) Theoretical trends of evolution of this sort, however, are of doubtful value because it seems evident that *parallel evolution* has played a major role in the formation of angiosperm inflorescences. In other words, comparative studies clearly suggest that not all "racemes," "umbels," "corymbs," or "heads," etc., are homologous; types which appear similar may have had different methods of origin during the evolution of the various taxa of the angiosperms.

An outstanding example of confusion in terminology is provided by the traditional recognition of two major groups of inflorescences: "racemose" and "cymose" types. The group of racemose inflorescences is commonly taken to include the raceme, panicle, and corymb in which, as represented in Fig. 19-59, there is presumed to be a more-or-less indefinite formation of flowers by the apex of the inflorescence. In contrast, the main axis of so-called cymose inflorescences ("cymes") — and the axes of the lateral flowering branches — are said to end in terminal flowers (Fig. 19-59). If defined in this way, the terms racemose and cymose are imprecise and inaccurate. Many simple racemes end in a functional terminal flower (Fig. 19-57, A); conversely some inflorescences, classified as cymose, are devoid of a terminal flower (Rickett, 1955, p. 425). In this connection, Troll (1964, p. 33) states the problem clearly by saying that although *cymose branching* may be found in various types of inflorescences, a "cymose inflorescence" does not exist as a separate morphological type.

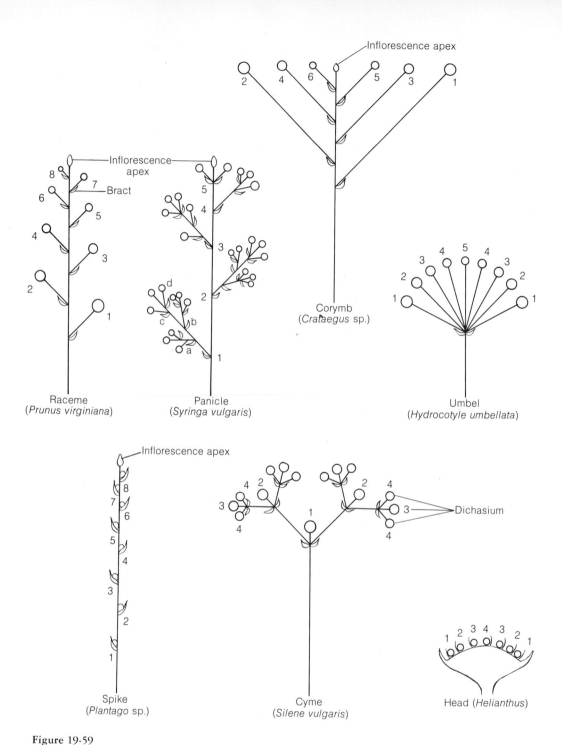

Figure 19-59

Diagrams of some of the common types of inflorescences in the angiosperms. Flowers are represented by circles and their order of development indicated in numerical sequence; in each inflorescence, flower 1 is the oldest in the group. The letters a–c, in the diagram of the panicle, depict the sequence of flower development in one of the lowest branches. See text for further explanation. [From *A Textbook of General Botany*, Ed. 2, by R. M. Holman and W. W. Robbins. Wiley, New York, 1928.]

When the various forms of inflorescences are precisely defined, as Rickett (1955) has attempted to do, important "characters" are provided for the descriptive needs of systematic botany. But the continued use of the often ambiguous morphological concepts of "racemose" and "cymose" to designate major categories of inflorescences, is highly questionable from either an ontogenetic or comparative point of view. (See Rickett, 1944, pp. 207–211.)

Unfortunately, there has been a conspicuous tendency in the past to treat inflorescences as static structures rather than as "modes of flowering." As Rickett (1944, p. 188) critically remarks, "the names (of inflorescences) have outlived the theories, to our ultimate confusion." The most recent attempt to take a new and broader view of the problem of inflorescence morphology has been made by Troll (1964, 1969) whose typological classification will now be briefly presented.

Typology of Inflorescences

According to Troll's (1935b, pp. 45–49) concept, the first objective of all morphological study should be the broadest possible comparison of the variant forms of a central "type" of organ or structure. To be specific, we should, for example, be fully informed regarding the *range* in form and organization of leaves throughout the angiosperms before we raise the question: what is *the* primitive "type" of leaf? Although typology in Troll's sense by no means excludes phylogenetic considerations, it should, according to him, have priority; once a typological series is recognized, delineation of phylogenetic series then becomes possible. In the light of this philosophy, Troll maintains that all the diverse forms of inflorescences can be grouped under two major types which he terms *polytelic* and *monotelic* (see Fig. 19-60, A, B). The suffix "telic," common to both

of these terms should be recognized as being derived from the Greek root *telos*, meaning an end or termination. Thus, the basic distinction between monotelic and polytelic depends upon the difference in the terminations of the axes of complex inflorescences, which Troll collectively calls "synflorescences."

Troll's typological analysis begins with relatively complex branched synflorescences, which he regards as composed of "systems of florescences." This starting point permits him to develop his ideas regarding (a) the derivation of the simpler forms of inflorescences (e.g., racemes, spikes, corymbs) as the result of the reduction of the lateral florescences to single flowers, as well as (b) the derivation of such complex forms as capita (heads) by the condensation and broadening of the end of the main inflorescence axis, and (c) the reduction of various types of inflorescences to a solitary terminal flower.

The salient features of a *polytelic synflorescence* are shown diagrammatically in Fig. 19-60, A. The main axis ends in a *group* of flowers termed the principal or "main" florescence (*mf* in Fig. 19-60, A). Below the main florescence there are lateral flowering shoots or "paracladia" (pc) that repeat the morphological organization of the terminal florescence and that are designated by Troll as "co-florescences" (cf in Fig. 19-60, A). Without exception, polytelic synflorescences are of the "open" type because the apices of *both* the main florescence and the co-florescences fail to produce terminal flowers, and as a consequence, *all* flowers terminate *secondary axes*.

The distinguishing feature of a *monotelic synflorescence*—regardless of the degree or type of branching—is the formation of a terminal flower at the tip of the main and lateral axes (Fig. 19-60, B–G, tf and tf[1]). Monotelic synflorescences therefore correspond to the "closed" type of inflorescence. Among monotelic inflorescences, however,

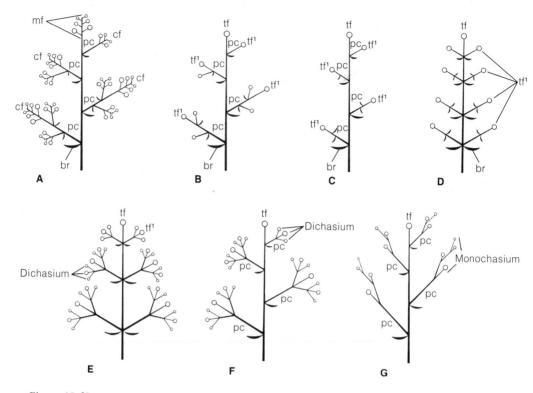

Figure 19-60

Schematic representations of the two principal types of synflorescences according to Troll. **A,** the polytelic synflorescence, in which the inflorescence axis, in place of forming a single terminal flower, ends in a many flowered main florescence (mf), the morphological organization of which is repeated by each of the "paracladia" (pc) that have formed lateral co-florescences (cf); **B,** the monotelic synflorescence, in which the main inflorescence axis ends with a terminal flower (tf), an organization which is repeated in each of the floral branches, where the terminal flowers are indicated as tf¹; **C, D,** simpler types of monotelic synflorescences; **E–G,** thyrsoid types of monotelic synflorescences, in which the paracladia are either dichasial (**E, F**) or monochasial (**G**). [From *Die Infloreszenzen* by W. Troll. Gustav Fischer, Stuttgart, 1964.]

may be found what appear to be "open" types. Troll maintains that this apparent inconsistency is the result of the abortion of the main terminal flower.

Because of the extraordinary range in the morphology of inflorescence types and the many forms that are transitional between them, we will only comment very briefly on the relation of Troll's typology to specific examples. Polytelic inflorescences are illustrated by *Capsella bursa-pastoris*, in which the florescences are simple racemes and by the more complex cymoid synflorescences of *Galeopsis* and other members of the Labiatae.

Monotelic inflorescences are exemplified by the panicle; in fact, Troll states that all panicles represent monotelic synflorescences. Under the broad category of monotelic synflorescences, Troll recognizes both paniculate (Fig. 19-60, B) and various types of thyrsoid inflorescences (Fig. 19-60, E–G). In the latter, the "units" of the inflorescence may be either *dichasia* (three-membered clusters, consisting of a terminal and two lateral flowers) or *monochasia* (in which each lateral branch continues the growth of a sympodial axis that ends with a terminal flower).

Both monotelic and polytelic inflorescences are widely distributed throughout the dicotyledons. The latter type prevails, according to Weberling's (1965, p. 220) summary, "without exception" in such families as Cruciferae, Leguminosae, Labiatae, Primulaceae, and Scrophulariaceae; monotelic inflorescences are characteristic of Caryophyllaceae, Oleaceae (e.g., *Syringa vulgaris*), Hydrophyllaceae, Solanaceae, and Boraginaceae. Troll (1964, p. 156) confesses that although it is a most difficult problem to decide whether the monotelic type has been derived from the polytelic, or *vice versa*, in general there is hardly any doubt, in a particular case, whether one is dealing with monotely or polytely. According to Weberling (1965, p. 220), "the polytelic type is more highly evolved than and perhaps de-

rived from the monotelic type," a possibility that Troll proposes to document in his future volumes on inflorescences.

In conclusion it should be emphasized that Troll's typology rests upon an exceptionally broad foundation. According to one of his former students (Weberling, 1965), Troll has studied the inflorescence morphology of more than 10,000 species distributed among 240 of the 342 families of angiosperms included in the last edition of Engler's (1964) *"Syllabus der Pflanzenfamilien"*! Regardless of future decision about the merits or defects of Troll's typology, his two books provide morphologists and taxonomists with the most comprehensive description and analysis of angiosperm inflorescences that have ever been attempted.

References

Andrews, H. N., Jr.
  1947. *Ancient Plants and the World They Lived In*. Comstock, Ithaca, N.Y.
Arber, A.
  1937. The interpretation of the flower: a study of some aspects of morphological thought. *Biol. Rev.* 12:157–184.
  1946. Goethe's botany. *Chron. Bot.* 10:67–124.
Arber, E. A. N., and J. Parkin
  1907. On the origin of angiosperms. *Jour. Linn. Soc. London Bot.* 38:29–80.
Arnott, H. J., and S. C. Tucker
  1963. Analysis of petal venation in *Ranunculus*. I. Anastomoses in *R. repens v. pleniflorus*. *Amer. Jour. Bot.* 50:821–830.
  1964. Analysis of petal venation in *Ranunculus*. II. Number and position of dichotomies in *R. repens v. pleniflorus*. *Bot. Gaz.* 125:13–26.
Axelrod, D. I.
  1952. A theory of angiosperm evolution. *Evolution* 6:29–60.

  1961. How old are the angiosperms? *Amer. Jour. Sci.* 259:447–459.
  1970. Mesozoic paleogeography and early angiosperm history. *Bot. Rev.* 36:277–319.
Ayensu, E. S.
  1972. *Anatomy of the Monocotyledons. VI. Dioscoreales*. Clarendon Press, Oxford.
Bailey, I. W.
  1944. The development of vessels in angiosperms and its significance in morphological research. *Amer. Jour. Bot.* 31:421–428.
  1953. Evolution of the tracheary tissue of land plants. *Amer. Jour. Bot.* 40:4–8.
  1954. *Contributions to Plant Anatomy*. Chronica Botanica, Waltham, Mass.
  1956. Nodal anatomy in retrospect. *Jour. Arnold Arboretum* 37:269–287.
  1957. Additional notes on the vesselless dicotyledon, *Amborella tricopoda* Baill. *Jour. Arnold Arboretum* 38:374–378.

Bailey, I. W., and C. G. Nast
  1943. The comparative morphology of the Winteraceae. I. Pollen and stamens. *Jour. Arnold Arboretum* 24:340–346.
  1945a. The comparative morphology of the Winteraceae. VII. Summary and conclusions. *Jour. Arnold Arboretum* 26:37–47.
  1945b. Morphology and relationships of *Trochodendron* and *Tetracentron*. I. Stem, root, and leaf. *Jour. Arnold Arboretum* 26:143–154.
Bailey, I. W., and B. G. L. Swamy
  1948. *Amborella trichopoda* Baill., a new morphological type of vesselless dicotyledon. *Jour. Arnold Arboretum* 29:245–254.
  1951. The conduplicate carpel of dicotyledons and its initial trends of specialization. *Amer. Jour. Bot.* 38:373–379.
Bailey, I. W., and W. P. Thompson
  1918. Additional notes upon the angiosperms *Tetracentron, Trochodendron,* and *Drimys,* in which vessels are absent from the wood. *Ann. Bot.* 32:503–512.
Baum, H.
  1948a. Über die postgenitale Verwachsung in Karpellen. *Osterr. Bot. Zeit.* 95:86–94.
  1948b. Postgenitale Verwachsung in und zwischen Karpell—und Staubblatt-kreisen. Sitzungsber. *Osterr. Akad. Wiss. Math.-Naturw.* Kl. Abt. I. 157:17–38.
Benzing, D. H.
  1967. Developmental patterns in stem primary xylem of woody Ranales. II. Species with trilacunar and multilacunar nodes. *Amer. Jour. Bot.* 54:813–820.
Bernier, G.
  1962. Evolution of the apical meristem of *Sinapis alba* L. (long-day plant) in long days, in short days, and during the transfer from short days to long days. *Caryologia* 15:303–325.
Bernier, G., J.-M. Kinet, and R. Bronchart
  1967. Cellular events at the meristem during floral induction in *Sinapis alba* L. *Physiol. Veg.* 5:311–324.
Blackman, E.
  1971. The morphology and development of cross veins in the leaves of Bread Wheat (*Triticum aestivum* L.). *Ann. Bot. n.s.* 35:653–665.

Boke, N. H.
  1940. Histogenesis and morphology of the phyllode in certain species of *Acacia*. *Amer. Jour. Bot.* 27:73–90.
  1947. Development of the adult shoot apex and floral initiation in *Vinca rosea* L. *Amer. Jour. Bot.* 34:433–439.
Brown, R. W.
  1956. *Palmlike Plants from the Dolores Formation (Triassic) in Southwestern Colorado.* U.S. Geol. Survey Prof. Paper. 274–H: 205–209.
Buvat, R.
  1952. Structure évolution et fonctionnement du méristème apical de quelques Dicotylédones. *Ann. Sci. Bot. Nat.* Ser. 11. 13:199–300.
Canright, J. E.
  1952. The comparative morphology and relationships of the Magnoliaceae. I. Trends of specialization in the stamens. *Amer. Jour. Bot.* 39:484–497.
  1953. The comparative morphology and relationships of the Magnoliaceae. II. Significance of the pollen. *Phytomorphology* 3:355–365.
  1955. The comparative morphology and relationships of the Magnoliaceae. IV. Wood and nodal anatomy. *Jour. Arnold Arboretum* 36:119–140.
Carlquist, S.
  1961. *Comparative Plant Anatomy.* Holt, Rinehart and Winston, New York.
Cheadle, V. I.
  1953. Independent origin of vessels in the monocotyledons and dicotyledons. *Phytomorphology* 3:23–44.
  1955. The taxonomic use of specialization of vessels in the metaxylem of Gramineae, Cyperaceae, Juncaceae, and Restionaceae. *Jour. Arnold Arboretum* 36:141–157.
  1956. Research on xylem and phloem. Progress in fifty years. *Amer. Jour. Bot.* 43:719–731.
Cheadle, V. I., and J. M. Tucker
  1961. Vessels and phylogeny of Monocotyledoneae. *Rec. Adv. Bot.* 1:161–165.
Chrtek, J.
  1962. Der Verlauf der Nervatur in den Kronblättern. bzw. Kronen der Dikotyledonen, p. 3–10. Novitates Bot. Horti. Bot. Univ. Carolineae Pragensis.

1963. Die Nervatur der Kronblätter bei den Vertretern der Ordnung Rosales s. l. Acta Horti. Bot. Pragensis. p. 13–29.

Conde, L. F., and D. E. Stone
1970. Seedling morphology in Juglandaceae, the cotyledonary node. *Jour. Arnold Arboretum* 51:463–477.

Corner, E. J. H.
1966. *The Natural History of Palms.* University of California Press, Berkeley and Los Angeles.

Cronquist, A.
1968. *The Evolution and Classification of Flowering Plants.* Houghton Mifflin, Boston.

Cutler, D. F.
1969. *Anatomy of the Monocotyledons. IV. Juncales.* Clarendon Press, Oxford.

Cutter, E. G.
1965. Recent experimental studies of the shoot apex and shoot morphogenesis. *Bot. Rev.* 31:7–113.

Dahl, A. D., and J. R. Rawley
1965. Pollen of *Degeneria. Jour. Arnold Arboretum* 46:308–323.

De Candolle, A. P.
1844. *Organographie végétale, Vol. 1.* Germer Bailliere, Paris.

Delevoryas, T.
1962. *Morphology and Evolution of Fossil Plants.* Holt, Rinehart and Winston, N. Y.

Dickison, W. C.
1969. Comparative morphological studies in Dilleniaceae IV. Anatomy of the node and vascularization of the leaf. *Jour. Arnold Arboretum* 50:384–400.

Eames, A. J.
1931. The vascular anatomy of the flower with refutation of the theory of carpel polymorphism. *Amer. Jour. Bot.* 18:147–188.
1951. Again: 'The New Morphology.' *New Phytol.* 50:17–35.
1953. Neglected morphology of the palm leaf. *Phytomorphology* 3:172–189.
1961. *Morphology of the Angiosperms.* McGraw-Hill, New York.

Eames, A. J., and L. H. MacDaniels
1947. *An Introduction to Plant Anatomy.* Ed. 2. McGraw-Hill, New York.

Engler, A.
1926. Angiospermae. *In* Engler and Prantl (eds.), *Die natürlichen Pflanzenfamilien,* Bd. 14a.

1964. *Syllabus der Pflanzenfamilien.* Ed. 12, Vol 2. Gebrüder Borntraeger, Berlin.

Erdtman, G.
1945. Pollen morphology and plant taxonomy. III. *Morina* L. with an addition on pollen morphological terminology. *Svensk Bot. Tidskr.* 39:187–191.
1963. Palynology. *In* Preston, R. D. (ed.), *Advances in Botanical Research,* Vol. 1. Academic Press, London and New York.
1969. *Handbook of Palynology. An Introduction to the Study of Pollen Grains and Spores.* Hafner, New York.

Esau, K.
1965. *Plant Anatomy,* Ed 2. Wiley, New York.
1967. Minor veins in *Beta* leaves: structure related to function. *Proc. Amer. Phil. Soc.* 111:219–233.

Faegri, K., and J. Iversen
1950. *Text-book of Modern Pollen Analysis.* Ejnar Munksgaard, Copenhagen.

Fagerlind, F.
1946. Strobilus und Blüte von *Gnetum* und die Moglichkeit, aus ihrer Struktur den Blütenbau der Angiospermen zu deuten. *Ark. Bot.* 33A(8):1–57.

Felix, C. J.
1961. An introduction to palynology. *In* Andrews, H. N., Jr. (ed.), *Studies in Paleobotany.* Wiley, New York.

Foster, A. S.
1928. Salient features of the problem of budscale morphology. *Biol. Rev.* 3:123–164.
1929. Investigations on the morphology and comparative history of development of foliar organs. I. The foliage leaves and cataphyllary structures in the horsechestnut (*Aesculus hippocastanum* L.). *Amer. Jour. Bot.* 16:441–501.
1931. Investigations on the morphology and comparative history of development of foliar organs. II. Cataphyll and foliage leaf form and organization in the black hickory (*Carya Buckleyi* var. *arkansana*). *Amer. Jour. Bot.* 18:864–887.
1932. Investigations on the morphology and comparative history of development of foliar organs. III. Cataphyll and foliage leaf ontogeny in the black hickory (*Carya Buckleyi* var *arkansana*). *Amer. Jour. Bot.* 19:75–99.

Foster, A. S. (*continued*)

1935. A histogenetic study of foliar determination in *Carya Buckleyi* var. *arkansana. Amer. Jour. Bot.* 22:88–147.

1950a. Morphology and venation of the leaf in *Quiina acutangula* Ducke. *Amer. Jour. Bot.* 37:159–171.

1950b. Venation and histology of the leaflets in *Touroulia guianensis* Aubl. and *Froesia tricarpa* Pires. *Amer. Jour. Bot.* 37:848–862.

1951. Heterophylly and foliar venation in *Lacunaria. Bull. Torrey Bot. Club* 78: 382–400.

1952. Foliar venation in angiosperms from an ontogenetic standpoint. *Amer. Jour. Bot.* 39:752–766.

1953. Venation patterns in the leaves of angiosperms, with special reference to the Quiinaceae. *Proc. Seventh Int. Bot. Congr.* (Stockholm, 1950) p. 380.

1955. Comparative morphology of the foliar sclereids in *Boronella* Baill. *Jour. Arnold Arboretum* 36:189–198.

1959. The morphological and taxonomic significance of dichotomous venation in *Kingdonia uniflora* Balfour f. et W. W. Smith. *Notes Bot. Gard. Edinb.* 23:1–12.

1961a. The floral morphology and relationships of *Kingdonia uniflora. Jour. Arnold Arboretum* 42:397–410.

1961b. The phylogenetic significance of dichotomous venation in angiosperms. *Rec. Adv. Bot.* 2:971–975.

1963. The morphology and relationships of *Circaeaster. Jour. Arnold Arboretum* 44: 299–321.

1966. Morphology of anastomoses in the dichotomous venation of *Circaeaster. Amer. Jour. Bot.* 53:588–599.

1968. Further morphological studies on anastomoses in the dichotomous venation of *Circaeaster. Jour. Arnold Arboretum* 49: 52–67.

1970. Types of blind vein-endings in the dichotomous venation of *Circaeaster. Jour. Arnold Arboretum* 51:70–80.

1971. Additional studies on the morphology of blind vein-endings in the leaf of *Circaeaster agrestis. Amer. Jour. Bot.* 58:263–272.

Foster, A. S., and H. J. Arnott

1960. Morphology and dichotomous vasculature of the leaf of *Kingdonia uniflora. Amer. Jour. Bot.* 47:684–698.

Foster, A. S., and F. A. Barkley

1933. Organization and development of foliar organs in *Paeonia officinalis. Amer. Jour. Bot.* 20:365–385.

Gifford, E. M., Jr., and G. E. Corson, Jr.

1971. The shoot apex in seed plants. *Bot. Rev.* 37:143–229.

Glück, H.

1919. Blatt-und blütenmorphologische Studien. Gustav Fischer, Jena.

Goebel, K.

1905. *Organography of Plants,* Pt. II. Eng. Ed. by I. B. Balfour. Clarendon Press, Oxford.

1930. *Organographie der Pflanzen.* 2 Ed., Pt. 2. p. 1145. G. Fischer, Jena.

Goethe, J. W. von

1790. *Versuch die Metamorphose der Pflanzen zu erklären.* Gotha.

Grégoire, V.

1938. La morphogénèse et l'autonomie morphologique de l'appareil floral. I. Le carpelle. *Cellule* 47:287–452.

Heslop-Harrison, J.

1968. Pollen wall development. *Science* 161: 230–237.

Hickey, L. J.

1973. Classification of the architecture of dicotyledonous leaves. *Amer. Jour. Bot.* 60:17–33.

Hiepko, P.

1965. Vergleichend-morphologische und entwicklungsgeschichtliche Untersuchungen über das Perianth bei den Polycarpicae. Teil I. *Bot. Jahrb.* 84:359–426.

Howard, R. A.

1970. Some observations on the nodes of woody plants with special reference to the problem of the 'split-lateral' versus the 'common gap.' *Bot. Jour. Linn. Soc.* (Suppl. 1) 63:195–214.

Hu, Z. H., K. Li, and C. L. Lee

1964. Distribution and general morphology of *Kingdonia uniflora. Acta Bot. Sinica* 12(4): 351–358.

Kaplan, D. R.
  1967. Floral morphology, organogenesis and interpretation of the inferior ovary in *Downingia bacigalupii*. *Amer. Jour. Bot.* 54:1274–1290.
  1970. Comparative foliar histogenesis in *Acorus calamus* and its bearing on the phyllode theory of monocotyledonous leaves. *Amer. Jour. Bot.* 57:331–361.

Kapp, R. O.
  1969. *Pollen and Spores*. W. C. Brown, Dubuque, Iowa.

Kasapligil, B.
  1951. Morphological and ontogenetic studies of *Umbellularia californica* Nutt. and *Laurus nobilis* L. *Univ. Calif. Publ. Bot.* 25:115–240.

Kerner, A. and F. W. Oliver
  1895. *The Natural History of Plants*. Blackie and Son, London.

Kosakai, H., M. F. Moseley, Jr., and V. I. Cheadle
  1970. Morphological studies of the Nymphaeaceae. V. Does *Nelumbo* have vessels? *Amer. Jour. Bot.* 57:487–494.

Lersten, N.
  1965. Histogenesis of leaf venation in *Trifolium wormskioldii* (Leguminosae). *Amer. Jour. Bot.* 52:767–774.

Mallory, T. E., Su-Hwa Chiang, E. G. Cutter, and E. M. Gifford, Jr.
  1970. Sequence and pattern of lateral root formation in five selected species. *Amer. Jour. Bot.* 57:800–809.

Marsden, M. P. F., and I. W. Bailey
  1955. A fourth type of nodal anatomy in dicotyledons, illustrated by *Clerodendron trichotomum* Thunb. *Jour. Arnold Arboretum* 36:1–50.

Mason, H. L.
  1957. The concept of the flower and the theory of homology. *Madroño* 14:81–95.

McCoy, R. W.
  1940. Floral organogenesis in *Frasera carolinensis*. *Amer. Jour. Bot.* 27:600–609.

Melville, R.
  1962. A new theory of the angiosperm flower. I. The gynoecium. *Kew Bull.* 16:1–50.
  1963. A new theory of the angiosperm flower. II. The androecium. *Kew Bull.* 17:1–63.

Metcalfe, C. R., and L. Chalk
  1950. *Anatomy of the Dicotyledons*. 2 v. Clarendon Press, Oxford.
  1960. *Anatomy of the Monocotyledons. I. Gramineae*. Clarendon Press, Oxford.
  1971. *Anatomy of the Monocotyledons. V. Cyperaceae*. Clarendon Press, Oxford.

Money, L. L., I. W. Bailey, and B. G. L. Swamy
  1950. The morphology and relationships of the Monimiaceae. *Jour. Arnold Arboretum* 31:372–404.

Moseley, M. F., Jr.
  1967. The value of the vascular system in the study of the flower. *Phytomorphology* 17:159–164.

Nougarède, A.
  1967. Experimental cytology of the shoot apical cells during vegetative growth and flowering. *Int. Rev. Cytol.* 21:203–351.

Ozenda, P.
  1952. Remarques sur quelques interprétations de l'étamine. *Phytomorphology* 2:225–231.

Pant, D. D., and B. Mehra
  1964. Nodal anatomy in retrospect. *Phytomorphology* 14:384–387.

Parkin, J.
  1914. The evolution of the inflorescence. *Jour. Linn. Soc. London Bot.* 42:511–562.
  1951. The protrusion of the connective beyond the anther and its bearing on the evolution of the stamen. *Phytomorphology* 1:1–8.

Periasamy K., and B. G. L. Swamy
  1956. The conduplicate carpel of *Cananga odorata*. *Jour. Arnold Arboretum* 37:366–372.

Philipson, W. R., and M. N. Philipson
  1968. Diverse nodal types in *Rhododendron*. *Jour. Arnold Arboretum* 49:193–217.

Post, D. M.
  1958. Studies in Gentianaceae. I. Nodal anatomy of *Frasera* and *Swertia perennis*. *Bot. Gaz.* 120:1–14.

Pray, T. R.
  1955. Foliar venation of angiosperms. III. Pattern and histology of the venation of *Hosta*. *Amer. Jour. Bot.* 42:611–618.
  1963. Origin of vein endings in angiosperm leaves. *Phytomorphology* 13:60–81.

Puri. V.
1951. The rôle of floral anatomy in the solution of morphological problems. *Bot. Rev.* 17:471–553.

Rickett, H. W.
1944. The classification of inflorescences. *Bot. Rev.* 10:187–231.
1955. Materials for a dictionary of botanical terms. III. Inflorescences. *Bull. Torrey Bot. Club* 82:419–445.

Riopel, J. L., and T. A. Steeves
1964. Studies on the roots of *Musa acuminata* cv. Gros Michel. I. The anatomy and development of main roots. *Ann. Bot.* n.s. 28:475–490.

Rüffle, L.
1968. Merkmalskomplex bei älteren Angiospermen—Blättern und die Kutikula von *Credneria* Zenker (Menispermaceae). *Palaeontographica* 123(B):132–143.
1969. Die Telomtheorie bei der Deutung von Angiospermen-organen und ihrer Herkunft. *Wiss. Zeit. Padagogische Hochschule Potsdam* 13:253–272.

Salisbury, F. G.
1963. *The Flowering Process*. Pergamon Press, Macmillan, New York.

Shobe, W. R., and N. R. Lersten
1967. A technique for clearing and staining gymnosperm leaves. *Bot. Gaz.* 128:150–152.

Sinnott, E. W.
1914. Investigations on the angiosperms. I. The anatomy of the node as an aid in the classification of angiosperms. *Amer. Jour. Bot.* 1:303–322.

Sinnott, E. W., and I. W. Bailey
1914. Investigations on the phylogeny of the angiosperms. III. Nodal anatomy and the morphology of stipules. *Amer. Jour. Bot.* 1:441–453.
1915. Investigations on the phylogeny of the angiosperms. V. Foliar evidence as to the ancestry and early climatic environment of the angiosperms. *Amer. Jour. Bot.* 2:1–22.

Solereder, H.
1908. *Systematic Anatomy of the Dicotyledons.* 2 v. Clarendon Press, Oxford.

Sporne, K. R.
1971. *The Mysterious Origin of Flowering Plants.* Oxford University Press, London.

Swamy, B. G. L.
1949. Further contributions to the morphology of the Degeneriaceae. *Jour. Arnold Arboretum* 30:10–38.
1953. The morphology and relationships of the Chloranthaceae. *Jour. Arnold Arboretum* 34:375–408.

Swamy, B. G. L., and I. W. Bailey
1950. *Sarcandra,* a vesselless genus of the Chloranthaceae. *Jour. Arnold Arboretum* 31:117–129.

Takhtajan, A.
1959. *Die Evolution der Angiospermen.* Gustav Fischer, Jena.
1969. *Flowering Plants. Origin and Dispersal.* Translated from the Russian by C. Jeffrey. Smithsonian Institution Press, Washington, D.C.

Tepfer, S. S.
1953. Floral anatomy and ontogeny in *Aquilegia formosa* var. *truncata* and *Ranunculus repens. Univ. Calif. Publ. Bot.* 25:513–648.

Thomas, H. H.
1955. Mesozoic Pteridosperms. *Phytomorphology* 5:177–185.

Thompson, J. McLean
1934. Studies in advancing sterility. VII. The state of flowering known as angiospermy (with special reference to placentation and the origin and nature of follicles and achenes). *Univ. Liverpool, Hartley Bot. Lab. Publ.* 12:5–47.

Tieghem, Ph. van
1875. Recherches sur la structure du pistil et sur l'anatomie comparée de la fleur. *Mem. Acad. Sci. Inst. Imp.* France 21:1–262.
1900. Sur les dicotylédones du groupe homoxylées. *Jour. Bot.* 14:259–297; 330–361.

Tomlinson, P. B.
1961. *Anatomy of the Monocotyledons. II. Palmae.* Clarendon Press, Oxford.
1969. *Anatomy of the Monocotyledons. III. Commelinales—Zingiberales.* Clarendon Press, Oxford.

Troll, W.

1935a. Vergleichende Morphologie der Fieder-blätter. *Nova Acta Leopoldina* n. f. 2:311–455.

1935b. *Vergleichende Morphologie der höheren Pflanzen*, Erster Bd. Lieferung I. Gebrüder Borntraeger, Berlin.

1937. *Vergleichende Morphologie der höheren Pflanzen*, Bd. I. Erster Teil, Lieferung 3. Gebrüder Borntraeger, Berlin.

1938. *Vergleichende Morphologie der höheren Pflanzen*, Bd. I. Zweiter Teil, Lieferung 1. Gebrüder Borntraeger, Berlin.

1939a. *Vergleichende Morphologie der höheren Pflanzen*, Bd. I. Zweiter Teil, Lieferung 2. Gebrüder Borntraeger, Berlin.

1939b. *Vergleichende Morphologie der höheren Pflanzen*, Bd. I. Zweiter Teil, Lieferung 4. Gebrüder Borntraeger, Berlin.

1949. Über die Grundbegriffe der Wurzel-morphologie. *Osterr. Bot. Zeit.* 96:444–452.

1964. *Die Infloreszenzen.* Bd. 1. Gustav Fischer, Stuttgart.

1967. *Vergleichende Morphologie der höheren Pflanzen.* Erster Band, Dritten Teil. (Authorized reprinting of original published in 1943.) Gebrüder Borntraeger, Berlin.

1969. *Die Infloreszenzen,* Bd. 2, Teil 1. Gustav Fischer, Stuttgart.

Vertrees, G. L.

1962. Plastochronic changes in the vegetative apex and the ontogeny of the floral apex of *Liriodendron tulipifera* L. M.S. thesis, University of California, Davis.

Von Ettinghausen, C. R.

1861. *Die Blatt-Skelette der Dicotyledonen mit besonderer Rücksicht auf die Untersuchung und Bestimmung der fossilen Pflanzen-reste.* Kais. Kön. Hof und Staatsdruck-erei, Wien.

Von Guttenberg, H.

1941. *Der primäre Bau der Gymnospermen-wurzel.* (Handbuch d. Pflanzenanatomie, Bd. 8, VIII. Lief. 41.) Gebrüder Born-traeger, Berlin.

1960. *Grundzüge der Histogenese höherer Pflan-zen. I. Die Angiospermen.* (Handbuch d. Pflanzenanatomie Bd. 8, Teil 3.) Ge-brüder Borntraeger, Berlin.

1968. *Die primäre Bau der Angiospermenwurzel,* Ed. 2, Bd. 8, Teil 5. Gebrüder Born-traeger, Berlin.

Walker, J. W.

1971. Pollen morphology, phytogeography and phylogeny of the Annonaceae. *Contrib. Gray Herb. Harvard Univ.* 202.

Wardlaw, C. W.

1965. The organization of the shoot apex. Pp. 966–1076 in *Handbuch der Pflanzen-physiologie,* Vol. 15, Pt. 1. Springer-Verlag, Berlin.

Weberling, F.

1965. Typology of inflorescences. *Jour. Linn. Soc. London Bot.* 59:215–221.

Willis, J. C.

1966. *A Dictionary of the Flowering Plants and Ferns,* Ed 7, revised by H. K. Airy Shaw. University Press, Cambridge.

Wilson, C. L.

1937. The phylogeny of the stamen. *Amer. Jour. Bot.* 24:686–699.

1942. The telome theory and the origin of the stamen. *Amer. Jour. Bot.* 29:759–764.

1953. The telome theory. *Bot. Rev.* 19:417–437.

Wilson, C. L., and T. Just

1939. The morphology of the flower. *Bot. Rev.* 5:97–131.

Wilson, T. K.

1964. Comparative morphology of the Canel-laceae. III. Pollen. *Bot. Gaz.* 125:192–197.

Wodehouse, R. P.

1935. *Pollen Grains.* McGraw-Hill, New York.

Zimmermann, M. H., and P. B. Tomlinson

1970. The vascular systems in the axis of *Dracaena fragrans.* 2. Distribution and development of secondary vascular tissue. *Jour. Arnold Arboretum* 51:478–491.

20

The Reproductive Cycle in Angiosperms

In the great majority of lower vascular plants the gametophyte generation, although comparatively small in size, is completely independent of the sporophyte and is a free-living plant. In marked contrast, the gametophyte generation of gymnosperms is not only reduced in size, but the female gametophyte is permanently enclosed within the ovule and thus entirely dependent for its nutrition on the sporophyte. The general evolutionary trend toward simplification and dependence reaches its culmination in the small, few-celled gametophytes of the angiosperms, the study of which, even today, with the best of technical methods, offers considerable difficulty to investigators.

Our present ontogenetic concepts of angiospermous gametophytes and our interpretation of the striking, novel aspects of sexual reproduction and seed formation in this group go back to the inspired and careful investigations that were made in Europe during the latter part of the nineteenth century. (See Maheshwari, 1950, for an excellent historical sketch of these early pioneering studies.) Modern studies on the details of the reproductive cycle in angiosperms are both voluminous and diversified, and they include systematic-phylogenetic surveys as well as the initiation of experimental work on the gametophytes and the developing seed. Some of these studies have resulted in clarification of the proper systematic disposition of certain genera or families. Others have greatly enriched our knowledge of the wide range in patterns of development of the endosperm tissue and the embryo. Hopefully, the continuous accumulation of knowledge in this area will eventually throw new and important light on the abiding mystery of the evolutionary relationships of flowering plants.

The publication of Maheshwari's invaluable book, *An Introduction to the Embryology of Angiosperms,* has provided morphologists with an able, coherent résumé of the salient aspects of sporogenesis, gametophyte development, double fertilization, endosperm formation, and embryogeny in flowering plants. In addition to this work, extensive discussions of the reproductive cycle have been prepared by Schnarf (1927–1929, 1931, 1941), Johansen (1950), Maheshwari (1948, 1949, 1950), Wardlaw (1955), and Davis (1966). These references will supplement the necessarily brief outline of the structures and processes in the reproductive cycle of angiosperms given in this chapter.

One of the characteristic features of sexual reproduction in angiosperms is the marked telescoping of the processes of sporogenesis and gametophyte development. Microsporogenesis, for example, is directly followed by only two mitoses, which yield the pollen grain with its typical, endosporic, three-celled male gametophyte. Megasporogenesis may involve, as in gymnosperms, the formation of a linear tetrad of morphologically discrete megaspores. But in many angiosperms, wall formation does not occur between the haploid megaspore nuclei, all four of which, by further divisions, contribute to the development of the female gametophyte. In short, the demarcation between the spore and the early phase in ontogeny of the gametophyte is frequently not sharp, and correct interpretation depends on a full appraisal of cytological details, especially the point at which meiosis occurs and the subsequent behavior and fate of the haploid nuclei.

In Chapter 19 the angiosperm flower was discussed from a broad morphological point of view, with particular emphasis on the interpretation of floral organs on the basis of their ontogeny and vasculation. In this chapter we are concerned with the rôle of the flower in reproduction, and we will attempt to present a coherent account of sporogenesis, the development of the gametophytes, pollination, fertilization, and the formation of endosperm and embryo in the seed. Because of the rapid accumulation in recent years of information on the life cycles of a wide range of angiosperms, it is now possible to present a synthetic description of the *basic features* common to all reproductive cycles in the angiosperms. This is done with full realization of the numerous—and possibly significant—departures in detail from the general situation as illustrated, for example, by the various types of embryo-sac development that are now recognized. To promote clarity in exposition, important deviations from what appears to be the typical condition will be reserved for brief discussion at the end of each topic.

Microsporangia

General Morphology

Our present general concepts of the structure and development of the microsporangia of angiosperms rest upon the classical studies of Warming (1873). He showed that typically four groups of pollen mother cells (i.e., microsporocytes) and the wall layers external to them, are derived from the periclinal divisions of cells of the hypodermal layer at the four "corners" of the differentiating anther. Warming emphasized that throughout all stages in development the epidermal cells external to each of the microsporangia, divide only in the anticlinal plane and that hence the initiation of *both* the sporogenous cells and the overlying layers of the "anther wall" are derived from a common hypodermal layer, which he termed the "mother layer."

The innumerable detailed studies on the differentiation of the angiosperm anther and its characteristic embedded (i.e., hypodermal) microsporangia that have been made since Warming's pioneering investigations fully support his general conclusions (Fig. 20-2). But from a comparative phylogenetic viewpoint, the significance of the embedded position of the microsporangia in angiosperms has raised problems and produced a somewhat confusing terminology in modern literature. The central morphological question at issue is whether angiosperm microsporangia are strictly homologous, in their mode of origin and structure, with the eusporangia of other vascular plants, including the gymnosperms.

Part of the problem has been generated by the use of different terms to designate *collectively* the layers of cells situated *between* the epidermis of the anther and the underlying areas of sporogenous cells (Fig. 20-1). Do these interposed cell layers collectively represent the wall of an embedded eusporangium or should they be considered as special wall layers of the anther itself? Warming (1873, p. 1) introduced the rather ambiguous term "anther wall" to designate the cell layers above each of the four masses of sporogenous cells and this term is still widely used in his sense in current morphological literature (see Schnarf, 1928–1929; Maheshwari, 1950; Esau, 1965; Eames, 1961; Davis, 1966; and Vasil, 1967). Unfortunately, the term "anther wall" has suggested the interpretation that the microsporangia of angiosperms are "wall-less" areas of sporogenous tissue. This view is strongly advocated by Eames (1961, p. 136) who states: "In the angiosperms there is no sporangium wall distinct from enclosing tissues of the sporophyll. The description of the microsporangium wall as 'many layered' is based on the interpretation of the anther-sac wall as the sporangium wall." From Eames' standpoint, the wall-less microsporangia "set the angiosperms well apart in method of microspore-bearing from other vascular plants."

Eames' theory of the "wall-less" micro-sporangium has been strongly criticized and rejected by Periasamy and Swamy (1964) who point out that "the chief attribute of a morphologically true sporangial wall is that it should arise from the same initial as the spore mother cell." As this requirement is fully supported by the facts of development, angiosperm microsporangia, aside from their hypodermal origin, do not differ fundamentally from typical eusporangia (see also Coulter and Chamberlain, 1912; Wettstein, 1924; Maheshwari and Kapil, 1966; and Chan and Hillson, 1971). In this connection, it should be emphasized that the *position* of the initial cells of eusporangia varies considerably within vascular plants. In *Lycopodium* and *Selaginella* (see Chapter 9) the initials are superficial cells in contrast to the consistently hypodermal location of the microsporangial initials in angiosperms. In some genera of the Coniferales, the position of the sporangial initials is superficial (*Pseudotsuga* and *Cedrus*) and in others it is hypodermal (*Cryptomeria, Cupressus, Chamaecyparis,* and *Taxus*) (see Chapter 17). If Eames' concept of the wall-less microsporangia were applied to the interpretation of microsporangial development in these gymnosperms, *Pseudotsuga* and *Cedrus* would have sporangia with "true" walls while the microsporangia of the other four genera would be "wall-less," like their "counterparts" in angiosperms! These morphological comparisons clearly demonstrate the artificiality of attempts to classify the eusporangia of vascular plants into "walled" and "wall-less" types. They also suggest that during the evolution of seed plants, the "center" of microsporangium initiation has shifted from superficial cells to hypodermal initials. As Fagerlind (1961) pointed out, future morphogenetic studies on the mode of initiation and development of sporangia in vascular plants at large may lead to the recognition of "a rather continuous typological series."

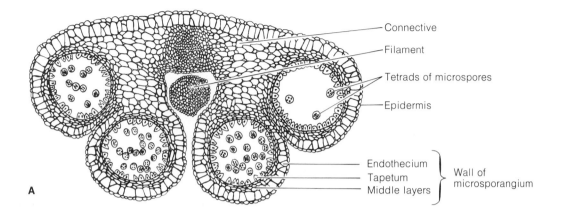

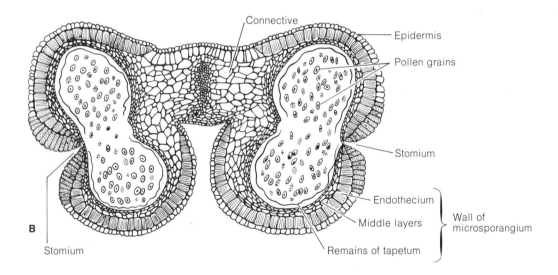

Figure 20-1
Transections of anthers of *Lilium*. **A,** anther with two pairs of nearly mature microsporangia, each sporangium containing tetrads of microspores. The cells of the endothecium have not yet acquired their characteristic wall thickenings; **B,** anther in later developmental stage, just prior to its dehiscence. The tissue between the members of each pair of sporangia has broken down, forming two pollen-containing sacs. Note wall thickenings in endothecium and the collapsed remains of the tapetum. The stomium at each side of the anther corresponds to the region of future longitudinal dehiscence. [Redrawn from *Handbuch der Systematischen Botanik* by R. Wettstein. Franz Deuticke, Leipzig u. Wien, 1924.]

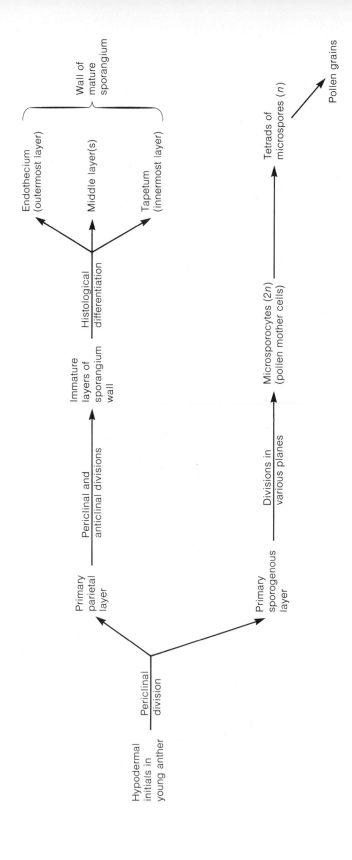

Figure 20-2
Diagram summarizing the general pattern of development of the microsporangium in the angiosperms.

Structure and Development

Although there is a wide range in the external form and size of angiosperm stamens, the anther, just prior to its dehiscence, most commonly contains four microsporangia arranged in pairs in the two anther lobes. Members of each pair of microsporangia are separated from each other by a plate of sterile tissue and the central tissue or *connective* of the anther is traversed by a strand of vascular tissue (Fig. 20-1, A). In some taxa, such as *Circaeaster* (Junell, 1931) and *Pulsatilla* (Bütow, 1955), only two microsporangia are developed, one in each lobe of the anther (see also Trapp, 1956). However, as Davis (1966, pp. 7–8) points out, some of the examples of bisporangiate anthers mentioned in the literature are open to question because they may actually represent the two locules, or pollen sacs, produced in many tetrasporangiate anthers by the breakdown of the sterile partitions between the members of each pair of microsporangia (compare A and B in Fig. 20-1).

Comparative studies on a wide range of angiosperm taxa have revealed considerable variation in the number and structure of the various layers of the microsporangium wall (see Davis, 1966, pp. 8–14). Before we describe specific examples which illustrate this variability, it is essential to indicate the general pattern of histogenesis which appears to predominate in angiosperms as a whole. This can be most easily accomplished by means of the schematic representation shown in Fig. 20-2. Many years ago Coulter and Chamberlain (1912) presented a lucid outline of the major steps in development of the microsporangium; the diagram shown in Fig. 20-2 is based upon their descriptions and terminologies.

In a typical tetrasporangiate stamen, each of the four microsporangia originates by the periclinal division of a group of hypodermal initials situated at the corners of the young anther. These periclinal divisions form a subepidermal *primary parietal layer* and an inner *primary sporogenous layer* (Fig. 20-2). The cells of the primary sporogenous layer may divide further prior to microsporogenesis, or else give rise directly to the diploid *microsporocytes*. When meiosis occurs in the sporocytes, tetrads of haploid microspores are formed, which, in turn, differentiate into pollen grains (Fig. 20-2). The cells of the primary parietal layer, by means of further periclinal and anticlinal divisions, form a variable number of concentrically arranged layers from which, by special histological differentiation, the various layers of the mature microsporangium wall take their origin.

As is shown diagrammatically in Fig. 20-2, the outermost layer of the wall is commonly termed the *endothecium*. Because of the characteristic thickenings, in the form of bars or networks, which develop on certain of the cell walls, the endothecium is termed by some morphologists the "fibrous layer." The endothecium appears to be a highly specialized wall layer concerned with the dehiscence of the anther (Fig. 20-1). Below the endothecium there is usually a layer— or several layers—of tabular thin-walled cells which collectively represent the *middle layers* of the sporangium wall. These middle layers become compressed or completely destroyed during the formation of the microspores (Fig. 20-1). The innermost layer of the sporangium wall is the *tapetum,* a glandular layer which, according to Carniel's (1963) detailed investigations, produces the enzymes, hormones, and nutritive materials used during the process of microsporogenesis. Further information on (1) the formation of tapetal cells at the *inner side* of the sporogenous tissue and (2) the morphological types of tapeta will be given later in this section of the chapter.

As already mentioned, Fig. 20-2 is intended only as a rather generalized histogenetic diagram which makes no attempt to show the wide variation in the *sequence* of periclinal divisions which may occur in the *derivatives* of the primary parietal layer. Recently, however, Davis (1966) has made a comprehensive systematic survey of the patterns of sporangium-wall development, which she maintains are consistent enough at the family level to justify classifying the methods of wall formation into four major types. She bases her recognition of each of the types on the behavior of the two "secondary parietal layers." These layers, according to her interpretation, are produced by the *first* periclinal division of the primary parietal layer.

In the Basic Type, *both* secondary parietal layers divide periclinally, yielding a four-layered wall consisting of an endothecium, two middle layers, and the tapetum. This type, illustrated in such families as the Winteraceae, Tiliaceae and Vitaceae, is regarded as the form from which the other three types have been derived. The Dicotyledonous Type, which is the most common, was found in half of all the families surveyed by Davis and is regarded as a derivative of the Basic Type because the *inner* of the two secondary parietal layers fails to divide further and differentiates directly into the tapetum. The periclinal division of the outer secondary parietal layer produces the endothecium and a single middle wall layer. In this way, a three-layered sporangium wall is formed. The third or Monocotyledonous Type, likewise characterized by a three-layered sporangium wall, differs from the preceding type by the suppression of periclinal divisions in the outer secondary parietal layer, which matures directly as the endothecium. Although this type is not restricted to monocotyledons, it is found, according to Davis, in 10 out of the 13 monocotyledonous families investigated. In the last or Reduced Type, *neither* of the secondary parietal layers undergoes periclinal division and the mature sporangium wall is two layered, consisting of the endothecium and the tapetum. In this type no

middle wall layers are formed. Examples of this probably highly advanced kind of sporangium wall are found in members of the Lemnaceae and Najadaceae.

Certain recent studies have given support for two of Davis' proposed types of sporangium-wall development. The Dicotyledonous Type, for example, is described in *Morina longifolia* (Vijayaraghavan and Sarveshwari, 1968) and *Centaurium ramosissimum* (Vijayaraghavan and Padmanaban, 1969); in *Nigella damascena,* the sporangium wall is described as developing according to the Basic Type (Vijayaraghavan and Marwah, 1969). But, on the other hand, we might seriously question whether, throughout the angiosperms, the sequences of periclinal divisions occur in such a precise and orderly fashion as is suggested in Davis' typology. It is evident, for example, from her own descriptions, that the number of middle layers of the mature wall—two in the Basic Type and only one in the Dicotyledonous Type—is by no means always consistent, as some of the cells of the "initial" middle layer or layers may undergo further periclinal divisions during wall development. Further evidence of the variability in the number of cell layers which develop *between* the tapetum and the anther-epidermis is provided by the investigations of Budell (1964). He found that in *Coleus* and *Inula* two such layers are produced: an outer endothecium and an inner middle layer. In contrast, four or five wall layers are formed during wall development in *Berberis* and *Rhododendron* and the number of layers in *Sassafras* varies from three to four. A particularly clear example of microsporangium wall development which does not readily "fit" into any of the types proposed by Davis is provided by Boke's (1949) study on *Vinca rosea* (Apocynaceae), which may now be described with the aid of Fig. 20-3.

In *Vinca,* the embryonic anther, as seen in transection, consists of a mass of ground meristem enclosed by a discrete protoderm

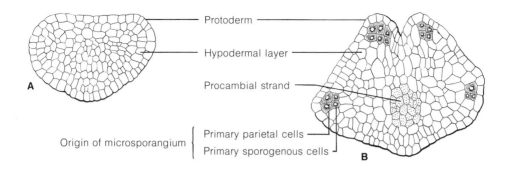

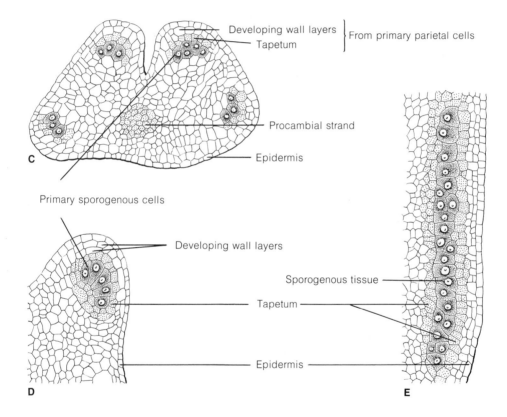

Figure 20-3

Origin and development of microsporangia in the stamen of *Vinca rosea*. **A–D**, transections of successively older stages in development of the anther. **A**, young anther with clearly defined hypodermal layer; **B**, later stage showing origin of microsporangia by periclinal divisions in four separate groups of hypodermal cells; **C**, stage illustrating the origin of a portion of the tapetum from the primary parietal cells; **D**, details of a later stage in development of a single microsporangium showing the developing wall layers and the group of sporogenous cells surrounded by the tapetum; **E**, longisection of portion of anther showing origin of inner portion of tapetum from cells adjacent to the sporogenous tissue (note the continued development of the wall layers below the epidermis). [Redrawn from Boke, *Amer. Jour. Bot.* 36:535, 1949.]

(Fig. 20-3, A). The protoderm develops into the anther-epidermis, and groups of 2–4 hypodermal cells, at the four corners of the differentiating anther, function as the initials of the microsporangia. Each initial group divides periclinally into an outer series of *primary parietal cells* and an inner layer of *primary sporogenous cells* (Fig. 20-3, B). The latter, with or without preliminary divisions, give rise to the microspore mother cells. The primary parietal cells at first may divide *either* anticlinally or periclinally (Fig. 20-3, C). After periclinal divisions have appeared, the inner cells become part of the future tapetum while the outer parietal cells divide periclinally once or twice and the derivative cells may then divide anticlinally. During this critical stage of wall development in *Vinca* there is apparently no fixed sequence in the plane of cell division and hence no evidence of the presence of definable "secondary parietal layers," in the sense of Davis' concept of the early ontogeny of either the Basic or Dicotyledonous types of sporangium wall formation (see Fig. 20-3, D, E).

As the microsporangia of *Vinca* approach maturity, the inner wall cells (including the tapetum) collapse so that only two wall layers are seen at the period of maturation of the pollen grains (see Boke, 1949, Fig. 32). It is not evident from Boke's account whether an endothecium, with characteristic secondary wall thickenings, is produced by the outermost layer of the microsporangium wall.

In *Vinca,* the innermost layer of the sporangium wall constitutes a considerable portion of the tapetum; the remainder of the tapetum originates, independently of the sporangium wall, from connective cells of the anther which lie adjacent to the inner boundary of the sporogenous tissue (Fig. 20-3, C, D). According to Boke's study, the inner part of the tapetum is frequently several layers in thickness in contrast to the predominantly one-layered structure of the outer tapetum which is a part of the sporangium wall proper (Fig. 20-3, E). A dual method of origin of tapetal cells is not confined to *Vinca* but has been observed in a number of other genera in the angiosperms (see Budell, 1964, and Periasamy and Parameswaran, 1965).

During the later stages of development of the microsporangium in angiosperms, the cells of the tapetum enlarge and, in many cases, become bi- or multinucleate as the result of the failure of cell-wall formation following the successive mitotic divisions. Recently it has also been shown that tapetal nuclei may increase in size and in chromosome number as a result of a process of *endomitosis*—i.e., the division of chromosomes within the nuclear membrane without the formation of a spindle. By this process, the conspicuously enlarged nuclei become polyploid to varying degrees. (See Carniel, 1963, for a detailed review of the extremely complex changes in nuclear and chromosome number which may accompany the maturation of tapetal cells in the angiosperms.)

From a structural as well as a physiological standpoint, it is interesting that, as in lower vascular plants, two major types of tapeta, distinguished on the basis of cell behavior during microsporogenesis, are recognized within the angiosperms: the *glandular* or *secretory tapetum,* the cells of which remain in their original position but finally become disorganized and obliterated (Fig. 20-1, A, B) and the *periplasmodial tapetum,* characterized by the preliminary breakdown of the cell walls and the fusion of the tapetal protoplasts to form a multinucleate periplasmodium which intrudes between the pollen mother cells (Fig. 20-4). Although an increase in the number of nuclei by means of normal mitosis takes place in the periplasmodial type, it is not known at present whether endomitosis also occurs (see Carniel, 1963, for additional information).

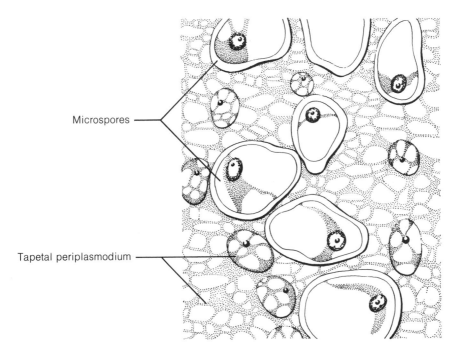

Microspores ────

Tapetal periplasmodium ────

Figure 20-4
Section of developing microsporangium of *Symphoricarpos racemosus* illustrating the structure
of the tapetal periplasmodium. [Adapted from *An Introduction to the Embryology of Angiosperms* by P. Maheshwari. McGraw-Hill, New York, 1950.]

Dehiscence

We may appropriately conclude the description of angiosperm microsporangia with a few brief comments on the various ways in which pollen is shed from the ripe anther and the possible role that the endothecial layer may play in the dehiscence process. In a comparatively few families (e.g., Ericaceae, Epacridaceae, Melastomaceae), the pollen is released from a small opening situated at one end (usually the distal end) of the anther. This is designated by the term poricidal dehiscence. More commonly, the anther dehisces *longitudinally* along the *stomium,* or furrow, situated at each side of the anther between the members of a pair of sporangia (Fig. 20-1, A, B). In the common type of anther represented in *Lilium,* the epidermal cells lining the base of each

stomium are very small and easily ruptured, causing the anther to split at two points, thus releasing the enclosed pollen grains.

Although it is commonly stated that the actual force which results in dehiscence is in some way related to the *endothecium,* the precise "mechanism" still awaits full explanation. Many investigators maintain that dehiscence is a hygroscopic phenomenon related to the peculiar pattern of secondary wall thickening of the endothecial cells. These thickenings may be restricted to the inner tangential and radial walls while the outer tangential walls remain thin. In an endothecium with this structure, the delicate outer walls are said to collapse more than the inner walls upon losing water and hence the two edges of the anther lobe curl, or reflex, in dry air. But it is also believed by certain investigators that *cohesion mechanisms* in the endothecium may play a role in anther dehiscence. For a more detailed discussion of the role of the endothecium in dehiscence, the student is referred to Haberlandt (1914, 1924).

Microsporogenesis

During the formation of the sporangium wall, the sporogenous cells may divide in various planes, the cells finally separating from one another and functioning as microsporocytes. The microsporocytes, or pollen mother cells (PMC), of angiosperms provide important and readily accessible material for the study of meiosis. Indeed, it is possible, under favorable or controlled conditions, to predict with considerable accuracy the stages in stamen development at which the various phases of the meiotic process will be most abundantly displayed. The laborious and time-consuming method of sectioning fixed anthers and subsequently staining the serial sections has been largely replaced by the process of "squashing" the anthers and then staining the masses of dividing pollen mother cells by the aceto-carmine, Feulgen, or crystal-violet techniques. (See Jensen, 1962, for detailed descriptions of the various procedures.) The use of these techniques also greatly facilitates the determination of the diploid and haploid number of chromosomes and the study of their form and structure.

Each functional microsporocyte, by means of meiosis and cytokinesis, gives rise to a tetrad of haploid microspores (Fig. 20-1, A). As is shown diagrammatically in Fig. 20-5, the arrangement of the four microspores in a tetrad varies widely and depends to some extent on the taxon in question. According to Maheshwari (1950) the *tetrahedral* and *isobilateral* types of microspore arrangement are very common but in some genera several kinds of arrangements may occur, even

in the same species. The *linear* arrangement —paralleled by the common arrangement of megaspores in angiosperm ovules—occurs in the family Asclepiadaceae but apparently from Davis' (1966) survey is a comparatively rare type.

At the time when meiosis begins, a thick *callose wall* is formed around each of the microsporocytes and later a wall of similar composition delimits each of the microspores in the tetrad (Waterkeyn, 1964). The process of cytokinesis that takes place during microspore formation has been intensively studied and has led to the recognition of two contrasted methods of wall formation.

In the so-called "successive type" illustrated by *Zea mays,* a *centrifugally* developing cell plate is formed at the end of meiosis I, dividing the microsporocyte into two cells (Fig. 20-6, A–C). Then the second meiotic division takes place in each of these two cells followed again by the centrifugal formation of cell plates (Fig. 20-6, D, E). In this way, an *isobilateral* type of microspore tetrad

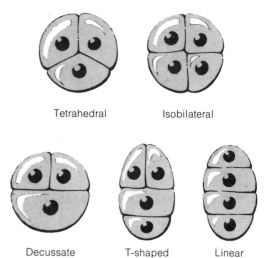

Tetrahedral   Isobilateral

Decussate   T-shaped   Linear

Figure 20-5
Various types of arrangement of microspores in a tetrad. [Adapted from *An Introduction to the Embryology of Angiosperms* by P. Maheshwari. McGraw-Hill, New York, 1950.]

is produced in *Zea*. This method of cytokinesis—the successive formation of two cell plates—is prevalent in the monocotyledons and has also been reported in some dicotyledonous families.

The distinguishing character of the so-called "simultaneous type" of cytokinesis, illustrated in Fig. 20-6, F–I, consists in the centripetal development of "constriction furrows," which usually first appear after the completion of meiosis. The furrows originate at the surface of the original wall of the microsporocyte and develop inwardly until they join in the center, resulting in the accompanying formation of walls that partition the sporocyte into four microspores. The simultaneous type of cytokinesis is found in members of both classes of the angiosperms; it is regarded by Maheshwari (1950) as the prevailing type within the dicotyledons.

Although it is evident from the extensive literature that microspore tetrads may be formed either by centrifugal cell-plate development or by centripetal furrowing, the widespread use of the terms "successive" and "simultaneous" to designate respectively these types of cytokinesis has recently been criticized by Sampson (1963, 1969). His review of the problem reveals that these two terms are, to some degree, misleading. In the genus *Schisandra* and in *Pseudowintera traversii,* for example, wall formation is delayed until the completion of meiosis II when a "simultaneous," rather than "successive" formation of centrifugal cell plates takes place (Kapil and Jalan, 1964; Sampson, 1970). Conversely, centripetal furrowing is not always "simultaneous" but, as in *Zygogynum, Annona,* and *Asimina,* may take place in two steps, beginning at the end of meiosis I (Swamy, 1952; Periasamy and Swamy, 1959). Such variations indicate that in drawing a distinction between the two types of cytokinesis, the major emphasis should be placed on the *contrasted processes* of cell-plate formation versus centripetal furrowing, rather than

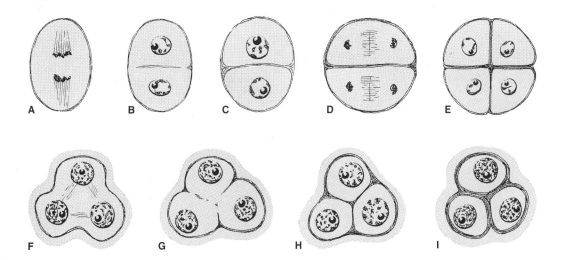

Figure 20-6
Types of cytokinesis in microsporocytes. **A–E,** the successive type in *Zea mays:* **F–I,** the simultaneous type in *Melilotus alba* (note centripetal direction of furrows separating the microspore protoplasts as shown in **G, H**). [Adapted from *An Introduction to the Embryology of Angiosperms* by P. Maheshwari. McGraw-Hill, New York, 1950.]

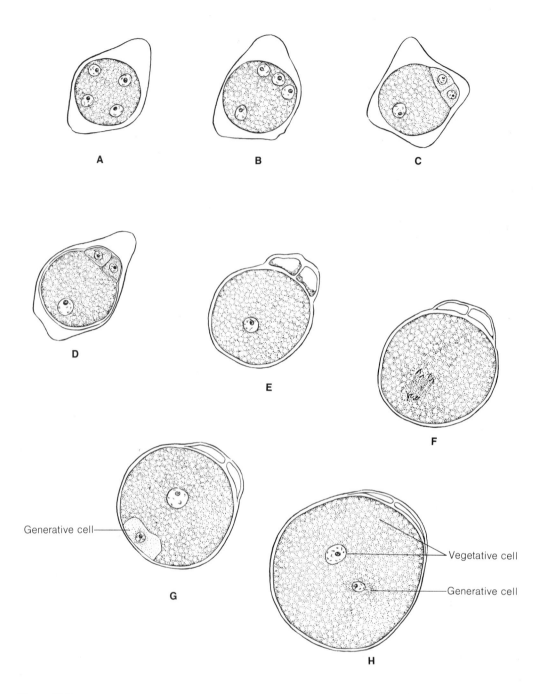

Figure 20-7
Pollen development in *Styphelia longifolia*. **A**, microsporocyte, immediately following meiosis, with four free microspore nuclei; **B**, migration of three microspore nuclei to one end of mother cell; **C**, formation of walls separating a large functional microspore from a cluster of three smaller microspores; **D–H**, death and progressive collapse of the three nonfunctional microspores; **F, G**, mitotic division of the nucleus of the functional microspore and formation of generative and vegetative cells of male gametophyte. [Redrawn from Smith-White, *Proc. Linn. Soc.* New South Wales 84:8, 1959.]

upon the times in the meiotic cycle when these processes take place. Future studies with the aid of the electron microscope may clarify the differences between the two methods of tetrad formation.

Remarkable patterns of microsporogenesis, characterized by the *absence of cytokinesis* during the meiotic divisions of the microsporocyte, have been found in members of two angiosperm families and deserve brief consideration at this point. In the family Epacridaceae, an essentially Australian group of heaths, an extraordinary diversity of pollen types has been described and classified (Smith-White, 1959; Venkata Rao, 1961). Besides the formation of "permanent" tetrads, each consisting of four normal functioning microspores, deviant types of tetrads occur, composed of three, two, or only one functional microspore. The most extreme pattern is illustrated in the genus *Styphelia* (a member of the Tribe Styphelieae), as is shown in Fig. 20-7. Following meiosis in the microsporocyte, the four free microspore nuclei become more-or-less uniformly spaced, either in a tetrahedral or quadrant arrangement (Fig. 20-7, A). Before partition of the mother cell, three of the nuclei migrate to one end of the cell and the fourth nucleus becomes situated at the opposite end (Fig. 20-7, B). Cell walls then arise and separate the three small microspores, which eventually become compressed and obliterated, from the large functional microspore (Fig. 20-7, C–E). Then the nucleus of the surviving microspore divides mitotically, forming the generative and vegetative cells of the young male gametophyte (Fig. 20-7, F–H). Thus the mature "tetrad" of *Styphelia,* which is termed a "monad" by Smith-White (1959), consists of one normal pollen grain and the degenerated remains of three microspores. For a detailed analysis of the morphogenetic, cytological, and genetic implications of the highly unusual pattern of microsporogenesis in *Styphelia* and other genera in the Epacri-

daceae, the student is referred to Smith-White's paper. Venkata Rao (1961) regards the "monad" type of pollen development of *Styphelia* as morphologically comparable with the monosporic type of embryo-sac development in angiosperms, in which three of the potential megaspores are eliminated and only a single megaspore is functional (see Fig. 20-15, E).

The only other known pattern of microsporogenesis similar to that in *Styphelia* is provided by certain members of the monocotyledonous family Cyperaceae. According to Shah's (1962) investigations, meiosis produces a four-nucleate mother cell in which only one of the microspore nuclei differentiates into a functional microspore. The other three nuclei behave somewhat as in *Styphelia:* a septum is formed between the functional spore and the three smaller nuclei; the latter become separated by delicate ephemeral membranes and finally degenerate and merge into an "irregular darkly stained mass." In the species studied by Shah, the "pseudomonad" is shed at the three-celled stage of development of the male gametophyte.

The phylogenetic significance of the highly aberrant and probably derivative types of microsporogenesis that occur in two obviously unrelated families—the Epacridaceae and the Cyperaceae—is obscure and for the present may be interpreted simply as a remarkable instance of parallel, or convergent, evolution. It is possible that additional examples in other taxa will be discovered as comparative studies of microsporogenesis in the angiosperms continue to be made.

The Male Gametophyte

After a resting period, which varies according to the taxon from a few hours to several months, each functional microspore develops into a *pollen grain* that consists of

648

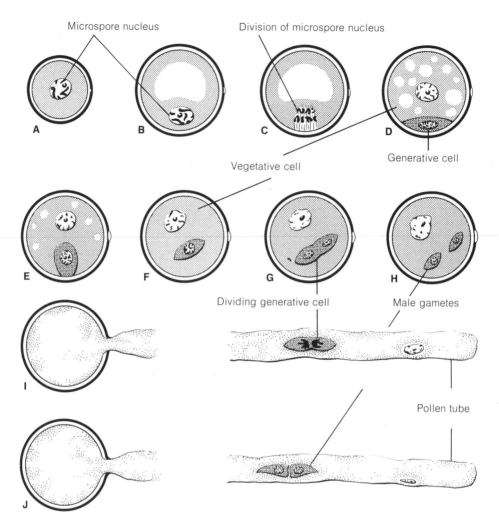

Figure 20-8
Development of the male gametophyte in angiosperms. See text for full discussion of this diagram.
[Redrawn from Maheshwari, *Bot. Rev.* 15:1, 1949.]

the young male gametophyte enclosed within the spore wall. Unlike the more complex and variable structure of the male gametophytes in gymnosperms, the male gametophyte in the angiosperms is relatively simple in organization and is further characterized by its uniform pattern of development. A total of two successive mitotic divisions yields a gametophyte that at maturity consists of a *vegetative cell* and

two male gametes (Fig. 20-8). From a phylogenetic standpoint, the morphologically simple male gametophyte, typical of modern angiosperms, seems to be the endproduct of profound evolutionary reduction, possibly from some hypothetical gymnosperm-like organization (Maheshwari, 1949, 1950; Battaglia, 1951).

Figure 20-8 presents schematically the main steps in the ontogeny of the male gametophyte in the angiosperms. Prior to the first mitotic division, the nucleus of the microspore becomes displaced from its original central position to a peripheral position where it divides to form two unequal cells: a smaller *generative cell* lying next to the spore wall and a much larger *vegetative cell* (Fig. 20-8, A–D). This first step in gametogenesis determines the polarity of the generative cell but its exact position, following mitosis, can only be determined accurately in taxa in which the pollen grains remain united in tetrads. In such taxa, the generative cell may form either at the proximal (i.e., inner) side of each pollen grain, as in *Pseudowintera* and *Zygogynum* (Fig. 20-9, G, H) or next to the distal (i.e., outer) face of each grain, as in *Degeneria, Cananga,* and certain members of the Orchidaceae (Fig. 20-10, B, C). These two locations appear sufficiently constant within taxa that they may be of systematic significance (Swamy, 1949b, 1952).

The generative cell, soon after its formation, moves away from its peripheral position and appears as an ellipsoidal, or lenticular, cell that lies free in the cytoplasm of the vegetative cell (Fig. 20-8, E, F). In apparently the majority of investigated angiosperms, the male gametophyte is in the two-celled stage of development when the pollen grains are shed from the anther and the division of the generative cell to form the two male gametes takes place in the pollen tube (Fig. 20-8, F, I, J). But in certain taxa, in both the monocotyledons and dicotyledons, the generative cell divides prior to the dehiscence of the anther and the pollen is shed at the three-celled stage in development of the gametophyte (Fig. 20-8, G, H).

Recently the systematic distribution and phylogenetic significance of the two-celled and three-celled types of angiosperm pollen grains have received detailed study by Brewbaker (1967). His comprehensive survey of descriptions of nearly 2,000 species of flowering plants included his own numerous cytological investigations of fresh pollen stained with acetocarmine, and the use of the data contained in the voluminous literature dealing with the male gametophyte. In approximately 70 percent of the species included in his survey, the pollen is released from the anther at the two-celled stage in development of the gametophyte. This type of pollen appears to predominate in all primitive families (e.g., those included under the Magnoliales, Ranales, Hamamelidales, and Nymphales) and is regarded by Brewbaker as the primitive form in the angiosperms as a whole. The three-celled type of pollen grain is thus a derivative type which, in his opinion, has originated independently many times during the evolution of the angiosperms. He concludes that the earliest angiosperms "were endowed with a unique binucleate pollen grain, distinguishing them from more primitive taxa." This type of pollen grain, in his view, played a role of great significance in the rapid expansion and diversification of angiosperms during the Cretaceous Period.

In concluding our discussion of the male gametophyte, a few comments are needed regarding the problematical role of the vegetative-cell nucleus. The assumption has commonly been made that the vegetative nucleus, also termed the tube nucleus, functions in some unexplained manner to "control" the direction of growth of the pollen tube. Maheshwari (1950, pp. 169–170) has presented evidence from the literature which casts serious doubt on the physiological importance of the vegetative nucleus. In many angiosperms, as shown diagrammatically in Fig. 20-8, G, H, the vegetative nucleus begins to exhibit signs of degeneration even before the emergence of the pollen tube (see also Brewbaker, 1967). Even when it can be recognized as a "normal" nucleus in the pollen tube, the

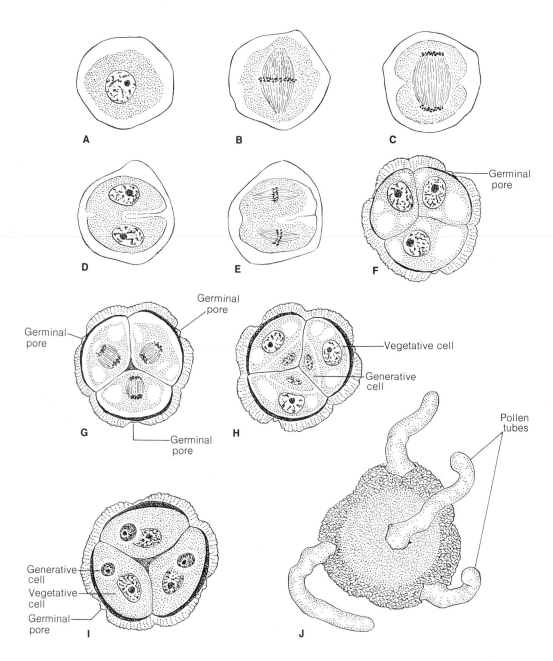

Figure 20-9
Microsporogenesis (**A–F**) and development of male gametophyte (**G–I**) in *Zygogynum bailloni* (Winteraceae).
A, microsporocyte; **B**, metaphase of meiosis I; **C**, late anaphase of meiosis I, showing preliminary equatorial constriction of cytoplasm; **D**, further progress in centripetal furrowing; **E**, meiosis II; **F**, tetrad of microspores; **G**, mitotic division of nucleus in each microspore; **H**, formation of generative cell next to proximal side of each pollen grain; **I**, tetrad of two-celled pollen grains at time of shedding; **J**, germination of each of the pollen grains of a tetrad removed from surface of the stigma. [Redrawn from Swamy, *Proc. Nat. Inst. Sci. India* B 18:399, 1952.]

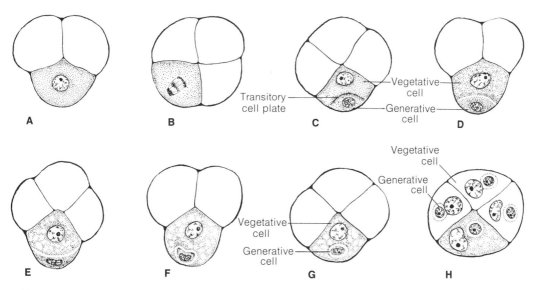

Figure 20-10

Development of male gametophyte in pollen grain of *Saccalobium filiforme* (Orchidaceae). **A**, tetrad of microspores; **B**, mitotic division of nucleus in a microspore; **C**, formation of generative cell next to outer wall and its separation from vegetative cell by a transitory cell plate; **D**, disappearance of cell plate; **E**, enlargement of vegetative cell, causing generative cell to appear crescentric in form; **F**, **G**, generative cell lying free in cytoplasm of vegetative cell; **H**, tetrad of two-celled pollen grains at time of shedding from anther. [Redrawn from Swamy, *Amer. Midland Natur.* 41:184, 1949.]

vegetative nucleus does not always lie ahead of the two male gametes but may lag considerably behind them. Such facts do not appear to indicate a vital role for the vegetative nucleus, and, as Maheshwari points out, its presumed function in "guiding" the growth of the pollen tube may be "discharged by the nucleus of the generative cell itself and later by the nuclei of the two male gametes formed by its division."

The Ovule

Since the ovule is the structure within which meiosis and megaspore formation take place, it corresponds *functionally* to a megasporangium and thus is the "promise" of a future seed. But, as in the gymnosperms, the angiosperm ovule, from a morphological standpoint, is not simply a spore case or sporangium as it usually consists of a nucellus invested by one or two integuments and a stalk or *funiculus* which is attached to the placenta of the ovary. Moreover, angiosperm ovules may develop a vascular system within the outer integument or, in a few taxa, there may be tracheids even in the lower part of the nucellus.

Before discussing megasporogenesis and the patterns of embryo-sac development, it will, therefore, be essential to summarize certain of the salient topographical and structural features of the ovule itself.

Placentation

Although ovules are attached, in most taxa, to the adaxial surface of the carpel, the position of the ovule-bearing regions or placentae vary widely in the angiosperms as a whole. In many types of free, simple carpels, the placentation is designated as submarginal, i.e., the ovule primordia arise very near the approximated or fused margins of the carpel (Fig. 19-44, A). However, in such primitive taxa as *Degeneria* and *Drimys piperita,* the two rows of ovules are situated a considerable distance away from the carpellary margins (Figs. 19-50, 19-51). This type of placentation is termed "laminar" by Eames (1961), who maintains that it represents "the primitive ovule position" from which the submarginal position has been derived by the "restriction of ovules to the near-marginal areas." Interpretation of the various types of placentation in syncarpous gynoecia is a very complex and controversial subject that cannot, however, be treated in this book (see Puri, 1952, and Eames, 1961, for detailed reviews of this problem). It will suffice here to note that in a certain widespread type of syncarpous gynoecium, which is found, for example, in *Lilium* and *Fritillaria,* the two rows of ovules in each locule of the ovary are attached to a central placental region formed by the fusion of the ventral margins of the carpels. This pattern is designated axile placentation (Fig. 19-41, K–M).

Types of Ovules

The ovules of angiosperms have long been classified into various types, depending upon whether the axis of the ovule is erect, with the micropyle facing away from the point of attachment of the funiculus, or curved, with the micropyle facing the placenta or the base of the funiculus. A brief description of some of the principal types of ovules classified from this standpoint will now be presented with the aid of the diagrams shown in Fig. 20-11.

The *orthotropous ovule,* which is found in members of such families as *Polygonaceae, Juglandaceae,* and *Najadaceae,* is erect and devoid of curvature, with the micropyle distal and directed away from the placenta (Fig. 20-11, A). In sharp contrast, the *ana-*

tropous ovule—a very common type throughout the angiosperms—is inverted in its orientation as a result of the approximately 180° curvature of the funiculus (Bersier 1960). The longitudinal axis of the nucellus, in this type, is parallel to the funicular axis and the micropyle thus faces down towards the placenta (Fig. 20-11, C). Very commonly, orthotropous and anatropous ovules are regarded as "basic types" but there is no convincing proof on which one may represent the ancestral form in the angiosperms as a whole. Eames (1961, p. 269), for example, on the basis of the prevalence of anatropous ovules in presumably primitive monocotyledons and dicotyledons, regards the orthotropous ovule as a deriva-

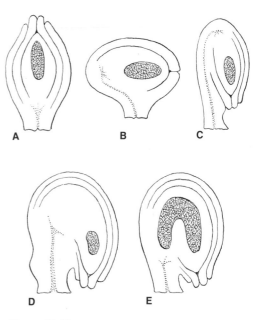

Figure 20-11
Diagrammatic longisectional views of the main types of ovules. The position of the embryo sac in each figure is shown by hatching and the course of the vascular strand by stippling. A, orthotropous; B, hemitropous; C, anatropous; D, campylotropous; E, amphitropous. [Redrawn from *Syllabus der Pflanzenfamilien,* Ed. 12, Vol. 2. by A. Engler. Gebrüder Borntraeger, Berlin 1964.]

tive type. On the other hand, Bocquet and Bersier (1960, p. 480) suggest that the orthotropous ovule may be more ancient. Support for this interpretation is in part derived from a study of ovule development, which shows that in many angiosperms, the primordium of an anatropous ovule is "orthotropous" prior to the initiation of the integuments; subsequent development is asymmetric and results in a curvature (of the upper region of the funiculus) and the final inverted orientation of the micropyle. A very clear example is provided by the development of the ovule in *Degeneria* in which according to Swamy (1949a), "the apex of the ovule undergoes a curvature of 90° . . . and by the time of sporogenesis, becomes bent on itself, thereby assuming a completely anatropous position" (Fig. 20-17). It should be noted further, that in the so-called *hemitropous* ovule, the degree of curvature is intermediate between the extremes which are represented by typical orthotropous and anatropous ovules (Fig. 20-11, B).

Two additional types of ovules remain to be considered: the *campylotropous* and the *amphitropous* (Fig. 20-11, D, E). According to the detailed analysis given by Bocquet and Bersier (1960), these are subordinate types which may appear *successively* as ontogenetic modifications of *either* the orthotropous or anatropous types of ovule. Figure 20-12 presents diagrammatically the orthotropous and anatropous series. In each series, the campylotropous ovule develops

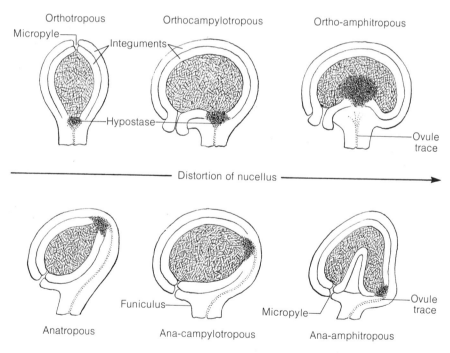

Figure 20-12
Diagram showing that both campylotropous and amphitropous ovules may arise by the successive ontogenetic modification of *either* an orthotropous or an anatropous ovule. Top row, the orthotropous series; lower row, the anatropous series. The hatched area in each ovule represents the embryo sac. [Adapted from Bocquet, *Phytomorphology* 9:222, 1959.]

a micropyle which is directed towards the base of the funiculus because of the ontogenetic curvature, or distortion, of the nucellus. A more exaggerated curvature in the development of the nucellus leads to the amphitropous type of ovule, the embryo sac of which becomes kidney-shaped. This excessive bending of the contour of nucellus (and embryo sac) is produced by the intrusion of a chalazal pad of specialized nucellar tissue—known as the *hypostase,* or basal body—into the main body of the ovule. A clear example of an amphitropous ovule is found in *Capsella* (Fig. 20-39); campylotropous ovules are characteristic of members of such families as Caryophyllaceae, Capparidaceae, and Leguminosae.

Bocquet's diagram (Fig. 20-12) shows that his basis for a decision about which of the two series a given ovule type should be assigned to depends in part on the course of the ovule trace. In the orthotropous series, regardless of the degree of curvature of the campylotropous and amphitropous derivatives, the ovule trace extends vertically upward through the funiculus and terminates below the weakly developed hypostase. On the other hand, in the anatropous series, because of the fundamental curvature of the prototype, the ovule trace in the derivative forms is likewise curved in its course before terminating at the base of the hypostase.

From the standpoint of the diagnostic value of ovule morphology as a taxonomic character, it is evident that the observations and conclusions of Bocquet and Bersier merit full consideration. If indeed a parallel, independent development of campylotropous and amphitropous ovules has taken place during the evolution of the angiosperms, this possibility must be fully recognized whenever the form, orientation, and vasculature of the ovule are used in an effort to determine the systematic relationships between genera or families.

Integuments

In the angiosperms, the ovule develops one or two integuments which completely enclose the nucellus except for the terminal *micropyle* (Fig. 20-15, J), which is a small opening in the apical portion of the integument and is usually the point of entrance of the pollen tube into the interior of the ovule.

There may be considerable variation in the number of integuments characteristic of the ovules of members of a genus, family, or order although, for example, *unitegmic* ovules (those with a single integument) appear to be almost "universal" in the Sympetalae and *bitegmic* ovules (those with two integuments) tend to predominate in the monocotyledons and many groups in the dicotyledons (Maheshwari, 1950; Kapil and Vasil, 1963). The taxonomic value of integument number, however, remains to be more fully explored because of (1) the occurrence of uni- and bitegmic ovules in the various genera of a single family, (2) the occurrence of interspecific variation within a single genus, such as in *Populus,* and (3) the present uncertainty about the original or primitive number of integuments in the remote ancestors of modern angiosperms. With reference to the third point, there seems to be evidence that at least in some taxa an apparently "single" integument has arisen by the congenital fusion of two integumentary primordia. But in other taxa, the unitegmic form may be the result of phylogenetic loss of one of the two ancestral integuments.

Apparently one of the few recent studies primarily concerned with the ontogeny of integuments is the detailed investigation by Roth (1957) on the bitegmic ovule of *Capsella bursa-pastoris.* Roth found that in this species, each integument originates from a *pair* of adjacent superposed cells of the protoderm layer; these integumentary initials are first evident on the side of the

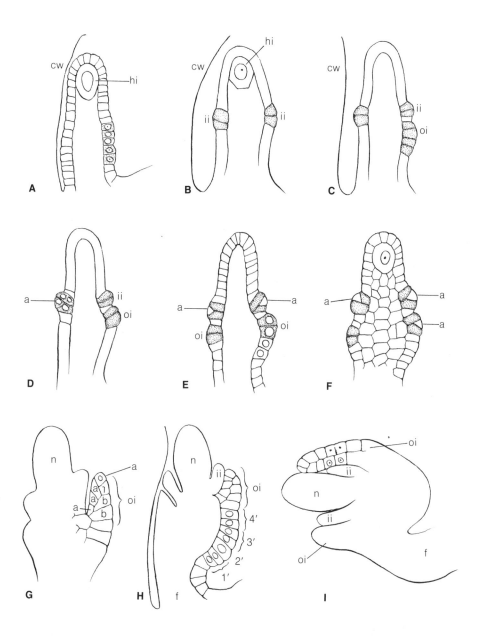

Figure 20-13

Initiation of integument primordia in ovule of *Capsella bursa-pastoris*. **A**, very early stage showing locus of integument initiation in protoderm of young ovule (stippled cells with nuclei); **B**, origin of inner integument from superposed pairs of protoderm cells; **C**, similar method of origin of outer integument; **D–F**, early differentiation of wedge-shaped "apical cells" in both integument primordia; **G**, later stage in development, showing a single derivative of apical cell (cell numbered 1) and several basal cells (a and b) formed by intercalary divisions in the outer integument primordium; **H**, early stage in asymmetrical curvature of ovule, resulting primarily from active division of protoderm cells below base of outer integument (cell groups 1'–4'); **I**, advanced stage in curvature of ovule, showing intercalary divisions in both cell layers of the outer integument. a, apical cell of integument primordium; cw, wall of carpel; f, funiculus; hi, hypodermal nucellar initial; ii, primordium of inner integument; n, nucellus; oi, primordium of outer integument. [Redrawn from Roth, *Flora* 145:212, 1957.]

ovule-primordium opposite (i.e., distal to) the wall of the carpel (Fig. 20-13, C). Commonly, the paired initial cells of the inner integument arise first but Roth also found examples of the "simultaneous" initiation of both integuments (compare parts B and C of Fig. 20-13). In either type, beginning with the original pair of initials, further differentiation of cells of the young integument spreads rapidly around the circumference of the ovule. In other words, the integument primordia soon became ring-shaped emergences near the base of the developing ovule. A distinctive feature of the very earliest stage in integument ontogeny is the formation, by means of obliquely oriented cell divisions, in either the upper or lower cell rows of the integumentary primordium, of a superficial series of wedge-shaped "apical cells" (Fig. 20-13, D–F). Although the divisions of these so-called apical cells contribute to the growth of the two-layered integument, intercalary cell divisions appear to play the dominant role in early histogenesis. Such intercalary divisions occur first in the cells located directly below the pyramidal apical initials of each integument primordium (Fig. 20-13, G). In later stages, intercalary cell divisions occur at more distal points in both layers of the young integument (Fig. 20-13, I).

During the early differentiation of the integuments, the ovule primordium of Capsella experiences a marked asymmetric curvature, and according to Roth, becomes "campylotropous." Two patterns of cell division operate to bring about this curvature; one is the result of periclinal divisions of cells of the hypodermal layer near the base of the outer integument; the other, and more important event, is the active anticlinal division of the protodermal cells situated directly below the base of the outer integument on the outer — i.e., convex — side of the ovule (Fig. 20-13, H). During the stages of ovule curvature, the growth of the outer integument on the inner concave side

of the ovule is markedly arrested and new cells in this part of the integument are primarily derived from the segmentation of the apical initials.

Additional studies, of the meticulous type illustrated by Roth's investigation on Capsella, are needed on a wide range of genera and indeed are essential in order to determine accurately the role of integuments in forming the very complex histology characteristic of the mature seed coat. (See Netolitzky, 1926, for an exhaustive treatment of seed-coat anatomy.) According to Roth's review of the literature, two-layered integuments, which arise as in Capsella from the protoderm, may later become multilayered as the result of periclinal divisions in the two original layers. But examples have also been noted, in the older literature, in which the subepidermal cells of the ovule-primordia may directly contribute to the initiation of the integuments. Comparative studies on the systematic occurrence of both kinds of integument formation would doubtless prove of considerable morphological and taxonomic interest (see Bouman 1971a, b).

There remain, for brief consideration, the occurrence and *patterns of venation* in the integuments of angiosperm ovules. Most commonly, ovules develop a comparatively simple type of vasculature: the ovule trace, usually a single bundle, is derived as a branch of one of the carpel veins (commonly a ventral carpel bundle) and extends through the funiculus, ending abruptly at the base of the chalaza. This simple pattern of vasculature is found — but not exclusively — in all the main types of ovules, as is shown in Fig. 20-11. The classical study of seed venation by LeMonnier (1872) and the later and very comprehensive survey made by Kühn (1928) have revealed that vascular strands, producing a wide range of venation patterns, may develop in the integuments of ovules. According to Kühn, vascularized integuments are found in more than 100 angiospermous genera distributed among 30 families, including such monocotyledonous families as Palmae, Amaryllidaceae, and Cannaceae. If two integuments are formed, the vasculature is usually restricted to the outer integument, but a few examples of ovules in which both of the integuments are vascularized have been discovered, particularly in members of the Euphorbiaceae (e.g., *Ricinus, Jatropha, Dalechampia,* and *Aleurites*).

Kühn (1928, pp. 353–356) classified the diversified patterns of integumentary venation into a series of "types" that she conveniently arranged under the three main categories of ovules — i.e., orthotropous, anatropous, and campylotropous. Figure 20-14, based on Kühn's figures and interpretations, depicts some of the extraordinary range in complexity of integumentary vascular systems in the angiosperms.

In the comparatively infrequent type of orthotropous ovule, the vascular bundle of the funiculus, upon reaching the chalazal region of the ovule, divides into numerous veins which extend, with or without branching, through the integument, ending blindly near the micropylar end (Fig. 20-14, A). Examples of this pattern are found in *Juglans, Myrica,* and *Sterculia.*

The patterns of integumentary venation in anatropous ovules are extremely diversified. Before considering them, however, it is necessary to explain the usage of the term *raphe.* As a result of the approximately 180° curvature which takes place during the ontogeny of an anatropous ovule, fusion may occur between the funiculus and the immediately adjacent portion of the outer integument (Bersier 1960). This type of adnation produces externally a ridge, termed the raphe, through which the upward continuation of the ovule trace extends in its course to the chalaza. The special term "raphe bundle" designates this vein and its position is indicated by "r" in Fig. 20-14, B–F.

In the simplest type of integumentary venation found in anatropous ovules, the raphe bundle extends as a curved and unbranched vein nearly to the micropylar region of the integument (Fig. 20-14, B). Kühn observed this type of vasculature in more than thirty widely scattered genera of angiosperms (e.g., *Anemone, Himantandra, Symphoricarpos,* and *Lonicera*). More complex, open venation patterns occur when the raphe bundle subdivides, near the chalazal end of the ovule, into a system of dichotomously branched veins which extend towards the micropylar end of the integument (Fig. 20-14, C, D). Examples of this pattern are found in *Thea* (Theaceae), *Castanea* (Fagaceae), and *Prunus* (Rosaceae).

Finally, a surprisingly complex reticulate type of venation is characteristic of the integuments of the ovule in several genera of the Cucurbitaceae (e.g., *Momordica, Cyclanthera,* and *Trichosanthes*). In this type, the raphe bundle continues beyond the chalaza almost entirely around the edge of the anatropous ovule and encloses an intricate *anastomosed* system of veinlets (Fig. 20-14, E).

According to Kühn, integumentary vasculature is not commonly developed in campylotropous ovules. Figure 20-14, F, shows the pattern of dichotomous-reticulate venation in the integument of *Lupinus luteus* (Leguminosae). A similar venation pattern was also observed in the integument of the ovule of the common cultivated legume, *Phaseolus vulgaris* (Kühn 1928, Taf XVI, Fig. 80).

What phylogenetic or taxonomic conclusions can be justifiably made on the basis of the extensive data provided by Kühn's survey of integumentary vasculature? From a phylogenetic viewpoint, it has frequently been maintained in the literature of plant morphology that when internal vasculature is present in angiosperm ovules, it is "vestigial" in character and represents the results of a profound reduction of the type of integumentary vascular system which characterized the seeds of the pteridosperms and which has persisted in the ovules of such

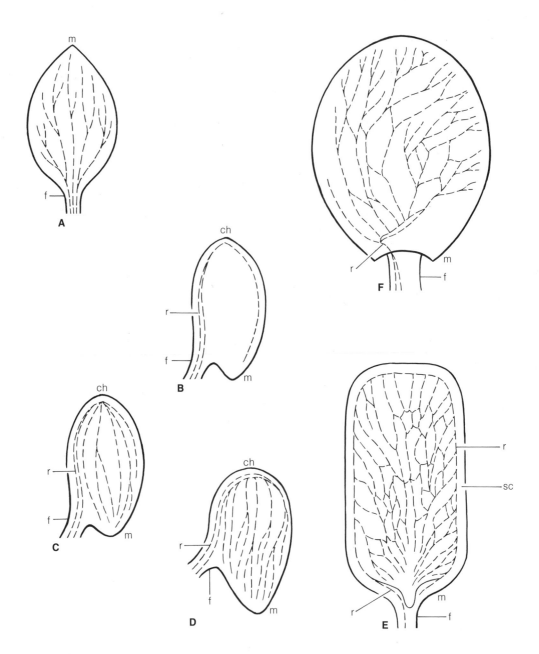

Figure 20-14
Types of venation patterns in the integument of the ovules of various angiosperms. **A,** tangential view (diagrammatic) of palmate-dichotomous venation in the orthotropous type of ovule; **B–E,** diagrams of venation patterns in anatropous ovules: **B,** simple venation, consisting of single unbranched vein, as seen in longisectional view of ovule; **C, D,** tangential views of integument, showing variations in point of origin of dichotomized lateral veins; **E,** tangential view of integument of *Momordica* showing complex reticulate pattern of venation; **F,** tangential view of dichotomous-reticulate venation in integument of campylotropous ovule of *Lupinus luteus*. Abbreviations for all figures: f, funiculus, ch, chalaza, m, micropyle, r, raphe bundle, sc, hard part of seed coat. [Redrawn from Kühn, *Bot. Jahrb.* 61:325, 1928.]

surviving gymnospermous "relics" as the cycads and *Ginkgo biloba* (Chapters 15 and 16). Kühn, however, rejected this idea and emphasized that one would seek "in vain" among either extinct or living gymnosperms for examples of the *manifold patterns* of integumentary venation developed by the ovules of many angiosperms. From a taxonomic viewpoint, Kühn maintains that the sporadic occurrence of similar patterns of integumentary venation within unrelated taxa negates their use in determining systematic affinities between genera or families. On the contrary, in her view, integumentary patterns of vasculature should be regarded as "new formations" which have developed independently of one another in both primitive and advanced taxa. Possibly the formation of veins in integuments may ultimately prove to have largely physiological significance during the maturation of an ovule into a seed, but there appears to be no experimental evidence to support this possibility.

In conclusion, it must be admitted that neither the method of initiation of integuments nor the presence of integumentary venation throw much light on the vexing question of the morphology of ovular integuments. While the superficial origin of the integuments from the protoderm and the presence of integumentary bundles may invite the conclusion that ovule envelopes are "foliar structures," this idea is wholly conjectural in view of our complete ignorance of the evolutionary origin of the angiosperms and the nature of the ovule in their remote ancestors (see Maheshwari, 1960). If Zimmermann's "Telome Theory" is applied to the elucidation of the problem, one might conclude that the integument represents a "syntelome," i.e., the result of the fusion of a ring of sterile telomes to form a cupule-like envelope surrounding a terminal sporangium (see Maheshwari, 1960, Boesewinkel and Bouman, 1967, and van Heel and Bouman, 1972, for discussions of

various phylogenetic theories about the origin and nature of the integument in angiosperm ovules).

Structure of Nucellus and Origin of Megasporocyte

Early in the ontogeny of the ovule, one or several *internal* cells near the apex of the nucellus enlarge and become different from the adjacent cells in structure and staining reactions. These distinctive cells are "potential" megasporocytes, as each of them, by meiotic divisions, can produce a tetrad of haploid megaspores. Most frequently only a single megasporocyte is formed in the developing nucellus (Fig. 20-15, B). But in the ovules of some angiosperms, a group of potential megasporocytes may arise, each of which theoretically can give rise to a tetrad of megaspores. A particularly striking example of the formation of "multiple megasporocytes" has been observed during the development of the massive nucellus of the ovule of *Paeonia californica*. In this species Walters (1962) found that 30–40 megasporocytes are differentiated and that many of them complete meiosis and form linear tetrads of megaspores. Even more remarkable is the fact that at the time of fertilization, several of the chalazal megaspores, from different tetrads, may have formed female gametophytes. During later ontogenetic stages, however, all but one of the megasporocytes and megagametophytes begin to degenerate and the mature seed contains only one embryo.

Because of the relatively infrequent—and apparently erratic—occurrence of multiple megasporocytes in the angiosperms as a whole, it is not possible at present to reach a conclusion about the phylogenetic significance of this condition. Perhaps, as Walters suggests, the development of multiple megasporocytes has occurred "independently

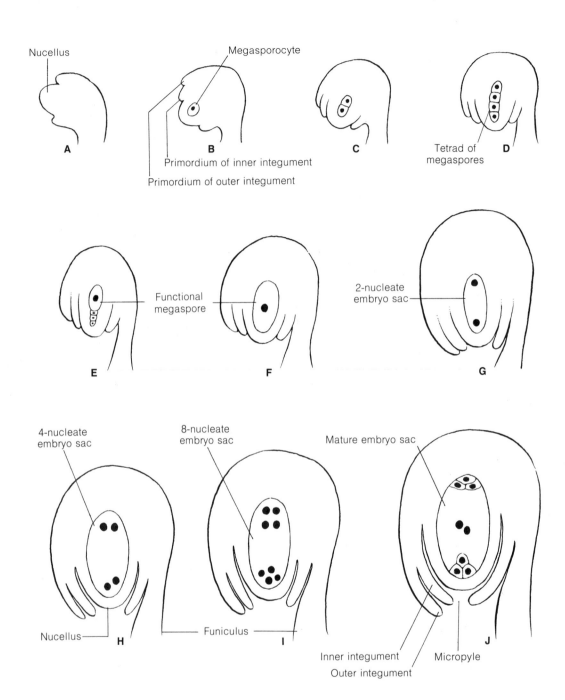

Figure 20-15
Development of an angiosperm ovule, beginning with the formation of the integuments and the single megasporocyte (**A, B**), continuing through the formation of megaspores (**C–E**), and concluding with the successive stages in development of the embryo sac (**F–J**). See text for detailed discussion of this figure. [Redrawn from *A Textbook of General Botany,* Ed. 4, by R. M. Holman and W. W. Robbins. Wiley, New York, 1951.]

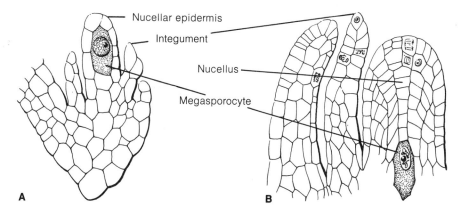

Figure 20-16
A, tenuinucellate type of ovule in *Orchis maculatus,* in which the megasporocyte is directly below the nucellar epidermis; **B,** crassinucellate type of ovule in *Quisqualis indica* (note deeply embedded position of megasporocyte and the active periclinal divisions in the nucellar epidermis). [Adapted from *An Introduction to the Embryology of Angiosperms* by P. Maheshwari. McGraw-Hill, New York, 1950.]

numerous times during the evolution of angiosperm families as well as gymnosperms."

More typically, in both monocotyledons and dicotyledons, only a single functional megasporocyte differentiates in the ovule and its *ultimate position* is usually correlated with the size and cellular organization of the nucellus. The terms *"crassinucellate"* and *"tenuinucellate",* originally introduced by Van Tieghem (1898), are widely used in current literature, to designate two contrasted types of nucellar organization. In a "typical" tenuinucellate ovule, the megasporocyte originates *directly* from a hypodermal cell located at the apical region of the relatively delicate nucellus, which consists of an epidermal layer and a comparatively few nonsporogenous internal cells (Fig. 20-16, A). On the other hand, in crassinucellate ovules, a more-or-less prominent hypodermal initial *first* divides periclinally into an outer *parietal cell* and an inner *sporogenous cell.* The latter functions as the megasporocyte while the parietal cell by means of further periclinal divisions forms one or many layers of cells. Thus, by this *indirect method of origin,* the megasporocyte

ultimately becomes deeply embedded within a relatively massive type of nucellus (Fig. 20-17, B–D). In some species, the division of parietal cells above the megasporocyte may be accompanied by periclinal divisions in the nucellar epidermis (Fig. 20-16, B). In such species, the deeply sunken position of the megasporocyte is very obvious and nucelli with this kind of epidermal proliferation tend to be particularly massive in structure.

In conclusion, it must be noted that the ovules of certain angiospermous genera cannot be rigidly classified as either crassinucellate or tenuinucellate if the presence or absence of parietal cells is used as a fundamental defining character. Dahlgren (1927), in his very comprehensive and critical treatise on the nucellus, has emphasized that parietal cells may be developed in ovules, which on the basis of their small size and weakly developed nucellus, would be classified as "tenuinucellate." Furthermore, there may even be variation between the ovules of the same species, some forming parietal cells and others being devoid of such cells. Conversely, not all "massive" ovules,

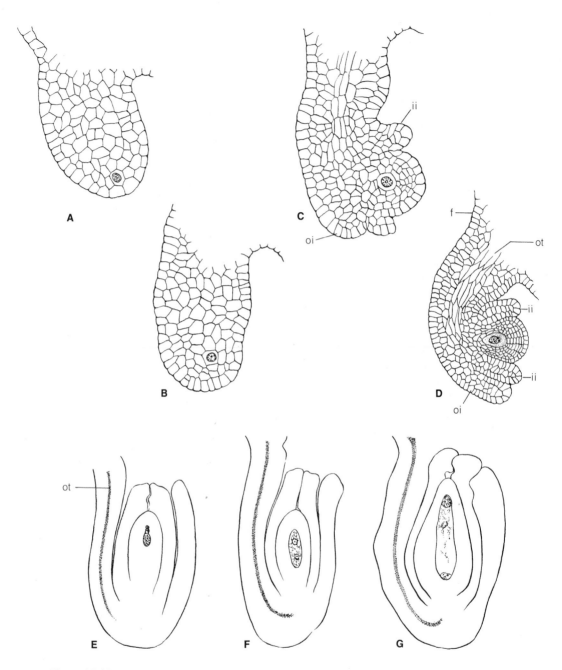

Figure 20-17
Ontogeny of the anatropous bitegmic, crassinucellate ovule of *Degeneria*. **A,** early stage in development of nucellus, showing prominent hypodermal initial; **B,** hypodermal initial has divided, forming megasporocyte (cell with nucleus) and several parietal cells; **C, D,** later stages, showing voluminous development of parietal tissue (radial rows of cells) and deeply embedded position of the megasporocyte; **E,** linear tetrad of megaspores; **F,** two-nucleate embryo sac; **G,** mature eight-nucleate embryo sac. f, funiculus; ii, inner integument; oi, outer integument; ot, ovule trace. [Redrawn from Swamy, *Jour. Arnold Arboretum* 30:10, 1949.]

which at maturity might be classified as crassinucellate, develop parietal cells during the differentiation of the megasporocyte. For example, in two species of *Calycanthus* investigated by Peter (1920), the nucellar epidermis produces a massive "nucellar cap" but the multiple megasporocytes originate *directly* from hypodermal cells without the formation of parietal cells. Davis (1966) proposed the new term "pseudocrassinucellar" for ovules with this type of nucellar organization. It should thus be evident that although it is generally useful to be able to classify ovules into two major types — i.e., crassinucellate and tenuinucellate — there are intergradations and deviations and these must be fully recognized whenever the organization of the nucellus is used as a taxonomic character.

During the development of the embryo sac, endosperm, and embryo in the ovule, most or all of the nucellar tissue becomes crushed and ultimately destroyed. In a few plants, however, such as various species of *Yucca,* a portion of the nucellus survives and forms a nutritive tissue, known as *perisperm,* in the mature seed (see Arnott, 1962, and Kapil and Vasil, 1963, for details).

Megasporogenesis

In lower heterosporous vascular plants and in most gymnosperms, the female gametophyte arises from the growth and division of a *single* haploid megaspore. There is a comparable method of origin for the megagametophyte, or embryo sac, according to Maheshwari (1950) "in at least 70 percent of the angiosperms now known." The salient features of this widespread type of megasporogenesis are illustrated in Fig. 20-15. Each of the two meiotic divisions is accompanied by wall formation resulting in a tetrad of megaspores. The first, or heterotypic, meiotic division, is always in the

transverse plane and produces two dyad cells (Fig. 20-15, C). Usually the second, or homotypic, meiotic division is followed by the formation of a transverse wall in each dyad cell. As a result, a file, or linear tetrad, of megaspores is produced (Fig. 20-15, D). In some cases, however, the plane of division in the micropylar dyad cell is longitudinal, resulting in a ⊥-shaped arrangement of the megaspores. If the division plane in the lower dyad cell is vertical, a T-shaped tetrad of spores is formed. Linear and T-shaped megaspore tetrads may, according to Maheshwari, arise in different ovules of the same ovary but the factors controlling this variability are at present unknown.

In the very common *Polygonum* type of embryo-sac origin, three megaspores of the linear tetrad degenerate while the chalazal megaspore enlarges and, by means of three successive mitotic divisions, gives rise to an eight-nucleate embryo sac (Fig. 20-15, E-I). Thus beginning with the megasporocyte, it takes a total of five nuclear divisions, of which two are meiotic, to produce the haploid female gametophyte (Fig. 20-15, J).

However, striking deviations from this common pattern of megasporogenesis and embryo-sac development have been discovered as the result of comparative studies on a wide range of angiosperm genera. One of the most remarkable deviations is the result of the participation of more than one megaspore nucleus in the formation of the embryo sac. This highly peculiar — and probably specialized condition — is the result of the partial or complete failure of wall formation during meiosis. In the most extreme expression, the completion of meiosis I and II results in four megaspore nuclei, which lie free in a cell that is termed the *coenomegaspore* (Maheshwari, 1950, footnote p. 106). Embryo sacs that arise from a coenomegaspore are thus *tetrasporic* and represent a type of female gametophyte origin that is without parallel in vascular plants, with the

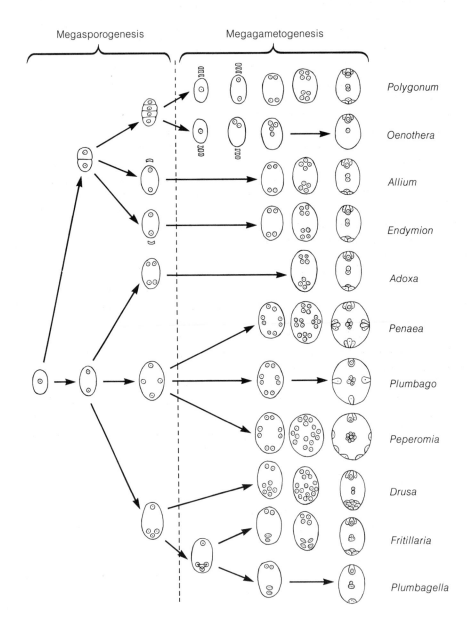

Figure 20-18
Diagrammatic comparison of the main types of megasporogenesis and megagametogenesis in the angiosperms. See text for detailed descriptions of selected examples of monosporic, bisporic, and tetrasporic patterns of development. [Redrawn and modified from Johri in *Recent Advances in the Embryology of Angiosperms*, p. 69, 1963.]

exception of two genera of gymnosperms, *Gnetum* and *Welwitschia* (see Waterkeyn, 1954, and Martens, 1962).

Types of Embryo-sac Development

Numerous studies in recent years have revealed a surprising diversity in the details of megasporogenesis and embryo-sac development within the angiosperms, and various typological classifications have been proposed (see Maheshwari, 1950, and Johri, 1963). The different types of embryo-sac development are distinguished on the basis of the following characters: (1) the number of megaspores or megaspore nuclei which participate in the formation of the embryo sac, (2) the total number of mitotic nuclear divisions during megagametogenesis, (3) the presence or absence of nuclear fusions, and (4) the number, arrangement, and chromosome number of the cells and free nuclei present in the fully mature embryo sac. The chart represented in Fig. 20-18 shows the principal types of megasporogenesis and embryo-sac development discussed in the recent review by Johri (1963). Although a detailed account of the eleven types represented in this diagram will not be given in this book, selected examples, under each of the three major categories of embryo-sac origin that are generally recognized, will be briefly described. These major categories are *monosporic development* (from a single megaspore), *bisporic development* (from two megaspore nuclei) and *tetrasporic development* (from four megaspore nuclei).

Monosporic Development

Under this category are grouped the *Polygonum* and *Oenothera* types of embryo sacs. Both types begin development from a single megaspore, and there are three successive mitotic divisions during megagametogenesis

in the *Polygonum* type but only two in the *Oenothera* type (Fig. 20-18).

In the *Polygonum* type — so named because of its discovery in this genus by Strasburger (1879) — the megaspore farthest from the micropyle is the functional cell and its first division yields two nuclei which move to the poles of the embryo sac (Fig. 20-19, D–G). Each of these nuclei then divides and a final division of the four nuclei produces a total of eight nuclei, arranged in quartets, at the micropylar and chalazal ends of the embryo sac. Three of the nuclei at the micropylar pole become differentiated as cells and constitute the *egg apparatus*; this consists of the female gamete, or *egg cell*, flanked by the two *synergids*. At the opposite end of the embryo sac, three of the four nuclei differentiate as the *antipodal cells*; in some plants, e.g., *Zea mays*, the original antipodals may increase in number by subsequent divisions. The two remaining nuclei — termed *polar nuclei* — migrate from the opposite ends of the sac into the *central cell* — i.e., the central region of the embryo sac (Fig. 20-15, J). The polar nuclei may remain separate until the discharge of the male gametes into the embryo sac or they may fuse, prior to fertilization, to form a diploid *secondary nucleus* (Fig. 20-19, I). The pattern of development that we have just described thus produces a seven-celled embryo sac which consists of a three-celled egg apparatus, three (or more) antipodal cells, and a central cell with either one or two nuclei.

The development of the *Oenothera* type of embryo sac, now recognized as characteristic of the family Onagraceae, presents a very interesting variant of the monosporic type of development. Usually the micropylar, rather than the chalazal spore of the tetrad, is the functional cell and a total of two (rather than three) nuclear divisions result in a very distinctive type of embryo sac consisting of a three-celled egg apparatus and a single polar nucleus (Fig. 20-18).

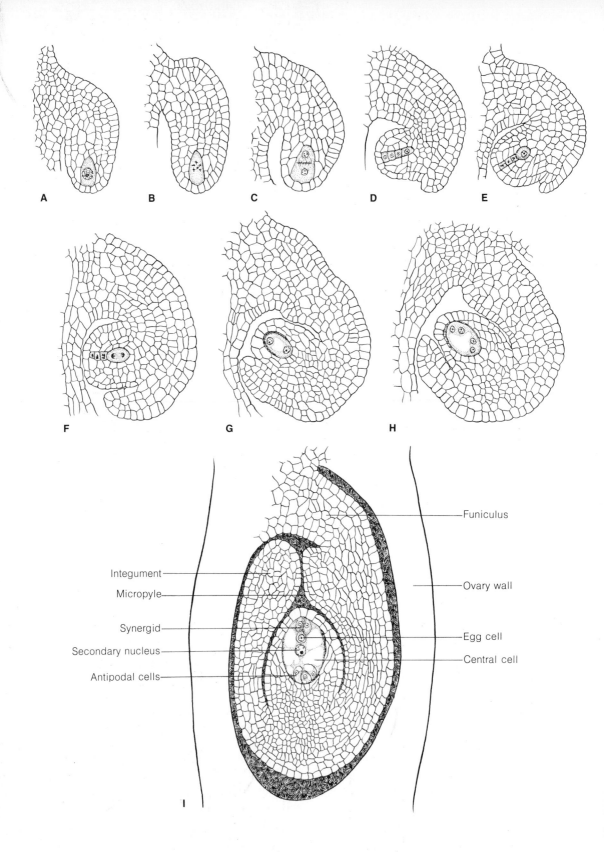

A

B

C

D

E

F

G

H

Funiculus

Integument

Micropyle

Ovary wall

Synergid

Secondary nucleus

Antipodal cells

Egg cell

Central cell

I

Figure 20-19 (*facing page*)
Megasporogenesis and development of the monosporic type of
embryo sac in *Anemone patens*. **A,** ovule primordium with
large hypodermal megasporocyte; **B, C,** meiosis I; **D,** linear
tetrad of megaspores following meiosis II; **E, F,** enlargement
and mitotic division of chalazal megaspore; **G,** binucleate
embryo sac; **H,** four-nucleate embryo sac; **I,** mature embryo
sac. [Redrawn from *Plant Morphology* by A. W. Haupt.
McGraw-Hill, New York, 1953.]

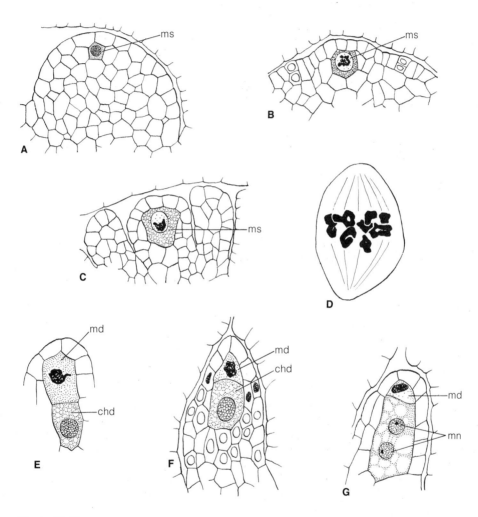

Figure 20-20
Origin and early development of bisporic type of embryo sac in *Allium cepa*. **A,** longisection
of nucellus showing hypodermal megasporocyte; **B, C,** prophase of meiosis I; **D,** heterotypic
division of megasporocyte; **E,** formation of the two dyad cells; **F,** chalazal dyad cell beginning
to enlarge, accompanied by abortion of micropylar dyad cell; **G,** formation of two megaspore
nuclei in chalazal dyad cell. chd, chalazal dyad cell; md, micropylar dyad cell; mn, megaspore
nuclei; ms, megasporocyte. [Redrawn from Jones and Emsweller, *Hilgardia* 10:415, 1936.]

Because of the smaller number of mitotic divisions during megagametogenesis, an additional polar nucleus and the antipodal cells are not formed. The *Oenothera* type of embryo sac is of particular morphological interest because, following double fertilization, the primary endosperm nucleus is diploid, rather than triploid as is the very widespread *Polygonum* type (Fig. 20-29).

Bisporic Development

In the bisporic type of embryo-sac development, one of the haploid binucleate dyad cells produced during meiosis gives rise to the mature female gametophyte (see Maheshwari, 1955). As the bisporic condition was first described for *Allium fistulosum* by Strasburger (1879), this method of embryo-sac origin is usually designated as the *Allium* type (Fig. 20-18).

The definitive features of the *Allium* type are illustrated by Jones and Emsweller's (1936) description of embryo-sac development in *Allium cepa* (Fig. 20-20). Following the heterotypic division in the megasporocyte (Fig. 20-20, D), two dyad cells, separated by a thin transverse wall, are produced (Fig. 20-20, E). The micropylar dyad cell soon aborts and Meiosis II, without the accompanying formation of a wall, produces two free haploid megaspore nuclei in the chalazal dyad cell (Fig. 20-20, F, G). Two successive mitotic divisions then take place in the functional dyad cell, yielding an eight-nucleate embryo sac which ultimately becomes organized into an egg apparatus, a group of antipodal cells, and two polar nuclei. (Contrast the *Allium* and *Polygonum* types of embryo-sac development shown diagrammatically in Fig. 20-18.)

An additional type of bisporic embryo-sac development was proposed by Battaglia (1958) and termed by him the *Endymion* type. In *Endymion,* a genus allied to *Scilla* (Liliaceae), the embryo sac originates from the upper binucleate dyad cell, rather than from the lower one (Fig. 20-18). According to Battaglia, the chalazal dyad cell, by means of two mitotic divisions, may reach a four-nucleate phase before it collapses.

In concluding this brief account of the bisporic type of embryo-sac formation, it is interesting to note that a type of megasporogenesis, which seems to represent a *transition* between the "typical" monosporic and bisporic patterns, has been observed in two other genera of the Liliaceae. In *Convallaria majalis,* according to Stenar (1941), meiosis results, as in monosporic types, first in two dyad cells and then in a tetrad of haploid megaspores (Fig. 20-21, a–c). But the walls between the members of *each pair* of spore nuclei are extremely delicate and soon disappear. As a result, the two original dyad cells, each of which now is binucleate, are "restored." The upper dyad cell finally degenerates and, as in *Allium,* an eight-nucleate embryo sac arises by the mitotic division of the megaspore nuclei in the chalazal dyad cell (Fig. 20-21, d–f). A similar sequence of events follows meiosis in *Smilacina racemosa,* but in this species the embryo sac arises from the micropylar rather than from the chalazal dyad cell (Fig. 20-21, a′–f′).

Tetrasporic Development

This category includes a series of complex and puzzling types of embryo-sac development, all of which, however, show two basic characteristics: (1) there is no wall formation during megasporogenesis and (2) all four megaspore nuclei participate in various ways in the formation of the mature embryo sac. Reference to Fig. 20-18 will show that, following meiosis, the megaspore nuclei exhibit three main types of polarity, or arrangement, in the coenomegaspore. In the *Adoxa* type, the megaspore nuclei show a 2 + 2 arrangement—i.e., one pair is situated

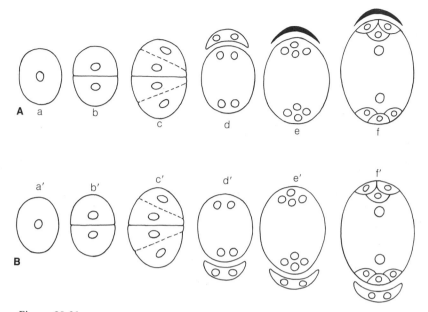

Figure 20-21
A, megasporogenesis and bisporic embryo-sac development in *Convallaria majalis:*
a, megasporocyte; b, dyad cells produced following meiosis I; c, tetrad of megaspores,
the pairs of megaspore nuclei separated by ephemeral walls represented by dotted
lines; d, four-nucleate embryo sac formed by the division of the two megaspore nuclei;
e, eight-nucleate embryo sac; f, mature embryo sac, organized into egg apparatus, three
antipodal cells, and two free polar nuclei. Note progressive degeneration of binucleate
micropylar dyad cell shown in d-f. **B**, similar sequences in megasporogenesis and
embryo-sac development in *Smilacina racemosa,* except that in this species, the upper
dyad cell is functional. [Redrawn from Stenar, *Bot. Notis.* pp. 123–128, 1941.]

at the micropylar end and the other pair at
the chalazal end of the coenomegaspore. A
tetrapolar, or 1 + 1 + 1 + 1, arrangement of
megaspore nuclei follows meiosis in the
Penaea, Plumbago, and *Peperomia* types of
embryo-sac development. In the 1 + 3
arrangement a single megaspore nucleus is
located at the micropylar end of the sac and
the other three nuclei are at the chalazal
pole. This peculiar kind of polarity characterizes the *Drusa, Fritillaria,* and *Plumbagella* types of embryo-sac development and,
in the two latter genera, the three chalazal
megaspore nuclei fuse later in development.
An example of each of the three main kinds
of megaspore arrangement and its subsequent relation to the process of megagametogenesis will now be briefly discussed. For

further details of all the types of tetrasporic
development, the student should consult
Maheshwari (1950) and Johri (1963) and
the extensive bibliographies which their
publications include.

ADOXA. In this type, the two meiotic
divisions of the megasporocyte yield four
megaspore nuclei arranged in pairs at the
opposite poles of the coenomegaspore.
Megagametogenesis entails a single mitotic
division of each megaspore nucleus, producing two quartets of nuclei which then
become differentiated to form a *Polygonum*
type of embryo-sac organization (Fig.
20-18). According to Maheshwari's (1946)
comprehensive review, continued research
has pronouncedly decreased the number of

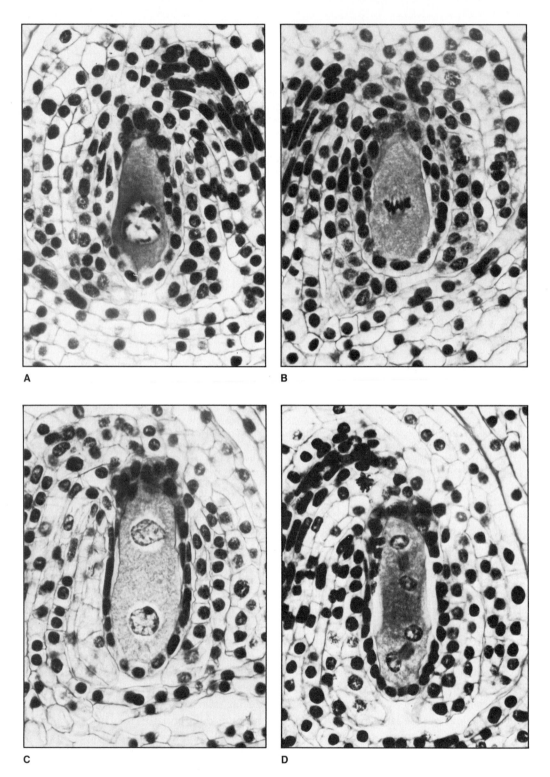

Figure 20-22
Megasporogenesis in *Fritillaria*. **A**, megasporocyte with nucleus in prophase; **B**, nucleus of megasporocyte in metaphase; **C**, two megaspore nuclei; **D**, four megaspore nuclei. [From slides prepared by Dr. F. V. Ranzoni.]

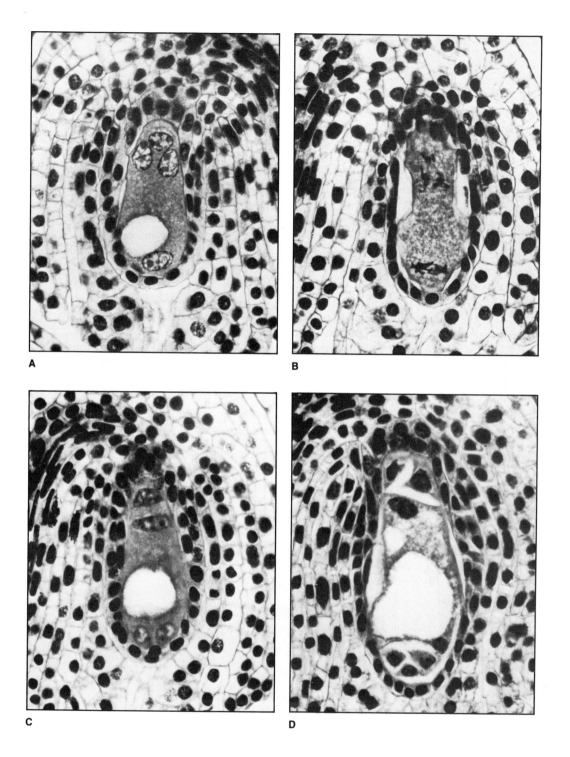

Figure 20-23

Development of embryo sac in *Fritillaria.* **A,** first four-nucleate stage consisting of a single micropylar and three chalazal megaspore nuclei; **B, C,** origin of second four-nucleate stage (note differences in size and shape of the micropylar and chalazal nuclei); **D,** eight-nucleate embryo sac consisting of a group of four haploid micropylar nuclei (lower end of sac) and a quartet of triploid chalazal nuclei. [From slides prepared by Dr. F. V. Ranzoni.]

[671]

plants considered to have a typical *Adoxa* type of embryo-sac development. Many previously cited examples have proved upon reexamination to belong to the *Fritillaria* type. Maheshwari (1950, p. 122) states that the *Adoxa* type is a regular feature of only five clearly unrelated genera: *Adoxa* and certain species of *Erythronium, Tulipa, Ulmus,* and *Sambucus.* The extremely clear-cut characteristics of the *Adoxa* type are very convincingly demonstrated and illustrated by Dolcher's (1951) investigation on embryo-sac development in *Sambucus nigra* L. var. *laciniata* and *typica.*

FRITILLARIA. The type of tetrasporic embryo-sac ontogeny illustrated in the genus *Fritillaria* is of particular interest because it occurs in a rather wide variety of genera including *Lilium,* which had long been used in botany courses to illustrate "typical" embryo-sac development and structure in the angiosperms. It was only after the careful investigations of Bambacioni (1928a, 1928b) and Cooper (1935) that the true sequence of events in the embryo-sac development of *Lilium* and *Fritillaria* was fully revealed (Fig. 20-18).

In *Lilium* and *Fritillaria,* the four megaspore nuclei resulting from meiosis behave in a very distinctive manner (Fig. 20-22). Three of the nuclei migrate to the chalazal end of the coenomegaspore while the remaining nucleus becomes situated at the micropylar pole (Fig. 20-23, A). This 3 + 1 arrangement of the megaspore nuclei represents the *first, four-nucleate* stage in the development of the embryo sac of both *Fritillaria* and *Lilium.* These four nuclei next undergo mitosis but the events which occur at the two poles of the coenomegaspore are strikingly dissimilar. The division of the micropylar nucleus is normal and yields two haploid nuclei. In contrast, the

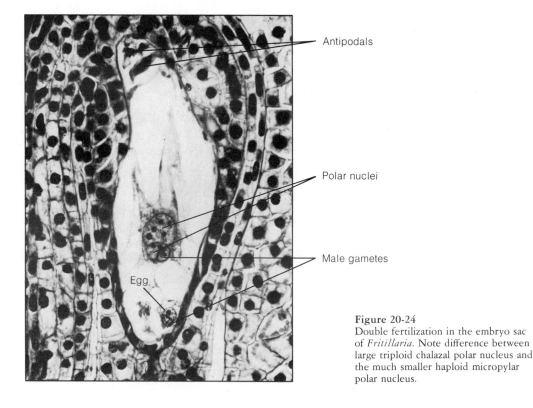

Antipodals

Polar nuclei

Male gametes

Egg

Figure 20-24
Double fertilization in the embryo sac of *Fritillaria.* Note difference between large triploid chalazal polar nucleus and the much smaller haploid micropylar polar nucleus.

spindles formed by the divisions of the three chalazal nuclei first unite to form a multipolar spindle. Then *all* the three sets of chromosomes become arranged on a bipolar spindle and the two nuclei which are reconstituted are therefore $3n$ in chromosome number. As a result of all of these events a *secondary four-nucleate* stage is produced which consists of two haploid micropylar nuclei and two larger triploid chalazal nuclei (Fig. 20-23, C). A final mitotic division produces a micropylar quartet of haploid nuclei and a chalazal quartet of triploid nuclei (Fig. 20-23, D). The organization of the mature embryo sac, with reference to the chromosome number of its cells and the two polar nuclei, is strikingly different from the *Polygonum* type. In *Fritillaria* (and *Lilium*), the egg apparatus is haploid, the antipodal cells are triploid, and one polar nucleus (derived from the micropylar quartet) is haploid while the other polar nucleus (derived from the chalazal quartet) is triploid. When double fertilization occurs, a normal diploid zygote is produced but the primary endosperm nucleus is $5n$ in chromosome number as the result of the union between a male gamete and the two polar nuclei (Figs. 20-24, 20-29).

On a comparative basis it is interesting to note that the ontogeny of the *Plumbagella* type of embryo sac follows the same pattern characteristic of *Fritillaria* up to the secondary four-nucleate stage. There are, however, no further nuclear divisions and the mature embryo sac of *Plumbagella* consists of an egg cell (unaccompanied by synergids), a single antipodal cell, and two polar nuclei, one haploid, the other triploid (Fig. 20-18). This peculiar type of embryo sac, at least from a morphological viewpoint, could be interpreted as a "reduced" *Fritillaria* type (see Maheshwari, 1950, and Johri, 1963).

PENAEA. As a last example of tetrasporic development, *Penaea* presents features of exceptional interest with reference to a full appreciation of the wide variation in the morphological organization of the embryo sac in angiosperms. In *Penaea,* the type genus, as well as in several members of other families such as the Malpighiaceae and Euphorbiaceae, the four megaspore nuclei are arranged in a tetrapolar—rather than a bipolar—pattern in the coenomegaspore (Fig. 20-18). A very clear example of the *Penaea* type is provided by Kapil's (1960) study of embryo-sac development in *Acalypha brachystachya* (Euphorbiaceae). In this species *each* megaspore nucleus, by means of two mitotic divisions, produces a quartet of nuclei and the young embryo sac as a result becomes sixteen-nucleate (Fig. 20-25, A–C). Four of the nuclei, one from each quartet, function as polar nuclei and fuse in the center of the sac to form a large, lobed $4n$ secondary nucleus while the other three nuclei in each group become organized as an egg apparatus (Fig. 20-25, D, E). According to Kapil's investigation, only the egg cell of the micropylar group enlarges and becomes functional, and the other three groups ultimately degenerate (Fig. 20-25, E). Despite this fact, the embryo sac in this species, as in *Penaea,* appears to be organized as though in fact it were a "compound" structure made up of four completely "fused" megagametophytes, each consisting of an egg apparatus and a single polar nucleus.

Concluding Remarks on Embryo-sac Morphology

Although it appears reasonable to suppose that the monosporic *Polygonum* type of embryo sac may represent the basic or primitive form in the angiosperms as a whole, the trends in evolutionary specialization which have led to the other types of embryo-sac development, particularly the tetrasporic patterns, are complex and obscure. The problem is further complicated by the observation—made rather frequently—that

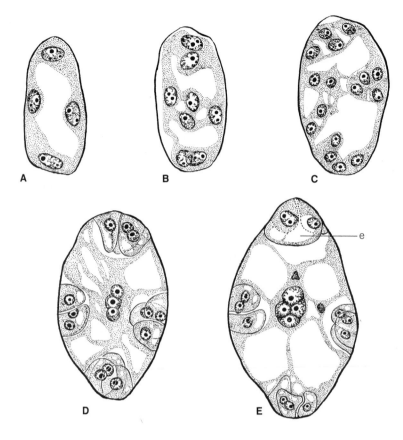

Figure 20-25
Embryo-sac development in *Acalypha brachystachya*. **A,** coenomegaspore showing
tetrapolar arrangement of the four megaspore nuclei; **B,** eight-nucleate stage;
C, 16-nucleate stage; **D,** organization of four egg apparati and migration of polar
nuclei (only three visible in this section) into center of sac; **E,** mature embryo
sac with enlarged egg cell (e) in micropylar egg apparatus and fusion of four
polar nuclei to form a lobed secondary nucleus. [Redrawn from Kapil, *Phytomorphology* 10:174, 1960.]

there may be more than one type of embryo-
sac in a single genus or even in one species.
A specific example is provided by the development of two types of ovules in the
carpel of *Clematis gauriana*. According to
Vijayaraghavan (1962) the embryo sac in
the single *functional* ovule is monosporic
and develops as in the *Polygonum* type. But
in the rudimentary, *sterile* ovules, *formed in
the same carpel,* the embryo sac develops
from a tetrasporic coenomegaspore in which
fusion of the megaspore nuclei in pairs
results in two diploid nuclei which then
divide to produce an embryo sac consisting

of an egg, two polar nuclei, and a single
antipodal cell. Hjelmquist (1964) has discussed numerous examples of the coexis-
tence of two or more types of embryo-sac
development in the various ovules of
one species. The student is referred to his
review for detailed comparisons and for an
analysis of the various factors which may condition variability in embryo-sac morphology.

If it is assumed that the *Polygonum* type
is primitive because of its monosporic
method of origin, the question arises
whether there is any convincing evidence to
link this type of megagametophyte struc-

ture with the types of female gametophytes developed in the majority of living gymnosperms. Several ingenious — but apparently invalid — theories have been proposed in an effort to demonstrate an homology between gymnosperm archegonia and certain cell groups in the angiosperm embryo sac. Maheshwari (1950) has criticized in detail the various "archegonial theories" which have been suggested and concludes that "it seems far more likely instead that the angiosperms have long passed the stage of archegonia or that they never had them at any time in their fossil history." A similar skeptical viewpoint was also expressed by Battaglia (1951), who maintains that "the reduction of the female gametophyte from gymnosperms to angiosperms is characterized by the disappearance of the archegonium."

To conclude, it seems that the phylogenetic origin of the embryo sac, like the mystery of the origin of double fertilization, represents at present a challenging but still unsolved aspect of the larger question of the origin of flowering plants and the morphology of their remote ancestors.

Pollination

In the preceding sections of this chapter, we have presented the information necessary for an understanding of sporogenesis and the ontogeny and structure of the male and female gametophytes in the angiosperms. But quite apart from the morphological details, it is essential at this juncture to recall that in angiosperms the male and female gametophytes are remote and isolated from each other because of the enclosure of the ovules within the ovarian cavity of the gynoecium. Unlike most gymnosperms, in which the ovule itself is directly pollinated, pollination in angiosperms involves the transferral of pollen grains — by such external agencies as wind or insects — from the opened anther of the stamen to the *receptive stigma* of the carpel. For convenience in the discussion which follows, pollination and fertilization will be treated separately although it should be realized that they are, respectively, the beginning and the endpoint of the process that constitutes sexual reproduction in flowering plants.

The agents or vectors responsible for pollen transfer in angiosperms have been grouped into two main categories: *abiotic agents* — inanimate forces in nature such as wind currents, gravity, and water — and *biotic agents* — various animal pollinators, with pollination by insects being termed *entomophily*, pollination by birds, *ornithophily*, and by bats, *cheirophily*. As the living gymnosperms, particularly the conifers, appear to be largely *anemophilous*, i.e., wind pollinated, it has been frequently argued that wind pollination may have been the original or primitive method in the angiosperms. However, the prevalence of anemophily in such specialized taxa as the Gramineae, Cyperaceae, and many members of the "Amentiferae" (e.g., some species of *Quercus*, Betulaceae, Juglandaceae) and the occurrence of entomophily in various primitive genera in the Ranales fail to support this speculation. The current viewpoint is that the ancient angiosperms were entomophilous and that their pollinators very probably were beetles. Among living angiosperms, pollination by beetles appears to be characteristic of a number of primitive types of dicotyledons such as *Eupomatia*, *Calycanthus*, *Paeonia*, and *Victoria* (Diels, 1916; Eames, 1961). According to Faegri and van der Pijl (1966, p. 84), beetles are one of the older groups of insects and "were already numerous at the time we assume the higher plants first came into existence during the Upper Jurassic or Lower Cretaceous whereas the higher hymenoptera and lepidoptera, so important in pollination today, had not yet developed." (See Grant, 1950; Faegri and van der Pijl, 1966; Baker and Hurd, 1968; and Carlquist, 1969, for further details on the phylogenetic and biological significance of beetle pollination in angiosperms.)

Among present-day angiosperms, various types of "attractants" seem primarily responsible for the establishment of direct relationships between flowers and their animal pollinators. One of the most evident attractants—and perhaps the oldest in a phylogenetic sense—is pollen, which constitutes the food either of an adult insect visitor or else is used by the insect to feed its brood of larvae. The outstanding examples of this relationship are provided by bees. Nectar, secreted from special glandular organs known as *nectaries,* is also an important (or even "primary") source of food for certain groups of insects and birds and doubtless plays a significant role in the syndrome of pollination in many angiosperms. (For descriptions of the varied structure and position of floral nectaries, see Eames, 1961, pp. 152–154, and Esau, 1965, pp. 311–313.)

The study of the fascinating and complex interrelationships between inflorescence- and flower-structure and animal pollinators, especially insects, constitutes an important aspect of modern plant biology. In some angiosperms, there appears to have developed an *obligatory* relationship in which a particular type of plant is dependent upon a particular type of insect for pollination. Classical examples are provided by the relationship between *Yucca* and its moth pollinator *Tegeticula* (=*Pronuba*), the orchid *Ophrys speculum* and its hymenopteran pollinator, and *Ficus* and its wasp pollinator *Blastophaga.* Other functional associations between specific pollinators and flower types have also been widely explored. Students interested in this rapidly developing field of research should consult the extensive

treatment of the intricate problems given by Faegri and van der Pijl (1966) and the more recent review of intrafloral ecology by Baker and Hurd (1968).

In conclusion it should be emphasized that a distinction must be drawn on the one hand between the *functional role* that animal pollinators (preeminently the insects) have played in floral evolution and diversification and, on the other hand, the unknown factors in the remote past which were responsible for the *phylogenetic origin* of the flower itself. "Adaptation" of flower structure to specific pollinators is a phenomenon which can be determined by direct field observations and experiments on modern angiosperms, but "adaptation" *per se* leaves unsolved the abiding mystery of the origin and early development of the angiosperm carpel. Perhaps the most significant biological role of the carpel in present-day flowering plants consists in the fact that it represents the primordium of the *fruit,* a structure which is of considerable importance in the protection and dispersal of the seeds (see Chapter 21).

Growth of the Pollen Tube

In the strict sense, the term "pollination" designates the actual transfer of pollen to some type of receptive surface of the carpel. "Fertilization," on the other hand, is a process of gametic fusion which takes place within the confines of the embryo sac. But as the ovules with their enclosed female gametophytes are situated within the ovary, there is inevitably an "intermediate phase" in the process of sexual reproduction which includes (1) the germination of the pollen grains on the stigma, and (2) the pollen tube's traversing a relatively long path through the style before finally penetrating the tissue of the ovule and entering the embryo sac.

Much attention has been given to the complex physiological and genetical aspects of pollen germination and pollen-tube growth but in this book we will only consider the problems entailed in a very brief and general fashion. Students wishing further information should consult the discussions and the reviews of the literature given by Maheshwari (1950), Steffen (1963), and Linskens (1969).

The germination of pollen grains and the emergence of the pollen tubes from the germ pores normally occur on a specialized "receptive" region of the carpel. Probably the most primitive type of receptive surface in living angiosperms is found in the remarkable carpels of *Degeneria* and *Drimys piperita*. In these ranalean plants, the pollen is deposited and held by a mat of interlocking glandular hairs on the unsealed margins of the "open" conduplicate carpels. When the pollen tubes emerge, they penetrate between the hairs into the papillate interior surface of the locule of the carpel, eventually reaching in their growth the ovules, which are remote from the unfused carpellary margins, or "stigmatic crests," which extend the full length of the carpel lamina (see Chapter 19). Bailey and Swamy (1951) have analyzed in detail the varied trends of carpel specialization that have led, in the Winteraceae, to the ultimate restriction of the stigmatic region to the terminal part of the carpel. (For detailed information on carpel morphogenesis in the Winteraceae, see Tucker and Gifford, 1966 and Sampson and Kaplan, 1970.)

In advanced angiosperm taxa, the pollen is deposited on a usually well-defined *terminal stigma* which may represent morphologically the "surviving" part of the type of stigmatic crests typical of *Degeneria* and various members of the Winteraceae. Although variable in form, the stigma functions as a secretory organ and the exudation of sugary solution from the epidermal cells

or stigmatic papillae induces the emergence of the pollen tube from one of the pores of the pollen grain. Soon after its emergence, the tip of the young pollen tube pushes its way between the cells of the stigma and thus begins its journey through the style to the ovary. A very important physiological and structural character of angiosperm carpels is the continuity which commonly exists between the glandular stigmatic tissue and the so-called "transmitting tissue" of the style and placental regions of the ovary. Esau (1965) collectively designates this stylar-ovarian tissue-complex as "*stigmatoid tissue,* on the basis of its apparent cytologic and physiologic similarity to the tissue of the stigma." In many monocotyledons and certain groups of dicotyledons the style is hollow and the transmitting tissue is represented by the glandular epidermis which lines the stylar canal. In such "open" styles, the growth of the pollen tube is superficial and the glandular epidermis serves both to nourish the elongating pollen tube and perhaps chemotrophically to guide the direction of its growth. More commonly, however, the style is more-or-less "solid" and the central region is composed of strands of transmitting tissue through which the pollen tube must pass on its way to the ovary. It is believed that the *intercellular growth* of the pollen tube in this type of style may involve a pectin-digesting enzyme which functions to loosen the cell walls that lie in the path of the advancing tube.

An important part of the physiology of pollen tubes which needs to be considered at this point is the synthesis, within the growing tube, of a complex polysaccharide known as *callose.* This carbohydrate, which has been most thoroughly studied with reference to its presence and presumable function in the sieve elements of the phloem, is readily demonstrated by the use of "resorcin blue," a dye which stains callose a brilliant blue (see Esau's 1969 monograph on

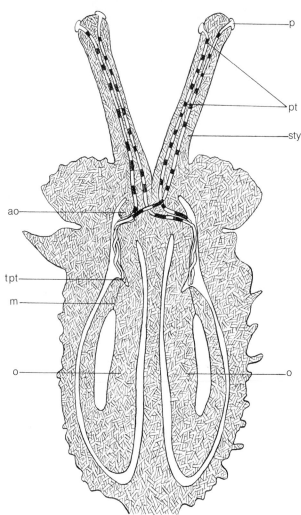

Figure 20-26
Diagram of median longisection of gynoecium of *Daucus carota* showing the course of the pollen tubes. The series of black areas in each of the pollen tubes represent callose plugs. ao, abortive ovule; m, micropyle; o, functional ovule; p, pollen grain on stigma; pt, pollen tubes with callose plugs (black areas); sty, style of bicarpellate gynoecium; tpt, tip of pollen tube. [Redrawn from Borthwick, *Bot. Gaz.* 92:23, 1931, University of Chicago Press.]

the phloem). In germinating pollen grains and in elongating pollen tubes, callose occurs not only in the innermost lamella of the intine but is locally deposited in the form of *callose plugs*. These remarkable structures have been observed in a wide variety of angiosperms, both in tubes growing naturally within the transmitting tissue of the style and in pollen tubes grown on artificial media. A good example of the abundant formation of callose plugs is provided by Borthwick's (1931) study on *Daucus carota* (Fig. 20-26). Using a combination of Heidenhain's haeme-

toxylin and resorcin blue as the stain, he found that plugs were most conspicuous in those portions of the tubes growing in the styles where the plugs are 3–10 μ in length and spaced at intervals of approximately 25 μ. Portions of the tubes which had reached the ovarian cavity or the micropyle of the ovule showed few or no callose plugs, but yet at this region the "entire wall of the tube is frequently much thickened with callose."

Accurate studies on the growth and structure of *complete* pollen tubes developing within the tissues of the gynoecium, however, are very difficult to make because of the frequently irregular, undulating course of the tubes. The most detailed and reliable information on the distribution of callose and the mode of development of callose plugs has been obtained by the study of pollen tubes grown in culture on artificial media composed of agar and varying concentrations of sucrose. (For details on the techniques of culturing pollen tubes see Iwanami, 1956, and Müller-Stoll and Lerch, 1957a.) As an example of the interesting results obtained from *in vitro* studies on pollen tubes, we refer to Fig. 20-27, which summarizes the main steps in callose formation revealed by the very thorough, detailed investigations of Müller-Stoll and Lerch (1957a, 1957b).

The *initial stage* in callose formation occurs during the swelling of the pollen grain and prior to the emergence of the young tube (Fig. 20-27, A). At this stage, a thin layer of callose is deposited on the inner surface of the intine and is particularly well developed near the germ pore. At the next stage, characterized by the emergence of the pollen tube, the wall (representing the protrusion of the intine layer) back of the growing tip consists of an outer pectin lamella and an inner layer composed of a mixture of cellulose and callose. The wall at the very tip, however, is devoid of callose, and consists only of pectin (Fig. 20-27, B, C). The salient feature of the stage during which the pollen tube attains its maximum length consists in the initiation and acropetal development of callose plugs. These structures are developed in two ways: (1) by the formation of concentric annular or cylindrical accumulations of callose which, like the closing of a diaphragm, grow inward from the wall until a complete septum is formed and (2) by the development of a peg-like protrusion of callose at one edge of the tube which, by continued growth, extends towards

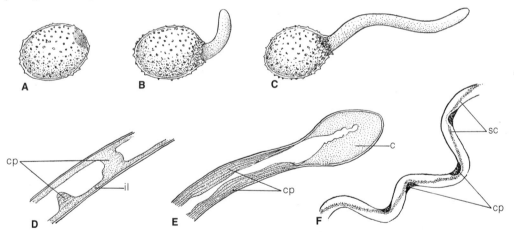

Figure 20-27
Diagrammatic summary of callose formation during successive stages of pollen tubes grown *in vitro:* **A,** swollen pollen grain showing marked accumulation of callose adjacent to germ pore; **B, C,** emergence and early development of pollen tube. Shaded area indicates callose-cellulose lamella, representing the innermost layer of pollen-tube wall; **D,** incomplete (lower) and continuous (upper) forms of callose plugs; **E,** terminal region of a senescent pollen tube showing patterns of callose deposition; **F,** portion of an undulate pollen tube showing regular distribution of callose plugs at inner edge of each bend and position of streaming cytoplasm. c, callose; cp, callose plug; il, callose-cellulose lamella; sc, streaming cytoplasm. [Redrawn from Müller-Stoll and Lerch, *Biol. Zentralbl.* 75:595, 1957b.]

and finally contacts the opposite part of the pollen tube wall (Fig. 20-27, D). In both types, additional deposits of callose may form at each side of the original plug, leading to highly bizarre shapes and contours (see Müller-Stoll and Lerch, 1957a, for details). In undulate or sinuous forms of pollen tubes, the callose plugs are initiated at the inner part of each of the bends and these regions of callose accumulation are the favored points of contact between the streaming cytoplasm and the wall of the tube (Fig. 20-27, F). Müller-Stoll and Lerch (1957a, Fig. 8) show, in one of their illustrations, the generative cell being carried by the streaming cytoplasm past a peg-shaped callose plug. At the terminal phase in growth of pollen tubes cultured *in vitro* there is a marked increase in the amount of callose that may accumulate either as centripetally developed lamellae near the end of the tube or as an amorphous mass of callose in the bulbous tip of the tube (Fig. 20-27, E).

The physiological significance of callose formation—particularly the formation of callose plugs—is problematical. It is evident that the formation of plugs is not limited to tubes grown *in vitro*, since callose plugs also form within tubes situated in the tissue of the style (Fig. 20-26). Brink (1924, p. 431) suggested that the formation of a succession of callose plugs isolates the terminal gamete-containing part of the pollen tube and thus serves to maintain its integrity and to *limit* "the region from which the vegetative cell absorbs nutrient materials to the less exhausted portions of the style." Other proposed "explanations" include the view that callose is either a reserve product or an ergastic substance. Esau (1969, p. 54) however concludes that "the role of callose in sieve elements and other locations has not been experimentally determined and statements about this matter are only conjectural."

After reaching the ovary, the pollen tube may enter an ovule by several possible

routes. Most commonly the tip of the pollen tube enters the micropyle and pushes through the nucellar tissue to the egg-apparatus end of the embryo sac. This type of ovule penetration is designated as *porogamy*. But some taxa (e.g., *Casuarina* and several members of the "Amentiferae") are said to be *chalazogamous* because the tip of the pollen tube penetrates the chalazal end of the ovule, and then continues its growth along the surface of the embryo sac before reaching the egg apparatus. Chalazogamy was formerly considered the primitive condition in angiosperms but at present the exact mode of ovule-penetration in a given taxon is believed to have physiological rather than phylogenetic significance. A highly peculiar—and apparently rare—mode of entry of the pollen tube is termed *mesogamy* and is illustrated in the monotypic *Circaeaster agrestis* (Junell 1931). In this plant, the pollen tube penetrates laterally the single integument of the ovule before reaching the egg apparatus of the embryo sac.

The final phase in the growth of the pollen tube consists in its entry into the micropylar end of the embryo sac. How this is accomplished and whether one of the synergids plays a significant role in the process are still unsettled, controversial problems. Aside from the inherent technical difficulties of securing a closely spaced series of critical stages between the entry of the pollen tube and the discharge of the male gametes, there appear to be real differences in the events that occur depending upon the taxon under consideration.

In many plants the pollen tube enters the embryo sac by passing either between one of the synergids and the egg or between a synergid and the embryo-sac wall (Maheshwari, 1950; Steffen, 1963). In such plants, one or both of the synergids are said to be destroyed by the intrusive growth of the pollen tube itself. Recently, however, intensive studies with the aid of modern

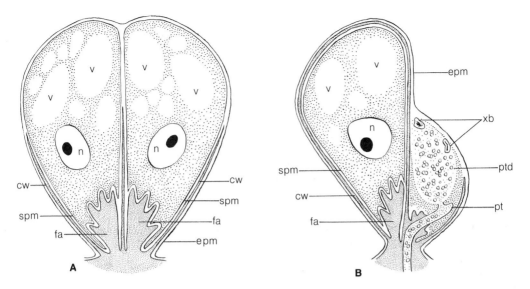

Figure 20-28
Diagrams summarizing the changes in the synergids of *Gossypium* before and after the discharge of the pollen tube. **A**, structure of the synergids in the embryo sac in an unpollinated flower; **B**, pollen-tube discharge in degenerating synergid (at right). cw, cell wall of synergid; epm, endosperm plasma membrane; fa, filiform apparatus; n, nucleus; pt, pollen tube with open pore near tip; ptd, pollen tube discharge; spm, synergid plasma membrane; v, vacuole; xb, X-bodies. [Redrawn from Jensen and Fisher, *Planta* 78:158, 1966.]

staining techniques and the use of the electron microscope indicate that in certain plants the synergid is not simply a "passive" cell, or buffer, but rather that it is a "receptive" cell into which the contents of the pollen tube are discharged (Jensen, 1973). In *Capsella,* according to Schulz and Jensen (1968a) "the pollen tube was never seen to enter the egg or grow through to the central cell" and the contents of the pollen tube were discharged in the synergid. This observation agrees with similar ones made on several other plants: *Torenia, Petunia, Gossypium,* and *Epidendrum* (for references, see Schulz and Jensen, 1968a).

Continued studies have revealed that in both cotton (*Gossypium*) and barley (*Hordeum*) the receptive synergid is a cell which shows signs of degeneration soon after pollination has taken place but before the pollen tube has reached the embryo sac (Jensen and Fisher, 1968; Cass and Jensen, 1970). In both of these plants, the pollen tube enters the degenerating synergid through the *filiform apparatus,* which is a struc-

turally modified part of the micropylar end of the synergid wall, and discharges its contents, including the two male gametes and the vegetative nucleus, into this cell. (Fig. 20-28 B, ptd). Jensen and Fisher also observed that discharge from the pollen tube of *Gossypium* is through a definite pore formed near the tip of the pollen tube; later the pore becomes occluded by the formation of a plug (Fig. 20-28, B). The important role of the degenerating synergid in regulating the site of pollen-tube entry and discharge in cotton is strongly supported by two instances in which Jensen and Fisher first observed the entry of a pollen tube into a normal, nondegenerative synergid, and then, however, the pollen tube "turned and grew through the wall separating the synergids and entered the degenerating one."

Following the discharge of the pollen tube into the synergid, the two male gametes, by some mechanism at present unknown, are liberated from the degenerating synergid and enter the egg and central cell of the embryo sac. Although the two male gametes in barley are liberated into the degenerating synergid as cells, each consisting of a nucleus and a cytoplasmic sheath, the role of the male cytoplasm in syngamy and triple fusion remains to be determined. Jensen and Fisher (1968) had suggested that the dark-staining X-bodies found associated with the material discharged into the synergid represented the remains of the cytoplasm of the two sperms. But in a subsequent investigation (Fisher and Jensen, 1969), the two X-bodies were identified as the remains of the nucleus of the degenerated synergid and the vegetative nucleus of the pollen tube (see also Jensen, 1973). This conclusion followed from the determination—made by means of autoradiography of sections treated with tritiated actinomycin D—that the X-bodies contain DNA. A similar interpretation of the X-bodies in the synergid of *Hordeum,* was proposed by Cass and Jensen (1970), who had used the same technique.

The studies that we have just briefly summarized clearly do not support the "classical" view that the entry of the pollen tube into the embryo sac entails the destruction of one of the synergids. But comparative studies on other species—using the techniques of both light- and electron-microscopy—are urgently needed in order to explore further the role of synergids in regulating the mode of entrance of the pollen tube and the site of discharge of the male gametes into the embryo sac of angiosperms.

Fertilization

One of the outstanding characteristics of angiosperms is the participation of *both* of the male gametes in an act of fusion: one unites with the egg to form a diploid zygote, from which the embryo originates, while the other gamete fuses with one or several polar nuclei (or with the secondary nucleus) to form the primary endosperm nucleus, from which the endosperm tissue takes its origin (Fig. 20-24). These complex events—the "dynamics" of which are still obscure—constitute the process now known as *double fertilization.* The male gametes are liberated either in one of the synergids or at a point external to the synergid; the mechanism by which these vital processes and the subsequent nuclear fusions take place have so far eluded convincing explanation. (For detailed discussions of various interpretations which have been suggested see Maheshwari, 1950; Steffen, 1963; Fisher and Jensen, 1969; and Jensen, 1973).

One of the many controversial aspects of the process of double fertilization concerns the mechanism of movement of the two male gametes. Are these two cells carried respectively to the egg and to the polar nuclei simply by means of the streaming movement of the cytoplasm (Jensen, 1973), or is "amoeboid movement" of the male gametes responsible for their divergent paths of movement? Steffen (1963) suggests that "it is also probable that after the sperm cells have entered the plasma stream by their own amoeboid movement, they can be transported further in a passive manner." In *Fritillaria,* according to Sax (1916), the events are indeed remarkable. After syngamy has occurred, one of the male gametes joins the upper polar nucleus and this pair of nuclei "migrates"—presumably by means of cytoplasmic streaming—to the chalazal end of the central cell where it fuses with the lower polar nucleus to form the primary endosperm nucleus.

Endosperm

The term "endosperm" designates the tissue, formed during the development of an angiosperm seed, which provides essential

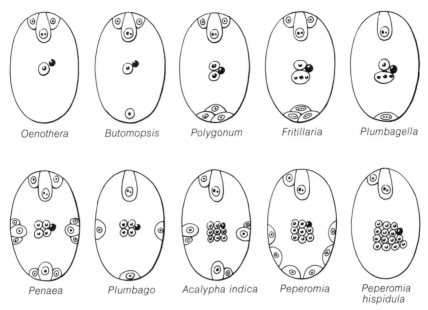

Oenothera *Butomopsis* *Polygonum* *Fritillaria* *Plumbagella*

Penaea *Plumbago* *Acalypha indica* *Peperomia* *Peperomia hispidula*

Figure 20-29
Embryo sacs. Note the wide variation in the number of polar nuclei that join with the sperm nucleus (shown in black) in the initiation of endosperm. [Adapted from *An Introduction to the Embryology of Angiosperms* by P. Maheshwari. McGraw-Hill, New York, 1950.]

food materials utilized in the growth of the embryo, and, in many cases, of the young seedling. Aside from the seeds of members of the Orchidaceae and Podostemonaceae, in which endosperm fails to develop or else degenerates early in ontogeny, the presence of endosperm is a salient and consistent feature of seed development in the angiosperms as a whole. In some angiosperms (e.g., beans, peas, and vetch) the endosperm is completely digested during embryogenesis —in such *exalbuminous* seeds, the embryo develops thick, fleshy, food-storing cotyledons, which provide nutrition to the seedling during germination (Fig. 21-7, E, F). Other angiosperms (e.g., castor bean, onion, palms, and corn) have *albuminous* seeds, in which copious amounts of endosperm tissue are present at the time of seed germination (Figs. 21-7, A–B; 21-8; 21-10; 21-11). The histological structure of endosperm and the types of reserve foods in its cells vary widely—in many plants the cells are densely

packed with starch grains, protein granules, or oils (Maheshwari, 1950, pp. 255–257; Chopra and Sachar, 1963, pp. 155–166). In certain palms (for example, the so-called "ivory-nut" palm) the endosperm cells develop very thick hard walls which are composed of hemicellulose, an important reserve-food material (Corner, 1966).

The initiation of endosperm normally requires an act of fusion between a male gamete and one or more of the polar nuclei in the embryo sac (Fig. 20-29). This process is one of the most definitive features of the reproductive cycle of angiosperms and indeed forms a most striking point of contrast with the type of seed ontogeny characteristic of the gymnosperms. In the latter, the nutritive tissue of the seed is already present *before fertilization* as the massive, haploid female gametophyte (Fig. 14-4, C). In other words, the nutritive tissues in gymnosperm and angiosperm seeds are *analogous* rather than *homologous*, from a morphological point of view. It seems necessary to emphasize this point because even in certain modern texts and research articles, the term "endosperm" is used in a loose sense for the reserve-food-containing gametophytic tissue of the seeds of gymnosperms.

Extensive comparative studies have revealed considerable variation in the pattern of development of endosperm and the following résumé presents the salient features of the three main ontogenetic types which are now recognized.

Types of Endosperm Development

NUCLEAR TYPE.   In this type, the division of the primary endosperm nucleus is followed by a variable number of subsequent free nuclear divisions; in some plants, as cited by Maheshwari (1950, p. 226), several hundred free nuclei may be produced. During this preliminary and definitive phase of development the center of the embryo sac is often occupied by a large vacuole, and the nuclei lie peripherally in the cytoplasm which lines the wall of the embryo sac (Fig. 20-39, A). Although the entire development of nuclear endosperm in some plants is limited to the production of free nuclei and the resultant formation of a multinucleate mass of protoplasm, in other plants there is a second phase in development that consists of the centripetal formation of cell walls. In such cases all or most of the endosperm is converted into a cellular tissue (Figs. 20-38, D; 20-39, B).

The young nuclear endosperm of the palm, *Cocos nucifera,* is known as "coconut milk," and contains mitochondria, free nuclei, protein granules, and oil droplets. In addition to these components, coconut milk contains certain important growth-promoting factors and has proved beneficial—or essential—as part of the media used for *in vitro* cultures of embryos and

excised plant tissue (Steward, 1968). Experimental studies that utilize coconut milk seek to discover the biochemical characteristics of endosperm and its role in morphogenesis (for additional information, see reviews of literature given by Maheshwari, 1950, pp. 389–390, and Chopra and Sachar, 1963, pp. 140–142 and pp. 158–159).

CELLULAR TYPE.   This type is well-demarcated from the nuclear type because the division of the primary endosperm nucleus is followed by the formation of either a longitudinally or transversely oriented wall (Fig. 20-30, A, E). When the plane of the first division is longitudinal, each of the two daughter cells divides, forming a vertical wall, with the result that four elongated endosperm cells are formed (Fig. 20-30, B). Then there are several repeated transverse divisions in each of these cells (Fig. 20-30, C); but soon the plane of subsequent divisions becomes irregular and the early tiered arrangement of the cells becomes obscured. Examples of this pattern of early wall formation in cellular endosperm are comparatively rare and are illustrated among dicotyledons by *Adoxa, Centranthus,* and *Circaeaster* (Junell, 1931). The more common pattern of wall formation consists in a series of preliminary transverse divisions which may yield a linear row of eight or more cells (Fig. 20-30, E–J). A striking example of this kind of regular development is provided by the early stages in ontogeny of the cellular endosperm of *Cercidiphyllum* (Swamy and Bailey 1949, p. 200, Figs. 40–43). In many plants, however, the early sequences of transverse and longitudinal divisions do not conform to a set pattern and the arrangement of cells is correspondingly variable and irregular (Fig. 20-33).

HELOBIAL TYPE.   This type of endosperm development was first designated as "helobial" by Schnarf (1929, p. 348), probably

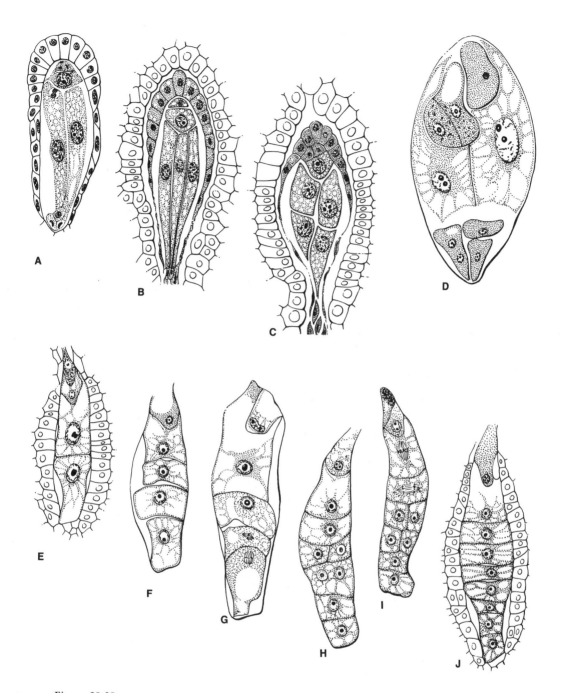

Figure 20-30
Early stages in development of the cellular type of endosperm in various angiosperms. In **A** and **D,** the plane of the first wall is vertical; in **E,** the first division wall is transverse. **A–C,** successive stages in *Adoxa moschatellina;* **D,** first longitudinal wall in *Centranthus macrosiphon;* **E–J,** successive stages in *Villarsia reniformis.* [Adapted from *An Introduction to the Embryology of Angiosperms* by P. Maheshwari. McGraw-Hill, New York, 1950.]

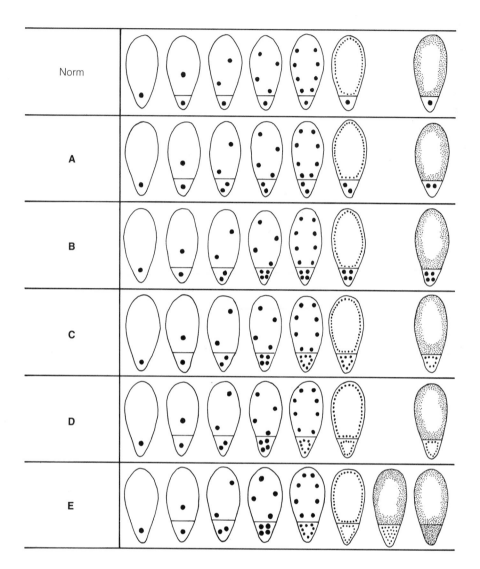

Figure 20-31
Diagrams showing the various patterns of development of helobial endosperm in the mono-cotyledons. In each developmental series, the formation of a transverse wall divides the embryo sac into a small lower (i.e., chalazal) chamber and a larger, upwardly oriented, micro-pylar chamber. See text for further explanations. [Adapted from Swamy and Parameswaran, *Biol. Rev.* 38:1, 1963.]

because of its prevalence among those monocotyledons classified under the order Helobiae (Helobiales). Swamy and Parameswaran (1963), on the basis of a very comprehensive survey, maintain that helobial endosperm is restricted to the monocotyledons and that the frequent reports of its presence in certain dicotyledonous families are due "to inadequate and often inaccurate observations, or to wrong interpretations, or to both." These authors conclude that helobial endosperm is not "intermediate" in pattern of development as was maintained by, for example, Schnarf (1929, p. 384), but rather is a type, coordinate with nuclear and cellular endosperm, which is "as distinctive a feature of monocotyledons as the other primary features which distinguish this group from the dicotyledons."

The mode of initiation and early method of formation of helobial endosperm appear as remarkably uniform processes throughout the monocotyledons. Swamy and Parameswaran (1963) recognize a basic type, or "norm," and five principal variants or deviations (Fig. 20-31). In the norm as well as in all the deviations from it, the primary endosperm nucleus is always found at the chalazal or basal end of the embryo sac. As a result, when the primary endosperm nucleus divides, a transverse wall is produced, dividing the embryo sac into a small chalazal cell, or chamber, and a much larger micropylar chamber. In the norm pattern, which has become stabilized in such families as the Aponogetonaceae, Hydrocharitaceae, and Potamogetonaceae, the nucleus in the chalazal cell does not divide further but becomes hypertrophied and this cell appears in later stages of development to function as an haustorium. In contrast, there are numerous free nuclear divisions in the larger micropylar cell. The free nuclei may later degenerate or wall formation may take place.

The deviations from the norm, shown by A–E in Fig. 20-31, fundamentally represent a series of increasing number of free nuclear divisions in the chalazal cell. If the number of free nuclear divisions is strictly limited—as it frequently is—only two or four nuclei are formed, and these subsequently become hypertrophied and ultimately degenerate (Fig. 20-31, A, B). Where there are a larger number of free nuclei formed in the chalazal cell, they are never as numerous as in the micropylar chamber and the successive divisions are not strictly synchronized in the two chambers (Fig. 20-31, C–E). A clear example of this is provided by *Eremurus himalaicus,* a member of the Liliaceae (Fig. 20-32). According to

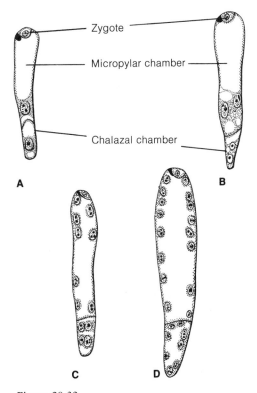

Figure 20-32
Development of the helobial type of endosperm in *Eremurus himalaicus.* [Adapted from *An Introduction to the Embryology of Angiosperms* by P. Maheshwari. McGraw-Hill, New York, 1950.]

Stenar's (1928) investigation on this species, the sequence of free nuclear divisions is *not synchronized* in the two chambers past the two-nucleate stage. Thus the four- and eight-nucleate stages in the chalazal chamber are accompanied by the formation of eight and sixteen nuclei in the micropylar compartment (Fig. 20-32, B–D). At much later stages in the development of helobial endosperm in *Eremurus,* Stenar found 30–32 nuclei in the basal cell while the upper cell contained a much larger number of free nuclei. Sections of the mature seed revealed that cell-wall formation ultimately takes place in the micropylar chamber. In this connection it should be mentioned that the survey made by Swamy and Parameswaran (1963) shows that cell-wall formation commonly occurs in the micropylar chamber of the norm type of development. If there is also wall formation in the chalazal chamber, its appearance may be conspicuously delayed (Fig. 20-31, E).

Phylogenetic Relationships Between the Types of Endosperm

The interpretation of the phylogeny of the nuclear, cellular, and helobial types of endosperm represents a most difficult and controversial problem in angiosperm morphology. Maheshwari (1950, pp. 245 and 252), for example, considers that the helobial type is "intermediate" in its developmental pattern between the nuclear and cellular types but maintains that "whether the series is to be read from the nuclear toward the cellular type or vice versa is not clear." Certain authors, such as Sporne (1954) and Swamy and Ganapathy (1957), writing at a later date than Maheshwari, have, however, taken a more "positive" position. Sporne, using statistical methods, found that nuclear endosperm is significantly correlated with a series of presumably primitive characters (e.g., woody habit, free petals, secretory

cells, ovules with two integuments) and concluded that in all probability the evolutionary development has been from the nuclear to the cellular type. On the other hand, Swamy and Ganapathy used the type of vessel perforation—i.e., whether scalariform or simple—as the basis for their efforts to obtain statistical correlation. They found that the nuclear type of endosperm is positively correlated with vessel members having simple perforations and that, conversely, the cellular type shows a negative correlation with this kind of perforation. Since there is excellent evidence (see Chapter 19) to support the idea that (1) vessel members originated from scalariformly pitted tracheids and (2) that the simple perforation plate is more specialized than the scalariform, Swamy and Ganapathy concluded that nuclear endosperm "is in all probability more advanced than cellular-type endosperm."

The divergent ideas that have just been discussed serve to emphasize the still uncertain status of the problem of endosperm typology in the angiosperms. The most recent and certainly the most thorough analysis of endosperm typology, from a morphogenetic as well as a phylogenetic standpoint, is the comprehensive monograph by Wunderlich (1959). The student is particularly referred to this work for a critical review of the literature and for a useful summary of the occurrence of the main types of endosperm in a wide range of families in both the monocotyledons and dicotyledons.

Wunderlich reached certain general conclusions about the phylogenetic development of endosperm on the premise that the type of endosperm is correlated, to a significant degree, with the morphology of the ovule itself. According to her theory, the starting point for the endosperm evolution is a crassinucellate ovule with two integuments, parietal cells, and cellular endosperm. As this kind of association between ovule

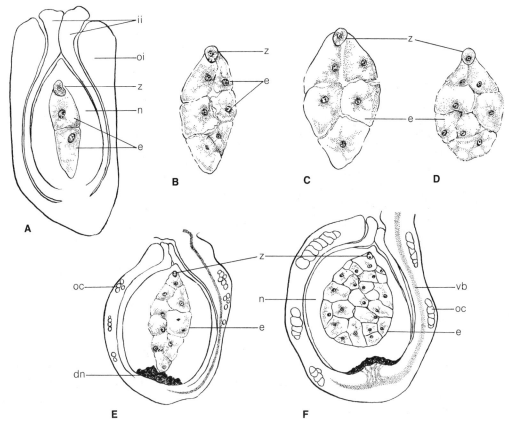

Figure 20-33
Ontogeny of the cellular type of endosperm during seed development in *Degeneria*. A, longisection of
ovule showing zygote and two-celled stage of the endosperm; **B-D,** later stages in endosperm development;
E, F, advanced stages in endosperm formation. Note that zygote has not yet divided. dn, pad of degenerating
nucellar tissue; e, endosperm; ii, inner integument; n, nucellus; oc, groups of oil-bearing cells; oi, outer
integument; vb, vascular bundle; z, zygote. [Redrawn from Swamy, *Jour. Arnold Arboretum* 30:10, 1949.]

and cellular endosperm is common—although not invariable—in various presumably primitive members of the Ranales, Wunderlich postulates that it represents the original form in the angiosperms. *Degeneria,* a primitive ranalean tree, provides a typical example of cellular endosperm that develops within a crassinucellate ovule (Fig. 20-33). However, Wunderlich's extensive survey reveals an apparent contradiction in that in the angiosperms *as a whole,* cellular endosperm is predominantly found in the advanced or tenuinucellate type of ovule (43 families) while in most families with

crassinucellate ovules, the prevailing type of endosperm is nuclear (121 families). To explain this apparent paradox, Wunderlich postulates that an important factor determining the type of endosperm is the "space relationships" within the nucellus. Thus, if in the course of phylogenetic development of a crassinucellate ovule the nucellus retained its original size—or perhaps even enlarged—then the relationship of space became "favorable" for a progressive development from cellular to nuclear endosperm. In other words, certain structural features of the ovule—such as a relatively

690

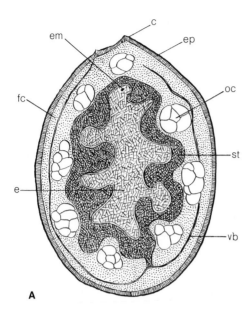

A

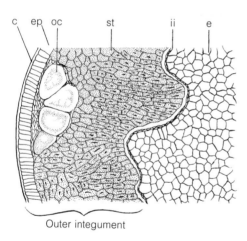

Outer integument

B

Figure 20-34
Structure of the mature seed of *Degeneria*. **A,** diagrammatic longisection showing minute embryo and ruminate endosperm; **B,** portion of mature seed, showing intrusion of stony layer of seed coat into the adjacent cellular endosperm. c, cuticle; e, endosperm; em, embryo; ep, epidermis of seed coat; fc, fleshy layer of seed coat; ii, inner integument; oc, oil-bearing cells; st, stony layer of seed coat; vb, vascular bundle. [Redrawn from Swamy, *Jour. Arnold Arboretum* 30:10, 1949.]

massive nucellus—may have retained their *original primitive* form while a phylogenetic advance was taking place in the development of the endosperm. The converse seems to have occurred, according to Wunderlich, in the trend towards reduction in size of many tenuinucellate ovules in which the resulting strict limitations of space seem to have conditioned the *persistence* of cellular endosperm and a *delay* in its phylogenetic development to nuclear endosperm.

To conclude this brief résumé of endosperm typology it should be emphasized that a great deal of survey work needs to be done on endosperm development and on the morphogenetic role of various factors within the ovule which may shape the pattern of its development. The endosperm formation of many angiospermous genera has not yet been adequately studied and there is a largely unexplored field for experimental studies on the biochemical factors which are necessary for the expression of the varied potentialities of this remarkable tissue when grown *in vitro* (see Chopra and Sachar, 1963, for a review of experimental studies on endosperm).

RUMINATE ENDOSPERM.   Among the other types of endosperm morphology that merit brief consideration is the peculiar one known as ruminate endosperm, which has been observed in members of at least 30 families of dicotyledons as well as in a number of genera of palms (Periasamy, 1962; Corner, 1966). In the mature seeds that form this type of endosperm, the outer surface of the endosperm tissue is irregular or furrowed to varying degrees. Ontogenetic studies have shown that the irregular contour of ruminate endosperm is the result of either the formation of outgrowths from the endosperm surface or the penetration, i.e., the "invasion," of the endosperm by portions of the integument of the ovule. A good example of the latter method of origin is found during seed development in *Degeneria*. According to Swamy (1949a) the ruminate endosperm of this plant is produced by localized patches of cells of the innermost layers of the outer integument that grow as wedges of tissues into the adjacent endosperm (Fig. 20-34, A). As the seed of *Degeneria* matures, the inner part of the outer integument and its protruding wedges differentiate into a hard sclerenchymatous tissue, which forms the stony inner seed coat (Fig. 20-34, B).

ENDOSPERM HAUSTORIA.   A truly bizarre characteristic of endosperm that has been observed in a number of angiosperms is the formation of more-or-less prominent *endosperm haustoria*. These remarkable outgrowths may arise at either or both ends of the developing endosperm and in some plants aggressively invade adjacent parts of the ovule such as the chalaza, the integuments, or even the funiculus. Endosperm haustoria are particularly prominent in a number of dicotyledons which develop the cellular type of endosperm. The haustoria formed by the cellular endosperm of *Downingia* (a genus in the Campanulaceae), arise at the micropylar and chalazal poles and their origin and aggressive growth into the tissues of the ovule have been studied in considerable detail by Kaplan (1969). A very striking example of an aggressive endosperm haustorium is found in *Impatiens roylei* (Dahlgren, 1934). In this species the uppermost cell at the micropylar pole of the endosperm forms a giant haustorium which extends through the micropyle of the ovule and then forms branches which penetrate the tissue of the funiculus (Fig. 20-35).

Although the invasion of various regions of the ovule by endosperm haustoria presumably contributes to the nutrition of the endosperm proper, evidence in support of this function is still circumstantial rather than experimental, as Kaplan (1969) has

already stated with reference to the role of endosperm haustoria in *Downingia*. The study of the varied types of endosperm haustoria from the standpoint of their ultrastructural organization and physiological role is surely a challenging problem for future research.

Interpretations of
Endosperm Morphology

Beginning with the classical studies of the past century and extending to the present, the morphological significance of the virtually universal formation of endosperm in angiosperms has been the subject of controversy and the most varied interpretations. Strasburger (1900) regarded the process of "triple fusion" as "vegetative fertilization" —in his view the union of one of the two male gametes with two polar nuclei constitutes a necessary "stimulus" for the rapid development of the nutritive endosperm. He interpreted the endosperm phylogenetically as equivalent to the tissue of the female gametophyte that has been delayed in its development. A similar interpretation was later adopted by Coulter and Chamberlain (1912). On the other hand, if triple fusion is regarded as an act of true fertilization, or syngamy, the endosperm might be regarded as a malformed or "unorganized" second embryo, the normal development of which was prevented by the participation of the chalazal polar nucleus in the formation of the triploid primary endosperm nucleus. This ingenious theory, proposed by Sargant (1900), obviously is not valid as an explanation of endosperm development in *Oenothera* (Onagraceae) and *Butomopsis,* in which the male gamete fuses with a single polar nucleus and hence the primary endosperm nucleus, like the zygote itself, is a diploid cell (Fig. 20-29). In these remarkable instances it would be highly interesting, with

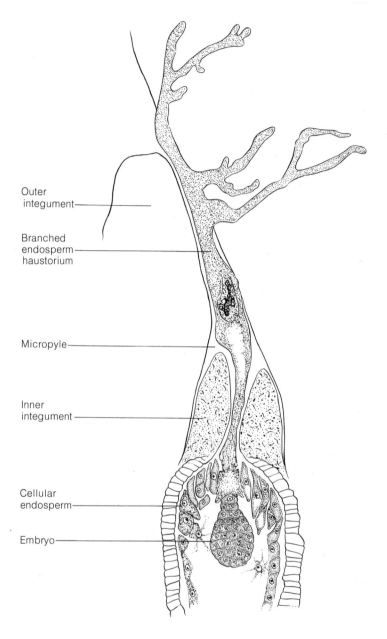

Outer
integument

Branched
endosperm
haustorium

Micropyle

Inner
integument

Cellular
endosperm

Embryo

Figure 20-35
Longisection through developing seed of *Impatiens roylei* showing giant,
branched endosperm haustorium. [Redrawn from Dahlgren, *Svensk Bot.
Tidskr.* 28:103, 1934.]

the use of modern *in vitro* culture techniques, to determine whether the diploid primary endosperm nucleus could experimentally be induced to form a normal embryo, comparable in development to the embryo formed by the zygote.

In more recent years, the significance of the fertilization of the polar nuclei has been interpreted from a genetical-physiological point of view by Brink and Cooper (1947). According to their hypothesis (1) triple fusion is essential for the inception and rapid development of endosperm and (2) as endosperm nuclei contain both maternal and paternal chromosomes, endosperm possesses *hybrid vigor* and hence is a physiologically aggressive tissue. But as Maheshwari (1950, pp. 427–428) has pointed out, Brink and Cooper's theory fails to explain why endosperm, which so commonly is triploid or polyploid, is sooner or later "digested" by the diploid embryo.

In the authors' opinion, none of the theories proposed for the interpretation of endosperm provides an explanation of the phylogenetic origin of this tissue or the remarkable variation in chromosome number in its cells, which ranges from $2n$ (*Oenothera*), $3n$ (*Polygonum*), and $5n$ (*Fritillaria*) to even $15n$ in, e.g., *Peperomia* (Fig. 20-29). As Coulter and Chamberlain (1912) stated so clearly many years ago, "the phylogeny of the endosperm must be traced and the place of triple fusion in its history determined before opinions cease to differ as to its morphological character."

Embryogeny

Origin and Development of the Proembryo

In its earliest stage of development the embryo of angiosperms is usually designated as the *proembryo*. The application of this term is, however, largely a matter of descriptive convenience because only an arbitrary boundary can be drawn between the "proembryo" and the subsequent "embryo proper." Johansen (1950, p. 297) defines the proembryo of angiosperms as "a more or less filamentous row of cells constituting the earliest phase of embryogenesis. It ends with the fourth cell generation when it consists of approximately sixteen cells." In contrast, the French embryologist Souèges, according to Crété (1963, p. 172), extended the concept of the proembryo to include those early stages of embryogeny prior to the appearance of the cotyledonary primordia. In view of the great variation in the number, arrangement, and future "destiny" of the cells of very young embryos, it would appear that Souèges' concept is more appropriate and flexible in the light of our present knowledge of comparative embryogenesis.

The extensive literature on embryogenesis in angiosperms indicates that the development of the endoscopic embryo begins usually with a transverse division of the zygote (Figs. 20-37, 20-41, and 20-43). Exceptions to the prevailing mode of embryo initiation have been observed and are usually classified as *irregular types*. In *Juglans regia,* for example, the plane of the first division of the zygote may be transverse *or* more-or-less conspicuously oblique (Nast, 1941). Additional examples of irregular types are provided by the vertical or oblique plane of the first division (or divisions) in the zygote of *Scabiosa* and members of the family Piperaceae (see Maheshwari, 1950; Johansen, 1950; and Crété, 1963). The morphogenetic or phylogenetic significance of vertical or oblique division planes in the initiation of an embryo is obscure at present, although Crété considers the peculiar embryogenesis of *Scabiosa* to be "primitive."

It was briefly emphasized in our general account of embryogeny in vascular plants

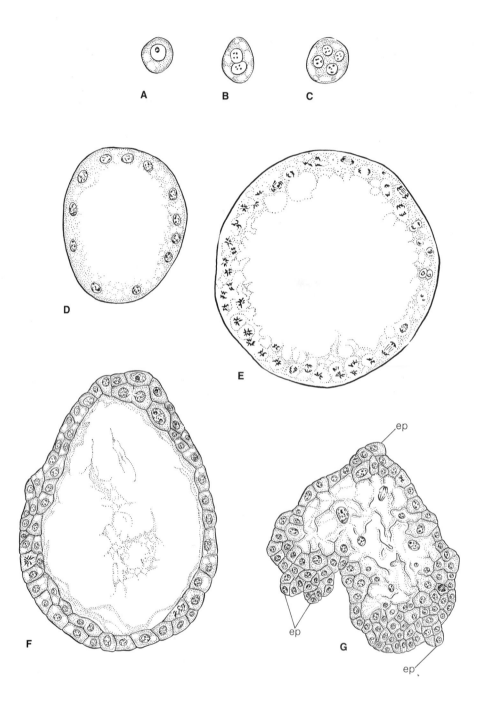

Figure 20-36
Development of the proembryo of *Paeonia*. **A**, zygote; **B–E**, successive stages in free-nuclear divisions in young proembryos; **F**, early stage in cell formation; **G**, later stage in development of cellular proembryo, showing several peripheral embryo primordia (ep). [Redrawn from Wunderlich, *Osterr. Bot. Zeit.* 113:395, 1966.]

(see Chapter 6) that the genus *Paeonia* exhibits a type of embryonic development remarkably different from that of the other angiosperms which have been investigated. The unique feature of *Paeonia* is a preliminary phase of *free nuclear divisions* during the early development of the proembryo (Fig. 20-36, A–E). The number of nuclei formed in the coenocytic proembryo varies, not only between different species but also within a single species. According to Cave et al. (1961), six or seven synchronous mitoses yield a proembryo of 64–128 free diploid nuclei. Carniel's (1967) investigations revealed that occasionally as many as 256 free nuclei may develop. Eventually, by means of wall formation, the multinucleate proembryo becomes differentiated into a cellular proembryo which, however, appears to lack the well-defined endoscopic polarity characteristic of the proembryo of other angiosperms and of those gymnosperms which show a free nuclear phase in their early embryogeny (Fig. 20-36, F). In *Paeonia,* numerous *embryo primordia* — as many as 25 in the species studied by Yakovlev (1967) — begin to differentiate at the periphery of the cellular proembryo (Fig. 20-36, G). Usually only one of these embryo primordia survives and gives rise to the single dicotyledonous type of embryo found in the maturing seed of *Paeonia* (see Cave et al., 1961, and Carniel, 1967, for details). Yakovlev (1967) lists *Paeonia* as an outstanding example among angiosperms of "proembryonal polyembryony."

The original discovery of the free nuclear phase in the development of the proembryo of *Paeonia* was made by Yakovlev and Yoffe (1957). Their observations have been fully confirmed by a series of more recent independent investigations (see Cave et al., 1961; Walters, 1962; Matthiessen, 1962; Wunderlich, 1966; and Carniel, 1967). The only disagreement in the interpretation of the facts was voiced by Murgai (1959, 1962) who maintained that (1) the first division of the zygote is accompanied by the formation of a transverse wall and (2) free nuclear divisions are restricted to the basal cell, termed the "suspensor haustorium," from which the cellular proembryo later takes its origin. However, the work of Cave et al., strongly indicates that a persistent synergid, appressed to the lower side of the zygote, may have been incorrectly interpreted by Murgai as the small apical cell of a two-celled proembryo.

Although Yakovlev and Yoffe (1957) believed that the coenocytic proembryo of *Paeonia* may provide a connecting link between the usually divergent embryogenesis of gymnosperms and angiosperms, this interpretation has been criticized by several other investigators. Cave et al. (1961), for example, regard the multinucleate phase of the proembryo of gymnosperms and *Paeonia* as an example of convergent evolution; they conclude that the peculiar free nuclear phase in the development of the proembryo of *Paeonia* is a derived rather than a primitive character. A similar type of conclusion was also reached by Matthiessen (1962) and Wunderlich (1966).

Comparative Embryogeny of Dicotyledons

During the present century, much attention has been paid to the early phases of development of the proembryo and elaborate and complex "embryonic" classifications have been devised, based on the origin, position, and histogenetic rôle of specific cells or of cell tiers. Particular emphasis has been placed on the plane in which the *terminal cell* divides during the formation of the four-celled proembryo, or "tetrad," as it is sometimes called. Most commonly, the terminal cell of the very young proembryo divides, forming either a vertical wall, as in *Capsella* (Fig. 20-37, E), or a transverse wall, as in *Nicotiana* (Fig. 20-41, B). However, embryological research also has revealed examples of dicotyledonous pro-embryos in which the plane of division of the terminal cell is clearly oblique. On the basis of these three possible planes of division of the terminal cell, Crété (1963, pp. 186–189) classified the four-celled stages of proembryos into three major series which form a significant feature of his embryogenic classification.

In the classification of embryonal types adopted by Johansen (1950) and Maheshwari (1950), considerable use is made, not only of the plane of division of the terminal cell but also of the degree to which the basal cell of the proembryo contributes to the formation of the embryo proper. In these schemes of classification, the name of each of the major types is derived from that of the family in which examples of that type are found. Thus Johansen (1950) recognizes six principal types of embryogeny: the piperad, onagrad, asterad, caryophyllad, solanad, and chenopodiad types.

A further critical review of the various—and often extremely complicated—classifications of embryonal types is beyond the scope and intention of the present text. Although the successive steps in *early* embryogeny of many of the species which have been studied in such detail by Souèges (see Crété, 1963, for a full listing of Souèges publications) seem remarkably precise, there is obviously considerable overlapping among the proposed types. A good example of the difficulties inherent in all efforts to assign fixed destinies to each cell or cell tier of the proembryo is provided by the concepts of the *hypophysis* and the *epiphysis,* which are frequently used in modern embryological descriptions and interpretations.

Hanstein (1870) in his classical study of angiosperm embryogeny, introduced the term hypophysis to designate a specific cell found at the juncture between the suspensor and the embryo proper. In the type of embryogenesis illustrated by *Capsella,* the hypophysis divides transversely. The lower of the two resulting cells initiates the cortex of the embryonic root, or radicle, and division of the upper cell forms the central portion of the root cap (Fig. 20-37, N–Q). In contrast, the hypophysis cell in *Nicotiana* arises at an earlier stage in the development of the proembryo and has a more limited histogenetic rôle as it gives rise to only a portion of the root cap and its calyptrogen layer (Fig. 20-41, D–I). The comparative studies of Souèges (1934) and Crété (1963), moreover, have resulted in expanding Hanstein's original concept of hypophysis to include a "group of cells belonging to different tiers of the proembryo." As both Souèges and von Guttenberg (1960) have indicated, the most consistent rôle of the hypophysis (whether a single cell or a group of cells) is the formation of the major portion of the embryonic root cap.

The apparent counterpart in structure and function of the hypophysis at the shoot pole of the embryo is the *epiphysis.* This term was devised by Souèges (1934) to designate a specific cell or cell group from which the epicotylar apex (i.e., the shoot apex) originates. In the simplest example, illustrated by *Papaver rhoeas* L., the terminal cell of the four-celled proembryo, by means of two successive vertical divisions, enters a quadrant stage; each of these four cells divides periclinally and thus are formed the first *epidermal* and *cortical* initials of the shoot apex (Souèges, 1934, p. 771, Fig. 28–32). In such plants as *Erodium, Geum,* and *Trifolium,* however, a wedge-shaped *apical cell,* which functions as the epiphysis, is produced by oblique divisions during the octant (i.e., eight-celled) stage of the proembryo. There are then periclinal and vertical divisions in the epiphysis and the derivative cells give rise to the primordium of the epicotyl. (See Souèges, 1923, 1934.) Instead of a single epiphysis cell, a terminal group of four cells, originating during either the quadrant or octant stage, may collectively function as the *epiphyseal tissue*; this is illustrated by the embryogeny of *Senecio vulgaris* L. and *Myosurus minimus* L. (Souèges, 1934, pp. 773–777).

Continued research is obviously needed in order to test Souèges' belief that a definable epiphysis is *consistently* differentiated during the development of the proembryo of dicotyledons. Doubtless in a good many plants, the cells from which the epicotyl originates are not clearly defined by their form or number from neighboring cells and in such plants the recognition of an epiphysis would be impossible or, at best, quite arbitrary.

In order to illustrate specifically some of the salient features of embryogenesis typical of certain dicotyledons, we have arbitrarily selected *Capsella bursa-pastoris* (an example of the onagrad type) and *Nicotiana* (an example of solanad type). For more detailed descriptions of these and numerous other types of dicotyledonous embryogeny, the student is referred to the work of Johansen (1950), Maheshwari (1950), Wardlaw (1955), and Crété (1963). Schulz and Jensen (1968b, 1968c) have provided the first information on the ultrastructure of the egg, zygote, and developing embryo of *Capsella*.

CAPSELLA BURSA-PASTORIS. The embryogeny of *Capsella* was first carefully investigated by Hanstein in 1870 and subsequently has served as an instructional model for the discussion of the dicotyledonous embryo. Souèges (1919) reinvestigated the successive steps in the developing embryo of this plant, and the essential events based on his study are shown in Fig. 20-37.

The first division of the zygote yields a terminal and basal cell (t and b, Fig. 20-37, B). Then the basal cell divides transversely and a *longitudinal division* follows in the terminal cell (Fig. 20-37, C–E). As a result a four-celled proembryo is developed. It will be necessary at first to trace separately the further division of the upper and lower pairs of cells. Each of the two terminal cells divides longitudinally, resulting in a quadrant stage (Fig. 20-37, F–J). There are then transverse divisions in each of these four cells, yielding the octant stage of the embryo (Figs. 20-37, K; 20-38, A). A critical histogenetic event then occurs: each of the eight cells divides periclinally into an outer dermatogen cell and an inner cell; as a consequence the young embryo proper now consists of eight external dermatogen cells, destined to produce by further anticlinal division the embryonic surface layer, or epidermis, and eight internal cells, from which the ground meristem and procambial system of the hypocotyl and cotyledons will gradually differentiate (Fig. 20-37, L–N). During the formation of the octant stage of the embryo proper the suspensor is developed from the two upper cells of the proembryo (Fig. 20-37, F–M). As is shown in this figure and in Fig. 20-39, A, the suspensor cell next to the micropylar end of the embryo sac usually fails to divide, but, instead, progressively enlarges to form a very conspicuous vesicular cell. A variable number of transverse divisions of the upper cell and its descendents produce 5–7 additional suspensor cells. The suspensor cell in contact with the base of the globular embryo represents the hypophysis (h in Fig. 20-37, N; see also Fig. 20-38, B). This cell, by transverse and longitudinal divisions, produces two four-celled tiers of cells (Figs. 20-37, N–Q; 20-38, C). Derivatives of the lower tier function as cortical initials of the embryonic root, or radicle, while derivatives of the tier next to the suspensor produce the central portion of the root cap.

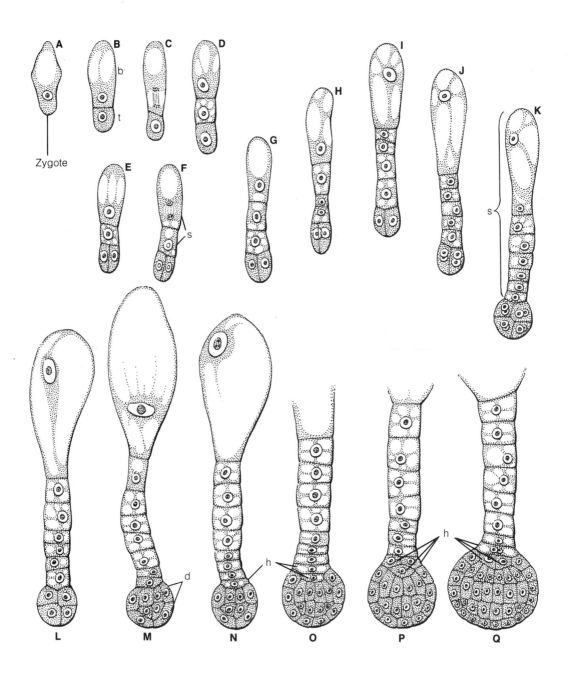

Figure 20-37

Early embryogeny in *Capsella bursa-pastoris*. **A, B,** division of zygote into terminal (t) and basal cell (b); **C, D,** development of three-celled proembryo; **E,** four-celled proembryo, or tetrad, formed by longitudinal division of terminal cell; **F–L,** initiation and development of suspensor (s), and division of terminal cells to yield quadrant stage (**J**) and octant stage (**K, L**) (note progressive enlargement of uppermost suspensor cell); **M,** origin of dermatogen (d), the surface cell layer or immature epidermis of the embryo; **N–Q,** origin of hypophysis (h), and derivation of cells from it, and continued increase in number of surface and internal cells in the globular embryo. [Redrawn from Souèges, *Ann. Sci. Nat. Bot.* X. 1:1, 1919.]

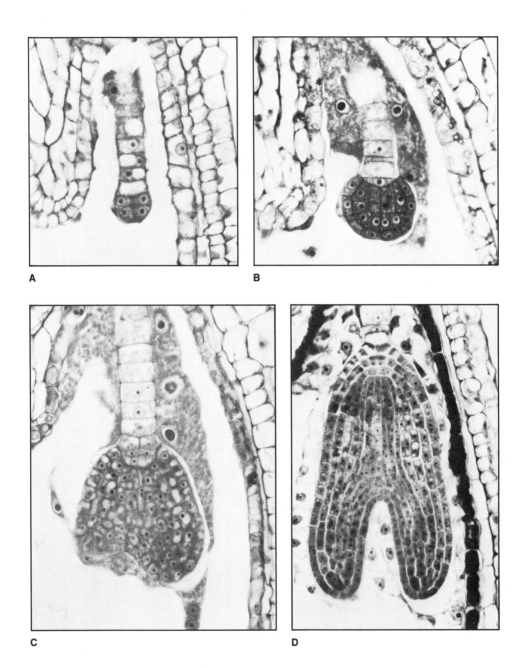

A

B

C

D

Figure 20-38
Embryogeny in *Capsella*. **A,** octant stage of proembryo, comparable with Fig. 20-37, L; **B,** later stage of proembryo with dermatogen layer comparable with 20-37, O (note lightly stained hypophysis cell at base of globular embryo); **C,** later stage, showing cell tiers derived from hypophysis and initiation of cotyledons; **D,** embryo with well-developed cotyledon primordia and procambium, comparable with Fig. 20-39, B. Free nuclear stage in endosperm development shown in **B, C;** cellular endosperm tissue in **D.**

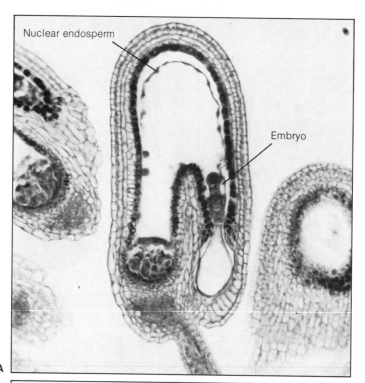

A

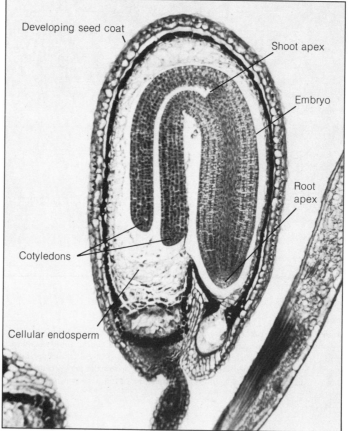

B

Figure 20-39 (*facing page*)
Longisections of developing seeds in *Capsella,* showing early embryogeny and free
nuclear endosperm in **A** and nearly mature embryo and cellular endosperm in **B**.

To recapitulate, the young embryo of
Capsella, as a result of a rather well-coordi-
nated sequence of cell divisions, consists
of a spherical group of cells attached to a
filamentous suspensor (Fig. 20-37, Q). The
paired cotyledons arise as two ridges of
tissue derived from the distal tier of the
embryo, and a few cells situated between
the bases of the cotyledon primordia remain
undifferentiated and constitute the future
shoot apex of the embryo (Figs. 20-38, C, D;
20-40). During cotyledon initiation, the division and differentiation of cells in the
lower tier of the embryo gradually produce
the young axis or hypocotyl of the embryo.
At this general stage the embryo is somewhat heart-shaped, as seen in longisectional
view. Continued enlargement of the hypocotyl and cotyledons results in a pronounced
curvature of the cotyledons which lie parallel to the axis of the embryo in the mature
seed (Figs. 20-39, B; 20-40).

NICOTIANA. We have already noted that
in the current schemes for classifying the
varied types of embryogeny considerable
importance is attached to the plane of cell
division of the terminal cell of the proembryo. In *Capsella,* the terminal cell divides,
forming a *longitudinal* wall, and it is primarily on this basis that the embryo is classified by Johansen (1950) as onagrad type. A
contrasted pattern of proembryo development is illustrated by *Nicotiana* because the
terminal and basal cell *both* divide *transversely* and thereby produce a linear four-
celled proembryo; on this basis Johansen
classifies *Nicotiana* as the solanad type.
Souèges (1920) carefully investigated the
distinctive development of the *Nicotiana*
proembryo and the following brief description as well as the schematic drawings in
Fig. 20-41 are taken from this paper.

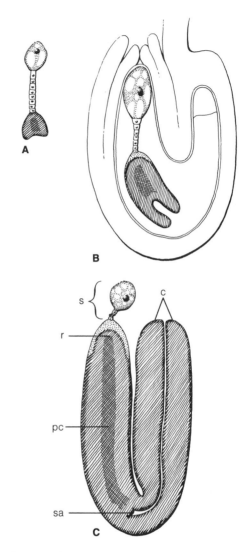

Figure 20-40
A, B, late stages in embryogeny of *Capsella.* **A,** the
cordate form of a longisection of an embryo at stage
of initiation of cotyledons; **B,** longisection of developing seed showing orientation and general structure
of an embryo with two cotyledons; **C,** longisection
of an embryo from a mature seed. c, cotyledons;
pc, procambium; r, tip of root; s, suspensor; sa, shoot
apex. [A and C redrawn from Schaffner, *Ohio Nat.*
7:1, 1906; B after Bergen and Caldwell and redrawn
from *A Textbook of General Botany,* Ed. 4, by R. M.
Holman and W. W. Robbins. Wiley, New York, 1951.]

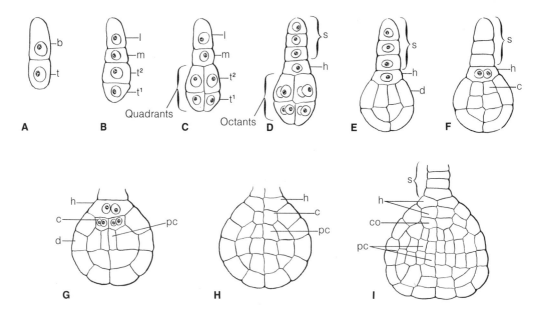

Figure 20-41

Development of the proembryo of *Nicotiana*. **A**, transverse division of zygote into terminal (t) and basal cell (b); **B**, transverse division of terminal and basal cell, forming the linear type of four-celled proembryo characteristic of *Nicotiana;* **C**, longitudinal divisions of cells t¹ and t² yield terminal cell quadrants; **D**, each quadrant cell then divides longitudinally, at right angles to the previous plane, forming terminal cell octants and the two cells m and l divide transversely to form the hypophysis (h) and the first cells of the developing suspensor (s); **E**, origin of dermatogen (d); **F, G**, origin of cortical initials of root (c) and early longitudinal subdivisions of hypophysis (h); **H, I**, formation of the two superposed cell tiers from hypophysis and differentiation of first procambium of future radicle (pc). [Redrawn from Souèges, *Compt. Rend. Acad. Sci.* (Paris) 170:1125, 1920.]

The terminal and basal cells, derived at the first stage of embryogenesis, *each* divide transversely and thus produce a filamentous four-celled proembryo (Fig. 20-41, A, B). The two lower cells (t¹ and t²) by successive longitudinal divisions, give rise first to the quadrant stage and then to the octant stage in proembryogenesis (Fig. 20-41, C, D). During this development, cell m, by transverse division, forms the hypophysis, h, and an adjacent suspensor cell and a similar division of cell l forms two additional suspensor cells, s (Fig. 20-41, D). Periclinal division of each of the octant cells separates the young dermatogen, d, from the internal cells of the proembryo (Fig. 20-41, E). Then the lower internal cells of the hypocotyledonary region, by means of transverse divisions, form the cortical initials of the future radicle (Fig. 20-41, F-I, c). Concomitantly,

the hypophysis cell, by longitudinal and transverse divisions, produces two four-celled tiers. (Fig. 20-41, F-I, h). The tier next to the suspensor, by means of radial divisions forms part of the external layer of the root cap; the lower tier, next to the embryo proper, divides tangentially and functions as part of the calyptrogen. As is shown in Fig. 20-41, I, the dermatogen cells immediately adjacent to the hypophyseal tissue divide periclinally and contribute to the development of the lateral portions of the root cap.

Comparative Embryogeny of Monocotyledons

The differences between the organography of the *mature embryos* of monocotyledons and dicotyledons are striking and have long been used to demarcate these two major classes in the angiosperms. With relatively few exceptions (e.g., the usual formation of three cotyledons in the embryo of *Degeneria*; see Swamy, 1949a) the embryo typical of dicotyledons develops a pair of *lateral cotyledons* between the bases of which is situated the small terminal shoot apex (Fig. 20-42). In contrast, according to the classical and still prevalent interpretation, the embryo in monocotyledons produces a single *terminal* cotyledon, below which is found the *lateral* shoot apex (Fig. 20-42).

One of the most interesting conclusions which has been reached through comparative studies is that the early phases of embryogenesis, prior to cotyledon initiation, are very similar in both dicotyledons and monocotyledons. Indeed, there are such close resemblances in the plane of division of the zygote, the steps leading to the formation of the four-celled proembryo, and the mode of origin of the terminal octants, that the proembryos of certain monocotyledonous and dicotyledonous genera have been classified under the same embryonic type (See Johansen's, 1950, pp. 121–123, and Crète's, 1963, pp. 192–195, schemes of classification.)

Beginning with the early investigations of Hanstein (1870), the embryogenesis of *Sagittaria* has come to be regarded as "typical" of the monocotyledons as a whole. A very detailed analysis of both early and late phases of embryogeny in *Sagittaria sagittaefolia* L. (Alismataceae) was made by Souèges (1931) and the following brief account and the accompanying illustrations are based on his investigation.

The embryogeny of *Sagittaria* begins with the usual transverse division of the zygote and the formation of a terminal and a basal cell (Fig. 20-43, A). However, the basal cell fails to contribute to the further development of the proembryo but instead becomes an enormous vesicular structure that persists into the final stages of embryo-

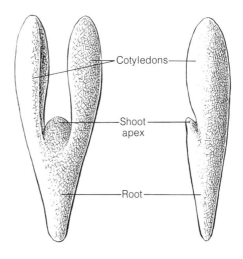

Dicotyledonous type of embryo

Monocotyledonous type of embryo

Figure 20-42
Differences, according to the classical interpretation, between the organography of the embryo in dicotyledons and monocotyledons.

genesis (Fig. 20-43, C–I). In contrast, the terminal cell of the two-celled proembryo functions as the initial from which the entire embryo (including the short suspensor) takes its origin. This peculiar and distinctive type of early demarcation between an inactive basal cell and an active terminal cell is also found during the preliminary stages of embryogenesis of certain dicotyledons (e.g., *Sagina,* a genus in the Caryophyllaceae). For this reason, the proembryo of *Sagittaria* has been classified by Johansen (1950, pp. 233–237) as one of the variations of the caryophyllad type of embryogeny.

Following the initial transverse division of the zygote, the terminal cell divides transversely, forming cells t' and t'' (Fig. 20-43, B, C). Then t' divides, forming a longitudinal wall, and cell t'' forms cells a and h. Disregarding the vesicular basal cell, the proembryo is now in the "tetrad," or four-celled stage of its development (Fig. 20-43, D). Souèges (1931, pp. 360–361) attaches great importance to the fact that in

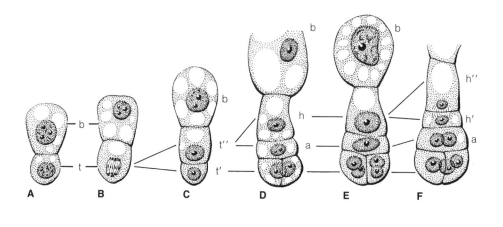

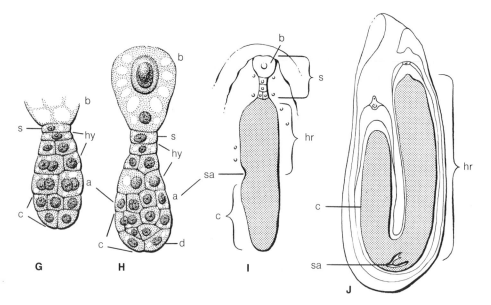

Figure 20-43

A–H, early embryogeny in *Sagittaria sagittaefolia* L. **A,** transverse division of zygote into terminal (t)
and basal (b) cells; note progressive enlargement of basal cell in later stages, **C–H; B, C,** transverse
division of terminal cells into cells t' and t''; **D,** longitudinal division of cell t', and transverse division
of cell t'' forming cells a and h; this is the four-celled stage of the proembryo; **E,** the two cells
derived from t' divide longitudinally, forming terminal cell quadrants; **F,** longitudinal and transverse
divisions, respectively, of cells a and h, result in an eight-celled proembryo; **G,** the four terminal
cells have divided and formed octants from which the terminal cotyledon (c) later develops (**H–J**);
H, origin of dermatogen (d); **I, J,** late stages in embryogeny of *Sagittaria variabilis:* **I,** longisection
of embryo with terminal cotyledons (c), a depression corresponding to future lateral shoot apex (sa),
hypocotyl root (hr), and suspensor (s); **J,** longisection of curved embryo in mature seed (labels as in
I). [A–H redrawn from Souèges, *Ann. Sci. Nat. Bot.* X. 13:353; I, J redrawn from Schaffner, *Bot.
Gaz.* 23:252, 1897.]

Sagittaria the proembryonic tetrad does *not* arise directly from the zygote (as it does in many angiosperms), but originates from the terminal cell of the bicellular proembryo.

After the four-celled stage in proembryo development, the juxtaposed terminal cells derived from cell t′ divide longitudinally, forming quadrants (Fig. 20-43, E). Then cell *a* divides longitudinally and cell h, by a transverse cleavage, produces cells h′ and h″. The proembryo consists at this point of eight cells, arranged in four tiers (Fig. 20-43, F). The lowermost quadrant tier, by means of transverse divisions, gives rise to octants which, according to Souèges' interpretation, collectively represent the primordium of the single *terminal* cotyledon, c (Fig. 20-43, G). The cell tiers that subsequently arise from the division of cell *a* initiate the *lateral* (i.e., subterminal) shoot apex; the upper part of the hypocotyl and cells h′ and h″ are the

points of origin, respectively, of the lower hypocotyl and radicle, and the short suspensor, s; the latter at maturity consists of 3–6 superposed cells. As in many dicotyledonous proembryos, the dermatogen in *Sagittaria* is initiated by the periclinal divisions of the octant cells (Fig. 20-43, H). During the late phases of embryogeny, the position of the shoot apex is evidenced by the development of a groove at one side of the embryo between the base of the cotyledon and the adjacent hypocotyl (Fig. 20-43, I). In the final stages of differentiation the embryo assumes a curved position in the seed and the shoot apex becomes surrounded by the sheathing base of the cotyledon (Fig. 20-43, J).

As very few embryogenetic studies have been made on the enormously diversified representatives of the monocotyledons (particularly the tropical genera), it would indeed be premature to generalize on the

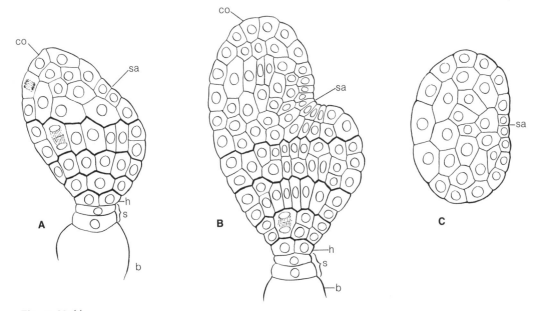

Figure 20-44
Young embryos of *Ottelia alismoides*. **A,** common origin of cotyledon primordium and shoot apex from terminal region (demarcated by heavy upper line) of proembryo; **B,** later stage, showing lateral displacement of shoot apex by growth of cotyledon; **C,** transverse section at level of shoot apex of stage represented in A. b, basal cell of proembryo; co, primordium of cotyledon; h, hypophysis; s, suspensor; sa, shoot apex. [Redrawn from Haccius, *Planta* 40:443, 1952.]

basis of the developmental pattern illustrated by *Sagittaria.* The need for reserving judgment is shown by Haccius' (1952a) study on *Ottelia alismoides,* a monocotyledon which, like *Sagittaria,* is classified as a member of the Helobiae. In *Ottelia,* both the cotyledon and the shoot apex are differentiated side-by-side from the terminal quadrants of the proembryo (20-44, A). In other words, the shoot apex is terminal in origin like that in a dicotyledonous embryo, and its subsequent "displacement" to one side results from the early aggressive growth of the single cotyledon (Fig. 20-44, B). It is thus clear that embryogenesis may be variable even within the same major taxon in the monoctyledons and that, from an ontogenetic standpoint, examples of both terminal (*Sagittaria*) and lateral (*Ottelia*) cotyledonary initiation do exist among monocotyledons (see Haccius and Lakshmanan, 1966). In this connection it should be mentioned that in several other monocotyledons — *Potamogeton indicus* (Potamogetonaceae), *Halophila ovata* (Hydrocharitaceae), and *Pistia stratiotes* (Araceae) — the single cotyledon and the shoot apex also jointly arise from the terminal cell tier of the proembryo as in *Ottelia,* but in *Lemna gibba* (Lemnaceae), the cotyledon is strictly terminal in method of origin (Swamy and Parameswaran, 1962; Swamy and Lakshmanan, 1962; Haccius and Lakshmanan, 1966).

For the present, it is necessary to conclude that the phylogenetic significance of the fluctuating position of the cotyledon in the embryo of the monocotyledons has not been satisfactorily explained. Souèges (1954), for example, gives a detailed defense of the "terminality" of the cotyledon while Haccius and Lakshmanan (1966) emphasize their conviction that within the monocotyledons "all" transitions occur between the clearly lateral method of cotyledon initiation and the so-called "terminal" method. In connection with this conclusion regarding monocotyledons it is interesting that in a number of examples of "monocotyledonous dicotyledons" that have recently been studied, the single cotyledon arises in a more-or-less obviously "lateral" position and there is no evidence of a "rudimentary" or "abortive" second cotyledon. Haccius (1954) found complete agreement between the relative position of the shoot apex and the single cotyledon in the young embryos of *Claytonia virginica* (a member of the dicotyledonous family Portulacaceae) and *Ottelia* (Hydrocharitaceae). Additional examples of monocotyledonous dicotyledons are provided by several species of *Pinguicula* (Lentibulariaceae), *Anemone apennina* (Ranunculaceae), certain geophilous members of the Umbelliferae, and *Cyclamen* (Primulaceae) (Haccius and Hartl-Baude, 1957; Haccius and Fischer, 1959; Haccius, 1952b; Haccius and Lakshmanan, 1967).

In addition to the problems posed by the position of the cotyledon in the embryos of monocotyledons, there still remains the question of the morphology of the cotyledon itself. Is this structure the homologue of a single foliar organ or is it in reality a "double" structure, equivalent to the fusion of a pair of cotyledons? Considerable effort was made by Sargant (1902) to demonstrate that the solitary cotyledon in monocotyledons consists of two congenitally united foliar organs. Arber (1925, p. 172), however, in a detailed analysis of Sargant's theory of syncotyly, concluded "that neither the consideration of external form, nor of internal structure, provides any convincing evidence of the bifoliar nature of the seed-leaf among Monocotyledons." A more widely held view regards the solitary cotyledon of monocotyledons as the surviving member of an ancestral pair of cotyledons (Eames, 1961, and Takhtajan, 1969). Although examples of "heterocotylous" embryos with one normal and one haustorial cotyledon have been found in certain dicotyledons (e.g., in some species of the genus *Peperomia*), comparable stages in the modification, reduction, or elimination of one of the two "original" cotyledons have been less convincingly demonstrated for the monocotyledons. In this connection it is interesting that the experimental treatment of the monocotyledonous embryo of *Cyclamen* with phenylboric acid does not inhibit development of the single cotyledon and is not "compensated by the development of a second cotyledon from the opposite side of the original one. Instead, the original cotyledonary zone bifurcates into two dissimilar cotyledonary stumps with united bases" (Haccius and Lakshmanan, 1967).

A highly original interpretation of the cotyledon in the monocotyledons has been proposed by Swamy (1962). In his view, the terminal octants in the proembryo of *both* dicotyledons and monocotyledons possess identical developmental potentialities: the initiation of two cotyledonary primordia and a *primary* terminal shoot apex. The distinctive feature of the embryogeny of monocotyledons is the *simultaneous* development of the shoot apex initials and the two cotyledonary initials, with the derivative cells soon becoming consolidated into a morphogenetically sterile structure which has been conventionally identified as the single terminal cotyledon. As, according to Swamy, the developing "cotyledon" incorporates the loci of two cotyledonary primordia as well as the initials of the epicotyl, a *new functional apex* arises from a lateral locus in a subterminal tier of the proembryo (Swamy and Lakshmanan, 1962, p. 244).

It is commonly held that the monocotyledons evolved from a dicotyledonous stock (Eames, 1961, pp. 362–363; Takhtajan, 1969). The acceptance of Swamy's hypothesis would not appear to lend substantial support to this belief or to shed useful light on the taxonomic interrelationships of the two classes of angiospermous plants. What really appears to be urgently needed at this juncture is a comprehensive series of studies on the initiation and development of the cotyledon and shoot apex during embryogenesis in a very wide range of living monocotyledons.

References

Arber, A.
　1925. *Monocotyledons. A Morphological Study.* University Press, Cambridge.
Arnott, H. J.
　1962. The seed, germination, and seedling of *Yucca. Univ. Calif. Publ. Bot.* 35:1–164.
Bailey, I. W., and B. G. L. Swamy
　1951. The conduplicate carpel of dicotyledons and its initial trends of specialization. *Amer. Jour. Bot.* 38:373–379.

Baker, H. G.
　1968. Intrafloral ecology. *Ann. Rev. Entomol.* 13:385–414.
Bambacioni, V.
　1928a. Ricerche sulla ecologia e sulla embriologia di *Fritillaria persica* L. *Ann. Bot.* 18:7–37.
　1928b. Contributo alla embriologia di *Lilium candidum* L. *Rend. Accad. Naz. Lincei* 8:612–618.

Battaglia, E.

1951. The male and female gametophytes of angiosperms—an interpretation. *Phytomorphology* 1:87–116.

1958. L'abolizione del tipo embriologico *Scilla* e la creazione dei nuovi tipi *Endymion* ed *Allium. Caryologia* 11: 247–252.

Bersier, J. D.

1960. L'ovule anatrope: Ranunculaceae. *Bull. Soc. Bot. Suisse* 70:171–176.

Bocquet, G.

1959. The campylotropous ovule. *Phytomorphology* 9:222–227.

Bocquet, G., and J. D. Bersier

1960. La valeur systématique de l'ovule: développements tératologiques. *Arch. Sci.* (Geneve) 13:475–496.

Boesewinkel, F. D., and F. Bouman

1967. Integument initiation in *Juglans* and *Pterocarya. Acta Bot. Neerl.* 16:86–101.

Boke, N. H.

1949. Development of the stamens and carpels in *Vinca rosea* L. *Amer. Jour. Bot.* 36:535–547.

Borthwick, H. A.

1931. Development of the macrogametophyte and embryo of *Daucus carota. Bot. Gaz.* 92:23–44.

Bouman, F.

1971a. The application of tegumentary studies to taxonomic and phylogenetic problems. *Ber. Deutsch. Bot. Ges.* 84:169–177.

1971b. Integumentary studies in the Polycarpicae. I. Lactoridaceae. *Acta. Bot. Neerl.* 20:565–569.

Brewbaker, J. L.

1967. The distribution and phylogenetic significance of binucleate and trinucleate pollen grains in the angiosperms. *Amer. Jour. Bot.* 54:1069–1083.

Brink, R. A.

1924. The physiology of pollen. *Amer. Jour. Bot.* 11:218–228; 283–294; 351–364; 417–436.

Brink, R. A., and D. C. Cooper

1947. The endosperm in seed development. *Bot. Rev.* 13:423–541.

Budell, B.

1964. Untersuchungen der Antherenentwicklung einiger Blütenpflanzen. *Zeit. Bot.* 52:1–28.

Bütow, R.

1955. Die Entwicklung der *Pulsatilla*—Anthere. *Zeit. Bot.* 43:423–449.

Carniel, K.

1963. Das Antherentapetum. Ein kritischer Überblick. *Osterr. Bot. Zeit.* 110:145–176.

1967. Über die Embryobildung in der Gattung *Paeonia. Osterr. Bot. Zeit.* 114: 4–19.

Carlquist, S.

1969. Toward acceptable evolutionary interpretations of floral anatomy. *Phytomorphology* 19:332–362.

Cass, D. D., and W. A. Jensen

1970. Fertilization in barley. *Amer. Jour. Bot.* 57:62–70.

Cave, M. S., H. J. Arnott, and S. A. Cook

1961. Embryogeny in the California peonies with reference to their taxonomic position. *Amer. Jour. Bot.* 48:397–404.

Chan, S. C. K., and C. J. Hillson

1971. Developmental morphology of the microsporangium in *Ipomoea reptans* Poir. *Bot. Gaz.* 132:224–229.

Chopra, R. N., and R. C. Sachar

1963. Endosperm. Pp. 135–170 *in* Maheshwari, P. (ed.), *Recent Advances in the Embryology of Angiosperms.* University of Delhi, Delhi.

Cooper, D. C.

1935. Macrosporogenesis and development of the embryo sac of *Lilium henryi. Bot. Gaz.* 97:346–355.

Corner, E. J. H.

1966. *The Natural History of Palms.* University of California Press, Berkeley and Los Angeles.

Coulter, J. M., and C. J. Chamberlain

1912. *Morphology of Angiosperms.* Appleton, New York.

Crété, P.

1963. Embryo. Pp. 171–220 *in* Maheshwari, P. (ed.), *Recent Advances in the Embryology of Angiosperms.* University of Delhi, Delhi.

Dahlgren, K. V. O.

1927. Die Morphologie des Nuzellus mit besonderer Berücksichitigung der deckzellosen Typen. *Jahr. Wiss. Bot.* 67: 347–426.

1934. Die Embryologie von *Impatiens roylei*. *Svensk. Bot. Tidskr.* 28:103–125.

Davis, G. L.
1966. *Systematic Embryology of the Angiosperms.* Wiley, New York.

Diels, L.
1916. Käferblumen bei den Ranales und ihre Bedeutung für die Phylogenie der Angiospermen. *Ber. Deutsch. Bot. Ges.* 34:758–774.

Dolcher, T.
1951. La meiosi e lo sviluppo del gametofito nell'ovulo di *Sambucus nigra* L. (Caprifoliaceae). *Carylogia* 4:58–76.

Eames, A. J.
1961. *Morphology of the Angiosperms.* McGraw-Hill, New York.

Esau, K.
1965. *Plant Anatomy.* Wiley, New York.
1969. *The Phloem.* (Handbuch d. Pflanzenanatomie, Band 5, Teil 2.) Gebrüder Borntraeger, Berlin.

Faegri, K., and L. van der Pijl
1966. *The Principles of Pollination Ecology.* Pergamon, London.

Fagerlind, F.
1961. The initiation and early development of the sporangium in vascular plants. *Svensk. Bot. Tidskr.* 55:299–312.

Fisher, D. B., and W. A. Jensen
1969. Cotton embryogenesis: The identification, as nuclei, of the X-bodies in the degenerated synergid. *Planta* 84:122–133.

Grant, V.
1950. The protection of the ovules in flowering plants. *Evolution* 4:179–201.

Guttenberg, H. von
1960. *Grundzüge der Histogenese höherer Pflanzen. I. Die Angiospermen.* (Handbuch d. Pflanzenanatomie, Band 8, Teil 3.) Gebrüder Borntraeger, Berlin.

Haberlandt. G.
1914. *Physiological Plant Anatomy.* Macmillan, London.
1924. *Physiologische Pflanzenanatomie,* Ed. 6. Wilhelm Engelmann, Leipzig.

Haccius, B.
1952a. Die Embryoentwicklung bei *Ottelia alismoides* und das Problem des terminalen Monokotylen–Keimblatts. *Planta* 40:443–460.

1952b. Verbreitung und Ausbildung der Einkeimblättrigkeit bei den Umbelliferen. *Osterr. Bot. Zeit.* 99:483–505.
1954. Embryologische und histogenetische Studien an "monokotylen Dikotylen." I. *Claytonia virginica. Osterr. Bot. Zeit.* 101:285–303.

Haccius, B., and E. Fischer
1959. Embryologische und histogenetische Studien an "monokotylen Dikotylen." III. *Anemone apennina* L. *Osterr. Bot. Zeit.* 106:373–389.

Haccius, B., and E. Hartl-Baude
1957. Embryologische und histogenetische Studien an "monokotylen Dikotylen." II. *Pinguicula vulgaris* L. und *P. alpina* L. *Osterr. Bot. Zeit.* 103:567–587.

Haccius, B., and K. K. Lakshmanan
1966. Vergleichende Untersuchung der Entwicklung von Kotyledon und Sprossscheitel bei *Pistia stratiotes* und *Lemna gibba,* ein Beitrag zum Problem der sogenannten terminalen Blattorgane. *Beitr. Biol. Pflanzen* 42:425–443.
1967. Experimental studies on monocotyledonous dicotyledons: Phenylboric acid-induced "dicotyledonous" embryos in *Cyclamen persicum. Phytomorphology* 17:488–494.

Hanstein, J.
1870. Entwicklungschichte der Keime der Monokotyle und Dikotyle. *Bot. Abhandl. Bonn* 1:1–112.

Hjelmquist, H.
1964. Variations in embryo sac development. *Phytomorphology* 14:186–196.

Iwanami, Y.
1956. Protoplasmic movement in pollen grains and tubes. *Phytomorphology* 6:288–295.

Jensen, W. A.
1962. *Botanical Histochemistry. Principles and Practice.* W. H. Freeman and Company, San Francisco.
1973. Fertilization in flowering plants. *BioScience* 23:21–27.

Jensen, W. A., and D. B. Fisher
1968. Cotton embryogenesis: The entrance and discharge of the pollen tube in the embryo sac. *Planta* 78:158–183.

Johansen, D. A.
1950. *Plant Embryology.* Chronica Botanica, Waltham, Massachusetts.

Johri, B. M.
   1963. Female gametophyte. Pp. 69–103 *in* Maheshwari, P. (ed.), *Recent Advances in the Embryology of Angiosperms*. University of Delhi, Delhi.
Jones, H. A., and S. L. Emsweller
   1936. Development of the flower and macrogametophyte of *Allium cepa. Hilgardia* 10:415–423.
Junell, S.
   1931. Die Entwicklungsgeschichte von *Circaeaster agrestis. Svensk Bot. Tidskr.* 25: 238–270.
Kapil, R. N.
   1960. Embryology of *Acalypha* Linn. *Phytomorphology* 10:174–184.
Kapil, R. N., and S. Jalan
   1964. *Schisandra michaux*—its embryology and systematic position. *Bot. Notis.* 117:285–306.
Kapil, R. N., and I. K. Vasil
   1963. Ovule. *In* Maheshwari, P. (ed.), *Recent Advances in the Embryology of Angiosperms*. University of Delhi, Delhi.
Kaplan, D. R.
   1969. Seed development in *Downingia. Phytomorphology* 19:253–278.
Kühn, G.
   1928. Beiträge zur Kenntnis der intraseminalen Leitbündel bei den Angiospermen. *Bot. Jahrb.* 61:325–379.
LeMonnier, G.
   1872. Recherches sur la nervation de la graine. *Ann. Sci. Nat. Bot.* Ser. 5. 16:233–305.
Linskens, H. F.
   1969. Fertilization mechanisms in higher plants. Pp. 189–253 *in* Metz, C. B., and A. Monnoy, eds, *Fertilization,* Vol. II. Academic, New York.
Maheshwari, P.
   1946. The *Adoxa* type of embryo sac: a critical review. *Lloydia* 9:73–113.
   1948. The angiosperm embryo sac. *Bot. Rev.* 14:1–56.
   1949. The male gametophyte of angiosperms. *Bot. Rev.* 15:1–75.
   1950. *An Introduction to the Embryology of Angiosperms*. McGraw-Hill, New York.
   1960. Evolution of the ovule. Pp. 1–13 in *A. C. Seward Memorial Lectures,* Ser. 7. Birbal Sahni Institute of Palaeobotany, Lucknow.

Maheshwari, P., and R. N. Kapil
   1966. Some Indian contributions to the embryology of angiosperms. *Phytomorphology* 16:239–291.
Maheshwari, S. C.
   1955. The occurrence of bisporic embryo sacs in angiosperms—a crtical review. *Phytomorphology* 5:67–99.
Martens, P.
   1962. Études sur les Gnétales. VI. Recherches sur *Welwitschia mirabilis.* III. L'ovule et le sac embryonnaire. *Cellule* 63:309–329.
Matthiessen, A.
   1962. A contribution to the embryogeny of *Paeonia. Acta Horti Bergiani* 20:57–61.
Müller-Stoll, W. R., and G. Lerch
   1957a. Über Nachweis, Enstehung und Eigenschaften der Kallosebildungen in Pollenschläuchen. *Flora* 144:297–334.
   1957b. Über den physiologischen Charakter der Kallosebildung in Siebröhren und Pollenschläuchen und die Beziehungen zu den Zellvorgängen. *Biol. Zentralbl.* 76:595–612.
Murgai, P.
   1959. The development of the embryo in *Paeonia. Phytomorphology* 9:275–277.
   1962. Embryology of *Paeonia* together with a discussion on its systematic position. Pp. 215–223 in *Plant Embryology: A Symposium.* Council of Scientific and Industrial Research, New Delhi.

Nast, C. G.
   1941. The embryogeny and seedling morphology of *Juglans regia* L. *Lilloa* 6:163–205.
Netolitzky, F.
   1926. *Anatomie der Angiospermen—Samen.* (Handbuch d. Pflanzenanatomie, Band 10, Lief. 14.) Gebrüder Borntraeger, Berlin.

Periasamy, K.
   1962. The ruminate endosperm. Development and types of rumination. Pp. 62–74 in *Plant Embryology: A Symposium.* Council of Scientific and Industrial Research, New Delhi.
Periasamy, K., and N. Parameswaran
   1965. A contribution to the floral morphology and embryology of *Tarenna asiatica. Beitr. Biol. Pflanzen* 41:123–138.

Periasamy, K., and B. G. L. Swamy
1959. Studies in the Anonaceae. I. Microsporogenesis in *Cananga odorata* and *Miliusa wightiana*. *Phytomorphology* 9: 251–263.
1964. Is the microsporangium of angiosperms wall-less? *Curr. Sci.* 33:735–738.
Peter, J.
1920. Zur Entwicklungsgeschichte einiger Calycanthaceen. *Beitr. Biol. Pflanzen* 14: 59–86.
Puri, V.
1952. Placentation in angiosperms. *Bot. Rev.* 18:603–651.
Roth, I.
1957. Die Histogenese der Integumente von *Capsella bursa-pastoris* und ihre morphologische Deutung. *Flora* 145:212–235.
Sampson, F. B.
1963. The floral morphology of *Pseudowintera*, the New Zealand member of the vesselless Winteraceae. *Phytomorphology* 13: 403–423.
1969. Cytokinesis in pollen mother cells of angiosperms, with emphasis on *Laurelia novae-zelandiae* (Monimiaceae). *Cytologia* 34:627–634.
1970. Unusual features of cytokinesis in meiosis of pollen mother cells of *Pseudowintera traversii* (Buchan.) Dandy (Winteraceae). *Beitr. Biol. Pflanzen* 47:71–77.
Sampson, F. B., and D. R. Kaplan
1970. Origin and development of the terminal carpel in *Pseudowintera traversii*. *Amer. Jour. Bot.* 57:1185–1196.
Sargant, E.
1900. Recent work on the results of fertilization in angiosperms. *Ann. Bot.* 14:689–712.
1902. The origin of the seed-leaf in monocotyledons. *New Phytol.* 1:107–113.
Sax, K.
1916. Fertilization in *Fritillaria pudica*. *Bull. Torrey Bot. Club* 43:505–522.
Schnarf, K.
1927–1928. *Embryologie der Angiospermen.* (Handbuch d. Pflanzenanatomie, Band 10.) Gebrüder Borntraeger, Berlin.
1929. *Embryologie der Angiospermen.* (Handbuch d. Pflanzenanatomie, Band 10, Teil 2.) Gebrüder Borntraeger, Berlin.
1931. *Vergleichende Embryologie der Angiospermen.* Gebrüder Borntraeger, Berlin.
1941. Contemporary understanding of embryo sac development among angiosperms. *Bot. Rev.* 2:565–585.
Schulz, Sister Richardis, and W. A. Jensen
1968a. *Capsella* embryogenesis: The synergids before and after fertilization. *Amer. Jour. Bot.* 55:541–552.
1968b. *Capsella* embryogenesis: The egg, zygote, and young embryo. *Amer. Jour. Bot.* 55:807–819.
1968c. *Capsella* embryogenesis: The early embryo. *Jour. Ultrastruct. Res.* 22:376–392.
Shah, C. K.
1962. Pollen development in some members of the Cyperaceae. Pp. 81–93 in *Plant Embryology: A Symposium.* Council of Scientific and Industrial Research, New Delhi.
Smith-White, S.
1959. Pollen development patterns in the Epacridaceae. A problem in cytoplasm-nucleus interaction. *Proc. Linn. Soc. New South Wales* 84:8–35.
Souèges, R.
1914. Nouvelles recherches sur le développement de l'embryon chez les Crucifères. *Ann. Sci. Nat. Bot.* Ser. 9. 19:311–339.
1919. Les premières divisions de l'oeuf et les différenciations du suspenseur chez le *Capsella bursa-pastoris. Ann. Sci. Nat. Bot.* Ser. 10. 1:1–28.
1920. Embryogénie des Solanacées. Développement de l'embryon chez les *Nicotiana. Compt. Rend. Acad. Sci.* (Paris) 170:1125–1127.
1923. Développement de l'embryon chez le *Geum urbanum* L. *Bull. Soc. Bot. France* 70:645–660.
1931. L'embryon chez le *Sagittaria sagittaefolia* L. Le cone végétatif de la tige et l'extrémité radiculaire chez les monocotylédones. *Ann. Sci. Nat. Bot.* Ser. 10. 13:353–402.
1934. L'hypophyse et l'épiphyse; les problèmes d'histogenèse qui leur sont liés. *Bull. Soc. Bot. France* 81:737–748; 769–778.
1954. L'origine du cone végétatif de la tige et la question de la "terminalité" du cotylédon des monocotylédones. *Ann. Sci. Nat. Bot.* Ser. 11. 15:1–20.

Sporne, K. R.
  1954. A note on nuclear endosperm as a primitive character among dicotyledons. *Phytomorphology* 4:275–278.
Steffen, K.
  1963. Fertilization. Pp. 105–133 *in* Maheshwari, P. (ed.), *Recent Advances in the Embryology of Angiosperms*. University of Delhi, Delhi.
Stenar, H.
  1928. Zur Embryologie der Asphodeline—Gruppe. Ein Beitrag zur systematischen Stellung der Gattungen *Bulbine* und *Paradisea*. *Svensk. Bot. Tidskr.* 22:145–159.
  1941. Über die Entwicklung des Embryosackes bei *Convallaria majalis* L. *Bot. Notis.* 1941, 123–128.
Steward, F. C.
  1968. *Growth and Organization in Plants*. Addison-Wesley, Reading, Mass.
Strasburger, E.
  1879. *Die Angiospermen und die Gymnospermen*. Jena.
  1900. Einige Bemerkungen zur Frage nach der "doppelten Befruchtung" bei Angiospermen. *Bot. Zeit.* II. 58:293–316.
Swamy, B. G. L.
  1949a. Further contributions to the morphology of the Degeneriaceae. *Jour. Arnold Arboretum* 30:10–38.
  1949b. Embryological studies in the Orchidaceae. *Amer. Midland Natur.* 41:184–201.
  1952. Some aspects in the embryology of *Zygogynum Bailloni* V. Tiegh. *Proc. Nat. Inst. Sci. India* 18:399–406.
  1962. The embryo of monocotyledons: a working hypothesis from a new approach. Pp. 113–123 in *Plant Embryology: A Symposium*. Council of Scientific and Industrial Research, New Delhi.
Swamy, B. G. L., and I. W. Bailey
  1949. The morphology and relationships of *Cercidiphyllum*. *Jour. Arnold Arboretum* 30:187–210.
Swamy, B. G. L., and P. M. Ganapathy
  1957. On endosperm in dicotyledons. *Bot. Gaz.* 119:47–50.

Swamy, B. G. L., and K. K. Lakshmanan
  1962. The origin of epicotylary meristem and cotyledon in *Halophila ovata* Gaudich. *Ann. Bot.* n.s. 26:243–249.
Swamy, B. G. L., and N. Parameswaran
  1962. On the origin of cotyledon and epicotyl in *Potamogeton indicus*. *Osterr. Bot Zeit.* 109:344–349.
  1963. The helobial endosperm. *Biol. Rev.* 38:1–50.
Takhtajan, A.
  1969. *Flowering Plants: Origin and Dispersal*. Translated from the Russian by C. Jeffrey. Oliver and Boyd, Edinburgh.
Trapp, A.
  1956. Zur Morphologie und Entwicklungsgeschichte der Staubblätter sympetaler Blüten. (Bot. Studien. Heft 5.) Gustav Fischer, Stuttgart.
Tucker, S. C., and E. M. Gifford, Jr.
  1966. Carpel development in *Drimys lanceolata*. *Amer. Jour. Bot.* 53:671–678.
Van Heel, W. A., and F. Bouman
  1972. Note on the early development of the integument in some Juglandaceae together with some general questions on the structure of angiosperm ovules. *Blumea* 20:155–159.
Van Tieghem, P.
  1898. Structure de quelques ovules et parti qu'on peut tirer pour améliorer la classification. *Jour. Bot.* 12:197–220.
Vasil, I. K.
  1967. Physiology and cytology of anther development. *Biol. Rev.* 42:327–373.
Venkata Rao, C.
  1961. Pollen types in the Epacridaceae. *Jour. Indian Bot. Soc.* 40:409–423.
Vijayaraghavan, M. R.
  1962. Studies in the family Ranunculaceae. II. The female gametophyte of *Clematis gauriana* Roxb. *Phytomorphology* 12:45–49.
Vijayaraghavan, M. R., and K. N. Marwah
  1969. Studies in the family Ranunculaceae—microsporangium, microsporogenesis and Ubisch granules in *Nigella damascena*. *Phyton Ann. Bot.* 13:203–209.

Vijayaraghavan, M. R., and U. Padmanaban
1969. Morphology and embryology of *Centuarium ramosissimum* Druce and affinities of the family Gentianaceae. *Beitr. Biol. Pflanzen* 46:15–37.

Vijayaraghavan, M. R., and G. S. Sarveshwari
1968. Embryology and systematic position of *Morina longifolia* Wall. *Bot. Notis.* 121:383–402.

Walters, J. L.
1962. Megasporogenesis and gametophyte selection in *Paeonia californica. Amer. Jour. Bot.* 49:787–794.

Wardlaw, C. W.
1955. *Embryogenesis in Plants.* Wiley, New York.

Warming, E.
1873. *Untersuchungen über Pollen bildende Phyllome und Kaulome.* (Bot. Abhand.) (Edited by J. Hanstein.) A. Marcus, Bonn.

Waterkeyn, L.
1954. Études sur les Gnétales. I. Le strobile femelle, l'ovule et la graine de *Gnetum africanum. Cellule* 56:105–146.
1964. Callose microsporocytaire et callose pollinque. Pp. 52–58 *in* Linskens, H. F. (ed.), *Pollen Physiology and Fertilization.* North Holland Publ. Co., Amsterdam.

Wettstein, R.
1924. *Handbuch der Systematischen Botanik,* Ed. 3. Franz Deuticke, Leipzig und Wien.

Wunderlich, R.
1959. Zur Frage der Phylogenie der Endospermtypen bei den Angiospermen. *Osterr. Bot. Zeit.* 106:203–293.
1966. Zur Deutung der eigenartigen Embryoentwicklung von *Paeonia. Osterr. Bot. Zeit.* 113:395–407.

Yakovlev, M. S.
1967. Polyembryony in higher plants and principles of its classification. *Phytomorphology* 17:278–282.

Yakovlev, M. S., and M. D. Yoffe
1957. On some peculiar features in the embryogeny of *Paeonia. Phytomorphology* 7:74–82.

21

Fruits, Seeds, and Seedlings

Fruits

In a strict morphological sense, a fruit is the seed-containing structure that originates from the enlargement and modified development of the gynoecium of a single flower. This definition seems particularly appropriate from the standpoint of etymology because the morphological unit of both an apocarpous and a syncarpous gynoecium is called a *carpel*, a term derived from the Greek word "karpos" meaning fruit (Fig. 21-1). Other parts of a flower, in addition to the modified gynoecium, may form a more-or-less conspicuous part of the mature fruit and in some texts, fruits of this sort are separately classified as "accessory fruits." Examples of so-called accessory fruits are found (1) in *Fragaria* (strawberry) in which

the fruit consists of the enlarged succulent receptacle of the flower with numerous small indehiscent fruitlets (achenes) embedded in its surface (Fig. 21-2) and (2) in *Malus* (apple), in which the outer fleshy edible portion — derived from the floral tube (i.e., the fused bases of the sepals, petals, and stamens) — surrounds the central core region derived from the inferior ovary of the gynoecium (see Fig. 21-3). Considerable ambiguity and confusion, however, are introduced in some textbooks by designating accessory fruits as "pseudocarps" or "false fruits" (see McLean and Ivimey-Cook, 1956). It must be emphasized, on the contrary, that it is impossible to draw a useful distinction between true fruits (derived exclusively from the gynoecium) and false or accessory fruits. The fruit of raspberry (*Rubus*) is a collection of small

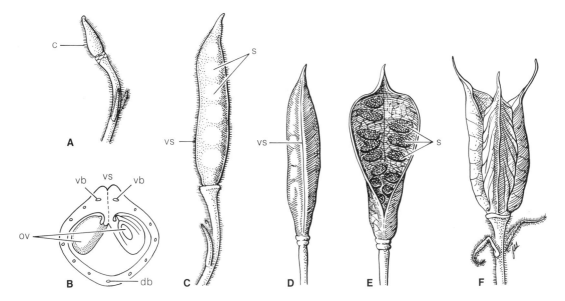

Figure 21-1
A–E, Development of single follicle of *Delphinium ajacis:* **A,** flower (after removal of perianth and stamens) showing the single carpel (c); **B,** transection of basal region of carpel showing two of the enclosed ovules (ov), the ventral suture (vs) where dehiscence will later occur, and the dorsal (db) and ventral (vb) carpel bundles; **C,** lateral view of ventral suture (vs) of young fruit with seeds (s); **D,** ventral view of nearly mature fruit; **E,** ripe follicle that has dehisced along the ventral suture (s, seeds). **F,** group of three dehiscent follicles of *Delphinium elatum.* [Redrawn from *Praktische Einführung in die Pflanzenmorphologie* by W Troll. Gustav Fischer, Jena, 1957.]

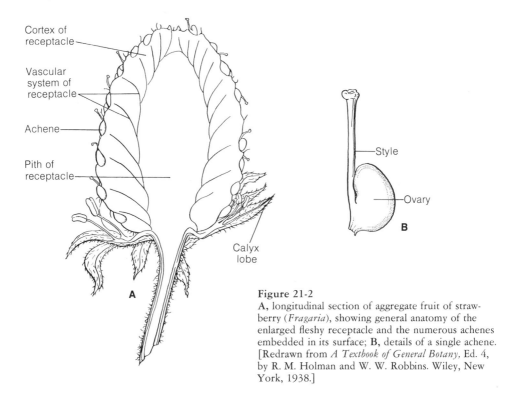

Figure 21-2
A, longitudinal section of aggregate fruit of strawberry (*Fragaria*), showing general anatomy of the enlarged fleshy receptacle and the numerous achenes embedded in its surface; **B,** details of a single achene. [Redrawn from *A Textbook of General Botany,* Ed. 4, by R. M. Holman and W. W. Robbins. Wiley, New York, 1938.]

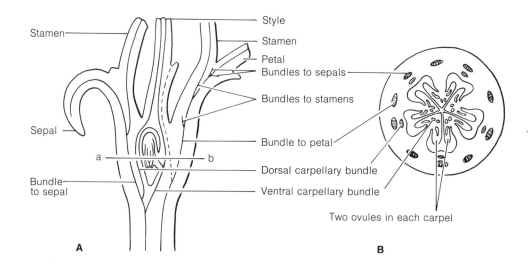

Stamen

Style
Stamen
Petal
Bundles to sepals
Bundles to stamens

Sepal

a — b

Bundle
to sepal

Bundle to petal

Dorsal carpellary bundle

Ventral carpellary bundle

Two ovules in each carpel

A

B

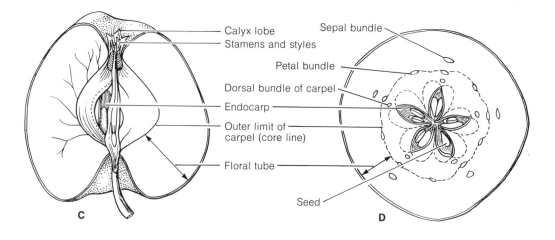

Calyx lobe
Stamens and styles

Sepal bundle

Petal bundle

Dorsal bundle of carpel

Endocarp

Outer limit of
carpel (core line)

Floral tube

Seed

C

D

Figure 21-3
The flower and fruit (a pome) of apple (*Malus sylvestris*). **A, B,** longitudinal and transverse sections, respectively, of flower, showing its vasculature and the inferior position of the five carpels; **C, D,** longitudinal and transverse sections, respectively, of the mature fruit, showing the outer fleshy region, derived from the floral tube, and the central core derived from the adnate five-carpellate gynoecium. [Redrawn from *Botany*, Ed. 3, by W. W. Robbins, T. E Weier, and C R. Stocking. Wiley, New York, 1964.]

fleshy drupes which, when ripe, separate as a unit-cluster from the receptacle (Fig. 21-4). But in the fruit of the closely allied blackberry (likewise a member of the genus *Rubus*) the receptacle becomes somewhat enlarged and fleshy and represents a significant part of the mature fruit.

Fruits are one of the definitive structures of the angiosperms and should not be confused with the analogous dispersal units of other groups of seed plants. Thus, for example, the large, fleshy, naked seed of *Ginkgo biloba* is sometimes erroneously referred to in botanical literature as a "fruit" (Rehder, 1927, p. 1). A similar misuse of the concept of fruit is illustrated when the fleshy seed cone of *Juniperus* (a member of the Coniferales) is described as "berry-like" (Sargent, 1922, p. 78).

Classification of Fruits

The great diversity in the organization of angiosperm flowers, especially the variation in the number, arrangement, degree of fusion and structure of the carpels which form the gynoecium, is in turn reflected in the wide range in the size, form, texture, and anatomy of fruits. We need only to realize that such familiar structures as watermelons, peaches, grapes, tomatoes, acorns, corn grains, and string beans are *all* fruits in the botanical sense, to understand fully the need for a scientific system, or classification, of fruits. But most of the classifications of fruits—especially those presented in elementary texts—are both artificial and confusing. A fundamental part of the difficulty, as will be discussed later, stems from the confusing concepts of aggregate and multiple fruits. Furthermore, there has also been a widespread tendency to overemphasize certain anatomical and biological characters of fruits, such as their texture (e.g., dry versus fleshy fruits) and their dehiscence or failure to dehisce when ripe. The use of such charac-

ters, while valid for purposes of taxonomic description and diagnosis, has led to much overlapping and duplication of subtypes in certain of the proposed fruit systems. An example that supports this criticism is the classification of the fruit of almond (*Prunus amygdalus*) as a drupe. Unlike, however, the "typical" fleshy indehiscent drupes of cherry and peach, the outer part of the fruit wall (pericarp) of the almond fruit *at maturity* is hard, dry, and dehiscent (see Robbins, Weir, and Stocking, 1964, p. 250, Fig. 15.1).

A detailed comparison of the various complex schemes of fruit classification cannot be undertaken in this book (see, for example, the fruit systems proposed by Winkler, 1939; Egler, 1943; and Baumann-Bodenheim, 1954). But many of the difficulties inherent in any effort to construct a "natural" system of fruits are apparent if we consider the current, popular classification of fruits into simple, aggregate, and multiple types. This classification appears to have originated nearly a century ago in Asa Gray's (1878) classical text and in essence is repeated today in many books on elementary botany (see, for example, Holman and Robbins, 1938; Robbins, Weir, and Stocking, 1964; Hill, Popp, and Grove, 1967; and Wilson and Loomis, 1967).

The morphological basis used for separating fruits into simple, aggregate, and multiple types is the ovary (or pistil) rather than the gynoecium as a whole. Thus a *simple fruit* is defined as one that originates from a *single ovary*, regardless of whether the ovary is unicarpellate or syncarpellate. During the maturation of simple fruits, the ovary wall differentiates into the *pericarp*, or fruit wall. There is great variation in the texture and histology of pericarps and much emphasis is placed in classifying simple fruits on whether the pericarp is fleshy or dry. A striking example of a fleshy fruit with a highly differentiated pericarp is provided by the *drupe*, illustrated by the fruits of cherry, peach, and

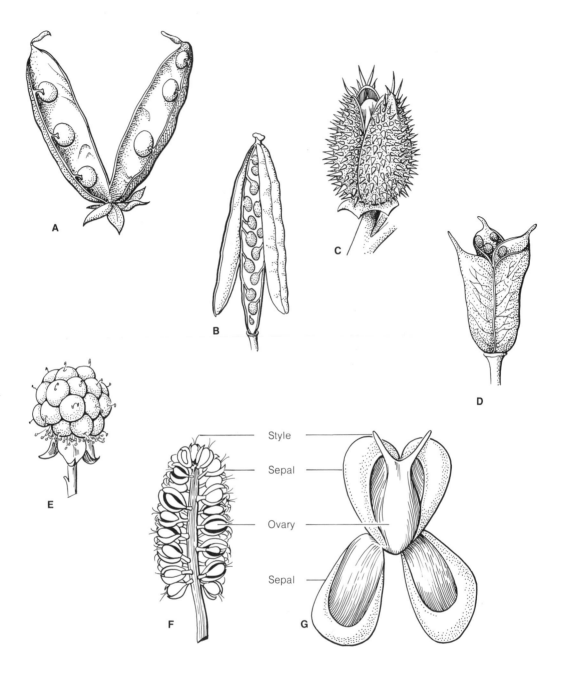

Figure 21-4

A–D, examples of various types of dry dehiscent fruits; **A,** legume of pea (*Pisum sativum*); **B,** silique of mustard (*Brassica*); **C,** capsule of Jimson weed (*Datura*); **D,** follicles of larkspur (*Delphinium*); **E,** aggregate fruit of raspberry (*Rubus*); **F,** longitudinal section of multiple fruit of mulberry (*Morus*); **G,** a single enlarged fruit of **F.** [**A–E** redrawn from *A Textbook of General Botany,* Ed. 4, by R. M. Holman and W. W. Robbins. Wiley, New York, 1951. **F, G** redrawn from *Botany,* Ed. 4, by J. B. Hill, H. W. Popp, and A. R. Grove, Jr. McGraw-Hill, New York, 1967.]

plum. In these fruits, the pericarp is clearly differentiated into an outer skin (exocarp), a thick fleshy middle portion (mesocarp), and a stony pit (endocarp) that encloses the seed. Simple dry fruits may either be indehiscent (*achene* of sunflower, *caryopsis* of *Zea*) or dehiscent (*legume* of pea and bean, *silique* of *Brassica, capsule* of *Datura,* and *follicle* of *Delphinium*). Figure 21-4, A–D, depicts these common examples of dry, dehiscent types of fruits. (For a detailed description of the developmental anatomy of the pericarp of various fruit types, see Esau, 1965, and Fahn, 1967.)

In contrast to the simple fruit, an *aggregate fruit* is said to consist of a collection of more-or-less separate simple fruits (or "fruitlets") derived from a corresponding number of separate ovaries or pistils of a single flower. Strawberries, raspberries, and blackberries are usually cited, in elementary texts, as typical examples of aggregate fruits (Figs. 21-2, 21-4, E).

Multiple fruits arise from the matured— and often coalesced—ovaries of the individual flowers of an inflorescence. The fruit of mulberry is commonly mentioned as an example of a multiple fruit (Fig. 21-4, F, G). Other examples of multiple fruits are the pineapple and the fig, in both of which the axis of the inflorescences constitutes a significant part of the ripe fruit.

The basic weakness of the classification of fruits into simple, aggregate, and multiple types is that it establishes categories that are not morphologically coordinate. Simple and aggregate fruits are defined with reference to the gynoecium of a single flower but the multiple "type" of fruit is set apart as a type derived from the closely crowded flowers of an inflorescence. This procedure obviously leads to confusion, since it attempts to compare nonhomologous structures. Some authors (e.g., Winkler, 1939, and Egler, 1943) have eliminated so-called multiple fruits from their schemes of classi-

fication. Another objection that can be raised against the classification we are discussing concerns the unfortunate choice of the term "simple" to include a bewildering array of fleshy and dry types of fruits (Figs. 21-1; 21-4, A–D). In this connection, the statement is frequently made that *both* aggregate and multiple fruits consist of clusters of simple fruits.

In an effort to devise a more "natural system" of fruits, Winkler (1939) proposed a classification that seems to eliminate many of the difficulties and inconsistencies we have been describing. His classification rests, in the first place, on his broad morphological concept that "a fruit is (that) which arises from the gynoecium (not simply the pistil) of a flower as the result of fertilization or of parthenocarpy." He maintains that if the *entire gynoecium* of the flower—rather than a single ovary or pistil—is used as the basis for defining and classifying fruits, all the past difficulties vanish and fruits can be naturally classified into two principal groups: *aggregate fruits*, which arise from the collection of ripe carpels of an *apocarpous gynoecium*, and *unit fruits*, which arise from a single *syncarpous gynoecium* (Fig. 21-5). It should be noted that although Winkler retains the old term "aggregate fruit," he uses it in a *collective sense* and thus avoids the confusion produced when *each* mature carpel is interpreted as a "simple fruit." He also makes the interesting point that if the separate ripe ovaries derived from an apocarpous gynoecium are called fruitlets, this term should not be used in a diminutive sense but in a partitive sense—i.e., in a manner analogous to the designation of the blades of a single compound leaf as leaflets. Winkler's *new* term unit fruit replaces the ambiguous and morphologically misleading older term simple fruit, a change in terminology which is both reasonable and welcome.

An important aspect of Winkler's classification of fruits consists in his using the superior versus the inferior position of the gynoecium of the flower as a basis for subdividing both aggregate and unit fruit types into "free fruits" and "cup fruits." Thus the fruit of *Ranunculus* – derived from a flower with a superior apocarpous gynoecium – is classified as an "aggregate free fruit" (Fig. 21-5, A). In contrast, the inferior gynoecium of the epigynous flower of *Rosa* gives rise to an "aggregate cup fruit" in which the ripe

carpels (fruitlets) are enclosed within the cup-like hypanthium or floral tube (Fig. 21-5, B). Parallel categories are recognized by Winkler for unit fruits. In the flower of poppy (*Papaver*), the superior syncarpous gynoecium develops into a "unit free fruit" (Fig. 21-5, C) and the "unit cup fruit" of the hawthorne (*Crataegus*) arises from an inferior gynoecium and is enclosed in the hollow floral tube (Fig. 21-5, D).

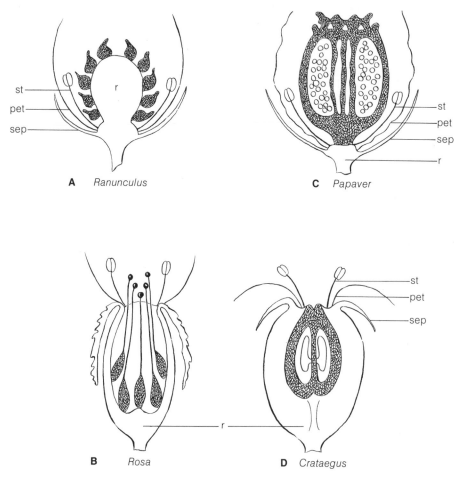

A   *Ranunculus*

C   *Papaver*

B   *Rosa*

D   *Crataegus*

Figure 21-5
Longisectional diagrams illustrating Winkler's (1939) morphological classification of fruits into aggregate (**A**, **B**) and unit (**C**, **D**) types. In all figures, the free or united carpels are shown by hatching and the other floral structures are labeled as follows: pet, petal; r, receptacle; sep, sepal; st, stamen. **A**, the aggregate free fruit; **B**, the aggregate cup fruit; **C**, the unit free fruit; **D**, the unit cup fruit. [Redrawn from Winkler, *Beitr. Biol. Pflanzen* 26:201, 1939.]

Although Winkler's classification primarily emphasizes the importance of the relative position and morphology of the gynoecium, he also employs secondary characters, such as texture, dehiscence versus indehiscence, and seed number, as additional features for his construction of an elaborate outline of fruit types (see Winkler, 1939, pp. 216–219). His conspectus clearly indicates the prevalence of parallel development in the evolution of angiosperm fruits and hence the difficulty at present of attaining a truly phylogenetic arrangement of fruit types. Eames (1961, pp. 379–380) maintains that the follicle—a many seeded fruit which dehisces along a single ventral suture—is "the primitive fruit type derived directly from the primitive carpel" (Fig. 21-1). This viewpoint is shared by several other morphologists (e.g., Juhnke and Winkler, 1938; Winkler, 1939, 1940; and Takhtajan, 1969) and may eventually provide the clue for a better understanding of the evolution of fruit types in the angiosperms.

Parthenocarpy

The growth of the gynoecium into a seed-containing fruit normally seems dependent on the stimulation provided by both pollination and the subsequent fertilization of the ovules. But some plants may develop seedless fruits without being pollinated or as the result of the abortion of the embryo in the seeds before fruit maturation. Fruits of this kind are termed *parthenocarpic* and have been divided by Leopold (1964, p. 263) into the following classes: (1) those which develop without pollination (e.g., the occasional parthenocarpic fruits of tomatoes, peppers, pumpkins, and cucumbers, as well as the consistently seedless fruits of banana and pineapple); (2) those in which pollination alone stimulates fruit development but in which fertilization of the ovules does not occur (e.g., certain orchids and the parthenocarpic species of *Poa*); and (3) those seedless fruits in which an abortion of the embryo in the developing seeds takes place *before* fruit maturation (e.g., the parthenocarpic fruits of some types of cherries, peaches, and grapes).

According to Leopold, "the physiological basis for parthenocarpy remains obscure," despite the considerable experimental and commercial attention which has been devoted to this problem. The artificial induction of parthenocarpy by the application of various auxins has been frequently tried and with tomatoes, for example, may result in seedless fruits even though fertilization of the ovules may have already taken place. For further details on the use of auxins and other chemicals in promoting fruit development and in controlling the commercially undesirable abscission of young fruits, the student is referred to the discussion of the subject given by Leopold (1964, Chapter 15).

Seeds

Our principal aim in the previous chapter was to describe those processes and structures that are essential to the final conversion of an ovule into a seed. Sporogenesis, the development of male and female gametophytes and pollination (including the growth of the pollen tube) "set the stage," so to speak, for the final act of double fertilization, which initiates the embryo and endosperm of the future seed. The complex changes that accompany the maturation of angiosperm seeds, however, cannot be considered in any detail in this book and the following brief résumé is intended only as a general outline of the subject. For detailed information on the anatomy of seeds, the student should consult Netolitzky's (1926) monograph and the discussions of seed development and anatomy given by Eames (1961), Esau (1965), and Fahn (1967).

One of the most interesting and important events in the maturation of a fertilized ovule into a seed concerns the "fate" of the endosperm. Endosperm is an important reserve-food tissue for the young developing embryo but in many dicotyledonous taxa the aggressive development of the embryo, prior to seed germination, destroys all or most of the endosperm. Seeds of this type are designated as *exalbuminous* and at maturity consist of a large embryo enclosed within the seed coat. Many legumes (e.g., pea, bean, and vetch) form exalbuminous seeds and the food reserves, used during germination, are stored in the thick fleshy cotyledons (Fig. 21-7, E). But in the very common type of *albuminous* seed, a large portion of the endosperm persists during embryogenesis and becomes a conspicuous tissue in the fully mature seed (Figs. 21-7, A; 21-8, A). The albuminous seeds of cereals, such as wheat, oats, corn, and rice, because of the rich food reserves in the endosperm, constitute one of the basic sources of food for man (Fig. 21-11, A).

During the development of the embryo sac, embryo, and endosperm, the nucellar tissue of the ovule is usually completely destroyed. In some taxa, however, the nucellus proliferates into a food-storing tissue known as *perisperm*. There may be both endosperm and perisperm in the same seed, as in the beet plant (*Beta*) or the perisperm may be the only food-storing tissue. As is shown in Fig. 21-9, a good example of copious perisperm development is provided by the seeds of various species of *Yucca*, a monocotyledonous genus that has received detailed investigation by Arnott (1962).

The development of the embryo and food-storing tissues (i.e., endosperm and perisperm) of angiosperm seeds is accompanied by the differentiation of the seed coat, or *testa*, from one or both integuments of the ovule. The *degree* to which integumentary tissues contribute to the formation of the mature testa is extremely varied and can

only be determined by careful histogenetic study (see Esau, 1965, and Fahn, 1967, for details). In the developing seed of *Asparagus officinalis*, for example, *both* integuments participate in forming the young seed coat. When the testa reaches its maximum thickness, two cuticles are evident, one situated at the boundary between the outer and the inner integument (Fig. 21-6, A, c[1]), the other between the inner integument and the adjacent nucellar tissue (Fig. 21-6, A, c[2]). As the seed approaches maturity, the two cell layers representing the inner integument become disintegrated and compressed and the two cuticles merge together. As a result, the testa of the ripe seed consists of a thick-walled epidermis overlying a zone of irregularly arranged thin-walled cells. Both of these tissues are derived from the outer integument, the inner boundary of which now is represented by the two closely appressed cuticles (Fig. 21-6, B, c[1], and c[2]). The nucellus (nu), recognizable during the early stages of seed development, is ultimately entirely absorbed (Fig. 21-6, A, B).

Comparative studies have revealed considerable diversity in the type and arrangement of the cells that compose the mature coat of angiosperm seeds (see the extensive monograph by Netolitzky, 1926). The histological complexity of the testa is due to (1) differences in the degree of destruction of portions of the integument during ontogeny, (2) the presence of a wide variety of cell types in the successive layers of the mature testa, (3) the pattern of the vascular system, which originates from the original venation of the integument, and (4) the formation of various types of specialized trichomes by the epidermis of the outer integument. Although a detailed treatment of these complex aspects of seed-coat anatomy cannot be undertaken in this book, two specific examples, which illustrate patterns of distribution of sclerenchyma in hard seed coats, deserve brief consideration.

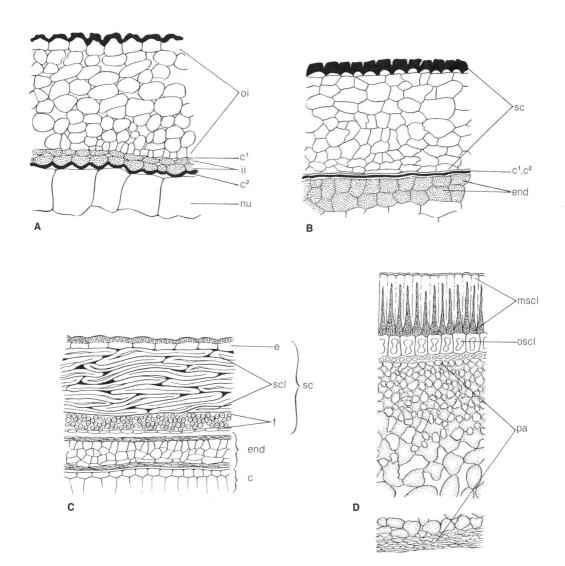

Figure 21-6

A, B, The seed coat in *Asparagus officinalis*. **A,** stage of maximum thickness, showing cells derived from both integuments of the ovule; **B,** mature seed coat after degeneration of the inner integument. **C,** the mature testa of the seed of apple; **D,** the mature testa of the seed of *Phaseolus multiflorus*. See text for further explanations. c, cotyledon of embryo; c^1, outer cuticle; c^2, inner cuticle; e, epidermis of seed coat; end, endosperm; f, zone of fibers; ii, inner integument; mscl, layer of macrosclereids; nu, nucellus; oi, outer integument; oscl, layer of osteosclereids; pa, parenchyma; sc, seed coat; scl, sclereids. [A, B redrawn from *Plant Anatomy* by K. Esau. Wiley, New York, 1965. C, D redrawn from *An Introduction to Plant Anatomy*, Ed. 2, by A. J. Eames and L. H. MacDaniels. McGraw-Hill, New York, 1947.]

In the seed of apple (*Malus*), there are two zones of sclerenchyma beneath the epidermis of the testa: an *outer zone*, composed of several layers of tightly joined sclereids, and an *inner zone* of fiber-like cells, the long axes of which are oriented at right angles to the longitudinal axes of the sclereids. (Fig. 21-6, C). A second and contrasted pattern of sclerenchyma development is illustrated by the seeds of some members of the family *Leguminosae*. Characteristically, as is shown in *Phaseolus multiflorus*, the outermost layer of the seed coat consists of a layer of columnar cells (derived from the epidermis of the integument (Fig. 21-6, D). The component cells—variously known as palisade cells, Malpighian cells, or macrosclereids—develop unevenly thickened walls and usually exhibit a highly refractive "light line" located in the walls near the outer ends of the cells. As is shown in Fig. 21-6, D, the hypodermal layer of the testa of *Phaseolus* consists of relatively short osteosclereids that because of the peculiar pattern of thickening of their walls, have been designated as "hour-glass cells." Beneath the sclerenchymatous hypodermal layer is an extensive zone of loosely arranged parenchyma cells (Fig. 21-6, D). A similar type of seed-coat anatomy is characteristic of *Pisum sativum* (see Reeve, 1946a, 1946b). But, as is shown by Corner's (1951) investigations, there is a great range of variation in the details of seed-coat histology in the seeds of the numerous genera of the Leguminosae that he examined, and the student is referred to his paper especially for information on the significance of seed-coat anatomy to systematic problems within the Leguminosae.

Seedlings

The behavior of angiosperm seeds, after their separation from the parent plant, varies widely. The embryo of some seeds is extremely small at the time of shedding and embryogenesis continues for some time before the actual rupture of the seed coat. In some plants (e.g., certain maples and oaks) the seeds promptly germinate if environmental conditions, such as light, temperature, and moisture are favorable; in others a more-or-less protracted period of seed dormancy passes before the seedling begins to develop. Because of its scientific as well as practical interest, much attention has been given to the physiological aspects of dormancy in seeds (Amen, 1963). In some plants, the seed coat seems relatively impervious to the absorption of the necessary water, and mechanical abrasion or rupture of the testa may be needed to induce germination. In other plants, however, it appears that the embryo itself is in a physiological state of dormancy that can artificially be broken only by relatively drastic measures, such as exposure of the seeds to very low temperatures or their treatment with special chemical reagents. In this connection it is of interest to note that dormant seeds, under natural conditions, may retain their viability for many years. A truly classical example is provided by the seeds of an Indian species of lotus (*Nelumbo*) that germinated after lying dormant in a peat bog for nearly two centuries!

Dicotyledons

In dicotyledons, the first stage in seed germination consists in the emergence of the radicle, or primary root, which ruptures the seed coat at the micropylar region and normally grows downward into the soil or substrate. Following this first stage, the two cotyledons and the epicotylar portion of the embryo may then emerge from the seed coat and, as a result of the elongation of the *hypocotyl,* become elevated above the level of the soil. This extremely common type of seed germination is termed *epigeal*

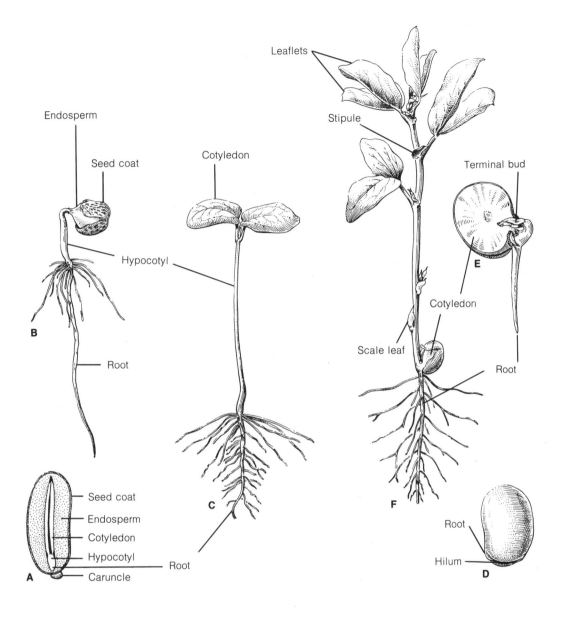

Leaflets

Stipule

Terminal bud

Endosperm

Seed coat

Cotyledon

Hypocotyl

Root

Cotyledon

Scale leaf

Root

E

Hypocotyl

Root

Root

Seed coat
Endosperm
Cotyledon
Hypocotyl
Caruncle

Root

Hilum

B

C

F

A

D

Figure 21-7
A–C, epigeal type of seed germination in castor bean (*Ricinus communis*). **A,** diagram of longisection of ripe seed showing the embryo surrounded by the copious endosperm tissue; **B,** young seedling showing the cotyledons still embedded in the endosperm; **C,** later stage (the elongation of the hypocotyl has elevated the two cotyledons and terminal bud above the surface of the ground).
D–F, hypogeal type of seed germination in vetch (*Vicia faba*). **D,** lateral view of seed showing position of embryonic root with reference to hilum; **E,** early stage in germination (seed coat removed) showing terminal bud of epicotyl, the two fleshy cotyledons, and the primary root; **F,** well-developed seedling, showing the elongated primary shoot (epicotyl) with scale leaves and stipulate foliage leaves (the cotyledons remain below ground because of the failure of growth in length of the hypocotyl). [Redrawn from *Vergleichende Morphologie der höheren Pflanzen* by W. Troll. Gebrüder Borntraeger, Berlin, 1935.]

and is illustrated by the seedlings of many everyday plants, such as radish, sunflower, pumpkin, and castor bean (Fig. 21-7, A–C). The cotyledons of seedlings with epigeal germination are extremely varied in size and form; many of them closely resemble small foliage leaves in their venation and in the development of buds in their axils (see Lubbock, 1892, for a monographic treatment of angiosperm seedlings). Cotyledons have various functions. The foliaceous types presumably carry on photosynthesis and in some annuals, such as *Circaeaster agrestis*, may persist as a pair of green appendages throughout the life of an individual plant (Foster, 1963, 1968). In some dicotyledons with albuminous seeds, such as *Ricinus*, the cotyledons may also function as organs which continue to absorb the reserve food material of the endosperm during the early phase of seedling development (Fig. 21-7, A, B). The cotyledons of some types of seedlings with epigeal germination, e.g., the common bean (*Phaseolus*), are thick, fleshy, food-storing organs which soon shrivel and fall away from the axis of the young seedling.

A contrasted type of germination, termed *hypogeal*, is found in a wide range of angiosperm genera. An example of an hypogeal type of dicotyledonous seedling is found in vetch (*Vicia*), in which the two food-storing cotyledons remain below ground within the seed coat and the portion of the seedling that emerges above the soil is the epicotyl (Fig. 21-7, D–F). Additional examples of dicotyledonous seedlings with hypogeal germination are found in various genera in the family Juglandaceae (Stone, 1970; Conde and Stone, 1970).

Monocotyledons

The development and organography of the seedlings in many monocotyledons present features of unusual interest, with particular reference to the structure and function of the single cotyledon. In some genera (especially, in those with exalbuminous seeds) the cotyledon consists of a basal sheath and a green photosynthetic laminar portion, and thus closely resembles a small foliage leaf (e.g., certain members of the families Alismataceae, Butomaceae, and Liliaceae). But in many monocotyledons, the cotyledon is morphologically and functionally highly specialized into three more-or-less clearly demarcated parts: (1) a *basal sheath*, which encloses the epicotylar bud during early phases of seed germination, (2) a slender, often conspicuously elongated *middle portion,* and (3) a distal tip, or *haustorium,* which remains within the seed (Fig. 21-9). The haustorium functions as a suctorial organ that absorbs the food reserves within the seed and transmits them, via the vascular system of the cotyledon, to the developing seedling. Comparative studies have revealed the wide range in size and form of the haustorial portion of the cotyledon and several examples will now be described to illustrate the extremes in variation.

A relatively simple illustration of a cotyledon with a specialized haustorial tip is provided by the seedling of *Allium cepa* (onion). When the seed germinates, the elongation of the middle portion of the cotyledon pushes the radicle, the epicotylar bud, and the lower sheathing part of the cotyledon out of the testa, leaving the coiled cotyledonary haustorium embedded within the endosperm of the seed (Fig. 21-8, A, B). As development continues, the elongating middle portion of the cotyledon acquires the form of a loop, or arch, which soon emerges above the surface of the soil (Fig. 21-8, C). Ultimately, the upper end of the cotyledon breaks away from the seed and appears as an erect green filiform organ with a withered tip and a conspicuous knee-like "joint" corresponding to the summit of the cotyledonary arch (Fig. 21-8, D). At this stage, the tip of the first foliage leaf begins to emerge through the slit-like opening of the cotyledonary sheath.

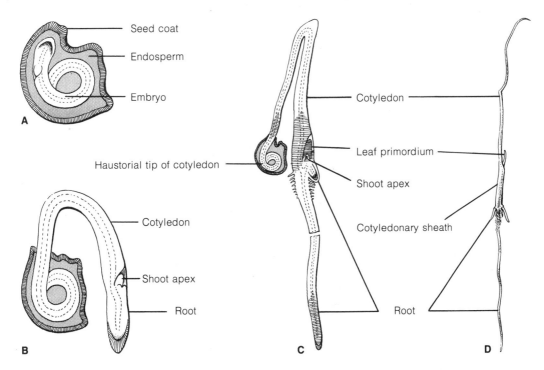

Figure 21-8
Seed germination and seedling morphology in *Allium cepa*. **A,** longisection of ripe seed showing coiled embryo surrounded by endosperm tissue; **B,** longisectional view of early stage in germination (the intercalary elongation of the single cotyledon has ruptured the seed coat and pushed out the root and shoot apex of the embryo; note that coiled haustorial tip of cotyledon remains in the seed); **C,** longisection of later stage in germination showing characteristic loop-like form of cotyledon, the sheathing base of which encloses the shoot apex and first leaf primordium; **D,** seedling in which the withered tip of cotyledon has separated from the seed (note emergence of first foliage leaf from cotyledonary sheath). [Redrawn from *Vergleichende Morphologie der höheren Pflanzen* by W. Troll. Gebrüder Borntraeger, Berlin, 1935.]

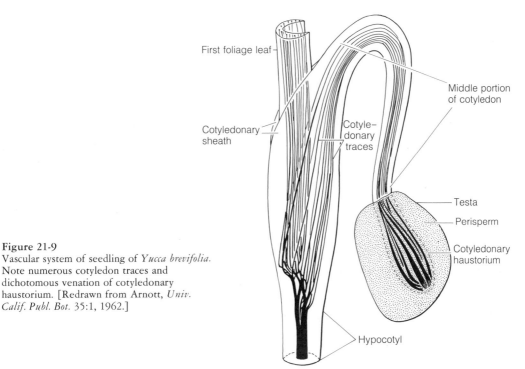

Figure 21-9
Vascular system of seedling of *Yucca brevifolia*. Note numerous cotyledon traces and dichotomous venation of cotyledonary haustorium. [Redrawn from Arnott, *Univ. Calif. Publ. Bot.* 35:1, 1962.]

The general morphology of the seedling and cotyledon in *Yucca* (Agavaceae) offers interesting points of similarity with the seedling of *Allium* (Fig. 21-9). According to Arnott's (1962) detailed study, the cotyledonary haustorium in *Yucca* increases significantly in size during the early phases of germination as a result of both cell enlargement and cell division. The reserve food materials (oil) in the perisperm, as well as the thick cell walls of this tissue, become digested and absorbed by the enlarged haustorium, which, in *Yucca brevifolia*, remains inside the seed "for nearly a month after germination." A striking point of contrast between *Allium* and *Yucca* concerns the difference in structure between the vascular systems in the mature cotyledons of these two genera. In *Allium cepa*, according to Arnott, the vasculature of the cotyledon is represented by a single unbranched strand of xylem extending from the hypocotyl to the haustorial tip of the cotyledon. The cotyledonary vascular system in *Yucca* is more complex and, depending upon the species, consists of two to twelve traces. In *Yucca brevifolia*, eight traces extend through the middle portion into the haustorium where, as in other species of this genus, the veins may dichotomize and, to varying degrees, anastomose (Fig. 21-9). The relatively profuse vasculature of the cotyledon of *Yucca* is of considerable anatomical and phylogenetic interest and also suggests that the haustorial type of cotyledon in this genus is an organ that is very well-equipped to function in the absorption and translocation of food materials to the developing seedling.

Without question the haustorium of the cotyledon in the monocotyledons reaches its most massive and bizarre development in the seedlings of certain palms. A very striking example — which has attracted much attention — is illustrated by the seedling of the coconut palm (*Cocos nucifera*). The "coconut" of commerce is not simply a large seed but on the contrary represents the inner nut-like part of a huge drupaceous type of fruit. Figure 21-10, A, represents diagrammatically a *complete fruit* as seen in longisectional view and shows the thick fibrous "husk" (the outer part of the pericarp, or fruit wall) and the enclosed "nut" (the hard sclerenchymatous shell which is the endocarp or inner part of the fruit wall). Within the endocarp is a single seed, with a greatly reduced testa and minute embryo, the latter embedded in the peripheral layer of cellular endosperm. The endosperm in ripe coconuts, yields the copra of commerce, important because of its rich content of oil. The interior of the seed is a large central cavity filled with liquid endosperm termed coconut milk. The coconut fruit is derived from a tricarpellate ovary and, when the husk has been removed, three "eyes" or shallow depressions can be seen at the lower or stem end of the nut (Fig. 21-10, B, C). These "eyes" are termed "germination pits" by Troll (1935, p. 142), and the largest one lies above the embryo and is the point where the shoot and radicle of the embryo break through the shell at the time of germination.

According to Troll (1935) the diminutive embryo of the coconut — prior to germination — consists of a short thick cotyledon at one end and the embryonic radicle and epicotyl at the opposite pole. Germination is a leisurely process and in nature occurs in nuts still enclosed within the husk. When the embryo begins to enlarge, the epicotyl and radicle rupture the endocarp below the germination pit and penetrate the husk of the fruit. As the process of germination proceeds, the shoot apex forms a series of scale leaves and additional roots arise from the nodes of these appendages as well as from the cotyledonary node (Fig. 29-10, D, E). During the formation of the primary shoot of the seedling, the cotyledon enlarges and becomes an enormous haustorium that grows into and absorbs the liquid endosperm and finally occupies the entire cavity

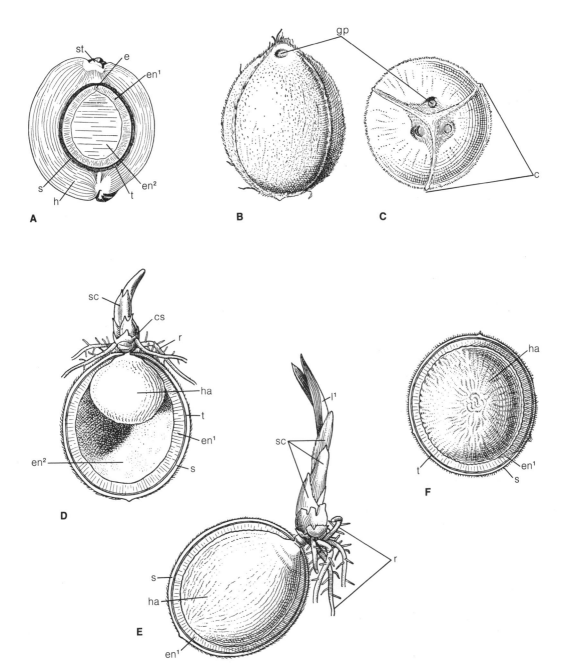

Figure 21-10

The fruit, seed, and seedling of the coconut palm (*Cocos nucifera*). **A**, diagram showing structure of the complete fruit, as seen in longisectional view; **B, C**, lateral and end views, respectively, of the surface of the nut, after removal of the husk (h), showing the three united carpel units (c) and the "germination pits" (gp); **D**, young seedling with lateral roots (r), developing shoot with its scale leaves (sc) and the greatly enlarged cotyledonary haustorium (ha) growing within the liquid endosperm (en²); **E**, later stage of seedling; the haustorium now occupies the entire central region of the nut; **F**, same stage in seedling development as in **E**, showing transectional view of cotyledonary haustorium. c, carpel unit; cs, cotyledonary sheath; e, embryo; en¹, cellular endosperm; en², liquid endosperm; gp, germination pit; h, husk of fruit; ha, cotyledonary haustorium; l¹, first foliage leaf; r, roots; s, shell (endocarp); sc, scale leaves; st, point of former attachment of fruit stalk; t, testa of seed. [Redrawn from *Vergleichende Morphologie der höheren Pflanzen* by W. Troll. Gebrüder Borntraeger, Berlin, 1935.]

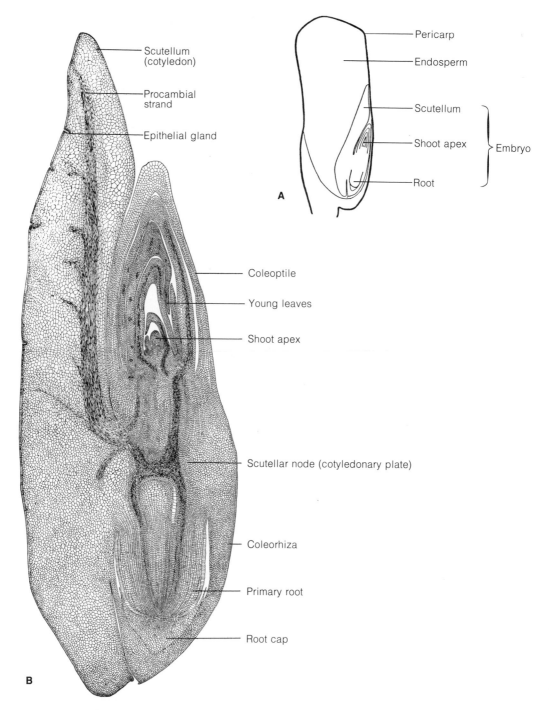

Figure 21-11
Structure of the embryo of *Zea mays*. **A,** diagram of median longitudinal section of corn grain (a caryopsis) showing the general orientation of the embryo with reference to the endosperm tissue; **B,** longitudinal section of embryo, removed from grain, showing position and cellular structure of the various organs and tissues. See text for further explanation. [Redrawn from Avery, *Bot. Gaz.* 89:1, 1930, University of Chicago Press.]

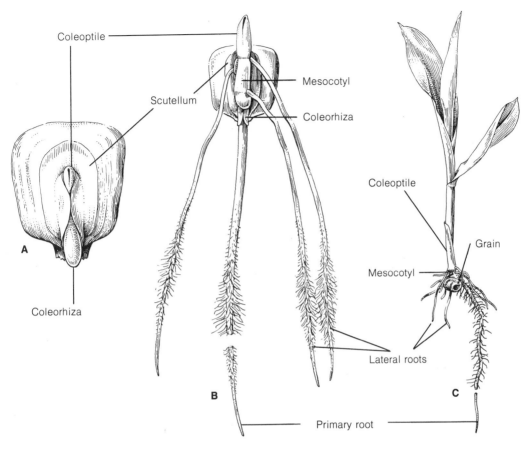

Figure 21-12
Seed germination and seedling morphology in corn (*Zea mays*). **A,** corn grain with pericarp removed showing the embryo with its coleorhiza, coleoptile, and scutellum attached to the endosperm tissue; **B,** early stage of germination showing elongation of primary root, the lateral roots developed from mesocotyl, and the sheath-like coleoptile; **C,** seedling showing that scutellum remains within grain below ground (note conspicuous sheath of the young foliage leaves and the persistent coleoptile). [Redrawn from *Vergleichende Morphologie der höheren Pflanzen* by W. Troll. Gebrüder Borntraeger, Berlin, 1935.]

of the nut (Fig. 29-10, D–F). Ultimately, the first foliage leaf emerges from the scaly terminal bud and the young palm seedling finally breaks its connection with the nut and begins its growth as an independent plant.

The most extreme and bizarre development of a cotyledonary haustorium occurs during the germination of the so-called "double coconut," or "coco-de-mer," which is the gigantic fruit of *Lodoicea maldivica*, a stately palm endemic to three islands of the Seychelles archipelago in the Indian Ocean (Bailey, 1942). The fruit of *Lodoicea* is usually bilobed (and is not actually a "pair" of fruits) and attains a length of nearly two feet. According to Corner (1966), the fruit may reach a weight of 40–50 pounds. The cotyledonary haustorium of *Lodoicea* completely destroys *all* the endosperm and at maturity is a colossal structure that, because of its two lobes and convoluted surface, was said by Thiselton-Dyer (1910) to resemble "a human

brain with the hemispheres drawn apart."
One of the most remarkable features of the
seedling of *Lodoicea* is the fact that the
middle portion of the cotyledon—the stalk
as it is often called—measures an inch in
thickness and during its enormous elongation
carries the plumule and radicle to a distance
of 3–12 feet from the parent nut.

In the examples of monocotyledonous
seedlings that we have just described, the
single cotyledon—although modified to
varying degrees with reference to its haustorial function—clearly appears to be a *unit
appendage* representing the first foliar organ
of the plant. The embryo and seedlings of
the grasses, however, have always proved
difficult to interpret and there are many
conflicting views about the morphology of
the cotyledon. A detailed discussion of the
various theories is beyond the scope of this
book but the nature of the problems involved
can be illustrated by a brief description of
the embryo and seedling of corn (*Zea mays*).

A corn grain is an indehiscent fruit, or
caryopsis, that contains within the pericarp
a single seed, consisting of a large embryo
which lies at one side of the massive endosperm tissue (Fig. 21-11, A). As seen in
median longisectional view, the embryo is
differentiated into the following main parts:
(1) a hood-like sheathing organ, termed the
coleoptile, which encloses the shoot apex
and several embryonic leaves; (2) a *primary
root*, enclosed within a sheath of tissue designated as the *coleorhiza*; and (3) a shield-
like organ, the *scutellum*, which lies in direct
contact with the adjacent endosperm tissue
of the grain (Fig. 21-11, B). When corn
germinates, the *scutellum* remains perma-

nently within the grain and functions as an
haustorial organ. The young emerging shoot
at first is completely enclosed within the
tubular sheathing coleoptile (Fig. 21-12,
A–B). Soon however, the epicotylar axis
begins to elongate and the young shoot,
with its foliage leaves, pushes through a
small slit-like opening at the tip of the
coleoptile, which remains for some time as
a sheath-like organ that becomes elevated
above its original position by the elongation
of a part of the seedling axis known as the
mesocotyl (Fig. 21-12, C). During the early
phases of germination, the primary root of
the embryo, together with additional lateral
roots, developing from the lower nodes of
the seedling axis, make their appearance
(Fig. 21-12, B, C).

Avery (1930), on the basis of his study of
the embryo and seedling of *Zea*, *Avena*, and
Triticum, concluded that the scutellum *alone*
represents the cotyledon while the coleoptile
is the second foliar appendage of the seedling. The divergent opinion, however, expressed by Goebel (1905), Arber (1925),
and Eames (1961), maintains that in grasses
the scutellum corresponds only to the cotyledonary haustorium and that the coleoptile
is homologous with the sheathing base of
the cotyledon. In criticizing this interpretation, Avery argues that the scutellum is a
lateral structure—comparable in position to
the cotyledon of other monocotyledons—
whereas the coleoptile, in its position and
mode of vascularization, is an appendage
of the epicotylar axis. In Avery's interpretation, the so-called "mesocotyl" represents
the *first internode* that separates the node of
the scutellum from the node of the coleoptile.

References

Amen, R. D.
  1963. The concept of seed dormancy. *Amer. Sci.* 51:408–424.
Arber, A.
  1925. *Monocotyledons, A Morphological Study.* University Press, Cambridge.
Arnott, H. J.
  1962. The seed, germination, and seedling of *Yucca. Univ. Calif. Publ. Bot.* 35:1–164.
Avery, G. S., Jr.
  1930. Comparative anatomy and morphology of embryos and seedlings of maize, oats, and wheat. *Bot. Gaz.* 89:1–39.
Bailey, L. H.
  1942. Palms of the Seychelles. *Gentes Herb.* 6(1):3–29.
Baumann-Bodenheim, M. G.
  1954. Prinzipien eines Fruchtsystems der Angiospermen. *Schweiz. Bot. Ges. Ber.* 64:94–112.
Conde, L. F., and D. E. Stone
  1970. Seedling morphology in Juglandaceae, the cotyledonary node. *Jour. Arnold Arboretum* 51:463–477.
Corner, E. J. H.
  1951. The leguminous seed. *Phytomorphology* 1:117–150.
  1966. *The Natural History of Palms.* University of California Press, Berkeley and Los Angeles.
Eames, A. J.
  1961. *Morphology of the Angiosperms.* McGraw-Hill, New York.
Egler, F. E.
  1943. The fructus and the fruit. *Chron. Bot.* 7(8):391–395.
Esau, K.
  1965. *Plant Anatomy,* Ed. 2. Wiley, New York.
Fahn, A.
  1967. *Plant Anatomy.* Pergamon, Oxford.
Foster, A. S.
  1963. The morphology and relationships of *Circaeaster. Jour. Arnold Arboretum* 44:299–321.

  1968. Further morphological studies on anastomoses in the dichotomous venation of *Circaeaster. Jour. Arnold Arboretum* 49:52–67.
Goebel, K.
  1905. *Organography of Plants,* Pt. 2. (English translation by I. B. Balfour.) Clarendon Press, Oxford.
Gray, A.
  1878. *Introduction to Structural and Systematic Botany and Vegetable Physiology.* Ivison, Blakeman, Tayler, New York.
Hill, J. B., H. W. Popp, and A. R. Grove, Jr.
  1967. *Botany. A textbook for Colleges,* Ed. 4. McGraw-Hill, New York.
Holman, R. M., and W. W. Robbins
  1938. *A Textbook of General Botany,* Ed. 4. Wiley, New York.
Juhnke, G., and H. Winkler
  1938. Der Balag als Grundelement des Angiospermengynaeceums. *Beitr. Biol. Pflanzen* 25:290–324.
Leopold, A. C.
  1964. *Plant Growth and Development.* McGraw-Hill, New York.
Lubbock, Sir John
  1892. *A Contribution to our Knowledge of Seedlings.* 2 v. Kegan Paul, Trench, Trübner, London.
McLean, R. C., and W. R. Ivimey-Cook
  1956. *Textbook of Theoretical Botany,* Vol. 2. Longmans, London.
Netolitzky, F.
  1926. *Anatomie der Angiospermen—Samen.* (Handbuch d. Pflanzenanatomie, Band 10, Lief. 14.) Gebrüder Borntraeger, Berlin.
Reeve, R. M.
  1946a. Structural composition of the sclereids in the integument of *Pisum sativum* L. *Amer. Jour. Bot.* 33:191–204.
  1946b. Ontogeny of the sclereids in the integument of *Pisum sativum* L. *Amer. Jour. Bot.* 33:806–816.

Rehder, A.
    1927. *Manual of Cultivated Trees and Shrubs Hardy in North America.* Macmillan, New York.

Robbins, W. W., T. E. Weir, and C. R. Stocking
    1964. *Botany. An Introduction to Plant Science,* Ed. 3. Wiley, New York.

Sargent, C. S.
    1922. *Manual of the Trees of North America*, Ed. 2. Houghton Mifflin, Boston and New York.

Stone, D. E.
    1970. Evolution of cotyledonary and nodal vasculature in the Juglandaceae. *Amer. Jour. Bot.* 57:1219–1225.

Takhtajan, A.
    1969. *Flowering Plants: Origin and Dispersal.* (Translated from the Russian by C. Jeffrey.) Oliver and Boyd, Edinburgh.

Thiselton-Dyer, W. T.
    1910. Morphological notes. XII. Germination of the double coco-nut. *Ann. Bot.* 24:223–230.

Troll, W.
    1935. *Vergleichende Morphologie der höheren Pflanzen,* Erster Band, Lieferung 1. Gebrüder Borntraeger, Berlin.

Wilson, C. L., and W. E. Loomis
    1967. *Botany,* Ed. 4. Holt, Rinehart and Winston, New York.

Winkler, H.
    1939. Versuch eines "natürliche" Systems der Früchte. *Beitr. Biol. Pflanzen* 26:201–220.
    1940. Zur Einigung und Weiterführung in der Frage des Fruchtsystems. *Beitr. Biol. Pflanzen* 27:92–130.

Index

Mai5